# SOCIOLOGY IN OUR TIMES

## the essentials

# about the AUTHOR

**Diana Kendall** is Professor of Sociology at Baylor University, where she was named an Outstanding University Professor. Dr. Kendall has taught a variety of courses, including Introduction to Sociology; Sociological Theory (undergraduate and graduate); Sociology of Medicine; Sociology of Law; and Race, Class, and Gender. Previously, she enjoyed many years of teaching sociology and serving as chair of the Social and Behavioral Science Division at Austin Community College.

Diana Kendall received her Ph.D. from the University of Texas at Austin, where she was invited to membership in Phi Kappa Phi Honor Society. Her areas of specialization and primary research interests are sociological theory and the sociology of medicine. In addition to *Sociology in Our Times: The Essentials,* tenth edition, she is the author of *Sociology in Our Times,* tenth edition; *The Power of Good Deeds: Privileged Women and the Social Reproduction of the Upper Class* (Rowman & Littlefield, 2002); *Members Only: Elite Clubs and the Process of Exclusion* (Rowman & Littlefield, 2008); and *Framing Class: Media Representations of Wealth and Poverty in America,* second edition (Rowman & Littlefield, 2012). Professor Kendall is a member of numerous sociological associations, including the American Sociological Association, the Society for the Study of Social Problems, and the Southwestern Sociological Association.

**10** EDITION

# SOCIOLOGY IN OUR TIMES

## the essentials

## Diana Kendall

Baylor University

CENGAGE
Learning®

Australia • Brazil • Mexico • Singapore • United Kingdom • United States

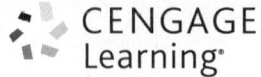

*Sociology in Our Times: The Essentials,*
**Tenth Edition**
Diana Kendall

Product Director: Marta Lee-Perriard

Product Manager: Jennifer Harrison

Content Developer: Lin Gaylord

Product Assistant: Julia Catalano

Media Developer: John Chell

Marketing Manager: Kara Kindstrom

Content Project Managers: Cheri Palmer
and Christy Frame

Art Director: Michelle Kunkler

Manufacturing Planner: Judy Inouye

Production Service: Greg Hubit Bookworks

Photo Researcher: Sundar Ananthapadmanabhan,
Lumina Datamatics

Text Researcher: Punitha Rajamohan,
Lumina Datamatics

Copy Editor: Donald Pharr

Proofreader: Debra Nichols

Text Designer: Diane Beasley

Cover Designer: Lee Friedman

Cover Image: Graham Crouch/Lonely Planet
Images/Getty Images

Compositor: MPS Limited

Library of Congress Control Number: 2014944052

ISBN: 978-1-305-09415-4

**Cengage Learning**
20 Channel Center Street
Boston, MA 02210
USA

Cengage Learning is a leading provider of customized learning solutions with office locations around the globe, including Singapore, the United Kingdom, Australia, Mexico, Brazil, and Japan. Locate your local office at **www.cengage.com/global**.

Cengage Learning products are represented in Canada by Nelson Education, Ltd.

To learn more about Cengage Learning Solutions, visit **www.cengage.com**.

Purchase any of our products at your local college store or at our preferred online store **www.cengagebrain.com**.

Printed in the United States of America
Print Number: 01    Print Year: 2014

# Brief Contents

# Contents

## PART 2 *Social Groups and Social Control*

### 4 Social Structure and Interaction in Everyday Life  103

### 5 Groups and Organizations  137

# PART 3 Social Inequality

# PART 4 *Social Institutions*

# PART 5 *Social Dynamics and Social Change*

# Features

## Census Profiles

FEATURES

# Preface

Welcome to the tenth edition of *Sociology in Our Times: The Essentials*! This edition marks a milestone in that this best-selling text has been extensively used for more than two decades in college and university classrooms across the United States, Canada, and other nations. However, *Sociology in Our Times: The Essentials* continues to live up to its name, remaining highly current and relevant to today's students and professors and reflecting the latest available data and new insights on what is going on in our nation and world from a sociological perspective.

This edition focuses on social change and ways in which media, particularly social media, and various other forms of technology inevitably bring about new ways of living, interacting with others, or doing some activity or task. For example, the theme of change and "going green" in large urban spaces is reflected by the tenth edition's livelier, exciting new cover and is carried throughout the text in updates to the interior design, artwork, and photos. In addition, new sections have been added throughout on social change, and the former "Media Framing" boxes have been expanded to include topics on social media and social change.

Like previous editions, the tenth edition highlights topics ranging from popular culture icons and social networking to far-more-serious issues of our times, such as the social effects of massive natural and human disasters, gun violence, terrorism, war, and the individual and social consequences of problems such as growing inequality between the wealthiest and the poorest people and nations, persistent unemployment, migration concerns worldwide, and other persistent issues and problems. In this edition even more attention is paid to the theme of social change and how it affects all of our lives.

The second decade of the twenty-first century offers unprecedented challenges and opportunities for each of us as individuals and for our larger society and world. In the United States, we can no longer take for granted the peace and economic prosperity that many—but far from all—people were able to enjoy in previous decades. However, even as some things change, others remain the same, and among the things that have not changed are the significance of education and the profound importance of understanding how and why people act the way they do. It is also important to analyze how societies grapple with issues such as economic hardship and the threat of terrorist attacks and war, and to gain a better understanding of why many of us seek stability in our social institutions—including family, religion, education, government, and media—even if we believe that some of these institutions might benefit from certain changes.

Like previous editions, the tenth edition of *Sociology in Our Times: The Essentials* highlights the relevance of sociology to help students connect with the subject and the full spectrum of topics and issues that it encompasses. It achieves this connection by providing a meaningful, concrete context within which to learn. Specifically, it presents the stories—the *lived experiences*—of real individuals and the social issues they face while discussing a diverse array of classical and contemporary theories and examining interesting and relevant research. The first-person commentaries that begin each chapter show students how sociology can help them understand the important questions and social issues that not only these other individuals face but that they themselves may face as well.

*Sociology in Our Times: The Essentials* includes the best work of classical and established contemporary sociologists, and it weaves an inclusive treatment of *all* people—across lines of race/ethnicity, class, gender, age, ability/disability, and other social attributes—into the examination of sociology in *all* chapters. It does not water down the treatment of sociology for students! *Sociology in Our Times: The Essentials* provides students with the most relevant information about sociological thinking and helps them to consider contemporary social issues through the lens of diversity. While helping students learn to appreciate how sociology can help them better understand the world, this text also encourages them to see themselves as *members of their communities* and shows them what can be done in responding to social issues. As a result, students learn how sociology is not only a collection of concepts and theories but also a field that can make a difference in their lives, their communities, and the world at large.

# What's New to the Tenth Edition?

The tenth edition builds on the best of previous editions but places more emphasis on social change, while offering new insights, learning tools, and opportunities to apply the content of each chapter to relevant sociological issues and major concerns of the twenty-first century. As it is my goal to make each edition better than the previous one, I have revised all the chapters to reflect the latest in sociological theory and research, and have updated examples throughout. Additionally, all statistics, such as data relating to crime, demographics, health, and the economy, are the latest available at the time of this writing.

To further emphasize the new theme on social change, the "Media Framing" box has been expanded to include issues and societal changes from the increased use of social media. In addition, the feature titled "Sociology and Social Policy" has been brought back to this edition. This feature discusses important social issues (such as the high rate of suicide in the military, the right of homeless people to occupy public spaces, and the extent to which employers should be able to spy on employees) to provide a systematic examination of how social policy and law may bring about social change or impede it.

To assist your students in learning about sociology and reflecting their knowledge on tests, I have continued to revise the learning objectives at the beginning of each chapter, provide integrated learning objective (LO) icons throughout the chapter, and offer students a study guide at the end of each chapter. The learning objectives have been carefully conceived to help the reader focus on the most crucial concepts of the chapter.

## Changes by Chapter

### Chapter 1, The Sociological Perspective and Research Process

- New chapter-opening lived experiences about the relationship among suicide, bullying, and social media
- Revised "Sociology and Everyday Life" quiz: "How Much Do You Know About Suicide?"
- Updated "Sociology Works!": "Durkheim's Classical Study of Suicide and Young People in India Today"
- Updated Figure 1.14, "Suicide Rates by Race and Sex"
- New "Sociology and Social Policy" box: "Establishing Policies to Help Prevent Military Suicides"
- Revised and updated "Understanding Data Presentations"

- Revised and updated Table 1.1: "Rates (per 100,000 U.S. Population) for Homicide, Suicide, and Firearm-Related Deaths of Youths Ages 15–19, by Gender, 2009"
- Updated Figure 1.20, "National Suicide Statistics at a Glance"
- New "Suicide in the Media" box: "Reporting on Bullying and Suicide: A Cause-and-Effect Relationship?"
- New discussion of Sudhir Venkatesh's *Floating City: A Rogue Sociologist Lost and Found in New York's Underground Economy* as an example of ethnography
- New second "Concept Quick Review": "Strengths and Weaknesses of Social Research Methods"

### Chapter 2, Culture

- New chapter-opening lived experience about the relationship between the food truck movement and cultural diversity
- Updated Figure 2.6, "States with Official English Laws"
- Updated "Census Profiles": "Languages Spoken in U.S. Households"
- Updated Figure 2.10, "Heterogeneity of U.S. Society"
- Updated Figure 2.11, "Cultural Diversity: A Nation of Immigrants"
- New "Sociology in Global Perspective" box: "What Do Cultural Norms Say About Drinking Behavior?"

### Chapter 3, Socialization

- Shortened discussion of social psychological theories of human development and added section on symbolic interactionist, functionalist, and conflict perspectives on human development
- Revised discussion of Mead's "I," "me," and the "self"
- Revised "Concept Quick Review," which now includes functionalist and conflict perspectives on socialization
- Revised and updated discussion on agents of socialization
- Revised "Sociology Works!": "The Issue of Excessive Praise"
- New Figure 3.4, "Types of Maltreatment Among Children Under Age 18"
- Updated "Sociology in Global Perspective": "Open Doors: Study Abroad and Global Socialization"

### Chapter 4, Social Structure and Interaction in Everyday Life

- Updated "Sociology and Everyday Life" quiz: "How Much Do You Know About Homeless

Persons and the Social Structure of Homelessness?"

- Revised "Homelessness in the Media" box
- Revised Figure 4.4, "Causes of Family Homelessness in 29 Cities"
- Revised Figure 4.8, "Who Are the Homeless?"
- New section: "Societies, Technology, and Sociocultural Change"
- Expanded discussion of Durkheim's mechanical and organic solidarity
- New Table 4.1, "Technocratic Bases of Society"
- New Table 4.2, "Comparing *Gemeinschaft* and *Gesellschaft* Societies"
- New examples of the self-fulfilling prophecy as it relates to homeless persons
- Expanded discussion of the dramaturgical perspective, with new concept of social script added
- Updated "Census Profiles": "Computer and Internet Access in U.S. Households"
- New "Sociology and Social Policy" box: "Homeless Rights Versus Public Space"

### Chapter 5, Groups and Organizations

- Updated "Community in the Media" reflecting how social media have affected this phenomenon
- Updated Figure 5.5, "Results of Milgram's Obedience Experiment"
- Revised "Sociology Works!": "Ingroups, Outgroups, and 'Members Only' Clubs"
- Updated information on Stanley Milgram's experiments on obedience to authority
- Added definition of *leadership* from a sociological viewpoint
- Expanded discussion and examples of iron law of oligarchy
- New "Sociology and Social Policy" box: "In a High-Tech World, Should Employers Be Allowed to 'Spy' on Their Workers?"
- Revised and updated "Organizations in the Future" to include "Socially Sustainable Organizations" and "Globalization, Technology, and 'Smart Working'"

### Chapter 6, Deviance and Crime

- New chapter-opening lived experience, "When the Unspeakable Happens," on the final police report on the shootings at Sandy Hook Elementary, Newtown, Connecticut
- New "Sociology and Everyday Life" quiz: "How Much Do You Know About Guns and Violence?"
- Updated crime statistics throughout chapter
- Revised "Sociology Works!": "Social Definitions of Deviance: Tracking Sasquatch"
- New section on "Internet Computer Crime"
- Updated Figure 6.8, "Distribution of Arrests by Type of Offense, 2012"

- Updated Figure 6.9, "The FBI Crime Clock, 2012"
- New Figure 6.10, "Top Reported Internet Crime Types"
- Updated Figure 6.13, "Arrest Rates by Gender, 2012 (Selected Offenses)"
- New Figure 6.14, "Arrest Rates by Race, 2012 (Selected Offenses)"
- Updated Figure 6.18, "Death Row Census, April 1, 2013"
- New "Sociology and Social Policy" box: "The Long War Over Gun Control"
- New "Media" box: "Violent Crime in the Media"
- Updated "U.S. Deviance and Crime in the Future" section
- Updated "Transnational Crime and the Global Criminal Economy" section

### Chapter 7, Class and Stratification in the United States

- New chapter-opening lived experience, "The Power of Class," on Annette Lareau's findings on the effects of "unequal childhoods" on long-term, class-based inequality
- Updated "Sociology and Everyday Life" quiz: "How Much Do You Know About Wealth, Poverty, and the American Dream?"
- Updated information and statistics on wealth, income, poverty, and health insurance coverage throughout chapter
- New "Sociology in Global Perspective": "A Day in Your Life: How Are You Touched by Modern Slavery?"
- Updated Figure 7.11, "Distribution of Pretax Income in the United States, 2011"
- Updated Figure 7.12, "Median Household Income in the United States"
- Updated Figure 7.13, "Median Household Income by Race/Ethnicity in the United States"
- Updated Figure 7.14, "Racial Divide in Net Worth, Shown in 2012 Dollars"
- Updated Figure 7.15, "Rate of Uninsurance by Household Income, 2012"
- Updated Figure 7.17, "Percentage of People in Poverty in the Past 12 Months by State, 2012"
- Updated Figure 7.18, "U.S. Poverty Rates by Age, 1959–2012"
- Expanded discussion of economic and structural sources of poverty
- Updated "U.S. Stratification in the Future" section

### Chapter 8, Global Stratification

- New chapter-opening lived experience, "Leaving the Snare of Poverty," on how impoverished girls have benefited from efforts to eradicate global poverty
- Updated statistics throughout

- Updated Figure 8.1, "Income Gap Between the World's Richest and Poorest People"
- New Figure 8.3 showing income amounts in "High-, Middle-, and Low-Income Economies in Global Perspective"
- Updated Figure 8.6, "Indicators of Human Development"
- Revised and updated "Child Labor Issues in the Media"
- Revised Concept Quick Review
- New "Sociology and Social Policy" box: "Fighting Extreme Poverty One Social Policy at a Time"
- Updated discussion of *maquiladora* plants in the twenty-first century
- Revised and updated: "Sociology Works!": "The Relationship Between Place and Global Poverty"

### Chapter 9, Race and Ethnicity

- Updated "Sociology and Everyday Life" quiz: "How Much Do You Know About Race, Ethnicity, and Sports?"
- Expanded discussion of race as a socially constructed reality
- Updated Figure 9.9, "U.S. Tribal Colleges and Universities"
- Revised and updated "Sociology in Global Perspective" box: "Racism in European Football"
- Expanded discussion of individual and institutional discrimination
- Revised and updated "Sociology Works!": "A Starting Point for Social Change"
- Revised and updated "Sports in the Media" box: "Media Coverage of Minorities in Sports"
- Updated information on all groups in "Racial and Ethnic Groups in the United States"
- Added sections on Arab Americans and Iranian (Persian) Americans under "Middle Eastern Americans"

### Chapter 10, Sex, Gender, and Sexuality

- New chapter-opening lived experience, "Child Beauty Pageants and the Sexualization of Girls," on child beauty pageants as related to body image and self-identity among girls and women
- Expanded and updated discussion of LGBT issues and statistics
- Revised and updated "Sociology Works!": "Institutional Discrimination: Women in a Locker-Room Culture"
- Expanded discussion of discrimination based on sexual orientation, including housing, health care, and marital rights
- Updated Figure 10.10, "The Wage Gap, 2012"
- Updated Figure 10.11, "Women's Wages as a Percentage of Men's in Each Racial–Ethnic Category, 2012"

- Updated Figure 10.12, "Women's Earnings as a Percentage of Men's by State and Puerto Rico, 2012"
- Updated Census Profile, "Single Mothers with Children Under 18"
- Updated Table 10.3, "Percentage of the Workforce Represented by Women, African Americans, Hispanics, and Asian Americans in Selected Occupations"
- Updated data on pay equity
- Revised and updated discussion on "Paid Work and Family Work," including new study by the Pew Research Center

### Chapter 11, Families and Intimate Relationships

- New chapter-opening lived experience, "Diverse Family Landscapes in the Twenty-First Century," about different patterns in contemporary families
- Updated "Census Profiles": "Household Composition, 1970 and 2012"
- Expanded discussion of cohabitation and domestic partnerships
- New section on same-sex marriages
- Updated statistics on marriage, housework, child care, birth and fertility rates, adoption, teen childbearing, single-parent households, two-parent households, remaining single, and divorce
- Revised discussion of children in foster care and family violence
- New Figure 11.6, "Estimated Number of Opposite-Sex Couples Cohabiting in the United States in Selected Years, 1996–2013"
- Updated Figure 11.11, "Living Arrangements of Children Under 18 Years Old for Selected Years, 1970–2012"
- Updated Figure 11.12, "Marital Status of U.S. Population Ages 18 and Over by Race/Ethnicity, 2012"

### Chapter 12, Education and Religion

- Chapter revised to provide latest available data on education and religion
- Updated "Sociology Works!": "Stopping Bullying: Character Building and Social Norms"
- New "Sociology and Social Policy" box: "Prayer in Public Schools? The Issue of Separation of Church and State"
- Revised discussion on tracking, detracking, and the hidden curriculum to show the continued relevance of these issues
- Updated Figure 12.6, "Percentage Distribution of Total Public Elementary–Secondary School System Revenue, 2012–2013"
- Updated Figure 12.7, "Status Dropout Rates for 16- to 24-Year-Olds, by Race/Ethnicity, Gender, and Region"

- Revised and updated discussion of racial segregation and resegregation
- Revised and updated discussion of school safety and school violence
- Revised and updated data on community colleges and four-year colleges and universities
- Updated information on lack of student and faculty diversity based on race/ethnicity and gender
- Expanded discussion on the future of education and Obama administration plans for funding and improving education
- Updated "Census Profiles": "Educational Achievement of Persons Ages 25 and Over"
- Updated Table 12.1, "Major World Religions," to show newer estimates of the number of current followers of world's religions
- Updated Table 12.3, "Top 25 U.S. Denominations That Self-Identify as Christian in 2012"
- New section under "The Secularization Debate" about the latest Pew study on people who do not affiliate with any religion
- Updated Figure 12.15, "U.S. Religious Traditions' Membership"

### Chapter 13, Politics and the Economy in Global Perspective

- New opening lived experience, "Experiencing Politics and History at a Presidential Inauguration," about college students' involvement in the 2013 presidential inauguration
- New "Sociology and Everyday Life" quiz: "How Much Do You Know About Politics and the Media?"
- Thoroughly revised and updated discussions on the 2012 presidential election
- New section on Super Pacs and the part they played in recent elections
- New Figure 13.4, "Influence of Super PACs in the 2012 Presidential Election"
- Revised "Sociology Works!": "C. Wright Mills's Ideas About the Media: Ahead of His Time?"
- Updated section "Discontent with the Current Political System and Parties," now including expanded discussions of the Tea Party and the Green Party
- Revised and updated discussion of voter participation and voter apathy in the 2012 presidential election
- New Figure 13.8, "Voter Participation in the 2012 Presidential Election by Race and Ethnicity"
- New Figure 13.9, "2012 Presidential Election: State by State"
- Updated discussion of federal bureaucracy
- Updated Figure 13.10, "The 'Typical' Federal Civilian Employee"
- Updated discussion of the military–industrial complex and defense spending
- New "Census Profiles" showing types of occupations by race and ethnicity of employees
- New "Politics in the Media" box: "'The Life of Julia': The Internet, Twitter, and Continuing Political Battles"
- Updated Figure 13.13, "The General Motors Board of Directors"
- Updated Table 13.1, "Revenues of the World's 20 Largest Public and Private Corporations, 2012"
- Updated Table 13.2, "The Music Industry's Big Three"

### Chapter 14, Health, Health Care, and Disability

- Updated statistics throughout on health in the United States and other nations
- Revised and updated discussion of lifestyle factors as they relate to health, including alcohol, nicotine, and sexually transmitted diseases
- Updated Figure 14.8, "Chlamydia: Rates by Age and Sex, United States, 2012"
- Updated Figure 14.9, "Adult Obesity in the United States, 2012"
- Updated Figure 14.11, "Increase in Cost of Health Care, 1993–2012"
- Updated Figure 14.12, "Uninsured Children by Poverty Status, Household Income, Age, Race, Hispanic Origin, and Nativity, 2012"
- Revised and updated discussion of Affordable Care Act implementation and possible consequences for new health insurance coverage and quality of health care

### Chapter 15, Population and Urbanization

- Updated Figure 15.1, "Growth in the World's Population, 2012"
- Updated information on rates of fertility, mortality, and migration
- Expanded and updated discussion on proposals regarding U.S. immigration law
- Updated Figure 15.12, "The World's Fourteen Largest Agglomerations"
- New "Immigration in the Media" box: "Language Matters: 'Illegal Immigrants' No Longer"

### Chapter 16, Collective Behavior, Social Movements, and Social Change

- New opening lived experience, "Many Voices Needed for Social Change"
- Updated discussion of Love Canal and ongoing activism by environmental groups
- Updated Table 16.1, "Top 10 Policy Priorities: 2014"

- Updated examples of panics, protest rallies, rumors, public opinion, and other forms of collective behavior
- New Figure 16.8, "Stages in Social Movements"

## Overview of the Text's Contents

*Sociology in Our Times: The Essentials,* tenth edition, contains sixteen high-interest, up-to-date, clearly organized chapters to introduce students to the best of sociological thinking. The length of the text makes full coverage of the book possible in the time typically allocated to the introductory course so that all students are purchasing a book that their instructors will have the time and desire to cover in its entirety.

*Sociology in Our Times: The Essentials* is divided into five parts.

**Part 1** establishes the foundation for studying society and social life. **Chapter 1** introduces students to the sociological imagination and traces the development of sociological thinking. The chapter sets forth the major theoretical perspectives used by sociologists in analyzing compelling social issues and shows students how sociologists conduct research. This chapter provides a thorough description of both quantitative and qualitative methods of sociological research, and shows how these approaches have been used from the era of Emile Durkheim to the present to study social concerns such as suicide. In **Chapter 2** culture is spotlighted as either a stabilizing force or a force that can generate discord, conflict, and even violence in societies. Cultural diversity is discussed as a contemporary issue, and unique coverage is given to popular culture and leisure and to divergent perspectives on popular culture. **Chapter 3** looks at the positive and negative aspects of socialization, including a lived experience of learning the socialization cues of medical school. This chapter presents an innovative analysis of gender and racial–ethnic socialization, and issues associated with recent immigration.

**Part 2** examines social groups and social control. **Chapter 4** applies the sociological imagination to an examination of society, social structure, and social interaction, using homelessness as a sustained example of the dynamic interplay of structure and interaction in society. Unique to this chapter are discussions of the sociology of emotions and of personal space as viewed through the lenses of race, class, gender, and age.

**Chapter 5** analyzes groups and organizations, including innovative forms of social organization and ways in which organizational structures may differentially affect people based on race, class, gender, and age. **Chapter 6** examines how deviance and crime emerge in societies, using diverse theoretical approaches to describe the nature of deviance, crime, and the criminal justice system. Key issues are dramatized for students through an analysis of recent research on peer cliques and gangs.

**Part 3** focuses on social differences and social inequality, looking at issues of class, race/ethnicity, and sex/gender, while also touching on issues relating to social inequality based on age. **Chapter 7** focuses on class and stratification in the United States, analyzing the causes and consequences of inequality and poverty, including a discussion of the ideology and accessibility of the American Dream. **Chapter 8** addresses the issue of global stratification and examines differences in wealth and poverty in rich and poor nations around the world. Explanations for these differences are discussed.

The focus of **Chapter 9** is race and ethnicity, which includes an illustration of the historical relationship (or lack of it) between sports and upward mobility by persons from diverse racial–ethnic groups. A thorough analysis of prejudice, discrimination, theoretical perspectives, and the experiences of diverse racial and ethnic groups is presented, along with global racial and ethnic issues. **Chapter 10** examines sex and gender, with special emphasis on gender stratification in historical perspective. Linkages between gender socialization and contemporary gender inequality are described and illustrated by lived experiences and perspectives on body image.

**Part 4** offers a systematic discussion of social institutions, building students' awareness of the importance of these foundational elements of society and showing how a problem in one often has a significant influence on others. Families and intimate relationships are explored in **Chapter 11**, which includes both U.S. and global perspectives on family relationships, a view of families throughout the life course, and a discussion of diversity in contemporary U.S. families. Education and religion are presented in **Chapter 12**, which highlights important sociological theories pertaining to these social institutions and integrates the theme of the influence of religion on education and life. In the process, the chapter highlights issues of race, class, and gender inequalities in current U.S. education. The chapter also provides a thorough discussion of religion in global perspective, including a survey of world religions and an analysis of how religious beliefs affect other aspects of social life. Current trends in U.S. religion are explored, including various sociological explanations of why people look to religion to find purpose and meaning in life.

**Chapter 13** discusses politics and the economy in global perspective, highlighting the international context in which contemporary political and economic systems operate and showing the intertwining nature of politics, the economy, and global media outlets. The chapter emphasizes the part that social media increasingly are playing in politics and the economy throughout the world.

**Chapter 14** analyzes health, health care, and disability from both U.S. and global perspectives. Among the topics included are social epidemiology, lifestyle factors influencing health and illness, health care organization in the United States and other nations, social implications of advanced medical technology, and holistic and alternative medicine. This chapter is unique in that it contains a thorough discussion of the sociological perspectives on disability and of social inequalities based on disability. It focuses on the health-related issues of families in global perspective and on the diversity found in U.S. families today.

**Part 5** surveys social dynamics and social change. **Chapter 15** examines population and urbanization, looking at demography, global population change, and the process and consequences of urbanization. Special attention is given to race- and class-based segregation in urban areas and the crisis in health care in central cities. **Chapter 16** concludes the text with an innovative analysis of collective behavior, social movements, and social change. Environmental activism is used as a sustained example to help students grasp the importance of collective behavior and social movements in producing social change.

# Distinctive, Classroom-Tested Features

The following special features are specifically designed to demonstrate the relevance of sociology in our lives, as well as to support students' learning. As the preceding overview of the book's contents shows, these features appear throughout the text, some in every chapter, others in selected chapters.

## Unparalleled Coverage of and Attention to Diversity

From its first edition, I have striven to integrate diversity in numerous ways throughout this book. The individuals portrayed and discussed in each chapter accurately mirror the diversity in society itself. As a result, this text speaks to a wide variety of students and captures their interest by taking into account

their concerns and perspectives. Moreover, the research used includes the best work of classical and established contemporary sociologists—including many white women and people of color—and it weaves an inclusive treatment of *all* people into the examination of sociology in *all* chapters. Therefore, this text helps students consider the significance of the interlocking nature of individuals' class, race, and gender (and, increasingly, age) in all aspects of social life.

## Personal Narratives That Highlight Issues and Serve as Chapter-Length Examples

Authentic first-person commentaries appear in the "Sociology and Everyday Life" features that open each chapter and personalize the issue that unifies the chapter's coverage. These lived experiences provide opportunities for students to examine social life beyond their own experiences and for instructors to systematically incorporate into lectures and discussions an array of interesting and relevant topics that help demonstrate to students the value of applying sociology to their everyday lives.

## Focus on the Relationship Between Sociology and Everyday Life

Each chapter has a brief quiz in the opening "Sociology and Everyday Life" feature that relates the sociological perspective to the pressing social issues presented in the vignette. (Answers are provided on a subsequent page.)

## Emphasis on the Importance of a Global Perspective

The global implications of all topics are examined throughout each chapter and in the "Sociology in Global Perspective" boxes, which highlight our interconnected world and reveal how the sociological imagination extends beyond national borders.

## Focus on Media

A significant benefit of a sociology course is encouraging critical thinking about such things as how the manner in which the media "package" news and entertainment influences our perception of social issues. This edition builds on the "Media" feature and expands coverage to include issues and societal changes related to the use of social media, as mentioned above.

## Census Profiles

The Census Profiles feature highlights current relevant data from the U.S. Census Bureau, providing students with further insight into the United States.

## Effective Study Aids

In addition to basic reading and study aids such as learning objectives, key terms, and a running glossary, *Sociology in Our Times* includes the following pedagogical aids to aid students' mastery of the course's content:

- **Concept Quick Review.** These tables categorize and contrast the major theories or perspectives on the specific topics presented in a chapter.
- **Questions for Critical Thinking.** Each chapter concludes with a set of questions to encourage students to reflect on important issues, to develop their own critical-thinking skills, and to highlight how ideas presented in one chapter often build on those developed previously.
- **Reflect & Analyze Questions.** From activating prior knowledge related to concepts and themes, to highlighting main ideas and reinforcing diverse perspectives, this text's questions encourage students to reflect on issues and to analyze content rather than to memorize and recall course content.
- **End-of-Chapter Summaries in Question-and-Answer Format.** Chapter summaries provide a built-in review for students by reexamining material covered in the chapter in an easy-to-read question-and-answer format to review, highlight, and reinforce the most important concepts and issues discussed in each chapter.

## Comprehensive Supplements Package
### Products for Blended and Online Courses

**MindTap™: The Personal Learning Experience**
*MindTap Sociology for Kendall's Sociology in Our Times: The Essentials,* tenth edition, Powered by Knewton from Cengage Learning, represents a new approach to a highly personalized, online learning platform. MindTap combines all of a student's learning tools—readings, multimedia, activities, and assessments—into a Learning Path that guides the student through the introduction to sociology course. Instructors personalize the experience by customizing the presentation of these learning tools to their students, even seamlessly introducing their own content into the Learning Path. Learn more at www.cengage.com/mindtap.

*MindTap Sociology for Kendall's Sociology in Our Times: The Essentials,* Powered by Knewton from Cengage Learning, is easy to use and saves instructors time by allowing them to

- Customize the course—from tools to text—and make adjustments "on the fly," making it possible to intertwine breaking news into their lessons and incorporate today's teachable moments
- Promote personalization by segmenting course content into movable objects, encourage interactivity, and ensure student engagement
- Integrate multimedia assets, in-context exercises, and supplements—student engagement will increase, leading to better student outcomes
- Track students' use, activities, and comprehension in real time, providing opportunities for early intervention to influence progress and outcomes
- Assess knowledge throughout each section: after readings, in activities, homework, and quizzes
- Automatically grade homework and quizzes

**Course Reader for Sociology.** CourseReader for Sociology allows you to create a fully customized online reader in minutes. You can access a rich collection of thousands of primary and secondary sources, readings, and audio and video selections from multiple disciplines. Each selection includes a descriptive introduction that puts it into context, and every selection is further supported by both critical-thinking and multiple-choice questions designed to reinforce key points. This easy-to-use solution allows you to select exactly the content you need for your courses, and it is loaded with convenient pedagogical features, such as highlighting, printing, note taking, and downloadable MP3 audio files for each reading. You have the freedom to assign and customize individualized content at an affordable price. CourseReader is the perfect complement to any class.

### Resources for Customizing Your Textbook

Cengage Learning is pleased to offer three modules that help you tailor *Sociology in Our Times: The Essentials,* tenth edition, to your course. In addition, you can choose to add your own materials or reorganize the table of contents. Work with your local Cengage Learning consultant to find out more.

***Careers in Sociology* Module.** Written by leading author Joan Ferrante, Northern Kentucky University, the *Careers in Sociology* module offers the most extensive and useful information on careers that is available. This module provides six career tracks, each of which has a "featured employer," a job description, and a letter of recommendation (written by a professor for a sociology student) or application (written by a sociology student). The module also includes résumé-building tips on how to make the most out of being a sociology major and offers specific course suggestions along with the transferable skills gained by taking these courses. As part of Cengage Learning's Add-a-Module Program, *Careers in Sociology* can be purchased separately, bundled, or customized with any of our introductory texts. The modules present topics not typically covered in most introductory texts but often requested by instructors.

*Sociology of Sports* **Module.** The *Sociology of Sports* module, authored by Jerry M. Lewis, Kent State University, examines why sociologists are interested in sports, mass media and sports, popular culture and sports (including feature-length films on sports), sports and religion, drugs and sports, and violence and sports. As part of Cengage Learning's Add-a-Module Program, *Sociology of Sports* can be purchased separately, bundled, or customized with any of our introductory texts. The modules present topics not typically covered in most introductory texts but often requested by instructors.

*Rural Sociology* **Module.** The *Rural Sociology* module, authored by Carol A. Jenkins, Glendale Community College–Arizona, presents the realities of life in rural America. Many people imagine a rural America characterized by farming, similar cultures, and close-knit communities. However, rural Americans and rural communities are extremely diverse—demographically, culturally, socially, economically, and environmentally. The module presents these characteristics of rural life in a comprehensive and accessible format for introductory sociology students. As part of Cengage Learning's Add-a-Module program, *Rural Sociology* can be purchased separately, bundled, or customized with any of our introductory sociology texts. The modules present topics not typically covered in most introductory texts but often requested by instructors.

## Teaching Aids for Instructors

A broad array of teaching aids is available to make course planning faster and easier, giving you more time to focus on your students. All of these resources can be accessed with a single account. Go to **login.cengage.com** to log in.

*Online Instructor's Resource Manual.* Prepared by Josh Packard of Northern Colorado University, this text's Online Instructor's Resource Manual is designed to maximize the effectiveness of your course preparation. Beginning with a list of What's New in Each Chapter and the Key Terms, it offers chapter outlines correlated to student learning objectives, creative lecture and teaching ideas, an extensive chapter outline with learning objectives correlated to headings, student active learning activities, Internet activities, and video suggestions.

**Online Test Bank.** Prepared by Josh Packard of Northern Colorado University, the tenth edition's test bank consists of revised and updated multiple-choice questions and true/false questions for each chapter of the text, along with an answer key and text references for each question. Each multiple-choice item has the question type (fact, concept, or application) indicated. Also included are short-answer and essay questions for each chapter. The author of the Test Bank has also keyed each test question to its related learning objective for the chapter.

**Online PowerPoint® Slides.** Helping you make your lectures more engaging while effectively reaching your more visually oriented students, these Microsoft® PowerPoint® slides present a classroom-ready presentation that includes graphics, key points, student polling questions, image-based Consider This questions, and a Quick Quiz. Prepared by Aurea Osgood of Winona State University, the PowerPoint slides are updated to reflect the content and organization of the new edition of the text.

**The Sociology Video Library Volumes I–IV.** These DVDs drive home the relevance of course topics through short, provocative clips of current and historical events. Perfect for enriching lectures and engaging students in discussion, many of the segments on these volumes have been gathered from BBC Motion Gallery. Ask your Cengage Learning representative for a list of contents.

**AIDS in Africa.** Expand your students' global perspective on HIV/AIDS with this award-winning documentary series on DVD, which focuses on controlling HIV/AIDS in southern Africa. Films focus on caregivers in the faith community; how young people share messages of hope through song and dance; the relationship of HIV/AIDS to gender, poverty, stigma, education, and justice; and the story of two HIV-positive women helping others.

**Cengage Learning Testing Powered by Cognero.** This is a flexible online system that allows you to

- Import, edit, and manipulate test bank content from the *Sociology in Our Times: The Essentials* test bank or elsewhere, including your own favorite test questions
- Create multiple test versions in an instant
- Deliver tests from your LMS, your classroom, or wherever you want

# Acknowledgments

*Sociology in Our Times: The Essentials,* tenth edition, would not have been possible without the insightful critiques of these colleagues, who have reviewed some or all of this text and its supplements or responded to a detailed survey. I extend my profound thanks to each one for engaging with me in the revision process.

Brenda Beach, Athens Technical College

Mark Beaudry, Northern Essex Community College

Jake Cummings, San Diego State University

Andrew Dzurisin, Middlesex County College

Rebecca Fahrlander, University of Nebraska Omaha

Pamela J. Forman, University of Wisconsin–Eau Claire

Malcolm Gold, Malone University

Jessica Halperin, Metropolitan Community College–Kansas City

Arthur Kimmel, Long Island University

Arnold Korotkin, Montclair State University

Chuck Kusselow, River Valley Community College

Elizabeth Lucal, Indiana University South Bend

Susan McCabe, Ohio University

Rafik Mohamed, Clayton State University

Kelly Mosel-Talavera, Texas State University

Rebecca Nees, Middle Georgia State College

Andrea Nichols, St. Louis Community College–Forest Park

Daniel Phillips, Lindsey Wilson College

Daniel Poole, University of Utah

David Rentler, University of Connecticut

Ruby Robinson, Purdue University

Mark Sherry, University of Toledo

Luceal Simon, Wayne State University

Pauli Taylorboyd, University of Wisconsin–Milwaukee

Shawn Trivette, Louisiana Tech University

Alan Turley

Connie Veldink, Everett Community College

Jessie Chatman Williams, Mineral Area College

I would also like to thank again the reviewers who helped throughout the revision of the previous edition: George Ansalone, St. Johns University; Stephanie Bucy, Columbia College; Kevin E. Early, University of Michigan–Dearborn; Obi N. I. Edde, University of Tennessee at Chattanooga; Carol Holdt, Portland State University; A. J. Jacobs, East Carolina University; Joan Luxenburg, University of Central Oklahoma; John McMullen, Frostburg State University; Rebecca Nees, Middle Georgia College; Diane Shinberg, Indiana University of Pennsylvania; and Rachel Stehle, Cuyahoga Community College.

I would also like to thank again the reviewers and survey respondents who helped throughout the revision of the previous editions: Cheryl Albers, Buffalo State University; Annette Allen, Troy University; Mel Barber, Flagler College; Amy Blackstone, University of Maine; Daniel Boudon, Hofstra University; Robin Brownfield, Rutgers University; Erica Burrell, Baton Rouge Community College; Suzi Calliham, Colby Community College; Margaret Choka, Pellissippi State Community College; J. Michael Cruz, Southern Methodist University; Karyn Daniels, Long Beach City College; Maria Flynn, Ocean County College; Mark Gaskill, State University of New York at Oneonta; Jeff Hahn, Mount Union College; Sara E. Hanna, Oakland Community College; Rebecca Hartzog, Milwaukee Area Technical College; Danielle Hedegard, University of Arizona; Jane Hellinghausen, Odessa College; Ursula Hines, Cuyamaca College; Karen Honeycutt, Keene State College; Connie Jamison, Volunteer State Community College; Chris Johnson, University of Louisiana at Monroe; Allison Jones, Ventura College; John C. Kilburn Jr., Texas A&M International University; Rosalind Kopfstein, Western Connecticut State University; Cindy J. Lahar, York County Community College; Jason J. Leiker, Utah State University; Chad Litton, University of Mary; Suzanne L. Maughan, University of Nebraska–Kearney; Nix McCray, Community College of Baltimore County; Christine Monnier, College of DuPage; Efren Padilla, California State University–East Bay; Jessica Pardee, University of Central Florida; Rita Sakitt, Suffolk County Community College; Donald P. Stewart, University of Nevada, Las Vegas; K. J. Stone, Los Angeles Southwest College; Kell Stone, El Camino College; Dawn Tawwater, Austin Community College; Pauli Taylorboyd, MATC and UWM; Robert Turley, College of the Canyons; Daniel Van Dussen, Youngstown State University; Connie Veldink, Everett Community College; Charles M. ViVona, Queensborough Community College; John White, Community College of Rhode Island; and Jessie M. Williams, Mineral Area College.

And for their many valuable contributions to earlier editions, I would also like to thank the following reviewers again: Jan Abu-Shakrah, Portland Community College; Ted Alleman, Penn State University; Ginna Babcock, University of Idaho; Kay Barches, Skyline College; Sampson Lee Blair, University of Oklahoma; Elizabeth Bodien, Northampton Community College; Barbara Bollmann, Community College of Denver; Gayle Gordon Bouzard, Texas State University, San Marcos; Joni Boye-Bearman, Wayne State College; John Brady, Texas Lutheran University; Valerie Brown, Case Western Reserve University; Deborah Carter, Johnson C. Smith University; Margaret Choka, Pellissippi State Community College; Dennis Cole, Owens Community College; Brooks Cowan, University of Vermont; Russell Curtis, University of Houston; Mary Ann Czarnezki, University of Wisconsin–Milwaukee; Lillian Daughaday, Murray State University; Kevin Delaney, Temple University; Judith Dilorio, Indiana University–Purdue University; Harold Dorton, Southwest Texas State University; Kevin Early, Oakland University; Don Ecklund, Lincoln Land Community College; Cindy Epperson, St. Louis Community College at Meramec; Julian Thomas Euell, Ithaca College; Grant Farr, Portland State University; Joe R. Feagin, University of Florida; Nancy Feather, West Virginia University; Kathryn M. Feltey, University of Akron; Leticia Fernandez, University of Texas at El Paso; William Finlay, University of Georgia; Andrea Fontana, UNLV; Mark Foster, Johnson County Community College; Howard Gabennesch, University of Southern Indiana; Richard Gale, University of Oregon; DeAnn K. Gauthier, University of Southwestern Louisiana; Robert B. Girvan, Clarion University; Tanya Gladney, University of Nebraska–Lincoln; Michael Goslin, Tallahassee Community College; Anna Hall, Delgado Community College; Lee Hamilton, New Mexico State University; Gary D. Hampe, University of Wyoming; Carole Hill, Shelton State Community College; Art Jipson, University of Dayton; Richard Kamei, Glendale Community College; William Kelly, University of Texas, Austin; Bud Khleif, University of New Hampshire; Joachim Kibirige, Missouri Western State University; Michael B. Kleiman, University of South Florida; Janet Koenigamen, College of St. Benedict; Rosalind Kopfstein, Western Connecticut State University; Amy Krull, Western Kentucky University; Monica Kumba, Buffalo State University; Yechiel Lahavy, Atlantic Community College; Gregory Leavitt, Idaho State University; Amy Leimer, Tennessee Tech University; Abraham Levine, El Camino College; Diane Levy, University of North Carolina, Wilmington; Stephen Lilley, Sacred Heart University; Michael Lovaglia, University of Iowa; Akbar Mahdo, Ohio Wesleyan University; Rebecca Matthews, University of Iowa; M. Cathey Maze, Franklin University; Jane McCandless, State University of West Georgia; Barbara Miller, Pasadena City College; John Molan, Alabama State University; Betty Morrow, Florida International University; Kelly Mosel-Talavera, Texas State University, San Marcos; Tina Mougouris, San Jacinto College; Mani Pande, Kansas State University; Dale Parent, Southeastern Louisiana University; Linda Petroff, Central Community College; Karen Polite, Harrisburg Area Community College–Lancaster Campus; J. Keith Price, West Texas A&M University; Barbara Richardson, Eastern Michigan University; Lois Rood, Tennessee Tech University; Ellen Rosengarten, Sinclair Community College; Phil Rutledge, University of North Carolina, Charlotte; Joseph Scimecca, George Mason University; Marcia Texler Segal, Indiana University Southeast; Dorothy Smith, State University of New York–Plattsburg; Teresa Sobieszczyk, University of Montana; Jackie Stanfield, Northern Colorado University; Mary Texeira, California State University–San Bernardino; Jamie Thompson, University of Central Oklahoma; Gary Tiedeman, Oregon State University; James Unnever, Radford University; Richard Valencia, Fresno City College; Steven Vasser, Mankato State University; Carl Wahlstrom, Genesee Community College; Henry Walker, Cornell University; Susan Waller, University of Central Oklahoma; Judith Ann Warner, Texas A&M International University; N. Ree Wells, Missouri Southern State University; Jerry Williams, Stephen Austin State; and John Zipp, University of Wisconsin–Milwaukee.

Finally, I deeply appreciate the energy, creativity, and dedication of the many people responsible for the development and production of *Sociology in Our Times: The Essentials.* I wish to thank Cengage Learning's Jennifer Harrison, Lin Gaylord, John Chell, Kara Kindstrom, Julia Catalano, Cheri Palmer, and Christy Frame for their enthusiasm and insights throughout the development of this text. Many other people worked hard on the production of the tenth edition of *Sociology in Our Times: The Essentials,* especially Greg Hubit, Donald Pharr, and Debra Nichols. I am extremely grateful to them.

I invite you to send your comments and suggestions about this book to me in care of:

Cengage Learning
    500 Terry A. Francois Blvd., 2nd Floor
    San Francisco, CA 94158

# SOCIOLOGY IN OUR TIMES

## the essentials

# 1 THE SOCIOLOGICAL PERSPECTIVE AND RESEARCH PROCESS

## LEARNING OBJECTIVES

**LO1** Explain what sociology can contribute to our understanding of social life.

**LO2** Discuss the sociological imagination and explain how we develop a global sociological imagination.

**LO3** Describe the historical context in which sociological thinking developed and state reasons why many early social thinkers were concerned with social order and stability.

**LO4** Identify reasons why many later social thinkers were concerned with social change.

**LO5** Describe key differences in functionalist, conflict, symbolic interactionist, and postmodern perspectives on contemporary social life.

**LO6** Explain why sociological research is necessary and point out how it may challenge our commonsense beliefs about pressing issues such as suicide.

**LO7** Distinguish between quantitative and qualitative research, and identify the steps in each method.

**LO8** Describe survey research and briefly discuss three types of surveys.

**LO9** Compare research methods used in surveys, secondary analysis of existing data, field research, and experiments.

**LO10** Discuss ethical issues in research and identify professional codes that protect research participants.

# The Sociology of Suicide Trends Today

**"Why are you alive?"**

**"You're ugly."**

Rebecca Ann Sedwick

Polaris/Newscom

**"You should die."**

**"Why don't you go kill yourself."**

**"Can u die please?"**

—After more than a year of receiving online bullying comments like these, Rebecca Ann Sedwick, a 12-year-old Tampa, Florida, girl, jumped from a concrete silo tower to her death at an abandoned cement plant. Prior to this terrible event, Rebecca changed her online screen name to "That Dead Girl." After her death, law enforcement officials stated that she was "absolutely terrorized on social media" (Alvarez, 2013), while psychologists pointed out that she had been a victim of the "cool to be cruel" cyber culture (Ng, 2013).

**" . . . go die, evry1 wud be happy."**

**"Drink bleach."**

—London, England, resident Hanna Smith, age 14, repeatedly had hostile comments like these posted about her on ASK.fm prior to her suicide (Dolan and Robinson, 2013).

**S**uicides committed by young people who have been the victims of online bullying deeply touch the lives of their families, friends, and others who have not met them. Although we will never know the full story of what happened to Rebecca Ann Sedwick and the others described, these tragic occurrences bring us to larger sociological questions: Why does anyone commit suicide? Is suicide purely an individual phenomenon, or is it related to our social interactions and the social environment and society in which we live? How have technologies such as smartphones and social media affected our communication—both positively and negatively—with others?

Social media use among teens and college students continues to grow rapidly. We are engulfed by smartphones, tablets, and computers. Facebook, Twitter, YouTube, and other Internet-based social networking sites are taken for granted. We enjoy the positive effects of social media, but the digital age may also produce harmful outcomes, particularly when some people harass others, causing psychological and physical harm and sometimes even bullying them into suicide.

Although suicide may seem like a "downer" for beginning our study of sociology, I have chosen this topic because it is one of the first social issues that early sociologists studied. These thinkers believed that identifying the *social causes* of such behavior sets sociology apart from psychology, philosophy, and other areas of inquiry.

In this chapter we examine how sociological theories and research can help us understand the *social* and *structural* origins of the seemingly *individualistic* act of taking one's own life. We will see how sociological theory and research methods might be used to answer complex questions, and we will wrestle with some of the difficulties that sociologists experience as they study human behavior.

## Putting Social Life into Perspective

**LO1** *Sociology* is the systematic study of human society and social interaction. It is a *systematic* study because sociologists apply

"You think you want to die, but in reality you just want to be saved."

—Hanna Smith posted a picture of this statement written in a spiral notebook to her Facebook page less than 24 hours prior to killing herself (Dolan and Robinson, 2013).

"Goodbye forever my good friends, goodbye, I regret nothing. I have chosen to go with

3 peoples advice and kill myself. I just wish it was faster."

—Prior to his suicide, Bart Palosz, a 15-year-old native of Poland who resided in Greenwich, Connecticut, for seven years, used his Google Plus page to express his suicidal musings during patterns of constant bullying by other boys at his school (Hussey and Leland, 2013).

## How Much Do You Know About Suicide?

**True  False**

T  F  1. Suicide is the number-one cause of injury-related death in the United States.

T  F  2. There is 1 suicide for every 25 attempted suicides annually in the United States.

T  F  3. Suicide rates are lower for young Native Americans (American Indians) and Alaska Natives between the ages of 15 to 34 than for persons in the same age bracket for any other racial–ethnic category in the United States.

T  F  4. In grades 9 through 12, significantly more Latina (Hispanic) female students report that they have attempted suicide in the last year than African American (black) non-Latin female students and white American non-Hispanic female students.

**True  False**

T  F  5. Females complete suicide (take their own life) at a much higher rate than that of males.

T  F  6. Firearms are the most commonly used method of suicide among males and females.

T  F  7. Alcohol intoxication is present in nearly one-fourth of all suicide deaths in the United States.

T  F  8. Government agencies such as the U.S. Centers for Disease Control and Prevention have developed public health programs to prevent suicidal behavior and now emphasize that promoting individual, family, and community connectedness is an important way to prevent such behavior.

Answers on page 6.

both theoretical perspectives and research methods (or orderly approaches) to examinations of social behavior. Sociologists study human societies and their social interactions in order to develop theories of how human behavior is shaped by group life and how, in turn, group life is affected by individuals.

## Why Should You Study Sociology?

Sociology helps us gain a better understanding of our social world and ourselves. It enables us to see how the groups to which we belong and the society in which we live largely shape behavior. A *society* **is a large social grouping that shares the same geographical territory and is subject to the same political authority and dominant cultural expectations,** such as the United States, Mexico, or Nigeria. Examining the world order helps us understand that each of us is affected by *global interdependence*—a relationship in which the lives of all people are intertwined closely and any one nation's problems are part of a larger global problem. Environmental

problems are an example: People throughout the world share the same biosphere. When environmental degradation, such as removing natural resources or polluting the air and water, takes place in one region, it may have an adverse effect on people around the globe.

Individuals can make use of sociology on a more personal level. Sociology enables us to move beyond established ways of thinking, thus allowing us to gain new insights into ourselves and to develop a greater awareness of the connection between our own "world" and that of other people. According to the sociologist Peter Berger (1963: 23), sociological inquiry helps us see that "things are not what they

**sociology** the systematic study of human society and social interaction.

**society** a large social grouping that shares the same geographical territory and is subject to the same political authority and dominant cultural expectations.

# ANSWERS to the Sociology Quiz on Suicide

**1. True.** Suicide became the number-one cause of injury-related death in the United States, surpassing car accidents, between 2000 and 2009 (the latest year for which data are available), when the death rate for suicide increased by 15 percent and decreased by 25 percent for car accidents.

**2. True.** According to the U.S. Centers for Disease Control and Prevention, 1 completed suicide occurs for every 25 persons who attempt suicide in the United States.

**3. False.** Native Americans (First Americans or American Indians) and Alaska Natives have the highest suicide rates per 100,000 in the United States. Among teens and young adults ages 15 to 34, the rate is 27.3 per 100,000, which is 2.4 times higher than the national average for that age group (11.4 per 100,000).

**4. True.** Significantly more Latina (Hispanic) female students (13.5 percent) reported attempting suicide in the last year than African American (black) non-Latin female students and white American non-Hispanic female students (7.9 percent).

**5. False.** Males *complete* suicide at a rate four times that of females. Male suicides account for 78.8 percent of all completed suicides in the United States; however, females are more likely than males to have had suicidal thoughts.

**6. False.** Although firearms are the most commonly used method of suicide among males (56.1 percent), poisoning is the most common method of suicide for females (38 percent).

**7. True.** Alcohol intoxication is present in 24 percent (nearly one-fourth) of all suicide deaths in the United States.

**8. True.** Government organizations such as the U.S. Centers for Disease Control and Prevention have developed these programs and now emphasize that promoting individual, family, and community connectedness is an important way to prevent such behavior.

Sources: Based on U.S. Centers for Disease Control and Prevention, 2012d, 2012e.

seem." Sociology provides new ways of approaching problems and making decisions in everyday life. For this reason, people with knowledge of sociology are employed in a variety of fields that apply sociological insights to everyday life (see Figure 1.1).

Sociology promotes understanding and tolerance by enabling each of us to look beyond intuition, common sense, and our personal experiences. Many of us rely on intuition or common sense gained from personal experience to help us understand our daily lives and other people's behavior. *Commonsense knowledge* guides ordinary conduct in everyday life. However, many commonsense notions are actually myths. A *myth* is a popular but false notion that may be used, either intentionally or unintentionally, to perpetuate certain beliefs or "theories" even in the light of conclusive evidence to the contrary. Before reading on, take the quiz in

| Health and Human Services | Business | Communication | Academia | Law |
|---|---|---|---|---|
| Counseling<br>Education<br>Medicine<br>Nursing<br>Social Work | Advertising<br>Labor Relations<br>Management<br>Marketing | Broadcasting<br>Public Relations<br>Journalism | Anthropology<br>Economics<br>Geography<br>History<br>Information Studies<br>Media Studies/<br>  Communication<br>Political Science<br>Psychology<br>Sociology | Law<br>Criminal Justice |

**FIGURE 1.1** FIELDS THAT USE SOCIAL SCIENCE RESEARCH

In many careers, including jobs in health and human services, business, communication, academia, and law, the ability to analyze social science research is an important need.

**6**

Source: Based on Katzer, Cook, and Crouch, 1991. © Cengage Learning.

the Sociology and Everyday Life box at the opening of the chapter.

By contrast, sociologists strive to use scientific standards, not popular myths or hearsay, in studying society and social interaction. They use systematic research techniques and are accountable to the scientific community for their methods and the presentation of their findings. Whereas some sociologists argue that sociology must be completely value free—free from distorting subjective (personal or emotional) bias—others do not think that total objectivity is an attainable or desirable goal when studying human behavior. However, all sociologists attempt to discover patterns or commonalities in human behavior. When they study suicide, for example, they look for recurring patterns of behavior even though *individual* people usually commit the acts and *other individuals* suffer as a result of these actions.

Consequently, sociologists seek out the multiple causes and effects of suicide or other social issues. They analyze the impact of the problem not only from the standpoint of the people directly involved but also from the standpoint of the effects of such behavior on all people.

## The Sociological Imagination

**LO2** Sociological reasoning is often referred to as the *sociological imagination—* **the ability to see the relationship between individual experiences and the larger society** (Mills, 1959b). The sociological imagination is important to each of us because having this awareness enables us to understand the link between our personal experiences and the social contexts in which they occur. Each of us lives in a society, and we live out a biography within some historical setting. Throughout our lives, we contribute to the shaping of society and to its history, even as we are made by society and the historical events that take place during our lifetime. The sociological imagination enables us to grasp the relationship between history at the societal level and our own biography at the individual level. It also helps us distinguish between personal troubles and social (or public) issues. C. Wright Mills (1959b) made this distinction between the two: *Personal troubles* are private problems that affect individuals and the networks of people with whom they associate regularly. As a result, individuals within their immediate social settings must solve those problems. For example, one person being unemployed may be a personal trouble. *Public issues* are problems that affect large numbers of people and often require solutions at the societal level. Widespread unemployment as a

result of economic changes such as plant closings is an example of a public issue. The sociological imagination helps us place seemingly personal troubles, such as losing one's job or feeling like committing suicide, into a larger social context, where we can distinguish whether and how personal troubles may be related to public issues.

**Suicide as a Personal Trouble** Often our individual experiences are largely beyond our own control. They are determined by society as a whole—by its historical development and its organization. However, in everyday life we frequently do not define personal experiences in these terms. If a person commits suicide, many people consider it to be the result of that individual's own personal problems.

**Suicide as a Public Issue** If we use the sociological imagination to look at the problem of suicide, we can see that it is often a public issue—a societal problem. Early sociologist Emile Durkheim refused to accept commonsense explanations of suicide. In what is probably the first sociological study to use scientific research methods, he related suicide to the issue of cohesiveness (or lack of it) in society instead of viewing suicide as an isolated act that could be understood only by studying individual personalities or inherited tendencies. In *Suicide* (1964b/1897), Durkheim documented his contention that a high suicide rate indicated large-scale societal problems.

## The Importance of a Global Sociological Imagination

Although existing sociological theory and research provide the foundation for sociological thinking, we must reach beyond studies that have focused primarily on the United States to develop a more comprehensive *global* approach for the future (see Figure 1.2). In the twenty-first century, we face important challenges in a rapidly changing nation and world.

The world's **high-income countries** are **nations with highly industrialized economies;**

**sociological imagination** C. Wright Mills's term for the ability to see the relationship between individual experiences and the larger society.

**high-income countries** nations with highly industrialized economies; technologically advanced industrial, administrative, and service occupations; and relatively high levels of national and personal income.

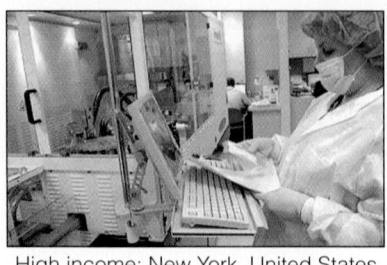
High income: New York, United States

Low income: Congo

Middle income: China

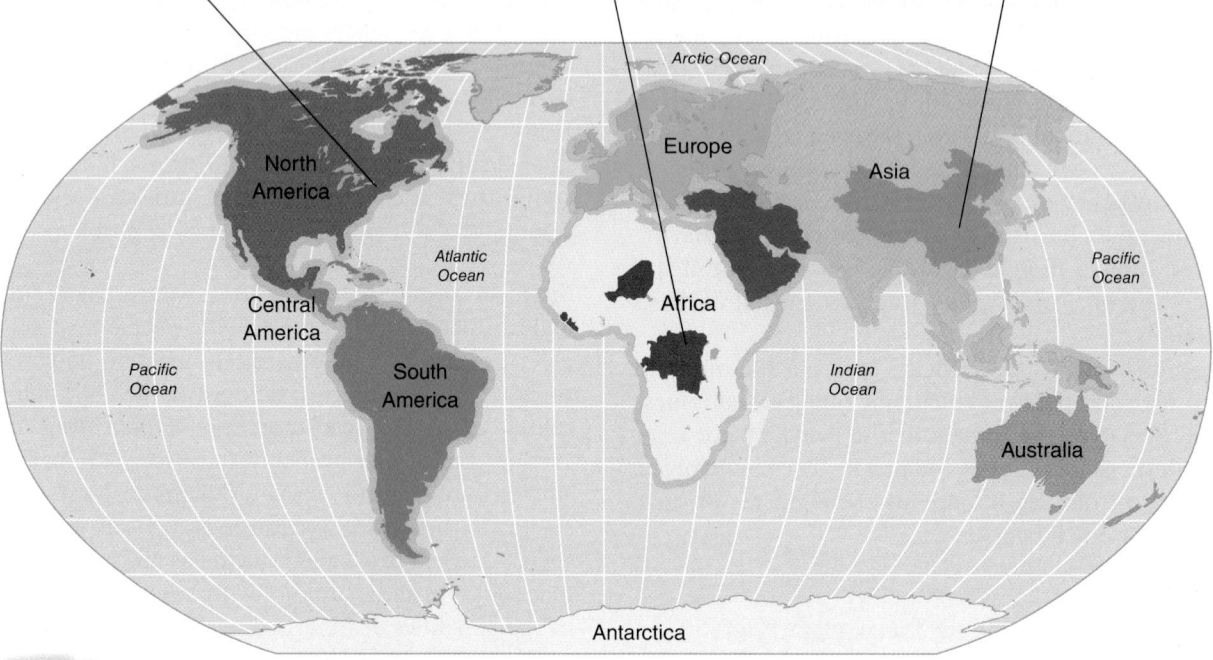

**FIGURE 1.2** THE WORLD'S ECONOMIES IN THE EARLY TWENTY-FIRST CENTURY
High-income, middle-income, and low-income countries.

© Cengage Learning. *Photos, left to right:* John Berry/Syracuse Newspapers/The Image Works; Gable/Alamy; philipbigg/Alamy.

**8**

**technologically advanced industrial, administrative, and service occupations; and relatively high levels of national and personal income.** Examples include the United States, Canada, Australia, New Zealand, Japan, and the countries of Western Europe.

As compared with other nations of the world, many high-income nations have a high standard of living and a lower death rate because of advances in nutrition and medical technology. However, everyone living in a so-called high-income country does not necessarily have these advantages. In contrast, *middle-income countries* **are nations with industrializing economies, particularly in urban areas, and moderate levels of national and personal income.** Examples of middle-income countries include the nations of Eastern Europe and many Latin American countries.

*Low-income countries* **are primarily agrarian nations with little industrialization and low levels of national and personal income.** Examples of low-income countries include many of the nations of Africa and Asia, particularly the People's Republic of China and India, where people typically work the land and are among the poorest in the world. However, generalizations are difficult to make because there are wide differences in income and standards of living within many nations (see Chapter 8, "Global Stratification"). Throughout this text, we will continue to develop our sociological imaginations by examining social life here and in other nations. The future of our nation is deeply intertwined with the future of all other nations of the world on economic, political, environmental, and humanitarian levels.

Whatever your race/ethnicity, class, sex, or age, are you able to include in your thinking the perspectives of people who are quite dissimilar in experiences and points of view? Before you answer this question, a few definitions are in order. *Race* is a term used by many people to specify groups of people distinguished by physical characteristics such as skin color. However, there are no "pure" racial types, and the concept of race is considered by most sociologists to be a social construction that people use to justify existing social inequalities. (When we say that something is a "social construction," we mean that race, ethnicity, class, and gender do not really mean anything apart from the meanings that people in societies give them.) When societies use these categories, the groupings privilege some categories of people over others. Unfortunately, the result is that many people act and respond as if these categories were a social reality rather than a construction.

*Ethnicity* refers to the cultural heritage or identity of a group and is based on factors such as language or country of origin. *Class* is the relative location of a person or group within the larger society, based on wealth, power, prestige, or other valued resources. *Sex* refers to the biological and anatomical differences between females and males. By contrast, *gender* refers to the meanings, beliefs, and practices associated with sex differences, referred to as *femininity* and *masculinity*.

# The Development of Sociological Thinking

**L03** Throughout history, social philosophers and religious authorities have made countless observations about human behavior. However, the idea of observing how people lived, finding out what they thought, and doing so in a systematic manner that could be verified did not take hold until the nineteenth century and the social upheaval brought about by industrialization and urbanization.

**Industrialization is the process by which societies are transformed from dependence on agriculture and handmade products to an emphasis on manufacturing and related industries.** This process occurred first during the Industrial Revolution in Britain between 1760 and 1850, and was soon repeated throughout Western Europe. By the mid-nineteenth century, industrialization was well under way in the United States (Figure 1.3). Massive economic, technological, and social changes occurred as machine technology and the factory system shifted the economic base of these nations from agriculture to manufacturing: textiles, iron smelting, and related industries. Many people who had labored on the land were forced to leave their tightly knit rural communities and sacrifice well-defined social relationships to seek employment as factory workers in the emerging cities, which became the centers of industrial work.

**Urbanization is the process by which an increasing proportion of a population lives in cities rather than in rural areas.** Although cities existed long before the Industrial Revolution, the development of the factory system led to a rapid

Everett Collection/Newscom

**FIGURE 1.3** As the Industrial Revolution swept through the United States beginning in the nineteenth century, children being employed in factories became increasingly common. Soon social thinkers began to explore such new social problems brought about by industrialization.

increase in both the number of cities and the size of their populations. People from very diverse backgrounds worked together in the same factory. At the same time, many people shifted from being *producers* to being *consumers*. For example, families living in the cities had to buy food with their wages because they could no longer grow their own crops to eat or to barter for other resources. Similarly, people had to pay rent for their lodging because they could no longer exchange their services for shelter.

These living and working conditions led to the development of new social problems: inadequate housing, crowding, unsanitary conditions, poverty, pollution, and crime. Wages were so low that entire

**middle-income countries** nations with industrializing economies, particularly in urban areas, and moderate levels of national and personal income.

**low-income countries** primarily agrarian nations with little industrialization and low levels of national and personal income.

**industrialization** the process by which societies are transformed from dependence on agriculture and handmade products to an emphasis on manufacturing and related industries.

**urbanization** the process by which an increasing proportion of a population lives in cities rather than in rural areas.

families—including very young children—were forced to work, often under hazardous conditions and with no job security. As these conditions became more visible, a new breed of social thinkers tried to understand why and how society was changing.

## The Development of Modern Sociology

At the same time that urban problems were growing worse, natural scientists had been using reason, or rational thinking, to discover the laws of physics and the movement of the planets. Social thinkers started to believe that by applying the methods developed by the natural sciences, they might discover the laws of human behavior and apply these laws to solve social problems. Historically, the time was ripe because the Age of Enlightenment had produced a belief in reason and humanity's ability to perfect itself.

### Early Thinkers: A Concern with Social Order and Stability

Early social thinkers—such as Auguste Comte, Harriet Martineau, Herbert Spencer, and Emile Durkheim—were interested in analyzing social order and stability, and many of their ideas have had a dramatic and long-lasting influence on modern sociology.

**Auguste Comte** The French philosopher Auguste Comte (1798–1857) coined the term *sociology* from the Latin *socius* ("social, being with others") and the Greek *logos* ("study of") to describe a new science that would engage in the study of society (Figure 1.4). Even though he never actually conducted sociological research, Comte is considered by some to be the "founder of sociology." Comte's theory that societies contain *social statics* (forces for social order and stability) and *social dynamics* (forces for conflict and change) continues to be used in contemporary sociology.

Comte stressed that the methods of the natural sciences should be applied to the objective study of society. He sought to unlock the secrets of society so that intellectuals like him could become the new secular (as contrasted with religious) "high priests" of society. For Comte, the best policies involved order and authority. He envisioned that a new consensus would emerge on social issues and that the new science of sociology would play a significant part in the reorganization of society.

**FIGURE 1.4** August Comte

Comte's philosophy became known as *positivism*—**a belief that the world can best be understood through scientific inquiry.** He believed that positivism had two dimensions: (1) methodological—the application of scientific knowledge to both physical and social phenomena—and (2) social and political—the use of such knowledge to predict the likely results of different policies so that the best one could be chosen.

Social analysts have praised Comte for his advocacy of sociology and his insights regarding linkages between the social structural elements of society (such as family, religion, and government) and social thinking in specific historical periods. However, a number of contemporary sociologists argue that Comte contributed to an overemphasis on the "natural science model" and focused on the experiences of a privileged few, to the exclusion of all others.

**Harriet Martineau** Comte's works were made more accessible for a wide variety of scholars through the efforts of the British sociologist Harriet Martineau (1802–1876) (Figure 1.5). Until recently, Martineau received no recognition in the field of sociology, partly because she was a woman in a male-dominated discipline and society. Not only did she translate and condense Comte's works, but

**FIGURE 1.5** Harriet Martineau

**FIGURE 1.6** Herbert Spencer

she was also an active sociologist in her own right. Martineau studied the social customs of Britain and the United States, analyzing the consequences of industrialization and capitalism. In *Society in America* (1962/1837), she examined religion, politics, child rearing, slavery, and immigration, paying special attention to social distinctions based on class, race, and gender. Her works explore the status of women, children, and "sufferers" (persons who are considered to be criminal, mentally ill, handicapped, poor, or alcoholic).

Martineau advocated racial and gender equality. She was also committed to creating a science of society that would be grounded in empirical observations and widely accessible to people. She argued that sociologists should be impartial in their assessment of society but that it is entirely appropriate to compare the existing state of society with the principles on which it was founded. Martineau believed that a better society would emerge if women and men were treated equally, enlightened reform occurred, and cooperation existed among people in all social classes (but led by the middle class).

**Herbert Spencer** Unlike Comte, who was strongly influenced by the upheavals of the French Revolution, the British social theorist Herbert Spencer (1820–1903) was born in a more peaceful and optimistic period in his country's history (Figure 1.6). Spencer's major contribution to sociology was an evolutionary perspective on social order and social change. Evolutionary theory helps to explain how organic and/or social change occurs in societies. According to Spencer's Theory of General Evolution, society, like a biological organism, has various interdependent parts (such as the family, the economy, and the government) that work to ensure the stability and survival of the entire society.

Spencer believed that societies develop through a process of "struggle" (for existence) and "fitness" (for survival), which he referred to as the "survival of the fittest." Because this phrase is often attributed to Charles Darwin, Spencer's view of society is known as *social Darwinism*—**the belief that those species of animals, including human beings, best adapted to their environment survive and prosper, whereas those poorly adapted die out.** Spencer equated this process of *natural selection* with progress because only the "fittest" members of society would survive the competition.

Critics believe that Spencer's ideas are flawed because societies are not the same as biological systems; people are able to create and transform the environment in which they live. Moreover, the notion of the survival of the fittest can easily

**positivism** a belief that the world can best be understood through scientific inquiry.

**social Darwinism** Herbert Spencer's belief that those species of animals, including human beings, best adapted to their environment survive and prosper, whereas those poorly adapted die out.

**11**

# Sociology WORKS!

## Durkheim's Classical Study of Suicide and Young People in India Today

The bond attaching [people] to life slackens because the bond which attaches [them] to society is itself slack.

—Emile Durkheim, *Suicide* (1964b/1897)

Although this statement describes social conditions accompanying the high rates of suicide found in late-nineteenth-century France, Durkheim's words ring true today as we look at contemporary suicide rates for young people in cities such as New Delhi, India. Suicide rates in India are highest in the 15–29 age category and are especially great among those living in the wealthier and more-educated regions of the nation (NDTV.com, 2012; *Lancet*, 2012).

Doesn't this seem unlikely? Many people think rural farmers facing poor harvests and high debt would have the highest risk of suicide; however, this has not proven true in India. At first glance, we might think that economic success and a good education would provide insurance against suicide because of the greater happiness and job satisfaction among individuals in cities such as New Delhi, as these individuals have gained new opportunities and higher salaries in recent years. However, this economic boom—including the more-open markets of India in the past twenty years—has not only created new opportunities for people; these changes have also contributed to rapid urbanization and weakened social ties. The result? Intensified job anxiety, higher expectations, and more pressure for individual achievement. Social bonds have been weakened or dissolved as people move away from their families and their community; ironically, newer technologies such as cell phones and social networking sites have contributed to the breakdown of traditional family units as communication has become more impersonal and fragmented.

In addition, life in the cities moves at a much faster pace than in the rural areas, and many individuals experience loneliness, sleep disorders, family discord, and major health risks such as heart disease and depression. In the words of Ramachandra Guha (2004), a social historian residing in India, Durkheim's sociology of suicide remains highly relevant to finding new answers to this challenging problem: "The rash of suicides in city and village is a qualitatively new development in our history. We sense that tragedies are as much social as they are individual. But we know very little of what lies behind them. What we now await, in sum, is an Indian Durkheim."

### reflect & analyze

**How does sociology help us to examine seemingly private acts such as suicide within a larger social context? Why are some people more inclined to commit suicide if they are not part of a strong social fabric and have, at the same time, high job anxiety and intensive pressure to achieve?**

---

be used to justify class, racial–ethnic, and gender inequalities.

**Emile Durkheim** French sociologist Emile Durkheim (1858–1917) stressed that people are the product of their social environment and that behavior cannot be understood fully in terms of *individual* biological and psychological traits. He believed that the limits of human potential are *socially* based, not *biologically* based.

In his work *The Rules of Sociological Method* (1964a/1895), Durkheim set forth one of his most important contributions to sociology: the idea that societies are built on social facts. **Social facts are patterned ways of acting, thinking, and feeling that exist outside any one individual but that exert social control over each person.** Durkheim believed that social facts must be explained by other social facts—by reference to the social structure rather than to individual attributes.

Durkheim observed that rapid social change and a more specialized division of labor produce *strains* in society. These strains lead to a breakdown in traditional organization, values, and authority and to a dramatic increase in *anomie*—**a condition in which social control becomes ineffective as a result of the loss of shared values and of a sense of purpose in society.** According to Durkheim, anomie is most likely to occur during a period of rapid social change. In *Suicide* (1964b/1897), he explored the relationship between anomic social conditions and suicide, a concept that remains important in the twenty-first century (see "Sociology Works!").

Durkheim's contributions are so significant that he is considered to be one of the crucial figures in the development of sociology as an academic area of study. He is one of the founding figures in the functionalist theoretical tradition, but he also made important contributions to other perspectives, particularly symbolic interactionism. Later in this chapter, we look at these theoretical approaches.

Although critics acknowledge Durkheim's important contributions, some argue that his emphasis on societal stability, or the "problem of order"—how society can establish and maintain social stability and cohesiveness—obscured the *subjective meanings* that individuals give to religion, work, and suicide. From this view, overemphasis on *structure* and the determining power of "society" resulted in a corresponding neglect of *agency* (the beliefs and actions of the actors involved) in much of Durkheim's theorizing.

## Differing Views on the Status Quo: Stability or Change?

**LO4** Together with Karl Marx, Max Weber, and Georg Simmel, Durkheim established the direction of modern sociology. We will look first at Marx's and Weber's divergent thoughts about conflict and social change in societies and then at Georg Simmel's microlevel analysis of society.

**Karl Marx** In sharp contrast to Durkheim's focus on the stability of society, German economist and philosopher Karl Marx (1818–1883) stressed that history is a continuous clash between conflicting ideas and forces. He believed that conflict—especially class conflict—is necessary in order to produce social change and a better society. For Marx, the most important changes are economic. He concluded that the capitalist economic system was responsible for the overwhelming poverty that he observed in London at the beginning of the Industrial Revolution (Marx and Engels, 1967/1848).

In the Marxian framework, *class conflict* is the struggle between the capitalist class and the working class. The capitalist class, or *bourgeoisie*, is those who own and control the means of production—the tools, land, factories, and money for investment that form the economic basis of a society. The working class, or *proletariat,* is those who must sell their labor because they have no other means to earn a livelihood. From Marx's viewpoint, the capitalist class controls and exploits the masses of struggling workers by paying less than the value of their labor. This exploitation results in workers' *alienation*—a feeling of powerlessness and estrangement from other people and from themselves. Marx predicted that the working class would become aware of its exploitation, overthrow the capitalists, and establish a free and classless society.

Marx is regarded as one of the most profound sociological thinkers; however, his social and economic analyses have also inspired heated debates among generations of social scientists. Central to his view was the belief that society should not just be studied but should also be changed, because the status quo involved the oppression of most of the population by a small group of wealthy people. Those who believe that sociology should be value free (see below) are uncomfortable with Marx's advocacy of what some perceive to be radical social change. As well, scholars who examine society through the lens of race, gender, and class believe that his analysis places too much emphasis on class relations.

**Max Weber** German social scientist Max Weber (pronounced VAY-ber) (1864–1920) was also concerned about the changes brought about by the Industrial Revolution. Although he disagreed with Marx's idea that economics is *the* central force in social change, Weber acknowledged that economic interests are important in shaping human action. Even so, he thought that economic systems were heavily influenced by other factors in a society.

Unlike many early analysts who believed that values could not be separated from the research process, Weber emphasized that sociology should be *value free*—research should be done scientifically, excluding the researcher's personal values and economic interests. However, Weber realized that social behavior cannot be analyzed by purely objective criteria. Although he recognized that sociologists cannot be totally value free, Weber stressed that they should employ *verstehen* (German for "understanding" or "insight") to gain the ability to see the world as others see it. In contemporary sociology, Weber's idea is incorporated into the concept of the sociological imagination (discussed earlier in this chapter).

Weber was also concerned that large-scale organizations (bureaucracies) were becoming increasingly oriented toward routine administration and a specialized division of labor, which he believed were destructive to human vitality and freedom. According to Weber, rational bureaucracy, rather than class struggle, is the most significant factor in determining the social relations between people in industrial societies. From this view, bureaucratic domination can be used to maintain powerful (capitalist) interests in society. As discussed in Chapter 5 ("Groups and Organizations"), Weber's work on bureaucracy has had a far-reaching impact.

**social facts** Emile Durkheim's term for patterned ways of acting, thinking, and feeling that exist outside any one individual but that exert social control over each person.

**anomie** Emile Durkheim's term for a condition in which social control becomes ineffective as a result of the loss of shared values and of a sense of purpose in society.

**FIGURE 1.7** Georg Simmel

the dynamics of communication and the overall interaction process. Simmel also developed *formal sociology,* an approach that focuses attention on the universal social forms that underlie social interaction. He referred to these forms as the "geometry of social life."

Like the other social thinkers of his day, Simmel analyzed the impact of industrialization and urbanization on people's lives. He concluded that class conflict was becoming more pronounced in modern industrial societies. He also linked the increase in individualism, as opposed to concern for the group, to the fact that people now had many cross-cutting "social spheres"—membership in a number of organizations and voluntary associations—rather than the singular community ties of the past.

Simmel's contributions to sociology are significant. He wrote more than thirty books and numerous essays on diverse topics, leading some critics to state that his work is fragmentary and piecemeal. However, his thinking has influenced a wide array of sociologists, including the members of the "Chicago School" in the United States.

## The Beginnings of Sociology in the United States

From Western Europe, sociology spread in the 1890s to the United States, where it thrived as a result of the intellectual climate and the rapid rate of social change. The first departments of sociology in the United States were located at the University of

Weber also provided important insights on the process of rationalization, bureaucracy, religion, and many other topics. In his writings, Weber was more aware of women's issues than many of the scholars of his day. Perhaps his awareness at least partially resulted from the fact that his wife, Marianne Weber, was an important figure in the women's movement in Germany.

**Georg Simmel** At about the same time that Durkheim was developing the field of sociology in France, the German sociologist Georg Simmel (pronounced ZIM-mel) (1858–1918) was theorizing about the importance of social change in his own country and elsewhere (Figure 1.7). Simmel was also focusing on how society is a web of patterned interactions among people (Figure 1.8). In *The Sociology of Georg Simmel* (1950/1902–1917), he described how social interactions are different based on the size of the social group. According to Simmel, interaction patterns differ between a *dyad* (a social group with two members) and a *triad* (a group with three members) because the presence of an additional person often changes

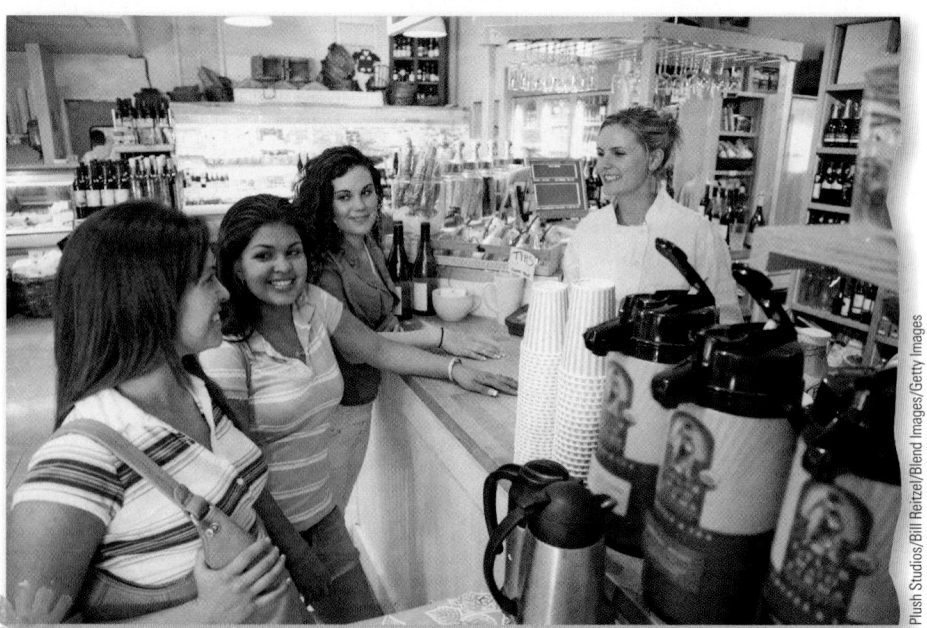

**FIGURE 1.8** According to the sociologist Georg Simmel, society is a web of patterned interactions among people. If we focus on the behavior of individuals only, we miss the underlying forms that make up the "geometry of social life."

Chicago and at Atlanta University, then an African American school.

**The Chicago School** The first department of sociology in the United States was established at the University of Chicago, where the faculty was instrumental in starting the American Sociological Society (now known as the American Sociological Association). Robert E. Park (1864–1944), a member of the Chicago faculty, asserted that urbanization has a disintegrating influence on social life by producing an increase in the crime rate and in racial and class antagonisms that contribute to the segregation and isolation of neighborhoods (Ross, 1991). George Herbert Mead (1863–1931), another member of the faculty at Chicago, founded the symbolic interaction perspective, which is discussed later in this chapter.

**Jane Addams** Jane Addams (1860–1935) is one of the best-known early women sociologists in the United States because she founded Hull House, one of the most famous settlement houses, in an impoverished area of Chicago (Figure 1.9). Throughout her career, she was actively engaged in sociological endeavors: She lectured at numerous colleges, was a charter member of the American Sociological Society, and published a number of articles and books. Addams was one of the authors of *Hull-House Maps and Papers,* a ground-breaking book that used a methodological technique employed by sociologists for the next forty years. She was also awarded a Nobel Prize for her assistance to the underprivileged.

FIGURE 1.10 W. E. B. Du Bois

Library of Congress Prints and Photographs Division [LC-DIG-ggbain-07435]

**W. E. B. Du Bois and Atlanta University** The second department of sociology in the United States was founded by W. E. B. Du Bois (1868–1963) at Atlanta University (Figure 1.10). He created a laboratory of sociology, instituted a program of systematic research, founded and conducted regular sociological conferences on research, founded two journals, and established a record of valuable publications. His classic work, *The Philadelphia Negro: A Social Study* (1967/1899), was based on his research into Philadelphia's African American community and stressed the strengths and weaknesses of a community wrestling with overwhelming social problems. Du Bois was one of the first scholars to note that a dual heritage creates conflict for people of color. He called this duality *double-consciousness*—the identity conflict of being both a black and an American. Du Bois pointed out that although people in this country espouse such values as democracy, freedom, and equality, they also accept racism and group discrimination. African Americans are the victims of these conflicting values and the actions that result from them.

AP Images

FIGURE 1.9 Jane Addams

CHAPTER 1 | THE SOCIOLOGICAL PERSPECTIVE AND RESEARCH PROCESS

15

# Contemporary Theoretical Perspectives

**L05** Given the many and varied ideas and trends that influenced the development of sociology, how do contemporary sociologists view society? Some see it as basically a stable and ongoing entity; others view it in terms of many groups competing for scarce resources; still others describe it based on the everyday, routine interactions among individuals. Each of these views represents a method of examining the same phenomena. Each is based on general ideas about how social life is organized and represents an effort to link specific observations in a meaningful way. Each uses a *theory*—**a set of logically interrelated statements that attempts to describe, explain, and (occasionally) predict social events.** Each theory helps interpret reality in a distinct way by providing a framework in which observations may be logically ordered. Sociologists refer to this theoretical framework as a *perspective*—an overall approach to or viewpoint on some subject. Three major theoretical perspectives have been predominant in U.S. sociology: the functionalist, conflict, and symbolic interactionist perspectives. Other perspectives, such as postmodernism, have emerged and gained acceptance among some social thinkers more recently. Before turning to the specifics of these perspectives, we should note that some theorists and theories do not fit neatly into any of these perspectives.

**FIGURE 1.11** Robert K. Merton

## Functionalist Perspectives

Also known as *functionalism* and *structural functionalism,* **functionalist perspectives are based on the assumption that society is a stable, orderly system.** This stable system is characterized by *societal consensus,* whereby the majority of members share a common set of values, beliefs, and behavioral expectations. According to this perspective, a society is composed of interrelated parts, each of which serves a function and (ideally) contributes to the overall stability of the society. Societies develop social structures, or institutions, that persist because they play a part in helping society survive. These institutions include the family, education, government, religion, and the economy. If anything adverse happens to one of these institutions or parts, all other parts are affected, and the system no longer functions properly.

### Talcott Parsons and Robert K. Merton

Talcott Parsons (1902–1979), perhaps the most influential contemporary advocate of the functionalist perspective, stressed that all societies must meet social needs in order to survive. Parsons (1955) suggested, for example, that a division of labor (distinct, specialized functions) between husband and wife is essential for family stability and social order. The husband/father performs the *instrumental tasks,* which involve leadership and decision-making responsibilities in the home and employment outside the home to support the family. The wife/mother is responsible for the *expressive tasks,* including housework, caring for the children, and providing emotional support for the entire family. Parsons believed that other institutions, including school, church, and government, must function to assist the family and that all institutions must work together to preserve the system over time (Parsons, 1955).

Functionalism was refined further by Robert K. Merton (1910–2003), who distinguished between manifest and latent functions of social institutions (Figure 1.11). **Manifest functions are intended and/or overtly recognized by the participants in a social unit.** In contrast, **latent functions are unintended functions that are hidden and remain unacknowledged by participants** (Figure 1.12). For example, a manifest function of education is the transmission of knowledge and skills from one generation to the next; a latent function is the establishment of social relations and networks. Merton noted that all features of a social system may not be functional at all times; *dysfunctions* are the undesirable consequences of any element of a society. A dysfunction of education in the United States is the perpetuation of gender, racial, and class inequalities. Such dysfunctions may threaten the capacity of a society to adapt and survive.

**FIGURE 1.12** Shopping malls are a reflection of a consumer society. A manifest function of a shopping mall is to sell goods and services to shoppers; however, a latent function may be to provide a communal area in which people can visit friends and enjoy an event.

St Petersburg Times/Zumapress/Newscom

**Applying a Functionalist Perspective to Suicide** Functionalism emphasizes the importance to a society of shared moral values and strong social bonds. It also highlights the significance of social support from others. When rapid social change or other disruptive conditions occur, moral values may erode, and people may become more uncertain about how to act and about whether or not their life has meaning. Social disruption and war are events that produce such feelings of anomie and suicidal ideation—suicidal thoughts or an unusual preoccupation with suicide.

One study that examined the functions of social support in reducing or preventing suicidal ideation in Air Force personnel during U.S. combat operations in Iraq and Afghanistan identified various forms of social support and found that all were not equally important protective factors against suicidal thoughts or actions on the part of military personnel. One form of social support is belonging, or companionship support, where we feel like we belong or fit in with others. Tangible support involves material issues, such as having someone willing to lend money or provide assistance with specific tasks. Appraisal support refers to someone providing useful information that helps us evaluate our situation and provide emotional validation for our thoughts and feelings. Esteem support involves having other people show a concern for our well-being or express

confidence in us and our ability to overcome the problems we face. Of these four types of social support, the researchers found that esteem support was the most important factor in whether the Air Force personnel in their study had experienced severe suicidal ideation (Bryan and Hernandez, 2013).

## Conflict Perspectives

According to *conflict perspectives,* **groups in society are engaged in a continuous power struggle for control of scarce resources.** Conflict may take the

**theory** a set of logically interrelated statements that attempts to describe, explain, and (occasionally) predict social events.

**functionalist perspectives** the sociological approach that views society as a stable, orderly system.

**manifest functions** functions that are intended and/or overtly recognized by the participants in a social unit.

**latent functions** unintended functions that are hidden and remain unacknowledged by participants.

**conflict perspectives** the sociological approach that views groups in society as engaged in a continuous power struggle for control of scarce resources.

**FIGURE 1.13** As one of the wealthiest and most-beloved entertainers in the world, Oprah Winfrey is an example of Max Weber's concept of prestige—a positive estimate of honor. In 2011 Ms. Winfrey moved beyond her very popular television show and started a cable channel, OWN.

form of politics, litigation, negotiations, or family discussions about financial matters. Simmel, Marx, and Weber contributed significantly to this perspective by focusing on the inevitability of clashes between social groups. Today, advocates of the conflict perspective view social life as a continuous power struggle among competing social groups.

**Max Weber and C. Wright Mills** As previously discussed, Karl Marx focused on the exploitation and oppression of the proletariat by the bourgeoisie. Max Weber recognized the importance of economic conditions in producing inequality and conflict in society, but he added *power* and *prestige* as other sources of inequality. Weber (1968/1922) defined *power* as the ability of a person within a social relationship to carry out his or her own will despite resistance from others, and *prestige* as a positive or negative social estimation of honor (Weber, 1968/1922) (Figure 1.13).

C. Wright Mills (1916–1962), a key figure in the development of contemporary conflict theory, encouraged sociologists to get involved in social reform. Mills encouraged everyone to look beneath everyday events in order to observe the major resource and power inequalities that exist in society. He believed that the most important decisions in the United States are made largely behind the scenes by the *power elite*—a small clique of top corporate, political, and military officials. Mills's power elite theory is discussed in Chapter 13 ("Politics and the Economy in Global Perspective").

The conflict perspective is not one unified theory but one with several branches. One branch is the neo-Marxist approach, which views struggle between the classes as inevitable and as a prime source of social change. A second branch focuses on racial–ethnic inequalities and the continued exploitation of members of some racial–ethnic groups. A third branch is the feminist perspective, which focuses on gender issues.

**The Feminist Approach** A feminist theoretical approach (or "feminism") directs attention to women's experiences and the importance of gender as an element of social structure. This approach is based on a belief in the equality of women and men and the idea that all people should be equally valued and have equal rights. According to feminist theorists, we live in a *patriarchy,* a system in which men dominate women and in which things that are considered to be "male" or "masculine" are more highly valued than those considered to be "female" or "feminine." The feminist perspective assumes that gender is socially created and that change is essential in order for people to achieve their human potential without limits based on gender. Some feminists argue that women's subordination can end only after the patriarchal system becomes obsolete. However, feminism is not one single, unified approach; there are several feminist perspectives, which are discussed in Chapter 10 ("Sex, Gender, and Sexuality").

**Applying Conflict Perspectives to Suicide**
How might we use a conflict approach to explain suicide?

**Social Class** Although many other factors may be present, social class pressures may affect rates of suicide among young people from lower-income families when they perceive that they have few educational or employment opportunities and little hope for the future. However, class-based inequality alone cannot explain suicides among young people.

**Gender** In North America, females are more likely to *attempt* suicide, whereas males are more likely to actually take their own life. Despite the fact that women's suicidal behavior has traditionally been attributed to problems in their interpersonal relationships, feminist analysts believe that we must examine social structural pressures on young women and how these may contribute to their behavior—for example, cultural assumptions about women and what their multiple roles should be in the family, education, and the workplace.

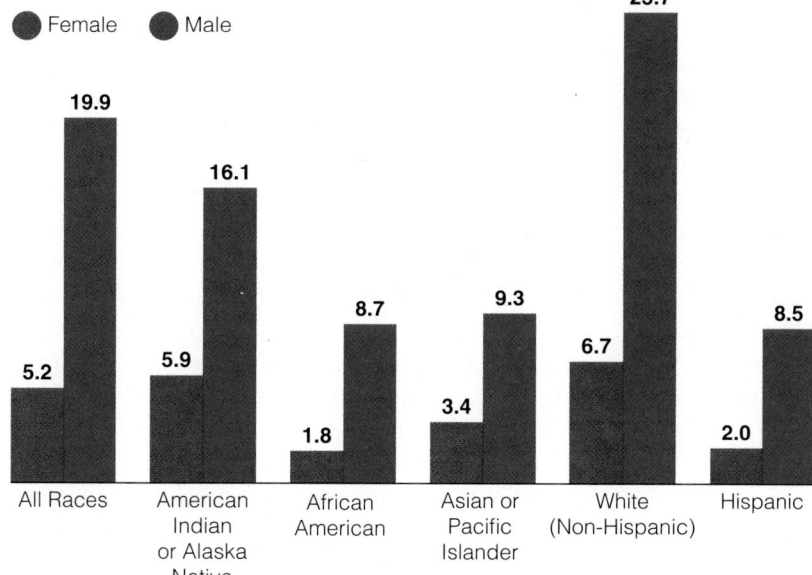

● Female ● Male

**FIGURE 1.14  SUICIDE RATES BY RACE AND SEX**
Rates are for U.S. suicides and indicate the number of deaths by suicide for every 100,000 people in each category for 2010 (the latest year for which data are available).

**Race/Ethnicity**  Racial and ethnic subordination may be a factor in some suicides. (Figure 1.14 displays U.S. suicides in terms of race and sex.) This fact is most glaringly reflected in the extremely high rate of suicide among Native Americans, who constitute about 1 percent of the U.S. population. On reservations in the Northern Plains (in states such as North Dakota, South Dakota, Nebraska, Montana, Wyoming, and Utah), parts of the Southwest, and in Alaska, Native American teens and young adults have suicide rates that are more than three times as high as those for other youths in the United States. Although some research has focused on individualistic reasons why young Native Americans commit suicide, sociologists focus on the effects of social inequalities and racial–ethnic discrimination on suicidal behavior. For example, many Native American young people reside in homes and communities characterized by extreme poverty, hunger, alcoholism, substance abuse, and family violence (Woodard, 2012).

## Symbolic Interactionist Perspectives

The conflict and functionalist perspectives have been criticized for focusing primarily on macrolevel analysis. A *macrolevel analysis* examines whole societies, large-scale social structures, and social systems instead of looking at important social dynamics in individuals' lives. Our third perspective, symbolic interactionism, fills this void by examining people's day-to-day interactions and their

behavior in groups. Thus, symbolic interactionist approaches are based on a *microlevel analysis,* which focuses on small groups rather than large-scale social structures.

We can trace the origins of this perspective to the Chicago School, especially George Herbert Mead and the sociologist Herbert Blumer (1900–1986), who is credited with coining the term *symbolic interactionism.* According to *symbolic interactionist perspectives,* society is the sum of the interactions of individuals and groups. Theorists using this perspective focus on the process of *interaction*—defined as immediate reciprocally oriented communication between two or more people—and the part that symbols play in communication. A *symbol* is anything that meaningfully represents something else. Examples include signs, gestures, written language, and shared values. Symbolic interaction occurs when people communicate through the use of symbols—for example, a ring to indicate a couple's engagement. But symbolic communication occurs in a variety of forms, including facial gestures, posture, tone of voice, and other symbolic gestures (such as a handshake or a clenched fist).

Symbols are instrumental in helping people derive meanings from social interactions. In social encounters, each person's interpretation or definition of a given situation becomes a *subjective reality* from that person's viewpoint. We often assume that what we consider to be "reality" is shared by others; however, this assumption is often incorrect. Subjective reality is acquired and shared through agreed-upon symbols, especially language. If a person shouts "Fire!" in a crowded movie theater, for example, that language produces the same response

**macrolevel analysis**  an approach that examines whole societies, large-scale social structures, and social systems instead of looking at important social dynamics in individuals' lives.

**microlevel analysis**  an approach that focuses on small groups rather than large-scale social structures.

**symbolic interactionist perspectives**  the sociological approach that views society as the sum of the interactions of individuals and groups.

**19**

(attempting to escape) in all of those who hear and understand it. When people in a group do not share the same meaning for a given symbol, however, confusion results: People who do not know the meaning of the word *fire* will not know what the commotion is about. How people *interpret* the messages they receive and the situations they encounter becomes their subjective reality and may strongly influence their behavior.

**Applying Symbolic Interactionist Perspectives to Suicide** Social analysts applying a symbolic interactionist framework to the study of suicide focus on a microlevel analysis of the people's face-to-face interactions and the roles they play in society. We define situations according to our own subjective reality. From this point of view, a suicide attempt may be a way of garnering attention—a call for help—rather than wanting to end one's life. People may attempt to communicate in such desperate ways because other forms of communication have failed.

## Postmodern Perspectives

According to *postmodern perspectives,* **existing theories have been unsuccessful in explaining social life in contemporary societies that are characterized by postindustrialization, consumerism, and global communications.** Postmodern social theorists reject the theoretical perspectives we have previously discussed, as well as how those theories were created (Ritzer, 2011).

Postmodern theories are based on the assumption that large-scale and rapid social change, globalization, and technology are central features in the postmodern era. Moreover, these conditions tend to have a harmful effect on people because they often result in ambiguity and chaos. One evident change is a significant decline in the influence of social institutions such as the family, religion, and education on people's lives. Those who live in postmodern societies typically pursue individual freedom and do not want the structural constraints that are imposed by social institutions. As social inequality and class differences increase, people are exposed to higher levels of stress that produce depression, fear, and ambivalence. Problems such as these are found in nations throughout the world.

Postmodern (or "postindustrial") societies are characterized by an *information explosion* and an economy in which large numbers of people either provide or apply information, or are employed in professional occupations (such as attorneys and physicians) or service jobs (such as fast-food servers and health care workers). There is a corresponding *rise of a consumer society* and the emergence of

a *global village* in which people around the world instantly communicate with one another.

Jean Baudrillard, a well-known French social theorist, has extensively explored how the shift from production of goods to consumption of information, services, and products has created a new form of social control. According to Baudrillard's approach, capitalists strive to control people's shopping habits, much like the output of factory workers in industrial economies, to enhance their profits and to keep everyday people from rebelling against social inequality (1998/1970). How does this work? When consumers are encouraged to purchase more than they need or can afford, they often sink deeper in debt and must keep working to meet their monthly payments. Consumption comes to be based on factors such as our "wants" and our need to distinguish ourselves from others. We will return to Baudrillard's general ideas on postmodern societies in Chapter 2 ("Culture").

Postmodern theory opens up broad new avenues of inquiry by challenging existing perspectives and questioning current belief systems. However, postmodern theory has also been criticized for raising more questions than it answers.

**Applying Postmodern Perspectives to Suicide** Although most postmodern social theorists have not addressed suicide as a social issue, some sociologists believe that postmodern theory can help us because it reminds us that social life is made up of real people with self-identities and lived experiences to share with others. Behind the groups, organizations, classes, and political parties that social scientists study are human beings who participate in the social construction of everyday life.

Looked at from this perspective, the relationship between suicide and race is important to consider. Although youths across ethnic categories share certain risk factors for suicide, young African American and Native American males appear to be at greater risk for suicide because people in this statistical category lack educational and employment opportunities. This problem is combined with other concerns, such as the systemic oppression that persons of color have experienced throughout U.S. history. In other words, the personal biographies of individuals are intertwined with the social worlds that they and others have helped create.

Each of the four sociological perspectives we have examined involves different assumptions. Consequently, each leads us to ask different questions and to view the world somewhat differently. Concept Quick Review 1 summarizes all four of these perspectives. Throughout this book, we will be using these perspectives as lenses through which to view our social world.

## The Major Theoretical Perspectives

| Perspective | Analysis Level | View of Society |
| --- | --- | --- |
| Functionalist | Macrolevel | Society is composed of interrelated parts that work together to maintain stability within society. This stability is threatened by dysfunctional acts and institutions. |
| Conflict | Macrolevel | Society is characterized by social inequality; social life is a struggle for scarce resources. Social arrangements benefit some groups at the expense of others. |
| Symbolic Interactionist | Microlevel | Society is the sum of the interactions of people and groups. Behavior is learned in interaction with other people; how people define a situation becomes the foundation for how they behave. |
| Postmodernist | Macrolevel/Microlevel | Societies characterized by postindustrialization, consumerism, and global communication bring into question existing assumptions about social life and the nature of reality. |

# The Sociological Research Process

**LO6** Most of us rely on our own experiences and personal knowledge to help us form ideas about what happens in everyday life and how the social world works. However, there are many occasions when this personal knowledge is not enough. This is why sociologists and other social scientists learn to question ordinary assumptions and to use specific research methods to find out more about the social world. As a result, sociology involves *debunking*—the unmasking of fallacies in everyday interpretations of social life—through research.

*Research* is the process of systematically collecting information for the purpose of testing an existing theory or generating a new one. What is the relationship between sociological theory and research? This relationship has been referred to as a continuous cycle, as shown in Figure 1.15 (Wallace, 1971).

Not all sociologists conduct research in the same manner. Some researchers primarily engage in quantitative research, whereas others engage in qualitative research. With **quantitative research, the goal is scientific objectivity, and the focus is on data that can be measured numerically.** Quantitative research typically emphasizes complex statistical techniques. Most sociological studies on suicide have used quantitative research. They have compared rates of suicide with almost every conceivable variable, including age, sex, race/ethnicity,

education, and even sports participation. For example, in one quantitative study, researchers found that Latinos/as (Hispanics) consistently had lower suicide rates than whites (non-Hispanics), particularly when they remained strongly attached to others in their own culture. Latinos/as who maintained shared belief systems, rituals, and social networks of friends and relatives that provided them with strong communities and intense feelings of group solidarity were much less likely to commit suicide, much as Durkheim had predicted (Wadsworth and Kubrin, 2007). However, the study also found that just as cultural assimilation (adopting another culture and leaving one's own culture behind) increases the rate of suicide for Hispanics, having higher income and social standing decreases the rate of suicide (Wadsworth and Kubrin, 2007). In quantitative research, these data are reported in a series of tables that summarize the findings of the researchers. ("Understanding Statistical Data Presentations"

**postmodern perspectives** the sociological approach that attempts to explain social life in contemporary societies that are characterized by postindustrialization, consumerism, and global communications.

**quantitative research** sociological research methods that are based on the goal of scientific objectivity and that focus on data that can be measured numerically.

**21**

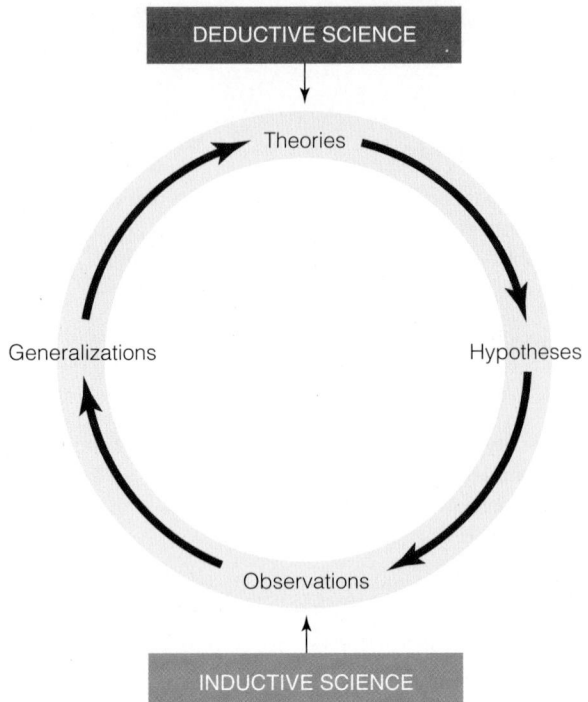

**FIGURE 1.15  THE THEORY AND RESEARCH CYCLE**
The theory and research cycle can be compared to a relay race; although all participants do not necessarily start or stop at the same point, they share a common goal—to examine all levels of social life.

explains how to read numerical tables and how to interpret the data and draw conclusions.)

With *qualitative research,* **interpretive descriptions (words) rather than statistics (numbers) are used to analyze underlying meanings and patterns of social relationships.** An example of qualitative research is a study in which the researchers systematically analyzed the contents of suicide notes to look for recurring themes (such as feelings of despair or failure) to determine if any patterns could be found that would help in understanding why people kill themselves (Leenaars, 1988).

## The Quantitative Research Model

**L07**  Research models are tailored to the specific problem being investigated and the focus of the researcher. Both quantitative research and qualitative research contribute to our knowledge of society and human social interaction, and involve a series of steps, as shown in Figure 1.16. We will now trace the steps in the "conventional" research model, which focuses on quantitative research. Then we will describe an alternative model that emphasizes qualitative research.

**FIGURE 1.16  STEPS IN SOCIOLOGICAL RESEARCH**

© Cengage Learning.

PART 1  |  STUDYING SOCIETY AND SOCIAL LIFE

# understanding
# STATISTICAL DATA PRESENTATIONS

Are young males or females more likely to die violently? How do homicide, suicide, and firearm death rates (per 100,000 population) compare for males and females ages 15 to 19 in the United States? Sociologists use statistical tables as a concise way to present data in a relatively small space; Table 1.1 gives an example. To understand a table, follow these steps:

1. **Read the title.** From the title, "Rates (per 100,000 U.S. Population) for Homicide, Suicide, and Firearm-Related Deaths of Youths Ages 15–19, by Gender, 2009," we learn that the table shows relationships between two variables: gender and three causes of violent deaths among young people in a specific age category.

2. **Check the source and other explanatory notes.** In this case, the sources are the U.S. Centers for Disease Control and Prevention, 2012e ("Web-Based Injury Statistics Query and Reporting System [WISQARS]"), a source that makes it possible for researchers to put in online inquiries for specific information from the CDC's database, and Child Trends Data Bank (2012), a data bank provided by a nonprofit research and policy center that reports on issues pertaining to children and young people. The explanatory note in this table states that firearm deaths, which constitute a majority of teen homicides and suicides, may also include *accidental* deaths that are firearm related. This distinction is made in Table 1.1 because it is possible for firearm-related deaths to occur accidentally; however, firearms were the method of death in 84 percent of teen homicides and 44 percent of teen suicides in 2009.

3. **Read the headings for each column and each row.** The main column headings in the table are "Method," "Males," and "Females." The columns present information (usually numbers) arranged vertically. The rows present information horizontally. Here, the row headings indicate homicide, suicide, and firearm-related death. Based on the explanation above regarding firearm-related deaths, we know that some overlap exists between the first two categories—homicide and suicide—and the third, deaths that are firearm related.

## table 1.1

Rates (per 100,000 U.S. Population) for Homicide, Suicide, and Firearm-Related Deaths of Youths Ages 15–19, by Gender, 2009

| Method | Males | Females |
|---|---|---|
| Homicide | 14.9 | 2.6 |
| Suicide | 12.0 | 3.3 |
| Firearm-Related Death[a] | 19.9 | 2.5 |

[a]Firearm deaths, which constitute a majority of teen homicides and suicides, also include accidental deaths that are firearm related.

Sources: U.S. Centers for Disease Control and Prevention, 2012e ("Web-Based Injury Statistics Query and Reporting System [WISQARS]"); and Child Trends Data Bank, 2012.

4. **Examine and compare the data.** To examine the data, determine what units of measurement have been used. In Table 1.1 the figures are rates per 100,000 males or females in a specific age category. For example, the suicide rate is 12.0 per 100,000 population of males between the ages of 15 and 19 as compared with only 3.3 per 100,000 population of females in the same age category.

5. **Draw conclusions.** By looking for patterns, some conclusions can be drawn from Table 1.1:

   a. *Determining differences by gender.* Males between the ages of 15 and 19 are about four times more likely than females to die from suicide (12.0 compared with 3.3 per 100,000). Males in this age category are almost six times more likely to die from homicide (14.9 compared with 2.6 per 100,000). And even more noteworthy, males are eight times more likely to die from a firearm-related incident (either intentional or unintentional) than females of this age. As shown in the table, 19.9 per 100,000 males ages 15 to 19 died by firearms in 2009, compared with 2.5 per 100,000 females in that same age category.

   b. *Drawing appropriate conclusions.* Although the rate of teen suicides has been relatively stable (between 7 and 8 per 100,000) over the past decade, important differences by gender exist. Males between the ages of 15 and 19 are much more likely than females in their age category to die violently, and many of those deaths are firearm related.

1. *Select and define the research problem.* Sometimes, a specific experience such as knowing someone who committed suicide can trigger your interest in a topic. Other times, you might select topics to fill gaps or challenge misconceptions in existing research or to test a specific theory (Babbie, 2013). Emile Durkheim selected suicide because he wanted to demonstrate the importance of *society* in situations that might appear to be arbitrary acts by individuals. Suicide was a suitable topic because it was widely believed that suicide was a uniquely individualistic act. However, Durkheim emphasized that *suicide rates* provide better explanations for suicide than do *individual* acts of suicide. He

**qualitative research** sociological research methods that use interpretive descriptions (words) rather than statistics (numbers) to analyze underlying meanings and patterns of social relationships.

reasoned that if suicide were purely an individual act, then the rate of suicide (the number of people who kill themselves in a given year) should be the same for every group regardless of culture and social structure. Moreover, Durkheim wanted to know why there were different rates of suicide—whether factors such as religion, marital status, sex, and age had an effect on social cohesion.

2. *Review previous research.* Before beginning the research, it is important to analyze what others have written about the topic. You should determine where gaps exist and note mistakes to avoid. When Durkheim began his study, very little sociological literature existed to review; however, he studied the works of several moral philosophers, including Henry Morselli (1975/1881).

3. *Formulate the hypothesis* (if applicable). You may formulate a ***hypothesis**—a statement of the expected relationship between two or more variables. A *variable* is any concept with measurable traits or characteristics that can change or vary from one person, time, situation, or society to another.* The most fundamental relationship in a hypothesis is between a dependent variable and one or more independent variables (see Figure 1.17). The ***independent variable** is presumed to be the cause of the relationship; the *dependent variable* is assumed to be caused by the independent variable(s)* (Babbie, 2013). Durkheim's hypothesis stated that the rate of suicide varies *inversely* with the degree of social integration. In other words, a low degree of

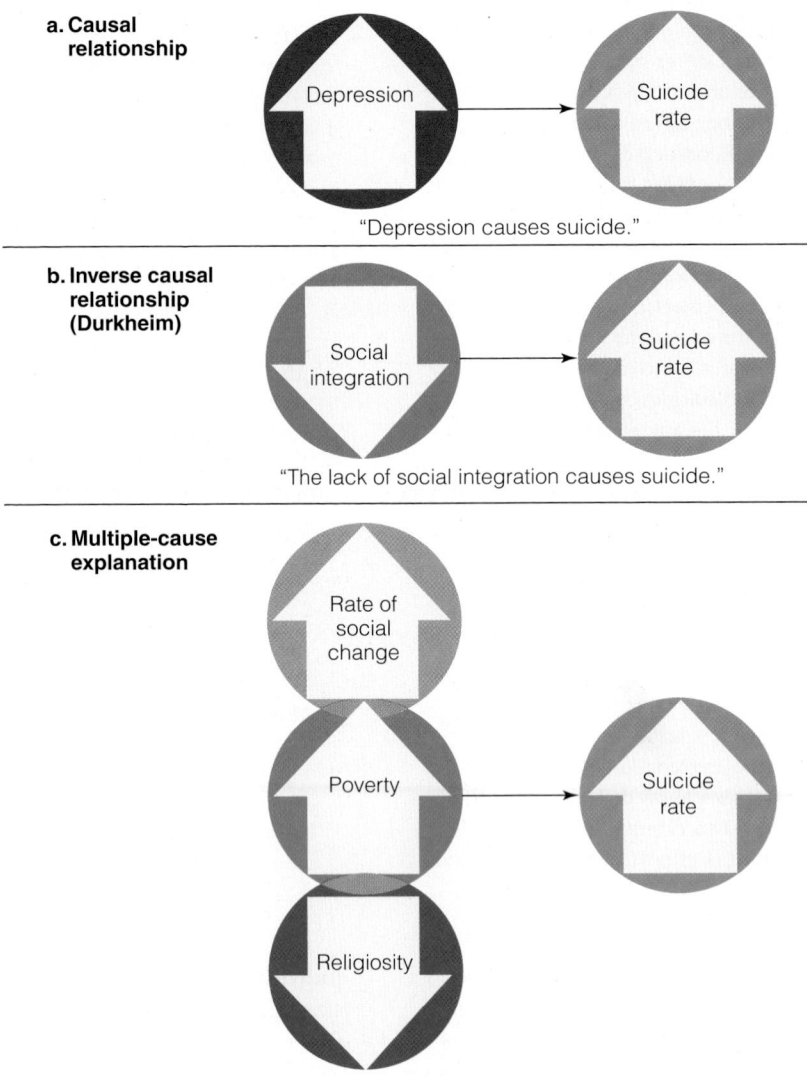

**a. Causal relationship**

Depression → Suicide rate

"Depression causes suicide."

**b. Inverse causal relationship (Durkheim)**

Social integration → Suicide rate

"The lack of social integration causes suicide."

**c. Multiple-cause explanation**

Rate of social change

Poverty → Suicide rate

Religiosity

"Many factors interact to cause suicide."

**FIGURE 1.17** HYPOTHESIZED RELATIONSHIPS BETWEEN VARIABLES
A causal hypothesis connects one or more independent (causal) variables with a dependent (affected) variable. The diagram illustrates three hypotheses about the causes of suicide. To test these hypotheses, social scientists would need to operationalize the variables (define them in measurable terms) and then investigate whether the data support the proposed explanation.

social integration (the independent variable) may "cause" or "be related to" a high rate of suicide (the dependent variable).

Not all social research uses hypotheses. If you plan to conduct an explanatory study (showing a cause-and-effect relationship), you will likely want to formulate one or more hypotheses to test theories. If you plan to conduct a descriptive study, however, you will be less likely to do so because you may desire only to describe social reality or provide facts.

4. *Develop the research design.* You must determine the unit of analysis to be used in the study. A *unit of analysis* is *what* or *whom* is being studied (Babbie, 2013). In social science research, individuals, social groups (such as families, cities, or geographic regions), organizations (such as clubs, labor unions, or political parties), and social artifacts (such as books, paintings, or weddings) may be units of analysis. As mentioned, Durkheim's unit of analysis was social groups, not individuals.

5. *Collect and analyze the data.* You must decide what population will be observed or questioned and then carefully select a sample. A *sample* is the people who are selected from the population to be studied; the sample should accurately represent the larger population. A *representative sample* is a selection from a larger population that has the essential characteristics of the total population. For example, if you interviewed five students selected haphazardly from your sociology class, they would not be representative of your school's student body. By contrast, if you selected five students from the student body by a random sample, they would be closer to being representative (although a random sample of five students would be too small to yield much useful data).

Validity and reliability may be problems in research. **Validity is the extent to which a study or research instrument accurately measures what it is supposed to measure.** A recurring issue in studies that analyze the relationship between religious beliefs and suicide is whether "church membership" is a valid indicator of a person's religious beliefs. In fact, one person may be very religious yet not belong to a specific church, whereas another person may be a member of a church yet not hold very deep religious beliefs. **Reliability is the extent to which a study or research instrument yields consistent results when applied to different individuals at one time or to the same individuals over time.** Sociologists have found that different interviewers get different answers from the people being interviewed. For example, how might the interviewers themselves

influence interviews with college students who have contemplated suicide?

Once you have collected your data, the data must be analyzed. *Analysis* is the process through which data are organized so that comparisons can be made and conclusions drawn. Sociologists use many techniques to analyze data. For his study of suicide, Durkheim collected data from vital statistics for approximately 26,000 suicides. He classified them separately according to age, sex, marital status, presence or absence of children in the family, religion, geographic location, calendar date, method of suicide, and a number of other variables. As Durkheim analyzed his data, four distinct categories of suicide emerged: egoistic, altruistic, anomic, and fatalistic. *Egoistic suicide* occurs among people who are isolated from any social group. For example, Durkheim concluded that suicide rates were relatively high in Protestant countries in Europe because Protestants believed in individualism and were more loosely tied to the church than were Catholics. Single people had proportionately higher suicide rates than married persons because they had a low degree of social integration, which contributed to their loneliness. In contrast, *altruistic suicide* occurs among individuals who are excessively integrated into society. An example is military leaders who kill themselves after defeat in battle because they have so strongly identified themselves with their cause that they believe they cannot live with defeat. Today, other factors such as extended periods of military service or lengthy wars may also contribute to relatively high rates of suicide among U.S. military personnel (see

**hypothesis** a statement of the expected relationship between two or more variables.

**variable** any concept with measurable traits or characteristics that can change or vary from one person, time, situation, or society to another.

**independent variable** in an experiment, the variable assumed to be the cause of the relationship between variables.

**dependent variable** in an experiment, the variable assumed to be caused by the independent variable(s).

**validity** the extent to which a study or research instrument accurately measures what it is supposed to measure.

**reliability** the extent to which a study or research instrument yields consistent results when applied to different individuals at one time or to the same individuals over time.

"Sociology and Social Policy"). According to Durkheim, people are more likely to commit suicide when social cohesion is either very weak or very strong, and/or when nations experience rapid social change.

6. *Draw conclusions and report the findings.* After analyzing the data, your first step in drawing conclusions is to return to your hypothesis or research objective to clarify how the data relate both to the hypothesis and to the larger issues being addressed. At this stage, you note the limitations of the study, such as problems with the sample, the influence of variables over which you had no control, or variables that your study was unable to measure.

Reporting the findings is the final stage. The report generally includes a review of each step taken in the research process in order to make the study available for *replication*—the repetition of the investigation in substantially the same way that it was originally conducted. Social scientists generally present their findings in papers at professional meetings and publish them in technical journals and books. In reporting his findings in *Suicide* (1964b/1897), Durkheim concluded that the suicide rate of a group is a social fact that cannot be explained in terms of the personality traits of individuals. Instead, his findings suggested that social conditions in a society are a more significant influence on rates of suicide.

We have traced the steps in the "conventional" research process (based on deduction and quantitative research). But what steps might be taken in an alternative approach based on induction and qualitative research?

### A Qualitative Research Model

Although the same underlying logic is involved in both quantitative and qualitative sociological research, the *styles* of these two models are very different. Qualitative research is more likely to be used when the research question does not easily lend itself to numbers and statistical methods. As compared to a quantitative model, a qualitative approach often involves a different type of research question and a smaller number of cases. Researchers using a qualitative approach may engage in *problem formulation* to clarify the research question and to formulate questions of concern and interest to people participating in the research.

In a qualitative approach, reviewing the literature and developing the research design often happen simultaneously. Typically, the next step is collecting and analyzing data to assess the validity of the starting proposition. Data gathering is the foundation of the research. Researchers pursuing a qualitative approach tend to gather data in natural settings, such as where the person lives or works, rather than in a laboratory or other research setting. Data collection and analysis frequently occur concurrently, and the analysis draws heavily on the language of the persons studied, not the researcher. Additional review of the literature may occur later in the process after data have been gathered and further insights are needed to help describe, explain, or make predictions from the data that have been analyzed. Finally, the researchers draw conclusions from their research and report their findings to others.

How would qualitative research work for the study of suicide? One Canadian study examined suicide among twenty-two older men who had previously experienced depression, and the study found important relationships between societal expectations about masculinity, depression, and suicide (Oliffe et al., 2011). To find out more about linkages between depression and suicide, researchers interviewed the men to learn more about how cumulative losses had contributed to their depression and lack of desire to live. The loss of social bonds, through deaths of family members and friends, was a major factor in their depression and thoughts about suicide. Other factors that contributed to their depression and suicidal tendencies were feelings that they were a failure as a breadwinner and/or beliefs that they had other shortcomings that their older age prevented them from overcoming. However, the qualitative study revealed that many of the participants would not commit suicide because of the stigma associated with this act and their guilt about the pain this act would bring to their family and friends (Oliffe et al., 2011).

From qualitative studies such as this one, we learn information that we might not find when using quantitative research. We find answers to questions such as "why" people might or might not engage in a specific behavior. Qualitative and quantitative research methods are often used in combination with each other in a research design to provide a more holistic view of the social world.

## Research Methods

**LO8** How do sociologists know which research method to use? Are some approaches best for a particular problem? **Research methods are specific strategies or techniques for systematically conducting research.** We will look at four of these methods: survey research, secondary analysis of existing data, field research, and experiments.

# SOCIOLOGY and Social Policy

## Establishing Policies to Help Prevent Military Suicides

Grandpa, I just wanted to give you my thanks for being a great influence in my life.

> —Marine Corporal Daniel O'Brien, who completed two tours of duty in Iraq, sent a note containing this statement to his grandfather prior to taking his own life (qtd. in Roberts, 2011).

We think that we are seeing a societal problem [in regard to suicides by military personnel], and frankly the Army is the canary in the mine shaft here.

> —General Ray Carpenter, commander of the Army National Guard, discussing how the incidence of suicide has continued to be a problem in most branches of the U.S. military (qtd. in Roberts, 2011)

**S**uicide is a problem nationwide, but it has grown as a special problem in the military, where it is now the number-two cause of death (after combat) of U.S. service members (Bumiller, 2012). Pentagon figures show that suicides among troops occur on average once a day and that there have been increases among active-duty members in all the services except the Navy. It is estimated that since 2001, more than 2,500 active-duty service members have taken their own lives (Zoroya, 2012).

Shocked by relatively high rates of both suicide and suicide attempts among military service members, the U.S. Department of Defense and the executive branch of the U.S. government have encouraged all branches of the military to learn more about the sociological causes of suicide and to develop comprehensive suicide-prevention initiatives to help reduce the problem and support military service members around the globe. A 2012 executive order, "Improving Access to Mental Health for Veterans, Service Members, and Military Families," set forth initiatives designed to improve mental health resources and intervention tools, such as increasing the number of crisis lines, peer-to-peer counselors, and mental health professionals available to service members and veterans. More recently, Facebook has become a tool for helping with posttraumatic stress, depression, and mental illness. The Durkheim Project (named for sociologist Emile Durkheim) combines Facebook, the Department of Defense, and the Department of Veterans Affairs in an effort to help veterans in distress. Those who voluntarily install an app on their digital devices get help from other military personnel if

Fabrizio Bensch/Reuters

**What unique social conditions do military personnel face that might contribute to suicide or other conditions such as depression and alcoholism?**

they post messages containing expressions of hopelessness, sleeplessness, or feelings of isolation (Basu, 2013).

Clearly, a variety of sociological issues are central in dealing with suicidal ideation and behavior in the military. In the words of a U.S. Army (2012: 51) report,

> Each potential suicide or attempted suicide is different with respect to contributing factors and triggering events. Each victim responds differently to pre-suicide stressors based on protective factors such as personal resilience, coping skills, and whether or not they are help-seeking. . . .

Studies are being conducted to gain a better understanding of why suicides are happening and what might be done to prevent this pressing problem. Although no single cause can be identified for suicide, factors such as financial worries, relationship issues, legal trouble, substance abuse, medical problems, and posttraumatic stress are all thought to be associated with suicidal behavior among military personnel. Perhaps greater awareness and new social policies being implemented by the military will help reduce this problem in the future.

### reflect & analyze

**How might lengthy wars and deployments away from home contribute to suicidal behavior among troops? Do you anticipate that military suicide rates might decrease in peacetime? Why or why not?**

## Survey Research

A *survey* is a poll in which the researcher gathers facts or attempts to determine the relationships among facts. Surveys are the most widely used research method in the social sciences because they make it possible to study things that are

> **research methods** specific strategies or techniques for systematically conducting research.
>
> **survey** a poll in which the researcher gathers facts or attempts to determine the relationships among facts.

not directly observable—such as people's attitudes and beliefs—and to describe a population too large to observe directly (Babbie, 2013) (Figure 1.18). Researchers frequently select a representative sample from a larger population to answer questions about attitudes, opinions, or behavior. *Respondents* are people who provide data for analysis through interviews or questionnaires. The Gallup and Harris polls are among the most widely known large-scale surveys; however, government agencies such as the U.S. Census Bureau conduct a variety of surveys as well.

Unlike many polls that use various methods of gaining a representative sample of the larger population, the Census Bureau attempts to gain information from all persons in the United States. The decennial census occurs every 10 years, in the years ending in "0." The purpose of this census is to count the population and housing units of the entire nation. The population count determines how seats in the U.S. House of Representatives are apportioned; however, census figures are also used in formulating public policy and in planning and decision making in the private sector. The Census Bureau attempts to survey the *entire* U.S. population by using two forms—a "short form" of questions asked of all respondents, and a "long form" that contains additional questions asked of a *representative sample* of about one in six respondents. Statistics from the Census Bureau provide information that sociologists use in their research. An example is shown in the Census Profiles feature: "How People in the United States Self-Identify Regarding Race." Note that because of recent changes in the methods used to collect data by the Census Bureau, information on race from the 2010 census is not directly comparable with data from earlier censuses.

**Types of Surveys** Survey data are collected by using self-administered questionnaires, face-to-face interviews, and telephone or computer surveys. A **questionnaire is a printed research instrument containing a series of items to which subjects respond.** Items are often in the form of statements with which

**FIGURE 1.18** Conducting surveys and polls is an important means of gathering data from respondents. Some surveys take place on street corners; increasingly, however, such surveys are done by telephone, Internet, and other means.

Paul Conklin/PhotoEdit

the respondent is asked to "agree" or "disagree." Questionnaires may be administered by interviewers in face-to-face encounters or by telephone, but the most commonly used technique is the *self-administered questionnaire,* which is either mailed to the respondent's home or administered to groups of respondents gathered at the same place at the same time. For example, in a now-classic study of suicide, race, and religion, the sociologist Kevin E. Early (1992) used survey data collected through questionnaires to test his hypothesis that suicide rates are lower among African Americans than among white Americans because of the influence of black churches. Data from questionnaires filled out by members of six African American churches in Florida supported Early's hypothesis that the church buffers some African Americans against harsh social forces—such as racism—that might otherwise lead to suicide.

Survey data may also be collected by interviews. An **interview is a data-collection encounter in which an interviewer asks the respondent questions and records the answers.** Survey research often uses *structured interviews,* in which the interviewer asks questions from a standardized questionnaire. Structured interviews tend to produce uniform or replicable data that can be elicited time after time by different interviews. For example, in addition to surveying congregation members, Early (1992) conducted interviews with pastors of African American churches to determine the pastors' opinions about the extent to which African American churches reinforce values and beliefs that discourage suicide.

Interviews have specific advantages such as being more effective in dealing with complicated issues and providing an opportunity for face-to-face communication between the interviewer and the respondent. Although interviews provide a wide variety of useful information, a major disadvantage is the cost and time involved in conducting the interviews and analyzing the results.

A quicker method of administering questionnaires is the *telephone* or *computer survey.* Telephone

## How People in the United States Self-Identify Regarding Race

Beginning with Census 2000 and continuing with Census 2010, the U.S. Census Bureau has made it possible for people responding to census questions regarding their race to mark more than one racial category. Although the vast majority of respondents select only one category (see below), the Census Bureau reports that in 2003, approximately 4.3 million people (1.48 percent of the population) in the United States self-identified as being of more than one race. As a result, if you look at the figures set forth below, they total more than 100 percent of the total population. How can this be? Simply stated, some individuals are counted at least twice, based on the number of racial categories they listed.

| Race | Percentage of Total Population |
|---|---|
| White alone or in combination with one or more other races | 74.8 |
| African American alone or in combination with one or more other races | 13.6 |
| Asian American alone or in combination with one or more other races | 5.6 |
| Native American and Alaska Native alone or in combination with one or more other races | 1.7 |
| Native Hawaiian and other Pacific Islander alone or in combination with one or more other races | 0.4 |
| Some other race alone or in combination with one or more other races | 7.0 |

Source: Humes, Jones, and Ramirez, 2011.

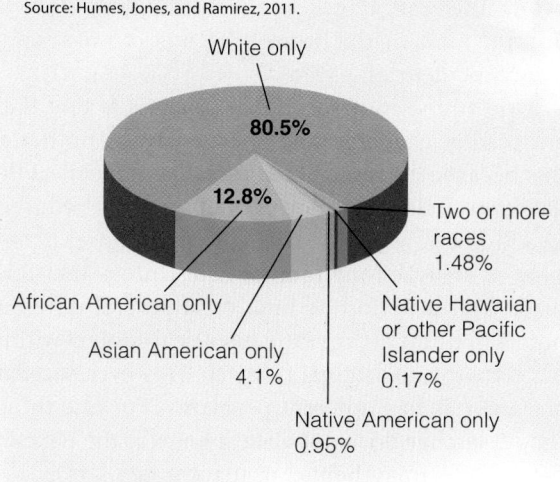

White only
80.5%

12.8%
African American only

Asian American only
4.1%

Native American only
0.95%

Native Hawaiian or other Pacific Islander only
0.17%

Two or more races
1.48%

and computer surveys give greater control over data collection and provide greater personal safety for respondents and researchers than do personal encounters (Figure 1.19). In *computer-assisted*

*telephone interviewing* (sometimes called CATI), the interviewer uses a computer to dial random telephone numbers, reads the questions shown on the video monitor to the respondent, and then types the responses into the computer terminal. The answers are immediately stored in the central computer, which automatically prepares them for data analysis. However, the respondent must answer the phone before the interview can take place, and many people screen their phone calls. In the past few years, online survey research has increased dramatically as software packages and online survey services have made this type of research easier to conduct. Online research makes it possible to study virtual communities, online relationships, and other types of computer-mediated communications networks around the world.

Survey research is useful in describing the characteristics of a large population without having to interview each person in that population. In recent years, computer technology has enhanced researchers' ability to do *multivariate analysis*—research involving more than two independent variables. For example, to assess the influence of religion on suicidal behavior among African Americans, a researcher might look at the effects of age, sex, income level, and other variables all at once to determine which of these independent variables influences suicide the most or least and how influential each variable is relative to the others. However, a weakness of survey research is the use of standardized questions; this approach tends to force respondents into categories in which they may or may not belong. Moreover, survey research relies on self-reported information, and some people may be less than truthful, particularly on emotionally charged issues such as suicide.

## Secondary Analysis of Existing Data

**LO9** In *secondary analysis*, researchers use existing material and analyze data that were originally collected by others. Existing data

**questionnaire** a printed research instrument containing a series of items to which subjects respond.

**interview** a data-collection encounter in which an interviewer asks the respondent questions and records the answers.

**secondary analysis** a research method in which researchers use existing material and analyze data that were originally collected by others.

**FIGURE 1.19** Computer-assisted telephone interviewing is an easy and cost-efficient method of conducting research. However, the widespread use of answering machines, voice mail, and caller ID may make this form of research more difficult in the twenty-first century.

sources include public records, official reports of organizations or government agencies, and *raw data* collected by other researchers. For example, Durkheim used vital statistics (death records) that were originally collected for other purposes to examine the relationship among variables such as age, marital status, and the circumstances surrounding the person's suicide. Today, many researchers studying suicide use data compiled by the U.S. Centers for Disease Control and Prevention (CDC) and the National Center for Injury Prevention and Control. For example, look at Figure 1.20, "National Suicide Statistics at a Glance," based on data compiled by the CDC, and try to develop several plausible sociological explanations for why suicide rates are the highest in the western and northwestern regions of the United States. Patterns of high suicide rates are also reported among some counties in the central Midwest, the South, and Central Florida. Can you provide an explanation of why rates might be higher in these areas?

Secondary analysis also includes ***content analysis*—the systematic examination of cultural artifacts or various forms of communication to extract thematic data and draw conclusions about social life.** Among the materials studied are written records (such as books, diaries, poems, and graffiti), narratives and visual texts (such as movies, television shows, advertisements, and greeting cards), and material culture (such as music, art, and even garbage). In content analysis, researchers look for regular patterns, such as the frequency of suicide as a topic on television talk shows. They may also examine subject matter to determine how it has been represented, such as how the mass media frame presentations of suicide (see "Suicide in the Media").

One strength of secondary analysis is that data are readily available and inexpensive. Another is that because the researcher often does not collect the data personally, the chances of bias may be reduced. In addition, the use of existing sources makes it possible to analyze longitudinal data (things that take place over a period of time or at several different points in time) to provide a historical context within which to locate original research. However, secondary analysis has inherent problems. For one thing, the researcher does not always know if the data are incomplete, unauthentic, or inaccurate.

## Field Research

*Field research* is the study of social life in its natural setting: observing and interviewing people where they live, work, and play. Some kinds of behavior

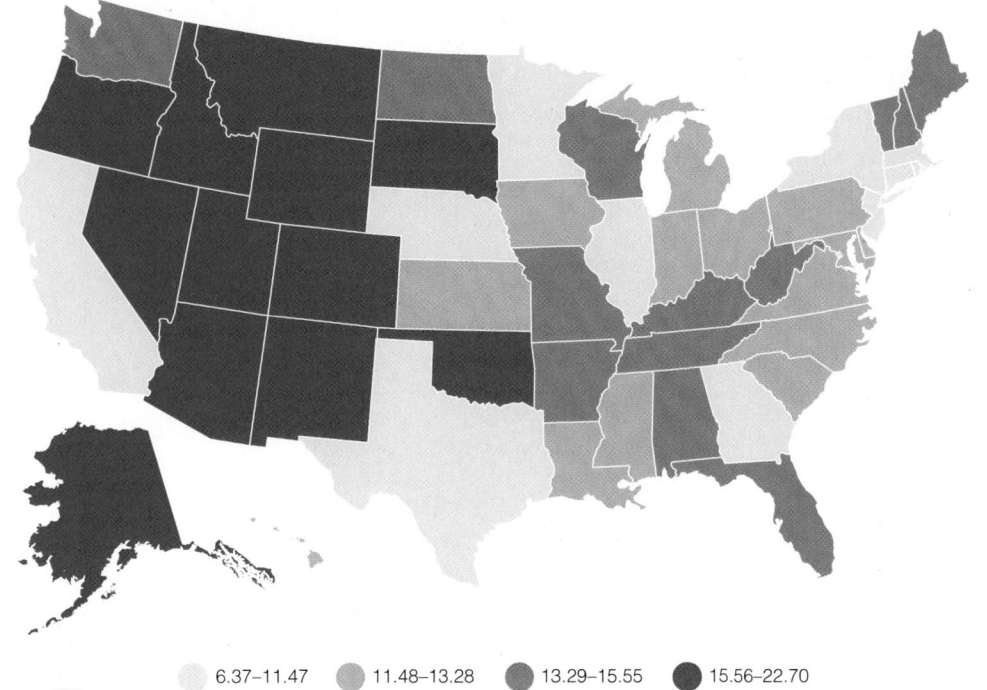

6.37–11.47    11.48–13.28    13.29–15.55    15.56–22.70

**FIGURE 1.20** NATIONAL SUICIDE STATISTICS AT A GLANCE

Age-adjusted suicide rate per 100,000 population per state, 2000–2006, United States.

© Cengage Learning®

can be best studied by "being there"; a fuller understanding can be developed through observations, face-to-face discussions, and participation in events. Researchers use these methods to generate qualitative data: observations that are best described verbally rather than numerically.

Sociologists who are interested in observing social interaction as it occurs may use *participant observation*—**the process of collecting systematic observations while being part of the activities of the group that the researcher is studying.** Participant observation generates more "inside" information than simply asking questions or observing from the outside. For example, to learn more about how coroners make a ruling of "suicide" in connection with a death and to analyze what (if any) effect such a ruling has on the accuracy of "official" suicide statistics, the sociologist Steve Taylor (1982) engaged in participant observation at a coroner's office over a six-month period. As he followed a number of cases from the initial report of death through the various stages of investigation, Taylor learned that it was important to "be around" so that he could listen to discussions and ask the coroners questions because intuition and guesswork play a large part in some decisions to rule a death as a suicide.

Another approach to field research is the *ethnography*—**a detailed study of the life and activities of a group of people by researchers who may live with that group over a period of years.** Unlike participant observation, ethnographic

studies usually take place over a longer period of time. For example, the sociologist Sudhir Venkatesh (2013) conducted more than a decade of research on New York City's underground economy before writing his book *Floating City: A Rogue Sociologist Lost and Found in New York's Underground Economy* (Figure 1.21 and Figure 1.22). He observed and interviewed a wide array of people, many of whom were involved in illegal transactions. Among his participants were poor immigrants, drug bosses and dealers, prostitutes, businesspeople, socialites, and academics, all of whom had their own stories to tell. Based on his extensive ethnographic research, Venkatesh was able to describe in great detail how a vast underground economy serves to unite the wealthy and poor in New York and other large cities.

**content analysis** the systematic examination of cultural artifacts or various forms of communication to extract thematic data and draw conclusions about social life.

**participant observation** a research method in which researchers collect systematic observations while being part of the activities of the group being studied.

**ethnography** a detailed study of the life and activities of a group of people by researchers who may live with that group over a period of years.

**31**

# Reporting on Bullying and Suicide: A Cause-and-Effect Relationship?

The media offer us different vantage points from which to view a given social event based on how they *frame* the information they provide to their audience. The term *media framing* refers to the process by which information and entertainment are packaged by the mass media (newspapers, magazines, radio and television stations and networks, the Internet, and various social media networks). For example, recurring media stories about suicides among young people have focused on individuals such as Rebecca Ann Sedwick (mentioned at the beginning of this chapter); Phoebe Prince, a 15-year-old girl who moved to Massachusetts from Ireland and killed herself after being bullied by high school classmates; and Amanda Cummings, another 15-year-old who committed suicide after what her family referred to as relentless bullying in New York. In the aftermath of these supposedly bullying-related suicides, media personnel often have framed extensive—and sometimes sensationalized—stories based on what they consider to be the "cause-and-effect" relationship between being bullied and committing suicide, especially among young people.

However, a question has been raised by mental health professionals and suicide-prevention workers: Is this the best way to inform media audiences about suicide, or is the framing of these stories far too simplistic? According to one specialist at the American Foundation for Suicide Prevention, "Bullying is so at the top of our consciousness that we're bending over backwards to get it into the story. Years and years of research has taught us that the overwhelming number of people who die by suicide had a diagnosable mental disorder at the time of their death" (qtd. in Bindley, 2012). From this perspective, other contributing factors, such as depression, family life, and the ending of a relationship, must also be considered when writing about suicidal behavior. Sometimes bullying becomes the "easy" answer, and we turn to it quickly because it is easy to understand, particularly when it is so often the focus of media reports on suicide by young people.

Thinking sociologically, what problems exist in how the media frame stories about a social issue such as bullying and suicide? As discussed in this chapter, the early sociologist Emile Durkheim believed that we should view even seemingly individual actions—such as suicide—from a *sociological perspective*

Mother Jones

**FIGURE 1.21** Sudhir Venkatesh

# FLOATING CITY

### A ROGUE SOCIOLOGIST LOST AND FOUND IN NEW YORK'S UNDERGROUND ECONOMY

## SUDHIR VENKATESH

AUTHOR of GANG LEADER FOR A DAY

The Penguin Press, 2013

**FIGURE 1.22** *Floating City: A Rogue Sociologist Lost and Found in New York's Underground Economy* by Sudhir Venkatesh that focuses on the part that social groups and societies play in patterns of behavior rather than focusing on individuals. If we use a sociological approach to analyzing how the media frame stories about suicide, here are three questions to consider:

- Do television and newspaper reports of suicides simplify the *reasons* for the suicide? The media often report a final precipitating situation (such as a fight with a boyfriend or girlfriend, losing one's job, or being bullied) that distressed the individual prior to suicide but do not inform audiences that this was not the *only* cause of this suicide.
- Are readers and viewers provided with *repetitive, ongoing, and excessive reporting* on a suicide? Repeated sensationalistic framing of stories about suicide may contribute to *suicide contagion.* Some people are more at risk because of age, stress, and/or other personal problems. For example, *suicide clusters* (suicides occurring close together in time and location) are most common among people 15 to 24 years of age who may have learned details from the media. Do the media use dramatic photographs (such as of the person committing suicide or of friends and family weeping and showing great emotion at the victim's funeral) primarily to sell papers or increase viewer ratings? Photographs and other sensational material are potentially most damaging for people who have long-standing mental health problems or who have limited social networks to provide them with hope, encouragement, and guidance.

The media frequently employ images when covering suicide. Do such images affect the way that you perceive the event?

In our mass-mediated culture, many sociologists agree that the media do much more than simply *mirror* society: The media help shape our society and our cultural perceptions on many issues, including seemingly individual behavior such as suicide.

## reflect & analyze

Are we influenced by how the media frame news and entertainment stories even when we think we are unaffected by such coverage? If so, how might this relate to media reporting of suicides?

---

## Experiments

An *experiment* **is a carefully designed situation in which the researcher studies the impact of certain variables on subjects' attitudes or behavior.** Experiments are designed to create "real-life" situations, ideally under controlled circumstances, in which the influence of different variables can be modified and measured. Conventional experiments require that subjects be divided into two groups: an experimental group and a control group. The *experimental group* **contains the subjects who are exposed to an independent variable (the experimental condition) to study its effect on them.** The *control group* **contains the subjects who are not exposed to the independent variable.** For example, the sociologist Arturo Biblarz and colleagues (1991) examined the effects of media violence and depictions of suicide on attitudes toward suicide by showing one group of subjects (an experimental group) a film about suicide, while a second (another experimental group) saw a film about violence, and a third (the control group) saw a film containing neither suicide nor violence. The research found some evidence that people exposed to suicidal acts or violence in the media may be more likely to demonstrate an emotional state favorable to suicidal behavior, particularly if they are already "at risk" for suicide. If we were able to replicate this study today, do you believe that we would find similar results? Why or why not?

Researchers may use experiments when they want to demonstrate that a cause-and-effect relationship exists between variables, such as whether violent video games cause an increase in violent tendencies in people who use them (Figure 1.23). In order to show that a change in one variable causes a change in another, three conditions must be satisfied: (1) a correlation between the two variables must be shown to exist (*correlation* **exists when two variables are associated more frequently than could be expected**

**experiment** a carefully designed situation in which the researcher studies the impact of certain variables on subjects' attitudes or behavior.

**experimental group** the group that contains the subjects who are exposed to an independent variable (the experimental condition) to study its effect on them.

**control group** the group that contains the subjects who are not exposed to the independent variable.

**correlation** a relationship that exists when two variables are associated more frequently than could be expected by chance.

*Jochen Tack/imageBROKER/AGE Fotostock*

**FIGURE 1.23** Do extremely violent video games cause an increase in violent tendencies in their users? Experiments are one way to test a researcher's hypothesis about this relationship.

by chance), (2) the independent variable must have occurred prior to the dependent variable, and (3) any change in the dependent variable must not have been due to an extraneous variable—one outside the stated hypothesis.

The major advantage of an experiment is the researcher's control over the environment and the ability to isolate the experimental variable. Because many experiments require relatively little time and money and can be conducted with limited numbers of subjects, it is possible for researchers to replicate an experiment several times by using different groups of subjects (Babbie, 2013). Perhaps the greatest limitation of experiments is that they are artificial: Social processes that are set up by researchers or that take place in a laboratory setting are often not the same as real-life occurrences (see Concept Quick Review 2).

## Ethical Issues in Sociological Research

**LO10** The study of people ("human subjects") raises vital questions about ethical concerns in sociological research. Researchers are required to obtain written "informed consent" statements from the persons they study—but what constitutes "informed consent"? And how do researchers protect the identity and confidentiality of their sources?

The American Sociological Association (ASA) *Code of Ethics* (1999) sets forth certain basic standards that sociologists must follow in conducting research. Among these standards are the following:

1. Researchers must endeavor to maintain objectivity and integrity in their research by disclosing their research findings in full and including all possible interpretations of the data (even when these interpretations do not support their own viewpoints).
2. Researchers must safeguard the participants' right to privacy and dignity while protecting them from harm.
3. Researchers must protect confidential information provided by participants, even when this information is not considered to be "privileged" (legally protected, as is the case between doctor and patient and between attorney and client) and legal pressure is applied to reveal this information.
4. Researchers must acknowledge research collaboration and assistance they receive from others and disclose all sources of financial support.

## Strengths and Weaknesses of Social Research Methods

| Research Method | Strengths | Weaknesses |
|---|---|---|
| Experiments (laboratory, field, natural) | Control over research | Artificial by nature |
| | Ability to isolate experimental factors | Frequent reliance on volunteers or captive audiences |
| | Relatively little time and money required | Ethical questions of deception |
| | Replication possible, except for natural experiments | |
| Survey research (questionnaire, interview, telephone survey) | Useful in describing features of a large population without interviewing everyone | Potentially forced answers |
| | Relatively large samples possible | Respondent untruthfulness on emotional issues |
| | Multivariate analysis possible | Data that are not always "hard facts" presented as such in statistical analyses |
| Secondary analysis of existing data (existing statistics, content analysis) | Data often readily available, inexpensive to collect | Difficulty in determining accuracy of some of the data |
| | Longitudinal and comparative studies | Failure of data gathered by others to meet goals of current research |
| | Replication possible | Questions of privacy when using diaries, other personal documents |
| Field research (participant observation, case study, ethnography, unstructured) | Opportunity to gain insider's view | Problems in generalizing results to a larger population |
| | Useful for studying attitudes and behavior in natural settings | Imprecise data measurements |
| | Longitudinal/comparative studies possible | Inability to demonstrate cause/effect relationships or test theories |
| | Documentation of important social problems of excluded groups possible | Difficult to make comparisons because of lack of structure |
| | Access to people's ideas in their words | Not a representative sample |
| | Forum for previously excluded groups | |
| | Documentation of need for social reform | |

Sociologists are obligated to adhere to this code and to protect research participants; however, many ethical issues still arise. A now-classic study in sociology illustrates this point. In the 1970s, sociologist William Zellner (1978) wanted to look at fatal single-occupant automobile accidents to determine if some drivers were actually committing suicide when they were involved in car crashes. To examine this issue further, he sought to interview the family, friends, and acquaintances of persons killed in single-car crashes to determine if the deaths were possibly intentional. To recruit respondents, Zellner told them that he hoped the research would reduce the number of automobile accidents in the future. He did not mention that he suspected "autocide" might have occurred in the case of their friend or loved one. From his data, Zellner concluded that at least 12 percent of the fatal single-occupant crashes were suicides—and that these crashes sometimes also killed or critically injured other people as well. However, Zellner's research raised important research questions: Was his research unethical? Did he misrepresent the reasons for his study?

In the twenty-first century, other researchers have called attention to the ethics of suicide research. These ethical problems include how to maintain the confidentiality of research participants, as well as how to respond in a sensitive and supportive manner to meet the needs of participants (Lakeman and Fitzgerald, 2009). When studies of suicides are conducted to provide better insights on how to reduce rates of suicide, these ethical concerns are important in all stages of the research process.

In this chapter we have looked at how theory and research work together to provide us with insights on human behavior. Theory and research are the "lifeblood" of sociology. Theory provides the framework for analysis; research provides opportunities for us to use our sociological imagination to generate new knowledge. Our challenge today is to find new ways of integrating knowledge and action and to include all people in the theory and research process in order to help fill the gaps in our existing knowledge about social life and how it is shaped by gender, race, class, age, and the broader social and cultural context in which everyday life occurs.

# chapter REVIEW

## LO1 What is sociology, and how can it help us understand others and ourselves?

Sociology is the systematic study of human society and social interaction. We study sociology to understand how human behavior is shaped by group life and, in turn, how individuals affect group life. Our culture tends to emphasize individualism, and sociology pushes us to consider more-complex connections between our personal lives and the larger world.

## LO2 What is meant by the sociological imagination, and how can we develop a global sociological imagination?

According to C. Wright Mills, the sociological imagination helps us understand how seemingly personal troubles, such as suicide, are actually related to larger social forces. It is the ability to see the relationship between individual experiences and the larger society. To develop a global sociological imagination, we must reach beyond past studies that have focused primarily on the United States and seek to develop a more comprehensive global approach. It is important to have a global sociological imagination because the future of this nation is deeply intertwined with the future of all nations of the world on economic, political, and humanitarian levels.

## LO3 What was the historical context in which sociological thinking developed, and why were many early social thinkers concerned with social order and stability?

The origins of sociological thinking as we know it today can be traced to the scientific revolution in the late-seventeenth and eighteenth centuries and to the Age of Enlightenment. Industrialization and urbanization increased rapidly in the late eighteenth century, and social thinkers began to examine the consequences of these powerful forces. As a result of rapid social change in societies, early thinkers—such as Auguste Comte, Harriet Martineau, Herbert Spencer, and Emile Durkheim—focused on social order and stability, and many of their ideas had a dramatic influence on modern sociology. Comte coined the term *sociology* to describe a new science that would engage in the study of society. Comte's works were made more accessible for a wide variety of scholars through the efforts of Martineau. Spencer's major contribution to sociology was an evolutionary perspective on social order and social change. Durkheim argued that rapid social change produces strains in society and that the loss of shared values and purpose can lead to a condition of anomie.

## LO4 Why were many later social thinkers concerned with social change?

In sharp contrast to Durkheim's focus on the stability of society, German economist and philosopher Karl Marx stressed that history is a continuous clash between conflicting ideas and forces. He believed that conflict—especially class conflict—is necessary in order to produce social change and a better society. Although he disagreed with Marx's idea that economics is *the* central force in social change, German social scientist Max Weber acknowledged that economic interests are important in shaping human action. Weber was particularly concerned with changes brought about by the Industrial Revolution and the influences these changes had on human behavior. In particular, Weber was concerned that large-scale organizations were becoming increasingly oriented toward routine administration and a specialized division of labor, which he believed were destructive to human vitality and freedom. Whereas other sociologists primarily focused on society as a whole, Simmel explored small social groups and argued that society was best seen as a web of patterned interactions among people.

## LO5 What are key differences in functionalist, conflict, symbolic interactionist, and postmodern perspectives on contemporary social life?

Functionalist perspectives assume that society is a stable, orderly system characterized by societal consensus. Conflict perspectives argue that society is a continuous power struggle among competing groups, often based on class, race, ethnicity, or gender. Symbolic interactionist perspectives focus on how people make sense of their everyday social interactions. From an alternative perspective, postmodern theorists believe that entirely new ways of examining social life are needed and that it is time to move beyond functionalist, conflict, and symbolic interactionist approaches.

## LO6 Why is sociological research necessary, and how does it challenge our commonsense beliefs about pressing social issues such as suicide?

Sociological research provides a factual and objective counterpoint to commonsense knowledge and ill-informed sources of information. It is based on an empirical approach that answers questions through a direct, systematic collection and analysis of data.

## LO7 How does quantitative research differ from qualitative research? What steps are involved in each method?

Quantitative research focuses on data that can be measured numerically (comparing rates of suicide, for example). Qualitative research focuses on interpretive description (words) rather than statistics to analyze underlying meanings and patterns of social relationships. Key steps in the quantitative research process include (1) selecting and defining the research problem; (2) reviewing previous research; (3) formulating the hypothesis, which involves constructing variables; (4) developing the research design; (5) collecting and analyzing the data; and (6) drawing conclusions and reporting the findings.

By contrast, key steps in the qualitative approach often include (1) reviewing the literature and formulating the problem to be studied instead of creating a hypothesis, (2) collecting and analyzing the data, and (3) reporting the results.

## LO8 What is survey research, and what are the three types of surveys?

The main types of research methods are surveys, secondary analysis of existing data, field research, and experiments. Surveys are polls used to gather facts about people's attitudes, opinions, or behaviors; a representative sample of respondents provides data through questionnaires or interviews. Survey data are collected by using self-administered questionnaires, face-to-face interviews, and telephone or computer surveys.

## LO9 How do the following research methods compare: surveys, secondary analysis of existing data, field research, and experiments?

Surveys are polls used to gather facts about people's attitudes, opinions, or behaviors; a representative sample of respondents provides data through questionnaires or interviews. In secondary analysis, researchers analyze existing data, such as a government census, or cultural artifacts, such as a diary. In field research, sociologists study social life in its natural setting through participant observation, case studies, unstructured interviews, and ethnography. Through experiments, researchers study the impact of certain variables on their subjects.

## LO10 What ethical issues are involved in sociological research, and what professional codes protect research participants?

Because sociology involves the study of people ("human subjects"), researchers are required to obtain the informed consent of the people they study; however, in some instances what constitutes "informed consent" may be difficult to determine. The American Sociological Association (ASA) *Code of Ethics* (1999) sets forth certain basic standards that sociologists must follow in conducting research.

# key TERMS

# questions for CRITICAL THINKING

1. Why does the concept of the sociological imagination help us see the relationship between our individual experiences and the larger society? How might this concept be applied to studying a current personal problem that you or one of your friends is experiencing?

2. As a sociologist, how would you remain objective yet still see the world as others see it? Would you make subjective decisions when trying to understand the perspectives of others?

3. With twenty-first-century digital technology, social media, and the rapid flow of global information, why are sociological theory and research still important in helping people understand social behavior and the larger social world?

# 2 CULTURE

## LEARNING OBJECTIVES

**LO1** Define *culture* and explain why it is important in helping people in their daily lives.

**LO2** Analyze material culture and nonmaterial culture, and give examples of each.

**LO3** Explain what is meant by the term *cultural universals*.

**LO4** Discuss how symbols and language reflect cultural values.

**LO5** Explain the differences among folkways, mores, and laws, and provide at least one example of each.

**LO6** Distinguish ways in which technological changes affect culture in a single nation and throughout the world.

**LO7** Define *culture shock, ethnocentrism*, and *cultural relativism*, and provide one example of each.

**LO8** Compare and contrast functionalist, conflict, symbolic interactionist, and postmodernist perspectives on society and culture.

41

# Spreading Culture Through Food Trucks?

**Food brings people** together in ways that words cannot, as it provides a universal outlet of creativity through which we can express ourselves. . . . We can pour culture and inspiration into our food, a subtle form of communication that is understood by all. The recent food truck craze in America has made this ability to bring people together through food even easier, and has indirectly spread cultural awareness with food as the medium.

—Veronica Werhane (2011), senior arts and culture editor for @neontommy, USC Annenberg School for Communication and Journalism, explaining how food trucks have become vehicles for spreading culture through food

How is the food that we consume linked to our identity and the larger culture of which we are a part? Do people who identify with more than one culture face more-complex issues when it comes to food preferences?

In a Yale University Introduction to Public Humanities class taught by Dean Ryan Brasseaux in fall 2013, students examined how New Haven, Connecticut, became a "food truck town" (*Yale News,* 2013). The students specifically wanted to find out how the local food culture was intertwined with New Haven's history, particularly in regard to racial and ethnic diversity. Using ethnographic filming, interactive mapping, and oral histories from the owners of food trucks ranging from the Cannoli Truck and Joe Grate's BBQ to Peking Edo Cart and Ricky D's Rib Shack,

**M**obile food trucks and street vendor carts have sold eats and drinks on the streets of U.S. cities for many years. Dating back to the 1960s, *loncheros,* or taco trucks, made daily stops at multiple locations in cities such as Los Angeles. The drivers slid open the shiny metal sides of their vehicles to display a variety of foods available for purchase at reasonable prices by workers from nearby offices or construction sites. By the 2000s, dramatic shifts had occurred in the mobile food industry as elaborately painted food trucks arrived on public streets and some college campuses throughout the nation. These newer trucks offered customers a wider variety of ethnic foods and sometimes even haute cuisine prepared from the menu of a famous chef. Many people had previously ridiculed food trucks, calling them "roach coaches," suggesting a lack of cleanliness, but now they referred to the elaborate trucks as a new cultural phenomenon and ordered foods such as Vietnamese bánh mi (a baguette with ingredients such as pork, cucumbers, fresh cilantro, sweet pickled carrots, daikon radish, cilantro, and chili peppers), Korean short-rib tacos and kimchi dogs, Brazilian barbeque, Greek sausages, Indian dosas, baklava (a popular dessert in Greece and Albania), and many other food offerings seemingly reflecting cultural diversity.

Although the trucks gained popularity partly because of the intense economic recession that began in 2007 and continued on for a number of years thereafter, economic problems were only one reason for the success of these mobile ventures. Other important cultural trends were also at play in the United States: The population was becoming more culturally diverse, younger generations were willing to experiment with a wider variety of tastes and textures in their food choices, and the explosion of social media made it possible for potential customers to keep up the students created a "food routes" project. The proprietors were asked questions, such as why they chose to enter the food truck business, how they determined their routes, what their relationship was to the cuisine they served, and where they saw themselves and their trucks fitting into the bigger picture of New Haven life. Then the students created maps that reflected patterns of cultural diversity and food consumption in New Haven. According to the students, the food truck movement helps to "illustrate the importance of migration to the city of New Haven and the various cultural and social networks that exist throughout the city" (*Yale News,* 2013).

What do you think you might learn about food and culture if you participated in a similar project in your own city? Has the food truck trend helped to spread cultural awareness? Why or why not?

## How Much Do You Know About Global Food and Culture?

| True | False | |
|------|-------|---|
| T | F | 1. Cheese is a universal food enjoyed by people of all nations and cultures. |
| T | F | 2. Giving round-shaped foods to the parents of new babies is considered to be lucky in some cultures. |
| T | F | 3. Wedding cakes are made of similar ingredients in all countries, regardless of culture or religion. |
| T | F | 4. Food is an important part of religious observance for many different faiths. |
| T | F | 5. In authentic Chinese cuisine, cooking methods are divided into "yin" and "yang" qualities. |

| True | False | |
|------|-------|---|
| T | F | 6. Because of the fast pace of life in the United States, virtually everyone relies on mixes and instant foods at home and fast foods when eating out. |
| T | F | 7. Potatoes are the most popular mainstay in the diet of first- and second-generation immigrants who have arrived in the United States over the past 40 years. |
| T | F | 8. According to sociologists, individuals may be offended when a person from another culture does not understand local food preferences or the cultural traditions associated with eating even if the person is obviously an "outsider" or a "tourist." |

Answers on page 45.

with the location and daily offerings of their favorite food trucks. Finally, people became more aware that food is intricately intertwined with culture and with our personal identities pertaining to race, ethnicity, religion, class, gender, age, and other factors.

For all of us, the food we consume is linked to our individual identity and to the larger culture of which we are a part. To some people, food consumption is nothing more than how we meet a basic biological need; however, many sociologists are interested in food and eating because of this subject's cultural significance in our lives.

In this chapter we examine society and culture, with special attention to how our material culture, including the food we eat, is related to our beliefs, values, and actions. We also analyze culture from functionalist, conflict, symbolic interactionist, and postmodern perspectives. Before reading on, test your knowledge of the relationship among food, its distribution, and the culture in which we live by answering the questions in the Sociology and Everyday Life box.

## Culture and Society in a Changing World

**LO1** What is culture? ***Culture* is the knowledge, language, values, customs, and material objects that are passed from person to person and from one generation to the next in a human group or society.** As previously defined, a

**culture** the knowledge, language, values, customs, and material objects that are passed from person to person and from one generation to the next in a human group or society.

*society* is a large social grouping that occupies the same geographic territory and is subject to the same political authority and dominant cultural expectations. Whereas a society is composed of people, a culture is composed of ideas, behavior, and material possessions. Society and culture are interdependent; neither could exist without the other.

How important is culture in determining how people think and act on a daily basis? Simply stated, culture is essential for our individual survival and our communication with other people. We rely on culture because we are not born with the information we need to survive. We do not know how to take care of ourselves, how to behave, how to dress, what to eat, which gods to worship, or how to make or spend money. We must learn about culture through interaction, observation, and imitation in order to participate as members of the group. Sharing a common culture with others simplifies day-to-day interactions. However, we must also understand other cultures and the worldviews therein.

Just as culture is essential for individuals, it is also fundamental for the survival of societies. Culture has been described as the common foundation or core that makes individuals understandable to the larger group of which they are a part. Some system of rule making and enforcing necessarily exists in all societies. What would happen, for example, if *all* rules and laws in the United States suddenly disappeared? At a basic level, we need rules in order to navigate our bicycles and cars through traffic. At a more abstract level, we need laws to establish and protect our rights.

In order to survive, societies need rules about civility and tolerance. We are not born knowing how to express certain kinds of feelings toward others. When a person shows kindness or hatred toward another individual, some people may say "Well, that's just human nature" when explaining this behavior. Such a statement is built on the assumption that what we do as human beings is determined by *nature* (our biological and genetic makeup) rather than *nurture* (our social environment)—in other words, that our behavior is instinctive. An *instinct* is an unlearned, biologically determined behavior pattern common to all members of a species that predictably occurs whenever certain environmental conditions exist. For example, spiders do not learn to build webs. They build webs because of instincts that are triggered by basic biological needs such as protection and reproduction.

Culture is similar to instincts in animals because it helps us deal with everyday life. Although people may have some instincts, what we most often think of as instinctive behavior can actually be attributed to reflexes and drives. A *reflex* is an unlearned, biologically determined involuntary response to some physical stimulus (such as a sneeze after breathing some pepper in through the nose or the blinking of an eye when a speck of dust gets in it). *Drives* are unlearned, biologically determined impulses common to all members of a species that satisfy needs such as those for sleep, food, water, or sexual gratification. Reflexes and drives do not determine how people will behave in human societies; even the expression of these biological characteristics is channeled by culture. For example, we may be taught that the "appropriate" way to sneeze (an involuntary response) is to use a tissue or turn our head away from others (a learned response). Similarly, we may learn to sleep on mats or in beds. Most contemporary sociologists agree that culture and social learning, not nature, account for most of our behavior patterns.

Because humans cannot rely on instincts in order to survive, culture is a "tool kit" for survival (Swidler, 1986). From this approach, culture serves as a tool kit full of abstract things such as our beliefs and rituals, symbols, personal narratives, and overall perspectives on the world. People use these in a variety of configurations to help solve the problems they face. The tools we choose vary according to our own personality and the situations we face. We are not puppets on a string; we make choices from among the items in our own "tool kit."

## Material Culture and Nonmaterial Culture

**LO2** Our cultural tool kit is divided into two major parts: material culture and nonmaterial culture. **Material culture consists of the physical or tangible creations that members of a society make, use, and share.** Initially, items of material culture begin as raw materials or resources such as ore, trees, and oil. Through technology, these raw materials are transformed into usable items, ranging from books and computers to guns and tanks. Technology is both concrete and abstract. For example, technology includes computers, smartphones, iPads and other tablets, and the knowledge and skills necessary to use them. At the most basic level, material culture is important because it is our buffer against the environment. For example, we create shelter to protect ourselves from the weather and to give ourselves privacy. Beyond the survival level, we make, use, and share objects that are interesting and important to us. Why are you wearing

# ANSWERS to the Sociology Quiz on Global Food and Culture

**1. False.** Although cheese is a popular food in many cultures, most of the people living in China find cheese distasteful and prefer delicacies such as duck's feet.

**2. True.** In some cultures, round foods such as pears, grapes, and moon cakes are given to celebrate the birth of babies because the shape of the food is believed to symbolize family unity.

**3. False.** Although wedding cakes are a tradition in virtually all nations and cultures, the ingredients of the cake—as well as other foods served at the celebration—vary widely. For example, the traditional wedding cake in Italy is made from biscuits.

**4. True.** Many faiths, including Christianity, Judaism, Islam, Hinduism, and Buddhism, have dietary rules and rituals that involve food; however, these practices and beliefs vary widely among individuals and communities.

**5. True.** Just as foods are divided into yin foods (e.g., bean sprouts and carrots) and yang foods (beef, chicken, and mushrooms), cooking methods are also referred to as having yin qualities (e.g., boiling and steaming) or yang qualities (roasting and stir-frying). For many Chinese Americans, yin and yang are complementary pairs that should be incorporated into all aspects of social life, including the ingredients and preparation of foods.

**6. False.** Although more people now rely on fast foods, some cultural and religious communities—such as the Amish of Ohio, Pennsylvania, and Indiana—encourage families to prepare their food from scratch and to preserve their own fruits, vegetables, and meats. Rural families are more likely to grow their own food or prepare it from scratch than are urban families.

**7. False.** Rice is a popular mainstay in the diets of people from diverse cultural backgrounds—from the Hmong and Vietnamese to Puerto Ricans and Mexican Americans—who have arrived in the United States over the past four decades. Among some in the younger generations, however, food choices have become increasingly Americanized, and items such as french fries and pizza have become popular.

**8. True.** Cultural diversity is a major issue in eating, and people in some cultures, religions, and nations expect that even an "outsider" will have a basic familiarity with, and respect for, their traditions and practices. However, social analysts also suggest that we should not generalize or imply that certain characteristics apply to *all* people in a cultural group or nation.

Sources: Based on Better Health Channel, 2007; Ohio State University, 2007; and PBS, 2005. © Cengage Learning.

the particular clothes you have on today? Perhaps you're communicating something about yourself, such as where you attend school, what kind of music you like, or where you went on vacation.

***Nonmaterial culture* consists of the abstract or intangible human creations of society that influence people's behavior.** Language, beliefs, values, rules of behavior, family patterns, and political systems are examples of nonmaterial culture. Even the gestures that we use in daily conversation are part of the nonmaterial culture in a society. As many international travelers and businesspeople have learned, it is important to know what gestures mean in various nations (see Figure 2.1). Although the "hook 'em Horns" sign—the pinky and index finger raised

up and the middle two fingers folded down—is used by fans to express their support for University of Texas at Austin sports teams, for millions of Italians the same gesture means "Your spouse is being unfaithful." In Argentina, rotating one's index

**material culture** the physical or tangible creations that members of a society make, use, and share.

**nonmaterial culture** the abstract or intangible human creations of society that influence people's behavior.

a. HORNS: "Hook 'em Horns" or "Your spouse is unfaithful."

b. CIRCLE: "OK" (absolutely fine) or "I'll kill you."

c. THUMBS UP: "Great" or an obscenity.

**FIGURE 2.1** HAND GESTURES WITH DIFFERENT MEANINGS

As international travelers and businesspeople have learned, hand gestures may have very different meanings in different cultures.

finger around the front of the ear means "You have a telephone call," but in the United States it usually suggests that a person is "crazy." Similarly, making a circle with your thumb and index finger indicates "OK" in the United States, but in Tunisia it means "I'll kill you!"

As the example of hand gestures shows, a central component of nonmaterial culture is **beliefs—the mental acceptance or conviction that certain things are true or real.** Beliefs may be based on tradition, faith, experience, scientific research, or some combination of these. Faith in a Supreme Being and trust in another person are examples of beliefs. We may also have a belief in items of material culture. When we travel by airplane, for instance, we believe that it is possible to fly at 33,000 feet and to arrive at our destination even though we know that we could not do this without the airplane itself.

## Cultural Universals

**LO3** Because all humans face the same basic needs (such as for food, clothing, and shelter), we engage in similar activities that contribute to our survival. Anthropologist George Murdock (1945: 124) compiled a list of more than seventy *cultural universals*—**customs and practices that occur across all societies.** His categories included appearance (such as bodily adornment and hairstyles), activities (such as sports, dancing, games, joking, and visiting), social institutions (such as family, law, and religion), and customary practices (such as cooking, folklore, gift giving, and hospitality). Whereas these general customs and practices may be *present* in all cultures, their specific *forms*

vary from one group to another and from one time to another within the same group. For example, although telling jokes may be a universal practice, what is considered to be a joke in one society may be an insult in another.

How do sociologists view cultural universals? In terms of their functions, cultural universals are useful because they ensure the smooth and continual operation of society. A society must meet basic human needs by providing food, shelter, and some degree of safety for its members so that they will survive (Figure 2.2). Children and other new members (such as immigrants) must be taught the ways of the group. A society must also settle disputes and deal with people's emotions. All the while, the self-interest of individuals must be balanced with the needs of society as a whole. Cultural universals help fulfill these important functions of society.

From another perspective, however, cultural universals are not the result of functional necessity; members of one society may have *imposed* these practices on members of another. Similar customs and practices do not necessarily constitute cultural universals. They may be an indication that a conquering nation used its power to enforce certain types of behavior on those who were defeated. Sociologists might ask questions such as "Who determines the dominant cultural patterns?" For example, although religion is a cultural universal, traditional religious practices of indigenous peoples (those who first live in an area) have often been repressed and even stamped out by subsequent settlers or conquerors who possess political and economic power over them. However, many people believe there is cause for optimism in the United

States because the democratic ideas of this nation provide more guarantees of religious freedom than do some other nations.

## Components of Culture

**LO4** Even though the specifics of individual cultures vary widely, all cultures have four common nonmaterial cultural components: symbols, language, values, and norms. These components contribute to both harmony and strife in a society.

### Symbols

A *symbol* **is anything that meaningfully represents something else.** Culture could not exist without symbols because there would be no shared meanings among people. Symbols can simultaneously produce loyalty and animosity, and love and hate. They help us communicate ideas such as love or patriotism because they express abstract concepts with visible objects (Figure 2.3). For example, flags can stand for patriotism, nationalism, school spirit, or religious beliefs held by members of a group or society. Symbols can stand for love (a heart on a valentine), peace (a dove), or hate (a Nazi swastika), just as words can be used to convey these meanings. Symbols can also transmit other types of ideas. A siren is a symbol that denotes an emergency situation and sends the message to clear the way immediately. Gestures are also a symbolic form of communication—a movement of the head, body, or hands can express our

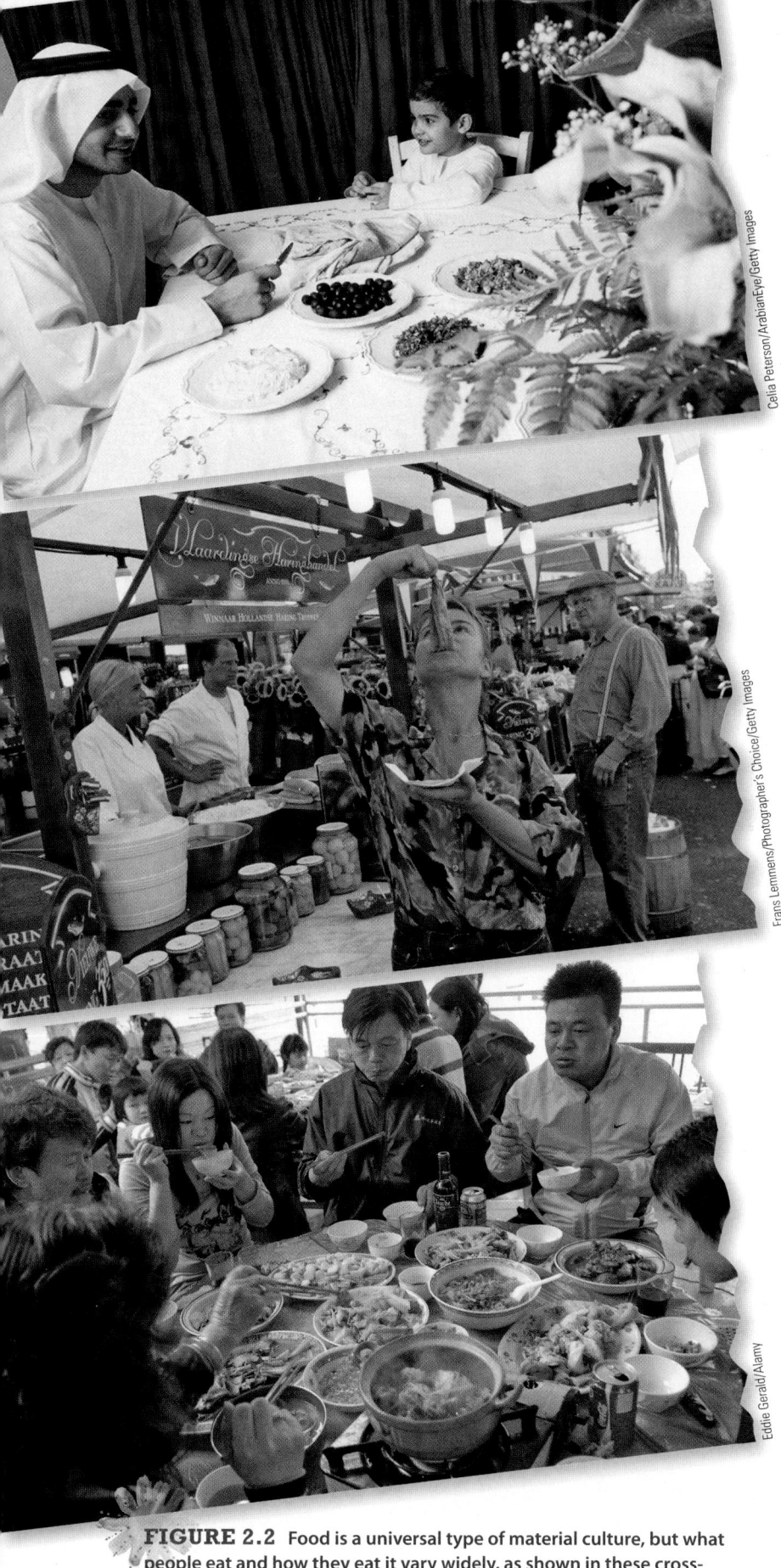

Celia Peterson/ArabianEye/Getty Images

Frans Lemmens/Photographer's Choice/Getty Images

Eddie Gerald/Alamy

**FIGURE 2.2** Food is a universal type of material culture, but what people eat and how they eat it vary widely, as shown in these cross-cultural examples from the United Arab Emirates (top), the Netherlands (middle), and China (bottom). What might be some reasons for the similarities and differences that you see in these photos?

**beliefs** the mental acceptance or conviction that certain things are true or real.

**cultural universals** customs and practices that occur across all societies.

**symbol** anything that meaningfully represents something else.

47

ideas or feelings to others. For example, in the United States, pointing toward your chest with your thumb or finger is a symbol for "me."

Symbols affect our thoughts about class. For example, how a person is dressed or the kind of car that he or she drives is often at least subconsciously used as a measure of that individual's economic standing or position. With regard to clothing, although many people wear casual clothes on a daily basis, where the clothing was purchased is sometimes used as a symbol of social status (Figure 2.4). Were the items purchased at Walmart, Old Navy, Abercrombie & Fitch, or Neiman Marcus? What indicators are there on the items of clothing—such as the Nike *swoosh*, some other logo, or a brand name—that indicate something about the status of the product? Automobiles and their logos are also symbols that have cultural meaning beyond the shopping environment in which they originate.

Finally, symbols may be specific to a given culture and have special meaning to individuals who share that culture but not necessarily to other people. Consider, for example, the use of certain foods to celebrate the Chinese New Year: Bamboo shoots and black-moss seaweed both represent wealth, peanuts and noodles symbolize a long life, and tangerines represent good luck. What foods in other cultures represent "good luck" or prosperity? In countries throughout the world, food and drink are powerful symbols of the history and cultural identity of people residing in the area.

## Language

**Language is a set of symbols that expresses ideas and enables people to think and communicate with one another.** Verbal (spoken) language and nonverbal (written or gestured) language help us describe reality. One of our most important human attributes is the ability to use language to share our experiences, feelings, and knowledge with others. Language can create visual images in our head, such as describing small white kittens as looking like "snowballs." Language also allows people to distinguish themselves from outsiders and maintain group boundaries and solidarity.

Language is not solely a human characteristic. Other animals use sounds, gestures, touch, and smell to communicate with one another, but they use signals with fixed

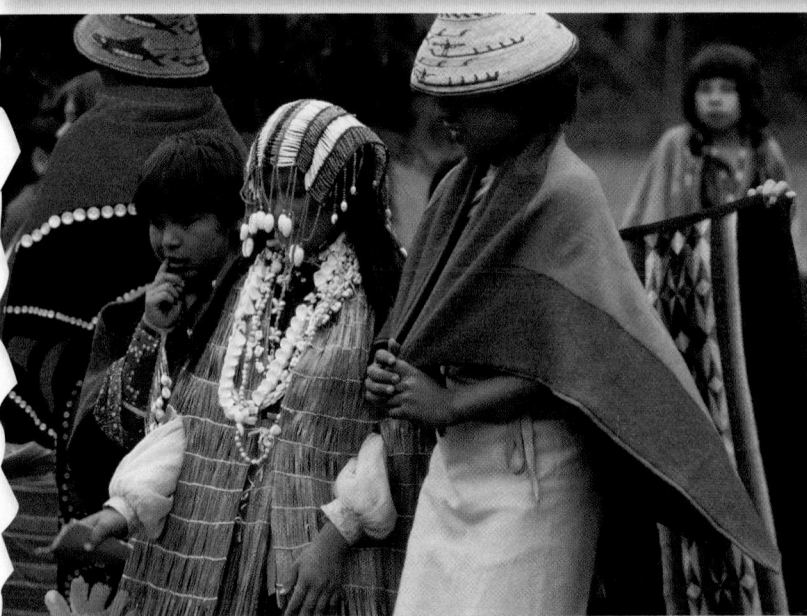

**FIGURE 2.3** The customs and rituals associated with weddings are one example of nonmaterial culture. What can you infer about beliefs and attitudes concerning marriage in the societies represented by these photographs?

meanings that are limited to the immediate situation (the present) and that cannot encompass past or future situations. For example, chimpanzees can use elements of American Sign Language and manipulate physical objects to make "sentences," but they are not physically endowed with the vocal apparatus needed to form the consonants required for oral language. As a result, non-human animals cannot transmit the more complex aspects of culture to their offspring. Humans have a unique ability to manipulate symbols to express abstract concepts and rules and thus to create and transmit culture from one generation to the next.

**Language and Social Reality** Does language *create* reality or simply *communicate* reality? Anthropological linguists Edward Sapir and Benjamin Whorf have suggested that language not only expresses our thoughts and perceptions but also influences our perception of reality. According to the **Sapir–Whorf hypothesis, language shapes the view of reality of its speakers** (Whorf, 1956; Sapir, 1961). If people are able to think only through language, then language must precede thought. If language actually shapes the reality we perceive and experience, then some aspects of the world are viewed as important and others are virtually neglected because people know the world only in terms of the vocabulary and grammar of their own language.

If language does create reality, does our language trap us? Many social scientists agree that the Sapir–Whorf hypothesis overstates the relationship between language and our thoughts and behavior patterns. Although they acknowledge that language has many subtle meanings and that words used by people reflect their central concerns, most sociologists contend that language may *influence* our behavior and interpretation of social reality but does not *determine* it.

**Language and Gender** How are language and gender related? What cultural assumptions about women and men does language reflect? Scholars have suggested several ways in which language and gender are intertwined:

- The English language ignores women by using the masculine form to refer to human beings in general. For example, the word *man* is used generically in words such as *chairman* and *mankind*, which allegedly include both men and women.
- Use of the pronouns *he* and *she* affects our thinking about gender. Pronouns show the gender of the person we *expect* to be in a particular

Justin Sullivan/Getty Images

**FIGURE 2.4** Would you expect the user of this device to be impoverished or affluent? What do possessions indicate about their owner's social class?

occupation. For instance, nurses, secretaries, and schoolteachers are usually referred to as *she,* but doctors, engineers, electricians, and presidents are usually referred to as *he.*

- Words have positive connotations when relating to male power, prestige, and leadership; when related to women, they carry negative overtones of weakness, inferiority, and immaturity. Table 2.1 shows how gender-based language reflects the traditional acceptance of men and women in certain positions, implying that the jobs are different when filled by women rather than men (Figure 2.5).
- A language-based predisposition to think about women in sexual terms reinforces the notion that women are sexual objects. Terms such as *fox, broad, bitch, babe,* and *doll* often describe women, which ascribe childlike or even pet-like characteristics to them. By contrast, men have performance pressures placed on them by being defined in terms of their sexual prowess, such as *dude, stud,* and *hunk.*

**language** a set of symbols that expresses ideas and enables people to think and communicate with one another.

**Sapir–Whorf hypothesis** the proposition that language shapes the view of reality of its speakers.

## table 2.1

### Language and Gender

| Male Term | Female Term | Neutral Term |
| --- | --- | --- |
| Teacher | Teacher | Teacher |
| Chairman | Chairwoman | Chair, chairperson |
| Congressman | Congresswoman | Representative |
| Policeman | Policewoman | Police officer |
| Fireman | Female fireman | Firefighter |
| Airline steward | Airline stewardess | Flight attendant |
| Race car driver | Female race car driver | Race car driver |
| Professor | Teacher/female professor | Professor |
| Doctor | Female doctor | Doctor |
| Bachelor | Bachelorette | Single person |
| Male prostitute | Prostitute | Prostitute |
| Welfare recipient | Welfare mother | Welfare recipient |
| Worker/employee | Working mother | Worker/employee |
| Janitor/maintenance man | Maid/cleaning lady | Custodial attendant |

Sources: Adapted from Korsmeyer, 1981: 122; and Miller and Swift, 1991.
© Cengage Learning.

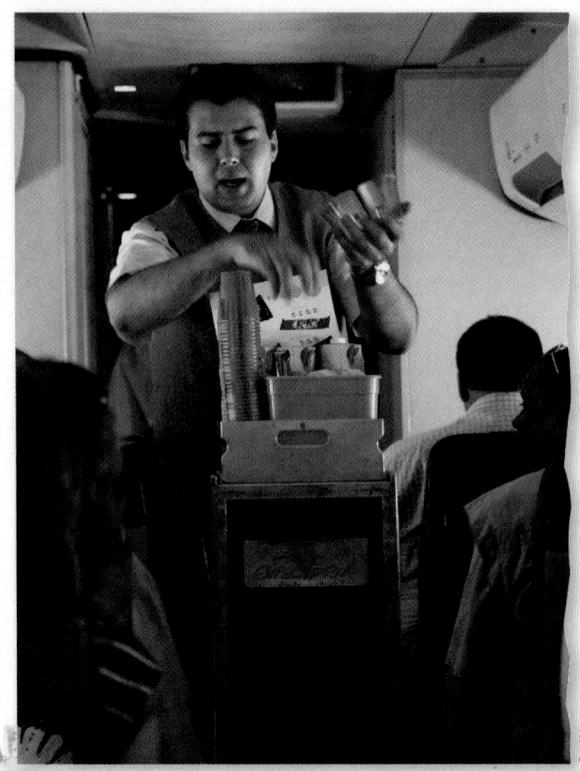

**FIGURE 2.5** Certain jobs are stereotypically considered to be "men's jobs"; others are "women's jobs." Is your perception of a male flight attendant the same as your perception of a female flight attendant?

Gender in language has been debated and studied extensively for many years now, and some changes have occurred. The preference of many women to be called *Ms.* (rather than *Miss* or *Mrs.* in reference to their marital status) has received a degree of acceptance in public life and the media. Many organizations and publications have established guidelines for the use of nonsexist language and have changed titles such as *chairman* to *chair* or *chairperson*. To develop a more inclusive and equitable society, many scholars suggest that a more inclusive language is needed.

**Language, Race, and Ethnicity** Language may create and reinforce our perceptions about race and ethnicity by transmitting preconceived ideas about the superiority of one category of people over another. Let's look at a few images conveyed by words in the English language in regard to race/ethnicity:

- Words may have more than one meaning and create and/or reinforce negative images. In the nineteenth and twentieth centuries, terms such as *black-hearted* (malevolent) and expressions such as a *black mark* (a detrimental fact) and *China-man's chance of success* (unlikely to succeed) were used to associate the words *black* and *Chinaman* with derogatory imagery. Although these terms are seldom used today, they are occasionally referenced in popular culture and film.

- Overtly derogatory terms such as *nigger, kike, gook, honky, chink, spic,* and other racial–ethnic slurs have been "popularized" in movies, music, comedy routines, and so on. Such derogatory terms are often used in conjunction with physical threats against persons.

- Words are frequently used to create or reinforce perceptions about a group. For example, Native Americans have been referred to as "savage" and "primitive," and African Americans have been described as "uncivilized," "cannibalistic," and "pagan."

- The "voice" of verbs may minimize or incorrectly identify the activities or achievements of people of color. For example, the use of the passive voice in the statement "African Americans *were given* the right to vote" ignores how African Americans *fought* for that right. Active-voice verbs may also inaccurately attribute achievements to people or groups. Some historians argue that cultural bias is shown by the very notion that "Columbus discovered America"—given that America was

Ort/Alamy

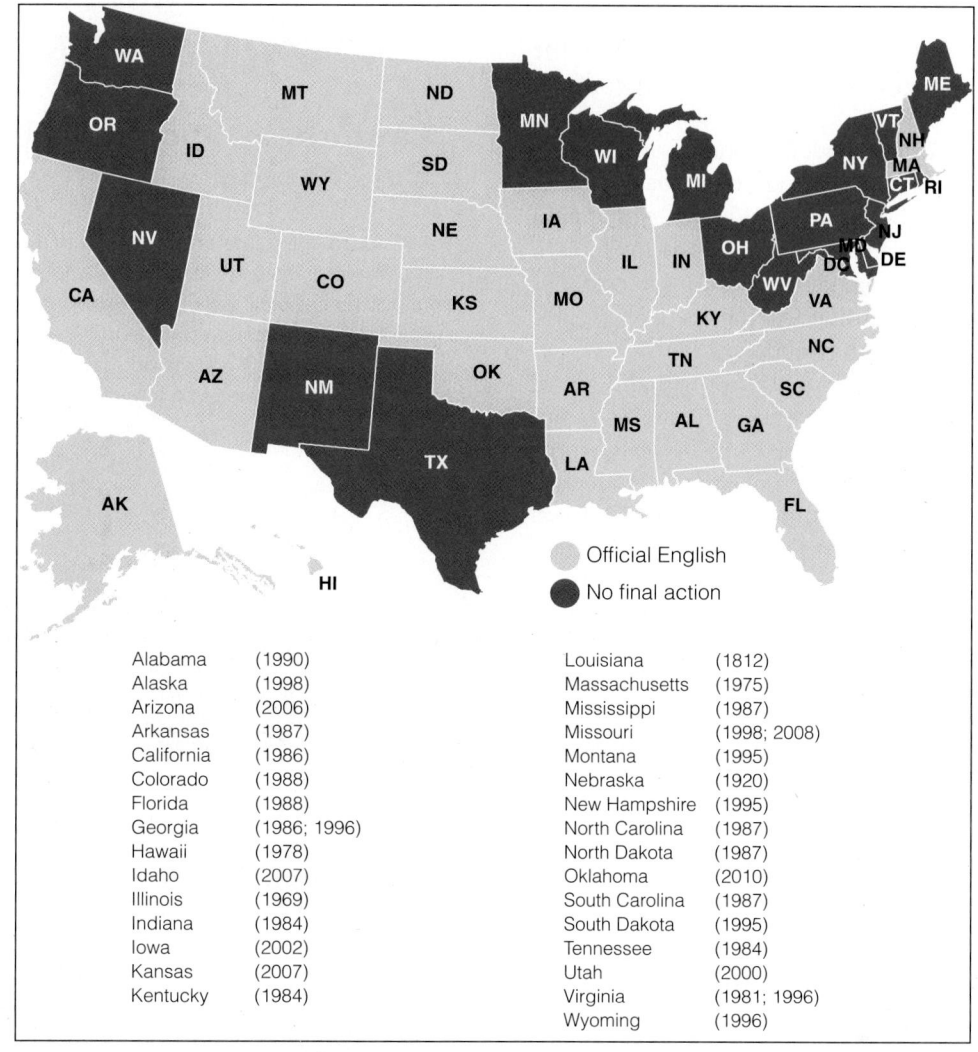

| | | | | |
|---|---|---|---|---|
| Alabama | (1990) | | Louisiana | (1812) |
| Alaska | (1998) | | Massachusetts | (1975) |
| Arizona | (2006) | | Mississippi | (1987) |
| Arkansas | (1987) | | Missouri | (1998; 2008) |
| California | (1986) | | Montana | (1995) |
| Colorado | (1988) | | Nebraska | (1920) |
| Florida | (1988) | | New Hampshire | (1995) |
| Georgia | (1986; 1996) | | North Carolina | (1987) |
| Hawaii | (1978) | | North Dakota | (1987) |
| Idaho | (2007) | | Oklahoma | (2010) |
| Illinois | (1969) | | South Carolina | (1987) |
| Indiana | (1984) | | South Dakota | (1995) |
| Iowa | (2002) | | Tennessee | (1984) |
| Kansas | (2007) | | Utah | (2000) |
| Kentucky | (1984) | | Virginia | (1981; 1996) |
| | | | Wyoming | (1996) |

**FIGURE 2.6** STATES WITH OFFICIAL ENGLISH LAWS

Why do some states have official English laws while others do not? How does the composition of the population in each state affect the passage of laws regarding language? Do you see any similarities in the states that have official English laws versus those that don't? What conclusions can you draw from this map?

Source: U.S. English, 2013

already inhabited by people who later became known as Native Americans.

In addition to these concerns about the English language, problems also occur when more than one language is involved. Across the nation, the question of whether the United States should have an "official" language continues to arise. Some people believe that there is no need to designate an official language; other people believe that English should be designated as the official language and that the use of any other language in official government business should be discouraged or negatively sanctioned. By the end of 2013, thirty-one states (see Figure 2.6) had passed laws requiring that all public documents, records, legislation, and regulations, as well as hearings, official ceremonies, and public meetings, be written or conducted solely in English.

Are deep-seated social and cultural issues embedded in social policy decisions such as these? Although the United States has always been a nation of immigrants, in recent decades this country has experienced rapid changes in population that have brought about greater diversity in languages and cultures. Recent data gathered by the U.S. Census Bureau (see "Census Profiles: Languages Spoken in U.S. Households") indicate that although 79 percent of the people in this country speak only English at home, 21 percent speak a language other than English. The largest portion of non-English speakers speak Spanish at home.

Language is an important means of cultural transmission. Through language, children learn about their cultural heritage and develop a sense of personal identity in relation to their group.

Latinos/as in New Mexico and South Texas use *dichos*—proverbs or sayings that are unique to the Spanish language—as a means of expressing themselves and as a reflection of their cultural heritage. Examples of *dichos* include *"Anda tu camino sin ayuda de vecino"* ("Walk your own road without the help of a neighbor") and *"Amor de lejos es para pendejos"* ("A long-distance romance is for fools"). *Dichos* are passed from generation to generation as a priceless verbal tradition whereby people can give advice or teach a lesson (Gandara, 1995).

Language is also a source of power and social control; language perpetuates inequalities between people and between groups because words are used (intentionally or not) to "keep people in their place." As the linguist Deborah Tannen (1993: B5) has suggested, "The devastating group hatreds that result in so much suffering in our own country and around the world are related in origin to the small intolerances in our everyday conversations—our readiness to attribute good intentions to ourselves and bad intentions to others." Language, then, is a reflection of our feelings and values.

## Values

**Values are collective ideas about what is right or wrong, good or bad, and desirable or undesirable in a particular culture.** Values do not dictate which behaviors are appropriate and which ones are not, but they provide us with the criteria by which we evaluate people, objects, and events. Values typically come in pairs of positive and negative values, such as being brave or cowardly, hardworking or lazy. Because we use values to justify our behavior, we tend to defend them staunchly.

**Core American Values** Do we have shared values in the United States? Sociologists disagree about the extent to which all people in this country share a core set of values. Functionalists tend to believe that shared values are essential for the maintenance of a society, and scholars using a functionalist approach have conducted most of the research on core values. Analysts who focus on the importance of core values maintain that the values identified by sociologist Robin M. Williams Jr., in 1970 are important to people in the United States but that they have changed somewhat in recent years. How important do you think the following ten values are?

1. *Individualism.* People in the United States are responsible for their own success or failure. Those who do not succeed have only themselves to

## census PROFILES

### Languages Spoken in U.S. Households

Among the categories of information gathered by the U.S. Census Bureau are data on the languages spoken in U.S. households. As shown below, English is the only language spoken at home in 79 percent of U.S. households. Some language other than English is the primary language spoken at home in 21 percent of U.S. households.

Source: U.S. Census Bureau, 2012b.

People who speak a language other than English at home are asked to indicate which other languages they speak. The principal languages other than English that are most frequently spoken at home are shown in the following chart. Do you think that changes in the languages spoken in this country will bring about other significant changes in U.S. culture? Why or why not?

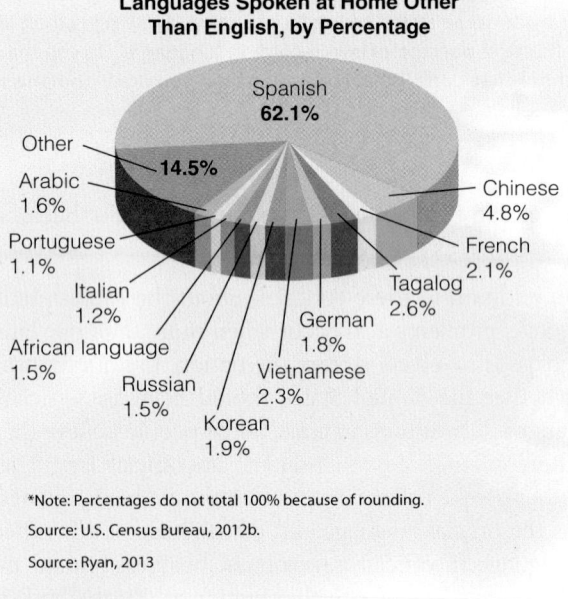

**Languages Spoken at Home Other Than English, by Percentage**

*Note: Percentages do not total 100% because of rounding.

Source: U.S. Census Bureau, 2012b.

Source: Ryan, 2013

blame because of their lack of ability, laziness, immorality, or other character defects.

2. *Achievement and success.* Personal achievement results from successful competition with others.

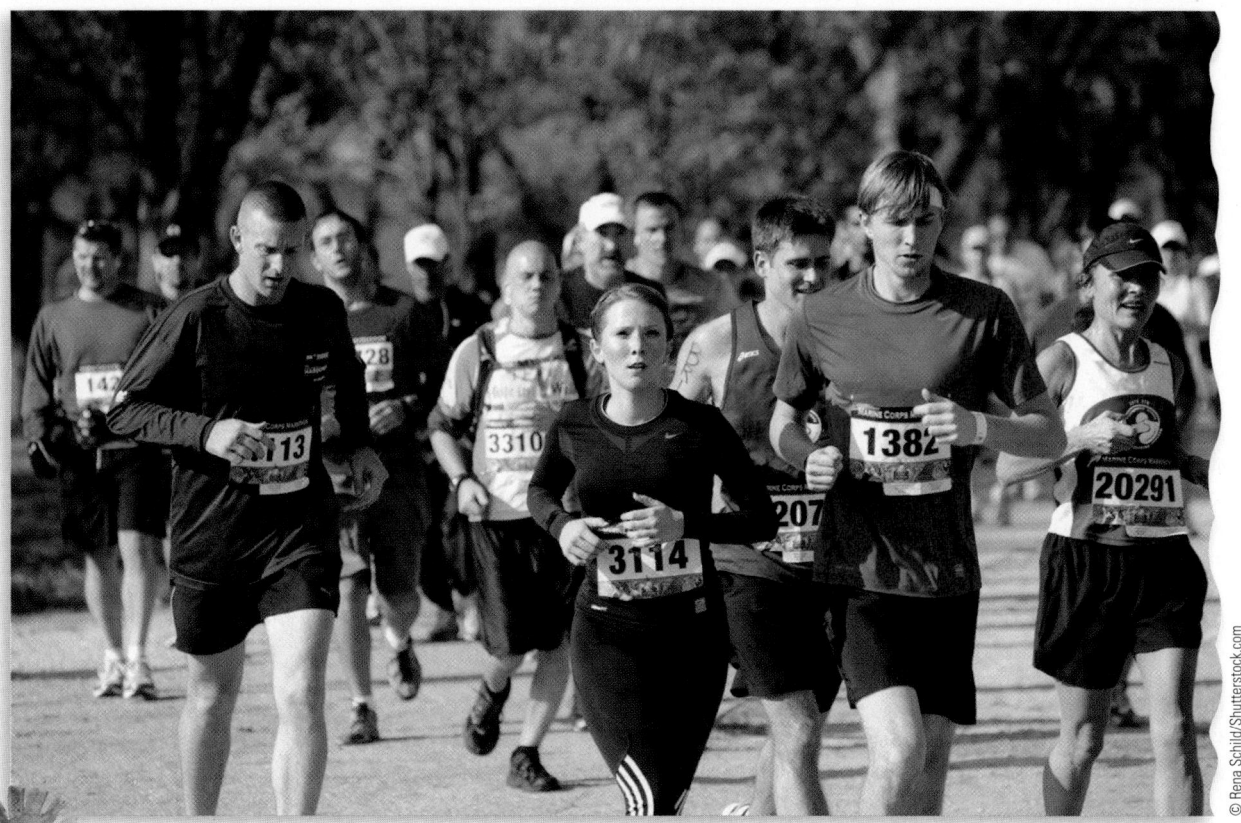

**FIGURE 2.7** Even during their leisure time, many people "work" in their play.

Individuals are encouraged to do better than others in school and to work in order to gain wealth, power, and prestige. Material possessions are seen as a sign of personal achievement.

3. *Activity and work.* People who are industrious are praised for their achievement; those perceived as lazy are ridiculed. From the time of the early Puritans, work has been viewed as important. Even during their leisure time, many people "work" in their play. For example, think of all the people who run in marathons, repair or restore cars, and so on in their spare time (Figure 2.7).

4. *Science and technology.* People in the United States have a great deal of faith in science and technology. They expect scientific and technological advances ultimately to control nature, the aging process, and even death.

5. *Progress and material comfort.* The material comforts of life include not only basic necessities (such as adequate shelter, nutrition, and medical care) but also the goods and services that make life easier and more pleasant.

6. *Efficiency and practicality.* People want things to be bigger, better, and faster. As a result, great value is placed on efficiency ("How well does it work?").

7. *Equality.* Since colonial times, overt class distinctions have been rejected in the United States. However, *equality* has been defined as "equality of *opportunity*"—an assumed equal chance to achieve success—not as "equality of *outcome*."

8. *Morality and humanitarianism.* Aiding others, especially following natural disasters (such as fires, hurricanes, and other natural disasters), is seen as a value. The notion of helping others was originally a part of religious teachings and tied to the idea of morality. Today, people engage in humanitarian acts without necessarily perceiving that it is the "moral" thing to do.

9. *Freedom and liberty.* Individual freedom is highly valued in the United States. The idea of freedom includes the right to private ownership of property, the ability to engage in private enterprise, freedom of the press, and other freedoms considered to be "basic" rights.

10. *Ethnocentrism and group superiority.* People value their own racial or ethnic group above all others. Such feelings of pride and superiority may lead to discrimination; in the past, slavery and segregation laws were an example. Many people

**values** collective ideas about what is right or wrong, good or bad, and desirable or undesirable in a particular culture.

**53**

# SOCIOLOGY *in Global Perspective*

## What Do Cultural Norms Say About Drinking Behavior?

In the United States, most of us are familiar with norms and rituals associated with alcohol consumption and/or drinking behavior. Some norms are formal, such as laws that set forth the legal age at which young people may consume alcoholic beverages. Other drinking norms are informal, such as whether champagne, wine, or beer should be served at a wedding celebration. Let's take a brief look at the Republic of Georgia, where highly formalized norms have been established regarding the consumption of wine at meals.

In the Republic of Georgia, people participate in celebratory and memorial occasions known as *supras*—feasts or banquets. More-formal *supras* take place at holidays or significant events, such as birthdays, weddings, baptisms, or funerals. Less-formal *supras* mark the gathering of friends and family to enjoy food, music, and dancing while engaging in ritualized wine drinking that marks one's personal and national identity, as well as his or her place in the social hierarchy.

The norms of eating are more informal at the *supra,* but the drinking norms are strictly formal, starting with the first toast, which sets the character, structure, and meaning of what is being celebrated or mourned (Muehlfried, 2007). Georgian wine cannot be consumed without first having an eloquent speech given by the *tamada,* or toastmaster, who must adhere to strict rules of etiquette to ensure that honor is appropriately distributed to the guests. The *tamada* must be able to give extensive toasts, share humorous stories, and consume large amounts of wine without showing any sign of fatigue or intoxication. The norms governing toasts emphasize the importance of whom the *tamada* addresses, when, how, in what order, and for how long. It is also important who drinks when and how much because such details demonstrate people's status. When a toast is given to women or the deceased, for example, women and children remain seated, but boys of a certain age must stand up to show that they have become men. Older men who no longer participate in

---

also believe in the superiority of this country and that "the American way of life" is best.

Are these values still important today? According to a report by the Pew Research Center Global Attitudes Project (2011), people in the United States remain more individualistic than people residing in Western Europe. However, Americans appear to be less inclined to view their culture and way of life as superior to others than they have in the past. Overall, it appears that core values are an important component of culture in all societies but that over time, they tend to shift based on economic conditions, social trends, religious beliefs, and other factors that arise in those nations and around the globe.

**Value Contradictions** All countries, including the United States, have value contradictions. *Value contradictions* **are values that conflict with one another or are mutually exclusive** (meaning that achieving one value makes it difficult, if not impossible, to achieve another). Core values of morality and humanitarianism may conflict with values of individual achievement and success. For example, humanitarian values reflected in welfare and other government-aid programs for people in need come

into conflict with values emphasizing hard work and personal achievement.

**Ideal Culture Versus Real Culture** What is the relationship between values and human behavior? Sociologists stress that a gap always exists between ideal culture and real culture in a society. *Ideal culture* refers to the values and standards of behavior that people in a society profess to hold. *Real culture* refers to the values and standards of behavior that people actually follow. For example, we may claim to be law-abiding (ideal cultural value) but may regularly drive over the speed limit (real cultural behavior) and still think of ourselves as "good citizens." We may believe in the value of honesty and, at the same time, engage in deception.

## Norms

**L05** Values provide ideals or beliefs about behavior but do not state explicitly how we should behave. Norms, on the other hand, do have specific behavioral expectations. *Norms* **are established rules of behavior or standards of conduct.** *Prescriptive norms* state what behavior is appropriate or acceptable. For example, persons making a certain amount of money are expected to drinking and toasting typically lose their status as head of the family (Muehlfried, 2007).

Why do Georgians maintain these cultural norms governing *supras* across generations? In the past, such norms may have helped Georgians hold on to their cultural identity when they were overtaken by other nations, such as Russia. Today, some social scientists believe that these norms and rituals continue to help people in Georgia maintain their identity and cultural heritage in the face of rapid social change and globalization (Muehlfried, 2007).

## reflect & analyze

Can you think of eating or drinking norms that are specific to your family's national, religious, or racial–ethnic traditions? What about those of your friends or acquaintances? Do these norms contribute to your identity and social interaction patterns?

**THE REPUBLIC OF GEORGIA**
© 2015 Cengage Learning.

file a tax return and pay any taxes they owe. Norms based on custom direct us to open a door for a person carrying a heavy load or give up our seat for an elderly person on a bus. By contrast, *proscriptive norms* state what behavior is inappropriate or unacceptable. Laws that prohibit us from driving over the speed limit and "good manners" that preclude you from texting during class are examples. Prescriptive and proscriptive norms operate at all levels of society, from our everyday actions to the formulation of laws.

**Formal and Informal Norms** Not all norms are of equal importance; those that are most crucial are formalized. *Formal norms* are written down and involve specific punishments for violators. Laws are the most common type of formal norms; they have been codified and may be enforced by sanctions. *Sanctions* **are rewards for appropriate behavior or penalties for inappropriate behavior.** Examples of *positive sanctions* include praise, honors, or medals for conformity to specific norms. *Negative sanctions* range from mild disapproval to the death penalty.

Norms that are considered to be less important are referred to as *informal norms*—unwritten standards of behavior understood by people who share a common identity. When individuals violate informal norms, other people may apply informal sanctions. *Informal sanctions* are not clearly defined and can be applied by any member of a group (such as frowning at someone or making a negative comment or gesture). Around the world, people of all nations have formal and informal norms that are unique to their culture and way of life (see "Sociology in Global Perspective").

**Folkways** Norms are also classified according to their relative social importance. *Folkways* **are informal norms or everyday customs that may be**

**value contradictions** values that conflict with one another or are mutually exclusive.

**norms** established rules of behavior or standards of conduct.

**sanctions** rewards for appropriate behavior or penalties for inappropriate behavior.

**folkways** informal norms or everyday customs that may be violated without serious consequences within a particular culture.

violated **without serious consequences within a particular culture** (Sumner, 1959/1906) (Figure 2.8). They provide rules for conduct but are not considered to be essential to society's survival. In the United States, folkways include using underarm deodorant, brushing our teeth, and wearing appropriate clothing for a specific occasion. Often, folkways are not enforced; when they are enforced, the resulting sanctions tend to be informal and relatively mild.

**Mores** Other norms are considered to be highly essential to the stability of society even though they are not laws. **Mores are strongly held norms with moral and ethical connotations that may not be violated without serious consequences in a particular culture.** Because mores (pronounced MOR-ays) are based on cultural values and are considered to be crucial to the well-being of the group, violators are subject to more-severe negative sanctions (such as ridicule or loss of employment) than are those who fail to adhere to folkways. The strongest mores are referred to as taboos. **Taboos are mores so strong that their violation is considered to be extremely offensive and even unmentionable.** Violation of taboos is punishable by the group or even, according to certain belief systems, by a supernatural force. The incest taboo, which prohibits sexual or marital relations between certain categories of kin, is an example of a nearly universal taboo. In the United States, mores are considered to be informal norms unless they are officially made into law.

**Laws** *Laws* **are formal, standardized norms that have been enacted by legislatures and are enforced by formal sanctions.** Laws may be either civil or criminal. *Civil law* deals with disputes among persons or groups. Persons who lose civil suits may encounter negative sanctions such as having to pay compensation to the other party or being ordered to stop certain conduct. *Criminal law,* on the other

hand, deals with public safety and well-being. When criminal laws are violated, fines and prison sentences are the most likely negative sanctions, although in some states the death penalty is handed down for certain major offenses.

**FIGURE 2.8** Folkways—such as how to behave in a crowded elevator—provide rules for conduct but are not considered to be essential to society's survival.

Nico Kai/The Image Bank/Getty Images

# Technology, Cultural Change, and Diversity

**L06** Cultures do not generally remain static. There are many forces working toward change and diversity. Some societies and individuals adapt to this change, whereas others suffer culture shock and succumb to ethnocentrism.

## Cultural Change

Societies continually experience cultural change at both material and nonmaterial levels. Changes in technology continue to shape the material culture of society. *Technology* **refers to the knowledge, techniques, and tools that allow people to transform resources into usable forms, and the knowledge and skills required to use them after they are developed.** Although most technological changes are primarily modifications of existing technology, *new technologies* refers to changes that make a significant difference in many people's lives. Examples of new technologies include the introduction of the printing press more than 500 years ago and the advent of computers and electronic communications in the twentieth century. The pace of technological change has increased rapidly in the past 150 years, as contrasted with the 4,000 years prior to that, during which humans advanced from digging sticks and hoes to the plow.

All parts of culture do not change at the same pace. When a change occurs in the material culture of a society, nonmaterial culture must adapt to that change. Frequently, this rate of change is uneven, resulting in a gap between the two. Sociologist William F. Ogburn (1966/1922) referred to this disparity as *cultural lag,* **which is a gap between the technical development of a society and its**

PhotoEdit/Alamy

**FIGURE 2.9** Facebook and other social networking companies are examples of cultural lag—a gap between technology and a society's morals and laws. Who owns what you have posted on your social networking page?

**moral and legal institutions** (Figure 2.9). In other words, cultural lag occurs when material culture changes faster than nonmaterial culture, thus creating a lag between the two cultural components. For example, at the material cultural level, the personal computer and electronic coding have made it possible to create a unique health identifier for each person in the United States. Based on available technology (material culture), it is possible to create a national data bank that would include everyone's individual medical records from birth to death. Using this identifier, health providers and insurance companies could rapidly transfer medical records around the globe, and researchers could access unlimited data on people's diseases, test results, and treatments. However, the availability of this technology does not mean that it will be accepted by people who believe (nonmaterial culture) that such a national data bank would constitute an invasion of privacy and could easily be abused by others. The failure of nonmaterial culture to keep pace with material culture is linked to social conflict and societal problems. As in the above example, such changes are often set in motion by discovery, invention, and diffusion.

*Discovery* is the process of learning about something previously unknown or unrecognized. Historically, discovery involved unearthing natural elements or existing realities, such as "discovering" fire or the true shape of the Earth. Today, discovery most often results from scientific research. For example, the discovery of a polio vaccine virtually eliminated one of the major childhood diseases. A future discovery of a cure for cancer or the common cold could result in longer and more productive lives for many people.

As more discoveries have occurred, people have been able to reconfigure existing material and nonmaterial cultural items through invention. *Invention* is the process of reshaping existing cultural items into a new form. Guns, airplanes,

TV sets, and digital devices are examples of inventions that positively or negatively affect our lives today.

When diverse groups of people come into contact, they begin to adapt one another's discoveries, inventions, and ideas for their own use. *Diffusion* is the transmission of cultural items or social practices from one group or society to another through such means as exploration, war, the media, tourism, and immigration. Today, cultural diffusion moves at a very rapid pace in the global economy.

## Cultural Diversity

*Cultural diversity* refers to the wide range of cultural differences found between and within nations. Cultural diversity between countries may be the result of natural circumstances (such as climate and geography) or social circumstances (such as level of technology and composition of the population). Some nations—such as Sweden—are referred to as *homogeneous societies,* meaning that they include people who share a common culture and who are typically from similar social, religious, political, and economic backgrounds (though this is changing in Sweden and other countries as they become more diverse). By contrast, other nations—including the United States—are referred to as *heterogeneous societies,* meaning that they include people who are dissimilar in regard to social characteristics such as religion, income, or race/ethnicity (see Figure 2.10).

Immigration contributes to cultural diversity in a society. Throughout its history, the United States

**mores** strongly held norms with moral and ethical connotations that may not be violated without serious consequences in a particular culture.

**taboos** mores so strong that their violation is considered to be extremely offensive and even unmentionable.

**laws** formal, standardized norms that have been enacted by legislatures and are enforced by formal sanctions.

**technology** the knowledge, techniques, and tools that allow people to transform resources into usable forms, and the knowledge and skills required to use them after they are developed.

**cultural lag** William Ogburn's term for a gap between the technical development of a society (material culture) and its moral and legal institutions (nonmaterial culture).

## Religious Affiliation

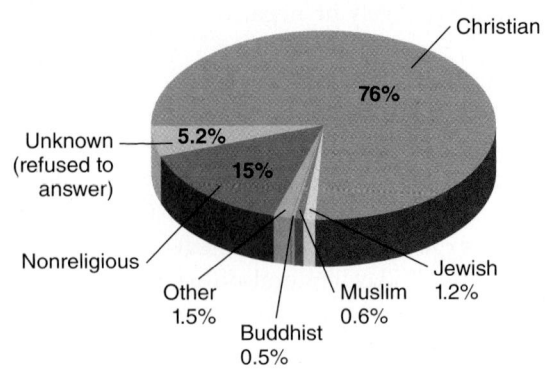

Christian 76%

Unknown (refused to answer) 5.2%

15%

Nonreligious

Other 1.5%

Buddhist 0.5%

Muslim 0.6%

Jewish 1.2%

## Household Income[a]

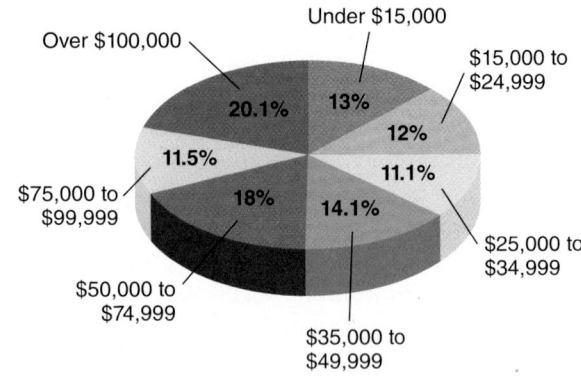

Over $100,000 — 20.1%

Under $15,000 — 13%

$15,000 to $24,999 — 12%

11.5%

$75,000 to $99,999 — 11.5%

18%

14.1%

11.1%

$25,000 to $34,999

$50,000 to $74,999 — 18%

$35,000 to $49,999

## Race and Ethnic Distribution*

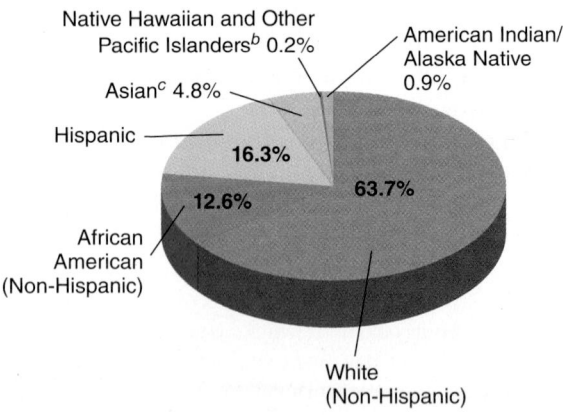

Native Hawaiian and Other Pacific Islanders[b] 0.2%

Asian[c] 4.8%

Hispanic — 16.3%

American Indian/ Alaska Native 0.9%

63.7%

African American (Non-Hispanic) 12.6%

White (Non-Hispanic)

*May not total to 100% due to rounding.

[a]In Census Bureau terminology, a household consists of people who occupy a housing unit.

[b]Includes Native Hawaiian, Guamanian or Chamarro, Samoan, and other Pacific Islanders.

[c]Includes Chinese, Filipino, Japanese, Asian, Indian, Korean, Vietnamese, and other Asians.

**FIGURE 2.10** HETEROGENEITY OF U.S. SOCIETY
Throughout history, the United States has been heterogeneous. Today, we represent a wide diversity of social categories, including our religious affiliations, income levels, and racial–ethnic categories.

has been a nation of immigrants (see Figure 2.11). Over the past 200 years, more than 60 million "documented" (legal) immigrants have arrived here; innumerable people have also entered the country as undocumented immigrants. Immigration can

cause feelings of frustration and hostility, especially in people who feel threatened by the changes that large numbers of immigrants may produce. Often, people are intolerant of those who are different from themselves. When societal tensions rise, people may look for others on whom they can place blame—or single out persons because they are the "other," the "outsider," the one who does not "belong."

Can people overcome these feelings in a culturally diverse society such as the United States? Some analysts believe that it is possible to communicate with others despite differences in race, ethnicity, national origin, age, sexual orientation, religion, social class, occupation, leisure pursuits, regionalism, and so on. People who differ from the dominant group may also find reassurance and social support in a subculture.

**Subcultures** A *subculture* **is a category of people who share distinguishing attributes, beliefs, values, and/or norms that set them apart in some significant manner from the dominant culture.** This concept has been applied to distinctions ranging from ethnic, religious, regional, and age-based categories to those categories presumed to be "deviant" or marginalized from the larger society. In the broadest use of the concept, members of thousands of categories of people residing in the United States might be classified as participants in one or more subcultures, including social media users, Native Americans, Lady Gaga fans, Muslims, and motorcycle enthusiasts. However, many sociological studies of subcultures have limited the scope of inquiry to more-visible, distinct subcultures such as the Old Order Amish and ethnic enclaves such as Little Saigon or Little Gaza to see how participants in subcultures interact with the dominant U.S. culture. Let's take a brief look at the Old Order Amish, a classic sociological example.

**The Old Order Amish** Having arrived in the United States in the early 1700s, members of the Old Order Amish have fought to maintain their distinct identity. Today, most of the Amish live in Pennsylvania, Ohio, and Indiana, where they practice their religious beliefs and remain in a relatively closed social network. According to sociologists, this religious community is a subculture because its members share values and norms that differ from those of people who primarily identify with the dominant culture. The Amish have a strong faith in God and reject worldly concerns. Their core values include the joy of work, the primacy of the home, faithfulness, thriftiness, tradition, and humility. The Amish hold a conservative view of the family, believing that women are subordinate

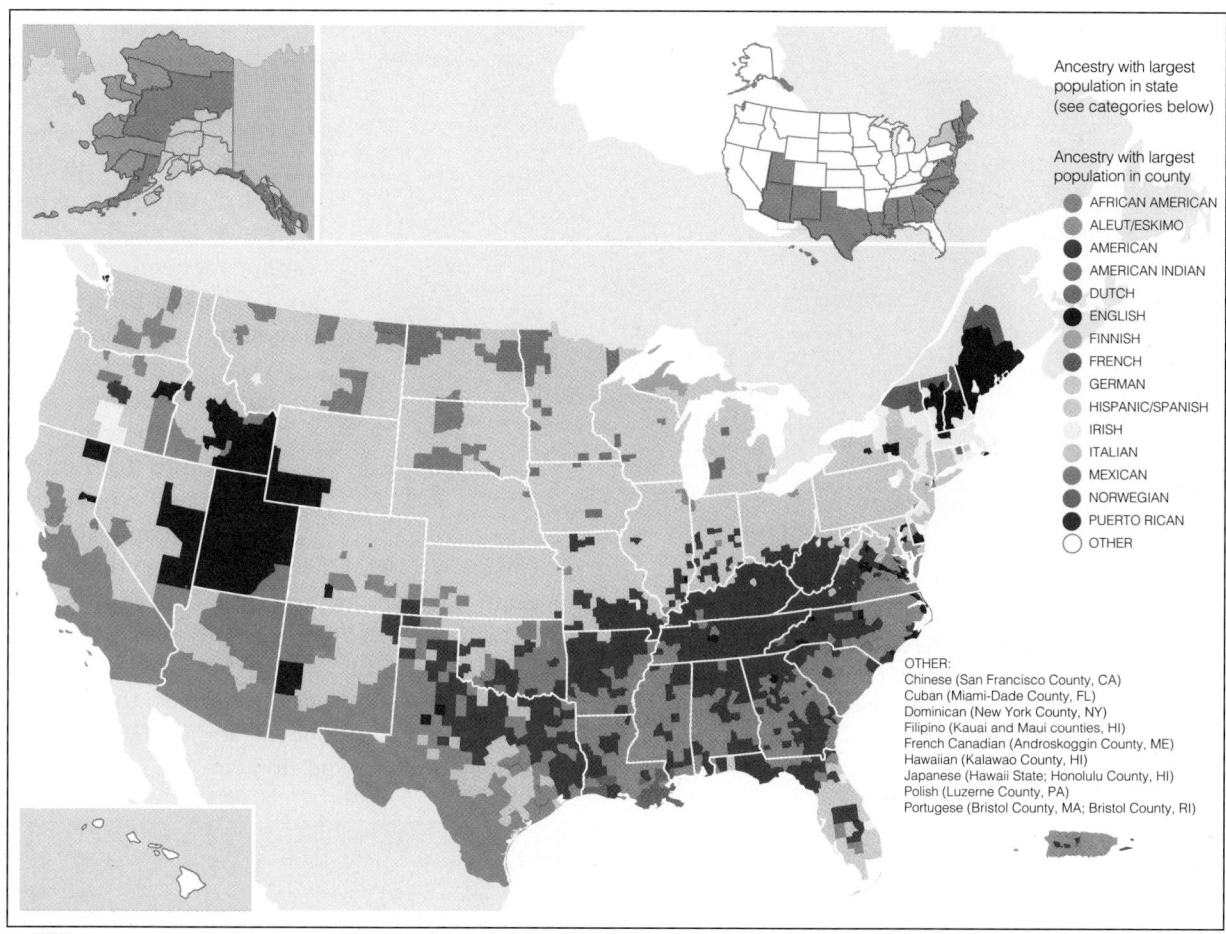

**FIGURE 2.11** CULTURAL DIVERSITY: A NATION OF IMMIGRANTS

Source: U.S. Census Bureau, 2012, tabulation by author.

to men, birth control is unacceptable, and wives should remain at home. Children (about seven per family) are cherished and seen as an economic asset: They help with the farming and other work. Many of the Old Order Amish speak Pennsylvania Dutch (a dialect of German) as well as English. They dress in traditional clothing, live on farms, and rely on the horse and buggy for transportation (Figure 2.12).

The Amish are aware that they share distinctive values and look different from other people; these differences provide them with a collective identity and make them feel close to one another. The belief system and group cohesiveness of the Amish remain strong despite the intrusion of corporations and tourists, the vanishing farmlands, and increasing levels of government regulation in their daily lives (Schaefer and Zellner, 2010).

**Ethnic Subcultures** Some people who have unique shared behaviors linked to a common racial, language, or nationality background identify themselves as members of a specific subculture, whereas others do not. Examples of ethnic subcultures include African Americans, Latinos/Latinas (Hispanic Americans), Asian Americans, and Native Americans. Some analysts include "white ethnics" such as Irish Americans, Italian Americans, and Polish Americans. Others also include Anglo Americans (Caucasians).

Although people in ethnic subcultures are dispersed throughout the United States, a concentration of members of some ethnic subcultures is visible in many larger communities and cities. For example, Chinatowns, located in cities such as San Francisco, Los Angeles, and New York, are one of the more visible ethnic subcultures in the United States. By living close to one another and retaining their original customs and language, first-generation immigrants can survive the abrupt changes they experience

**subculture** a category of people who share distinguishing attributes, beliefs, values, and/or norms that set them apart in some significant manner from the dominant culture.

**59**

FIGURE 2.12 Although modernization and consumerism have changed the way of life of some subcultures, groups such as the Old Order Amish have preserved some of their historical practices, including traveling by horse-drawn carriage.

## Culture Shock

**LO7** *Culture shock* is the disorientation that people feel when they encounter cultures radically different from their own and believe they cannot depend on their own taken-for-granted assumptions about life. When people travel to another society, they may not know how to respond to that setting. For example, in a now-classical study, Napoleon Chagnon (1992) described his initial shock at seeing the Yanomamö (pronounced yahnoh-MAH-mah) tribe of South America on his first trip in 1964 (Figure 2.13).

The Yanomamö (also referred to as the "Yanomami") are a tribe of about 20,000 South American Indians who live in the rain forest.

in material and nonmaterial cultural patterns. In New York City, for example, Korean Americans and Puerto Rican Americans constitute distinctive subcultures, each with its own food, music, and personal style. In San Antonio, Mexican Americans enjoy different food and music than do Puerto Rican Americans or other groups.

Subcultures provide opportunities for the expression of distinctive lifestyles, as well as sometimes helping people adapt to abrupt cultural change. Subcultures can also serve as a buffer against the discrimination experienced by many ethnic or religious groups in the United States. However, some people may be forced by economic or social disadvantage to remain in such ethnic enclaves.

**Countercultures** A *counterculture* **is a group that strongly rejects dominant societal values and norms and seeks alternative lifestyles.** Young people are most likely to join countercultural groups, perhaps because younger persons generally have less invested in the existing culture. Examples of countercultures include the beatniks of the 1950s, the flower children of the 1960s, the drug enthusiasts of the 1970s, and contemporary members of nonmainstream religious sects or cults. Occupy Wall Street and its counterparts throughout the United States and other nations began as a counterculture because participants took a stand against dominant cultural values of wealth, power, and political privilege.

FIGURE 2.13 Even as global travel and the media make us more aware of people around the world, the distinctiveness of the Yanomamö in South America remains apparent. Are people today more or less likely than those in the past to experience culture shock upon encountering diverse groups of people such as these Yanomamö?

Although Chagnon traveled in a small aluminum motorboat for three days to reach these people, he was not prepared for the sight that met his eyes when he arrived:

> I looked up and gasped to see a dozen burly, naked, sweaty, hideous men staring at us down the shafts of their drawn arrows. Immense wads of green tobacco were stuck between their lower teeth and lips, making them look even more hideous, and strands of dark-green slime dripped from their nostrils—strands so long that they reached down to their pectoral muscles or drizzled down their chins and stuck to their chests and bellies. We arrived as the men were blowing ebene, a hallucinogenic drug, up their noses. . . . I was horrified. What kind of welcome was this for someone who had come to live with these people and learn their way of life—to become friends with them? But when they recognized Barker [a guide], they put their weapons down and returned to their chanting, while keeping a nervous eye on the village entrances. (Chagnon, 1992: 12–14)

The Yanomamö have no written language, system of numbers, or calendar. They lead a nomadic lifestyle, carrying everything they own on their backs. They wear no clothes and paint their bodies; the women insert slender sticks through holes in the lower lip and through the pierced nasal septum. In other words, the Yanomamö—like the members of thousands of other cultures around the world—live in a culture very different from that of the United States.

Andrea Magugliani/Alamy

**FIGURE 2.14** These children in Norway wave their country's flag on May 17 (Norwegian Constitution Day), displaying a form of positive ethnocentrism.

## Ethnocentrism and Cultural Relativism

When observing people from other cultures, many of us use our own culture as the yardstick by which we judge their behavior. Sociologists refer to this approach as *ethnocentrism*—**the practice of judging all other cultures by one's own culture.** Ethnocentrism is based on the assumption that one's own way of life is superior to all others. For example, most schoolchildren are taught that their own school and country are the best (Figure 2.14). The school song, the pledge to the flag, and the national anthem are forms of *positive ethnocentrism.* However, *negative ethnocentrism* can also result from constant emphasis on the superiority of one's own group or nation. Negative ethnocentrism is manifested in derogatory stereotypes that ridicule recent immigrants whose customs, dress, eating habits, or religious beliefs are markedly different from those of dominant-group members. Long-term U.S. residents who are members of racial and ethnic minority groups, such as Native Americans, African Americans, and Latinas/os, have also been the target of ethnocentric practices by other groups.

An alternative to ethnocentrism is *cultural relativism*—**the belief that the behaviors and**

**counterculture** a group that strongly rejects dominant societal values and norms and seeks alternative lifestyles.

**culture shock** the disorientation that people feel when they encounter cultures radically different from their own and believe they cannot depend on their own taken-for-granted assumptions about life.

**ethnocentrism** the practice of judging all other cultures by one's own culture.

**cultural relativism** the belief that the behaviors and customs of any culture must be viewed and analyzed by the culture's own standards.

**61**

customs of any culture must be viewed and analyzed by the culture's own standards. For example, the anthropologist Marvin Harris (1974, 1985) uses cultural relativism to explain why cattle, which are viewed as sacred, are not killed and eaten in India, where widespread hunger and malnutrition exist. From an ethnocentric viewpoint, we might conclude that cow worship is the cause of the hunger and poverty in India. However, according to Harris, the Hindu taboo against killing cattle is very important to the Indian economic system. Live cows are more valuable than dead ones because they have more important uses than as a direct source of food. As part of the ecological system, cows consume grasses of little value to humans. Then they produce some valuable resources—oxen (the neutered offspring of cows) to power farmers' plows and manure (for fuel and fertilizer)—as well as milk for children. As Harris's study reveals, culture must be viewed from the standpoint of those who live in a particular society.

Cultural relativism also has a downside. It may be used to excuse customs and behavior (such as cannibalism) that may violate basic human rights. Cultural relativism is a part of the sociological imagination; researchers must be aware of the customs and norms of the society they are studying and then spell out their background assumptions so that others can spot possible biases in their studies. However, according to some social scientists, issues surrounding ethnocentrism and cultural relativism may become less distinct in the future as people around the globe increasingly share a common popular culture. Others, of course, disagree with this perspective. Let's see what you think.

# A Global Popular Culture?

Before taking this course, what was the first thing you thought about when you heard the term *culture*? In everyday life, culture is often used to describe the fine arts, literature, and classical music. When people say that a person is "cultured," they may mean that the individual has a highly developed sense of style or aesthetic appreciation of the "finer" things.

## High Culture and Popular Culture

Some sociologists use the concepts of high culture and popular culture to distinguish between different cultural forms. **High culture consists** of classical music, opera, ballet, live theater, and other activities usually patronized by elite audiences, composed primarily of members of the upper-middle and upper classes, who have the time, money, and knowledge assumed to be necessary for its appreciation. In the United States, high culture is often viewed as being international in scope, arriving in this country through the process of diffusion, because many art forms originated in European nations or other countries of the world.

By contrast, much of U.S. popular culture is often thought of as "homegrown" in this country. *Popular culture* consists of activities, products, and services that are assumed to appeal primarily to members of the middle and working classes. These include rock concerts, spectator sports, movies, and television soap operas and situation comedies. Although we will distinguish between "high" and "popular" culture in our discussion, it is important to note that some social analysts believe that the rise of a consumer society in which luxury items have become more widely accessible to the masses has greatly reduced the great divide between activities and possessions associated with wealthy people or a social elite.

However, most sociological examinations of high culture and popular culture focus primarily on the link between culture and social class. French sociologist Pierre Bourdieu's (1984) *cultural capital theory* views high culture as a device used by the dominant class to exclude the subordinate classes. According to Bourdieu, people must be trained to appreciate and understand high culture. Individuals learn about high culture in upper-middle-class and upper-class families and in elite education systems, especially higher education. Once they acquire this trained capacity, they possess a form of *cultural capital*. Persons from poor and working-class backgrounds typically do not acquire this cultural capital. Because knowledge and appreciation of high culture are considered a prerequisite for access to the dominant class, its members can use their cultural capital to deny access to subordinate-group members and thus preserve and reproduce the existing class structure.

## Forms of Popular Culture

Three prevalent forms of popular culture are fads, fashions, and leisure activities. A *fad* is a temporary but widely copied activity followed enthusiastically by large numbers of people. Most fads are short-lived novelties. Fads can be divided into four major categories. First, *object fads* are items that people purchase despite the fact that they have little use value, such as wristbands that make a statement or support a cause. Second, *activity fads* include pursuits such as body piercing or flash mobs. Third are *idea fads*, such as New Age ideologies. Fourth are *personality fads*—for example, those surrounding celebrities such as Lady Gaga, Justin Bieber, and Miley Cyrus (based on Lofland, 1993).

A *fashion* is a currently valued style of behavior, thinking, or appearance that is longer lasting and more widespread than a fad. Examples of fashion are found in many areas, including child rearing, education, arts, clothing, music, and sports. Soccer is an example of a fashion in sports. Until fairly recently, only schoolchildren played soccer in the United States. Now it has become a popular sport, perhaps in part because of immigration from Latin America and other areas of the world where soccer is widely played.

Like soccer, other forms of popular culture move across nations. In fact, popular culture is one of the United States' largest exports to other nations. In turn, people in this country continue to be strongly influenced by popular culture from other nations. Will the spread of popular culture produce a homogeneous global culture? Critics argue that the world is not developing a global culture; rather, other cultures are becoming Westernized. Political and religious leaders in some nations oppose this process, which they view as *cultural imperialism*—**the extensive infusion of one nation's culture into other nations.** For example, some view the widespread infusion of the English language into countries that speak other languages as a form of cultural imperialism. On the other hand, the concept of cultural imperialism may fail to take into account various cross-cultural influences. For example, cultural diffusion of literature, music, clothing, and food has occurred on a global scale. A global culture, if it comes into existence, will most likely include components from many societies and cultures.

# Sociological Analysis of Culture

**LO8** Sociologists regard culture as a central ingredient in human behavior. Although all sociologists share a similar purpose, they typically see culture through somewhat different lenses as different theoretical perspectives in their research guide them. What do these perspectives tell us about culture?

## Functionalist Perspectives

Functionalist perspectives are based on the assumption that society is a stable, orderly system with interrelated parts that serve specific functions. Anthropologist Bronislaw Malinowski (1922) suggested that culture helps people meet their *biological needs* (including food and procreation), *instrumental needs* (including law and education), and *integrative needs* (including religion and art). Societies in which people share a common language and core values are more likely to have consensus and harmony.

How might functionalist analysts view popular culture? According to many functionalist theorists, popular culture serves a significant function in society in that it may be the "glue" which holds society together. Regardless of race, class, sex, age, or other characteristics, many people are brought together (at least in spirit) to cheer teams competing in major sporting events such as the Super Bowl and the Olympics. Television, the Internet, and social media help integrate recent immigrants into the mainstream culture, whereas longer-term residents may become more homogenized as a result of seeing the same images and being exposed to the same beliefs and values. However, functionalists acknowledge that all societies have dysfunctions, which produce a variety of societal problems. When a society contains numerous subcultures, discord may result from a lack of consensus about values and social norms. In fact, popular culture may undermine cultural values rather than reinforce them. For example, popular culture may glorify crime, rather than hard work, as the quickest way to get ahead. According to some analysts, excessive violence in music, video games, movies, and television shows

**high culture** classical music, opera, ballet, live theater, and other activities usually patronized by elite audiences.

**popular culture** activities, products, and services that are assumed to appeal primarily to members of the middle and working classes.

**cultural imperialism** the extensive infusion of one nation's culture into other nations.

may be harmful to children and young people. From this perspective, popular culture can be a factor in antisocial behavior such as hate crimes and fatal shootings.

A strength of the functionalist perspective on culture is its focus on the needs of society and the fact that stability is essential for society's continued survival. A shortcoming is its overemphasis on harmony and cooperation. This approach also fails to fully account for factors embedded in the structure of society—such as class-based inequalities, racism, and sexism—that may contribute to conflict among people in the United States or to global strife.

## Conflict Perspectives

Conflict perspectives are based on the assumption that social life is a continuous struggle in which members of powerful groups seek to control scarce resources. According to this approach, values and norms help create and sustain the privileged position of the powerful in society while excluding others. As early conflict theorist Karl Marx stressed, ideas are *cultural creations* of a society's most powerful members. Thus, it is possible for political, economic, and social leaders to use *ideology*—an integrated system of ideas that is external to, and coercive of, people—to maintain their positions of dominance in a society. As Marx stated,

> The ideas of the ruling class are in every epoch the ruling ideas, i.e., the class which is the ruling material force in society, is at the same time, its ruling intellectual force. The class, which has the means of material production at its disposal, has control at the same time over the means of mental production. . . . The ruling ideas are nothing more than the ideal expression of the dominant material relationships, the dominant material relationships grasped as ideas. (Marx and Engels, 1970/1845–1846: 64)

Many contemporary conflict theorists agree with Marx's assertion that ideas, a nonmaterial component of culture, are used by agents of the ruling class to affect the thoughts and actions of members of other classes. The role of the mass media in influencing people's thinking about the foods that they should—or should not—eat is an example of ideological control (see "Culture in the Media" on the facing page).

How might conflict theorists view popular culture? Some conflict theorists believe that popular culture, which originated with everyday people, has been largely removed from their domain and has become nothing more than a part of the capitalist economy in the United States. From this approach, media conglomerates such as Time Warner and ABC/Disney create popular culture, such as films, television shows, and amusement parks, in the same way that they would produce any other product or service. Creating new popular culture also promotes consumption of *commodities*—objects outside ourselves that we purchase to satisfy our human needs or wants. According to contemporary social analysts, consumption—even of things that we do not necessarily need—has become prevalent at all social levels, and some middle- and lower-income individuals and families now use as their frame of reference the lifestyles of the more affluent in their communities.

AP Images/Ann Johansson

**FIGURE 2.15** Rapid changes in language and culture in the United States are reflected in this sign at a shopping center. How do functionalist and conflict theorists' views regarding language differ?

# How We Become Opinionated Eaters

Media Headlines: "Agra Business Use of Pesticides Linked to Health Hazards"

"Big Agra Exploits Cheap Labor"

The term *media framing* refers to the process by which information and entertainment are packaged by the mass media (newspapers, magazines, radio and television stations and networks, the Internet, and various social media networks). Therefore, a media *frame* constitutes a story line or an unfolding narrative about an issue. Such narratives wield power because they influence how we make sense of the world (Kendall, 2011)—emphasizing some beliefs and values over others and directing people's attention to some ideas while ignoring others. For example, how headlines and media stories about food are framed—including the story's exposure, delivered tone, and accompanying visuals—affects how we think about food and what we eat. If a food product undergoes a much-publicized recall, for example, many stop buying it for some time, even after the recall has ended.

How does media framing influence the way in which we conceive of the production of our food? Consider how the media have influenced the way you feel about companies that produce organic and nonorganic foods, for example. "Big Agra" is often used in the media to describe the major corporations that grow and market nonorganic food. These megacorporations own giant cultivated tracts that use procedures intended to maximize the crop yield, harvest that crop (whether plants or animals) at the lowest price possible, and distribute the crop to markets in the United States and other countries. Stories in the media often emphasize how such maximizing of crop yield involves the use of chemicals and pesticides—and cheap labor. "Big Agra" obviously has a stake in the battle over how the media frame stories, especially those that compare its foods negatively with organic foods; the organic-food industry also has a stake in the battle. Accordingly, both sides attempt to influence how stories about their products are framed in the media because that can make a big difference in their respective profits.

You may ask "What does media framing of food production have to do with culture?" The answer is simple: Food is very much a part of all cultures. What we eat and how it is grown and prepared are a product of the culture of the society in which we live. Whether it is fast food or fresh spinach, companies that produce food want to have an effect on what you buy and where you buy it.

UPPA/ZUMAPRESS/Newscom

Has media framing affected your view of genetically modified organisms (GMOs) and food?

## reflect & analyze
**Have the media influenced your perception of large companies that produce food? How so?**

As a result, many families live on credit in order to purchase the goods and services that they would like to have or that keep them on the competitive edge with their friends, neighbors, and coworkers.

A strength of the conflict perspective is that it stresses how cultural values and norms may perpetuate social inequalities. It also highlights the inevitability of change and the constant tension between those who want to maintain the status quo and those who desire change (Figure 2.15). A limitation is its focus on societal discord and the divisiveness of culture.

## Symbolic Interactionist Perspectives

Unlike functionalists and conflict theorists, who focus primarily on macrolevel concerns, symbolic interactionists engage in a microlevel analysis that views society as the sum of all people's interactions. From this perspective, people create, maintain, and modify culture as they go about their everyday activities. Symbols make communication with others possible because symbols provide us with shared meanings.

According to some symbolic interactionists, people continually negotiate their social realities. Values and norms are not independent realities that automatically determine our behavior. Instead, we reinterpret them in each social situation we encounter. However, the classical sociologist Georg Simmel warned that the larger cultural world—including both material culture and nonmaterial culture—eventually takes on a life of its own apart from the actors who daily re-create social life. As a result, individuals may be more controlled by culture than they realize.

Simmel (1990/1907) suggested that money is an example of how people may be controlled by their culture. According to Simmel, people initially create money as a means of exchange, but then money acquires a social meaning that extends beyond its purely economic function. Money becomes an end in itself, rather than a means to an end. Today, we are aware of the relative "worth" not only of objects but also of individuals. Many people revere wealthy entrepreneurs and highly paid celebrities, entertainers, and sports figures for the amount of money they make, not for their intrinsic qualities. According to Simmel (1990/1907), money makes it possible for us to *relativize* everything, including our relationships with other people. When social life can be reduced to money, people become cynical, believing that anything—including people, objects, beauty, and truth—can be bought if we can pay the price. Although Simmel acknowledged the positive functions of money, he believed that the social interpretations people give to money often produce individual feelings of cynicism and isolation.

A symbolic interactionist approach highlights how people maintain and change culture through their interactions with others. However, interactionism does not provide a systematic framework for analyzing how we shape culture and how it, in turn, shapes us. It also does not provide insight into how shared meanings are developed among people, and it does not take into account the many situations in which there is disagreement on meanings. Whereas the functional and conflict approaches tend to overemphasize the macrolevel workings of society, the interactionist viewpoint often fails to take these larger social structures into account.

## Postmodernist Perspectives

Postmodernist theorists believe that much of what has been written about culture in the Western world is Eurocentric—that it is based on the uncritical assumption that European culture (including its dispersed versions in countries such as the United States, Australia, and South Africa) is the true, universal culture in which all the world's people ought to believe. By contrast, postmodernists believe that we should speak of *cultures,* rather than *culture.*

However, Jean Baudrillard, one of the best-known French social theorists and key figures in postmodern theory, believes that the world of culture today is based on *simulation,* not reality. According to Baudrillard, social life is much more a spectacle that simulates reality than it is reality itself. Many U.S. children, upon entering school for the first time, have already watched more hours of television than the total number of hours of classroom instruction they will encounter in their entire school careers. Add to this the number of hours that some children will have spent playing computer games or using the Internet, where they often find that it is more interesting to deal with imaginary heroes and villains than to interact with "real people" in real life. Baudrillard refers to this social creation as *hyperreality*—a situation in which the *simulation* of reality is more real than experiencing the event itself and having any actual connection with what is taking place. For Baudrillard, everyday life has been captured by the signs and symbols generated to represent it, and we ultimately relate to simulations and models as if they were reality.

Baudrillard (1983) uses Disneyland as an example of a simulation—one that conceals the reality that exists outside rather than inside the boundaries of the artificial perimeter. According to Baudrillard, Disney-like theme parks constitute a form of seduction that substitutes symbolic (seductive) power for real power, particularly the ability to bring about social change. From this perspective, amusement park "guests" may feel like "survivors" after enduring the rapid speed and gravity-defying movements of the roller-coaster rides or see themselves as "winners" after surviving fights with hideous cartoon villains on the "dark rides." In reality, they have been made

## Analysis of Culture

| Components of culture | Symbol | Anything that meaningfully represents something else. |
|---|---|---|
| | Language | A set of symbols that expresses ideas and enables people to think and communicate with one another. |
| | Values | Collective ideas about what is right or wrong, good or bad, and desirable or undesirable in a particular culture. |
| | Norms | Established rules of behavior or standards of conduct. |
| Sociological analysis of culture | Functionalist perspectives | Culture helps people meet their biological, instrumental, and expressive needs. |
| | Conflict perspectives | Ideas are a cultural creation of society's most powerful members and can be used by the ruling class to affect the thoughts and actions of members of other classes. |
| | Symbolic interactionist perspectives | People create, maintain, and modify culture during their everyday activities; however, cultural creations can take on a life of their own and end up controlling people. |
| | Postmodern perspectives | Much of culture today is based on simulation of reality (e.g., what we see on television) rather than reality itself. |

to *appear* to have power, but they do not actually possess any real power.

In their examination of culture, postmodernist social theorists thus make us aware of the fact that no single perspective can grasp the complexity and diversity of the social world. There is no one, single, universal culture. They also make us aware that reality may not be what it seems. According to the postmodernist view, no one authority can claim to know social reality, and we should deconstruct—take apart and subject to intense critical scrutiny—existing beliefs and theories about culture in hopes of gaining new insights.

Although postmodern theories of culture have been criticized on a number of grounds, we will mention only three. One criticism is postmodernism's lack of a clear conceptualization of ideas. Another is the tendency to critique other perspectives as being "grand narratives," whereas postmodernists offer their own varieties of such narratives. Finally, some analysts believe that postmodern analyses of culture lead to profound pessimism about the future.

This chapter's Concept Quick Review summarizes the components of culture as well as how the four major perspectives view it.

## Culture in the Future

As we have discussed in this chapter, many changes are occurring in the United States. Increasing cultural diversity can either cause long-simmering racial and ethnic antagonisms to come closer to a boiling point or result in the creation of a truly "rainbow culture" in which diversity is respected and encouraged (Figure 2.16).

In the future the issue of cultural diversity will increase in importance, especially in schools. Multicultural education that focuses on the contributions of a wide variety of people from different backgrounds will continue to be an issue of controversy from kindergarten through college. In schools throughout the nation, students speak many different languages and dialects. It is estimated that 325 languages are spoken in U.S. schools today. Consequently, schools continue to face the challenge of embracing widespread cultural diversity while conveying a sense of

# Sociology WORKS!

## Schools as Laboratories for Getting Along: Having Lunch Together

Sociology makes us aware of the importance of culture in daily life. Research has shown that schools and friendship groups can expose children and young people to cultures that are different from their own. Studies have also shown that it may be easier for children to set aside their differences and get to know one another than it is for adults to do so, particularly if they are able to spend time together at lunch or talking in other informal settings. Consider what happened among some children at the International Community School in Decatur, Georgia. Some students were born in the United States, but some were refugees from as many as 40 war-torn countries. This school became a "laboratory for getting along," particularly as some children took the initiative to befriend and help others by doing such things as eating lunch with them (St. John, 2007). An excellent example is the friendship that developed between nine-year-old Dante Ramirez and Soung Oo Hlaing, an eleven-year-old Burmese refugee who spoke no English:

> The two boys met on the first day of school this year. Despite the language barrier, Dante managed to invite the newcomer to sit with him at lunch.
>
> "I didn't think he'd make friends at the beginning because he didn't speak that much English," Dante said. "So I thought I should be his friend."

In the next weeks, the boys had a sleepover. They trick-or-treated on Soung's first Halloween. Soung, a gifted artist, gave Dante pointers on how to draw. And Dante helped Soung with his English. "I use simple words that are easy to know and sometimes hand movements," Dante explained. "For 'huge,' I would make my hands bigger. And for 'big,' I would make my hands smaller than for huge." (St. John, 2007: A14)

Over time, as the boys got to know each other better, their mothers also developed a friendship and began to celebrate ethnic holidays together even though they largely relied on gestures (a form of nonverbal communication) to communicate with each other. Could "laboratories" such as this help more people from diverse cultures get along? From a sociological perspective, a personal or community effort, such as arranging lunches or other meals together, is clearly a right start.

### reflect & analyze

**What examples can you provide that show how eating together or having other social gatherings may create new bonds across diverse cultural groups on your college campus or in the community where you reside?**

AP Images/John Raoux

David McNew/Getty Images News/Getty Images

**FIGURE 2.16** In recent years there has been a significant increase in the number of immigrants who have become U.S. citizens. However, an upsurge in anti-immigrant sentiment has put pressure on the Border Patrol and the U.S. Citizenship and Immigration Services, which are charged with enforcing immigration laws.

community and national identity to students (see "Sociology Works!").

Technology will continue to have a profound effect on culture. Television and radio, films and videos, and electronic communications will continue to accelerate the flow of information and expand cultural diffusion throughout the world. Global communication devices will move images of people's lives, behavior, and fashions instantaneously among almost all nations. Increasingly, computers, the Internet, and social media are people's window on the world, resulting in either greater integration or in fragmentation. Integration occurs when there is a widespread acceptance of ideas and items—such as democracy or the latest trend in music or clothing—across cultures. By contrast, fragmentation occurs when people in one culture disdain the beliefs and actions of other cultures. As a force for both cultural integration and fragmentation, technology continues to revolutionize communications, but most of the world's population is only slowly, if at all, becoming part of this revolution.

From a sociological perspective, the study of culture helps us not only understand our own "tool kit" of symbols, stories, rituals, and worldviews but also expand our insights to include those of other people of the world, who also seek strategies for enhancing their own lives. If we understand how culture is used by people, how cultural elements constrain or further certain patterns of action, what aspects of our cultural heritage have enduring effects on our actions, and what specific historical changes undermine the validity of some cultural patterns and give rise to others, we can apply our sociological imagination not only to our own society but to the entire world as well.

# chapter REVIEW

## LO1 What is culture, and why is it important in helping people in their daily lives?

Culture is the knowledge, language, values, customs, and material objects that are passed from person to person and from one generation to the next in a human group or society. Culture is essential for our individual survival and our communication with other people.

## LO2 What is material culture, and what are the four nonmaterial components of culture that are common to all societies?

Material culture consists of physical and tangible creations that members of society make, use, and share. Changes in technology continue to shape the material culture of society. Nonmaterial components of culture are symbols, language, values, and norms.

## LO3 What are cultural universals?

Cultural universals are customs and practices that exist in all societies and include activities and institutions such as storytelling, families, and laws. However, specific forms of these universals vary from one cultural group to another.

## LO4 How do symbols and language reflect cultural values?

Symbols express shared meanings; through them, groups communicate cultural ideas and abstract concepts. Language is a set of symbols through which groups communicate. One of our most important human attributes is the ability to use language to share our experiences, feelings, and knowledge with others.

## LO5 What are the differences among folkways, mores, and laws?

Folkways are norms that express the everyday customs of a group, whereas mores are norms with strong moral and ethical connotations, and are essential to the stability of a culture. Laws are formal, standardized norms that are enforced by formal sanctions.

## LO6 How can technological changes affect the culture of a nation and the world?

All parts of culture do not change at the same pace. However, technological change widens a cultural lag between a technical development in a society and the society's moral and legal institutions. The pace of technology modifies people's daily lives, as evident in the spread of information via smart technologies, such as handheld devices and automated jobs.

## LO7 What are culture shock, ethnocentrism, and cultural relativism?

Culture shock refers to the anxiety that people experience when they encounter cultures radically different from their own. Ethnocentrism is the assumption that one's own culture is superior to others. Cultural relativism views and analyzes another culture in terms of that culture's own values and standards.

## LO8 How do the major sociological perspectives view society and culture?

A functionalist analysis of culture assumes that a common language and shared values help produce consensus and harmony. According to some conflict theorists, certain groups may use culture to maintain their privilege and exclude others from society's benefits. Symbolic interactionists suggest that people create, maintain, and modify culture as they go about their everyday activities. Postmodern thinkers believe that there are many cultures within the United States alone. In order to grasp a better understanding of how popular culture may simulate reality rather than be reality, postmodernists believe that we need a new way of conceptualizing culture and society.

## key TERMS

beliefs 46

counterculture 60

cultural imperialism 63

cultural lag 56

cultural relativism 61

cultural universals 46

culture 43

culture shock 60

ethnocentrism 61

folkways 55

high culture 62

language 48

laws 56

material culture 44

mores 56

nonmaterial culture 45

norms 54

popular culture 62

sanctions 55

Sapir–Whorf hypothesis 49

subculture 58

symbol 47

taboos 56

technology 56

value contradictions 54

values 52

## questions for CRITICAL THINKING

1. Would it be possible today to live in a totally separate culture in the United States? Could you avoid all influences from the mainstream popular culture or from the values and norms of other cultures? How would you be able to avoid any change in your culture?

2. Consider a wide variety of fads and fashions: musical styles, video games and social media, literature, and political, social, and religious ideas. Do fads and fashions reflect and reinforce or challenge and change the values and norms of a society?

3. You are doing a survey analysis of recent immigrants to the United States to determine the effects of popular culture on their views and behavior. What are some of the questions that you would use in your survey?

# 3 SOCIALIZATION

## LEARNING OBJECTIVES

**LO1** Debate the extent to which people would become human beings without adequate socialization.

**LO2** Discuss the sociological perspective on human development, emphasizing the contributions of Charles Horton Cooley and George Herbert Mead.

**LO3** Contrast functionalist and conflict theorists' perspectives on the roles that families play in the socialization process.

**LO4** Describe how schools socialize children in both formal and informal ways.

**LO5** Explain the role that peer groups and media play in socialization now, and predict the part that these agents will play in the future.

**LO6** Identify ways in which gender socialization and racial–ethnic socialization occur.

**LO7** Discuss the stages in the life course and demonstrate why the process of socialization is important in each stage.

**LO8** Distinguish between voluntary and involuntary resocialization, and give examples of each.

# Socialization in Higher Education

Nothing could have gone more wrong for me on my first day [as a college student]. . . . I hadn't bought any of my books yet because I thought they would be passed out to me as was customary in high school, and, worst of all, I had never really read a map on my own before so finding the classrooms in the huge main campus was really hard. Avoid looking like an idiot on your first day of higher education by following a few pieces of advice, all of which revolve around the two most important things I learned about being a college student: stay organized and make friends whenever possible. . . . Good luck on your first day!

—As a student, Amelie (2010) found that she was unprepared for the first day of college, so she now helps other students with the socialization process by providing tips on how to survive in higher education.

Life is a continuous socialization process, whether we are visiting with others in an informal setting or learning in a formal medical classroom. What recurring patterns of socialization can you identify in this picture?

AP Images/Matt Rourke

The white coat ceremony . . . was intended to herald our introduction into the [Harvard Medical School] community on our first day of medical school. While not the long coat of a physician or resident, the white coat signaled our medical affiliation and differentiated us from the civilian visitors and volunteers.

This was not an affiliation I was ready to claim as a first year medical student. Over the course of the year, after taking courses in anatomy, pharmacology, physiology, genetics, and embryology, I was more deeply impressed by how little I knew than by how much I had learned. Yet every Monday in our Patient–Doctor course I found myself in my white coat interviewing still another patient.

**W**hat do the comments of Amelie and Dr. Ellen Lerner Rothman have in common? Sociologically speaking, their statements express concern about the socialization process and how we learn to adapt when we join a new social organization. Many of us experience stress when we take on new and seemingly unfamiliar roles and find that we must learn the appropriate norms regarding how people in a specific role should think, act, and communicate with others. This is especially true when we start college or a graduate program and find ourselves in a new setting surrounded by many people we do not know.

How does socialization work in college? Look around in your classes at the beginning of each semester, and you will probably see other students trying to find out what is going to be expected of them in a particular course. What is the course going to cover? What are the instructor's requirements? How should students communicate with the instructor and other students in the class? Some information of this type is learned through formal instruction, such as in the syllabus or classroom, but much of what we know about school is learned informally through our observations of other people, by listening to what they say when we are in their physical presence, or through interacting with them by cell phone, e-mail, or text messaging when we are apart. Sociologists use the term *socialization* to refer to both the formal and informal processes by which people learn a new role and find out how to be a part of a group or organization. Moreover, the process of socialization continues throughout an individual's life.

In this chapter we examine the process of socialization and identify reasons why socialization is crucial to the well-being of individuals, groups, and societies. We discuss both sociological and social psychological theories of human development. We look at the dynamics of socialization—how it occurs and what shapes it. We also focus on positive and

Despite the uncertainty of my place in the medical world, my white coat ushered me into the foreign world of the patient–doctor dynamic. . . . These weekly interviews as part of our Patient–Doctor course were about learning the important questions, the right mannerisms, and the appropriate responses to our patients. Our instructors taught us to take a careful, methodical history, which I more or less skillfully replicated every week with a different patient. Although the goal of these weekly patient interactions was to discover a person's experience with illness, these interviews were more about my learning process than about the patient's story. . . . [I]n the medical world my white coat did not offer the solace of anonymity but forced me to take on power that I was not ready to accept.

—Ellen Lerner Rothman, M.D. (1999: 2–3), describing the professional socialization that students encounter in the early years of medical training.

## How Much Do You Know About Socialization and the College Experience?

| True | False | |
|---|---|---|
| T | F | 1. Professors are the primary agents of socialization for college students. |
| T | F | 2. Researchers have found that few students spend time studying with other students. |
| T | F | 3. Many students find that college courses are stressful because the classes are an abrupt change from those found in high school. |
| T | F | 4. Law and medical students often report high levels of academic pressure because they know that their classmates were top students during their undergraduate years. |
| T | F | 5. Academic stress may be positive for students: It does not necessarily trigger psychological stress. |

| True | False | |
|---|---|---|
| T | F | 6. College students typically find the socialization process in higher education to be less stressful than the professional socialization process they experience when they enter an occupation or profession. |
| T | F | 7. Students who hold jobs outside of school experience higher levels of stress than students who are not employed during their college years. |
| T | F | 8. Getting good grades and completing schoolwork are the top sources of stress reported by college students. |

Answers on page 77.

negative aspects of the socialization process, including the daily stresses that may be involved in this process.

Before reading on, test your knowledge about socialization and the college experience by taking the "Sociology and Everyday Life" quiz.

## Why Is Socialization Important Around the Globe?

**LO1** *Socialization* is the lifelong process of social interaction through which individuals acquire a self-identity and the physical, mental, and social skills needed for survival in society. It is the essential link between the individual and society because it helps us to become aware of ourselves as members of the larger groups and organizations of which we are a part. Socialization also helps us to learn how to communicate with other people and to have knowledge of how other people expect us to behave in a variety of social settings. Briefly stated, socialization enables us to develop our human potential and to learn the ways of thinking, talking, and acting that are necessary for social living.

Socialization is most crucial during childhood because it is essential for the individual's survival and for human development (Figure 3.1). The many people who met the early material and social needs of each of us were central to our establishing our own identity. During the first three years of our life, we begin to develop both a unique identity and the

**socialization** the lifelong process of social interaction through which individuals acquire a self-identity and the physical, mental, and social skills needed for survival in society.

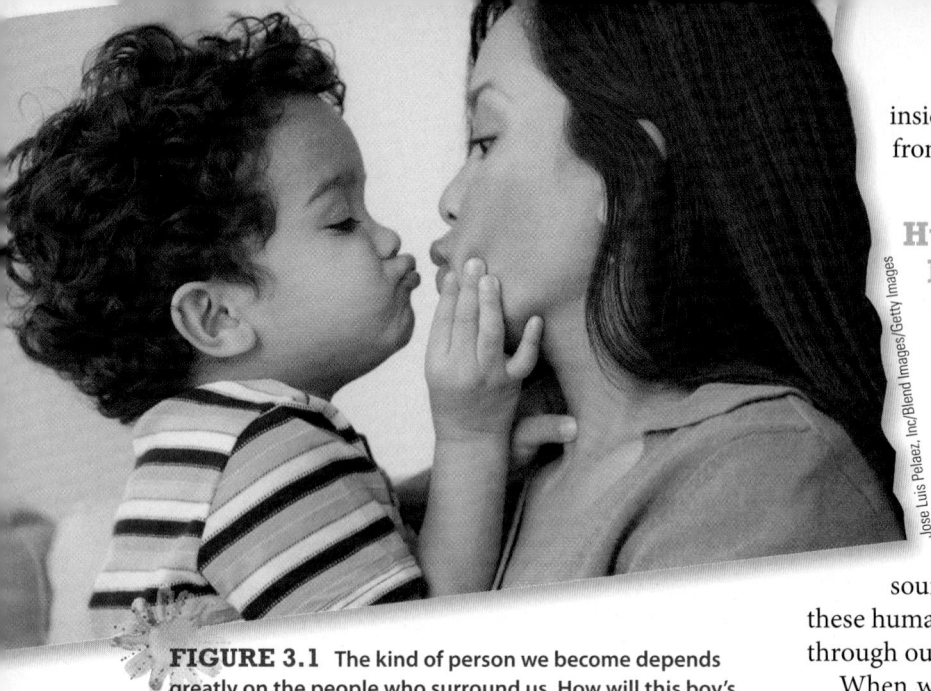

**FIGURE 3.1** The kind of person we become depends greatly on the people who surround us. How will this boy's life be shaped by his close and warm relationship with his mother?

Jose Luis Pelaez, Inc/Blend Images/Getty Images

ability to manipulate things and to walk. We acquire sophisticated cognitive tools for thinking and for analyzing a wide variety of situations, and we learn effective communication skills. In the process we begin a socialization experience that takes place throughout our lives and through which we also have an effect on the people who watch us.

Socialization is also essential for the survival and stability of society. Members of a society must be socialized to support and maintain the existing social structure. From a functionalist perspective, individual conformity to existing norms is not taken for granted; rather, basic individual needs and desires must be balanced against the needs of the social structure. The socialization process is most effective when people conform to the norms of society because they believe that doing so is the best course of action. Socialization enables a society to "reproduce" itself by passing on its culture from one generation to the next.

Although the techniques used to teach newcomers the beliefs, values, and rules of behavior are somewhat similar in many nations, the *content* of socialization differs greatly from society to society. How people walk, talk, eat, make love, and wage war are all functions of the culture in which they are raised. At the same time, we are also influenced by our exposure to subcultures of class, race, ethnicity, religion, and gender. In addition, each of us has unique experiences in our family and friendship groupings. The kind of human being that we become depends greatly on the particular society and social groups that surround us at birth and during early childhood. What we believe about ourselves, our society, and the world does not spring full-blown from inside ourselves; rather, we learn these things from our interactions with others.

## Human Development: Biology and Society

What does it mean to be "human"? To be human includes being conscious of ourselves as individuals, with unique identities, personalities, and relationships with others. As humans, we have ideas, emotions, and values. We have the capacity to think and to make rational decisions. But what is the source of "humanness"? Are we born with these human characteristics, or do we develop them through our interactions with others?

When we are born, we are totally dependent on others for our survival. We cannot turn ourselves over, speak, reason, plan, or do many of the things that are associated with being human. Although we can nurse, wet, and cry, most small mammals can also do those things. As discussed in Chapter 2, we humans differ from nonhuman animals because we lack instincts and must rely on learning for our survival. Human infants have the potential to develop human characteristics if they are exposed to an adequate socialization process.

Every human being is a product of biology, society, and personal experiences—that is, of heredity and environment or, in even more basic terms, "nature" and "nurture." How much of our development can be explained by socialization? How much by our genetic heritage? Sociologists focus on how humans design their own culture and transmit it from generation to generation through socialization. By contrast, sociobiologists assert that nature, in the form of our genetic makeup, is a major factor in shaping human behavior. **Sociobiology** is the systematic study of "social behavior from a biological perspective" (Wilson and Wilson, 2007: 328). According to the zoologist Edward O. Wilson, who pioneered sociobiology, genetic inheritance underlies many forms of social behavior, such as war and peace, envy of and concern for others, and competition and cooperation. Most sociologists disagree with the notion that biological principles can be used to explain all human behavior. Obviously, however, some aspects of our physical makeup—such as eye color, hair color, height, and weight—are largely determined by our heredity.

How important is social influence ("nurture") in human development? There is hardly a single behavior that is not influenced socially. Except for simple reflexes, most human actions are social, either in their causes or in their consequences. Even solitary actions such as crying or brushing our teeth are ultimately

# ANSWERS to the Sociology Quiz on Socialization and the College Experience

**1. False.** Studies have concluded that although professors are important in helping students learn about the academic side of the college experience, our friends and acquaintances help us adapt to higher education.

**2. False.** Slightly more than 85 percent of first-year students at four-year colleges report that they have studied with other students. Similar data are not available for students at two-year schools. How might data for this group differ?

**3. True.** The college environment is stressful for many students, who find that it is an abrupt change from high school because workloads increase, students are expected to manage their time independently, and grades are increasingly important for a person's future endeavors.

**4. True.** The competitive nature of the admission process in law schools and medical schools virtually guarantees that new students will be surrounded by classmates who were exceptional students during their undergraduate years. However, this level of achievement may be a source of stimulation for some students rather than a source of stress.

**5. True.** Some amount of academic stress may be positive in helping students reach their academic and career goals; however, excessive academic stress may be detrimental if it results in high levels of psychological stress or problematic behaviors such as alcohol abuse.

**6. False.** Studies have found that stress levels among college students are higher than those of people entering a new occupation or profession. For this reason, students are encouraged to develop coping skills and build support networks of friends, family, and others.

**7. False.** Most research has not shown a significant relationship between the number of hours worked and levels of stress among students. Earning money for school and personal expenses appears to offset additional time and responsibility in the workplace.

**8. True.** The top stressors most frequently reported in college are getting good grades and completing schoolwork. However, first-year college students also report that changes in eating and sleeping habits, increased workloads and new responsibilities, and going home for holidays and other breaks are major sources of stress for them.

Sources: *Campus Times*, 2008; *Chronicle of Higher Education*, 2011; Messenger, 2009; and Reuters News Service, 2008. © Cengage Learning.

---

social. We cry because someone has hurt us. We brush our teeth because our parents (or dentist) told us it was important. Social environment probably has a greater effect than heredity on the way we develop and the way we act. However, heredity does provide the basic material from which other people help to mold an individual's human characteristics.

Our biological and emotional needs are related in a complex equation. Children whose needs are met in settings characterized by affection, warmth, and closeness see the world as a safe and comfortable place and see other people as trustworthy and helpful. By contrast, infants and children who receive less-than-adequate care or who are emotionally rejected or abused often view the world as hostile and have feelings of suspicion and fear.

## Problems Associated with Social Isolation and Maltreatment

Social environment, then, is a crucial part of an individual's socialization. Even nonhuman primates

such as monkeys and chimpanzees need social contact with others of their species in order to develop properly. As we will see, appropriate social contact is even more important for humans.

**Isolation and Nonhuman Primates** Researchers have attempted to demonstrate the effects of social isolation on nonhuman primates raised without contact with others of their own species. In a series of laboratory experiments, the psychologists Harry and Margaret Harlow (1962, 1977) took infant rhesus monkeys from their mothers and isolated them in separate cages. Each cage contained two nonliving "mother substitutes" made of wire, one with a feeding bottle attached and the other covered with soft terry cloth but without a bottle

**sociobiology** the systematic study of "social behavior from a biological perspective."

(Figure 3.2). The infant monkeys instinctively clung to the cloth "mother" and would not abandon it until hunger drove them to the bottle attached to the wire "mother." As soon as they were full, they went back the to the cloth "mother" seeking warmth, affection, and physical comfort.

The Harlows' experiments show the detrimental effects of isolation on nonhuman primates. When the young monkeys were later introduced to other members of their species, they cringed in the corner. Having been deprived of social contact during their first six months of life, they never learned how to relate to other monkeys or to become well-adjusted adults—they were fearful of or hostile toward other monkeys (Harlow and Harlow, 1962, 1977).

**FIGURE 3.2** As Harry and Margaret Harlow discovered, humans are not the only primates that need contact with others. Deprived of its mother, this infant monkey found a substitute.

Because humans rely more heavily on social learning than do monkeys, the process of socialization is even more important for us.

**Isolated Children** Of course, sociologists would never place children in isolated circumstances so that they could observe what happened to them. However, some cases have arisen in which parents or other caregivers failed to fulfill their responsibilities, leaving children alone or placing them in isolated circumstances. From analysis of these situations, social scientists have documented cases in which children were deliberately raised in isolation. A look at the lives of two children who suffered such emotional abuse provides important insights into the importance of a positive socialization process and the negative effects of social isolation.

**Anna** Born in 1932 in Pennsylvania to an unmarried, mentally impaired woman, Anna was an unwanted child. She was kept in an attic-like room in her grandfather's house. Her mother, who worked on the farm all day and often went out at night, gave Anna just enough care to keep her alive; she received no other care. Sociologist Kingsley Davis (1940) described Anna's condition when she was found in 1938:

> [Anna] had no glimmering of speech, absolutely no ability to walk, no sense of gesture, not the least capacity to feed herself even when the food was put in front of her, and no comprehension of cleanliness. She was so apathetic that it was hard to tell whether or not she could hear. And all of this at the age of nearly six years.

When she was placed in a special school and given the necessary care, Anna slowly learned to walk, talk, and care for herself. Just before her death at the age of ten, Anna reportedly could follow directions, talk in phrases, wash her hands, brush her teeth, and try to help other children (Davis, 1940).

**Genie** About three decades later, Genie was found in 1970 at the age of thirteen (Figure 3.3). She had been locked in a bedroom alone, alternately strapped down to a child's potty chair or straitjacketed into a sleeping bag, since she was twenty months old. She had been fed baby food and beaten with a wooden paddle when she whimpered. She had not heard the sounds of human speech because no one talked to her and there was no television or radio in her room (Curtiss, 1977; Pines, 1981). Genie was placed in a pediatric hospital, where one of the psychologists described her condition:

> At the time of her admission she was virtually unsocialized. She could not stand erect, salivated continuously, had never been toilet-trained and had no control over her urinary or bowel functions. She was unable to chew solid food and had the weight, height and appearance of a child half her age. (Rigler, 1993: 35)

Martin Rogers/The Image Bank/Getty Images

**FIGURE 3.3** A victim of extreme child abuse, Genie was isolated from human contact and tortured until she was rescued at the age of thirteen. What are the consequences to children of isolation and physical abuse, as contrasted with social interaction and parental affection? Sociologists emphasize that the social environment is a crucial part of an individual's socialization.

In addition to her physical condition, Genie showed psychological traits associated with neglect, as described by one of her psychiatrists:

> If you gave [Genie] a toy, she would reach out and touch it, hold it, caress it with her fingertips, as though she didn't trust her eyes. She would rub it against her cheek to feel it. So when I met her and she began to notice me standing beside her bed, I held my hand out and she reached out and took my hand and carefully felt my thumb and fingers individually, and then put my hand against her cheek. She was exactly like a blind child. (Rymer, 1993: 45)

Extensive therapy was used in an attempt to socialize Genie and develop her language abilities (Curtiss, 1977; Pines, 1981). These efforts met with limited success: In the 1990s, Genie was living in a board-and-care home for retarded adults (see Angier, 1993; Rigler, 1993; Rymer, 1993).

Why do we discuss children who have been the victims of maltreatment in a chapter that looks at the socialization process? The answer lies in the fact that such cases are important to our understanding of the socialization process because they show the importance of this process and reflect how detrimental social isolation and neglect can be to the well-being of people.

**Child Maltreatment** What do the terms *child maltreatment* and *child abuse* mean to you? When asked what constitutes child maltreatment, many people first think of cases that involve severe physical injuries or sexual abuse. However, neglect is the most frequent form of child maltreatment (Mattingly and Walsh, 2010). Child neglect occurs when children's basic needs—including emotional warmth and security, adequate shelter, food, health care, education, clothing, and protection—are not met, regardless of cause (Mattingly and Walsh, 2010). Neglect often involves acts of omission (where parents or caregivers fail to provide adequate physical or emotional care for children) rather than acts of commission (such as physical or sexual abuse). Neglect is the most common type of maltreatment among children under age eighteen (see Figure 3.4). Of course, what constitutes child maltreatment differs from society to society.

# Social Psychological Theories of Human Development

Over the past hundred years, a variety of psychological and sociological theories have been developed not only to explain child abuse but also to describe how a positive process of socialization occurs. Although these are not sociological theories, it is important to be aware of the contributions of Freud, Piaget, Kohlberg, and Gilligan because knowing about these theories provides us with a framework for comparing various perspectives on human development.

## Freud and the Psychoanalytic Perspective

The basic assumption in Sigmund Freud's (1924) psychoanalytic approach is that human behavior and personality originate from unconscious

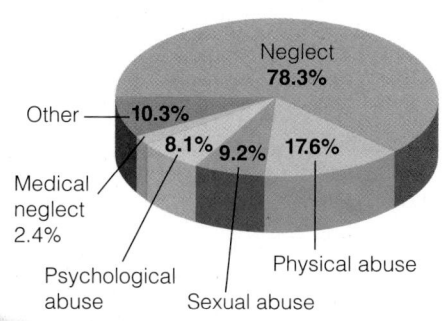

**FIGURE 3.4** TYPES OF MALTREATMENT AMONG CHILDREN UNDER AGE 18*

Source: U.S. Department of Health and Human Services, 2013.

*A child may have suffered from multiple forms of maltreatment and was counted once for each maltreatment type.

**FIGURE 3.5** Sigmund Freud, founder of the psychoanalytic perspective.

Mary Evans/The Image Works

the personality, which he referred to as the *id, ego,* and *superego.* The **id is the component of personality that includes all of the individual's basic biological drives and needs that demand immediate gratification.** For Freud, the newborn child's personality is all id, and from birth the child finds that urges for self-gratification—such as wanting to be held, fed, or changed—are not going to be satisfied immediately. However, id remains with people throughout their life in the form of *psychic energy,* the urges and desires that account for behavior.

By contrast, the second level of personality—the *ego*—develops as infants discover that their most basic desires are not always going to be met by others. The **ego is the rational, reality-oriented component of personality that imposes restrictions on the innate pleasure-seeking drives of the id.** The ego channels the desire of the id for immediate gratification into the most advantageous direction for the individual. The third level of personality—the superego—is in opposition to both the id and the ego. The **superego, or conscience, consists of the moral and ethical aspects of personality.** It is first expressed as the recognition of parental control and eventually matures as the child learns that parental control is a reflection of the values and moral demands of the larger society. When a person is well adjusted, the ego successfully manages the opposing forces of the id and the superego. Figure 3.6 illustrates Freud's theory of personality.

forces within individuals. Freud (1856–1939), who is known as the founder of psychoanalytic theory, developed his major theories in the Victorian era, when biological explanations of human behavior were prevalent (Figure 3.5). For example, Freud based his ideas on the belief that people have two basic tendencies: the urge to survive and the urge to procreate.

According to Freud (1924), human development occurs in three states that reflect different levels of

**FIGURE 3.6** FREUD'S THEORY OF PERSONALITY

This illustration shows how Freud might picture a person's internal conflict over whether to commit an antisocial act such as stealing a candy bar. In addition to dividing personality into three components, Freud theorized that our personalities are largely unconscious—hidden from our normal awareness. To dramatize his point, Freud compared conscious awareness (portions of the ego and superego) to the visible tip of an iceberg. Most of personality—including the id, with its raw desires and impulses—lies submerged in our subconscious.

**80**

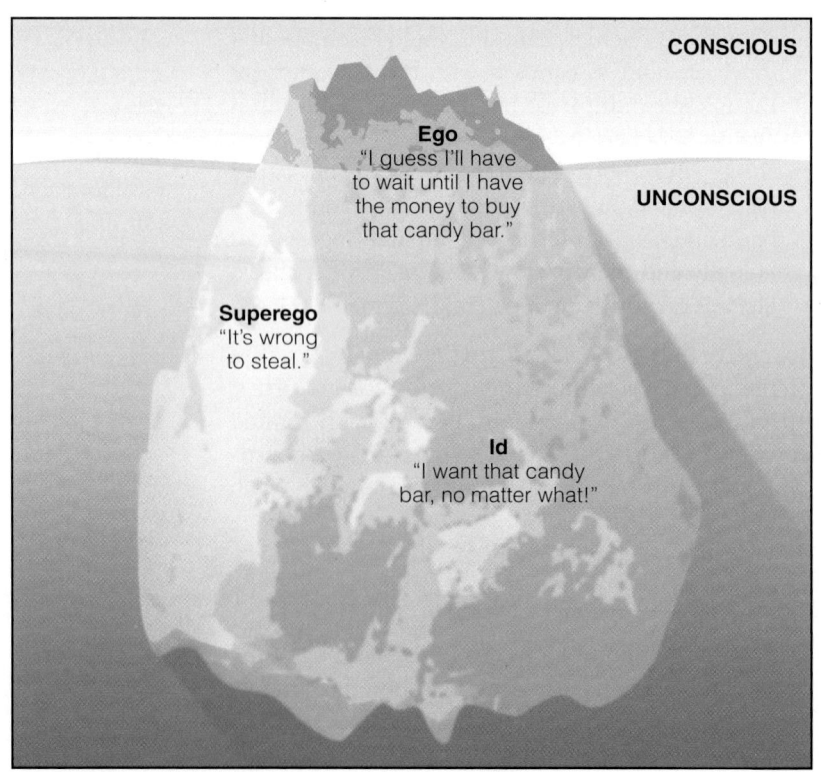

CONSCIOUS

**Ego**
"I guess I'll have to wait until I have the money to buy that candy bar."

UNCONSCIOUS

**Superego**
"It's wrong to steal."

**Id**
"I want that candy bar, no matter what!"

**FIGURE 3.7** Jean Piaget, a pioneer in the field of cognitive development.

development are organized around specific tasks that, when mastered, lead to the acquisition of new mental capacities, which then serve as the basis for the next level of development. Piaget emphasized that all children must go through each stage in sequence before moving on to the next one, although some children move through them faster than others.

1. *Sensorimotor stage* (birth to age two). During this period, children understand the world only through sensory contact and immediate action because they cannot engage in symbolic thought or use language. Toward the end of the second year, children comprehend *object permanence*; in other words, they start to realize that objects continue to exist even when the items are out of sight.

2. *Preoperational stage* (age two to seven). In this stage, children begin to use words as mental symbols and to form mental images. However, they are still limited in their ability to use logic to solve problems or to realize that physical objects may change in shape or appearance while still retaining their physical properties (see Figure 3.8).

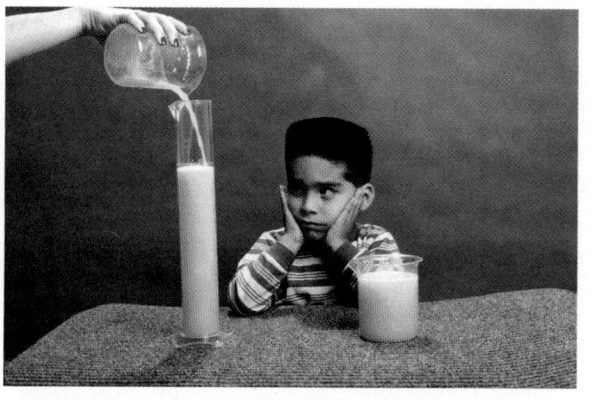

**FIGURE 3.8** Psychologist Jean Piaget identified four stages of cognitive development, including the preoperational stage, in which children have limited ability to realize that physical objects may change in shape or appearance. Piaget showed children two identical beakers filled with the same amount of water. After the children agreed that both beakers held the same amount of water, Piaget poured the water from one beaker into a taller, narrower beaker and then asked them about the amounts of water in each beaker. Those still in the preoperational stage believed that the taller beaker held more water because the water line was higher than in the shorter, wider beaker.

## Piaget and Cognitive Development

Jean Piaget (1896–1980), a Swiss psychologist, was a pioneer in the field of cognitive (intellectual) development (Figure 3.7). Cognitive theorists are interested in how people obtain, process, and use information—that is, in how we think. Cognitive development relates to changes over time in how we think.

Piaget (1954) believed that in each stage of development (from birth through adolescence), children's activities are governed by their perception of the world around them. His four stages of cognitive

**id** Sigmund Freud's term for the component of personality that includes all of the individual's basic biological drives and needs that demand immediate gratification.

**ego** Sigmund Freud's term for the rational, reality-oriented component of personality that imposes restrictions on the innate pleasure-seeking drives of the id.

**superego** Sigmund Freud's term for the conscience, consisting of the moral and ethical aspects of personality.

3. *Concrete operational stage* (age seven to eleven). During this stage, children think in terms of tangible objects and actual events. They can draw conclusions about the likely physical consequences of an action without always having to try the action out. Children begin to take the role of others and start to empathize with others' viewpoints.
4. *Formal operational stage* (age twelve through adolescence). By this stage, adolescents are able to engage in highly abstract thought and understand places, things, and events they have never seen. They can think about the future and evaluate different options or courses of action.

### Kohlberg and the Stages of Moral Development

Lawrence Kohlberg (1927–1987) elaborated on Piaget's theories of cognitive reasoning by conducting a series of studies in which children, adolescents, and adults were presented with moral dilemmas that took the form of stories. Based on his findings, Kohlberg (1969, 1981) classified moral reasoning into three sequential levels:

1. *Preconventional level* (age seven to ten). Children's perceptions are based on punishment and obedience. Evil behavior is that which is likely to be punished; good conduct is based on obedience and avoidance of unwanted consequences.
2. *Conventional level* (age ten through adulthood). People are most concerned with how they are perceived by their peers and with how one conforms to rules.
3. *Postconventional level* (few adults reach this stage). People view morality in terms of individual rights; "moral conduct" is judged by principles based on human rights that transcend government and laws.

### Gilligan's View on Gender and Moral Development

Psychologist Carol Gilligan (b. 1936) noted that both Piaget and Kohlberg did not take into account how gender affects the process of social and moral development. According to Gilligan (1982), Kohlberg's model was developed solely on the basis of research with male respondents, who often have different views from women on morality. Gilligan believes that men become more concerned with law and order but that women tend to analyze social relationships and the social consequences of behavior. Gilligan argues that men are more likely to use *abstract standards* of right and wrong when making moral decisions, whereas women are more likely to be concerned about the *consequences* of behavior. Does this constitute a "moral deficiency" on the part of either women or men? Not according to Gilligan, who believes that people make moral decisions according to both abstract principles of justice and principles of compassion and care.

# Sociological Theories of Human Development

**LO2** Although social scientists acknowledge the contributions of social–psychological explanations of human development, sociologists believe that it is important to bring a sociological perspective to bear on how people develop an awareness of self and learn about the culture in which they live. Let's look at symbolic interactionist, functional, and conflict approaches to describing the socialization process and its outcomes.

### Symbolic Interactionist Perspectives on Socialization

According to a symbolic interactionist approach to socialization, we cannot form a sense of self or personal identity without intense social contact with others. How do we develop ideas about who we are? How do we gain a sense of self? The self represents the sum total of perceptions and feelings that an individual has of being a distinct, unique person—a sense of who and what one is. When we speak of the "self," we typically use words such as *I, me, my, mine,* and *myself* (Cooley, 1998/1902). This sense of self (also referred to *self-concept*) is not present at birth; it arises in the process of social experience. **Self-concept is the totality of our beliefs and feelings about ourselves.** Four components make up our self-concept: (1) the physical self ("I am tall"), (2) the active self ("I am good at soccer"), (3) the social self ("I am nice to others"), and (4) the psychological self ("I believe in world peace"). Between early and late childhood, a child's focus tends to shift from the physical and active dimensions of self toward the social and psychological aspects. Self-concept is the foundation for communication with others; it continues to develop and change throughout our lives.

Our *self-identity* is our perception about what kind of person we are and our awareness of our unique identity. Self-identity emerges when we ask the question "Who am I?" Factors such as individuality, uniqueness, and personal characteristics and personality are components of self-identity. As we have seen, socially isolated children do not have typical self-identities because they have had no experience of "humanness."

According to symbolic interactionists, we do not know who we are until we see ourselves as we believe that others see us. The perspectives of symbolic interactionists Charles Horton Cooley and George Herbert Mead help us understand how our self-identity is developed through our interactions with others.

### Cooley: The Looking-Glass Self

Charles Horton Cooley (1864–1929) was one of the first U.S. sociologists to describe how we learn about ourselves through social interaction with other people. Cooley used the concept of the *looking-glass self* to describe how the self emerges. The **looking-glass self refers to the way in which a person's sense of self is derived from the perceptions of others.** Our looking-glass self is based on our perception of *how* other people think of us (Cooley, 1998/1902). As Figure 3.9 shows, the looking-glass self is a self-concept derived from a three-step process:

We imagine how we appear to other people.

We imagine how other people judge the appearance that we think we present.

If we think the evaluation is favorable, our self-concept is enhanced.

If we think the evaluation is unfavorable, our self-concept is diminished.

**FIGURE 3.9** HOW THE LOOKING-GLASS SELF WORKS
© Cengage Learning

1. We imagine how our personality and appearance will look to other people.
2. We imagine how other people judge the appearance and personality that we think we present.
3. We develop a self-concept. If we think the evaluation of others is favorable, our self-concept is enhanced. If we think the evaluation is unfavorable, our self-concept is diminished. (Cooley, 1998/1902)

Because the looking-glass self is based on how we *imagine* other people view us, we may develop self-concepts based on an inaccurate perception of what other individuals think about us. Consider, for example, the individual who believes that other people see him or her as "fat" when, in actuality, he or she is a person of average height, weight, and build. The consequences of such a false perception may lead to excessive dieting or health problems such as anorexia, bulimia, and other eating disorders.

### Mead: Role-Taking and Stages of the Self

George Herbert Mead (1863–1931) extended Cooley's insights by linking the idea of self-concept to *role-taking*—**the process by which a person mentally assumes the role of another person or group in order to understand the world from that person's or group's point of view.** Role-taking often occurs through play and games, as children try out different roles (such as being mommy, daddy, doctor, or teacher) and gain an appreciation of them. First, people come to take the role of the other (role-taking).

By taking the roles of others, the individual hopes to ascertain the intention or direction of the acts of others. Then the person begins to construct his or her own roles (role-making) and to anticipate other individuals' responses. Finally, the person plays at her or his particular role (role-playing).

According to Mead (1934), children in the early months of life do not realize that they are separate from others. However, they do begin early on to see a mirrored image of themselves in others. Shortly after birth, infants start to notice the faces of those around them, especially the significant others, whose faces start to have meaning because they are associated with experiences such as feeding and cuddling. **Significant others are those persons whose care, affection, and approval are especially**

**self-concept** the totality of our beliefs and feelings about ourselves.

**looking-glass self** Charles Horton Cooley's term for the way in which a person's sense of self is derived from the perceptions of others.

**role-taking** the process by which a person mentally assumes the role of another person or group in order to understand the world from that person's or group's point of view.

**significant others** those persons whose care, affection, and approval are especially desired and who are most important in the development of the self.

**desired and who are most important in the development of the self.** Gradually, we distinguish ourselves from our caregivers and begin to perceive ourselves in contrast to them. As we develop language skills and learn to understand symbols, we begin to develop a self-concept. When we can represent ourselves in our minds as objects distinct from everything else, our self has been formed.

According to Mead (1934), the self has two sides, which he divided into the "me" and the "I." The "me" is what is learned by interaction with others in the larger social environment; it is the organized set of attitudes of others that an individual assumes. The "me" is the objective element of the self, which represents an internalization of the expectations and attitudes of others and the individual's awareness of those demands. By contrast, the "I" is the person's individuality—it is the response of the person to the attitudes of other individuals; we might think of the "me" as the social self and the "I" as the response to the "me." According to Mead, the "I" develops first, and the "me" takes form during the three stages of self-development (Figure 3.10):

1. During the *preparatory stage,* up to about age three, interactions lack meaning, and children largely imitate the people around them, particularly parents and other family members. At this stage, children are preparing for role-taking.

2. In the *play stage,* from about age three to five, children learn to use language and other symbols, thus enabling them to pretend to take the roles of specific people. At this stage, they begin to see themselves in relation to others, but they do not see role-taking as something they have to do.

3. During the *game stage,* which begins in the early school years,

**FIGURE 3.10** According to sociologist George Herbert Mead, the self develops through three stages. In the preparatory stage, children imitate others; in the play stage, children pretend to take the roles of specific people; and in the game stage, children become aware of the "rules of the game" and the expectations of others.

# Sociology WORKS!

## The Issue of Excessive Praise

I sometimes say that praise is fine "when praise is due." We get into the habit of praising when it isn't praise that is appropriate but encouragement. For example, we're always saying to young children: "Oh, what a beautiful picture," even when their pictures aren't necessarily beautiful. So why not really look at each picture? Maybe a child has painted a picture with many wonderful colors. Why don't we comment on that—on the reality of the picture? (Docia Zavitkovsky, qtd. in *Scholastic Parent & Child*, 2007)

Child development specialist Docia Zavitkovsky points out that excessive praise or unearned compliments may be problematic for children because, rather than bolstering their self-esteem, such praise may increase a child's dependence on adults. As children increasingly rely on constant praise and on significant others to identify what is good or bad about their performance, they may not develop the ability to make meaningful judgments about what they have done. Excessively praising children may make it more difficult for them to develop a positive self-concept and visualize an accurate picture of what is expected of them as they grow into young adulthood.

Zavitkovsky's ideas remind us of the sociological insights of the early sociologist George Herbert Mead, who described how children learn to take into account the expectations of the larger society and to balance the "I" (the subjective element of the self: the spontaneous and unique traits of each person) with the "me" (the objective element of the self: the internalized attitudes and demands of other members of society and the individual's awareness of those demands).

Does this mean that children should not be praised? Definitely not! As noted above, what children may need sometimes is not praise but *encouragement*. From this perspective, positive feedback can have a very important influence on a child's self-esteem because he or she can learn how to do a "good job" when engaging in a specific activity or accomplishing a task rather than simply being praised for any effort expended. Mead's concept of the generalized other makes us aware of the importance of other people's actions in how self-concept develops.

### reflect & analyze

**What effect does receiving praise when we are young have on us when we are college students? Also, when we are dealing with our peers, how might we thoughtfully use the phrase "Good job!" without making it into an overworked expression?**

---

children understand not only their own social position but also the positions of others around them. In contrast to play, games are structured by rules, are often competitive, and involve a number of other "players." At this time, children become concerned about the demands and expectations of others and of the larger society.

Mead's concept of the ***generalized other* refers to the child's awareness of the demands and expectations of the society as a whole or of the child's subculture.** According to Mead, the generalized other is evident when a person takes into account other people and groups when he or she speaks or acts. In sum, both the "I" and the "me" are needed to form the social self. The unity of the two (the "generalized other") constitutes the full development of the individual and a more thorough understanding of the social world.

How useful are symbolic interactionist perspectives in enhancing our understanding of the socialization process? This approach has made a positive contribution to our understanding of how the self develops (see "Sociology Works!").

**More-Recent Symbolic Interactionist Perspectives** The symbolic interactionist approach emphasizes that socialization is a collective process in which children are active and creative agents, not just passive recipients of the socialization process. From this view, childhood is a *socially constructed* category. As children acquire language skills and interact with other people, they begin to construct their own shared meanings. Sociologist William A. Corsaro (2011) refers to this as the "orb web model," whereby the cultural knowledge that children possess consists not only of beliefs found in the adult world but also of unique interpretations from the children's own peer culture. According to Corsaro, children create and share a *peer culture,* which is an established set of activities, routines, and beliefs that are in some ways different from adult culture. This peer culture emerges through interactions as children "borrow" from the adult culture but transform it so that it fits their own situation. In fact, according to Corsaro, peer culture is the most significant arena in which children and young people acquire cultural knowledge.

**generalized other** George Herbert Mead's term for the child's awareness of the demands and expectations of the society as a whole or of the child's subculture.

**FIGURE 3.11** Some theorists identify three stages of socialization: primary, secondary, and tertiary. At what stage might socialization be occurring for the people volunteering together in this photo?

## Functionalist Perspectives on Socialization

As discussed in Chapter 1, functionalist theorists such as Talcott Parsons and Robert Merton saw socialization as the process by which individuals internalize social norms and values. They believed that socialization is important to societies as well as to individuals because social institutions must be maintained and preserved for a nation to survive. For these institutions to be efficient, individuals must play their roles appropriately, or dysfunctions will occur. Simply stated, the socialization process plays an integral part in teaching the next generation, as well as new arrivals, about how to conform to the rules of the game, and this keeps the society functioning properly. As a result of adequate socialization, people come to support a society that is stable and orderly. Individuals learn to accept the values, beliefs, and behavioral expectations that keep society, and sometimes the larger global community, functioning effectively.

Some functionalist theorists identify three stages of socialization (Figure 3.11): ***Primary socialization* refers to the process of learning that begins at birth and occurs in the home and family;** by contrast, ***secondary socialization* refers to the process of learning that takes place outside the home—in settings such as schools, religious organizations, and the workplace—and helps individuals learn how to act in appropriate ways in various situations.** Secondary socialization often occurs when we are teenagers and young adults. ***Tertiary socialization* takes place when adults move into new settings where they must accept certain ideas or engage in specific behaviors that are appropriate to that specific setting.** For example, older persons entering a retirement community often have to internalize new social norms and values that are appropriate to the setting in which they now reside. From a functionalist approach, problems in the socialization process contribute not only to individual concerns but also to larger societal issues, such as high rates of crime and poverty, school dropouts and failures, and family discord.

## Conflict Perspectives on Socialization

Based on an assumption that groups in society are engaged in a continuous power struggle for control of scarce resources, conflict theorists stress that socialization contributes to "false consciousness"—a lack of awareness and a distorted perception of the reality of class as it affects all aspects of social life. As a result, socialization reaffirms and reproduces the class structure in the next generation rather than challenging existing conditions. For example, children in low-income families may be unintentionally socialized to believe that acquiring an education and aspiring to lofty ambitions are pointless because of existing economic conditions in the family. By contrast, middle- and upper-income families typically instill ideas of monetary and social success in children. As discussed later, schools may also provide different experiences to children depending on their gender, social class, racial–ethnic background, and other factors. This chapter's Concept Quick Review summarizes the major theories of human development and socialization.

## Agents of Socialization

*Agents of socialization* **are the persons, groups, or institutions that teach us what we need to know in order to participate in society.** We are exposed to many agents of socialization throughout our lifetime; in turn, we have an influence on those socializing agents and organizations. In this section we look at the most pervasive agents of socialization in childhood—the family, the school, peer groups, and the mass media.

### The Family

**LO3** The family is the most important agent of socialization in all societies. From our infancy onward, our families transmit cultural and social values to us. As discussed later in this book, families vary in size and structure. Some families consist of two parents and their biological children, whereas others consist of a single parent and one or more children. Still other families reflect changing patterns of divorce and remarriage, and an increasing number are made up of same-sex partners and their children. Over time, patterns have changed in some two-parent families so that fathers, rather than mothers, are the primary daytime agents of socialization for their young children.

Theorists using a functionalist perspective emphasize that families serve important functions in society because they are the primary locus for the procreation and socialization of children (Figure 3.12). Most of us form an emerging sense of self and acquire most of our beliefs and values within the family context. We also learn about the larger dominant culture (including language, attitudes, beliefs, values, and norms) and the primary subcultures to which our parents and other relatives belong.

Families are also the primary source of emotional support. Ideally, people receive love, understanding,

Comstock / Jupiterimages

**FIGURE 3.12** As this birthday celebration attended by several generations of family members illustrates, socialization enables society to "reproduce" itself.

security, acceptance, intimacy, and companionship within families. The role of the family is especially significant because young children have little social experience beyond the family's boundaries; they have no basis for comparing or evaluating how they are treated by their own family.

To a large extent, the family is where we acquire our specific social position in society. From birth, we are a part of the specific racial, ethnic, class, religious, and regional subcultural grouping of our family. Many parents socialize their children somewhat differently based on race, ethnicity, and class. Some families instruct their children about the unique

**primary socialization** the process of learning that begins at birth and occurs in the home and family.

**secondary socialization** the process of learning that takes place outside the home—in settings such as schools, religious organizations, and the workplace—and helps individuals learn how to act in appropriate ways in various situations.

**tertiary socialization** the process of learning that takes place when adults move into new settings where they must accept certain ideas or engage in specific behaviors that are appropriate to that specific setting.

**agents of socialization** the persons, groups, or institutions that teach us what we need to know in order to participate in society.

## Psychological and Sociological Theories of Human Development and Socialization

| | | |
|---|---|---|
| Social psychological theories | Freud's psychoanalytic perspective | |
| | Piaget's cognitive development | Children go through four stages of cognitive (intellectual) development, moving from understanding only through sensory contact to engaging in highly abstract thought. |
| | Kohlberg's stages of moral development | People go through three stages of moral development, from avoidance of unwanted consequences to viewing morality based on human rights. |
| | Gilligan: gender and moral development | Women go through stages of moral development from personal wants to the greatest good for themselves and others. |
| Symbolic interactionist theories | Cooley's looking-glass self | A person's sense of self is derived from his or her perception of how others view him or her. |
| | Mead's three stages of self-development | In the preparatory stage, children prepare for role-taking. In the play stage, they pretend to take the roles of specific people. In the game stage, they learn to take into account the demands and expectations of the larger society and to develop a generalized other. |
| Functionalist theories | Parsons's and Merton's views | The socialization process serves a central function for both individuals and society by helping people learn the appropriate norms, values, and behaviors that support social institutions and the larger social group. |
| Conflict theories | Based on Marx's work | Socialization contributes to false consciousness and reproduces inequalities in the class structure in the next generation as well as ignoring crucial differences based on gender, race/ethnicity, and other factors. |

racial–ethnic and/or cultural backgrounds of their parents and grandparents so that they will have a better appreciation of their heritage. Other families teach their children primarily about the dominant, mainstream culture in hopes that this will help their children get ahead in life.

In terms of class, some upper-class parents focus on teaching their children about the importance of wealth, power, and privilege; however, many downplay this aspect of life and want their children to make their own way in life, fearing that "spoiling them" will not be in their best interest. Middle-class parents have typically focused on academic achievement and the importance of hard work to achieve

the American Dream. However, with the recent economic recession, some optimism that the middle class once passed on to their children may have diminished. Parents in the working-class and lower-income levels often have to struggle to keep shelter overhead and food on the table, and they have little time to help their children learn about things that will help them succeed in school and life (Kendall, 2002, 2011). This problem contributes to and reinforces social inequality, and this is one reason why conflict theorists are concerned about the long-term effects of the socialization process.

However, we should note that socialization is a bidirectional process in which children and young people socialize their agents of socialization, including parents, teachers, and others, as well as receiving socialization from these important agents. ***Reciprocal socialization is the process by which the feelings, thoughts, appearance, and behavior of individuals who are undergoing socialization also have a direct effect on those agents of socialization who are attempting to influence them.*** Examples of this process include parents whose preferences in music, hairstyles, and clothing are influenced by their children, and teachers whose choice of words ("like, you know," "LOL," and other slang terms) is similar to that of their students.

## The School

**LO4** As the amount of specialized technical and scientific knowledge has expanded rapidly and as the amount of time that children are in educational settings has increased, schools continue to play an enormous role in the socialization of young people. For many people, the formal education process is an undertaking that lasts up to twenty years.

As the number of one-parent families and families in which both parents work outside the home has increased dramatically, the number of children in day-care and preschool programs has also grown rapidly. Currently, about 60 percent of all U.S. preschool children are in day care, either in private homes or institutional settings, and this percentage continues to climb (Children's Defense Fund, 2012b). Generally, studies have found that quality day-care and preschool programs have a positive effect on the overall socialization of children. These programs provide children with the opportunity to have frequent interactions with teachers and to learn how to build their language and literacy skills. High-quality programs also have a positive effect on the academic performance of children, particularly those from low-income families. Today, however, the cost of child-care programs has become a major concern for many families.

Although schools teach specific knowledge and skills, they also have a profound effect on children's self-image, beliefs, and values (Figure 3.13). As children enter school for the first time, they are evaluated and systematically compared with one another by the teacher. A permanent, official record is kept of each child's personal behavior and academic activities. From a functionalist perspective, schools are responsible for (1) socialization, or teaching students to be productive members of society; (2) transmission of culture; (3) social control and

Jamie Grill/Getty Images

**FIGURE 3.13** Students are sent to school to be educated. However, what else will they learn in school beyond the academic curriculum? Sociologists differ in their responses to this question.

personal development; and (4) the selection, training, and placement of individuals on different rungs in the society (Ballantine and Hammack, 2012).

In contrast, conflict theorists assert that students have different experiences in the school system depending on their social class, their racial–ethnic background, the neighborhood in which they live, their gender, and other factors. For example, Langhout and Mitchell (2008), after investigating the "hidden curriculum" in a low-income elementary school, concluded that African American and Latino boys were disproportionately punished for violating the rules (e.g., raising your hand to speak) when compared to their white and female counterparts. Thus, schools do not socialize children for their own well-being but rather for their later roles in the workforce, where it is important to be well behaved and "know your place." Students who are destined for leadership or elite positions acquire different skills and knowledge than those who will enter working-class and middle-class occupations.

**reciprocal socialization** the process by which the feelings, thoughts, appearance, and behavior of individuals who are undergoing socialization also have a direct effect on those agents of socialization who are attempting to influence them.

**FIGURE 3.14** The pleasure of participating in activities with friends is one of the many attractions of adolescent peer groups. What groups have contributed the most to your sense of belonging and self-worth?

## Peer Groups

**LO5** As soon as we are old enough to have acquaintances outside the home, most of us begin to rely heavily on peer groups as a source of information and approval about social behavior (Figure 3.14). A *peer group* **is a group of people who are linked by common interests, equal social position, and (usually) similar age.** In early childhood, peer groups are often composed of classmates in day care, preschool, and elementary school. Preadolescence—the latter part of the elementary school years—is an age period in which children's peer culture has an important effect on how children perceive themselves and how they internalize society's expectations (Robnett and Susskind, 2010). For example, boys who have a large proportion of same-gender friends are more likely to reject "feminine" traits, which they associate with girls. As a result, this may play a part in socializing them to have negative attitudes toward femininity that they display later in life (Robnett and Susskind, 2010). In adolescence, peer groups are typically made up of people with similar interests and social activities. As adults, we continue to participate in peer groups of people with whom we share common interests and comparable occupations, income, and/or social position.

Peer groups function as agents of socialization by contributing to our sense of "belonging" and our feelings of self-worth. As early as the preschool years, peer groups provide children with an opportunity for successful adaptation to situations such as gaining access to ongoing play, protecting shared activities from intruders, and building solidarity and mutual trust during ongoing activities (Corsaro, 2011). Unlike families and schools, peer groups provide children and adolescents with some degree of freedom from parents and other authority figures. They also teach cultural norms such as what constitutes "acceptable" behavior in a specific situation. Peer groups simultaneously reflect the larger culture and serve as a conduit for passing on culture to young people. As a result, the peer group is both a product of culture and one of its major transmitters.

Is there such a thing as "peer pressure"? Individuals must earn their acceptance with their peers by conforming to a given group's norms, attitudes, speech patterns, and dress codes. When we conform to our peer group's expectations, we are rewarded; if we do not conform, we may be ridiculed or even expelled from the group. Conforming to the demands of peers frequently places children and adolescents at cross-purposes with their parents. For example, young people are frequently under pressure to obtain certain valued material possessions (such as toys, clothing, athletic shoes, or cell phones); they then pass the pressure on to their parents through emotional pleas to purchase the desired items.

## Mass Media

An agent of socialization that has a profound impact on both children and adults is the *mass media,* **composed of large-scale organizations that use print or electronic means (such as radio, television, film, and the Internet) to communicate with large numbers of people.** Today, the term *media* also includes the many forms of Web-based and mobile technologies that we refer to as "social media," such as Facebook, Twitter, and YouTube. For many years, the media have functioned as socializing agents in several ways: (1) they inform us about events; (2) they introduce us to a wide variety of people; (3) they provide an array of viewpoints on current issues; (4) they make us aware of products

# COLLEGE LIFE IN THE MEDIA

## Learning About College from TV Shows

*Community* is a classic culture clash comedy. Familiar archetypes—a snooty girl; a spindly nerd; a brash, heavyset black woman; a cocky former high school jock; a neurotic drop out; an eccentric retiree—are forced to interact in a Spanish study group. (Stanley, 2009)

[In *Greek*,] after years of living on the straight and narrow, nerdy freshman Rusty enters college determined to catch up on all the partying he's missed by joining a fraternity. His big sister, Casey, is a member of the best sorority on campus. Through them and other students at Cyprus Rhodes University, viewers see the social minefield that is the Greek system. (ABC Family, 2010)

Community and *Greek* are two TV series, now available in syndication and on DVD, that claim to reflect college life, or at least some aspect of it. *Greek* supposedly shows life in the sorority and fraternity system at a four-year university, while *Community* seeks to entertain audiences by showing a somewhat dysfunctional group of students at Greendale Community College in Colorado.

How realistic are programs such as these in socializing students for college life? And what forms of media framing are most prevalent? Here are a few of the many media frames that are apparent in *Community* and *Greek*:

- *Antithesis framing*: Professors are *not* like us! Rather, professors are portrayed as odd individuals who think and dress peculiarly and have little understanding of what today's students are like. For example, in an episode of *Community*, Betty White, an over-ninety actor, plays the role of an anthropology instructor whose antics include using a blowgun to shoot an arrow at a student and claiming that she is drinking urine from an antique instrument.

- *Socialization framing*: College is all about fun, romance, and heartbreaks, not about studying and gaining new insights. For example, *Greek* portrays college life as a series of ongoing social interactions, romantic conquests, and theme-based drinking parties at sorority and fraternity houses. Classrooms with students and professors are occasionally shown but primarily as a backdrop so that characters can set up a new romantic liaison or have a public confrontation with a member of a rival Greek organization.

- *Post-racism and post-homophobia framing*: College in the past was laden with racial and ethnic tensions, as well as homophobia, but supposedly we can now laugh about these issues. For example, in *Greek*, Calvin is both gay and African American, so his membership in a traditional, white, homophobic Greek fraternity produces two-dimensional tensions that are typically swept away with comments which suggest that racial differences and prejudice against gays are all in the past and no longer have to be confronted or challenged by dominant-group members.

Although *Community* and *Greek* are only two of many TV series that have focused on college life, they borrow heavily from stereotypes that have been popular for many years in media representations of college students. Clearly, these programs are created for entertainment purposes and are not true representations of the socialization process that occurs at college or of how students truly live and interact.

## reflect & analyze

**What part did your family, peers, and the media play in forming your own expectations about what college life would be like? What media frames can you identify from TV shows and films you have viewed about college students and academic life?**

and services that, if we purchase them, will supposedly help us to be accepted by others; and (5) they entertain us by providing the opportunity to live vicariously (through other people's experiences).

Although most of us take for granted that the media play an important part in contemporary socialization, we frequently underestimate the enormous influence that this agent of socialization may have on our attitudes and behavior. If we look at media framing of entertainment programs about a social setting with which we are very familiar, such as college life, it is possible to see how television and films can affect how we view students and the academic life in general (see "College Life in the Media").

As previously noted, the use of social media such as Facebook and Twitter has grown exponentially in recent years. Today, 93 percent of teens report that they use the Internet, and most indicate that they use it to interact with friends, share their artistic creations, and tell stories through online journals and blogs (Purcell, 2012). Although boys are more likely to play video games, girls lead the charge in the teen blogosphere. Social networking has been added as an additional layer on top of existing layers of media use, particularly among young people.

> **peer group** a group of people who are linked by common interests, equal social position, and (usually) similar age.
>
> **mass media** large-scale organizations that use print or electronic means (such as radio, television, film, and the Internet) to communicate with large numbers of people.

**FIGURE 3.15** As part of the socialization process in the past, our parents and grandparents might have spent time learning how to read a paper map to help them when they wanted to go somewhere. Today many of us simply use the GPS on our smartphone or tablet to guide us in the right direction. How does this process differ from earlier learning experiences? For example, a paper map cannot talk to the walker in this photo.

Consider children, for example. Research has shown that daily media use among U.S. children between the ages of eight and eighteen has increased dramatically. Studies indicate that children in this age range spend an average of 7 hours and 38 minutes a day (which adds up to more than 53 hours a week) using entertainment media. Cell phones, iPads, and other mobile media make it possible for young people to have access to media 24 hours per day, 7 days per week (Figure 3.15), even when they are seemingly away from everyone else (Rideout, Foehr, and Roberts, 2010).

Exactly how the process of socialization has changed as media use has grown is unknown, but it is known that African American and Latino/a children are spending far more time with media than are white American (non-Hispanic) children. A major study by the Kaiser Family Foundation found black and Hispanic children spending almost 4.5 more hours involved in daily media use than their white American counterparts. Even when factors such as the education level of the children's parents, whether they live in a one- or two-parent household, and the children's ages are taken into account, the amount of media use still varies by race/ethnicity (Rideout, Foehr, and Roberts, 2010). Does this make a significant difference in childhood socialization? Future studies will no doubt continue to examine new media's effects on children and how increased use relates to grades, family interaction patterns, social networks, and other important issues in reaching maturity.

Parents, educators, social scientists, and public officials have widely debated the consequences of young people watching violence on television and elsewhere. In addition to concerns about violence in television programming, motion pictures, and electronic games, television shows have been criticized for projecting negative images of women and people of color. Although the mass media have changed some of the roles that they depict women as playing, some newer characters tend to reinforce existing stereotypes of women as sex objects even when they are in professional roles such as doctors or lawyers. Often, women are shown in traditional roles as homemakers or caretakers who do not receive respect from other family members unless they demand it (Collins, 2011). What effect does this have on children and young people as they develop their own ideas about the "adult world"? Throughout this text, we will look at additional examples of how the media socialize us in ways that we may or may not realize.

## Gender Socialization

**LO6** *Gender socialization* **is the aspect of socialization that contains specific messages and practices concerning the nature of being female or male in a specific group or society.** Through the process of gender socialization we learn which attitudes and behaviors are considered to be appropriate for girls and boys, men and women, in a particular society. Different sets of gender norms are appropriate for females and males in the United States and in most other nations.

One of the primary agents of gender socialization is the family. In some families this process begins even before the child's birth. Parents who learn the sex of the fetus through ultrasound or amniocentesis often purchase color-specific and gender-typed clothes, toys, and nursery decorations in anticipation of their daughter's or son's arrival. After the child's birth, parents may respond differently toward male and female infants; they often play more roughly with boys and talk more lovingly to girls. Throughout childhood and adolescence, boys and girls are typically assigned different household chores and given different privileges, such as boys being given more latitude to play farther away from home than girls and being allowed to stay out later at night (Lareau, 2011).

In regard to gender-socialization practices among various racial–ethnic groups, some sociologists have found that children typically are not taught to think of gender strictly in "male–female" terms. Both daughters and sons are socialized toward autonomy, independence, self-confidence, and nurturance of children. Sociologist Patricia Hill Collins (2000) has suggested that "othermothers" (women other than a child's biological mother) play an important part in the gender socialization and motivation of African

PART 1 | STUDYING SOCIETY AND SOCIAL LIFE

American children, especially girls. Othermothers often serve as gender-role models and encourage women to become activists on behalf of their children and community (Collins, 2000). By contrast, in the past, Korean American and Latino/a families have typically engaged in more-traditional gender socialization, but evidence in the 2010s suggests that this pattern has continued to change as young women are spending more time away from older family members and are gaining greater freedom of expression at school and in the workplace.

Like the family, schools, peer groups, and the media also contribute to our gender socialization. From kindergarten through college, teachers and peers reward gender-appropriate attitudes and behavior. Sports reinforce traditional gender roles through a rigid division of events into male and female categories. The media are also a powerful source of gender socialization; starting very early in childhood, children's books, television programs, movies, and music provide subtle and not-so-subtle messages about how boys and girls should act (see Chapter 10, "Sex, Gender, and Sexuality").

## Racial–Ethnic Socialization

In addition to gender-role socialization, we receive racial socialization throughout our lives. ***Racial socialization is the aspect of socialization that contains specific messages and practices concerning the nature of one's racial or ethnic status as it relates to our identity, interpersonal relationships, and location in the social hierarchy.*** Racial socialization includes direct statements regarding race, modeling behavior (wherein a child imitates the behavior of a parent or other caregiver), and indirect activities such as exposure to an environment that conveys a specific message about a racial or ethnic group ("We are better than they are," for example).

The most important aspects of racial identity and attitudes toward other racial–ethnic groups are passed down in families from generation to generation (Figure 3.16). As the sociologist Martin Marger (1994: 97) notes, "Fear of, dislike for, and antipathy toward one group or another is learned in much the same way that people learn to eat with a knife or fork rather than with their bare hands or to respect others' privacy in personal matters." These beliefs can be transmitted in subtle and largely unconscious ways; they do not have to be taught directly or intentionally. Scholars have found that ethnic values and attitudes begin to crystallize among children as young as age four (Van Ausdale and Feagin, 2001). By this age, the society's ethnic hierarchy has become apparent to the child. Some minority parents feel that racial socialization is essential because it provides

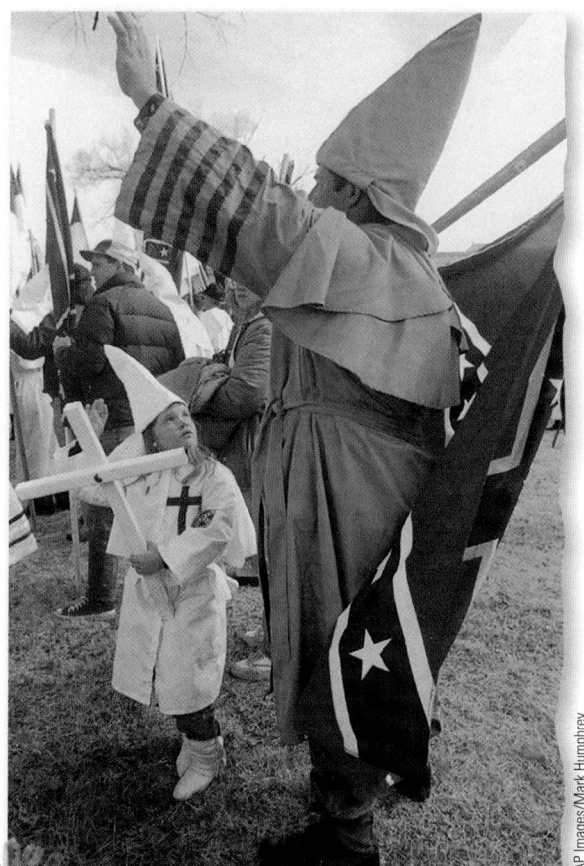

AP Images/Mark Humphrey

**FIGURE 3.16** Do you believe that what this child is learning here will have an influence on her actions in the future? What other childhood experiences might offset early negative racial socialization?

children with the skills and abilities they will need to survive in the larger society.

## Socialization Through the Life Course

**LO7** Why is socialization a lifelong process? Throughout our lives, we continue to learn. Each time we experience a change in status (such as becoming a college student or getting married), we learn a new set of rules, roles, and relationships. Even before we achieve a new status, we often

**gender socialization** the aspect of socialization that contains specific messages and practices concerning the nature of being female or male in a specific group or society.

**racial socialization** the aspect of socialization that contains specific messages and practices concerning the nature of one's racial or ethnic status as it relates to identity, interpersonal relationships, and location in the social hierarchy.

93

**FIGURE 3.17** An important rite of passage for many Latinas is the *quinceañera*—a celebration of their fifteenth birthday and their passage into womanhood. Can you see how this occasion might also be a form of anticipatory socialization?

participate in *anticipatory socialization*—**the process by which knowledge and skills are learned for future roles.** Many societies organize social activities according to age and gather data regarding the age composition of the people who live in that society. Some societies have distinct *rites of passage,* based on age or other factors, that publicly dramatize and validate changes in a person's status. In the United States and other industrialized societies, the most common categories of age are childhood, adolescence, and adulthood (often subdivided into young adulthood, middle adulthood, and older adulthood).

### Childhood

Some social scientists believe that a child's sense of self is formed at an early age and that it is difficult to change this self-perception later in life. Symbolic interactionists emphasize that during infancy and early childhood, family support and guidance are crucial to a child's developing self-concept. In some families, children are provided with emotional warmth, feelings of mutual trust, and a sense of security. These families come closer to our ideal cultural belief that childhood should be a time of carefree play, safety, and freedom from economic, political, and sexual responsibilities. However, other families reflect the discrepancy between cultural ideals and reality— children grow up in a setting characterized by fear, danger, and risks that are created by parental neglect, emotional maltreatment, or premature economic and sexual demands. Abused and neglected children often experience physical consequences, such as damage to their growing brains, which can lead to cognitive delays or emotional difficulties. Psychological problems can also occur that involve high-risk behavior such as smoking, alcohol or drug abuse, or similar activities. Other psychological problems manifest as low self-esteem, an inability to trust others, feelings of isolation and powerlessness, and denial of one's feelings (Child Welfare Information Gateway, 2008).

### Adolescence

In industrialized societies the adolescent (or teenage) years represent a buffer between childhood and adulthood. It is a time during which young people pursue their own routes to self-identity and adulthood. Anticipatory socialization is often associated with adolescence, with many young people spending time planning or being educated for future roles they hope to occupy. Although no specific rites of passage exist in the United States to mark *every* child's transition between childhood and adolescence or between adolescence and adulthood, some rites of passage are observed. For example, a celebration known as a Bar Mitzvah is held for some Jewish boys on their thirteenth birthday, and a Bat Mitzvah is held for some Jewish girls on their twelfth birthday; these events mark the occasion upon which young people accept moral responsibility for their own actions and the fact that they are now old enough to own personal property. Similarly, some Latinas are honored with the *quinceañera*—a celebration of their fifteenth birthday that marks their passage into young womanhood (Figure 3.17). Although it is not officially designated as a rite of passage, many of us think of the time when we get our first driver's license or graduate from high school as another way in which we mark the transition from one period of our life to the next.

Adolescence is often characterized by emotional and social unrest. In the process of developing their own identities, some young people come into conflict with parents, teachers, and other authority figures who attempt to restrict their freedom. Adolescents may also find themselves caught between the demands of adulthood and their own lack of financial independence and experience in the job market.

**EARLY FALL** ▶

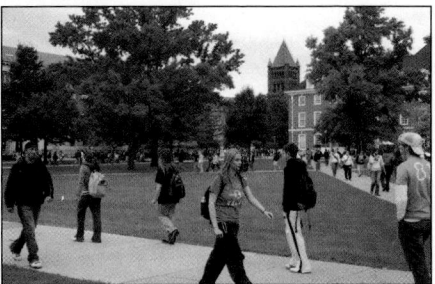

- Adapting to new people and new situations

- Anticipation and excitement about studying in a new setting

- Insecurity about academic demands

- Homesickness

- If employed, trying to balance school and work life

**MID FALL** ▶

- Social pressures from others: What would my parents think?

- Anticipation (and dread) of midterm exams and major papers

- Time-management problems between school and social life

- Intense need for a break

- Concerns about role conflict between school and work

**LATE FALL** ▶

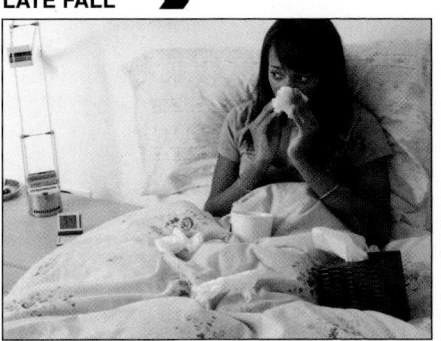

- Positive or negative assessment of grades so far

- Pre-final studying and jitters

- Making up for lost time and procrastination

- First college illnesses likely to occur because of late hours, poor eating habits, and proximity to others who become ill

- Potential problems with roommates or others who make excessive demands on one's time and/or personal space

**END OF TERM** ◆

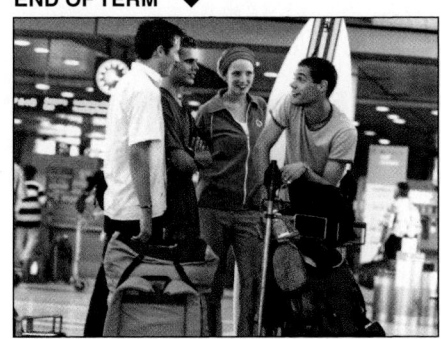

- Final exams: late nights, extra effort, and stress

- Concerns about leaving new friends and college setting for winter break

- Anticipation (and tension) associated with going home for break for those who have been away

- Reassessment of college choice, major, and career options: Am I on the right track?

- Acknowledgment that growth has occurred and much has been learned, both academically and otherwise, during the first college term

**FIGURE 3.18** TIME LINE FOR FIRST-SEMESTER COLLEGE SOCIALIZATION

Sources: Based on the author's general observations of college student life and on Kansas State University, 2014. © Cengage Learning

The experiences of individuals during adolescence vary according to race, class, and gender. Based on their family's economic situation and personal choices, some young people leave high school and move directly into the world of work, whereas others pursue a college education and may continue to receive advice and financial support from their parents. Others are involved in both the world of work and the world of higher education as they seek to support themselves and to acquire more years of formal education or vocational/career training. Whether or not a student works while in college may affect the process of adjusting to college life (see Figure 3.18). In the second decade of the twenty-first century,

**anticipatory socialization** the process by which knowledge and skills are learned for future roles.

# SOCIOLOGY *in Global Perspective*

## Open Doors: Study-Abroad and Global Socialization

The first month or so of the study-abroad experience feels like a vacation in that everything is exciting and new. After this "honeymoon" period, the experience becomes something other than merely a vacation or fleeting visit. You start to relate to the people, the culture, and life in that country not from the eyes of a tourist passing through, but progressively from the eyes of those around you—the citizens who were born and raised there. That is the perspective which is unattainable without actually living in another country, and a perspective which I have come to appreciate and understand more fully as I settle back into life here back at home.

> —John R. R. Howie (2010), then a Boston College economics and Mandarin Chinese major, explaining what studying abroad at Peking University, in Beijing, meant to him. Howie has since graduated and is now employed as a financial analyst.

Studying abroad is an important part of the college socialization process for preparing to live and work in an interconnected world. Here are a few interesting facts about studying abroad (adapted from Institute of International Education, 2011):

- About 274,000 U.S. students participated in study-abroad programs for credit in 2010–2011, and this number continues to increase each year.
- The United Kingdom, Italy, Spain, France, and China are the top destinations for study-abroad; however, Latin America, Asia, Oceania—which includes Australia, New Zealand, and the South Pacific islands—and Africa are also popular destinations.

more college students are exploring international study programs as part of their adult socialization in order to help them gain new insights into divergent cultures and the larger world of which they are a part (see "Sociology in Global Perspective").

## Adulthood

One of the major differences between child socialization and adult socialization is the degree of freedom of choice. If young adults are able to support themselves financially, they gain the ability to make more choices about their own lives. In early adulthood (usually until about age forty), people work toward their own goals of creating relationships with others, finding employment, and seeking personal fulfillment. Of course, young adults continue to be socialized by their parents, teachers, peers, and the media, but they also learn new attitudes and behaviors. For example, when we marry or have children, we learn new roles as partners or parents.

*Workplace (occupational) socialization* is one of the most important types of early adult socialization. This type of socialization tends to be most intense immediately after a person makes the transition from school to the workplace; however, many people experience continuous workplace socialization as a result of having more than one career in their lifetime.

In middle adulthood—between the ages of forty and sixty-five—people begin to compare their accomplishments with their earlier expectations. This is the point at which people either decide that they have reached their goals or recognize that they have attained as much as they are likely to achieve.

Late adulthood may be divided into three categories: (1) the "young-old" (ages sixty-five to seventy-four), (2) the "old-old" (ages seventy-five to eighty-five), and (3) the "oldest-old" (over age eighty-five). Although these are somewhat arbitrary divisions, the "young-old" are less likely to suffer from disabling illnesses, whereas some of the "old-old" are more likely to suffer such illnesses. Increasingly, studies in gerontology and the sociology of medicine have come to question these arbitrary categories and show that some persons defy the expectations of their age grouping based on individual genetic makeup, lifestyle choices, and a zest for living. Perhaps "old age" is what we make it!

## Late Adulthood and Ageism

In older adulthood, some people are quite happy and content (Figure 3.19); others are not. Erik Erikson noted that difficult changes in adult attitudes and behavior occur in the last years of life, when people experience decreased physical ability, lower prestige, and the prospect of death. Older

- The top fields of study for U.S. study-abroad students are social sciences, business/management, humanities, fine or applied arts, physical/life sciences, and foreign languages.
- More than half of study-abroad students remain in their host country for a short-term stay (a summer or eight weeks or less during the academic year).

Sociologists are interested in studying the profile of U.S. study-abroad students because the data provide interesting insights on differences in students' participation by classification, gender, race, and class. Based on the latest figures available (2010/2011), most students participating in study-abroad programs are classified as juniors (35.8 percent) or seniors (23.4 percent). Women make up 64.4 percent of all study-abroad students, and men make up 35.6 percent. White students make up the vast majority of study-abroad students (77.8 percent). Other groups include Asian or Pacific Islander (7.9 percent), Hispanic or Latino(a) (6.9 percent), and black or African American (4.8 percent) (Institute of International Education, 2011).

Socialization for life in the global community is necessary for all students because of the increasing significance of international understanding and the need to learn how to live and work in a diversified nation and world. Even more important may be the opportunity for each student to gain direction and meaning in his or her own life. Do you think that studying abroad might make an important contribution to your own socialization while in college? Why or why not?

### reflect & analyze

**What are the positive aspects of study-abroad programs in the college socialization process? What are the limitations of such programs? If you are unable to participate in a study-abroad program, what other methods and resources might you use to gain "global socialization," which could be beneficial in helping you meet your goals for the future?**

Sandy Huffaker/Corbis

**FIGURE 3.19** Throughout life, our self-concept is influenced by our interactions with others. How might the self-concept of these women be influenced by one another? By society in general?

adults in industrialized societies may experience *social devaluation*—**wherein a person or group is considered to have less social value than other persons or groups.** Social devaluation is especially acute when people are leaving roles that have defined their sense of social identity and provided them with meaningful activity.

Negative images regarding older persons reinforce *ageism*—**prejudice and discrimination against people on the basis of age, particularly against older persons.** Ageism is reinforced by stereotypes, whereby people have narrow, fixed images of certain groups. Older persons are often stereotyped as thinking and moving slowly; as being bound to themselves and their past, unable to change and grow; as being unable to move forward and often moving backward.

Negative images also contribute to the view that women are "old" ten or fifteen years sooner than men (Lauzen and Dozier, 2005). In popular films, male characters increase in leadership roles and powerful positions as they grow older; women are either moved into the background, where they have few, if any, goals, or they disappear from sight altogether (Lauzen and Dozier, 2005). Similarly, the multibillion-dollar cosmetics industry helps perpetuate the myth that age reduces the "sexual value" of women but increases it for men. Men's sexual value is defined more in terms of personality, intelligence, and earning power than by physical appearance. For women, however, sexual attractiveness is based on youthful appearance. By idealizing this "youthful" image of women and playing up the fear of growing older, sponsors sell thousands of products and services that claim to prevent or fix the "ravages" of aging.

Although not all people act on appearances alone, Patricia Moore, an industrial designer, found that many do. At age twenty-seven, Moore disguised herself as an eighty-five-year-old woman by donning age-appropriate clothing and placing baby oil in her eyes to create the appearance of cataracts. With the help of a makeup artist, Moore supplemented the "aging process" with latex wrinkles, stained teeth, and a gray wig. For three years, "Old Pat Moore" went to various locations, including a grocery store, to see how people responded to her:

> When I did my grocery shopping while in character, I learned quickly that the Old Pat Moore behaved—and was treated—differently from the Young Pat Moore. When I was 85, people were more likely to jockey ahead of me in the checkout line. And even more interesting, I found that when it happened, I didn't say anything to the offender, as I certainly would at age 27. It seemed somehow, even to me, that it was okay for them to do this to the Old Pat Moore, since they were undoubtedly busier than I was anyway. And further,

they apparently thought it was okay, too! After all, little old ladies have plenty of time, don't they? And then when I did get to the checkout counter, the clerk might start yelling, assuming I was deaf, or becoming immediately testy, assuming I would take a long time to get my money out, or would ask to have the price repeated, or somehow become confused about the transaction. What it all added up to was that people feared I would be trouble, so they tried to have as little to do with me as possible. And the amazing thing is that I began almost to believe it myself. . . . I think perhaps the worst thing about aging may be the overwhelming sense that everything around you is letting you know that you are not terribly important any more. (Moore with Conn, 1985: 75–76)

If we apply our sociological imagination to Moore's study, we find that "Old Pat Moore's" experiences reflect what many older persons already know—it is other people's *reactions* to their age, not their age itself, that place them at a disadvantage.

Many older people buffer themselves against ageism by continuing to view themselves as being in middle adulthood long after their actual chronological age would suggest otherwise. Other people begin a process of resocialization to redefine their own identity as mature adults.

# Resocialization

**LO8** *Resocialization* **is the process of learning a new and different set of attitudes, values, and behaviors from those in one's background and previous experience.** Resocialization may be voluntary or involuntary. In either case, people undergo changes that are much more rapid and pervasive than the gradual adaptations that socialization usually involves.

## Voluntary Resocialization

Resocialization is voluntary when we assume a new status (such as becoming a student, an employee, or a retiree) of our own free will. Sometimes, voluntary resocialization involves medical or psychological treatment or religious conversion, in which case the person's existing attitudes, beliefs, and behaviors must undergo strenuous modification to a new regime and a new way of life. For example, resocialization for adult survivors of emotional/physical child abuse includes extensive therapy in order to form new patterns of thinking and action, somewhat like Alcoholics Anonymous and its twelve-step program, which has become the basis for many other programs dealing with addictive behavior.

**FIGURE 3.20** New inmates are taught how to order their meals. Two fingers raised means two portions. There is no talking in line. Inmates must eat all their food. This "ceremony" suggests how much freedom and dignity an inmate loses when beginning the resocialization process.

*Journal Courier/The Image Works*

## Involuntary Resocialization

Involuntary resocialization occurs against a person's wishes and generally takes place within a *total institution*—**a place where people are isolated from the rest of society for a set period of time and come under the control of the officials who run the institution** (Goffman, 1961a) (Figure 3.20). Military boot camps, jails and prisons, concentration camps, and some mental hospitals are considered total institutions. Involuntary resocialization is a two-step process. First, people are stripped of their former selves—or depersonalized—through a degradation ceremony (Goffman, 1961a). For example, inmates entering prison are required to strip, shower, and wear assigned institutional clothing. In the process, they are searched, weighed, fingerprinted, photographed, and given no privacy even in showers and restrooms. Their official identification becomes not a name but a number. In this abrupt break from their former existence, they must leave behind their personal possessions and their family and friends. The depersonalization process continues as they are required to obey rigid rules and to conform to their new environment.

The second step in the resocialization process occurs when the staff at an institution attempt to build a more compliant person. A system of rewards and punishments (such as providing or withholding television or exercise privileges) encourages conformity to institutional norms.

Individuals respond to involuntary resocialization in different ways. Some people are rehabilitated; others become angry and hostile toward the system that has taken away their freedom. Although the assumed purpose of involuntary resocialization is to reform people so that they will conform to societal standards of conduct after their release, the ability of total institutions to modify offenders' behavior in a meaningful manner has been widely questioned. In many prisons, for example, inmates may conform to the norms of the prison or of other inmates but have little respect for the norms and the laws of the larger society.

**social devaluation** a situation in which a person or group is considered to have less social value than other individuals or groups.

**ageism** prejudice and discrimination against people on the basis of age, particularly against older persons.

**resocialization** the process of learning a new and different set of attitudes, values, and behaviors from those in one's background and previous experience.

**total institution** Erving Goffman's term for a place where people are isolated from the rest of society for a set period of time and come under the control of the officials who run the institution.

## Socialization in the Future

What will socialization be like in the future? The family is likely to remain the institution that most fundamentally shapes and nurtures people's personal values and self-identity. However, other institutions, including education, religion, and the media, will continue to exert a profound influence on individuals of all ages. A central value-oriented issue facing parents and teachers as they attempt to socialize children is the growing dominance of television, the Internet, and social media, which make it possible for children and young people to experience many things outside their homes and schools and to communicate routinely with people around the world.

The socialization process in colleges and universities will become more diverse as students have an even wider array of options in higher education, including attending traditional classes in brick-and-mortar buildings, taking independent-study courses, enrolling in online courses and degree programs, participating in study-abroad programs, and choosing from options that are unknown at this time. However, it remains to be seen whether newer approaches to socialization in higher education will be more stressful or less stressful than current methods. One thing remains clear: The socialization process will continue to be a dynamic and important part of our life as we assume various roles throughout our life span.

# chapter REVIEW

### LO1 What is the extent to which people would become human beings without adequate socialization?

Socialization is the lifelong process through which individuals acquire their self-identity and learn the physical, mental, and social skills needed for survival in society. The kind of person we become depends greatly on what we learn during our formative years from our surrounding social groups and social environment. Social contact is essential in developing a self, or self-concept, which represents an individual's perceptions and feelings of being a distinct or separate person. Much of what we think about ourselves is gained from our interactions with others and from what we perceive that others think of us.

### LO2 What is the sociological perspective on human development?

According to Charles Horton Cooley's concept of the looking-glass self, we develop a self-concept as we see ourselves through the perceptions of others. Our initial sense of self is typically based on how our families perceive and treat us. George Herbert Mead suggested that we develop a self-concept through role-taking and learning the rules of social interaction. According to Mead, the self is divided into the "I" and the "me." The "I" represents the spontaneous and unique traits of each person. The "me" represents the internalized attitudes and demands of other members of society.

### LO3 How do functionalist and conflict theorists' perspectives differ on the roles that families play in the socialization process?

Theorists using a functionalist perspective emphasize that families serve important functions in society because they are the primary locus for the socialization of children. The family influences an emerging sense of self and the acquisition of beliefs and values. Families are also the primary source of emotional support. Ideally, people receive love, understanding, security, acceptance, intimacy, and companionship within families. On the other hand, conflict theorists stress that socialization contributes to false consciousness—a lack of awareness and a distorted perception of the reality of class as it affects all aspects of social life. As a result, socialization reaffirms and reproduces the class structure of the next generation rather than challenging the conditions that presently exist.

### LO4 How do schools socialize children in both formal and informal ways?

Schools continue to play an enormous role in the socialization of young people. Schools primarily teach knowledge and skills but also have a profound influence on the self-image, beliefs, and values of children.

### LO5 What role do peer groups and media play in socialization now, and what part might these agents play in the future?

Peer groups contribute to our sense of belonging and self-worth, and are a key source of information about acceptable behavior. Peer groups simultaneously reflect the larger culture and serve as a conduit for passing on culture to young people. The media function as socializing agents by (1) informing us about world events; (2) introducing us to a wide variety of people; (3) providing an array of viewpoints on current issues; (4) making us aware of products and services that, if we purchase them, will supposedly help us to be accepted by others; and (5) providing an opportunity to live vicariously through other people's experiences.

# LO6 What are ways in which gender socialization and racial–ethnic socialization occur?

Through the process of gender socialization, we learn about what attitudes and behaviors are considered to be appropriate for girls and boys, men and women, in a particular society. One of the primary agents of gender socialization is the family. Racial socialization includes direct statements regarding race, modeling behavior, and indirect activities such as exposure to an environment that conveys a specific message about a racial or ethnic group.

# LO7 What are the stages in the life course, and why is the process of socialization important in each stage?

Socialization is ongoing throughout the life course—from childhood to adolescence to adulthood and old age. We learn knowledge and skills for future roles through anticipatory socialization. Throughout childhood, we are socialized by our parents, schools, peers, and other groups. Workplace (occupational) socialization is one of the most important types of early adult socialization.

# LO8 What is the difference between voluntary and involuntary resocialization?

Resocialization is the process of learning new attitudes, values, and behaviors, either voluntarily or involuntarily. Resocialization is voluntary when we assume a new status (such as becoming a student) of our own free will. Involuntary resocialization occurs against a person's wishes and generally takes place within a total institution, such as a jail or prison.

# key TERMS

ageism 98

agents of socialization 87

anticipatory socialization 94

ego 80

gender socialization 92

generalized other 85

id 80

looking-glass self 83

mass media 90

peer group 90

primary socialization 86

racial socialization 93

reciprocal socialization 89

resocialization 98

role-taking 83

secondary socialization 86

self-concept 82

significant others 83

social devaluation 98

socialization 75

sociobiology 76

superego 80

tertiary socialization 86

total institution 99

# questions for CRITICAL THINKING

1. Consider the concept of the looking-glass self. How do you think others perceive you? Do you think most people perceive you correctly?
2. What are your "I" traits? What are your "me" traits? Which ones are stronger?
3. What are some different ways that you might study the effect of toys on the socialization of children? How could you isolate the toy variable from other variables that influence children's socialization?
4. Is the attempted rehabilitation of criminal offenders—through boot camp programs, for example—a form of socialization or resocialization?

# 4 SOCIAL STRUCTURE AND INTERACTION IN EVERYDAY LIFE

## LEARNING OBJECTIVES

**LO1** Explain why social structure is important in our interaction with others.

**LO2** Distinguish among ascribed, achieved, and master statuses, and give examples of each.

**LO3** Explain each of these terms: role, role expectation, role performance, role conflict, role strain, and role exit.

**LO4** Compare functionalist and conflict views on social institutions.

**LO5** Explain how social change occurs in preindustrial, industrial, and postindustrial societies.

**LO6** Discuss symbolic interactionist views on the social construction of reality and the self-fulfilling prophecy.

**LO7** Compare ethnomethodology and dramaturgical analysis as two research methods for observing how people deal with everyday life.

**LO8** Explain how the sociology of emotions and the study of nonverbal communication add to our understanding of human behavior.

# The Art of Diving for Dinner

I began Dumpster diving [scavenging in a large garbage bin] about a year before I became homeless . . . . The area I frequent is

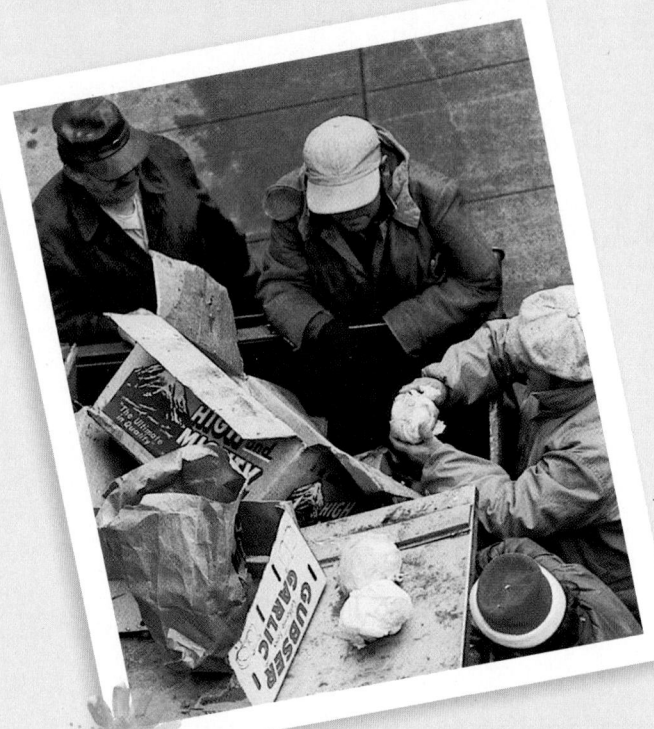

All activities in life—including scavenging in garbage bins and living "on the streets"—are social in nature.

inhabited by many affluent college students. I am not here by chance; the Dumpsters in this area are very rich. Students throw out many good things, including food. In particular they tend to throw everything out when they move at the end of a semester, before and after breaks, and around midterm, when many of them despair of college. So I find it advantageous to keep an eye on the academic calendar. I learned to scavenge gradually, on my own. Since then I have initiated several companions into the trade. I have learned that there is a predictable series of stages a person goes through in learning to scavenge.

At first the new scavenger is filled with disgust and self-loathing. He is ashamed of being seen and may lurk around, trying to duck behind things, or he may dive at night. (In fact, most people instinctively look away from a scavenger. By skulking around, the novice calls attention to himself and arouses suspicion. Diving at night is ineffective and needlessly messy.) . . . That

Bob Collins/The Image Works

ighner's "diving" activities reflect a specific pattern of social behavior. All activities in life—including scavenging in garbage bins and living "on the streets"—are social in nature. Homeless persons and domiciled persons (those with homes) participate in a social world that has predictable patterns. This is true for the rest of us as well. In this chapter we look at the relationship between social structure and social interaction in everyday life. In the process, homelessness is used as an example of how social problems may occur and how they may be reduced or perpetuated within social structures and patterns of interaction in communities and nations.

Let's start by defining social interaction and social structure. Although we frequently are not aware of it, our daily interactions with others and the larger patterns found in the social world of which we are a part are important ingredients

in the framework of our individual daily lives. *Social interaction* **is the process by which people act toward or respond to other people and is the foundation for all relationships and groups in society.** As discussed in Chapter 3, we learn virtually all of what we know from our interactions with other people.

Socialization is a small-scale process, whereas social structure is a much more encompassing framework. *Social structure* **is the complex framework of societal institutions (such as the economy, politics, and religion) and the social practices (such as rules and social roles) that make up a society and that organize and establish limits on people's behavior.** This structure is essential for the survival of society and for the well-being of individuals because it provides a social web of familial support and social relationships that connects each of us to the larger society. Many homeless people have lost stage passes with experience. The scavenger finds a pair of running shoes that fit and look and smell brand-new. . . . He begins to understand: People throw away perfectly good stuff, a lot of perfectly good stuff.

At this stage, Dumpster shyness begins to dissipate. The diver, after all, has the last laugh. He is finding all manner of good things that are his for the taking. Those who disparage his profession are the fools, not he.

—Author Lars Eighner recalls his experiences as a Dumpster diver while living under a shower curtain in a stand of bamboo in a public park. Eighner became homeless when he was evicted from his "shack" after being unemployed for about a year (Eighner, 1993: 111–119).

## How Much Do You Know About Homeless People and the Social Structure of Homelessness?

| True | False | |
|------|-------|---|
| T | F | 1. Local, state, and federal assistance to homeless people has shrunk in recent years. |
| T | F | 2. A majority of people who are counted as homeless live on the streets or in cars, abandoned buildings, or other places not intended for human habitation. |
| T | F | 3. Many homeless people have full-time employment. |
| T | F | 4. Homelessness is affected by both income and the affordability of available housing. |

| True | False | |
|------|-------|---|
| T | F | 5. Homeless people typically panhandle (beg for money) so that they can buy alcohol or drugs. |
| T | F | 6. Shelters for the homeless consistently have clients who sleep on overflow cots, in chairs, in hallways, and in other nonstandard sleeping arrangements. |
| T | F | 7. There have always been homeless people throughout the history of the United States. |
| T | F | 8. "Doubled-up" populations (people who live with friends, family, or other nonrelatives for economic reasons) have decreased in recent years. |

Answers on page 107.

this vital linkage. As a result, they often experience a loss of personal dignity and a sense of moral worth because of their "homeless" condition. Before reading on, learn more about homeless people and how the pressing national problem of homelessness is related to social structure and interaction by taking the "Sociology and Everyday Life" quiz.

## Social Structure: The Macrolevel Perspective

**LO1** Social structure provides the framework within which we interact with others. This framework is an orderly, fixed arrangement of parts that together make up the whole group or society (see Figure 4.1). At the macrolevel, the social structure of a society has several essential elements: social institutions, groups, statuses, roles, and norms.

Functional theorists emphasize that social structure is essential because it creates order and predictability in a society. Social structure is also important for our human development. As discussed in Chapter 3, we develop a self-concept as we learn the attitudes, values, and behaviors of the people around us. When these attitudes and values are

**social interaction** the process by which people act toward or respond to other people; the foundation for all relationships and groups in society.

**social structure** the complex framework of societal institutions (such as the economy, politics, and religion) and the social practices (such as rules and social roles) that make up a society and that organize and establish limits on people's behavior.

**FIGURE 4.1** SOCIAL STRUCTURE FRAMEWORK

© Cengage Learning

part of a predictable structure, it is easier to develop that self-concept.

Social structure gives us the ability to interpret the social situations we encounter. For example, we expect our families to care for us, our schools to educate us, and our police to protect us. When our circumstances change dramatically, most of us feel an acute sense of anxiety because we do not know what to expect or what is expected of us. For example, newly homeless individuals may feel disoriented because they do not know how to function in their new setting. The person is likely to ask questions: "How will I survive on the streets?" "Where do I go to get help?" "Should I stay at a shelter?" "Where can I get a job?" Social structure helps people make sense out of their environment even when they find themselves on the streets.

However, conflict theorists maintain that there is more to social structure than is readily visible and that we must explore the deeper, underlying structures that determine social relations in a society. For example, Karl Marx suggested that the way economic production is organized is the most important structural aspect of any society. In capitalistic societies, where a few people control the labor of many, the social structure reflects a system of relationships of domination among categories of people (for example, owner–worker and employer–employee).

Social structure creates boundaries that define which persons or groups will be the "insiders" and which will be the "outsiders." *Social marginality* is the state of being part insider and part outsider in the social structure. Sociologist Robert Park (1928) coined this term to refer to persons (such as immigrants) who simultaneously share the life and traditions of two distinct groups. Social marginality results in stigmatization. A *stigma* is any physical or social attribute or sign that so devalues a person's

social identity that it disqualifies that person from full social acceptance (Goffman, 1963b). A convicted criminal wearing a prison uniform is an example of a person who has been stigmatized; the uniform says that the person has done something wrong and should not be allowed unsupervised outside the prison walls.

# Components of Social Structure

**LO2** The social structure of a society includes its social positions, the relationships among those positions, and the kinds of resources attached to each of the positions. Social structure also includes all the groups that make up society and the relationships among those groups (Smelser, 1988). We begin by examining the social positions that are closest to the individual.

## Status

A *status* is a socially defined position in a group or society characterized by certain expectations, rights, and duties. Statuses exist independently of the specific people occupying them; the statuses of professional athlete, rock musician, professor, college student, and homeless person all exist exclusive of the individuals who occupy these social positions. For example, although thousands of new students arrive on college campuses each year to occupy the status of first-year student, the status of college student and the expectations attached to that position have remained relatively unchanged for the past century.

Does the term *status* refer only to high-level positions in society? Although many people

# ANSWERS to the Sociology Quiz on Homeless Persons and the Social Structure of Homelessness

**1. True.** Debt and deficit reduction at the federal level, combined with fiscal crises at the local and state levels, has reduced funds for assistance available to homeless people and the organizations that assist them.

**2. False.** A majority of people who are counted as homeless are found in emergency shelters or transitional housing programs, but these organizations offer only a temporary break from the larger problem of homelessness.

**3. True.** Many homeless people do have full-time employment, but they are among the working poor. The minimum-wage jobs they hold do not pay enough for them to support their families and pay the high rents that are typical in many cities.

**4. True.** Although many people think of homelessness as being based solely on lack of income to pay for housing, another significant factor is the cost of housing. In some cities and regions, available housing simply is not affordable to people who have limited financial resources.

**5. False.** Many homeless people panhandle to pay for food, a bed at a shelter, or other survival needs.

**6. True.** Overcrowded shelters throughout the nation often attempt to accommodate as many homeless people as possible on a given night, particularly when the weather is bad. As a result, any available spaces—including offices, closets, and hallways—are used as sleeping areas until the individuals can find another location or weather conditions improve.

**7. True.** Scholars have found that homelessness has always existed in the United States. However, the number of homeless people has increased or decreased with fluctuations in the national economy.

**8. False.** The "doubled-up" population in the United States increased by more than 50 percent from 2005 to 2010 as many people faced difficult economic times during the Great Recession. Many people lost their homes to foreclosure, others were unable to pay rent because of losing their jobs, and still others found themselves in a downward economic spiral.

Source: United States Conference of Mayors, 2013.

---

equate the term with high levels of prestige, sociologists use it to refer to all socially defined positions—high rank and low rank. For example, both the position of director of the Department of Health and Human Services in Washington, D.C., and that of a homeless person who is paid about five dollars a week (plus bed and board) to clean up the dining room at a homeless shelter are social statuses.

Take a moment to answer the question "Who am I?" To determine who you are, you must think about your social identity, which is derived from the statuses you occupy and is based on your status set. A *status set* **comprises all the statuses that a person occupies at a given time.** For example, Maria may be a psychologist, a professor, a wife, a mother, a Catholic, a school volunteer, a Texas resident, and a Mexican American. All of these socially defined positions constitute her status set.

**Ascribed Status and Achieved Status** Statuses are distinguished by the manner in which we acquire them. An *ascribed status* **is a social position conferred at birth or received involuntarily later in life, based on attributes over which the individual has little or no control, such as race/ethnicity, age, and gender.** For example, Maria is a female born to Mexican American parents; she was assigned these statuses at birth. She is an adult and—if she lives long enough—will someday become an "older adult," which is an ascribed status received involuntarily later in life.

**status** a socially defined position in a group or society characterized by certain expectations, rights, and duties.

**status set** all the statuses that a person occupies at a given time.

**ascribed status** a social position conferred at birth or received involuntarily later in life, based on attributes over which the individual has little or no control, such as race/ethnicity, age, and gender.

An *achieved status* is a social position that a person assumes voluntarily as a result of personal choice, merit, or direct effort. Achieved statuses (such as occupation, education, and income) are thought to be gained as a result of personal ability or successful competition (Figure 4.2). Most occupational positions in modern societies are achieved statuses. For instance, Maria voluntarily assumed the statuses of psychologist, professor, wife, mother, and school volunteer. However, not all achieved statuses are positions most people would want to attain; for example, being a criminal, a drug addict, or a homeless person is a negative achieved status.

Ascribed statuses have a significant influence on the achieved statuses that we occupy. Race/ethnicity, gender, and age affect each person's opportunity to acquire certain achieved statuses. Those who are privileged by their positive ascribed statuses are more likely to achieve the more prestigious positions in a society. Those who are disadvantaged by their ascribed statuses may more easily acquire negative achieved statuses.

**Master Status** If we occupy many different statuses, how can we determine which is the most important? Sociologist Everett Hughes has stated that societies resolve this ambiguity by determining master statuses. A *master status* is the most important status that a person occupies; it dominates all the individual's other statuses and is the overriding ingredient in determining a person's general social position (Hughes, 1945). Being poor or rich is a master status that influences many other areas of life, including health, education, and life opportunities (Figure 4.3). For men, occupation has usually been the most important status, although occupation is increasingly a master status for many women as well. "What do you do?" is one of the first questions most people ask when meeting another. Occupation provides important clues to a person's educational level, income, and family background. An individual's race/ethnicity may

also constitute a master status in a society in which dominant-group members single out members of other groups as "inferior" on the basis of real or alleged physical, cultural, or nationality characteristics.

Master statuses confer high or low levels of personal worth and dignity on people. These are not characteristics that we inherently possess; they are derived from the statuses we occupy. For those who have no residence, being a homeless person readily becomes a master status regardless of the person's other attributes. Homelessness is a stigmatized master status that confers disrepute on its occupant because domiciled people often believe that a homeless person has a "character flaw." Some-

**FIGURE 4.2** In the past, a person's status was primarily linked to his or her family background, education, occupation, and other sociological attributes. Today, some sociologists believe that celebrity status has overtaken the more traditional social indicators of status. Miley Cyrus, shown here performing at a concert, is an example of celebrity status.

times this assumption is supported by how the media frame stories about homeless people (see "Homelessness in the Media"). The circumstances under which someone becomes homeless determine the extent to which that person is stigmatized (see Figure 4.4). In a now-classic study, the sociologists David Snow and Leon Anderson (1993: 199) described the effects of homelessness as a master status:

It was late afternoon, and the homeless were congregated in front of [the Salvation Army shelter] for dinner. A school bus approached that was packed with Anglo junior high school students being bused from an eastside barrio school to their upper-middle and upper-class homes in the city's northwest neighborhoods. As the bus rolled by, a fusillade of coins came flying out the windows, as the students made obscene gestures and shouted, "Get a job." Some of the homeless gestured back, some scrambled for the scattered coins—mostly pennies—others

**FIGURE 4.3** Sociologists believe that being rich or poor may be a master status in the United States. How do the lifestyles of these two men differ based on their master statuses?

angrily threw the coins at the bus, and a few seemed oblivious to the encounter. For the passing junior high schoolers, the exchange was harmless fun, a way to work off the restless energy built up in school; but for the homeless it was a stark reminder of their stigmatized status and of the extent to which they are the objects of negative attention.

**Status Symbols** When people are proud of a particular social status that they occupy, they often choose to use visible means to let others know about

their position. ***Status symbols are material signs that inform others of a person's specific status.*** For example, just as wearing a wedding ring proclaims that a person is married, owning a Rolls-Royce announces that one has "made it." As we saw in Chapter 3, achievement and success are core U.S. values. For this reason, people who have "made it" tend to want to display symbols to inform others of their accomplishments.

Status symbols for the domiciled and for the homeless may have different meanings. Among affluent persons, a full shopping cart in the grocery store and bags of merchandise from expensive department stores indicate a lofty financial position. By contrast, among the homeless, bulging shopping bags and overloaded grocery carts suggest a completely different status. Carts and bags are essential to street life; there is no other place to keep things, as shown by this description of Darian, a homeless woman in New York City:

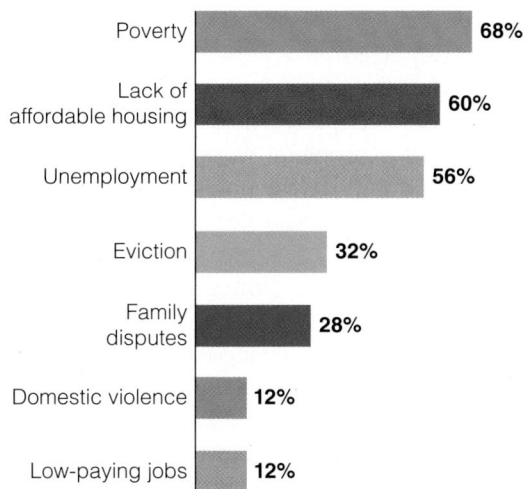

Note: Survey cities were asked to identify the three main causes of homelessness among families with children.

**FIGURE 4.4** CAUSES OF FAMILY HOMELESSNESS IN TWENTY-FIVE CITIES

Source: U.S. Conference of Mayors, 2013.

**achieved status** a social position that a person assumes voluntarily as a result of personal choice, merit, or direct effort.

**master status** the most important status that a person occupies.

**status symbol** a material sign that informs others of a person's specific status.

The possessions in her postal cart consist of a whole house full of things, from pots and pans to books, shoes, magazines, toilet articles, personal papers and clothing, most of which she made herself. . . .

Because of its weight and size, Darian cannot get the cart up over the curb. She keeps it in the street near the cars. This means that as she pushes it slowly up and down the street all day long, she is living almost her entire life directly in traffic. She stops off along her route to sit or sleep for a while and to be both stared at as a spectacle and to stare back. Every aspect of her life including sleeping, eating, and going to the bathroom is constantly in public view. . . . [S]he has no space to call her own and she never has a moment's privacy. Her privacy, her home, is her cart with all its possessions. (Rousseau, 1981: 141)

Although this description is more than thirty years old, homeless persons today can still be spotted in large urban centers, such as New York, San Francisco, and Seattle, and smaller cities, such as Waco, Texas, where they live with all of their possessions in a shopping cart or other portable conveyance.

For homeless women and men, possessions are not status symbols as much as they are a link with the past, a hope for the future, and a potential source of immediate cash. As Snow and Anderson (1993: 147) note, selling personal possessions is not uncommon among most social classes; members of the working and middle classes hold garage sales, and those in the upper classes have estate sales. However, when homeless persons sell their personal possessions, they do so to meet their immediate needs, not because they want to "clean house."

## Role

**L03** A *role* is a set of behavioral expectations associated with a given status. For example, a carpenter (employee) hired to remodel a kitchen is not expected to sit down uninvited and join the family (employer) for dinner. A role is the dynamic aspect of a status. Whereas we occupy a status, we play a role.

*Role expectation* is a group's or society's definition of the way that a specific role ought to be played. By contrast, *role performance* is how a person actually plays the role. Role performance does not always match role expectation. Some statuses have role expectations that are highly specific, such as that of surgeon or college professor. Other statuses, such as friend or significant other, have less structured expectations. The role expectations tied to the status of student are more specific than those

of being a friend. Role expectations are typically based on a range of acceptable behavior rather than on strictly defined standards.

Our roles are relational (or complementary); that is, they are defined in the context of roles performed by others. We can play the role of student because someone else fulfills the role of professor. Conversely, to perform the role of professor, the teacher must have one or more students.

*Role ambiguity* occurs when the expectations associated with a role are unclear. For example, it is not always clear when the provider–dependent aspect of the parent–child relationship ends. Should it end at age eighteen or twenty-one? When a person is no longer in school? People will answer these questions differently depending on their experiences and socialization, as well as on the parents' financial capability and psychological willingness to continue contributing to the welfare of their adult children.

**Role Conflict and Role Strain** Most people occupy a number of statuses, each of which has numerous role expectations attached. For example, Charles is a student who attends morning classes at the university, and he is an employee at a fast-food restaurant, where he works from 3:00 to 10:00 P.M. He is also Stephanie's boyfriend, and she would like to see him more often. On December 7, Charles has a final exam at 7:00 P.M., when he is supposed to be working. Meanwhile, Stephanie is pressuring him to take her to a movie. To top it off, his mother calls, asking him to fly home because his father is going to have emergency surgery. How can Charles be in all these places at once? Such experiences of role conflict can be overwhelming.

*Role conflict* **occurs when incompatible role demands are placed on a person by two or more statuses held at the same time.** When role conflict occurs, we may feel pulled in different directions. To deal with this problem, we may prioritize our roles and first complete the one we consider to be most important. Or we may compartmentalize our lives and separate our roles from one another. That is, we may perform the activities linked to one role for part of the day and then engage in the activities associated with another role in some other time period or elsewhere. For example, under routine circumstances, Charles would fulfill his student role for part of the day and his employee role for another part of the day. In his current situation, however, he is unable to compartmentalize his roles.

Role conflict may occur as a result of changing statuses and roles in society. Research has found that women who engage in behavior that is gender-typed as "masculine" tend to have higher

# HOMELESSNESS IN THE MEDIA

## Are They Statistics, or Are They People?

"It is distressingly common."

"It's tough to get your head around that such a rich country can have such impoverished folks, and that they're so disenfranchised that they don't see hope that things can get better for them."

These are comments made by people in a newspaper article describing homeless people living in tent cities west of downtown Erie, Pennsylvania. The tent city community is located behind big-box stores and near the railroad tracks, where it remains largely invisible to the general public (Guerriero, 2012). As the article suggests, governmental agencies and organizations such as the National Alliance to End Homelessness do not keep statistics on tent cities such as this all over the United States. Like other homeless populations, these tent cities are an indication that the homeless are undercounted in this country, and sometimes media sources are the primary way we learn that such homeless persons exist.

Most media framing of newspaper articles and television reports about the problem of homelessness can be classified into one of two major categories: thematic framing and episodic framing. *Thematic framing* refers to news stories that focus primarily on statistics about the homeless population and recent trends in homelessness (Kendall, 2011). Examples include stories about changes in homeless rates or poverty rates in the United States. This type of article is abstract and impersonal, as it presents data and an expert's interpretation of what those data mean.

By contrast, *episodic framing* presents public issues such as poverty and homelessness as concrete events, showing them to be specific instances that occur more or less in isolation (Kendall, 2011). For example, a news article may focus on the problems of one homeless family, describing how the parents and kids live in a car and eat meals from a soup kitchen. Often, what is not included is the *macro view of homelessness*, including causes and possible solutions for the "homeless problem" in the United States.

Media framing of stories about the poor and homeless represent the less fortunate in various ways. If homelessness is shown as nothing more than statistical data, then we are unable to make a balanced assessment of the larger social problems involved. On the other hand, if the media make us aware of situations—such as tent cities—that we otherwise would overlook, we may develop a better awareness of the pervasive nature of the problem and the need for broader social intervention.

### reflect & analyze

How are the poor and homeless represented in the media that you follow? Do you think that social media will increase public awareness of social problems such as homelessness? Why or why not?

---

rates of role conflict than those who engage in traditional "feminine" behavior. Role conflict may sometimes be attributed not to the roles themselves but to the pressures that people feel when they do not fit into culturally prescribed roles. A study of women athletes in college sports programs found that female gymnasts and softball players faced a "female/athlete paradox" which caused them to construct images that were based on femininity on some occasions and images that were based on athleticism on other occasions. It appears that these young women had constructed their own approach for dealing with conflict inherent in playing the female student-athlete role (Ross and Shinew, 2008).

Whereas role conflict occurs between two or more statuses, role strain takes place within one status. *Role strain* **occurs when incompatible demands are built into a single status that a person occupies.** For example, married or cohabiting women may experience more role strain than married or cohabiting men because many of them experience work overload, meaning that they work for wages outside the household but are also responsible for most of the parenting and household responsibilities within the family.

Recent social and economic changes in society may have increased role strain for men. In some families, men's traditional position of dominance has eroded as more women have entered the paid

**role** a set of behavioral expectations associated with a given status.

**role expectation** a group's or society's definition of the way that a specific role ought to be played.

**role performance** how a person actually plays a role.

**role conflict** a situation in which incompatible role demands are placed on a person by two or more statuses held at the same time.

**role strain** a condition that occurs when incompatible demands are built into a single status that a person occupies.

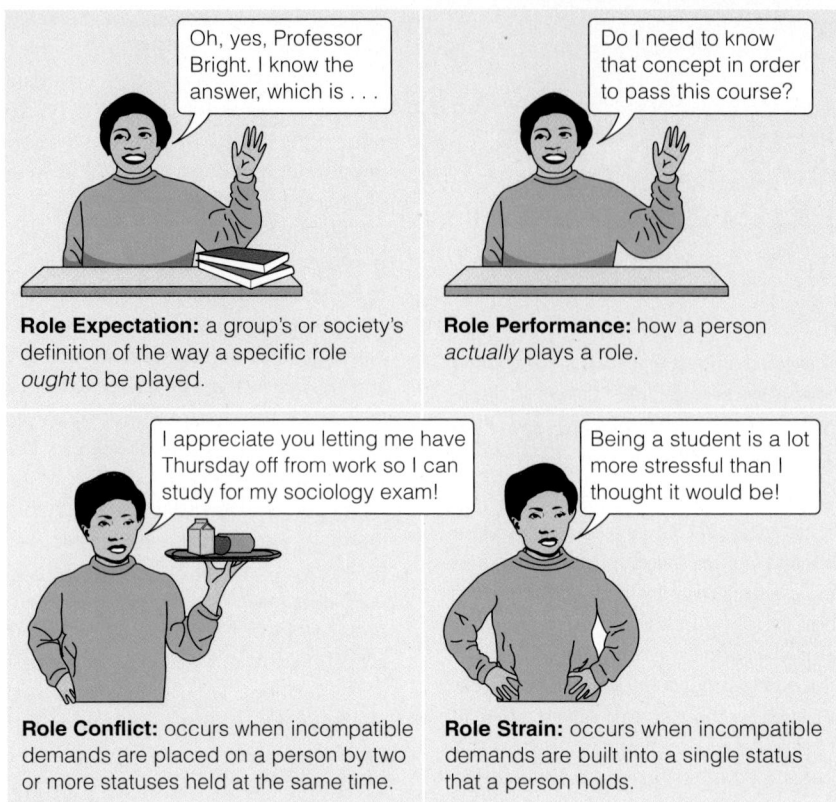

**Role Expectation:** a group's or society's definition of the way a specific role *ought* to be played.

**Role Performance:** how a person *actually* plays a role.

**Role Conflict:** occurs when incompatible demands are placed on a person by two or more statuses held at the same time.

**Role Strain:** occurs when incompatible demands are built into a single status that a person holds.

**FIGURE 4.5** ROLE EXPECTATION, PERFORMANCE, CONFLICT, AND STRAIN
When playing the role of "student," do you sometimes personally encounter these concepts?

© Cengage Learning

labor force and, in more cases, become the primary or sole breadwinner for the family. The concepts of role expectation, role performance, role conflict, and role strain are illustrated in Figure 4.5.

Individuals frequently distance themselves from a role they find extremely stressful or otherwise problematic. *Role distancing* occurs when people consciously foster the impression of a lack of commitment or attachment to a particular role and merely go through the motions of role performance (Goffman, 1961b). People use distancing techniques when they do not want others to take them as the "self" implied in a particular role, especially if they think the role is "beneath them." While Charles is working in the fast-food restaurant, for example, he does not want people to think of him as a "loser in a dead-end job." He wants them to view him as a college student who is working there just to "pick up a few bucks" until he graduates. When customers from the university come in, Charles talks to them about what courses they are taking, what they are majoring in, and what professors they have. He does not discuss whether the bacon cheeseburger is better than the chili burger. When Charles is really involved in role distancing, he tells his friends that he "works there but wouldn't eat there."

**Role Exit** *Role exit occurs when people disengage from social roles that have been central to their self-identity* (Figure 4.6). Sociologist Helen Rose Fuchs Ebaugh (1988) studied this process by interviewing ex-convicts, ex-nuns, retirees, divorced men and women, and others who had exited voluntarily from significant social roles. According to Ebaugh, role exit occurs in four stages. The first stage is doubt, in which people experience frustration or burnout when they reflect on their existing roles. The second stage involves a search for alternatives; here, people may take a leave of absence from their work or temporarily separate from their marriage partner. The third stage is the turning point at which people realize that they must take some final action, such as quitting their job or getting a divorce. The fourth and final stage involves the creation of a new identity. Consider, for example, attempting to exit the "homeless" role: This is a very difficult process because the longer an individual remains on the streets, the more that person's personal resources diminish, and his or her work experience and skills become outdated and unmarketable.

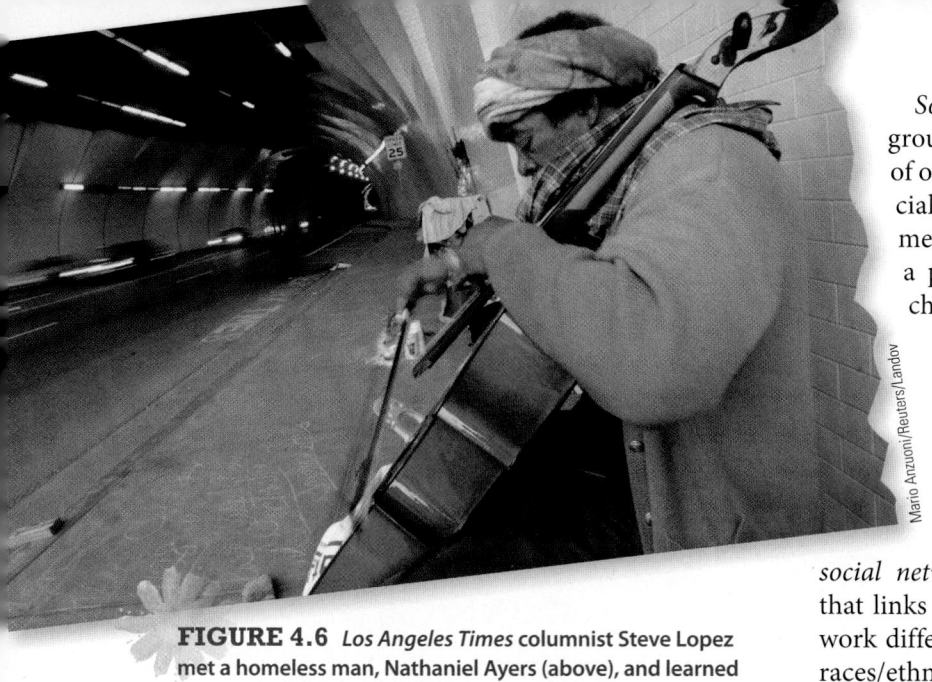

**FIGURE 4.6** *Los Angeles Times* columnist Steve Lopez met a homeless man, Nathaniel Ayers (above), and learned that he had been a promising musician studying at the Juilliard School who had dropped out because of his struggle with mental illness. In his book *The Soloist*, Lopez chronicles the relationship that he developed with Ayers and how he eventually helped get Ayers off the street and treated for his schizophrenia. This story is an example of role exit, and you can see it in the movie version of *The Soloist*.

## Groups

Groups are another important component of social structure. To sociologists, a *social group* **consists of two or more people who interact frequently and share a common identity and a feeling of interdependence.** Throughout our lives, most of us participate in groups, from our families and childhood friends, to our college classes, to our work and community organizations, and even to society.

Primary and secondary groups are the two basic types of social groups. A *primary group* **is a small, less specialized group in which members engage in face-to-face, emotion-based interactions over an extended period of time.** Primary groups include our family, close friends, and school- or work-related peer groups. By contrast, a *secondary group* **is a larger, more specialized group in which members engage in more-impersonal, goal-oriented relationships for a limited period of time.** Schools, churches, and corporations are examples of secondary groups. In secondary groups, people have few, if any, emotional ties to one another. Instead, they come together for some specific, practical purpose, such as getting a degree or a paycheck. Secondary groups are more specialized than primary ones; individuals relate to one another in terms of specific roles (such as professor and student) and more-limited activities (such as course-related endeavors). Primary and secondary groups are further discussed in Chapter 5 ("Groups and Organizations").

*Social solidarity*, or cohesion, refers to a group's ability to maintain itself in the face of obstacles. Social solidarity exists when social bonds, attractions, or other forces hold members of a group in interaction over a period of time. For example, if a local church is destroyed by fire and congregation members still worship together in a makeshift setting, then they have a high degree of social solidarity.

Many of us build social networks that involve our personal friends in primary groups and our acquaintances in secondary groups (Figure 4.7). A *social network* is a series of social relationships that links an individual to others. Social networks work differently for men and women, for different races/ethnicities, and for members of different social classes. Traditionally, people of color and white women have been excluded from powerful "old-boy" social networks. At the middle- and upper-class levels, individuals tap social networks to find employment, make business deals, and win political elections. However, social networks typically do not work effectively for poor and homeless individuals. Snow and Anderson (1993) found that homeless men have fragile social networks that are plagued with instability. Most of the avenues for exiting the homeless role and acquiring housing are intertwined with the large-scale, secondary groups that sociologists refer to as *formal organizations*.

A *formal organization* **is a highly structured group formed for the purpose of completing certain tasks or achieving specific goals.** Many of us spend most of our time in formal organizations such as colleges, corporations, or the government.

**role exit** a situation in which people disengage from social roles that have been central to their self-identity.

**social group** a group that consists of two or more people who interact frequently and share a common identity and a feeling of interdependence.

**primary group** a small, less specialized group in which members engage in face-to-face, emotion-based interactions over an extended period of time.

**secondary group** a larger, more specialized group in which members engage in more-impersonal, goal-oriented relationships for a limited period of time.

**formal organization** a highly structured group formed for the purpose of completing certain tasks or achieving specific goals.

**FIGURE 4.7** For many years, powerful "old-boy" social networks have dominated capitalism. Informal discussions, such as the one shown here, often close major business deals.

In Chapter 5 we analyze the characteristics of bureaucratic organizations; however, at this point we should note that these organizations are a very important component of social structure in all industrialized societies. We expect such organizations to educate us, solve our social problems (such as crime and homelessness), and provide us work opportunities.

Today, formal organizations such as the U.S. Conference of Mayors and the National Law Center on Homelessness and Poverty work with groups around the country to make people aware that homelessness must be viewed within the larger context of poverty and to educate the public on the nature and extent of homelessness among various categories of people in the United States (see Figure 4.8 for statistics on homelessness).

## Social Institutions

**L04** At the macrolevel of all societies, certain basic activities routinely occur—children are born and socialized, goods and services are produced and distributed, order is preserved, and a sense of purpose is maintained. Social institutions are the means by which these basic needs are met.

A *social institution* **is a set of organized beliefs and rules that establishes how a society will attempt to meet its basic social needs.** In the past, these needs centered around five basic social institutions:

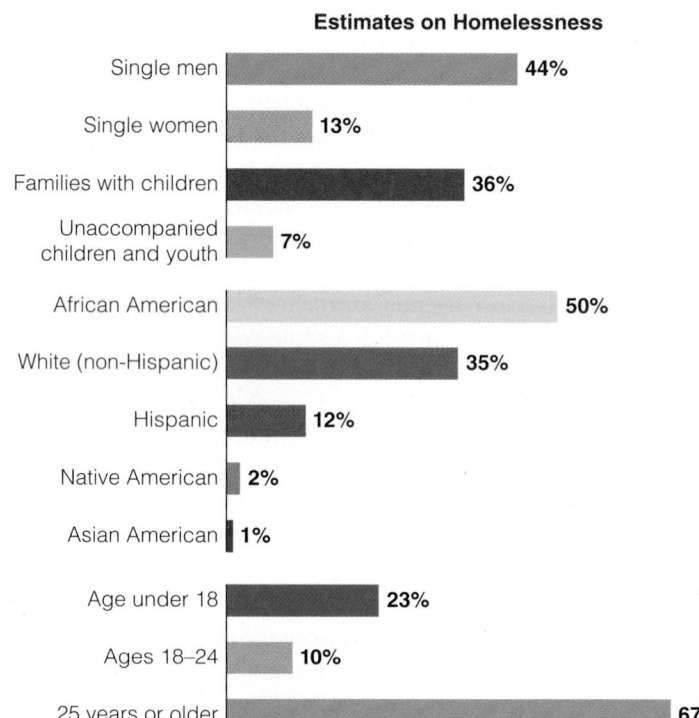

**Estimates on Homelessness**

| | |
|---|---|
| Single men | 44% |
| Single women | 13% |
| Families with children | 36% |
| Unaccompanied children and youth | 7% |
| African American | 50% |
| White (non-Hispanic) | 35% |
| Hispanic | 12% |
| Native American | 2% |
| Asian American | 1% |
| Age under 18 | 23% |
| Ages 18–24 | 10% |
| 25 years or older | 67% |

**FIGURE 4.8** WHO ARE THE HOMELESS?

Source: U.S. Department of Housing and Urban Development, 2013.

the family, religion, education, the economy, and the government or politics. Today, mass media, sports, science and medicine, and the military are also considered to be social institutions.

What is the difference between a group and a social institution? A group is composed of specific, identifiable people; an institution is a standardized way of doing something. The concept of "family" helps to distinguish between the two. When we talk about "your family" or "my family," we are referring to a specific family. When we refer to the family as a social institution, we are talking about ideologies and standardized patterns of behavior that organize family life. For example, the family as a social institution contains certain statuses organized into well-defined relationships, such as husband–wife, parent–child, and brother–sister. Specific families do not always conform to these ideologies and behavior patterns.

### Functionalist Views on Social Institutions

Functional theorists emphasize that social institutions exist because they perform five essential tasks:

1. *Replacing members.* Societies and groups must have socially approved ways of replacing members who move away or die.
2. *Teaching new members.* People who are born into a society or move into it must learn the group's values and customs.
3. *Producing, distributing, and consuming goods and services.* All societies must provide and distribute goods and services for their members.
4. *Preserving order.* Every group or society must preserve order within its boundaries and protect itself from attack by outsiders.
5. *Providing and maintaining a sense of purpose.* In order to motivate people to cooperate with one another, a sense of purpose is needed.

Although this list of functional prerequisites is shared by all societies, the institutions in each society perform these tasks in somewhat different ways depending on their specific cultural values and norms.

### Conflict Views on Social Institutions

Conflict theorists agree with functionalists that social institutions are originally organized to meet basic social needs. However, they do not believe that social institutions work for the common good of everyone. For example, some conflict theorists might point out that families may be a source of problems (rather than solutions) for young people. Some children are abused or neglected; others have arguments with their parents or other authority figures in the household that contribute to a decision to run away from home and try their luck living on the streets. Traumatic incidents in families may trigger fear, anxiety, and dread that contribute to homelessness among young people.

# Societies, Technology, and Sociocultural Change

As we think about homeless people today, it is difficult to realize that for people in some societies, being without a place of residence is a way of life. Where people live and the mode(s) of production they use to generate a food supply are related to *subsistence technology*—the methods and tools that are available for acquiring the basic needs of daily life. Social scientists have identified five types of societies based on various levels of subsistence technology: hunting and gathering, horticultural and pastoral, agrarian, industrial, and postindustrial societies. These types of societies are said to have different *technoeconomic bases* related to the technology that is available and the economic structure of the society. (The features of the different types of societies, distinguished by technoeconomic base, are summarized in Table 4.1.) The first three of these societies—hunting and gathering, horticultural and pastoral, and agrarian—are also referred to as preindustrial societies. According to social scientist Gerhard Lenski, societies change over time through the process of *sociocultural evolution,* the changes that occur as a society gains new technology (see Nolan and Lenski, 2010).

## Hunting and Gathering Societies

From the origins of human existence (several million years ago) until about 10,000 years ago, hunting and gathering societies were the only type of human society that existed. **Hunting and gathering societies use simple technology for hunting animals and gathering vegetation.** The technology in these societies is limited to tools and weapons that are used for basic subsistence, including spears, bows and arrows, nets, traps for hunting, and digging sticks for plant collecting. All tools and weapons are made of natural materials such as stone, bone, and wood.

**social institution** a set of organized beliefs and rules that establishes how a society will attempt to meet its basic social needs.

**hunting and gathering societies** societies that use simple technology for hunting animals and gathering vegetation.

## table 4.1

### Technoeconomic Bases of Society

| | Hunting and Gathering | Horticultural and Pastoral | Agrarian | Industrial | Postindustrial |
|---|---|---|---|---|---|
| Change from Prior Society | — | Use of hand tools, such as digging stick and hoe | Use of animal-drawn plows and equipment | Invention of steam engine | Invention of computer and development of "high-tech" society |
| Economic Characteristics | Hunting game, gathering roots and berries | Planting crops, domestication of animals for food | Labor-intensive farming | Mechanized production of goods | Information and service economy |
| Control of Surplus | None | Men begin to control societies | Men own land or herds | Men own means of production | Corporate shareholders and high-tech entrepreneurs |
| Inheritance | None | Shared—patrilineal and matrilineal | Patrilineal | Bilateral | Bilateral |
| Control Over Procreation | None | Increasingly by men | Men—to ensure legitimacy of heirs | Men—but less so in later stages | Mixed |
| Women's Status | Relative equality | Decreasing in move to pastoralism | Low | Low | Varies by class, race, and age |

Source: Adapted from Lorber, 1994: 140.

In hunting and gathering societies, the basic social unit is the kinship group or family. People do not have private households or residences as we think of them. Instead, they live in small groups of about twenty-five to forty people. Kinship ties constitute the basic economic unit through which food is acquired and distributed. With no stable food supply, hunters and gatherers continually search for wild animals and edible plants. As a result, they remain on the move and seldom establish a permanent settlement (Nolan and Lenski, 2010).

Hunting and gathering societies are relatively egalitarian. Because it is impossible to accumulate a surplus of food, there are few resources upon which individuals or groups can build a power base. Some specialization occurs, primarily based on age and sex. Young children and older people are expected to contribute what they can to securing the food supply, but healthy adults of both sexes are expected to obtain most of the food. In some societies, men hunt for animals, and women gather plants; in others, both women and men gather plants and hunt for wild game, with women more actively participating when smaller animals are nearby (Figure 4.9).

Contemporary hunting and gathering societies are located in relatively isolated geographical areas. However, some analysts predict that these groups will soon cease to exist, as food producers with more-dominating technologies usurp the geographic areas from which these groups have derived their food supply (Nolan and Lenski, 2010).

## Horticultural and Pastoral Societies

The period between 13,000 and 7,000 B.C.E. marks the beginning of horticultural and pastoral societies. During this period, there was a gradual shift from *collecting* food to *producing* food, a change that has been attributed to three factors: (1) the depletion of the supply of large game animals as a source of food, (2) an increase in the size of the human population to feed, and (3) dramatic weather and environmental changes that probably occurred by the end of the Ice Age.

Why did some societies become horticultural while others became pastoral? Water supply, terrain, and soils are three critical factors in whether horticultural activities or pastoral activities became a society's primary mode of food production. **Pastoral societies are based on technology that supports the domestication of large animals to provide food** and emerged in mountainous regions and areas with low amounts of annual rainfall. Pastoralists—people in pastoral societies—typically remain nomadic as they seek new grazing lands and water sources for their animals. **Horticultural societies are based on technology that supports the cultivation of plants to provide food.** These societies emerged in more-fertile areas that were better suited for growing plants through the use of hand tools.

The family is the basic unit in horticultural and pastoral societies. Because they typically do not move as often as hunter-gatherers or pastoralists,

**FIGURE 4.9** In contemporary hunting and gathering societies, women contribute to the food supply by gathering plants and sometimes hunting small animals. These women of the Kalahari in Botswana gather and share edible roots.

horticulturalists establish more-permanent family ties and create complex systems for tracing family lineage. Some social analysts believe that the invention of a hoe with a metal blade was a contributing factor to the less nomadic lifestyle of the horticulturalists because this made planting more efficient and productive. As a result, people become more *sedentary,* remaining settled for longer periods in the same location.

Unless there are fires, floods, droughts, or environmental problems, herding animals and farming are more reliable sources of food than hunting and gathering. When food is no longer in short supply, more infants are born, and children have a greater likelihood of surviving. When people are no longer nomadic, children are viewed as an economic asset: They can cultivate crops, tend flocks, or care for younger siblings.

Division of labor increases in horticultural and pastoral societies. As the food supply grows, not everyone needs to be engaged in food production. Some pursue activities such as weaving cloth or carpets, crafting jewelry, serving as priests, or creating the tools needed for building the society's structure. Horticultural and pastoral societies are less egalitarian than hunter-gatherers, and the idea of property rights emerges as people establish more-permanent settlements. At this stage, families with the largest surpluses have an economic advantage and gain prestige and power.

In contemporary horticultural societies, women do most of the farming while men hunt game, clear land, work with arts and crafts, make tools,

participate in religious and ceremonial activities, and engage in war. Gender inequality is greater in pastoral societies because men herd the large animals and women contribute relatively little to subsistence production. In some herding societies, women's primary value is their ability to produce male offspring so the family lineage is preserved and a sufficient number of males are available to protect against enemy attack.

## Agrarian Societies

About five to six thousand years ago, agrarian (or agricultural) societies emerged, first in Mesopotamia and Egypt and slightly later in China. *Agrarian societies* **use the technology of large-scale farming, including animal-drawn or energy-powered plows and equipment, to produce their food supply** (Figure 4.10). Farming made it possible for people to spend their entire lives in the same location, and food surpluses made it possible for people to live in cities, where they were not directly involved in food production. The use of animals to pull plows made it possible for people to generate a large surplus of food. The land can be used more or less continuously because the plow turns the topsoil, thus returning more nutrients to the soil. In some cases, farmers reap several harvests each year from the same plot of land.

In agrarian societies, social inequality is the highest of all preindustrial societies in terms of both class and gender. The two major classes are the landlords and the peasants. The landlords own the fields and the harvests produced by the peasants. Inheritance becomes important as families of wealthy landlords own the same land for generations. By contrast, the landless peasants enter into an agreement with the

**pastoral societies** societies based on technology that supports the domestication of large animals to provide food.

**horticultural societies** societies based on technology that supports the cultivation of plants to provide food.

**agrarian societies** societies that use the technology of large-scale farming, including animal-drawn or energy-powered plows and equipment, to produce their food supply.

Ariadne Van Zandbergen/Alamy

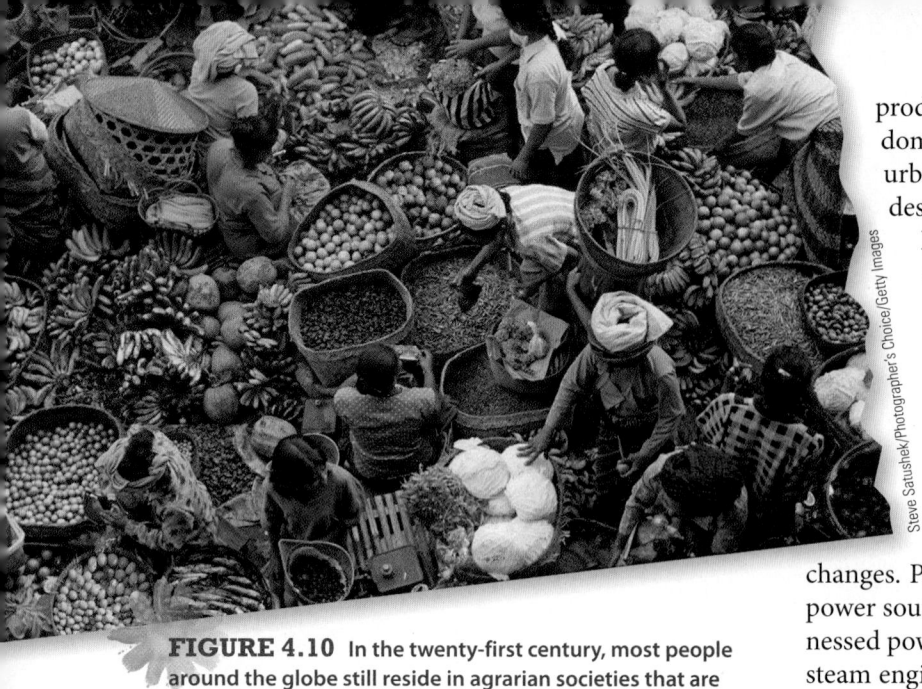

**FIGURE 4.10** In the twenty-first century, most people around the globe still reside in agrarian societies that are in various stages of industrialization. Open-air markets such as this one in Bali, where people barter or buy their food from one another, are a common sight in agrarian societies.

landowners to live on and cultivate a parcel of land in exchange for part of the harvest or other economic incentives. Over time, the landlords grow increasingly wealthy and powerful as they extract labor, rent, and taxation from the landless workers. Politics is based on a feudal system controlled by a political–economic elite made up of the ruler, his royal family, and members of the landowning class. Peasants have no political power and may be suppressed through the use of force.

Gender-based inequality grows dramatically in agrarian societies. Men gain control over both the disposition of the food surplus and the kinship system. Because agrarian tasks require more labor and greater physical strength than horticultural ones, men become more involved in food production. Women may be excluded from these tasks because they are seen as too weak for the work or because it is believed that their child-care responsibilities are incompatible with the full-time labor that the tasks require. Today, gender inequality continues in agrarian societies; the division of labor between women and men is very distinct in areas such as parts of the Middle East. Here, women's work takes place in the private sphere (inside the home), and men's work occurs in the public sphere, providing men with more recognition and greater formal status.

## Industrial Societies

***Industrial societies* are based on technology that mechanizes production.** Originating in England during the Industrial Revolution, this mode of production dramatically transformed predominantly rural and agrarian societies into urban and industrial societies. Chapter 1 describes how the revolution first began in Britain and then spread to other countries, including the United States.

Industrialism involves the application of scientific knowledge to the technology of production, thus making it possible for machines to do the work previously done by people or animals. New technologies, such as the invention of the steam engine and fuel-powered machinery, stimulated many changes. Previously, machines were run by natural power sources (such as wind or water mills) or harnessed power (either human or animal power). The steam engine made it possible to produce goods by machines powered by fuels rather than undependable natural sources or physical labor.

As inventions and discoveries build upon one another, the rate of social and technological change increases. For example, the invention of the steam engine brought about new types of transportation, including trains and steamships. Inventions such as electric lights made it possible for people to work around the clock without regard to whether it was daylight or dark outside. Industrialism changes the nature of subsistence production. In countries such as the United States, large-scale agribusinesses have practically replaced small, family-owned farms and ranches. However, large-scale agriculture has produced many environmental problems while providing solutions to the problem of food supply.

In industrial societies a large proportion of the population lives in or near cities. Large corporations and government bureaucracies grow in size and complexity. The nature of social life changes as people come to know one another more as statuses than as individuals. In fact, a person's occupation becomes a key defining characteristic in industrial societies, whereas his or her kinship ties are most important in preindustrial societies.

Social institutions are transformed by industrialism. The family diminishes in significance as the economy, education, and political institutions grow in size and complexity. The family is now a consumption unit, not a production unit. Although the influence of traditional religion is diminished in industrial societies, religion remains a powerful institution. Religious organizations are important in determining what moral issues will be brought to the forefront (e.g., unapproved drugs, abortion, and violence and sex in the media) and in trying to influence lawmakers to pass laws

Steve Satushek/Photographer's Choice/Getty Images

**FIGURE 4.11** In postindustrial economies, the focus is often on service- and information-based jobs.

regulating people's conduct. Politics in industrial societies is usually based on a democratic form of government. As nations such as South Korea, the People's Republic of China, and Mexico have become more industrialized, many people in these nations have intensified their demands for political participation.

Although the standard of living rises in industrial societies, social inequality remains a pressing problem. As societies industrialize, the status of women tends to decline further. For example, after industrialization occurred in the United States, the division of labor between men and women in the middle and upper classes became much more distinct: Men were responsible for being "breadwinners"; women were seen as "homemakers." This gendered division of labor increased the economic and political subordination of women. In short, industrial societies have brought about some of the greatest innovations in all of human history, but they have also maintained and perpetuated some of the greatest problems.

## Postindustrial Societies

A *postindustrial society* is one in which technology supports a service- and information-based economy. As discussed in Chapter 1, postmodern (or "postindustrial") societies are characterized by an *information explosion* and an economy in which large numbers of people either provide or apply information (IT specialists, for example) (Figure 4.11) or are employed in service jobs (such as fast-food servers or health care workers). For example, banking, law, and the travel industry are characteristic forms of employment in postindustrial societies, whereas producing steel or automobiles is representative of employment in industrial societies. In fact, some analysts refer to postindustrial societies as "service economies" because many workers provide services for others. However, most of the new service occupations pay relatively low wages and offer limited opportunities for advancement.

Postindustrial societies produce knowledge that becomes a commodity. This knowledge can be leased or sold to others, or it can be used to generate goods, services, or more knowledge. In the previous types of societies we have examined, machinery or raw materials are crucial to how the

**industrial societies** societies based on technology that mechanizes production.

**postindustrial societies** societies in which technology supports a service- and information-based economy.

**119**

## Computer and Internet Access in U.S. Households

The U.S. Census Bureau and other agencies such as the U.S. Department of Commerce collect data on information such as the use of computers and the level of Internet access in U.S. households. Since 1984, the first year in which the Census Bureau collected data on computer ownership and use, there has been a vast increase in the percentage of households with computers. Starting in 1998, the Census Bureau included Internet access in its data. As you can see, in 2011, the most recent year available, more than 70 percent of all U.S. households in the Census data had Internet access, and more than 75 percent reported Internet use at home.

**Computers and Internet Access in the Home: 1984 to 2011**
(civilian noninstitutional population)

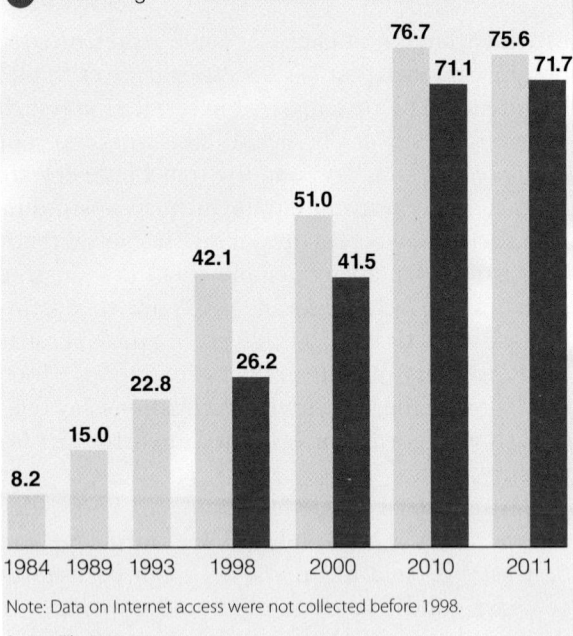

○ Percentage of households with a computer

● Percentage of households with Internet access

Note: Data on Internet access were not collected before 1998.

Source: File, 2013.

dial-up or broadband service (File, 2013). However, this still leaves slightly less than 30 percent of households without Internet access (see "Census Profiles: Computer and Internet Access in U.S. Households").

Previous forms of production, including agriculture and manufacturing, do not disappear in postindustrial societies. Instead, they become more efficient through computerization and other technological innovations. Work that relies on manual labor is often shifted to less technologically advanced societies, where workers are paid low wages to produce profits for corporations based in industrial and postindustrial societies.

Knowledge is viewed as the basic source of innovation and policy formulation in postindustrial societies. As a result, formal education and other sources of information become crucial to the success of individuals and organizations. Scientific research becomes institutionalized, and newer industries—such as computer manufacturing and software development—come into existence that would not have been possible without the new knowledge and technological strategies.

## Sociological Perspectives on Stability and Change in Society

**LO5** Changes in social structure have a dramatic impact on individuals, groups, and societies. Social arrangements in contemporary societies have grown more complex with the introduction of new technology, changes in values and norms, and the rapidly shrinking "global village." How do societies maintain some degree of social solidarity in the face of such changes? Sociologists Emile Durkheim and Ferdinand Tönnies developed typologies to explain the processes of stability and change in the social structure of societies. A *typology* is a classification scheme containing two or more mutually exclusive categories that are used to compare different kinds of behavior or types of societies.

### Durkheim: Mechanical and Organic Solidarity

Emile Durkheim (1933/1893) was concerned with the question "How do societies manage to hold together?" According to Durkheim, social solidarity derives from a society's social structure, which, in turn, is based on the society's division of labor. *Division of labor* **refers to how the various tasks of**

economy operates. In postindustrial societies, the economy is based on involvement with people and communications technologies such as the mass media, computers, and the Web. For example, data from the U.S. Department of Commerce indicate that more than 70 percent of homes in the United States have Internet access either through a society are divided up and performed. People in diverse societies (or in the same society at different points in time) divide their tasks somewhat differently, based on their own history, physical environment, and level of technological development. Durkheim claimed that preindustrial societies are held together by strong traditions and by the members' shared moral beliefs and values. As societies industrialized and developed more-specialized economic activities, social solidarity came to be rooted in the members' shared dependence on one another.

To explain social change, Durkheim categorized societies as having either mechanical or organic solidarity. **Mechanical solidarity refers to the social cohesion of preindustrial societies, in which there is minimal division of labor and people feel united by shared values and common social bonds.** Durkheim used the term *mechanical solidarity* because he believed that people in such preindustrial societies feel a more or less automatic sense of belonging. Social cohesion comes from the similarity of individuals who feel connected because they engage in the same kind of work and have similar education, religious beliefs, and lifestyles. Social interaction is characterized by face-to-face, intimate, primary-group relationships. Because everyone is engaged in similar work, little specialization is found in the division of labor. In societies of this kind, the focus is on the group, not the individual, and social interaction is much more personal.

*Organic solidarity* **refers to the social cohesion found in industrial (and perhaps postindustrial) societies, in which people perform very specialized tasks and feel united by their mutual dependence.** Durkheim chose the term *organic solidarity* because he believed that individuals in industrial societies come to rely on one another in much the same way that the organs of the human body function interdependently. Social interaction is less personal, more status oriented, and more focused on specific goals and objectives. People no longer rely on morality or shared values for social solidarity; instead, they are bound together by interdependence and practical considerations.

## Tönnies: *Gemeinschaft* and *Gesellschaft*

German sociologist Ferdinand Tönnies (1855–1936) used the terms *Gemeinschaft* and *Gesellschaft* to characterize the degree of social solidarity and social control found in societies. He was especially concerned about what happens to social solidarity in a society when a "loss of community" occurs.

The *Gemeinschaft* **(guh-MINE-shoft) is a traditional society in which social relationships are based on personal bonds of friendship and kinship and on intergenerational stability.** Tönnies (1963/1887) used the German term *Gemeinschaft* because it means "commune" or "community"; social solidarity and social control are maintained by the community. In this kind of society, relationships are based on ascribed (from birth) status rather than achieved (acquired) status. For example, the child of a farmer is likely to become, and remain, a farmer as well. In the *Gemeinschaft*, people have a commitment to the entire group and feel a sense of togetherness: They tend to focus more on the needs and interests of the group rather than their own self-interest. Members have a strong sense of belonging, but they also have very limited privacy. External social control is seldom needed because control is maintained through informal means such as persuasion or gossip.

By contrast, the *Gesellschaft* **(guh-ZELL-shoft) is a large, urban society in which social bonds are based on impersonal and specialized relationships, with little long-term commitment to the group or consensus on values.** Tönnies (1963/1887) selected the German term *Gesellschaft* because it means "association"; relationships are based on achieved statuses, and interactions among people are both rational and calculated. For example, achieved status might be based on education

**division of labor** how the various tasks of a society are divided up and performed.

**mechanical solidarity** Emile Durkheim's term for the social cohesion of preindustrial societies, in which there is minimal division of labor and people feel united by shared values and common social bonds.

**organic solidarity** Emile Durkheim's term for the social cohesion found in industrial (and perhaps postindustrial) societies, in which people perform very specialized tasks and feel united by their mutual dependence.

*Gemeinschaft* (guh-MINE-shoft) a traditional society in which social relationships are based on personal bonds of friendship and kinship and on intergenerational stability.

*Gesellschaft* (guh-ZELL-shoft) a large urban society in which social bonds are based on impersonal and specialized relationships, with little long-term commitment to the group or consensus on values.

**121**

# SOCIOLOGY and Social Policy

## Homeless Rights Versus Public Space

I had a bit of a disturbing experience yesterday as I was running errands downtown.... When I reached the corner of Queen and Bay [in Toronto, Canada], I saw some police officers and city workers "taking action on sidewalk clearance." They were clearing a homeless person's worldly belongings off the sidewalk. Using shovels. And a pickup truck....

I think what I saw yesterday is unacceptable. Sure, the situation is complicated. Yes, there are a lot of stakeholders and stories to appreciate. But it's unfairness I want to see shoveled out of public space. Not people. Not blankets. Not kindness. And I hope I'm not alone. (Sandals, 2007)

Mark Ludak/The Image Works

Sidewalk clearance and public space protection are controversial topics in cities where law enforcement officials have been instructed to remove homeless individuals and their possessions from public spaces. What are the central issues in this social policy debate? Why should this problem be of concern to each of us?

"**P**ublic space protection" has become an issue in many cities, both in the United States and elsewhere. Record numbers of homeless individuals and families seek refuge on the streets and in public parks because they have nowhere else to go. However, this seemingly individualistic problem is actually linked to larger social concerns, including long-term unemployment, lack of education and affordable housing, and cutbacks in government and social service budgets. The problem of homelessness also raises significant social policy issues, including the extent to which cities can make it illegal for people to remain for extended periods of time in public spaces.

Should homeless people be allowed to sleep on sidewalks, in parks, and in other public areas? This issue has been the source of controversy. As cities have sought to improve their downtown areas and public spaces, they have taken measures to enforce city ordinances controlling loitering (standing around or sleeping in public spaces), "aggressive panhandling," and disorderly conduct. Advocates for the homeless and civil liberties groups have filed lawsuits claiming that the rights of the homeless are being violated by the enforcement of these laws. The lawsuits assert that the homeless have a right to sleep in parks because no affordable housing is available for them. Advocates also argue that panhandling is a legitimate means of livelihood for some of the homeless and is protected speech under the First Amendment. In addition, they accuse public and law enforcement officials of seeking to

Source: Based on National Coalition for the Homeless, 2012.

punish the homeless on the basis of their "status," a cruel and unusual punishment prohibited by the Eighth Amendment to the U.S. Constitution.

The "homeless problem" is not a new one for city governments. Of the limited public funding that is designated for the homeless, most has been spent on shelters that are frequently overcrowded and otherwise inadequate. Officials in some cities have given homeless people a one-way ticket to another city. Still others have routinely run them out of public spaces or tried to relocate them to marginalized areas of the city.

What responsibility does society have for the homeless? Are laws restricting the hours that public areas or parks are open to the public unfair to homeless persons? These questions highlight pressing social policy concerns because affordable housing and job opportunities are not available for large numbers of people, and many homeless persons are left with nowhere to go.

### reflect & analyze

What responsibility do communities have to the homeless? Is it possible to maintain public safety and not criminalize homelessness? What do you think should be done?

level or the kind of work that people do, rather than the family into which they were born. In such societies, most people are "strangers" who perceive that they have very little in common with most other people. Consequently, self-interest dominates, and little consensus exists regarding values. Table 4.2 compares the characteristics of *Gemeinschaft* and *Gesellschaft* societies.

## table 4.2

**Comparing *Gemeinschaft* and *Gesellschaft* Societies**

| *Gemeinschaft* | *Gesellschaft* |
| --- | --- |
| Characterized by rural life | Characterized by urban life |
| Sense of community based on similarity | Lack of feeling of community |
| Intimate, face-to-face social interactions | Impersonal and task-oriented relationships |
| Primary focus on personal relationships | Primary focus on tasks or goals to be accomplished |
| Social control on an informal basis | Formal social control |
| Ascribed statuses most important | Achieved statuses most important |
| Limited social change | Social change more prevalent |

Source: © Cengage Learning 2015.

### Social Structure and Homelessness

In *Gesellschaft* societies such as the United States, a prevailing value is that people should be able to take care of themselves. Consequently, some politicians and everyday people argue that social agencies and institutions have little or no responsibility to assist homeless persons. Others believe that the social structure of societies should provide a safety net for everyone regardless of their economic status. Clearly, there is no simple answer to questions about what should be done to help homeless persons. Nor, as discussed in "Sociology and Social Policy," is there any consensus on what rights the homeless have to occupy public spaces such as parks and city sidewalks. The answers we derive as a society and as individuals are often based on our social construction of the reality of life for the homeless.

## Social Interaction: The Microlevel Perspective

**LO6** So far in this chapter, we have focused on society and social structure from a macrolevel perspective, seeing how the structure of society affects the statuses we occupy, the roles we play, and the groups and organizations to which we belong. Functionalist and conflict perspectives provide a macrosociological overview because they concentrate on large-scale events and broad social features. By contrast, the symbolic interactionist perspective takes a microsociological approach, asking how social institutions affect our daily lives.

### Social Interaction and Meaning

When you are with other people, do you often wonder what they think of you? If so, you are not alone! Because most of us are concerned about the meanings that others ascribe to our behavior, we try to interpret their words and actions so that we can plan how we will react toward them. We know that others have expectations of us. We also have certain expectations about them. For example, if we get on a bus or a subway that has only one other person on it, we do not expect that individual to confront us and stare into our eyes. As a matter of fact, we would be quite upset if the person did so (Figure 4.12).

Social interaction within a given society has certain shared meanings across situations. For instance, our reaction would be the same regardless of *which* elevator we rode in *which* building. Sociologist Erving Goffman (1961b) described these shared meanings in his observation about two pedestrians approaching each other on a public sidewalk. He noted that each will tend to look at the other just long enough to acknowledge the other's presence. By the time they are about eight feet away from each other, both individuals will tend to look downward. Goffman referred to this behavior as *civil inattention*—the ways in which an individual shows an awareness that another is present without making this person the object of particular attention. The fact that people engage in civil inattention demonstrates that interaction does have a pattern, or *interaction order,* which regulates the form and processes (but not the content) of social interaction.

Does everyone interpret social interaction rituals in the same way? No. Race/ethnicity, gender, and social class play a part in the meanings we give to our interactions with others, including chance encounters on elevators or the street. Our perceptions about the meaning of a situation vary widely based on the statuses we occupy and our unique personal experiences. For example, women often do not perceive street encounters, particularly at night, to be "routine" rituals. They fear for their personal safety and try to avoid comments and propositions that are sexual in nature. Persons of color may also feel uncomfortable in street encounters. A middle-class African American college student described his experiences walking home at night from a campus job:

> So, even if you wanted to, it's difficult just to live a life where you don't come into conflict with others. . . . Every day that you live as a black person you're reminded how you're perceived in society. You walk the streets at night;

**FIGURE 4.12** How do our expectations of social interaction affect how we behave in a crowded bus or subway?

white people cross the streets. I've seen white couples and individuals dart in front of cars to not be on the same side of the street. Just the other day, I was walking down the street, and this white female with a child, I saw her pass a young white male about 20 yards ahead. When she saw me, she quickly dragged the child and herself across the busy street. . . . [When I pass,] white men tighten their grip on their women. I've seen people turn around and seem like they're going to take blows from me. . . . So, every day you realize [you're black]. Even though you're not doing anything wrong; you're just existing. You're just a person. But you're a black person perceived in an unblack world. (qtd. in Feagin, 1991: 111–112)

Although this statement was made more than twenty years ago, some current students of color reading it can still relate to the experiences described here. As the passage indicates, social encounters have different meanings for men and women, whites and people of color, individuals from different social classes, and sometimes people from different areas of the country. Members of the dominant classes regard the poor, unemployed, and working class as less worthy of attention, frequently subjecting them to subtle yet systematic "attention deprivation" (Derber, 1983). The same can

certainly be said about how members of the dominant classes "interact" with the homeless. There are some who say that all of this has changed since the time that some of this research was originally conducted, but, unfortunately, violent incidents are periodically reported in the media, renewing the concerns that we have a long way to go in protecting the safety and security of all individuals in our nation.

## The Social Construction of Reality

If we interpret other people's actions so subjectively, can we have a shared social reality? Some symbolic interaction theorists believe that there is very little shared reality beyond that which is socially created. Symbolic interactionists refer to this as the *social construction of reality*—**the process by which our perception of reality is largely shaped by the subjective meaning that we give to an experience** (Berger and Luckmann, 1967). This meaning strongly influences what we "see" and how we respond to situations.

As discussed previously, our perceptions and behavior are influenced by how we initially define situations: We act on reality as we see it. Sociologists describe this process as the *definition of the situation,* meaning that we analyze a social context in which we find ourselves, determine what is in our best interest, and adjust our attitudes and actions accordingly. This process can result in a *self-fulfilling prophecy—a false belief or prediction that produces behavior that makes the originally false belief come true* (Merton, 1968). An example would be a person who has been told repeatedly that she or he is not a good student; eventually, this person might come to believe it to be true, stop studying, and receive failing grades. Dominant-group members with prestigious statuses may have the ability to establish how other people define "reality" (Berger and Luckmann, 1967: 109).

An example of the self-fulfilling prophecy is a study of homeless persons in the United Kingdom where extensive interviews were conducted with eight homeless individuals to learn about their experiences with health-related social services. When the study was conducted, early in this century, approximately 400,000 persons were considered to be "hidden homeless" in the United Kingdom, or approximately more than two in every thousand people in that nation. According to the researchers, homeless people in the UK were trapped between the *homed* system—the social service system, which offered formal help but was quick to label and stigmatize them, while enforcing extensive rules on them—and the *homeless* system of informal help, where other homeless individuals were supportive but often intensified their problems by offering alcohol, drugs, or other "solutions" that did not help their overall situation. The researchers concluded that a self-fulfilling prophecy occurs when homeless persons who have been labeled as "sofa surfers," "homeless," or similar terms come to view the label as central to how they see themselves and then decide to embed themselves within the homeless system, where they can have the support of other homeless people who also live without establishment-ordered rules and routines (Ogden and Avades, 2011). From the choices these individuals made, the self-fulfilling prophecy became a reality because their homeless situation was unchanged, and social organizations and institutions within the community were unable to meet their needs.

## Ethnomethodology

**LO7** How do we know how to interact in a given situation? What rules do we follow? Ethnomethodologists are interested in the answers to these questions. *Ethnomethodology is the study of the commonsense knowledge that people use to understand the situations in which they find themselves* (Heritage, 1984: 4). Sociologist Harold Garfinkel (1967) initiated this approach and coined the term: *ethno* for "people" or "folk" and *methodology* for a "system of methods." Garfinkel was critical of mainstream sociology for not recognizing the ongoing ways in which people create reality and produce their own world. Consequently, ethnomethodologists examine existing patterns of conventional behavior in order to uncover people's background expectancies—that is, their shared interpretation of objects and events, as well as their resulting actions. According to ethnomethodologists, interaction is based on assumptions of shared expectancies. For example, when you are talking with someone, what expectations do you have that you will take turns? Based on your background expectancies, would you be surprised if the other person talked for an hour and never gave you a chance to speak?

To uncover people's background expectancies, ethnomethodologists frequently break "rules" or act as though they do not understand some basic rule of social life so that they can observe other people's responses. In a series of *breaching experiments,* Garfinkel assigned different activities to his students to see how breaking the unspoken rules of behavior created confusion.

The ethnomethodological approach contributes to our knowledge of social interaction by making us aware of subconscious social realities in our daily lives. However, a number of sociologists regard ethnomethodology as a frivolous approach to studying human behavior because it does not examine the impact of macrolevel social institutions—such as the economy and education—on people's expectancies. Some scholars suggest that ethnomethodologists fail to do what they claim to do: look at how social realities are created. Rather, they take ascribed statuses (such as race, class, gender, and age) as "givens," not as *socially created* realities.

**social construction of reality** the process by which our perception of reality is largely shaped by the subjective meaning that we give to an experience.

**self-fulfilling prophecy** a situation in which a false belief or prediction produces behavior that makes the originally false belief come true.

**ethnomethodology** the study of the commonsense knowledge that people use to understand the situations in which they find themselves.

## Dramaturgical Analysis

How is everyday life like watching a dramatic presentation? Erving Goffman suggested that day-to-day interactions have much in common with being on stage or in a dramatic production. ***Dramaturgical analysis* is the study of social interaction that compares everyday life to a theatrical presentation.** In this presentation, there is a stage, actors, and an audience to observe and analyze the social interactions of the actors. The actors have a ***social script*—a playbook that the actors use to guide their verbal replies and overall performance to achieve the desired goal of the conversation or fulfill the role they are playing.** Although most of us do not have scripted conversations, we have a good idea how a social exchange will occur. For example, when someone asks us how we are doing, we expect to reply, "Fine, thanks." If we take our vehicle to the drive-up order station at a fast-food restaurant, we expect to hear a voice say, "May I take your order, please?" We do not expect to reply, "Do you know what the high temperature will be today?"

Because we are familiar with most scripts in our daily lives, we know what to expect; however, there is often more than one way to interpret a script, leading to confusion and sometimes to conflict. According to Goffman (1959, 1963a), members of our "audience" judge our performance and are aware that we may slip and reveal our true character. Consequently, most of us attempt to play our role as well as possible and to create and sustain favorable impressions. ***Impression management (presentation of self)* refers to people's efforts to present themselves to others in ways that are most favorable to their own interests or image.**

For example, suppose that a professor has returned graded exams to your class. Will you discuss the exam and your grade with others in the class? If you are like most people, you probably play your student role differently depending on whom you are talking to and what grade you received on the exam. Your "presentation" may vary depending on the grade earned by the other person (your "audience"). In one study, students who all received high grades ("Ace–Ace encounters") willingly talked with one another about their grades and sometimes engaged in a little bragging about how they had "aced" the test. However, encounters between students who had received high grades and those who had received low or failing grades ("Ace–Bomber encounters") were uncomfortable. The Aces felt as if they had to minimize their own grade. Consequently, they tended to attribute their success to "luck" and were quick to offer the Bombers words of encouragement. On the other hand,

the Bombers believed that they had to praise the Aces and hide their own feelings of frustration and disappointment. Students who received low or failing grades ("Bomber–Bomber encounters") were more comfortable when they talked with one another because they could share their negative emotions. They often indulged in self-pity and relied on face-saving excuses (such as an illness or an unfair exam) for their poor performances (Albas and Albas, 1988, 2011).

In Goffman's terminology, ***face-saving behavior* refers to the strategies people use to rescue their performance when they experience a potential or actual loss of face.** When the Bombers made excuses for their low scores, they were engaged in face-saving; the Aces attempted to help them save face by asserting that the test was unfair or that it was only a small part of the final grade. Why would the Aces and Bombers both participate in face-saving behavior? In most social interactions, all role players have an interest in keeping the "play" going so that they can maintain their overall definition of the situation in which they perform their roles.

Goffman noted that people consciously participate in *studied nonobservance,* a face-saving technique in which one role player ignores the flaws in another's performance to avoid embarrassment for everyone involved. Most of us remember times when we have failed in our role and know that it is likely to happen again; thus, we may be more forgiving of the role failures of others.

Social interaction, like a theater, has a front stage and a back stage. The *front stage* is the area where a player performs a specific role before an audience (Figure 4.13). The *back stage* is the area where a player is not required to perform a specific role because it is out of view of a given audience. For example, when the Aces and Bombers were talking with one another at school, they were on the "front stage." When they were in the privacy of their own residences, they were in "back stage" settings—they no longer had to perform the Ace and Bomber roles and could be themselves.

The need for impression management is most intense when role players have widely divergent or devalued statuses. As we have seen with the Aces and Bombers, the participants often play different roles under different circumstances and keep their various audiences separated from one another. If one audience becomes aware of other roles that a person plays, the impression being given at that time may be ruined. For example, people facing or experiencing homelessness are not only stigmatized but may also find that they lose the opportunity to get a job if their homelessness becomes known (see "Sociology Works!"). However, many homeless individuals and not the underlying substance. This approach may not place enough emphasis on the ways in which our everyday interactions with other people are influenced by occurrences within the larger society. For example, if some political leaders or social elites in a community deride homeless people by saying they are "lazy" or "unwilling to work," it may become easier for everyday people walking down a street to treat homeless individuals poorly. Overall, however, Goffman's dramaturgical analysis has been highly influential in the development of the sociology of emotions, an important area of contemporary theory and research.

## The Sociology of Emotions

**LO8** Why do we laugh, cry, or become angry? Are these emotional expressions biological or social in nature? To some extent, emotions are a biologically given sense (like hearing, smell, and touch), but they are also social in origin. We are socialized to feel certain emotions, and we learn how and when to express (or not express) those emotions (Hochschild, 1983).

How do we know which emotions are appropriate for a given role? Sociologist Arlie Hochschild (1983) suggests that we acquire a set of *feeling rules* that shapes the appropriate emotions for a given role or specific situation (Figure 4.14). These rules include how, where, when, and with whom an emotion should be expressed. For example, for the role of a mourner at a funeral,

**FIGURE 4.13** Erving Goffman believed that people spend a great amount of time and effort managing the impressions that they present. How do politicians use impression management as they seek to accomplish their goal of being elected to public office?

do not passively accept the roles into which they are cast. For the most part, they attempt—as we all do—to engage in impression management in their everyday lives.

The dramaturgical approach helps us think about the roles we play and the audiences who judge our presentation of self; however, this perspective has also been criticized for focusing on appearances

**dramaturgical analysis** Erving Goffman's term for the study of social interaction that compares everyday life to a theatrical presentation.

**social script** a "playbook" that "actors" use to guide their verbal replies and overall performance to achieve the desired goal of the conversation or fulfill the role they are playing.

**impression management (presentation of self)** Erving Goffman's term for people's efforts to present themselves to others in ways that are most favorable to their own interests or image.

**face-saving behavior** Erving Goffman's term for the strategies people use to rescue their performance when they experience a potential or actual loss of face.

# Sociology WORKS!

## Goffman's Stigmatization Theory and Contemporary Homelessness

As we have seen in this chapter, homelessness carries a stigma like the ones that sociologist Erving Goffman described in his book *Stigma: Notes on the Management of Spoiled Identity*, which was originally published in 1963. According to Goffman, there are two types of stigma: (1) *discredited stigmas* are obvious to other people because the source of the stigma is visible (such as a person with a missing nose or other physical impairment), and (2) *discreditable stigmas* are not obvious to others and are not known or perceived by them.

People who are visibly homeless are more likely to be stigmatized than those who make up the invisible homeless. When homelessness is highly visible, homeless individuals must manage the tension that occurs because others know of their problem. When homeless is not highly visible (for example, homeless persons who live in their cars or with relatives), the central concern of these individuals is managing information so that others do not find out about their problem (Goffman, 1963b).

As a result of the stigma that is attached to homelessness, it is difficult for homeless people to reestablish a routine in everyday life. Some individuals who are homeless try to "pass" by acting like they have a place to live and by disassociating themselves with other homeless people. Others respond differently to stigma and instead embrace homelessness by becoming part of a subculture comprising homeless individuals who provide mutual support for one another (Johnson, 2006).

As we look at Goffman's ideas on stigma today, we find rich new opportunities for application of classical sociological insights to our understanding of pressing social problems such

Bruce Ayres/Getty Images

Contrary to a popular myth that most homeless people are single drifters, an increasing number of families are now homeless. How might this family's visible homelessness affect the degree to which they are stigmatized?

as homelessness. We can also hope that political and social leaders will seek to provide new pathways out of homelessness through innovative public policy rather than continuing to reinforce social structures that contribute to long-term homelessness and stigmatization for many people.

### reflect & analyze

How might you apply Goffman's ideas about stigma to problems at your college or university? For example, would his analysis be useful in studying issues such as eating disorders or alcohol abuse among college students?

---

feeling rules tell us which emotions are required (sadness and grief, for example), which are acceptable (a sense of relief that the deceased no longer has to suffer), and which are unacceptable (enjoyment of the occasion expressed by laughing out loud) (see Hochschild, 1983: 63–68).

Feeling rules also apply to our occupational roles. For example, the truck driver who handles explosive cargos must be able to suppress fear. Although all jobs place some burden on our feelings, *emotional labor* occurs only in jobs that require personal contact with the public or the production of a state of mind (such as hope, desire, or fear) in others (Hochschild, 1983). With emotional labor, employees must display only certain carefully selected emotions. For example, flight attendants are required to

act friendly toward passengers, to be helpful and open to requests, and to maintain an "omnipresent smile" in order to enhance the customers' status. By contrast, bill collectors are encouraged to show anger and make threats to customers, thereby supposedly deflating the customers' status and wearing down their presumed resistance to paying past-due bills. In both jobs, the employees are expected to show feelings that are often not their true ones (Hochschild, 1983).

Social class and race are determinants in managed expression and emotion management. Emotional labor is emphasized in middle- and upper-class families. Because middle- and upper-class parents often work with people, they are more likely to teach their children the importance of emotional

**FIGURE 4.14** Are there different gender-based expectations in the United States about the kinds of emotions that men, as compared with women, are supposed to show? What feeling rules shape the emotions of the man in this photo?

labor in their own careers than are working-class parents, who tend to work with things, not people (Hochschild, 1983). Race is also an important factor in emotional labor. People of color spend much of their life engaged in emotional labor because racist attitudes and discrimination make it continually necessary to manage one's feelings.

Emotional labor may produce feelings of estrangement from one's "true" self. C. Wright Mills (1956) suggested that when we "sell our personality" in the course of selling goods or services, we engage in a seriously self-alienating process. In other words, the "commercialization" of our feelings may dehumanize our work role performance and create alienation and contempt that spill over into other aspects of our life.

Hochschild (2012) also conducted research to demonstrate how many middle- and upper-income individuals "outsource" emotional labor in the more-intimate aspects of their lives, such as having other people professionally plan their family birthday parties and weddings, selecting names for their children, overseeing the daily lives of their children, and assuming major caregiving responsibilities for their elderly parents. Hochschild (2012) uses the term "outsourced self" to refer to what happens when individuals defer most of their emotional labor to others, particularly as the market continues to invade private life.

## Nonverbal Communication

In a typical stage drama, the players not only speak their lines but also use nonverbal communication to convey information. ***Nonverbal communication is the transfer of information between persons without the use of words*** (Figure 4.15). It includes not only visual cues (gestures, appearances) but also vocal features (inflection, volume, pitch) and environmental factors (use of space, position) that affect meanings. Facial expressions, head movements, body positions, and other gestures carry as much of the total meaning of our communication with others as our spoken words do.

### Functions of Nonverbal Communication

We obtain first impressions of others from various kinds of nonverbal communication, such as the clothing they wear and their body positions. Head and facial movements may provide us with information about other people's emotional states, and others receive similar information from us. Through our body posture and eye contact, we signal that we do or do not wish to speak to someone. For example, we may look down at the sidewalk or off into the distance when we pass homeless persons who look as if they are going to ask for money.

Nonverbal communication establishes the relationship among people in terms of their responsiveness to and power over one another. For example, we show that we are responsive toward or like another person by maintaining eye contact and attentive body posture and perhaps by touching and standing close. Studies of communications in the doctor–patient relationship confirm that trust is vital for quality health care outcomes and that the best communications are established through actions more than words. Nonverbal communication, such as good eye contact and attentive listening posture, show that the doctor is paying attention to what the patient is saying, and this helps to build rapport and establish the agenda of the professional in relation to that of the patient (Brown et al., 2011).

**nonverbal communication** the transfer of information between persons without the use of words.

**FIGURE 4.15** Nonverbal communication may be thought of as an international language. What message do you receive from the facial expression, body position, and gestures of each of these people? Is it possible to misinterpret these messages?

Goffman (1956) suggested that *demeanor* (how we behave or conduct ourselves) is relative to social power. People in positions of dominance are allowed a wider range of permissible actions than are their subordinates, who are expected to show deference. *Deference* is the symbolic means by which subordinates give a required permissive response to those in power; it confirms the existence of inequality and reaffirms each person's relationship to the other.

**Facial Expression, Eye Contact, and Touching** Deference behavior is important in regard to facial expression, eye contact, and touching. This type of nonverbal communication is symbolic of our relationships with others. Who smiles? Who stares? Who makes and sustains eye contact? Who touches whom? All these questions relate to demeanor and deference; the key issue is the status of the person who is doing the smiling, staring, or touching relative to the status of the recipient (Goffman, 1967).

Facial expressions, especially smiles and eye contact, also reflect gender-based patterns of dominance and subordination in society. Typically, women have been socialized to smile and frequently do so even when they are not actually happy (LaFrance and Hecht, 2000). Jobs held predominantly by women (including flight attendant, secretary and administrative assistant, elementary schoolteacher, and nurse) are more closely associated with being pleasant and smiling than are "men's jobs." By contrast, men tend to display less emotion through smiles or other facial expressions and instead seek to show that they are reserved and in control. Even as women have entered previously male-dominated professions, such as medicine, law, and college teaching and administration, an expectation that they will be more personable, more understanding, and more nurturing than their male counterparts has remained in the minds of some patients, clients, customers, colleagues, students, supervisors, and others who come into daily contact with the woman professional.

Women are also more likely to sustain eye contact during conversations (but not otherwise) as a means of showing their interest in and involvement with others. By contrast, men are less likely to maintain prolonged eye contact during conversations but are more likely to stare at other people (especially men) in order to challenge them and assert their own status (Hall, Carter, and Horgan, 2000).

Eye contact can be a sign of domination or deference. For example, in a classic participant-observation study of domestic (household) workers and their employers, the sociologist Judith Rollins (1985) found that household workers were supposed to show deference by averting their eyes when they talked to their employers. Deference also required that they present an "exaggeratedly subservient demeanor" by standing less erect and walking tentatively. This kind of behavior is depicted in the best-selling book and movie *The Help.* More-recent relationships of domination and deference in household relationships among immigrant workers and their affluent employers in the United States are discussed in the research of the sociologist Pierrette Hondagneu-Sotelo (2007). Other contemporary examples of using eye contact as a means to express domination and deference are when a teacher or principal is correcting a student and when a detainee is in a juvenile facility or prison.

Touching is another form of nonverbal behavior that has many different shades of meaning. Although touching is a universal aspect of people's communication with each other, it varies greatly by gender, age, and culture. Gender and power differences are evident in tactile communication from birth: Boys are touched more roughly and playfully, whereas girls are handled more gently and protectively. This pattern continues into adulthood, with women touched more frequently than men. Clearly, touching has a different meaning to women than to men. Women may hug and touch others to indicate affection and emotional support, but men are more likely to touch others to give directions, assert power, and express sexual interest.

Age is also a factor in touching: Different patterns of touching have been identified in people under thirty years of age as compared with those of older adults. Younger men behave more possessively and women more submissively in regard to touching behavior. Touching behavior also involves large cultural differences in personal space, such as the handshake as a preferred means of personal greeting in the United States as compared to extensive hugs and kisses in some other countries. Let's look more closely at the concept of personal space.

**Personal Space** *Personal space* **is the immediate area surrounding a person that the person claims as private.** Our personal space is contained within an invisible boundary surrounding our body, much like a snail's shell or an invisible bubble or zone around a person. This space is fluctuating, and it is part of a communication style. When others invade our space, we may retreat, stand our ground, or even lash out, depending on our cultural background.

Anthropologist Edward Hall (1959, 1966) first described the concept of personal space and identified four dimensions of this space among people in the United States: *Intimate distance* (or *intimate zone*) involves a high level of intimacy between two persons, including touching; *personal distance* (or *personal zone*) is the distance between two persons who know each other with a relative intimacy, such as friends, brothers, sisters, or other relatives; and *social distance* (or *social zone*) is the more impersonal form of communication or business interaction that takes place among individuals at a social gathering or employees in a business environment. The fourth dimension is *public distance* (or *public zone*), in which no intimacy exists between a speaker on a platform and an audience where the distance is greater than thirteen feet and may be as much as twenty-seven or more feet. Age, gender, and cultural differences are important factors in the allocation of personal space. With regard to age, adults generally do not hesitate to enter the personal space of a child. Women tend to interact at closer distances than men in the United States, and this appears to remain relatively consistent across age categories. Cross-cultural studies have confirmed that people in the United States are more comfortable with larger zones of personal space than are individuals from Latin American countries. Less personal space is also required among South Americans, Southern and Eastern Europeans, and Arabs, while larger amounts of personal space are required among Asians and Northern Europeans (Beaulieu, 2004). Figure 4.16 illustrates differences in social distance rules between two contrasting cultures, North America and Latin America, which are known for having different social distance rules.

In sum, all forms of nonverbal communication are influenced by gender, race, social class, and the personal contexts in which they occur. Although it is difficult to generalize about people's nonverbal behavior, we still need to think about

**personal space** the immediate area surrounding a person that the person claims as private.

## Social Interaction: The Microlevel Perspective

| | |
|---|---|
| Social interaction and meaning | In a given society, forms of social interaction have shared meanings, although these may vary to some extent based on race/ethnicity, gender, and social class. |
| Social construction of reality | The process by which our perception of reality is largely shaped by the subjective meaning that we give to an experience. |
| Ethnomethodology | Studying the commonsense knowledge that people use to understand the situations in which they find themselves makes us aware of subconscious social realities in daily life. |
| Dramaturgical analysis | The study of social interaction that compares everyday life to a theatrical presentation. This approach includes impression management (people's efforts to present themselves favorably to others). |
| Sociology of emotions | We are socialized to feel certain emotions, and we learn how and when to express (or not express) them. |
| Nonverbal communication | The transfer of information between persons without the use of speech, such as by facial expressions, head movements, and gestures. |

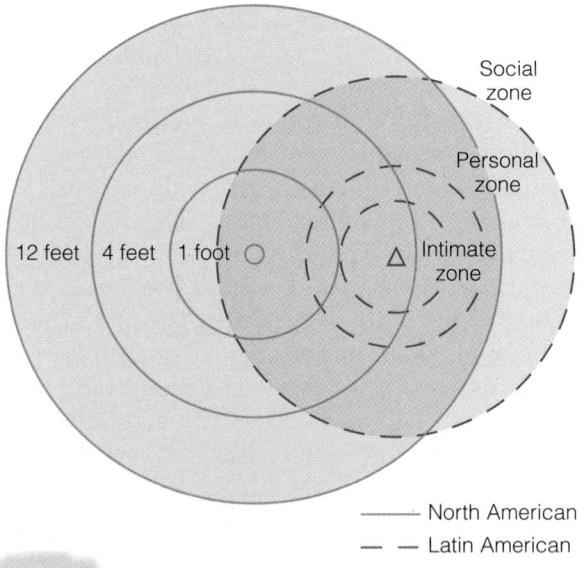

**FIGURE 4.16** NORTH AMERICAN AND LATIN AMERICAN SOCIAL DISTANCE RULES

Source: From *Cultural Anthropology* by Paul Hiebert. Copyright © 1983 by Baker Academic, a division of Baker Publishing Group. Used by permission.

our own nonverbal communication patterns. Recognizing that differences in social interaction exist is important. Learning to understand and respect alternative styles of social interaction enhances our personal effectiveness by increasing the range of options we have for communicating with different people in diverse contexts and for varied reasons. (The Concept Quick Review summarizes the microlevel approach to social interaction.)

# Changing Social Structure and Interaction in the Future

The social structure in the United States has been changing rapidly in recent decades, and we can no longer think of our future as separate from the larger world of which we are a part. Currently, there are more possible statuses for persons to occupy and roles to play than at any other time in history. Although achieved statuses are considered very important, ascribed statuses still have a significant effect on people's options and opportunities. National issues are now fused with global concerns regarding economic crises, actual and potential war and terrorism, more-frequent occurrences of natural disasters, and other problems that affect people around the world.

Ironically, at a time when we have more ability to communicate than ever before, more technological capability, more leisure activities and types of entertainment, and more quantities of material goods available for consumption, many people experience problems that are beyond their individual ability to resolve, such as chronic unemployment and homelessness. As the sociological imagination suggests, these individuals are dealing not only with personal troubles but also with public issues that affect large numbers of people and require solutions at societal or global levels. The future of our country rests on our collective ability to deal with major social problems at both the macrolevel and the microlevel of the social world

# chapter REVIEW

## LO1 Why is social structure important in our interaction with others?

The stable patterns of social relationships within a particular society make up its social structure. Social structure is a macrolevel influence because it shapes and determines the overall patterns in which social interaction occurs. Social structure provides an ordered framework for society and for our interactions with others. Social structure comprises statuses, roles, groups, and social institutions.

## LO2 What are the differences among ascribed, achieved, and master statuses?

A status is a specific position in a group or society and is characterized by certain expectations, rights, and duties. Ascribed statuses, such as gender, class, and race/ethnicity, are acquired at birth or involuntarily later in life. Achieved statuses, such as education and occupation, are assumed voluntarily as a result of personal choice, merit, or direct effort. A master status is the most important status a person occupies. For some, occupation is the chief indicator of their status. Occupation provides important clues to a person's educational level, income, and family background. Master statuses confer high or low levels of personal worth and dignity on people.

## LO3 How are role, role expectation, role performance, role conflict, role strain, and role exit alike or different?

A role is a set of behavioral expectations associated with a given status. Role expectation is a group's definition of the way that a specific role ought to be played, whereas role performance is how a person actually plays the role. When role conflict occurs, we may feel pulled in different directions. To deal with this problem, we may prioritize our roles and first complete the one we consider to be most important. Role conflict may occur as a result of changing statuses and roles in society. Role strain occurs when incompatible demands are built into a single status that a person occupies. For example, married women might feel role strain when they have the majority of responsibility to work full time, manage household duties, and take care of the family. Role exit occurs when people disengage from social roles that have been central to their self-identity. Role exit is a four-stage process, ending with the creation of a new identity.

## LO4 What are the functionalist and conflict perspectives on social institutions?

According to functionalist theorists, social institutions perform several prerequisites of all societies: replace members; teach new members; produce, distribute, and consume goods and services; preserve order; and provide and maintain a sense of purpose. Conflict theorists suggest that social institutions do not work for the common good of all individuals: Institutions may enhance and uphold the power of some groups but exclude others, such as the homeless.

## LO5 How does social change occur in preindustrial, industrial, and postindustrial societies?

According to Emile Durkheim, although changes in social structure may dramatically affect individuals and groups, societies manage to maintain some degree of stability. According to Durkheim, social solidarity derives from a society's social structure, which, in turn, is based on the society's division of labor. People in preindustrial societies are united by mechanical solidarity because they have shared values and common social bonds. As societies industrialized and developed more-specialized economic activities, social solidarity came to be rooted in the members' shared dependence on one another. Industrial societies are characterized by organic solidarity, which refers to the cohesion that results when people perform specialized tasks and are united by mutual dependence.

## LO6 What are the symbolic interactionist views on the social construction of reality and the self-fulfilling prophecy?

Symbolic interactionists refer to the social construction of reality as the process by which our perception of reality is largely shaped by the subjective meaning that we give to an experience. We analyze a social context in which we find ourselves, determine what is in our best interest, and adjust our attitudes and actions accordingly. This process can result in a self-fulfilling prophecy—a false belief or prediction that produces behavior that makes the originally false belief come true.

## LO7 What are dramaturgical analysis and ethnomethodology?

According to Erving Goffman's dramaturgical analysis, our daily interactions are similar to dramatic productions. Presentation of self refers to efforts to present our own self to others in ways that are most favorable to our interests or self-image. Ethnomethodology is the study of the common-sense knowledge that people use to understand the situations in which they find themselves. Ethnomethodologists frequently break "rules" or act as though they do not understand some basic rule of social life so that they can observe other people's responses.

**LO8 How do the sociology of emotions and the study of nonverbal communication add to our understanding of human behavior?**

Our emotions are not always private, and specific emotions may be demanded of us on certain occasions. Feeling rules shape the appropriate emotions for a given role or specific situation. Nonverbal communication is the transfer of information between persons without the use of words. It establishes the relationship among people in terms of their responsiveness to and power over one another.

# key TERMS

achieved status 108

agrarian society 117

ascribed status 107

division of labor 120

dramaturgical analysis 126

ethnomethodology 125

face-saving behavior 126

formal organization 113

*Gemeinschaft* 121

*Gesellschaft* 121

horticultural society 116

hunting and gathering society 115

impression management (presentation of self) 126

industrial society 118

master status 108

mechanical solidarity 121

nonverbal communication 129

organic solidarity 121

pastoral society 116

personal space 131

postindustrial society 119

primary group 113

role 110

role conflict 110

role exit 112

role expectation 110

role performance 110

role strain 111

secondary group 113

self-fulfilling prophecy 125

social construction of reality 124

social group 113

social institution 114

social interaction 104

social script 126

social structure 104

status 106

status set 107

status symbol 109

# questions for CRITICAL THINKING

1. Think of a person you know well who often irritates you or whose behavior grates on your nerves. First, list that person's statuses and roles. Then analyze the person's possible role expectations, role performance, role conflicts, and role strains. Does anything you find in your analysis help to explain the irritating behavior? How helpful are the concepts of social structure in analyzing individual behavior?

2. You are conducting field research on gender differences in nonverbal communication styles. How are you going to account for variations among age, race, and social class?

3. When communicating with other genders, races, and ages, is it better to express and acknowledge different styles or to develop a common, uniform style?

# 5 GROUPS AND ORGANIZATIONS

## LEARNING OBJECTIVES

**LO1** Explain what constitutes a social group as opposed to an aggregate or a category.

**LO2** Distinguish among ingroups, outgroups, and reference groups, and give an example of each.

**LO3** Discuss how group size shapes members' communication, leadership styles, and pressures to conform.

**LO4** Applying the concept of groupthink, describe how people often respond differently in a group context than they might if they were alone.

**LO5** Identify the three categories of formal organizations and state how they differ in membership.

**LO6** Debate the strengths and weaknesses of bureaucracies in contemporary nations such as the United States.

**LO7** Define the iron law of oligarchy and apply the concept to a brief analysis of the U.S. government.

**LO8** Identify alternative forms of organization that exist today in nations such as Japan.

Marmaduke St. John/Alamy

137

# Social Media and the Classroom

**At my university, professors are divided about whether they should meddle** [with students who bring smartphones, iPads, and computers to class]. Our students,

Although books are still an integral part of higher education, computers and other digital technologies are rapidly changing the social and learning environments of today's colleges and universities.

some say, are grown-ups. It is not for us to dictate how they take notes or to get involved if they let their attention wander from class-related materials. But when I stand in back of our Wi-Fi enabled lecture halls, students are on Facebook and YouTube, and they are shopping, mostly for music. I want to engage my students in conversation. I don't think they should use class time for any other purpose. One year, I raised the topic for general discussion and suggested using notebooks (the paper kind) for note taking. Some of my students claimed to be relieved. "Now I won't be tempted by Facebook messages," said one sophomore. Others were annoyed, almost surly. . . . I maintained my resolve, but the following year, I bowed to common practice and allowed students to do what they wished.

—Sherry Turkle (2011), a professor at MIT, describing her feelings about students' use of digital technology in classrooms where professors are attempting to form groups and build community among students

A ccording to sociologists, we need groups and organizations—just as we need culture and socialization—to live and participate in a society. Historically, the basic premise of groups and organizations was that individuals engage in face-to-face interactions in order to be part of such a group; however, millions of people today communicate with others through the Internet, cell phones, and other forms of information technology that make it possible for them to "talk" with individuals they have never met and who may live thousands of miles away. A variety of networking websites, including Facebook, LinkedIn, Twitter, and Google+, now compete with or—in some cases—replace live, person-to-person communications. For many college students, Facebook has become an enjoyable way to get to know other people, to join online groups with similar interests or activities, and to plan "real-life" encounters. Despite the wealth of information and opportunities for new social connections that such websites offer, many of our daily activities require that we participate in social groups and formal organizations where *face time*—time spent interacting with others on a face-to-face basis, rather than via Internet or cell phone—is necessary.

What do social groups and formal organizations mean to us in an age of rapid telecommunications? What is the relationship between information and social organizations in societies such as ours? How can we balance the information that we provide to other people about us with our own right to privacy and need for security? These questions are of interest to sociologists who seek to apply the sociological imagination to their studies of social groups, bureaucratic organizations, social networking, and virtual communities. Before we take a closer look at groups and organizations, work through the "Sociology and Everyday Life" quiz on issues pertaining to personal privacy in groups and organizations.

## Social Groups

**LO1** Three strangers are standing at a street corner waiting for a traffic light to change. Do they constitute a group? Five hundred

# How Much Do You Know About Privacy in Groups and Organizations?

| True | False | |
|------|-------|---|
| T | F | 1. A college student's privacy is protected when using a school-owned computer as long as he or she deletes from the computer all e-mails or other documents he or she has worked on and thus prevents anyone else from examining those documents. |
| T | F | 2. Parents of students at all U.S. colleges and universities are entitled to obtain a transcript of their children's college grades, regardless of the student's age. |
| T | F | 3. If you work for a business that monitors phone calls with a pen register (an electronic device that records information about calls to or from a particular phone extension), your employer has the right to maintain and examine a list of phone numbers dialed by your extension and how long each call lasted. |
| T | F | 4. Members of a high school football team can be required to submit to periodic, unannounced drug testing. |

| True | False | |
|------|-------|---|
| T | F | 5. A company has the right to keep its employees under video surveillance anywhere at the company's place of business—even in the restrooms. |
| T | F | 6. A professor can legally post students' grades in public, using each student's Social Security number as an identifier, as long as the student's name does not appear with the number. |
| T | F | 7. Students at a church youth-group meeting who hear one member of the group confess to an illegal act can be required to divulge what that member said. |
| T | F | 8. If you apply for a job at a company that has more than 25 employees, your employer can require that you provide a history of your medical background or take a physical examination prior to offering you a job. |

Answers on page 140.

women and men are first-year graduate students at a university. Do they constitute a group? In everyday usage, we use the word *group* to mean any collection of people. According to sociologists, however, the answer to these questions is no; individuals who happen to share a common feature or to be in the same place at the same time do not constitute social groups.

## Groups, Aggregates, and Categories

As we saw in Chapter 4, a *social group* is a collection of two or more people who interact frequently with one another, share a sense of belonging, and have a feeling of interdependence. Several people waiting for a traffic light to change constitute an **aggregate—a collection of people who happen to be in the same place at the same time but share little else in common.** Shoppers in a department store and passengers on an airplane flight are also examples of aggregates. People in aggregates share a

common purpose (such as purchasing items or arriving at their destination) but generally do not interact with one another, except perhaps briefly. The first-year graduate students, at least initially, constitute a *category*—**a number of people who may never have met one another but share a similar characteristic, such as education level, age, race, or gender.** Men and women make up categories, as do Native Americans and Latinos/as, and victims of sexual or racial harassment. Categories are not social groups because the people in them do not

**aggregate** a collection of people who happen to be in the same place at the same time but share little else in common.

**category** a number of people who may never have met one another but share a similar characteristic, such as education level, age, race, or gender.

**139**

# ANSWERS to the Sociology Quiz on Privacy

**1. False.** Deleting an e-mail or other document from a computer does not actually remove it from the computer's memory. Until other files are entered that write over the space where the document was located, experts can retrieve the document that was deleted.

**2. False.** The Family Educational Right to Privacy Act, which allows parents of a student under age 18 to obtain their child's grades, requires the student's consent once he or she has attained age 18. However, that law applies only to institutions that receive federal educational funds.

**3. True.** Telephone numbers called from a company's phone extensions can be recorded on a pen register, and this information can be used by the employer in evaluating the amount of time employees have spent talking with clients—or with other people. However, personal cell phones now provide employees with a way to talk to friends, family, and others without being detected by their employer.

**4. True.** The U.S. Supreme Court has ruled that schools may require students to submit to random drug testing as a condition of participating in extracurricular activities.

**5. False.** An employer may not engage in video surveillance of its employees in situations where they have a reasonable right of privacy. At least in the absence of a sign warning of such surveillance, employees have this right in company restrooms.

**6. False.** The Federal Educational Rights and Privacy Act states that Social Security numbers are "personally identifiable information" that may not be released without written consent from the student. Posting grades by Social Security number violates this provision unless the student has consented to the number being disclosed.

**7. True.** Although confidential communications made privately to a minister, priest, rabbi, or other religious leader (or to an individual the person reasonably believes to hold such a position) generally cannot be divulged without the consent of the person making the communication, this does not apply when other people are present who are likely to hear the statement.

**8. False.** The Americans with Disabilities Act prohibits employers in companies with more than 25 employees from asking job applicants about medical information or requiring a physical examination prior to employment.

---

usually create a social structure or have anything in common other than a particular trait.

Occasionally, people in aggregates and categories form social groups. For instance, people within the category known as "graduate students" may become an aggregate when they get together for orientation to graduate school. Some of them may form social groups as they interact with one another in classes and seminars, find that they have mutual interests and concerns, and develop a sense of belonging to the group. Information technology raises new and interesting questions about what constitutes a group. For example, some people question whether we can form a social group on the Internet (see "'Community' in the Media").

## Types of Groups

**LO2** As you will recall from Chapter 4, groups have varying degrees of social solidarity and structure. This structure is flexible in some groups and more rigid in others. Some groups are small and personal; others are large

and impersonal. We more closely identify with the members of some groups than we do with others.

### Cooley's Primary and Secondary Groups

Sociologist Charles H. Cooley (1963/1909) used the term *primary group* to describe a small, less specialized group in which members engage in face-to-face, emotion-based interactions over an extended period of time. We have primary relationships with other individuals in our primary groups—that is, with our *significant others,* who frequently serve as role models.

In contrast, as you will recall, a *secondary group* is a larger, more specialized group in which the members engage in more-impersonal, goal-oriented relationships for a limited period of time. The size of a secondary group may vary. Twelve students in a graduate seminar may start out as a secondary group but eventually become a primary group as they get to know one another and communicate on a more personal basis. Formal organizations are secondary groups, but they also contain many primary groups

# "COMMUNITY" IN THE MEDIA

## Can You Create a Community on the Internet?

Do you have friends you've never met? Many people believe that they can make friends and establish a community online. Although chat groups are one of the oldest forms of communication in the online community, they remain popular all around the Internet, from search engines to social networking sites (PerfSpot.com, 2011). With chat groups, you can have a real-time conversation with another person, and chat groups now provide people with an option to have a seemingly face-to-face conversation by using a webcam. Many popular chat groups are dedicated to a specific topic, such as support groups for people with certain types of chronic illnesses or people who like to talk about a special interest or hobby.

Although chat groups are often framed as a way to make friends, get dates, and establish a community, as you study sociology you might ask whether this form of "community" is actually a true community. Sociologists define a *social group* as a collection of two or more people who interact frequently with one another, share a sense of belonging, and have a feeling of interdependence. Are we able to form social groups and establish communities with people whom we have never actually met?

Some social scientists believe that virtual communities established on the Internet constitute true communities (see Wellman, 2001). However, sociologists Robyn Bateman Driskell and Larry Lyon think otherwise. According to Driskell and Lyon, although the Internet provides us with the opportunity to share interests with others whom we have not met and to communicate with people we already know, the original concept of community, which "emphasized local place, common ties, and social interaction that is intimate, holistic, and all-encompassing," is lacking (Driskell and Lyon, 2002: 6).

Driskell and Lyon argue that virtual communities on the Internet do not have geographic and social boundaries, are limited in their scope to specific areas of interest, are psychologically detached from close interpersonal ties, and have only limited concern for their "members" (Driskell and Lyon, 2002). In fact, they say, if we spend many hours in social isolation doing impersonal searches for information, the Internet may reduce community rather than enhance it. Even so, Driskell and Lyon believe it is possible that the Internet will create a "weak community replacement" for people based on a virtual community of specialized ties developed by e-mail correspondence and chatroom discussions (Driskell and Lyon, 2002). What do you think?

Colin Young-Wolff/PhotoEdit

Chat groups and other forms of communication on the Internet are extremely popular with millions of people; however, some sociologists question whether we can actually form social groups and true communities on the Internet. Is cyber chat really different from our face-to-face interactions with others?

## reflect & analyze

**Have you had opportunities to gain new friends and build "community" on the Internet? What are the strengths and limitations of virtual communications?**

within them. For example, how many primary groups do you think there are within the secondary-group setting of your college?

**Sumner's Ingroups and Outgroups** All groups set boundaries by distinguishing between insiders who are members and outsiders who are not. Sociologist William Graham Sumner (1959/1906) coined the terms *ingroup* and *outgroup* to describe people's feelings toward members of their own and other groups. An *ingroup* **is a group to which a person belongs and with which the person feels a sense of identity.** An *outgroup* is a group to which **a person does not belong and toward which the person may feel a sense of competitiveness or hostility.** Distinguishing between our ingroups and our outgroups helps us establish our individual identity

**ingroup** a group to which a person belongs and with which the person feels a sense of identity.

**outgroup** a group to which a person does not belong and toward which the person may feel a sense of competitiveness or hostility.

and self-worth. Likewise, groups are solidified by ingroup and outgroup distinctions; the presence of an enemy or a hostile group binds members more closely together.

Group boundaries may be formal, with clearly defined criteria for membership. For example, a country club that requires an applicant for membership to be recommended by four current members and to pay a $50,000 initiation fee has clearly set requirements for its members (see "Sociology Works!"). Club members often develop *consciousness of kind*—a term used by sociologists to describe the awareness that individuals may have when they believe that they share important commonalities with certain other people. Consciousness of kind is strengthened by membership in clubs ranging from country clubs to college sororities, fraternities, and other by-invitation-only college or university social clubs.

Most of us are aware that our ingroups are very important to us: They provide us with a unique sense of identity, but they also give us the ability to exclude those individuals whom we do not want in our inner circle of friends and acquaintances. The early sociologist Max Weber captured this idea in his description of the *closed relationship*—a setting in which the "participation of certain persons is excluded, limited, or subjected to conditions" (Gerth and Mills, 1946: 139). However, friendship groups usually do not have clear guidelines for membership; rather, the boundaries tend to be very informal and vaguely defined.

Ingroup and outgroup distinctions may encourage social cohesion among members, but they may also promote classism, racism, sexism, and ageism. Ingroup members typically view themselves positively and members of outgroups negatively. These feelings of group superiority, or *ethnocentrism,* are somewhat inevitable. Some group members may never act on these beliefs of superiority and inferiority because the larger organization of which they are a part actively discourages ethnocentric beliefs and discriminatory actions. However, other organizations may covertly foster ethnocentrism and negative ingroup/outgroup distinctions by denying that these beliefs exist among group members or by failing to take action when misconduct occurs that is rooted in racism, sexism, and/or ageism. An example is a college Greek letter organization whose national leadership strongly opposes theme parties with racist or sexist overtones sponsored on local campuses, but some affiliates continue to hold social gatherings with decorations, clothing, music, and slogans that ridicule subordinate-group members such as persons of color, older individuals, persons with a disability, or women who have been turned into sex objects. Although campus social organizations often promote social cohesion among

members by making them feel like they are the "in group" and everyone else is in the "out group," such beliefs and practices may also promote classism, racism, sexism, and/or ageism.

**Reference Groups** Ingroups provide us not only with a source of identity but also with a point of reference. A *reference group* **is a group that strongly influences a person's behavior and social attitudes, regardless of whether that individual is an actual member.** When we attempt to evaluate our appearance, ideas, or goals, we automatically refer to the standards of some group. Sometimes, we will refer to our membership groups, such as family or friends. Other times, we will rely on groups to which we do not currently belong but that we might wish to join in the future, such as a social club or a profession.

Reference groups help explain why our behavior and attitudes sometimes differ from those of our membership groups. We may accept the values and norms of a group with which we identify rather than one to which we belong. We may also act more like members of a group that we want to join than members of groups to which we already belong. In this case, reference groups are a source of anticipatory socialization. For most of us, our reference-group attachments change many times during our life course, especially when we acquire a new status in a formal organization.

**Networks** A *network* **is a web of social relationships that links one person with other people and, through them, with other people they know.** Frequently, networks connect people who share common interests but who otherwise might not identify and interact with one another. For example, if A is tied to B, and B is tied to C, then a network is more likely to be formed among individuals A, B, and C. Today, the term *networking* is widely used to describe the contacts that people make to find jobs or other opportunities; however, sociologists have studied social networks for many years in an effort to learn more about the linkages between individuals and their group memberships.

What are your networks? For a start, your networks consist of all the people linked to you by primary ties, including your relatives and close friends. Your networks also include your secondary ties, such as acquaintances, classmates, professors, and—if you are employed—your supervisor and coworkers. However, your networks actually extend far beyond these ties to include not only the people that you *know* but also the people that you *know of*—and who know of you— through your primary and secondary ties. In fact, your networks potentially include a pool of between 500 and 2,500 acquaintances if you count the connections of everyone in your networks (Milgram, 1967).

# Sociology WORKS!

## Ingroups, Outgroups, and "Members Only" Clubs

In this country we have a God-given right to associate with whomever we please. And frankly, this includes my right to *not* associate with people I don't want to. If I don't want to be around somebody, why should I have to let them in my club? Let them go start their own club.

　—Phil, a white, male attorney who is a member of several prestigious private clubs, explaining why he believes he has the right to establish his own ingroup through private club memberships (qtd. in Kendall, 2008)

A key characteristic of the city clubs and country clubs where Phil is a member is that each organization has formal group boundaries, with people becoming members "by invitation only." In other words, prospective members must be nominated by current members and be voted into the club: They cannot simply decide to join the organization. For this reason, people who are invited to join typically feel special (like "insiders") because they know that club membership is not available to everyone. Such exclusive clubs also typically have signs posted on gates, fences, or buildings that state "Members Only." These organizations do not welcome outsiders within their walls, and members are often pledged to loyalty and secrecy about their club's activities.

Similarly, many college fraternities and sororities thrive on rituals, secrecy, and the importance of what it means to pledge—to have accepted a bid to join but not having yet been initiated into—the group of one's choice (Robbins, 2004: 342). For many students, joining a "members only" sorority or fraternity is a means of establishing an ingroup to which they will belong throughout their years in school.

Will such "members only" clubs remain relevant in the future? Recent studies on private clubs and exclusive college social organizations show that the sociological concepts of "ingroup" and "outgroup" remain highly relevant today when we conduct research on the processes of inclusion and exclusion to learn more about how such activities affect individuals and groups (see Kendall, 2008).

### reflect & analyze

**What areas of sociological research or personal interest can you think of that might benefit from applying the ingroup/outgroup concept to your analysis? How might these concepts be applied to other areas of college life besides invitational social organizations?**

**Sometimes, the distinction between what constitutes an ingroup and an outgroup is subtle. Other times, it is not subtle at all. Would you feel comfortable entering or joining this club?**

## The Purpose of Groups: Multiple Perspectives

What purpose do groups serve? Why are individuals willing to relinquish some of their freedom to participate in groups? According to functionalists, people form groups to meet instrumental and expressive needs. *Instrumental*, or task-oriented, needs cannot always be met by one person, so the group works cooperatively to fulfill a specific goal.

**reference group** a group that strongly influences a person's behavior and social attitudes, regardless of whether that individual is an actual member.

**network** a web of social relationships that links one person with other people and, through them, with other people they know.

**FIGURE 5.1** According to the sociologist Georg Simmel, interaction patterns change when a third person joins a dyad—a group composed of two members. How might the conversation between these two women have changed when the man arrived to talk to them?

Groups help members do jobs that are impossible to do alone or that would be very difficult and time-consuming at best. For example, think of how hard it would be to function as a one-person football team or to single-handedly build an apartment complex. In addition to instrumental needs, groups also help people meet their *expressive,* or emotional, needs, especially those involving self-expression and support from family, friends, and peers.

Although not disputing that groups ideally perform such functions, conflict theorists suggest that groups also involve a series of power relationships through which the needs of individual members may not be equally served. Symbolic interactionists focus on how the size of a group influences the kind of interaction that takes place among members. To many postmodernists, groups and organizations—like other aspects of postmodern societies—are generally characterized by superficiality and depthlessness in social relationships. One postmodern thinker who focuses on this issue is the literary theorist Fredric Jameson, who believes that people experience a waning of emotion in organizations where fragmentation and superficiality are a way of life. For example, fast-food restaurant employees and customers interact in extremely superficial ways that are largely scripted: The employees follow scripts in taking and filling customers' orders ("Would you like fries and a drink with that?"), and the customers respond with their own routine reactions. What examples can you think of that fit this description?

## Group Characteristics and Dynamics

**LO3** We will now look at certain characteristics of groups, such as how size affects group dynamics.

### Group Size

The size of a group is one of its most important features. Interactions are more personal and intense in a *small group,* **a collectivity small enough for all members to be acquainted with one another and to interact simultaneously.**

Sociologist Georg Simmel (1950/1902–1917) suggested that small groups have distinctive interaction patterns that do not exist in larger groups. According to Simmel, in a *dyad*—**a group composed of two members**—the active participation of both members is crucial to the group's survival. If one member withdraws from interaction or "quits," the group ceases to exist. Examples of dyads include best friends, married couples, and domestic partnerships. Dyads provide members with an intense bond and a sense of unity not found in most larger groups.

When a third person is added to a dyad, a *triad,* **a group composed of three members,** is formed. The nature of the relationship and interaction patterns changes with the addition of the third person (Figure 5.1). In a triad, even if one member ignores another or declines to participate, the group can still

**144**

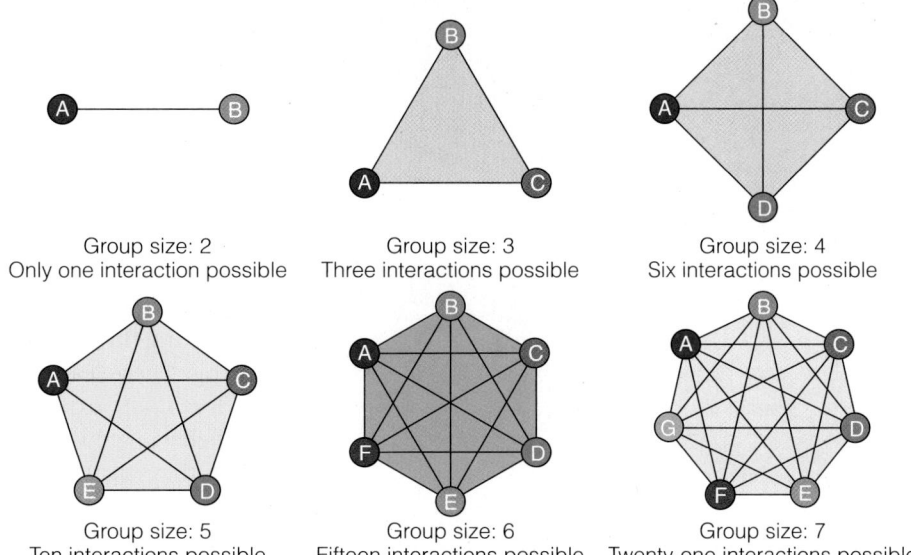

**FIGURE 5.2** GROWTH OF POSSIBLE SOCIAL INTERACTION BASED ON GROUP SIZE
© Cengage Learning

function. In addition, two members may unite to create a coalition that can subject the third member to pressure to conform. A *coalition* is an alliance created in an attempt to reach a shared objective or goal. If two members form a coalition, the other member may be seen as an outsider or intruder.

As the size of a group increases beyond three people, members tend to specialize in different tasks, and everyday communication patterns change. For instance, in groups of more than six or seven people, it becomes increasingly difficult for everyone to take part in the same conversation; therefore, several conversations will probably take place simultaneously. Members are also likely to take sides on issues and form a number of coalitions. In groups of more than ten or twelve people, it becomes virtually impossible for all members to participate in a single conversation unless one person serves as moderator and guides the discussion. As shown in Figure 5.2, when the size of the group increases, the number of possible social interactions also increases.

Although large groups typically have less social solidarity than small ones, they may have more power. However, the relationship between size and power is more complicated than it might initially seem. The power relationship depends on both a group's *absolute* size and its *relative* size (Simmel, 1950/1902–1917; Merton, 1968). The absolute size is the number of members the group actually has; the relative size is the number of potential members. For example, suppose that 300 people band together to "march on Washington" and demand enactment of a law on some issue that they feel is important. Although 300 people is a large number in some contexts, opponents of this group would argue that the low turnout (compared with the number of people in this country) demonstrates that most people don't believe the issue is important. At the same time, the power of a small group to demand change may be based on a "strength in numbers" factor if the group is seen as speaking on behalf of a large number of other people (who are also voters).

Larger groups typically have more-formalized leadership structures, and their leaders are expected to perform a variety of roles, some related to the internal workings of the group and others related to external relationships with other groups.

## Group Leadership

What role do leaders play in groups? **Leadership refers to the ability to influence what goes on in a group or social system.** Leaders are responsible for directing plans and activities so that the group completes its task or fulfills its goals. Primary groups generally have informal leadership. For example, most of us do not elect or appoint leaders in our own families. Various family members may assume a leadership

**small group** a collectivity small enough for all members to be acquainted with one another and to interact simultaneously.

**dyad** a group composed of two members.

**triad** a group composed of three members.

**leadership** the ability to influence what goes on in a group or social system.

**FIGURE 5.3** Organizations have different leadership styles based on the purpose of the group. How do leadership styles in the military differ from those on college and university campuses and in office workplaces, for example?

role at various times or act as leaders for specific tasks. In traditional families, the father or eldest male is usually the leader. However, in today's more-diverse families, leadership and power are frequently in question, and power relationships may be quite different, as discussed later in this text. By comparison, larger groups typically have more-formalized leadership structures. For example, leadership in secondary groups (such as colleges, governmental agencies, and corporations) involves a clearly defined chain of command, with written responsibilities assigned to each position in the organizational structure.

**Leadership Functions** Both primary and secondary groups have some type of leadership or positions that enable certain people to be leaders, or at least to wield power over others. From a functionalist perspective, if groups exist to meet the instrumental and expressive needs of their members, then leaders are responsible for helping the group meet those needs. *Instrumental leadership* **is goal or task oriented;** this type of leadership is most appropriate when the group's purpose is to complete a task or reach a particular goal. *Expressive leadership* **provides emotional support for members;** this type of leadership is most appropriate when the group is dealing with emotional issues and when harmony, solidarity, and high morale are needed. Both kinds of leadership are needed for groups to work effectively.

**Leadership Styles** Three major styles of leadership exist in groups: authoritarian, democratic, and laissez-faire. *Authoritarian leaders* **make all major group decisions and assign tasks to members.** These leaders focus on the instrumental tasks of

the group and demand compliance from others. In times of crisis, such as a war or natural disaster, authoritarian leaders may be commended for their decisive actions (Figure 5.3). In other situations, however, they may be criticized for being dictatorial and for fostering intergroup hostility. By contrast, *democratic leaders* **encourage group discussion and decision making through consensus building.** These leaders may be praised for their expressive, supportive behavior toward group members, but they may also be blamed for being indecisive in times of crisis.

*Laissez-faire* literally means "to leave alone." *Laissez-faire leaders* **are only minimally involved in decision making and encourage group members to make their own decisions.** On the one hand, laissez-faire leaders may be viewed positively by group members because they do not flaunt their power or position. On the other hand, a group that needs active leadership is not likely to find it with this style of leadership, which does not work vigorously to promote group goals.

Studies of kinds of leadership and decision-making styles have certain inherent limitations. They tend to focus on leadership that is imposed externally on a group (such as bosses or political leaders) rather than leadership that arises within a group. Different decision-making styles may be more effective in one setting than another. For example, imagine attending a college class in which the professor asked the students to determine what should be covered in the course, what the course requirements should be, and how students should be graded. It would be a difficult and cumbersome way to start the semester; students might spend the entire term negotiating these matters and never actually learn anything.

## Group Conformity

To what extent do groups exert a powerful influence on our lives? Groups have a significant amount of influence on our values, attitudes, and behavior. In order to gain and then retain our membership in groups, most of us are willing to exhibit a high level of conformity to the wishes of other group members. *Conformity* **is the process of maintaining or changing behavior to comply with the norms established by a society, subculture, or other group.** We often experience powerful pressure from other group members to conform. In some situations, this pressure may be almost overwhelming.

Researchers have found that the pressure to conform may cause group members to say they see something that is contradictory to what they are actually seeing or to do something that they would otherwise be unwilling to do. Conforming to group pressure begins as early as preschool age (Haun and Tomasello, 2011). As we look at two classic studies of group conformity (which would be impossible to conduct today for ethical reasons), ask yourself what you might have done if you had been involved in this research.

**Asch's Research** Pressure to conform is especially strong in small groups in which members want to fit in with the group. In a series of experiments conducted by Solomon Asch (1955, 1956), the pressure toward group conformity was so great that participants were willing to contradict their own best judgment if the rest of the group disagreed with them.

One of Asch's experiments involved groups of undergraduate men (seven in each group) who were allegedly recruited for a study of visual perception. All the men were seated in chairs. However, the person in the sixth chair did not know that he was the only actual subject; all the others were assisting the researcher. The participants were first shown a large card with a vertical line on it and then a second card with three vertical lines (see Figure 5.4). Each of the seven participants was asked to indicate which of the three lines on the second card was identical in length to the "standard line" on the first card.

In the first test with each group, all seven men selected the correct matching line. In the second trial, all seven still answered correctly. In the third trial, however, the actual subject became very uncomfortable when all the others selected the incorrect line. The subject could not understand what was happening and became even more confused as the others

continued to give incorrect responses on eleven out of the next fifteen trials.

Asch (1955) found that about one-third of all subjects chose to conform by giving the same (incorrect) responses as Asch's assistants. In discussing the experiment afterward, most of the subjects who gave incorrect responses indicated that they had known the answers were wrong but decided to go along with the group in order to avoid ridicule or ostracism.

Asch concluded that the size of the group and the degree of social cohesion felt by participants were important influences on the extent to which individuals respond to group pressure. If you had been in the position of the subject, how would you have responded? Would you have continued to give the correct answer, or would you have been swayed by the others?

**Milgram's Research** How willing are we to do something because someone in a position of authority has told us to do it? How far are we willing to go to follow the demands of that individual? Approximately fifty years ago, Stanley Milgram (1963, 1974) conducted a series of controversial experiments to find answers to these questions about people's obedience to authority. *Obedience* is a form of compliance in which people follow direct orders from someone in a position of authority.

Milgram's subjects were men who had responded to an advertisement for participants in an experiment. When the first (actual) subject arrived at the Yale University lab, he was told that the study concerned the effects of punishment on learning. After the second subject (an assistant of Milgram's) arrived, the two men were instructed to draw slips of paper from a hat to get their assignments as either the "teacher" or the "learner." Because the drawing was rigged, the actual subject always became

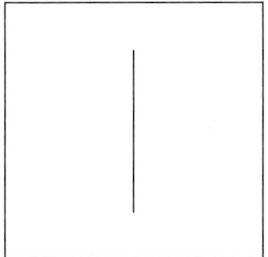

**FIGURE 5.4** ASCH'S CARDS
Although Line 2 is clearly the same length as the line in the lower card, Solomon Asch's research assistants tried to influence "actual" participants by deliberately picking Line 1 or Line 3 as the correct match. Many of the participants went along rather than risking the opposition of the "group."

Source: Asch, 1955. © Cengage Learning

**instrumental leadership** goal- or task-oriented leadership.

**expressive leadership** leadership that provides emotional support for members.

**authoritarian leaders** leaders who make all major group decisions and assign tasks to members.

**democratic leaders** leaders who encourage group discussion and decision making through consensus building.

**laissez-faire leaders** leaders who are only minimally involved in decision making and encourage group members to make their own decisions.

**conformity** the process of maintaining or changing behavior to comply with the norms established by a society, subculture, or other group.

the teacher, and the assistant the learner. Next, the learner was strapped into a chair with protruding electrodes that looked something like an electric chair. The teacher was placed in an adjoining room and given a realistic-looking but nonoperative shock generator. The "generator's" control panel showed levels that went from "Slight shock" (15 volts) on the left, to "Intense shock" (255 volts) in the middle, to "Danger: severe shock" (375 volts), and finally "XXX" (450 volts) on the right.

The teacher was instructed to read aloud a pair of words and then repeat the first of the two words. At that time, the learner was supposed to respond with the second of the two words. If the learner could not provide the second word, the teacher was instructed to press the lever on the shock generator so that the learner would be punished for forgetting the word. Each time the learner gave an incorrect response, the teacher was supposed to increase the shock level by 15 volts. The alleged purpose of the shock was to determine whether punishment improves a person's memory.

What was the maximum level of shock that a "teacher" was willing to inflict on a "learner"? The learner had been instructed (in advance) to beat on the wall between him and the teacher as the experiment continued, pretending that he was in intense pain. The teacher was told that the shocks might be "extremely painful" but that they would cause no permanent damage. At about 300 volts, when the learner quit responding at all to questions, the teacher often turned to the experimenter to see what he should do next. When the experimenter indicated that the teacher should give increasingly painful shocks,

65 percent of the teachers administered shocks all the way up to the "XXX" (450-volt) level (see Figure 5.5). By this point in the process, the teachers were frequently sweating, stuttering, or biting on their lip. According to Milgram, the teachers (who were free to leave whenever they wanted to) continued in the experiment because they were being given directions by a person in a position of authority (a university scientist wearing a white coat).

What can we learn from Milgram's study? The study provides evidence that obedience to authority may be more common than most of us would like to believe. None of the "teachers" challenged the process before they had applied 300 volts. Almost two-thirds went all the way to what could have been a deadly jolt of electricity if the shock generator had been real. For many years, Milgram's findings have been found to be consistent in studies conducted both in the United States and in other countries and with variations in the research design (Burger, 2009; Tavris, 2013).

This research once again raises some questions concerning research ethics. As was true of Asch's research, Milgram's subjects were deceived about the nature of the study in which they were asked to participate. Many of them found the experiment extremely stressful, and subjects were not informed that they were not actually administering escalating levels of shock (Perry, 2013). Such conditions cannot be ignored by social scientists because subjects may receive lasting emotional scars from this kind of research. Today, it would be virtually impossible to obtain permission to replicate this experiment in a university setting.

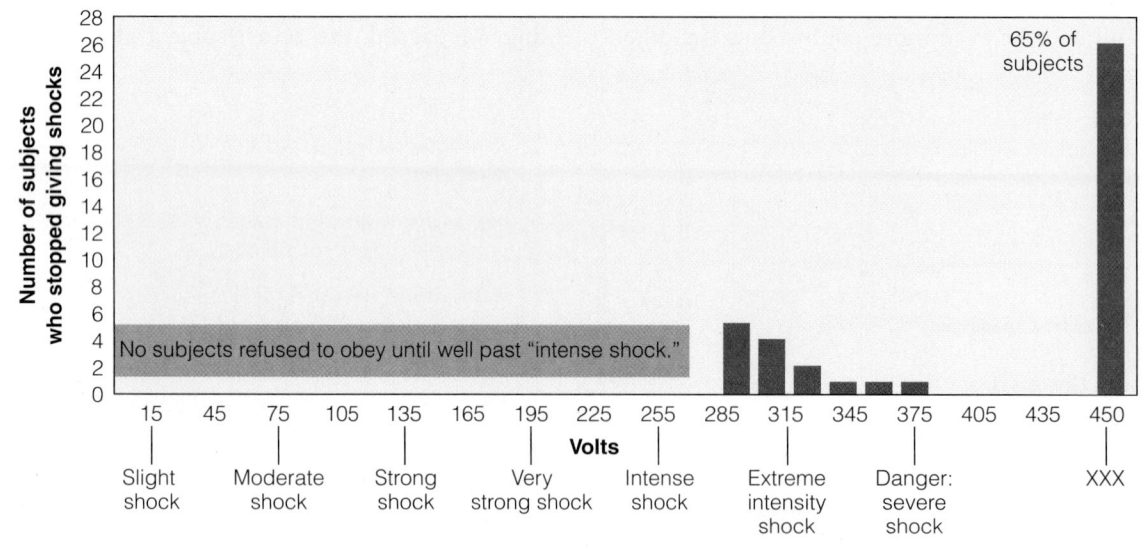

**FIGURE 5.5** RESULTS OF MILGRAM'S OBEDIENCE EXPERIMENT
Even Milgram was surprised by subjects' willingness to administer what they thought were severely painful and even dangerous shocks to a helpless "learner."

Source: Milgram, 1963. © Cengage Learning

**Process of Groupthink**　　　　　**Example: Deepwater Horizon Explosion**

| PRIOR CONDITIONS | |
|---|---|
| Isolated, cohesive, homogeneous decision-making group<br><br>Lack of impartial leadership<br><br>High stress | Millions of dollars had been spent on production of mobile offshore drilling unit; BP was running behind schedule and was under pressure to complete work despite reports of a leak in the rig's blowout preventer. |
| **SYMPTOMS OF GROUPTHINK** | |
| Closed-mindedness<br><br>Rationalization<br><br>Squelching of dissent<br><br>"Mindguards"<br><br>Feelings of righteousness and invulnerability | Although rig workers reported pieces of rubber seal coming loose, superiors stated this happened often. Superiors closed off debate by saying blowout-preventer problem would be resolved if something went wrong. |
| **DEFECTIVE DECISION MAKING** | |
| Incomplete examination of alternatives<br><br>Failure to examine risks and contingencies<br><br>Incomplete search for information | In March 2010, when rig workers informed superiors of actual leaks in the gasket on blowout preventer on the rig, no decision was made to repair the rubber seal or to stop work. Superiors began to hide bad news, and decisions were made without a clear sense of what risks were involved. |
| **CONSEQUENCES** | |
| Poor decisions | No one stopped production. The BP oil rig exploded, killing 11 and injuring 17. The spill was the largest of its kind in the history of the petroleum industry. BP, Transocean, and others were blamed for making a series of bad decisions based on money, time pressures, and too many people thinking alike. |

**FIGURE 5.6** JANIS'S DESCRIPTION OF GROUPTHINK

In Janis's model, prior conditions such as a highly homogeneous group with committed leadership can lead to potentially disastrous "groupthink," which short-circuits careful and impartial deliberation. Events leading up to the tragic 2010 explosion of the BP oil rig have been cited as an example of this process.

Source: Mackin, 2010. © Cengage Learning

AP Images/Anonymous/US Coast Guard

## Groupthink

**LO4** As we have seen, individuals often respond differently in a group context than they might if they were alone. Social psychologist Irving Janis (1972, 1989) examined group decision making among political experts and found that major blunders in U.S. history can be attributed to pressure toward group conformity. To describe this phenomenon, he coined the term *groupthink*—**the process by which members of a cohesive group arrive at a decision that many individual members privately believe is unwise.** Why not speak up at the time? Members usually want to be "team players." They may not want to be the ones who undermine

**groupthink** the process by which members of a cohesive group arrive at a decision that many individual members privately believe is unwise.

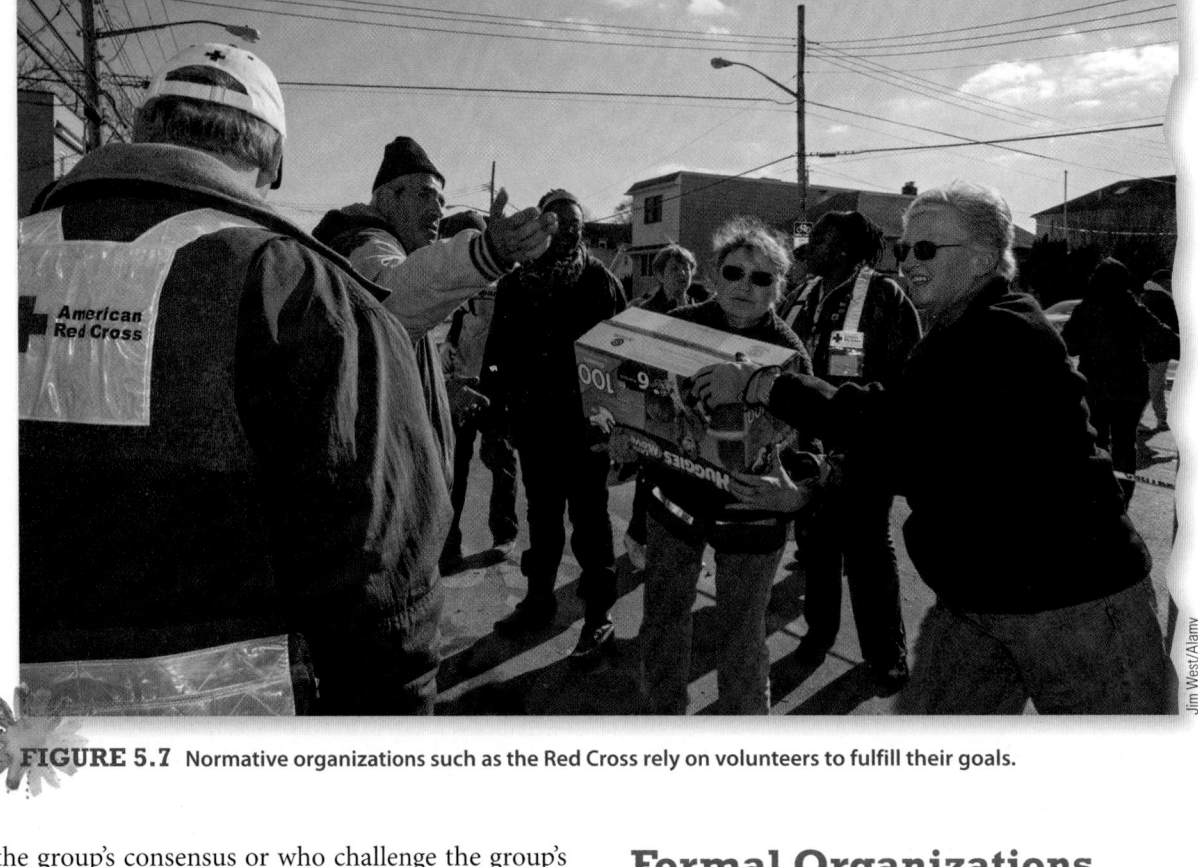

**FIGURE 5.7** Normative organizations such as the Red Cross rely on volunteers to fulfill their goals.

the group's consensus or who challenge the group's leaders. Consequently, members often limit or withhold their opinions and focus on consensus rather than on exploring all of the options and determining the best course of action. Figure 5.6 summarizes the dynamics and results of groupthink.

The tragic 2010 explosion of the BP Deepwater Horizon oil rig, owned by British Petroleum and located in the Gulf of Mexico, is an example of this process. Errors in decision making contributed to one of the worst oil spills and marine and wildlife disasters in U.S. history. Eleven people were killed and seventeen were injured in the rig explosion, and it is impossible to estimate the full extent of the damage done to the Gulf Coast and the fishing and tourism industries because of this massive accident. Why is this disaster an example of groupthink? Because officials for BP, Transocean, and Halliburton, the major transnational corporations responsible for this error in decision making, closed off their discussions about safety and hid bad news from one another and public officials; because they began to think alike in their assumptions about safety, namely that a blowout preventer would keep such a massive disaster from occurring; and because their companies were already behind schedule, had put millions of dollars into production, and did not want to stop to check out reports that a rubber safety seal was broken.

# Formal Organizations in Global Perspective

**LO5** Over the past century, the number of formal organizations has increased dramatically in the United States and other industrialized nations. Previously, everyday life was centered in small, informal, primary groups, such as the family and the village. With the advent of industrialization and urbanization (as discussed in Chapter 1), people's lives became increasingly dominated by large, formal, secondary organizations. A *formal organization,* you will recall, is a highly structured secondary group formed for the purpose of achieving specific goals in the most efficient manner. Formal organizations (such as corporations, schools, and government agencies) usually keep their basic structure for many years in order to meet their specific goals.

## Types of Formal Organizations

We join some organizations voluntarily and others out of necessity. Sociologist Amitai Etzioni (1975) classified formal organizations into three categories—normative, coercive, and utilitarian—based on the nature of membership in each.

**Normative Organizations** We voluntarily join *normative organizations* when we want to pursue

## Characteristics of Groups and Organizations

| | | |
|---|---|---|
| **Types of social groups** | Primary group | Small, less specialized group in which members engage in face-to-face, emotion-based interaction over an extended period of time |
| | Secondary group | Larger, more specialized group in which members engage in more-impersonal, goal-oriented relationships for a limited period of time |
| | Ingroup | A group to which a person belongs and with which the person feels a sense of identity |
| | Outgroup | A group to which a person does not belong and toward which the person may feel a sense of competitiveness or hostility |
| | Reference group | A group that strongly influences a person's behavior and social attitudes, regardless of whether the person is actually a member |
| **Group size** | Dyad | A group composed of two members |
| | Triad | A group composed of three members |
| | Formal organization | A highly structured secondary group formed for the purpose of achieving specific goals |
| **Types of formal organizations** | Normative | Organizations that we join voluntarily to pursue some common interest or gain personal satisfaction or prestige by joining |
| | Coercive | Associations that people are forced to join (total institutions such as boot camps and prisons are examples) |
| | Utilitarian | Organizations that we join voluntarily when they can provide us with a material reward that we seek |

some common interest or gain personal satisfaction or prestige from being a member (Figure 5.7). Political parties, ecological activist groups, religious organizations, parent–teacher associations, and college sororities and fraternities are examples of normative, or voluntary, associations.

Class, gender, and race are important determinants of a person's participation in a normative association. Class is the most significant predictor of whether a person will participate in mainstream normative organizations; membership costs may exclude some from joining. Those with higher socioeconomic status are more likely to be not only members but also active participants in these groups. Gender is also an important determinant. Historically, all-male voluntary organizations have had a higher level of prestige than many women's organizations. In the twenty-first century, some of these patterns have changed.

Throughout history, people of all racial–ethnic categories have participated in voluntary organizations to bring about racial equality and social justice. Women have often taken leadership roles in these movements. African American women were actively involved in antislavery societies in the nineteenth century and in the civil rights movement in the twentieth century. Similarly, Native American women participated in the American Indian Movement, a group organized to fight problems ranging from police brutality to housing and employment discrimination. Mexican American women have held a wide range of leadership positions in La Raza Unida Party and the League of United Latin American Citizens, organizations oriented toward civic activities and protests against injustices.

**Coercive Organizations** People do not voluntarily become members of *coercive organizations*—associations that people are forced to join. Total institutions, such as boot camps, prisons, and some mental hospitals, are examples of coercive organizations. As discussed in Chapter 3, the assumed goal of total institutions is to resocialize people through incarceration. These environments are characterized by restrictive barriers (such as locks, bars, and security guards) that make it impossible for people to leave freely. When people leave without being officially dismissed, their exit is referred to as an "escape."

**Utilitarian Organizations** We voluntarily join *utilitarian organizations* when they can provide us with a material reward that we seek. To make a living or earn a college degree, we must participate in organizations that can provide us these opportunities. Although we have some choice regarding where we work or attend school, utilitarian organizations are not always completely voluntary. For example, most people must continue to work even if the conditions of their employment are less than ideal. (This chapter's Concept Quick Review summarizes the types of groups, sizes of groups, and types of formal organizations.)

## Bureaucracies

**LO6** The bureaucratic model of organization remains the most universal organizational form in government, business, education, and religion. A *bureaucracy* **is an organizational model characterized by a hierarchy of authority, a clear division of labor, explicit rules and procedures, and impersonality in personnel matters.**

Sociologist Max Weber (1968/1922) was interested in the historical trend toward bureaucratization that accelerated during the Industrial Revolution. To Weber, bureaucracy was the most "rational" and efficient means of attaining organizational goals because it contributed to coordination and control. According to Weber, *rationality* **is the process by which traditional methods of social organization, characterized by informality and spontaneity, are gradually replaced by efficiently administered formal rules and procedures.** Bureaucracy can be seen in all aspects of our lives, from small colleges with perhaps a thousand students to multinational corporations employing many thousands of workers worldwide.

In his study of bureaucracies, Weber relied on an ideal-type analysis, which he adapted from the field of economics. An *ideal type* **is an abstract model that describes the recurring characteristics of some phenomenon** (such as bureaucracy). To develop this ideal type, Weber abstracted the most characteristic bureaucratic aspects of religious, educational, political, and business organizations. Weber acknowledged that no existing organization would exactly fit his ideal type of bureaucracy.

**Ideal Characteristics of Bureaucracy** Weber set forth several ideal-type characteristics of bureaucratic organizations. His model (see Figure 5.8) highlights the organizational efficiency and productivity that bureaucracies strive for in these five central elements of the ideal organization:

- *Division of labor.* Bureaucratic organizations are characterized by specialization, and each member has highly specialized tasks to fulfill.
- *Hierarchy of authority.* In a bureaucracy, each lower office is under the control and supervision of a higher one. Those few individuals at the top of the hierarchy have more power and exercise more control than do the many at the lower levels. Those who are lower in the hierarchy report to (and often take orders from) those above them in the organizational pyramid. Persons at the upper levels are responsible not only for their own actions but also for those of the individuals they supervise.

| Characteristics | Effects |
|---|---|
| • Division of labor | • Inefficiency and rigidity |
| • Hierarchy of authority | • Resistance to change |
| • Rules and regulations | • Perpetuation of race, class, and |
| • Qualification-based employment |   gender inequalities |
| • Impersonality | |

**FIGURE 5.8** CHARACTERISTICS AND EFFECTS OF BUREAUCRACY
The very characteristics that define Weber's idealized bureaucracy can create or exacerbate the problems that many people associate with this type of organization. Can you apply this model to an organization with which you are familiar?

- *Rules and regulations.* Rules and regulations establish authority within an organization. These rules are typically standardized and provided to members in a written format. In theory, written rules and regulations offer clear-cut standards for determining satisfactory performance so that each new member does not have to reinvent the rules.
- *Qualification-based employment.* Bureaucracies require competence and hire staff members and professional employees based on specific qualifications. Individual performance is evaluated against specific standards, and promotions are based on merit as spelled out in personnel policies.
- *Impersonality.* Bureaucracies require that everyone must play by the same rules and be treated the same. Personal feelings should not interfere with organizational decisions.

**Contemporary Applications of Weber's Theory** How well do Weber's theory of rationality and his ideal-type characteristics of bureaucracy withstand the test of time? More than 100 years later, many organizational theorists still apply Weber's perspective. For example, the sociologist George Ritzer used Weber's theories to examine fast-food restaurants such as McDonald's. According to Ritzer, the process of "McDonaldization" has become a global phenomenon that can be seen in fast-food restaurants and other "speedy" or "jiffy" businesses (such as Sir Speedy Printing and Jiffy Lube). *McDonaldization* is the term coined by Ritzer to describe the process of rationalization, which takes a task and breaks it down into smaller tasks. This process is repeated until all tasks have been broken down into the smallest possible level. The resulting tasks are then rationalized to find the single most efficient method for completing each task. The result is an efficient, logical sequence of methods that can be completed the same way every time to produce the desired outcome. Ritzer (2013) identifies four dimensions of formal rationality (McDonaldization) found in fast-food restaurants:

- *Efficiency* means the search for the best means to the end; the drive-through window is a good example of heightening the efficiency of obtaining a meal.
- *Predictability* means a world of no surprises; the Big Mac in Los Angeles is indistinguishable from the one in New York; similarly, the one we consume tomorrow or next year will be just like the one we eat today.
- *Emphasis on quantity rather than quality.* The Big Mac is a good example of this emphasis on quantity rather than quality.

- *Control through nonhuman technologies* is exemplified by unskilled cooks following detailed directions and assembly-line methods applied to the cooking and serving of food.

Finally, such a formally rational system brings with it various irrationalities, most notably the dehumanization of the dining experience. For example, people in a McDonaldized world may become more enthusiastic about quickly purchasing extremely large portions of relatively inexpensive foods than with having a "slow" dining experience where people bond over the experience of cooking and eating a more nutritious (not mass-produced) meal with their friends or family. How applicable are some of Weber's ideas today? While still useful, Weber's ideal type largely failed to take into account the informal side of bureaucracy.

**The Informal Side of Bureaucracy** When we look at an organizational chart, the official, formal structure of a bureaucracy is readily apparent. In practice, however, a bureaucracy has patterns of activities and interactions that cannot be accounted for by its organizational chart. These have been referred to as *bureaucracy's other face* (Page, 1946).

The ***informal side of a bureaucracy* is composed of those aspects of participants' day-to-day activities and interactions that ignore, bypass, or do not correspond with the official rules and procedures of the bureaucracy.** An example is an informal "grapevine" that spreads information (with varying degrees of accuracy) much faster than do official channels of communication, which tend to be slow and unresponsive. The informal structure has also been referred to as *work culture* because it includes the ideology and practices of workers on the job. Workers create this work culture in order to confront, resist, or adapt to the constraints of their jobs,

**bureaucracy** an organizational model characterized by a hierarchy of authority, a clear division of labor, explicit rules and procedures, and impersonality in personnel matters.

**rationality** the process by which traditional methods of social organization, characterized by informality and spontaneity, are gradually replaced by efficiently administered formal rules and procedures.

**ideal type** an abstract model that describes the recurring characteristics of some phenomenon.

**informal side of a bureaucracy** those aspects of participants' day-to-day activities and interactions that ignore, bypass, or do not correspond with the official rules and procedures of the bureaucracy.

as well as to guide and interpret social relations on the job. Today, computer networks and e-mail offer additional opportunities for workers to enhance or degrade their work culture. Some organizations have sought to control offensive communications so that workers will not be exposed to a hostile work environment brought about by colleagues, but such control has raised significant privacy issues (see "Sociology and Social Policy").

Is the informal side of bureaucracy good or bad? Should it be controlled or encouraged? Two schools of thought have emerged with regard to these questions. One approach emphasizes control of informal groups in order to ensure greater worker productivity. By contrast, the other school of thought asserts that informal groups should be nurtured because such networks may serve as a means of communication and cohesion among individuals (Figure 5.9). Large organizations would be unable to function without strong informal norms and relations among participants.

Informal networks continue to thrive in contemporary organizations because e-mail and the Internet have made it possible for people to communicate throughout the day without ever having to engage in face-to-face interaction. The need to meet at the water fountain or the copy machine in order to exchange information is long gone: Workers now have an opportunity to tell one another—and higher-ups, as well—what they think.

## Problems of Bureaucracies

The characteristics that make up Weber's "rational" model of bureaucracy have a dark side that has frequently given this type of organization a bad name. Three of the major problems of bureaucracies are (1) inefficiency and rigidity, (2) resistance to change, and (3) perpetuation of race, class, and gender inequalities.

**FIGURE 5.9** How do people use this informal "grapevine" to spread information? Is it faster than the organization's official channels of communication? Is it more or less accurate than official channels?

© iStockphoto.com/Milenko Bokan

**Inefficiency and Rigidity** Bureaucracies experience inefficiency and rigidity at both the upper and lower levels of the organization. The self-protective behavior of officials at the top may render the organization inefficient. One type of self-protective behavior is the monopolization of information in order to maintain control over subordinates and outsiders. Information is a valuable commodity in organizations, and those persons in positions of authority guard information because it is a source of power for them—others cannot "second-guess" their decisions without access to relevant (and often "confidential") information.

When those at the top tend to use their power and authority to monopolize information, they also fail to communicate with workers at the lower levels. As a result, they are often unaware of potential problems facing the organization and of high levels of worker frustration. Bureaucratic regulations are written in far greater detail than is necessary in order to ensure that almost all conceivable situations are covered. *Goal displacement* **occurs when the rules become an end in themselves rather than a means to an end, and organizational survival becomes more important than achievement of goals** (Merton, 1968).

Inefficiency and rigidity occur at the lower levels of the organization as well. Workers often engage in *ritualism;* that is, they become most concerned with "going through the motions" and "following the rules." According to Robert Merton (1968), the term *bureaucratic personality* **describes those workers who are more concerned with following correct procedures than they are with getting the job done correctly.** Such workers are usually able to handle routine situations effectively but are frequently incapable of handling a unique problem or an emergency. Thorstein Veblen (1967/1899) used the term *trained incapacity* to characterize situations

# SOCIOLOGY and Social Policy

## In a High-Tech World, Should Employers Be Allowed to "Spy" on Their Workers?

I know that when I'm at work, my time belongs to the company, but somehow I still feel like it's an invasion of my privacy for some computer to monitor every single thing I do. I mean, I work for a company that makes cardboard boxes, not the CIA!
—a student in one of the author's classes, expressing her irritation over computer surveillance at work

**D**o employers really have the right to monitor everything that their employees do on company-owned computers? Generally speaking, the answer is yes, and the practice is widespread, with more than three-quarters of U.S. companies monitoring employee Internet use (Petrecca, 2010). Employers assert not only that they have the right to engage in such surveillance but also that it may be necessary for them to do so for their own protection. Employers state that they own the computer network and the monitors, pay for the Internet service, and pay the employee to spend time on company business. Employers also argue that they may be held legally responsible for costly harassment suits from other individuals who may be exposed to offensive content sent on company computers and that surveillance is the only way to protect against such liability. Unchecked Internet activity can also expose a company's network and systems to malware and other intrusions that the company otherwise might not encounter. As a result, many employers take the position that First Amendment (privacy) rights are left at the office door when a person agrees to work for a private employer.

What is the legal position on this issue? In most instances, courts have upheld monitoring by employers (see, for example, *Bourke v. Nissan, Smyth v. Pillsbury,* and *Shoars v. Epson*). Yet there are valid arguments against computer surveillance, and invasion of a worker's privacy is certainly one of them. When an employee makes a personal phone call while at work, the employee usually has a reasonable expectation of privacy—a reasonable belief that neither fellow workers nor his or her employer is eavesdropping on that call. If the employer is going to read an employee's e-mail or track the person's

B Busco/Getty Images

Do you think that employers should have the right to monitor everything that their employees do on company-owned computers?

Internet activities, shouldn't the employer at least have to make sure that its workers are aware of that policy? And this is exactly what most *endpoint security* in businesses now suggests: Be forthright with people, and let everyone know what is being tracked and why. A simple statement in the employee handbook and in orientation sessions that tells employees that everything they do on company computers, including individual keystrokes, can and will be tracked lets workers know that their behavior is being monitored (Strohmeyer, 2011).

### reflect & analyze

**Are you concerned about computer privacy in your own life? Should businesses and colleges have the right to monitor our digital communication? If so, how should they go about this process?**

---

in which workers have become so highly specialized or have been given such fragmented jobs to do that they are unable to come up with creative solutions to problems. Workers who have reached this point also tend to experience bureaucratic alienation—they really do not care what is happening around them.

**Resistance to Change** Once bureaucratic organizations are created, they tend to resist change. This resistance not only makes bureaucracies

**goal displacement** a process that occurs in organizations when the rules become an end in themselves rather than a means to an end, and organizational survival becomes more important than achievement of goals.

**bureaucratic personality** a psychological construct that describes those workers who are more concerned with following correct procedures than they are with getting the job done correctly.

**FIGURE 5.10** According to conflict theorists, members of the capitalist class benefit from the work of laborers such as the people shown here, who are harvesting onions on a farm in the Texas Rio Grande Valley. How do low wages and lack of job security contribute to class-based inequalities in the United States?

members by hierarchical bureaucratic structures and exclusion from informal networks.

Like racial inequality, social-class divisions may be perpetuated in bureaucracies. The theory of a "dual labor market" has been developed to explain how social-class distinctions are perpetuated through different types of employment. Middle- and upper-middle-class employees are more likely to have careers characterized by higher wages, more job security, and opportunities for advancement. By contrast, poor and working-class employees work in occupations characterized by low wages, lack of job security, and few opportunities for promotion. The "dual economy" not only reflects but may also perpetuate people's current class position. Conflict theorists point out that persons in the lowest-wage and highest-potential-for-injury jobs, such as agricultural harvesters and other seasonal laborers, are among the workers most harmed by the presence of a dual economy and its role in perpetuating race-, gender-, and class-based inequalities in the United States and other nations (Figure 5.10).

Gender inequalities are also perpetuated in bureaucracies. Women in traditionally male organizations may feel more visible and experience greater performance pressure. They may also find it harder to gain credibility in management positions.

Inequality in organizations has many consequences. People who lack opportunities for integration and advancement tend to be pessimistic and to have lower self-esteem. Believing that they have few opportunities, they may resign themselves to staying put and surviving at that level. By contrast, those who enjoy full access to organizational opportunities tend to have high aspirations and high self-esteem. They often feel loyalty to the organization and typically see their job as a means for mobility and growth.

virtually impossible to eliminate but also contributes to bureaucratic enlargement. Because of the assumed relationship between size and importance, officials tend to press for larger budgets and more staff and office space. To justify growth, administrators and managers must come up with more tasks for workers to perform.

Resistance to change may also lead to incompetence. Based on organizational policy, bureaucracies tend to promote people from within the organization. As a consequence, a person who performs satisfactorily in one position is promoted to a higher level in the organization. Eventually, people reach a level that is beyond their knowledge, experience, and capabilities.

**Perpetuation of Race, Class, and Gender Inequalities** Some bureaucracies perpetuate inequalities of race, class, and gender because this form of organizational structure creates a specific type of work or learning environment. This structure was typically created for middle-class and upper-middle-class white men, who for many years were the predominant organizational participants.

For people of color, *entry* into dominant white bureaucratic organizations does not equal actual *integration*. Instead, many have experienced an internal conflict between the bureaucratic ideals of equal opportunity and fairness and the prevailing norms of discrimination and hostility that exist in many organizations. Research has found that people of color are more adversely affected than dominant-group

## Bureaucracy and Oligarchy

**LO7** Why do a small number of leaders at the top make all the important organizational decisions? According to the German political sociologist Robert Michels (1949/1911), all organizations encounter the ***iron law of oligarchy—the tendency to become a bureaucracy ruled by the few.*** His central idea was that those who control bureaucracies not only wield power but also have an interest in retaining their power. For example, formal and informal political party leaders often do not want to relinquish their control over the party because they are able to influence who runs for public office and how campaigns are conducted. Officials elected to Congress frequently choose to serve multiple terms in office because it provides them with the opportunity to become more involved not only in service to their country but also in bureaucratic power. Some members of Congress have served more than half a century as elected officials (Manning, 2011).

Michels found that the hierarchical structures of bureaucracies and oligarchies go hand in hand. On the one hand, power may be concentrated in a few people because rank-and-file members must inevitably delegate a certain amount of decision-making authority to their leaders. Leaders have access to information that other members do not have, and they have "clout," which they may use to protect their own interests. On the other hand, oligarchy may result when individuals have certain outstanding qualities that make it possible for them to manage, if not control, others. The members choose to look to their leaders for direction; the leaders are strongly motivated to maintain the power and privileges that go with their leadership positions.

Are there limits to the iron law of oligarchy? The leaders in most organizations do not have unlimited power. Divergent groups within a large-scale organization often compete for power, and informal networks can be used to "go behind the backs" of leaders. In addition, members routinely challenge leaders' decisions, and sometimes they (or the organization's governing board) can remove leaders when they are not pleased with their actions.

# Alternative Forms of Organization

**LO8** Many organizations have sought new and innovative ways to organize work more efficiently than the traditional hierarchical model.

## Humanizing Bureaucracy

In the early 1980s there was a movement in the United States to *humanize bureaucracy*—to establish an organizational environment that develops rather than impedes human resources. More-humane bureaucracies are characterized by (1) less rigid hierarchical structures and greater sharing of power and responsibility by all participants, (2) encouragement of participants to share their ideas and try new approaches to problem solving, and (3) efforts to reduce the number of people in dead-end jobs, train people in needed skills and competencies, and help

people meet outside family responsibilities while still receiving equal treatment inside the organization (Kanter, 1983, 1985, 1993/1977). However, this movement has been overshadowed by globalization and the perceived strengths of systems of organizing work in other nations, such as Japan.

## Organizational Structure in Japan, Russia, and India

For several decades the Japanese model of organization was widely praised for its innovative structure because it focused on lifetime employment and company loyalty. Although the practice of lifetime employment has largely been replaced by the concept of long-term employment, many workers in Japan have higher levels of job security than do U.S. workers. According to advocates of the Japanese system, this model encourages worker loyalty and a high level of productivity. Managers move through various parts of the organization and acquire technical knowledge about the workings of many aspects of the corporation, unlike their U.S. counterparts, who tend to become highly specialized. Unlike top managers in the United States who have given themselves pay raises and bonuses even when their companies were financially strapped, many Japanese managers have taken pay cuts under similar circumstances. Japanese management is characterized as being people oriented, taking a long-term view, and having a culture that focuses on *how* work gets done rather than on the result alone.

In the twenty-first century the Japanese organization is often based on a management style where information flows from the bottom to the top. As a result, senior managers serve in a supervisory capacity, rather than taking a "hands-on" approach, and policies usually originate at middle organizational levels and then are passed upward for senior managers' approval. According to analysts, this approach is beneficial because the same persons responsible for implementing policies are the ones who have an active role in initially shaping the rules, policies, and procedures (Bizshifts-Trends, 2011).

In the Japanese organization, managers are expected to be "father figures" and create an environment in which groups can succeed and goals can be met (Figure 5.11). Effective leadership is not based on individual personality or a dictatorial manner,

**iron law of oligarchy** according to Robert Michels, the tendency of bureaucracies to be ruled by a few people.

**FIGURE 5.11** The Japanese model of organization—including planned group-exercise sessions for employees—has become a part of the workplace in many nations. Would it be a positive change if more workplace settings, such as the one shown here, were viewed as an extension of the family? Why or why not?

and there is disapproval for those who appear to be overly ambitious.

Unlike Japanese organizational structure and management style, organizations in Russia and India are more likely to be hierarchical, centralized, and highly directive. Most organizations also have a "top-down" approach in which chief executives or the highest leaders issue orders for subordinates to follow, and very little consultation takes place with persons in the lower sectors. Leaders who allow too much participation in organizational decision making are often viewed as weak and indecisive. However, middle managers who have privileged access to top elites often become more powerful managers than managers who lack such access. Looking specifically at India, many organizations are family-owned businesses that are tightly controlled across generations; however, there are indications that Western management styles have become more prevalent in that nation as the children and grand-children of company founders increasingly have been educated in universities in the United States or other high-income nations.

What can we learn by examining alternative organizational structures in other countries? We can see that all organizations are not established on the same premises about how leadership should operate and how decisions should be made. We can also see that different types of leadership affect how

organizations will go about their tasks. Some leadership styles are more democratic (managers delegate authority to subordinates in the decision-making process), others are more autocratic (decisions are made solely by those at the top of the hierarchy), and others are more participatory. Finally, we can see that cultural differences do have an important effect on how organizations operate and how leaders think and act (Bizshifts-Trends, 2011).

# Organizations in the Future

What is the best organizational structure for the future? Of course, this question is difficult to answer because it requires the ability to predict economic, political, and social conditions. Nevertheless, we can make several observations.

## Socially Sustainable Organizations

First, organizations have been affected by growing social inequality in the United States and other nations because of heightening differences between high- and low-income segments of populations. Having *socially sustainable organizations* is of increasing importance because television, the Internet, and international travel have made people more aware of the wide disparities in the resources and power of "haves" and "have-nots" both within a single country and across nations.

The term *socially sustainable organizations* is used here to refer to those organizations that take into account the social effects of organizational activities on workers and other persons in the community, the nation, and sometimes the world. Researchers have shown how organizations interact with their physical environments and may produce problems such as pollution and environmental degradation. But the focus of the socially sustainable organizational approach is more on the human and social environment and what organizations can do to sustain and sometimes enhance those aspects of the environment that are not strictly physical or biological. As a result of emphasizing the social sustainability factor, organizations must be developed that are both economically efficient and as equitable as possible.

Some organizational and management analysts suggest that more attention must be paid to the "stakeholders" of an organization. Stakeholder theory is based on the assumption that organizations and their managers must focus on morals and values in goal-setting and decision-making processes. For example, at a college or university, stakeholders would include (but not be limited to) students, faculty and staff, administrators, alumni, major contributors, boards of regents, suppliers, the community where the school is located, and the society as a whole. The management structure and the morals and values of the institution should reflect the interests of those constituent groups. The goals of the organization should be based on taking into account the interests of these various stakeholders and working toward organizational goals and outcomes that will not only ensure organizational success but also provide the greatest good for the greatest number of stakeholders. Although academic success, winning sports teams, and college financial stability are important in higher education, other criteria should also be used in assessing the effectiveness and overall output of the college community. In other types of organizations, similar stakeholders can be identified and goals established to meet the needs of various constituencies.

## Globalization, Technology, and "Smart Working"

Second, *globalization* is the key word for management and change in many organizations, and the use of technology is intricately linked with

William Perlman/Star Ledger/Corbis

**FIGURE 5.12** "Smart working" is based on the assumption that innovation is crucial and that people should have flexibility in where, when, and how they work. Do nontraditional office spaces like this one reflect the idea of "smart working"?

performing flexible, mobile work anywhere in the world. Based on the assumption that organizations must respond to a rapidly changing environment or they will not thrive, several twenty-first-century organizational models are based on the need to relegate traditional organizational structure to dinosaur status and move ahead with structures that fully use technology. A central emphasis is on the need to communicate more effectively. As the pace of communication has increased dramatically and information overload has become prevalent, the leaders of organizations are seeking new ways in which to more efficiently manage their organizations and to be ahead of change, rather than merely adapting to change after it occurs.

One recent approach is referred to as "smart working," which is based on the assumption that innovation is crucial and that organizational leaders must be able to use the talents and energies of the people who work with them. At one level, "smart working" refers to "anytime, anywhere" ways of work that have become prevalent because of communications technologies such as smartphones and computers. However, another level focuses on the ways in which smart working makes it possible for people to have flexibility and autonomy in where, when, and how they work (chiefexecutive.com, 2010) (Figure 5.12). According to one management specialist,

> It turns out that the sort of collaborative, challenging work with potential for learning and personal development that people find satisfying is exactly the sort of work needed to adapt to current turbulent global operating conditions.

**159**

Smart working is an outcome of designing organizational systems that are good for business and good for people. (chiefexecutive.com, 2010)

From this perspective, organizations must adapt to change; empower all organizational participants to become involved in collaboration, problem solving, and innovation; and create a work environment that people find engaging and that inspires them to give their best to the organization.

Exactly how these organizations might look is not fully clear, although some analysts suggest that corporations such as Google, Microsoft, and other high-tech companies have actively sought to redefine organizational culture and environment by being responsive to employees, customers, and other shareholders. Although management continues to exist, the distinction between managers and the managed becomes less prevalent, and the idea that management knowledge will be everyone's responsibility becomes more predominant. Emphasis is also placed on the importance of improving communication and on acquiring the latest technologies to make this process even more fast, secure, and efficient. Overall, there is a focus on change and the assumption that people in an organization should be change agents, not individuals who merely respond to change after it occurs.

Ultimately, everyone has a stake in seeing that organizations operate in an effective, humane manner and that opportunities are widely available to all people regardless of race, gender, class, or age. Workers and students alike can benefit from organizational environments that make it possible for people to explore their joint interests without fear of losing their privacy or being pitted against one another in a competitive struggle for advantage.

# chapter REVIEW

## LO1 How do sociologists distinguish among social groups, aggregates, and categories?

Sociologists define a social group as a collection of two or more people who interact frequently, share a sense of belonging, and depend on one another. People who happen to be in the same place at the same time are considered an aggregate. Those who share a similar characteristic are considered a category. Neither aggregates nor categories are considered social groups.

## LO2 How do sociologists distinguish among ingroups, outgroups, and reference groups?

Sociologists distinguish between primary groups and secondary groups. Primary groups are small and personal, and members engage in emotion-based interactions over an extended period. Secondary groups are larger and more specialized, and members have less personal and more-formal, goal-oriented relationships. Sociologists also divide groups into ingroups, outgroups, and reference groups. Ingroups are groups to which we belong and with which we identify. Outgroups are groups we do not belong to or perhaps feel hostile toward. Reference groups are groups that strongly influence people's behavior whether or not they are actually members.

## LO3 How does the size of a group shape its members' communication, leadership styles, and pressures to conform?

In small groups, all members know one another and interact simultaneously. In groups with more than three members, communication dynamics change, and members tend to assume specialized tasks. Leadership may be authoritarian, democratic, or laissez-faire. Authoritarian leaders make major decisions and assign tasks to individual members. Democratic leaders encourage discussion and collaborative decision making. Laissez-faire leaders are minimally involved and encourage members to make their own decisions. Groups may have significant influence on members' values, attitudes, and behaviors.

## LO4 In terms of the concept of groupthink, how do people often respond differently in a group context than they might if they were alone?

Groupthink is the process by which members of a cohesive group arrive at a decision that many individual members privately believe is unwise. In order to maintain ties with a group, many members are willing to conform to norms established and reinforced by group members.

**LO5** **What are the three types of formal organizations, and how do they differ in membership?**

Normative, coercive, and utilitarian organizations are formal organizations. We voluntarily join normative organizations when we want to pursue some common interest or gain personal satisfaction or prestige from being a member. People do not voluntarily become members of coercive organizations—associations that people are forced to join. We voluntarily join utilitarian organizations when they can provide us with a material reward that we seek.

**LO6** **What are the strengths and weaknesses of bureaucracies in contemporary nations such as the United States?**

A bureaucracy is a formal organization characterized by hierarchical authority, division of labor, explicit procedures, and impersonality. According to Max Weber, bureaucracy supplies a rational means of attaining organizational goals because it contributes to coordination and control. A bureaucracy also has an informal structure, which includes the daily activities and interactions that bypass the official rules and procedures. The informal structure may enhance productivity or may be counterproductive to the organization. A bureaucracy may be inefficient, resistant to change, and a vehicle for perpetuating class, race, and gender inequalities.

**LO7** **What is the iron law of oligarchy, and how does the concept apply to the U.S. government?**

The iron law of oligarchy is the tendency to become a bureaucracy ruled by the few. Those who control bureaucracies not only wield power but also have an interest in retaining their power. For example, officials elected to the U.S. Congress frequently choose to serve multiple terms in office because it provides them with the opportunity to become more involved not only in service to their country but also in bureaucratic power.

**LO8** **What alternative forms of organization exist today in nations such as Japan?**

Some organizations have adopted Japanese management techniques based on long-term employment and company loyalty as alternative forms of bureaucratic structures. Unlike Japanese organizational structure and management style, organizations in Russia and India are more likely to be hierarchical, centralized, and highly directive. More recently, having socially sustainable organizations is becoming increasingly important.

# key TERMS

aggregate 139
authoritarian leaders 146
bureaucracy 152
bureaucratic personality 154
category 139
conformity 146
democratic leaders 146
dyad 144

expressive leadership 146
goal displacement 154
groupthink 149
ideal type 152
informal side of a bureaucracy 153
ingroup 141
instrumental leadership 146
iron law of oligarchy 156

laissez-faire leaders 146
leadership 145
network 142
outgroup 141
rationality 152
reference group 142
small group 144
triad 144

# questions for CRITICAL THINKING

1. Who might be more likely to conform in a bureaucracy, those with power or those wanting more power?
2. Do the insights gained from Milgram's research on obedience outweigh the elements of deception and stress that were forced on its subjects?

3. How would you organize a large-scale organization or company for the second decade of the twenty-first century?

# 6 DEVIANCE AND CRIME

## LEARNING OBJECTIVES

**LO1** Define deviance and explain when deviant behavior is considered a crime.

**LO2** Identify and compare the key functionalist perspectives on deviance.

**LO3** State the key ideas of conflict explanations of deviance and crime that focus on power relations, capitalism, feminism, and the intersection of race, class, and gender.

**LO4** Explain these symbolic interactionist perspectives on deviance: differential association theory, social bond theory, and labeling theory.

**LO5** Describe how postmodern perspectives on deviance differ from other theoretical approaches, specifically identifying Michel Foucault's contributions to the study of deviance and social control.

**LO6** Define the following types of crime: violent crime, property crime, public order crime, occupational crime, corporate crime, organized crime, and political crime.

**LO7** Explain why official statistics may not be a good indicator of how many crimes are committed, particularly with regard to factors such as age, race, gender, and class.

**LO8** Identify the components of the criminal justice system and list the goals of punishment.

AP Images/Daniel Luna

163

# When the Unspeakable Happens

STATE OF CONNECTICUT
DEPARTMENT OF EMERGENCY SERVICES
 AND PUBLIC PROTECTION
OFFICE OF THE COMMISSIONER
December 27, 2013

(Cover Letter, Final Report on the Sandy Hook Elementary School Killings)

First and foremost, my thoughts and prayers . . . continue to be with the families of the victims, the survivors, and the entire Newtown community. On December 14, 2012, your lives were changed forever in a matter of moments. The horror and unspeakable sadness of that day and the lives that were lost will never be forgotten.

In the midst of the darkness of that day, we also saw remarkable heroism and glimpses of grace. We saw Sandy Hook Elementary School faculty and staff doing everything in their power to protect their charges. We saw law enforcement officers from dozens of local, state, and federal agencies running as one into harm's way at enormous personal risk. We saw dispatchers calmly and efficiently coordinating resources and directing personnel while at the very same time compassionately reassuring

Mass shootings have become an all-too-common occurrence in the United States. Some of the most traumatic shootings have been in elementary or secondary schools and at colleges or universities. Violence at Columbine High School in Littleton, Colorado, and at Virginia Tech in Blacksburg, Virginia, were well-documented, and then came the tragic occurrence at Sandy Hook Elementary School in Newtown, Connecticut, where an armed shooter killed twenty young children, six teachers, and his own mother (in another location) prior to committing suicide.

People worldwide were left reeling by the Newtown deaths and turned to multiple explanations, none satisfactory, of why such a terrible event had occurred. Blame was placed on lax gun control policies, mental illness, divorced parents, violent video games, media glorification of violence and killing, and a myriad of other possible "causes" of this horrific event. The final report, issued one year later, did not provide a definitive answer to pressing questions about what triggered the Newtown murder/suicide rampage. Although this is often the case, we cannot stop looking for the causes of violent crime if we have any hope of reducing violent attacks and other forms of criminal conduct in the future.

For this reason, crimes such as mass shootings are of great interest and concern to all who teach and study sociology because of the harm that such incidents inflict not only on victims and their families but also on entire communities, the nation, and the larger global population linked through instantaneous communications and social media networks.

In this chapter we look at deviance and crime to learn more about sociological perspectives on deviance, the nature and extent of crime in the United States, and how the criminal justice system operates. Before reading on, take the "Sociology and Everyday Life" quiz on violence and guns in the United States to see how much you know about the subject.

## What Is Deviance?

**LO1** *Deviance* is any behavior, belief, or condition that violates significant social norms in the society or group in which it victims. We saw fire and EMS personnel from all over the region offering their services, state and municipal employees and contractors working countless hours in innumerable ways to support the first responders and the community at large, and thousands of volunteers and donors from across the globe reaching out in any way they could. For everything that was done by all of these and more, I am profoundly grateful. . . .

I hope that the release of this [final] report, though painful, will allow those who have been affected by it to continue their personal process of healing, and will provide helpful information that can be put to use to prevent such tragedies in the future.

Sincerely,
Reuben F. Bradford
Commissioner

## How Much Do You Know About Violence and Guns in the United States?

| True | False | |
|------|-------|---|
| T | F | 1. The total number of reported violent crimes increases each year in the United States. |
| T | F | 2. Aggravated assaults account for the highest number of violent crimes in this country. |
| T | F | 3. Firearms are used in more than half of all murders in the United States. |
| T | F | 4. Despite extensive media coverage about mass killings, shooting sprees are rare in the United States. |

| True | False | |
|------|-------|---|
| T | F | 5. The West is the most violent region in the United States. |
| T | F | 6. The presence of more guns tends to indicate a likelihood of more homicides. |
| T | F | 7. Residents in urban areas experience the highest rates of violence. |
| T | F | 8. Each year about 50 percent of violent victimizations are reported to the police. |

Answers on page 167.

**occurs.** We are most familiar with *behavioral* deviance, based on a person's intentional or inadvertent actions. For example, a person may engage in intentional deviance by drinking too much or robbing a bank, or in inadvertent deviance by losing money in a casino or laughing at a funeral.

Although we usually think of deviance as a type of behavior, people may be regarded as deviant if they express a radical or unusual *belief system.* Members of far-right-wing or far-left-wing political groups may be considered deviant when their religious or political beliefs become known to people with more-conventional cultural beliefs. However, individuals who are considered to be "deviant" by one category of people may be seen as conformists by another group. For example, adolescents in some peer cliques and youth gangs may shun mainstream cultural beliefs and values but routinely conform to subcultural codes of dress, attitude (such as defiant individualism), and behavior.

In addition to their behavior and beliefs, individuals may also be regarded as deviant because they possess a specific *condition* or *characteristic.* A wide range of conditions and characteristics have been identified as "deviant," including being obese, having excessive tattoos or body piercings, and being diagnosed with AIDS (Weitz, 2004) (Figure 6.1). For example, in a now-classic study that remains valid today, sociologist Rose Weitz (2004) found that persons with AIDS live with a stigma that affects their relationships with other people, including family members, friends, lovers, colleagues, and health care workers. Chapter 4 defines a *stigma* as any physical or social attribute or sign that so devalues a person's social identity that it disqualifies the person from full social acceptance (Goffman, 1963b). Based on this definition, the stigmatized person has a "spoiled identity" as a result of being negatively evaluated by others (Goffman, 1963b).

**deviance** any behavior, belief, or condition that violates significant social norms in the society or group in which it occurs.

**165**

To avoid or reduce stigma, many people seek to conceal the characteristic or condition that might lead to stigmatization.

## Who Defines Deviance?

Are some behaviors, beliefs, and conditions inherently deviant? In commonsense thinking, deviance is often viewed as inherent in certain kinds of behavior or people. For sociologists, however, deviance is related to social situations and social structures rather than to the behavior of individual actors. As the sociologist Kai T. Erikson (1964: 11) explains,

> Deviance is not a property inherent in certain forms of behavior; it is a property conferred upon these forms by the audiences which directly or indirectly witness them. *The critical variable in the study of deviance, then, is the social audience rather than the individual actor, since it is the audience which eventually determines whether or not any episode of behavior or any class of episodes is labeled deviant* [emphasis added].

Based on this statement, we can conclude that deviance is *relative*—that is, an act becomes deviant when it is socially defined as such. Definitions of deviance vary widely from place to place, from time to time, and from group to group (see "Sociology Works!"). Clothing styles are an example: What some people wear in public today might have landed them in jail in their grandparents' or great-grandparents' day.

Deviant behavior also varies in *degree of seriousness,* ranging from mild transgressions of folkways to more-serious infringements of mores to quite serious violations of the law. Have you skipped class or pretended that you were sick so that you would have more time to complete a homework assignment or study for an exam? If so, you have violated a folkway. Others probably view your infraction as relatively minor; at most, you might receive a lower grade. Violations of mores—such as falsifying a

**FIGURE 6.1** Do you consider this man deviant?

© Andrey Arkusha/Shutterstock.com

college application or cheating on an examination—are viewed as more-serious infractions and are punishable by stronger sanctions, such as academic probation or expulsion. Some forms of deviant behavior violate the criminal law, which defines the behaviors that society labels as criminal. A *crime* is **behavior that violates criminal law and is punishable with fines, jail terms, and/or other negative sanctions.** Crimes range from minor offenses (such as traffic violations) to major offenses (such as murder). A subcategory, *juvenile delinquency,* **refers to a violation of law or the commission of a status offense by young people.** Note that the legal concept of juvenile delinquency includes not only crimes but also *status offenses,* which are illegal only when committed by younger people (such as cutting school or running away from home).

## What Is Social Control?

Societies not only have norms and laws that govern acceptable behavior; they also have various mechanisms to control people's behavior. *Social control* **refers to the systematic practices that social groups develop in order to encourage conformity to norms, rules, and laws and to discourage deviance.** Social control mechanisms may be either internal or external. *Internal social control* takes place through the socialization process: Individuals *internalize* societal norms and values that prescribe how people should behave and then follow those norms and values in their everyday lives. By contrast, *external social control* involves the use of negative sanctions that proscribe certain behaviors and set forth the punishments for rule breakers and nonconformists. In contemporary societies the criminal justice system, which includes the police, the courts, and the prisons, is the primary mechanism of external social control.

If most actions deemed deviant do little or no direct harm to society or its members, why is social

## ANSWERS to the Sociology Quiz on Violence and Guns in the United States

**1. False.** Looking at recent trends, the total number of reported violent crimes decreased by about 15 percent between 2007 and 2011.

**2. True.** Aggravated assaults in recent years accounted for the highest number (62.4 percent) of violent crimes reported to law enforcement officials.

**3. True.** FBI data show that firearms are used in 67.7 percent of the nation's murders.

**4. False.** Over the past three decades, at least 61 mass murders have been carried out with firearms in the United States. These killings have taken place in 30 different states, ranging from Massachusetts to Hawaii.

**5. False.** The South is the most violent region in the United States.

**6. True.** The Harvard Injury Control Research Center found evidence that links higher rates of gun ownership with more murders in various U.S. states as well as in different countries.

**7. True.** Residents in urban areas experience the highest rates of violent crime when compared to suburban and rural residents.

**8. True.** About 50 percent of violent victimizations are reported to the police annually. Crimes such as domestic violence, serious violent crimes involving weapons, and serious violent crimes involving injury are the most likely to be reported.

Sources: Based on Klein, 2012; FBI, 2012; and Truman and Planty, 2012.

control so important to groups and societies? Why is the same belief or action punished in one group or society and not in another? These questions pose interesting theoretical concerns and research topics for sociologists and criminologists who examine issues pertaining to law, social control, and the criminal justice system. *Criminology* **is the systematic study of crime and the criminal justice system, including the police, courts, and prisons.**

The primary interest of sociologists and criminologists is not questions of how crime and criminals can best be controlled but rather social control as a social product. Sociologists do not judge certain kinds of behavior or people as being "good" or "bad." Instead, they attempt to determine what types of behavior are defined as deviant, who does the defining, how and why people become deviants, and how society deals with deviants. Although sociologists have developed a number of theories to explain deviance and crime, no one perspective is a comprehensive explanation of all deviance. Each theory provides a different lens through which we can examine aspects of deviant behavior.

## Functionalist Perspectives on Deviance

**LO2** As we have seen in previous chapters, functionalists focus on societal stability and the ways in which various parts of society contribute to the whole. According to functionalists,

a certain amount of deviance contributes to the smooth functioning of society.

## What Causes Deviance, and Why Is It Functional for Society?

Sociologist Emile Durkheim believed that deviance is rooted in societal factors such as rapid social change and lack of social integration among people. As you will recall, Durkheim attributed the social upheaval he saw at the end of the nineteenth century to the shift from mechanical to organic solidarity, which was brought about by rapid industrialization and urbanization. Although many people continued to follow the dominant morality (norms, values, and laws) as best they could, rapid social change

**crime** behavior that violates criminal law and is punishable with fines, jail terms, and/or other negative sanctions.

**juvenile delinquency** a violation of law or the commission of a status offense by young people.

**social control** systematic practices that social groups develop in order to encourage conformity to norms, rules, and laws and to discourage deviance.

**criminology** the systematic study of crime and the criminal justice system, including the police, courts, and prisons.

contributed to *anomie*—a social condition in which people experience a sense of futility because social norms are weak, absent, or conflicting. According to Durkheim, as social integration (bonding and community involvement) decreased, deviance and crime increased. However, from his perspective, this was not altogether bad because he believed that deviance has positive social functions in terms of its consequences. For Durkheim (1964a/1895), deviance is a natural and inevitable part of all societies. Likewise, contemporary functionalist theorists suggest that deviance is universal because it serves three important functions:

1. *Deviance clarifies rules.* By punishing deviant behavior, society reaffirms its commitment to the rules and clarifies their meaning.
2. *Deviance unites a group.* When deviant behavior is seen as a threat to group solidarity and people unite in opposition to that behavior, their loyalties to society are reinforced.
3. *Deviance promotes social change.* Deviants may violate norms in order to get them changed. For example, acts of *civil disobedience*—including lunch counter sit-ins and bus boycotts—were used to protest and eventually correct injustices such as segregated buses and lunch counters in the South. More recently, this is what organizers of groups in the "Occupy movement," such as Occupy Wall Street, hoped to accomplish, but their goal of redistribution of some wealth from the richest 1 percent to the bottom 99 percent constituted a very complex issue. Although some analysts have since declared the Occupy movement dead, other Occupy organizers, such as those found in Portland, Oregon, are now committed to more-specific issues such as fighting for the rights of homeless persons (Davidson, 2013).

Functionalists acknowledge that deviance may also be dysfunctional for society. If too many people violate the norms, everyday existence may become unpredictable, chaotic, and even violent. If even a few people commit acts that are so violent that they threaten the survival of a society, then deviant acts move into the realm of the criminal and even the unthinkable. One example that stands out in everyone's mind is terrorist attacks around the world and the fear that remains constantly present as a result.

Although there is a wide array of contemporary functionalist theories regarding deviance and crime, many of these theories focus on social structure. For this reason, the first theory we will discuss is referred to as a structural functionalist approach. It describes the relationship between the society's economic structure and why people might engage in various forms of deviant behavior.

## Strain Theory: Goals and Means to Achieve Them

Modifying Durkheim's (1964a/1895) concept of *anomie,* the sociologist Robert Merton (1938, 1968) developed strain theory. According to **strain theory, people feel strain when they are exposed to cultural goals that they are unable to obtain because they do not have access to culturally approved means of achieving those goals.** The goals may be material possessions and money; the approved means may include an education and a job. When denied legitimate access to these goals, some people seek access through deviant means.

Merton identified five ways in which people adapt to cultural goals and approved ways of achieving them: conformity, innovation, ritualism, retreatism, and rebellion (see Table 6.1). According to Merton, *conformity* occurs when people accept culturally approved goals and pursue them through approved means. Persons who want to achieve success through conformity work hard, save their

### table 6.1

Merton's Strain Theory of Deviance

| Mode of Adaptation | Method of Adaptation | Seeks Culture's Goals | Follows Culture's Approved Ways |
|---|---|---|---|
| Conformity | Accepts culturally approved goals; pursues them through culturally approved means | Yes | Yes |
| Innovation | Accepts culturally approved goals; adopts disapproved means of achieving them | Yes | No |
| Ritualism | Abandons society's goals but continues to conform to approved means | No | Yes |
| Retreatism | Abandons both approved goals and the approved means to achieve them | No | No |
| Rebellion | Challenges both the approved goals and the approved means to achieve them | No—seeks to replace | No—seeks to replace |

# Sociology WORKS!

## Social Definitions of Deviance: Tracking Sasquatch?

Sociologist Carson Mencken sat perched in a tree in the Sam Houston National Forest. It was 1 a.m. Mencken had accompanied a group of Bigfoot (also known as Sasquatch) hunters as they set out to do "call blasting" in their quest for Bigfoot, of which some 800 to 1,000 specimens are thought to roam about East Texas. Giant tape recorders and speakers were set up in the trees to assist in luring the legendary creature.

The Bigfoot hunters used prerecorded audio samples to "drive" the Bigfoot out into the open, Mencken said. "As soon as that howl went off, every animal in those woods freaked out, as you can well imagine. . . . And I started to think, I really hope that there's not a Bigfoot, because if that's what it sounds like, we're in a lot of trouble." (Aydelotte, 2010)

Sociology contributes to our thinking about conformity and deviance by making us aware that the people we are around help us define what we think of as "normal" beliefs and actions. When individuals who believe in Bigfoot are surrounded with other like-minded individuals, they think of their beliefs as normal and gain a personal sense of belonging. For this reason, some join groups such as the Bigfoot Field Researchers Organization so that they can share their outings, compare field notes on recent sightings, and feel that they are part of an important group or a clique. Bigfoot followers treat one another with respect in accordance to their documentation, rather than receiving blank stares or skeptical commentary by nonbelievers (Wischnowsky, 2005).

Looking at the seemingly deviant behavior of going out looking for Bigfoot from a sociological perspective, researchers place these actions within a larger social context. One context for studying people's fascination with Bigfoot sightings and other paranormal occurrences is the sociology of religion. Many people who believe in Bigfoot "believe that it is a matter of faith" (Weiss, 2004). This faith may be intensified by use of the Internet, where true believers may easily report their sightings without fear of ridicule or being identified as deviant by outsiders.

### reflect & analyze

At your college or university, what beliefs and actions of individuals and groups might be classified as conformity by some people but identified as deviance by others?

---

money, and so on. Even people who find that they are blocked from achieving a high level of education or a lucrative career may take a lower-paying job and attend school part time, join the military, or seek alternative (but legal) avenues, such as playing the lottery to "strike it rich."

Merton classified the remaining four types of adaptation as deviance:

- *Innovation* occurs when people accept society's goals but adopt disapproved means for achieving them. Innovations for acquiring material possessions or money cover a wide variety of illegal activities, including theft and drug dealing.
- *Ritualism* occurs when people give up on societal goals but still adhere to the socially approved means for achieving them. Ritualism is the opposite of innovation; persons who cannot obtain expensive material possessions or wealth may nevertheless seek to maintain the respect of others by being a "hard worker" or "good citizen."
- *Retreatism* occurs when people abandon both the approved goals and the approved means of achieving them. Merton included persons such as skid-row alcoholics and drug addicts in this category; however, not all retreatists are destitute. Some may be middle- or upper-income individuals who see themselves as rejecting the conventional trappings of success or the means necessary to acquire them.
- *Rebellion* occurs when people challenge both the approved goals and the approved means for achieving them and advocate an alternative set of goals or means. To achieve their alternative goals, rebels may use violence (such as rioting) or may register their displeasure with society through acts of vandalism or graffiti.

## Opportunity Theory: Access to Illegitimate Opportunities

Expanding on Merton's strain theory, sociologists Richard Cloward and Lloyd Ohlin (1960) suggested that for deviance to occur, people must have

> **strain theory** the proposition that people feel strain when they are exposed to cultural goals that they are unable to obtain because they do not have access to culturally approved means of achieving those goals.

**FIGURE 6.2** Members of the California group known as the Culver City Boyz typify how gang members use items of clothing and gang signs made with their hands to assert their membership in the group and solidarity with one another. Researchers have found that some gang members may have insufficient legitimate means to achieve conventional goals of status and wealth but have illegitimate opportunity structures through which they can achieve these goals.

A. Ramey/PhotoEdit

access to ***illegitimate opportunity structures—circumstances that provide an opportunity for people to acquire through illegitimate activities what they cannot achieve through legitimate channels.*** For example, in studies of juvenile gangs, researchers have found that gang members may have insufficient legitimate means to achieve conventional goals of status and wealth but have illegitimate opportunity structures—such as theft, drug dealing, or robbery—through which they can achieve these goals (Figure 6.2). In his now-classic sociological study of the "Diamonds," a Chicago street gang whose members are second-generation Puerto Rican youths, sociologist Felix M. Padilla (1993) found that gang membership was linked to the members' belief that they might reach their aspirations by transforming the gang into a business enterprise. Coco, one of the Diamonds, explains the importance of sticking together in the gang's income-generating organization:

> We are a group, a community, a family—we have to learn to live together. If we separate, we will never have a chance. We need each other even to make sure that we have a spot for selling our supply [of drugs]. You know, there is people around here, like some opposition, that want to take over your *negocio* [business]. And they think

that they can do this very easy. So we stick together, and that makes other people think twice about trying to take over what is yours. In our case, the opposition has never tried messing with our hood, and that's because they know it's protected real good by us fellas. (qtd. in Padilla, 1993: 104)

Although Padilla's study is more than two decades old, his findings continue to be supported by research, popular culture, and media accounts of how gangs stick together in income-generating business organizations. In the latest available study of national youth gangs, law enforcement agencies estimate that 29,400 gangs operate across the United States and that about 756,000 individuals are gang members. Some gangs are believed to be involved in criminal activities such as drug trafficking, weapon smuggling, and extortion (Egley and Howell, 2012).

Other early research by Cloward and Ohlin (1960) identified three basic gang types—criminal, conflict, and retreatist—that emerge on the basis of what type of illegitimate opportunity structure is available in a specific area. The *criminal gang* is devoted to theft, extortion, and other illegal means of securing an income. For young men who grow up in a criminal gang, running drug houses and selling drugs on street corners make it possible for them to support themselves and their families as well as purchase material possessions to impress others. By contrast, *conflict gangs* emerge in communities that do not provide either legitimate or illegitimate opportunities. Members of conflict gangs seek to acquire a "rep" (reputation) by fighting over "turf" (territory) and adopting a value system of toughness, courage, and similar qualities. On some Native American reservations, for example, homegrown gangs routinely fight their rivals, often over a minor incident or slight, and engage in thefts, assaults, and property crimes in some of the nation's poorest, most neglected places, including the Pine Ridge Indian Reservation (Eckholm, 2009). Unlike criminal and conflict gangs, members of *retreatist gangs* are unable to gain success through legitimate means and are unwilling to do so through illegal ones. As a result, the consumption of drugs is stressed, and addiction is prevalent.

How useful are social structural approaches such as opportunity theory and strain theory in explaining deviant behavior? Although there are weaknesses in these approaches, they focus our attention on one crucial issue: the close association between certain forms of deviance and social class position. If we view gangs as a microcosm of the larger society, we can see connections between poverty and inequality and larger patterns of crime in the United States.

# Conflict Perspectives on Deviance

**LO3** Who determines what kinds of behavior are deviant or criminal? Different branches of conflict theory offer somewhat divergent answers to this question. One branch emphasizes power as the central factor in defining deviance and crime: People in positions of power maintain their advantage by using the law to protect their interests. Another branch emphasizes the relationship between deviance and capitalism, whereas a third focuses on feminist perspectives and the confluence of race, class, and gender issues in regard to deviance and crime.

## Deviance and Power Relations

Conflict theorists who focus on power relations in society suggest that the lifestyles considered deviant by political and economic elites are often defined as illegal. According to this approach, norms and laws are established for the benefit of those in power and do not reflect any absolute standard of right and wrong. As a result, the activities of poor and lower-income individuals are more likely to be defined as criminal than those of persons from middle- and upper-income backgrounds. The media often contribute to this perception by frequent reporting of African American perpetrators that portrays them as criminal through both the context of the story and through the social structural context in which the news stories are reported (Bjornstrom et al., 2010).

Moreover, the criminal justice system is more focused on, and is less forgiving of, deviant and criminal behavior engaged in by people in specific categories. For example, research shows that young, single, urban males are more likely to be perceived as criminals and receive stricter sentences in courts (Rehavi and Starr, 2012). One study of racial disparity in federal criminal sentencing found that African Americans receive sentences that are almost 10 percent longer than those given to white Americans arrested for the same offenses (Rehavi and Starr, 2012). This finding may partly be attributed to the fact that prosecutors are almost twice as likely to file charges against African Americans that carry a mandatory-minimum sentence. This is especially true in drug cases. By contrast, another study found that black–white disparity in sentencing of offenders is less frequent in courts in those counties where more African American lawyers practice, suggesting that having representation by a person from one's own racial or ethnic category may bring greater scrutiny to criminal proceedings (King, Johnson, and McGeever, 2010).

## Deviance and Capitalism

A second branch of conflict theory—Marxist/critical theory—views deviance and crime as a function of the capitalist economic system. Although the early economist and social thinker Karl Marx wrote very little about deviance and crime, many of his ideas are found in a critical approach that has emerged from earlier Marxist and radical perspectives on criminology. The critical approach is based on the assumption that the laws and the criminal justice system protect the power and privilege of the capitalist class. As you may recall from Chapter 1, Marx based his critique of capitalism on the inherent conflict that he believed existed between the capitalists (bourgeoisie) and the working class (proletariat). In a capitalist society, social institutions (such as law, politics, and education, which make up the superstructure) legitimize existing class inequalities and maintain the capitalists' superior position in the class structure. According to Marx, capitalism produces haves and have-nots, who engage in different forms of deviance and crime.

According to the sociologist Richard Quinney (2001/1974), people with economic and political power define as criminal any behavior that threatens their own interests. The powerful use laws to control those who are without power. For example, drug laws enacted early in the twentieth century were actively enforced in an effort to control immigrant workers, especially the Chinese, who were being exploited by the railroads and other industries. By contrast, antitrust legislation passed at about the same time was seldom enforced against large corporations owned by prominent families such as the Rockefellers, Carnegies, and Mellons. Having antitrust laws on the books merely shored up the government's legitimacy by making it appear responsive to public concerns about big business.

In sum, the Marxist/critical approach argues that criminal law protects the interests of the affluent

**illegitimate opportunity structures** circumstances that provide an opportunity for people to acquire through illegitimate activities what they cannot achieve through legitimate channels.

and powerful. The way that laws are written and enforced benefits the capitalist class by ensuring that individuals at the bottom of the social class structure do not infringe on the property or threaten the safety of those at the top (Reiman and Leighton, 2010). However, others assert that critical theorists have not shown that powerful economic and political elites actually manipulate lawmaking and law enforcement for their own benefit. Rather, people of all classes share a consensus about the criminality of certain acts. For example, laws that prohibit murder, rape, and armed robbery protect not only middle- and upper-income people but also low-income people, who are frequently the victims of such violent crimes.

## Feminist Approaches

Can theories developed to explain male behavior be used to understand female deviance and crime? According to feminist scholars, the answer is no. A new interest in women and deviance developed in 1975 when two books—Freda Adler's *Sisters in Crime* and Rita James Simons's *Women and Crime*—declared that women's crime rates were going to increase significantly as a result of the women's liberation movement. Although this so-called *emancipation theory* of female crime has been refuted by subsequent analysts, Adler's and Simons's works encouraged feminist scholars (both women and men) to examine more closely the relationship among gender, deviance, and crime. For example, it is important to understand gender-specific differences in crimes committed by men and women (Van Gundy and Kappeler, 2013). Consequently, feminist scholars have developed theories and conducted research to fill the void in our knowledge about gender and crime. For example, in a study by Janet Davidson and Meda Chesney-Lind (2009), the authors conclude that socio-biographical variables such as a history of physical and sexual abuse are more predictive of female criminality than of male criminality. Although there is no single feminist perspective on deviance and crime, three schools of thought have emerged.

Why do women engage in deviant behavior and commit crimes? According to the *liberal feminist approach,* women's deviance and crime are a rational response to the gender discrimination that women experience in families and the workplace. From this view, lower-income and minority women typically have fewer opportunities not only for education and good jobs but also for "high-end" criminal endeavors.

By contrast, the *radical feminist approach* views the cause of women's crime as originating in patriarchy (male domination over females). From this view, arrests and prosecution for crimes such as prostitution reflect our society's sexual double standard whereby it is acceptable for a man to pay for sex but unacceptable for a woman to accept money for such services. Although state laws usually view both the female prostitute and the male customer as violating the law, in most states the woman is far more likely than the man to be arrested, brought to trial, convicted, and sentenced.

The third school of feminist thought, the *Marxist (socialist) feminist approach,* is based on the assumption that women are exploited by both capitalism and patriarchy. Because many females have relatively low-wage jobs (if any) and few economic resources, crimes such as prostitution and shoplifting become a means to earn money or acquire consumer goods. However, instead of freeing women from their problems, prostitution institutionalizes women's dependence on men and makes women more vulnerable to criminal prosecution. Lower-income women are further victimized by the fact that they are often the targets of violent acts by lower-class males, who perceive themselves as being powerless in the capitalist economic system.

These approaches make a contribution to our understanding of deviance and crime by focusing on gender as a central concern; however, some theories neglect the centrality of race, ethnicity, class, and sexual orientation in deviance, crime, and the criminal justice system (Figure 6.3).

## Approaches Focusing on the Interaction of Race, Class, and Gender

Some studies have focused on the simultaneous effects of race, class, and gender on deviant behavior and crime. Multiracial feminist approaches have examined how the intersecting systems of race, class, and gender act as "structuring forces" that affect how people act, what opportunities they have available, and how their behavior is socially defined. Examples include how the legal system responds to individual offenders based on their social location in hierarchies based on race, class, and/or gender. Some studies have found, for instance, that young unemployed African American (black) and Hispanic men are treated differentially in sentencing decisions by judges and juries based on their location at the margins of race, age, and gender systems (see Burgess-Proctor, 2006). Other work has also been conducted using an intersectional approach to learn more about domestic violence and other crimes in society.

**FIGURE 6.3** In March of 2013, New Orleans police arrest a young sex worker for soliciting a pedestrian in the French Quarter. Which of the feminist theories of women's crime best explains this young women's offense?

# Symbolic Interactionist Perspectives on Deviance

**LO4** Symbolic interactionists focus on *social processes,* such as how people develop a self-concept and learn conforming behavior through socialization. According to this approach, deviance is learned in the same way as conformity—through interaction with others. Although there are a number of symbolic interactionist perspectives on deviance, we will examine four major approaches—differential association and differential reinforcement theories, rational choice theory, control theory, and labeling theory.

## Differential Association Theory and Differential Reinforcement Theory

How do people learn deviant behavior through their interactions with others? According to the sociologist Edwin Sutherland (1939), people learn the necessary techniques and the motives, drives, rationalizations, and attitudes of deviant behavior from people with whom they associate. *Differential association theory* **states that people have a greater tendency to deviate from societal norms when** **they frequently associate with individuals who are more favorable toward deviance than conformity.** From this approach, criminal behavior is learned within intimate personal groups such as one's family and peer groups.

Differential association theory contributes to our knowledge of how deviant behavior reflects the individual's learned techniques, values, attitudes, motives, and rationalizations. It calls attention to the fact that criminal activity is more likely to occur when a person has frequent, intense, and long-lasting interactions with others who violate the law. However, it does not explain why many individuals who have been heavily exposed to people who violate the law still engage in conventional behavior most of the time.

Criminologist Ronald Akers (1998) has combined differential association theory with elements

**differential association theory** the proposition that people have a greater tendency to deviate from societal norms when they frequently associate with persons who are more favorable toward deviance than conformity.

**FIGURE 6.4** Is this example of graffiti likely to be the work of isolated artists or of gang members? In what ways do gangs reinforce such behavior?

of psychological learning theory to create *differential reinforcement theory,* which suggests that both deviant behavior and conventional behavior are learned through the same social processes. Akers starts with the fact that people learn to evaluate their own behavior through interactions with significant others. If the persons and groups that a particular individual considers most significant in his or her life define deviant behavior as being "right," the individual is more likely to engage in deviant behavior; likewise, if the person's most significant friends and groups define deviant behavior as "wrong," the person is less likely to engage in that behavior. This approach helps explain not only juvenile gang behavior but also how peer cliques on high school campuses have such a powerful influence on people's behavior.

### Rational Choice Theory

Another approach to studying deviance is rational choice theory, which suggests that people weigh the rewards and risks involved in certain types of behavior and then decide which course of action to follow. Rational choice theory is based on the assumption that when people are faced with several courses of action, they will usually do what they believe is likely to have the best overall outcome. The *rational choice theory of deviance* states that deviant behavior occurs when a person weighs the costs and benefits of nonconventional or criminal behavior and determines that the benefits will outweigh the risks involved in such actions. Rational choice approaches suggest that most people who commit crimes do not engage in random acts of antisocial behavior. Instead, they make careful decisions based on weighing the available information regarding *situational factors,* such as the place of the crime, suitable targets, and the availability of people to deter the behavior, and *personal factors,* such as what rewards they may gain from their criminal behavior (Figure 6.4).

How useful is rational choice theory in explaining deviance and crime? A major strength of this theory is that it explains why high-risk youths do not constantly engage in delinquent acts: They have learned to balance risk against the potential for criminal gain in each situation. Moreover, rational choice theory is not limited by the underlying assumption of most social structural theories, which is that the primary participants in deviant and criminal behaviors are people in the lower classes. Rational choice theory also has important policy implications regarding crime reduction or prevention, suggesting that people must be taught that the risks of engaging in criminal behavior far outweigh any benefits they may gain from their actions. Thus, people should be taught *not* to engage in crime.

### Control Theory: Social Bonding

Control theory is based on the assumption that conformity is associated with a person's bonds to other people. According to the sociologist Walter Reckless (1967), society produces pushes and pulls that move people toward criminal behavior; however, some people "insulate" themselves from such pressures by having positive self-esteem and good group cohesion. Reckless suggests that many people do not resort to deviance because of *inner containments*—such as self-control, a sense of responsibility, and resistance to diversions—and *outer containments*—such as supportive family and friends, reasonable social expectations, and supervision by others (Figure 6.5). Those with the strongest containment mechanisms are able to withstand external pressures that might cause them to participate in deviant behavior. As you can see, control/social bonding theories have elements of functionalist and symbolic interactionist perspectives embedded within them because they legitimacy of conventional values and norms. Later, Michael R. Gottfredson and Hirshi (1990) modified the earlier theory that strong social bonds minimize criminal conduct and focused instead on the importance of self-control as a determinate of who will be likely to commit crime. According to Gottfredson and Hirshi (1990), high self-control is related to an individual's likelihood of conforming to norms and laws; low self-control can help explain a person's propensity to commit or refrain from committing crimes. From this perspective, young children who are adequately socialized and have behavioral problems are more likely to grow into delinquents than into adult offenders. Social bond theory and control theory are both rooted in a functionalist assumption about the division of labor in families between men and women, with women being primarily responsible for how children are socialized and whether they become conformists or deviants.

**FIGURE 6.5** According to control theory, strong bonds—including close family ties—are a factor in explaining why many people do not engage in deviant behavior. Why do some sociologists believe that quality family time is more important in discouraging delinquent behavior than is time spent with other young people?

focus on both social control and the bonds that tie people together.

Extending Reckless's containment theory, sociologist Travis Hirschi's (1969) social control theory is based on the assumption that deviant behavior is minimized when people have strong bonds that bind them to families, schools, peers, churches, and other social institutions. *Social bond theory holds that the probability of deviant behavior increases when a person's ties to society are weakened or broken.* According to Hirschi, social bonding consists of (1) *attachment* to other people, (2) *commitment* to conformity, (3) *involvement* in conventional activities, and (4) *belief* in the

## Labeling Theory

*Labeling theory states that deviance is a socially constructed process in which social control agencies designate certain people as deviants and they, in turn, come to accept the label placed upon them and begin to act accordingly.* Based on the symbolic interactionist theory of Charles H. Cooley

**rational choice theory of deviance** the proposition that deviant behavior occurs when a person weighs the costs and benefits of nonconventional or criminal behavior and determines that the benefits will outweigh the risks involved in such actions.

**social bond theory** the proposition that the probability of deviant behavior increases when a person's ties to society are weakened or broken.

**labeling theory** the proposition that deviance is a socially constructed process in which social control agencies designate certain people as deviants and they, in turn, come to accept the label placed upon them and begin to act accordingly.

175

| Primary deviance | | Secondary deviance | | Tertiary deviance | **FIGURE 6.6** A CLOSER |
|---|---|---|---|---|---|
| Initial rule breaking | ➡ | New identity accepted, deviance continues | ➡ | Individual relabels behavior as nondeviant | LOOK AT LABELING THEORY |
| | | | | | © Cengage Learning |

and George H. Mead, labeling theory focuses on the variety of symbolic labels that people are given in their interactions with others.

How does the process of labeling occur? The act of fixing a person with a negative identity, such as "criminal" or "mentally ill," is directly related to the power and status of those persons who *do* the labeling and those who are *being labeled*. Behavior, then, is not deviant in and of itself; it is defined as such by a social audience (Erikson, 1962). According to the sociologist Howard Becker (1963), *moral entrepreneurs* are often the ones who create the rules about what constitutes deviant or conventional behavior. Becker believes that moral entrepreneurs use their own perspectives on "right" and "wrong" to establish the rules by which they expect other people to live. They also label others as deviant. Often these rules are enforced on persons with less power than the moral entrepreneurs. Becker (1963: 9) concludes that the deviant is "one to whom the label has successfully been applied; deviant behavior is behavior that people so label."

As the definition of labeling theory suggests, several stages may occur in the labeling process (see Figure 6.6). **Primary deviance refers to the initial act of rule breaking** (Lemert, 1951). However, if individuals accept the negative label that has been applied to them as a result of the primary deviance, they are more likely to continue to participate in the type of behavior that the label was initially meant to control. *Secondary deviance* **occurs when a person who has been labeled a deviant accepts that new identity and continues the deviant behavior.** For example, a person may shoplift an item of clothing from a department store but not be apprehended or labeled as a deviant. The person may subsequently decide to forgo such behavior in the future. However, if the person shoplifts the item, is apprehended, is labeled as a "thief," and subsequently accepts that label, then the person may shoplift items from stores on numerous occasions. A few people engage in *tertiary deviance,* **which occurs when a person who has been labeled a deviant seeks to normalize the behavior by relabeling it as nondeviant** (Kitsuse, 1980). An example would be drug users who believe that using marijuana or other illegal drugs is no more deviant than drinking alcoholic beverages and therefore should not be stigmatized.

How can labeling theory be used in research on deviance? In a now-classic study that is often used to show how labeling theory works, the sociologist

William Chambliss (1973) studied two groups of adolescent boys in a high school: the "Saints" and the "Roughnecks." Members of both groups were constantly involved in acts of truancy, drinking, wild parties, petty theft, and vandalism. Although the Saints committed more offenses than the Roughnecks, the Roughnecks were the ones who were labeled as "troublemakers" and arrested by law enforcement officials. By contrast, the Saints were described as being the "most likely to succeed," and none of the Saints were ever arrested. According to Chambliss (1973), the Roughnecks were more likely to be labeled as deviants because they came from lower-income families, did poorly in school, and were generally viewed negatively, whereas the Saints came from "good families," did well in school, and were generally viewed positively. Although both groups engaged in similar behavior, only the Roughnecks were stigmatized by a deviant label.

How successful is labeling theory in explaining deviance and social control? One contribution of labeling theory is that it calls attention to the way in which social control and personal identity are intertwined: Labeling may contribute to the acceptance of deviant roles and self-images. However, labeling theory does not explain what caused the original acts that constituted primary deviance, nor does it provide insight into why some people accept deviant labels but others do not.

## Postmodernist Perspectives on Deviance

**LO5** Departing from other theoretical perspectives on deviance, some postmodern theorists emphasize that the study of deviance reveals how the powerful exert control over the powerless by taking away their free will to think and act as they might choose. From this approach, institutions such as schools, prisons, and mental hospitals use knowledge, norms, and values to categorize people into "deviant" subgroups such as slow learners, convicted felons, or criminally insane individuals, and then to control them through specific patterns of discipline.

An example of this idea is found in social theorist Michel Foucault's *Discipline and Punish* (1979), in which Foucault examines the intertwining nature of power, knowledge, and social control. In this study of prisons from the mid-1800s to the early 1900s,

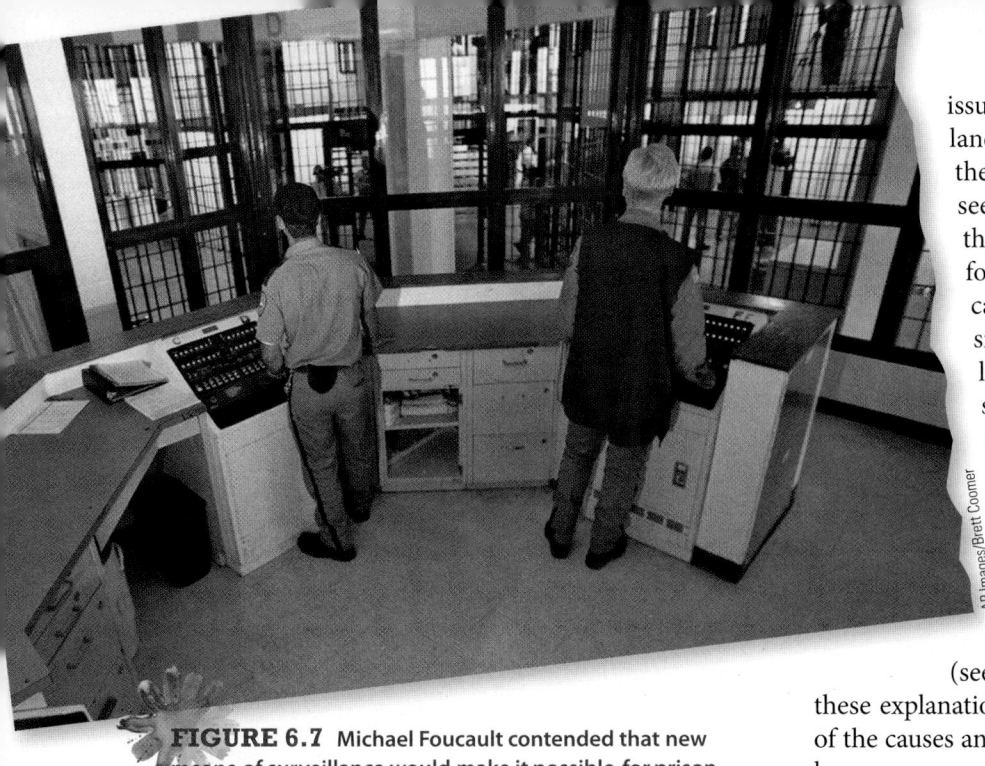

**FIGURE 6.7** Michael Foucault contended that new means of surveillance would make it possible for prison officials to use their knowledge of prisoners' activities as a form of power over the inmates. These guards are able to monitor the activities of many prisoners without ever leaving their station.

Foucault found that many penal institutions ceased torturing prisoners who disobeyed the rules and began using new surveillance techniques to maintain social control. Although the prisons appeared to be more humane in the post-torture era, Foucault contends that the new means of surveillance impinged more on prisoners and brought greater power to prison officials. To explain, he described the *Panoptican*—a structure that gives prison officials the possibility of complete observation of criminals at all times. Typically, the Panoptican was a tower located in the center of a circular prison from which guards could see all the cells (Figure 6.7). Although the prisoners knew they could be observed at any time, they did not actually know when their behavior was being scrutinized. As a result, prison officials were able to use their knowledge as a form of power over the inmates. Eventually, the guards did not even have to be present all the time because prisoners believed that they were under constant scrutiny by officials in the observation post. If we think of this in contemporary terms, we can see how cameras, computers, and other devices have made continual surveillance quite easy in virtually all institutions. In such cases, social control and discipline are based on the use of knowledge, power, and technology.

Foucault's view on deviance and social control has influenced other social analysts, including researchers who have looked at a variety of social issues, ranging from workplace surveillance to squelching music pirating on the Internet. These analyses typically see the world as a modern Panoptican that gives supervisors and law enforcement officials virtually unlimited capabilities for surveillance. Today, smartphones and social media outlets provide new opportunities for surveillance by government officials, corporate supervisors, and others who are not visible to the individuals being watched.

We have examined functionalist, conflict, symbolic interactionist, and postmodernist perspectives on social control, deviance, and crime (see the Concept Quick Review). All of these explanations contribute to our understanding of the causes and consequences of deviant behavior; however, we now turn to the subject of crime itself.

## Crime Classifications and Statistics

**LO6** Crime in the United States can be divided into different categories. We will look first at the legal classifications of crime and then at categories typically used by sociologists and criminologists.

### How the Law Classifies Crime

Crimes are divided into felonies and misdemeanors. The distinction between the two is based on the seriousness of the crime. A *felony* is a serious crime such as rape, homicide, or aggravated assault, for which punishment typically ranges from more than a year's imprisonment to death. A *misdemeanor* is a minor crime that is typically punished by less than one year in jail. In either event, a fine may be part of the sanction as well. Actions that constitute felonies and misdemeanors are determined by the legislatures in the various states; thus, their definitions vary from jurisdiction to jurisdiction.

**primary deviance** the initial act of rule breaking.

**secondary deviance** deviance that occurs when a person who has been labeled a deviant accepts that new identity and continues the deviant behavior.

**tertiary deviance** deviance that occurs when a person who has been labeled a deviant seeks to normalize the behavior by relabeling it as nondeviant.

## Theoretical Perspectives on Deviance

| | Theory | Key Elements |
|---|---|---|
| **Functionalist Perspectives** | | |
| Robert Merton | Strain theory | Deviance occurs when access to the approved means of reaching culturally approved goals is blocked. Innovation, ritualism, retreatism, or rebellion may result. |
| Richard Cloward/Lloyd Ohlin | Opportunity theory | Lower-class delinquents subscribe to middle-class values but cannot attain them. As a result, they form gangs to gain social status and may achieve their goals through illegitimate means. |
| **Conflict Perspectives** | | |
| Karl Marx Richard Quinney | Critical approach | The powerful use law and the criminal justice system to protect their own class interests. |
| Kathleen Daly and Meda Chesney-Lind | Feminist approach | Historically, women have been ignored in research on crime. Liberal feminism views women's deviance as arising from gender discrimination, radical feminism focuses on patriarchy, and socialist feminism emphasizes the effects of capitalism and patriarchy on women's deviance. |
| **Symbolic Interactionist Perspectives** | | |
| Edwin Sutherland | Differential association | Deviant behavior is learned in interaction with others. A person becomes delinquent when exposure to law-breaking attitudes is more extensive than exposure to law-abiding attitudes. |
| Travis Hirschi | Social control/social bonding | Social bonds keep people from becoming criminals. When ties to family, friends, and others become weak, an individual is most likely to engage in criminal behavior. |
| Howard Becker | Labeling theory | Acts are deviant or criminal because they have been labeled as such. Powerful groups often label less-powerful individuals. |
| Edwin Lemert | Primary/secondary deviance | Primary deviance is the initial act. Secondary deviance occurs when a person accepts the label of "deviant" and continues to engage in the behavior that initially produced the label. |
| **Postmodernist Perspective** | | |
| Michel Foucault | Knowledge as power | Power, knowledge, and social control are intertwined. In prisons, for example, new means of surveillance that make prisoners think they are being watched all the time give officials knowledge that inmates do not have. Thus, the officials have a form of power over the inmates. |

## Other Crime Categories

The *Uniform Crime Report (UCR)* is the major source of information on crimes reported in the United States. The UCR has been compiled since 1930 by the Federal Bureau of Investigation based on information filed by law enforcement agencies throughout the country. When we read that the rate of certain types of crimes has increased or decreased when compared with prior years, for example, this information is usually based on UCR data.

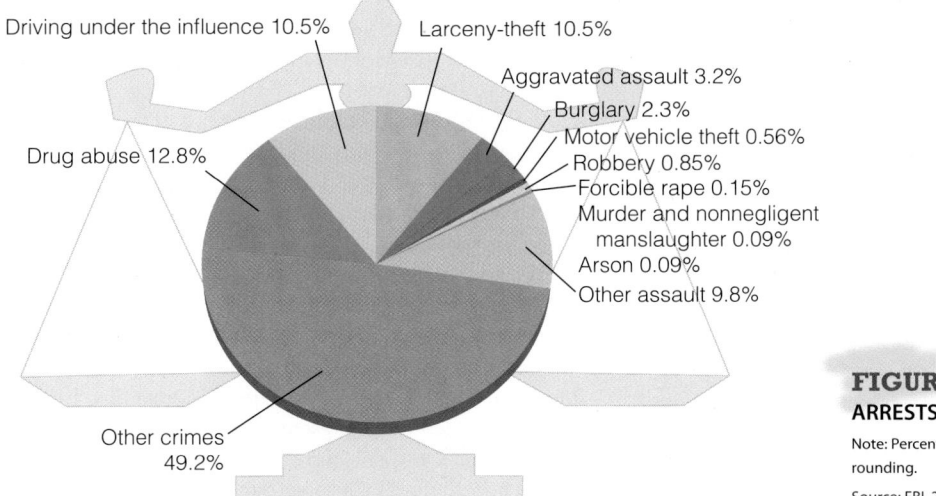

Driving under the influence 10.5%
Larceny-theft 10.5%
Aggravated assault 3.2%
Burglary 2.3%
Motor vehicle theft 0.56%
Robbery 0.85%
Forcible rape 0.15%
Murder and nonnegligent manslaughter 0.09%
Arson 0.09%
Other assault 9.8%
Drug abuse 12.8%
Other crimes 49.2%

**FIGURE 6.8** DISTRIBUTION OF ARRESTS BY TYPE OF OFFENSE, 2012

Note: Percentages may not equal 100% because of rounding.

Source: FBI, 2013.

The UCR focuses on violent crime and property crime (which, prior to 2004, were jointly referred to in that report as "index crimes") but also contains data on other types of crime (see Figure 6.8). In 2012 an estimated 12,196,959 arrests were made for all criminal infractions (excluding traffic violations) in the United States. Although the UCR gives some indication of crime, the figures do not reflect the *actual* number and kinds of crimes, as will be discussed later.

**Violent Crime** *Violent crime* consists of actions—murder, rape, robbery, and aggravated assault—involving force or the threat of force against others. Although only 4.3 percent of all arrests in the United States in 2012 were for violent crimes, this category is probably the most anxiety-provoking of all criminal behavior: Most of us know someone who has been a victim of violent crime, or we have been so ourselves. Victims are often physically injured or even lose their lives; the psychological trauma may last for years after the event. Violent crime also receives the most sustained attention from the media (see "Violent Crime in the Media").

**Property Crime** *Property crimes* include burglary (breaking into private property to commit a serious crime), motor vehicle theft, larceny-theft (theft of property worth $50 or more), and arson. In the United States a property crime occurs, on average, once every 3.5 seconds; a violent crime occurs, on average, once every 26 seconds (see Figure 6.9). In most property crimes the primary motive is to obtain money or some other desired valuable.

**Public Order Crime** *Public order crimes* (sometimes referred to as "morals" crimes) involve an illegal action voluntarily engaged in by the participants, such as prostitution, illegal gambling, the private use of illegal drugs, and illegal pornography. Many people assert that such conduct should not be labeled as a crime; these offenses are often referred to as *victimless crimes* because they **involve a willing exchange of illegal goods or services among adults.** However, public order crimes can include children and adolescents as well as adults. Young children and adolescents may unwillingly become child pornography "stars" or prostitutes.

**Occupational and Corporate Crime** Although the sociologist Edwin Sutherland (1949) developed the theory of white-collar crime more than sixty years ago, it was not until the 1980s that the public became fully aware of its nature. *Occupational (white-collar) crime* comprises illegal activities committed by people in the course of their employment or financial affairs.

In addition to acting for their own financial benefit, some white-collar offenders become involved in criminal conspiracies designed to improve the

**violent crime** actions—murder, rape, robbery, and aggravated assault—involving force or the threat of force against others.

**property crimes** burglary (breaking into private property to commit a serious crime), motor vehicle theft, larceny-theft (theft of property worth $50 or more), and arson.

**victimless crimes** crimes involving a willing exchange of illegal goods or services among adults.

**occupational (white-collar) crime** illegal activities committed by people in the course of their employment or financial affairs.

# VIOLENT CRIME IN THE MEDIA

## Causing More Harm Than Good?

If you have people bringing dolls or flowers to makeshift memorials and they're crying, that's a powerful image, it's part of this story, it's part of our history right now, and we have to deal with it.
  —Wolf Blitzer, television anchor for CNN, describing why reporters sometimes remain on the scene of a tragic killing rampage for days and even weeks after a tragedy such as the Sandy Hook School shootings occurs (qtd. in Applebome and Stelter, 2012)

In the aftermath of every U.S. mass shooting in recent years, communities have been overrun with worldwide media outlets bringing in broadcasting trucks, bright lights, cameras, and hundreds of reporters and commentators to continually describe, analyze, and reanalyze what might have happened to cause such a tragic event. Some information provided by the media later proves to be accurate; however, other statements are found to be flawed or completely inaccurate. In the Newtown tragedy, for example, some initial media reports misidentified the shooter, provided inaccurate information about the shooter's mother (who was killed), and misstated how the shooter initially got into the school building where the violence occurred (Applebome and Stelter, 2012).

From a sociological perspective, it is important to ask "Does media framing of stories about violence produce more harm than good in informing the public?" Let's consider both sides of this issue: On the one hand, people learn details about crime and violence that they otherwise might not know about or experience firsthand. This information may help them form an opinion on issues such as school safety, gun control, and the nature and extent of violence in the United States. Stories that focus on those who lost their life, such as personal stories about the children and teachers who died at Newtown, give voice to the grief of relatives and friends while helping strangers know more about those who were killed and offering survivors a way to honor a person and remember his or her life.

On the other hand, continual media reports about a tragedy provide us with endless details about the crime and the fear that it has produced in survivors, families, neighbors, and others in the community. Many media presentations of crime prey on fear, disbelief, and distrust. Without a systematic framework for thinking about serious violent crimes, media audiences are left with representations of crime as the product of "bad people" and evil in the world—ideas which divert attention from important linkages that these crimes have to the larger social structure or culture (Reiner, 2007). For people who are already fearful of violent crime, media representations confirm these concerns and make individuals more afraid to leave home and interact with others. Violence leaves millions worrying about their safety and that of their children and other family members, even if the statistical odds of a similar occurrence in their own community are small. Recently, extensive discussions have taken place about the effects of media reporting on individuals' perception of violent crime, particularly those occurrences that are the most devastating. Will such discussions affect how various forms of media report on mass murders and other horrendous incidents? Or will people go back to "business as usual" until the next dreadful event occurs?

### reflect & analyze

When should media representatives say "Enough is enough!" and leave grieving families and communities alone to work through their shared losses? Do some journalists provide too many graphic details about gruesome events such as mass shootings, or does the public have a right to know all of the details?

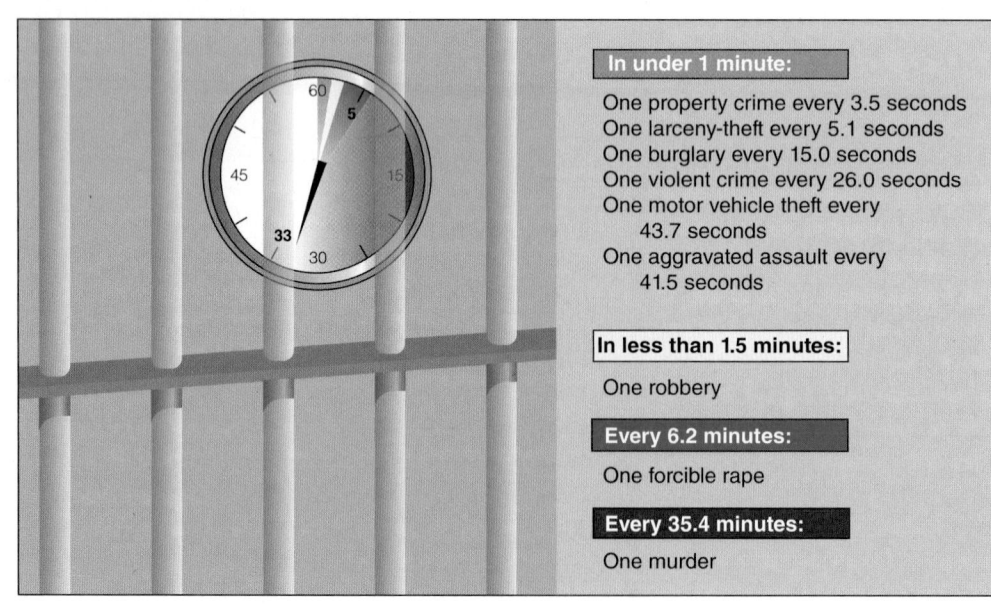

In under 1 minute:
One property crime every 3.5 seconds
One larceny-theft every 5.1 seconds
One burglary every 15.0 seconds
One violent crime every 26.0 seconds
One motor vehicle theft every
    43.7 seconds
One aggravated assault every
    41.5 seconds

In less than 1.5 minutes:
One robbery

Every 6.2 minutes:
One forcible rape

Every 35.4 minutes:
One murder

FIGURE 6.9 THE FBI CRIME CLOCK, 2012

Source: FBI, 2012.

market share or profitability of their companies. This is known as *corporate crime*—**illegal acts committed by corporate employees on behalf of the corporation and with its support.** Examples include antitrust violations, tax evasion, misrepresentations in advertising, price fixing, financial fraud, and infringements on patents, copyrights, and trademarks. These crimes are a result of deliberate decisions made by corporate personnel to enhance resources or profits at the expense of competitors, consumers, and the general public.

Although people who commit occupational and corporate crimes can be arrested, fined, and sent to prison, some people are less likely to regard such behavior as "criminal." In many cases, punishment for such offenses has been a fine and a relatively brief prison sentence in settings sometimes referred to as "country club prisons" because of their amenities.

For many years, public concern and media attention focused primarily on the street crimes disproportionately committed by persons who were poor, powerless, and nonwhite. Today, however, part of our focus has shifted to crimes committed by top banking officials in corporate suites, such as fraud, tax evasion, and insider trading by executives at large and well-known corporations. For example, Occupy Wall Street and similar protest movements in the second decade of the twenty-first century have highlighted how the collapse of the housing and financial markets, partly because of the greed of wealthy individuals and corporate entities, has seriously harmed the lifestyles and opportunities of millions of Americans. In one of the most notorious of these instances, Bernard Madoff, the former chairperson of NASDAQ, admitted to defrauding his clients of up to $50 billion in a massive scheme that took place over a number of years. Madoff was sentenced to life in prison in 2008. However, he was not an isolated example of such criminal endeavors: Numerous occupational and corporate criminals have engaged in activities that have cost other people billions of dollars and, in some cases, their life savings.

Corporate crimes are often more costly in terms of money and lives lost than street crimes. Thousands of jobs and billions of dollars have been lost annually as a result of corporate crime (Reiman and Leighton, 2010). Deaths resulting from corporate crimes such as polluting the air and water, manufacturing defective products, and selling unsafe foods and drugs far exceed the number of deaths caused by homicide each year. Other costs include the effect on the moral climate of society (Simon, 2012). Throughout the United States the confidence of everyday people in the nation's economy has been shaken badly by the greedy and illegal behavior of corporate insiders.

**Internet Crime** *Internet crime* consists of FBI-related scams, identity theft, advance fee fraud, nonauction/nondelivery of merchandise, and overpayment fraud. The proliferation of computers and Internet access worldwide has contributed to the growth of lucrative crimes in which the victim never meets the perpetrator. Auto fraud is one of the most frequently reported Internet crimes, in which perpetrators attempt to sell vehicles they do not own. Typically, they advertise vehicles for sale on the Internet at prices well below market value and claim that they must sell the vehicles quickly because of personal issues, such as relocation for work, deployment by the military, or a family crisis. Victims are instructed to wire full or partial payment to a third-party agent without meeting the seller or inspecting the vehicle prior to purchase. Because victims think they are getting a good deal, they contact their bank for a wire transfer of the money. After the third-party agent receives the victim's money, the perpetrator pockets the cash but does not deliver the vehicle.

Another frequently reported Internet crime is the FBI-related scam, in which a perpetrator poses as a high-ranking government official or as the FBI to defraud victims (see Figure 6.10). The perpetrator sends an e-mail that looks legitimate because it contains an "official" government letterhead or the FBI seal, and the message contains information about money or property that the recipient allegedly inherited, bogus lottery winning notifications, or occasionally extortion threats. However, the recipient must contribute a specific amount of money to gain the inheritance, lottery, or other prize. Although no monetary losses are incurred in some cases, public trust is still undermined, and these scams pose a viable threat to national security (Internet Crime Complaint Center, 2013).

Real estate fraud occurs when perpetrators find websites that list homes for sale and take legitimate information and misuse it with their own e-mail addresses on Craigslist (without consent) under the housing rental category. Victims are instructed to send money for the first and last month's rent via a wire transfer to an overseas account. By the time

**corporate crime** illegal acts committed by corporate employees on behalf of the corporation and with its support.

**Internet crime** illegal acts committed on the Internet, including FBI-related scams, identity theft, advance fee fraud, nonauction/nondelivery of merchandise, and overpayment fraud.

Romance scams
4,476

Real-estate fraud
14,432

Scareware/ransomware
1,969

Hit-man scams
1,354

Intimidation/
extortion scams
8,324

Auto fraud
17,159

FBI impersonation
e-mail scams
14,141

**FIGURE 6.10** TOP REPORTED INTERNET CRIME TYPES

Source: Internet Crime Complaint Center, 2013.

they find out that this is a scam, their money is long gone, and they have no legitimate claim on the house that they hoped to occupy for a below-market rental rate. Other forms of real estate fraud include time-share marketing schemes and loan-modification scams in which a bogus loan company contacts a homeowner who is having financial problems and offers a loan-modification plan in which the homeowner stops making mortgage payments and starts sending the money, along with additional fees, to the bogus loan company instead.

Other frequently reported Internet crimes include the intimidation/extortion scam, such as payday loans or grandparent scams in which older individuals are targeted by fraudsters who claim to be a relative in a legal or financial crisis and make an urgent plea for money to be sent to help them. Using another twist, romance scams are perpetrated by individuals who promise love and romance to online victims they meet in chat rooms, dating sites, and social media networks. Romance scams have grown in popularity because it is relatively easy to target individuals who seek companionship or romance online. The victim believes that he or she is becoming acquainted with, or dating, a real person online because scammers sometimes write poetry or send flowers and gifts to victims until they persuade them to part with their money. Romance scams alone accounted for $55 million lost in 2012.

Scareware and ransomware schemes are hoaxes pulled on computer users through pop-up messages which alert individuals that their computers are infected with viruses and bait them into purchasing nonexistent or useless software that allegedly will remove the viruses from their computer. One of the most frightening of the Internet schemes is the Hit Man Scam, in which scammers send an e-mail claiming to be a hit man hired to kill the victim.

Recipients must pay money to make sure that the hit man does not carry out the death contract. More recently, scammers using this approach have gained personal information about the alleged victim based on what he or she has posted on social media sites, giving more credibility to their threat.

Internet crime continues to grow because perpetrators can so easily change their tactics. According to the FBI, the total loss involving known Internet crimes is more than $525 million annually (Internet Crime Complaint Center, 2013).

**Organized Crime** *Organized crime* **is a business operation that supplies illegal goods and services for profit.** Premeditated, continuous illegal activities of organized crime include drug trafficking, prostitution, loan-sharking, money laundering, and large-scale theft such as truck hijackings (Simon, 2012). No single organization controls all organized crime; rather, many groups operate at all levels of society. In recent decades, organized crime in the United States has become increasingly transnational in nature. The globalization of the economy and the introduction of better communications technology have made it possible for groups around the world to operate in the United States and other nations. The FBI has identified a number of major categories of organized crime threats in the United States, as shown in Figure 6.11.

Organized crime thrives because there is great demand for illegal goods and services. Criminal organizations initially gain control of illegal activities by combining threats and promises. For example, small-time operators running drug or prostitution rings may be threatened with violence if they compete with organized crime or fail to make required payoffs.

Apart from their illegal enterprises, organized crime groups have infiltrated the world of legitimate

## FIGURE 6.11 ORGANIZED CRIME THREATS

*Includes groups born in or with families from the former Soviet Union or Central Europe.

Source: www.fbi.gov/hq/cid/orgcrime/ocshome.htm.

FIGURE 6.12 Australian Julian Assange, founder of Wikileaks, engaged in political crime when he published classified material provided by U.S. Army Private First Class Bradley Manning.

business. Known linkages between legitimate businesses and organized crime exist in banking, hotels and motels, real estate, garbage collection, vending machines, construction, delivery and long-distance hauling, garment manufacture, insurance, stocks and bonds, vacation resorts, and funeral homes. In addition, some law enforcement and government officials are corrupted through bribery, campaign contributions, and favors intended to buy them off.

In the future, organized crime groups may further intensify their illegal activities in global drug and sex trafficking, identity theft, Internet scams, and other forms of cybercrime that are even more difficult to detect and can produce millions of dollars in a short period of time.

**Political Crime** The term *political crime* **refers to illegal or unethical acts involving the usurpation of power by government officials or illegal/unethical acts perpetrated against the government by outsiders seeking to make a political statement, undermine the government, or overthrow it.** Government officials may use their authority unethically or illegally for the purpose of material gain or political power (Simon, 2012). They may engage in graft (taking advantage of political position to gain money or property) through bribery, kickbacks, or "insider" deals that financially benefit them.

Some acts committed by agents of the government against persons and groups believed to be threats to national security are also classified as political crimes. Four types of political deviance have been attributed to some officials: (1) secrecy and deception designed to manipulate public opinion, (2) abuse of power, (3) prosecution of individuals because of their political activities, and (4) official violence, such as police brutality against people of color or the use of citizens as unwilling guinea pigs in scientific research (Simon, 2012).

Outsiders may also engage in political crime. For example, Australian Julian Assange, founder of WikiLeaks—a website that disseminates information from whistleblowers worldwide—published classified material provided by U.S. Army Private First Class Bradley Manning (Figure 6.12). In 2010 WikiLeaks published U.S. diplomatic cables about U.S. involvement in the wars in Iraq and Afghanistan, including many classified documents that contained the names of people who had spoken confidentially to officials. The release of this information might have resulted in their being fired, prosecuted, physically injured, or even killed. This type of political crime is possible because newer technologies

**organized crime** a business operation that supplies illegal goods and services for profit.

**political crime** illegal or unethical acts involving the usurpation of power by government officials or illegal/unethical acts perpetrated against the government by outsiders seeking to make a political statement, undermine the government, or overthrow it.

have made it easy for information to be gathered and stored beyond the reach of any one government and for large amounts of data to be released instantaneously and globally without going through processes such as the "declassification" of information, which typically takes months or years to accomplish.

## Crime Statistics

**LO7** How useful are crime statistics as a source of information about crime? As mentioned previously, official crime statistics provide important information on crime; however, the data reflect only those crimes that have been reported to the police.

Why are some crimes not reported? People are more likely to report crime when they believe that something can be done about it (apprehension of the perpetrator or retrieval of their property, for example). About half of all assault and robbery victims do not report the crime because they may be embarrassed or fear reprisal by the perpetrator. Thus, the number of crimes reported to police represents only the proverbial "tip of the iceberg" when compared with all offenses actually committed. Official statistics are problematic in social science research because of these limitations.

The *National Crime Victimization Survey* was developed by the Bureau of Justice Statistics as an alternative means of collecting crime statistics. Known as the NCVS, this annual survey collects information on nonfatal crimes reported and not reported to the police against persons age 12 or older from a nationally representative sample of U.S. households. In 2012 members of 92,390 households were interviewed in person to determine if they had been the victims of crimes even if they had not reported the crime to law enforcement officials. Between 2011 and 2012, the number of violent crimes, known and unknown to the police, reported in the NCVS increased by 17.7 percent, as compared to an increase of only 0.7 percent in violent crimes as shown in the Uniform Crime Report and thus known to the police. Similarly, the number of property crimes reported in the NCVS increased by 15 percent, as compared to a decline of 0.9 percent in the number of property crimes reported in the UCR. These figures suggest that underreporting occurs in both violent and property crimes, and that having data from the NCVS provides additional information about the nature and extent of criminal activity in the United States (Truman, Langton, and Planty, 2013).

Studies based on anonymous self-reports of criminal behavior also reveal much higher rates of crime than those found in official statistics. For example, self-reports tend to indicate that adolescents of all classes violate criminal laws. However, not all children who commit juvenile offenses are apprehended and referred to court. Some children from white, affluent families have their cases handled outside the juvenile justice system (for example, a youth may be sent to a private school or hospital rather than to a juvenile court and a public correctional facility).

Some crimes committed by persons of higher socioeconomic status in the course of business are handled by an administrative or quasi-judicial body, such as the Securities and Exchange Commission or the Federal Trade Commission, or by civil courts. As a result, many elite crimes are never classified as "crimes," nor are the businesspeople who commit them labeled as "criminals."

## Terrorism and Crime

In the twenty-first century the United States and other nations are confronted with a difficult prospect: how to deal with terrorism. **Terrorism is the calculated, unlawful use of physical force or threats of violence against persons or property in order to intimidate or coerce a government, organization, or individual for the purpose of gaining some political, religious, economic, or social objective.** An example of terrorism in the United States was the 2013 bombings at the Boston Marathon, when multiple bombs were used to target innocent civilians who were participants in and spectators at this racing event. The most devastating act of terrorism in the United States took place on September 11, 2001 (the "9/11 attacks"), when nearly 3,000 people, including 19 airplane hijackers, lost their lives in attacks on the World Trade Center in New York and the Pentagon in Washington, D.C. In the aftermath of this tragic event, national security in the United States was revamped and intensified in an effort to prevent the recurrence of these tragedies.

How are sociologists and criminologists to explain world terrorism, which may have its origins in more than one nation and include diverse "cells" of terrorists who operate in a somewhat gang-like manner but are believed to be following directives from leaders elsewhere? In order to deal with the aftermath of terrorist attacks, government officials typically focus on "known enemies" such as Osama bin Laden, who was killed by U.S. military forces in 2011. The nebulous nature of the "enemy" and the problems faced by any one government trying to identify and apprehend the perpetrators of acts of terrorism have resulted in a global "war on terror."

Social scientists who use a rational choice approach suggest that terrorists are rational actors who constantly calculate the gains and losses of

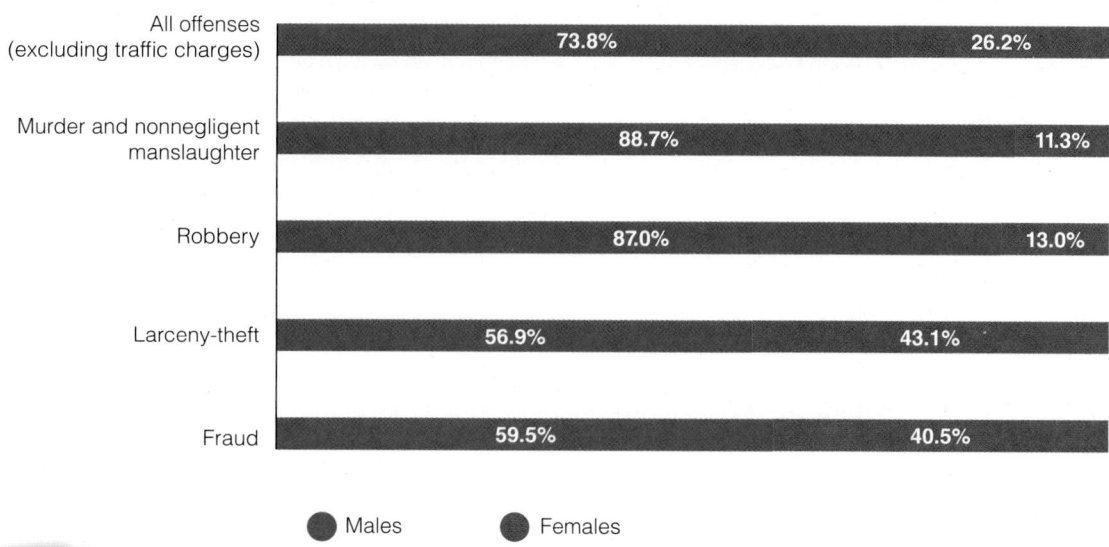

**FIGURE 6.13** ARREST RATES BY GENDER, 2012 (SELECTED OFFENSES)

participation in violent—and sometimes suicidal—acts against others. For example, terrorists created a climate of fear prior to the 2014 Winter Olympic Games, hosted by Russia, through a series of deadly terror attacks by suicide bombers prior to the games. Major sporting events, including the Boston Marathon, have been the targets of terrorists who act when the whole world is watching and seek to put their beliefs or causes out for everyone to see. Unfortunately, the bombings of buses, schools, and other public accommodations become frequent occurrences and provide a venue for persons to kill or threaten to kill many innocent civilians and to destroy millions of dollars in property. *Rational choice* seems an odd term for behavior such as this.

## Street Crimes and Criminals

Given the limitations of official statistics, is it possible to determine who commits crimes? We have much more information available about conventional (street) crime than elite crime; therefore, statistics concerning street crime do not show who commits all types of crime. Gender, age, class, and race are important factors in official statistics pertaining to street crime.

**Gender and Crime** There is a gender gap in crime statistics: Males are arrested for significantly more crimes than females. In 2012 about 74 percent of all persons arrested nationwide were male. Males made up slightly more than 80 percent of persons arrested for violent crimes and almost 63 percent of all persons arrested for property crimes (FBI, 2013). Females have higher arrests rates than males for

prostitution, commercial vice, and embezzlement, but in all other categories, males have higher rates.

Before further consideration of differences in crime rates by males and females, three similarities should be noted. First, the most common arrest categories for both men and women are driving under the influence of alcohol, drug abuse violations, larceny, and minor or criminal mischief types of offenses. These categories account for about 47 percent of all male arrests and about 49 percent of all female arrests. Second, liquor-law violations (such as underage drinking), simple assault, and disorderly conduct are middle-range offenses for both men and women. Third, the rate of arrests for murder, arson, and embezzlement is relatively low for both men and women.

The most important gender differences in arrest rates are reflected in the proportionately greater involvement of men in major property crimes (such as robbery and larceny-theft) and violent crime, as shown in Figure 6.13. In 2012 men accounted for almost 88 percent of robberies and murders and almost 57 percent of all larceny-theft arrests in the United States. Property crimes for which women are most frequently arrested are nonviolent in nature, including shoplifting, theft of services, passing bad checks, credit card fraud, and employee pilferage.

**terrorism** the calculated, unlawful use of physical force or threats of violence against persons or property in order to intimidate or coerce a government, organization, or individual for the purpose of gaining some political, religious, economic, or social objective.

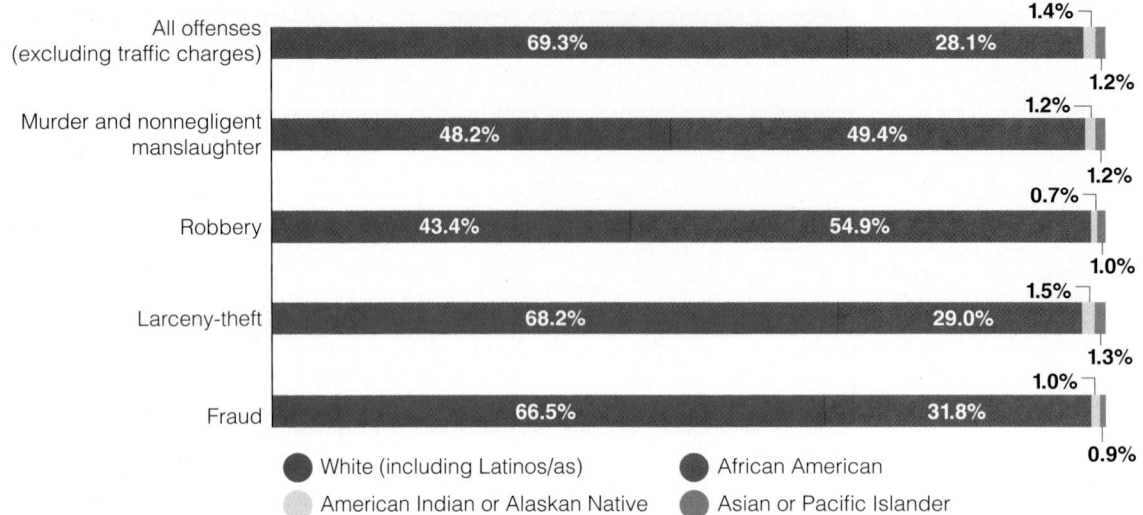

**FIGURE 6.14** ARREST RATES BY RACE, 2012 (SELECTED OFFENSES)

Note: Classifications as used in the Uniform Crime Report.

Source: FBI, 2013.

Often when women are arrested for serious violent and property crimes, they are seen as accomplices of the men who planned the crime and instigated its commission; however, this assumption frequently does not prove true today. Some studies have found that some women play an active role in planning and carrying out robberies and other major crimes.

**Age, Class, and Crime** Of all factors associated with crime, the age of the offender is one of the most significant. Arrest rates for violent crime and property crime are highest for people between the ages of 13 and 25, with the peak being between ages 16 and 17. In 2012 individuals under age 18 accounted for nearly 12 percent of all arrests for violent crime and 18 percent of all arrests for property crime. Persons under age 18 were arrested for 26 percent of all vandalism and 21 percent of all disorderly conduct charges (FBI, 2013).

The median age of those arrested for aggravated assault and homicide is somewhat older, generally in the late twenties. Typically, white-collar criminals are even older because it takes time to acquire both a high-ranking position and the skills needed to commit this particular type of crime. At every age and for nearly all offenses, rates of arrest remain higher for males than females. This female-to-male ratio remains fairly constant across all age categories. The most significant gender difference in the age curve is for prostitution (a nonviolent crime). In 2012 nearly 60 percent of all women arrested for prostitution were under age 30. For individuals over age 45, many more men than women are arrested for sex-related offenses (including procuring the services of a prostitute). More stringent enforcement

of prostitution statutes often exists when young females are involved, and opportunities for prostitution are typically greater for younger women.

Individuals from all classes commit crimes; they simply commit different kinds of crimes. Persons from lower socioeconomic backgrounds are more likely to be arrested for violent and property crimes. By contrast, persons from the upper part of the class structure generally commit white-collar or elite crimes, although only a small proportion of these individuals will be arrested or convicted of a crime.

**Race and Crime** Are people from some racial–ethnic categories more likely to be arrested for committing a crime? In 2012 whites (including Hispanics or Latinos/as) accounted for about 69 percent of *all arrests*. When these data were gathered, it was impossible to separate out arrest rates for white (non-Hispanic) offenders and Hispanic (Latino/a) offenders. However, the FBI started collecting nationwide data on ethnicity in 2014 to bring UCR data closer in line with U.S. Census Bureau statistics. In 2012 the total estimated arrest rate for African Americans was 28 percent, as shown in Figure 6.14. The total arrest rate for American Indians or Alaskan Natives was 1.4 percent, slightly above the 1.2 percent for Asians or Pacific Islanders. African Americans made up more than 49 percent of all arrests for murder and nonnegligent manslaughter. By contrast, arrest rates for whites were higher for nonviolent property crimes, including more than 68 percent of all arrests for larceny-theft and slightly more than 66 percent for fraud (FBI, 2013).

Although official arrest records reveal certain trends, these data tell us very little about the actual dynamics of crime by racial–ethnic category. According to official statistics, African Americans are overrepresented in arrest data. African Americans made up over 13 percent of the U.S. population in 2012 but accounted for 28 percent of all arrests. In some areas, African Americans were the vast majority of those arrested for certain categories of crime. For example, African Americans made up 67 percent of arrests for gambling and almost 55 percent of arrests for robbery. Likewise, African Americans are more likely than people in other racial or ethnic classifications to become crime victims (FBI, 2013). What sociological factors might account for this disparity in crime statistics?

In 2012 Native Americans (designated in the UCR as "American Indians" or "Alaskan Natives") accounted for 1.4 percent of all arrests, although their highest-ranking crimes were drunkenness, liquor-law violations, vagrancy, disorderly conduct, and offenses against the family (FBI, 2013). In that same year, 1.2 percent of all arrests were of Asians or Pacific Islanders. Among the higher percentages of arrests for members of this category were prostitution, commercialized vice, gambling, and embezzlement (FBI, 2013).

Are arrest statistics a true reflection of crimes committed? According to criminologists Jeffrey Reiman and Paul Leighton (2010), arrest statistics reflect the UCR's focus on violent and property crimes, especially property crimes, which are committed primarily by low-income people. This emphasis draws attention away from the white-collar and elite crimes committed by middle- and upper-income people (Figure 6.15). Police may also demonstrate bias and racism in their decisions regarding whom to question, detain, or arrest under certain circumstances (Reiman and Leighton, 2010).

Statistics may also show that a disproportionate number of people of color are arrested because law enforcement officials may focus on certain types of crime and direct their attention to specific "high-crime" neighborhoods in which deviance is believed to be more prevalent. As discussed previously, many poor, young, central-city males turn to criminal activity because of lack of education, job skills, and employment opportunities, giving them little or no chance of earning a living wage through legitimate employment. Law enforcement has focused on drug-related offenses, particularly among young people of color, and arrest rates have been high for this type of crime.

Nick Koudis/Photodisc/Getty Images

**FIGURE 6.15** According to Karl Marx, capitalism produces haves and have-nots, and each group engages in different types of crime. Statistically, the man being arrested here is much more likely to be suspected of a financial crime than a violent crime.

Finally, arrests should not be equated with guilt: Being arrested does not necessarily mean that a person is guilty of the crime with which he or she has been charged. In the United States, individuals accused of crimes are, at least theoretically, "innocent until proven guilty."

## Crime Victims

How can we learn more about crime victims? As previously discussed, the National Crime Victimization Survey (NCVS), compiled by the U.S. Census Bureau for the Bureau of Justice Statistics, provides annual data about crimes (reported and not reported to the police) from which we can find out more about who is actually victimized by crime. Based on this data, researchers found that the rate of violent victimization increased from 22.6 per 1,000 persons age 12 or older in 2011 to 26.1 in 2012. An increase in simple assault accounted for most of the rise in total violence between 2011 and 2012 (Truman, Langton, and Planty, 2013). During the same time period, the rate of property crime increased from 138.7 per 1,000 households to 155.8. The majority of the increase in property crime has been attributed to thefts.

Sex is important in looking at victimization studies. Males had a higher rate of total violent victimization than females. The rate of violent victimization for males ages 12 and older was 29.1 per 1,000 in 2012, up from 25.5 per 1,000 in 2011. The rate for female victimization was 23.3 per 1,000, up from 19.8 in 2011. Violent victimization includes these categories: rape/sexual assault, robbery, total assault, aggravated assault, and simple assault. As might be expected, robbery victimization was higher for males; victimization for rape or sexual assault was higher for females.

Race and age are also important factors in studying victimization. In terms of violent victimization, American Indians/Alaska Natives had the highest rates (46.9 per 1,000), followed by persons of two or more races (42.8 per 1,000), when compared to whites (non-Hispanics), black/African Americans, Hispanics/Latinos, and Asians/Native Hawaiians/other Pacific Islanders. Asians and Pacific Islanders have the lowest rates of both violent and property crime victimizations. Persons ages 12 to 17 had higher violent victimization rates (48.4 per 1,000) when compared to those between the ages of 25 and 34 (34.2 per 1,000) and those who are older (Truman, Langton, and Planty, 2013).

What can we learn from victimization studies? Because all crimes are not reported, we can find out more about the nature and extent to which people in specific regions of the country, income categories, racial or ethnic groupings, ages, and other demographic characteristics are victimized by violent and property crimes in the United States. These data can be compared with official crime statistics to see if arrest rates and convictions are an accurate reflection of crime in the United States. The NCVS particularly helps us to learn about the types and number of offenses that go unreported to police or other officials who are part of the criminal justice system.

# The Criminal Justice System

**LO8** Of all the agencies of social control (including families, schools, and churches) in contemporary societies, only the criminal justice system has the power to control crime and punish those who are convicted of criminal conduct. The *criminal justice system* **refers to the local, state, and federal agencies that enforce laws, adjudicate crimes, and treat and rehabilitate criminals.** The system includes the police, the courts, the correctional facilities, and the people in myriads of police agencies, courts, prosecutorial agencies, correctional institutions, and probation and parole departments.

The term *criminal justice system* is somewhat misleading because it implies that law enforcement agencies, courts, and correctional facilities constitute one large, integrated system when, in reality, the criminal justice system is made up of many bureaucracies that have considerable discretion in how decisions are made. *Discretion* refers to the use of personal judgment by police officers, prosecutors, judges, and other criminal justice system officials regarding whether and how to proceed in a given situation (see Figure 6.16). The police are a prime example of discretionary processes because they have the power to selectively enforce the law and have on many occasions been accused of being too harsh or too lenient on alleged offenders.

## The Police

The role of the police in the criminal justice system continues to expand. The police are responsible for crime control and maintenance of order, but local police departments now provide numerous other human-service functions, including improving community relations, resolving family disputes, and helping people during emergencies. Not all "police officers" are employed by local police departments; they are also employed in governmental agencies ranging from local jurisdictions to federal levels. However, we will focus primarily on metropolitan

**Police**

- Enforce specific laws
- Investigate specific crimes
- Search people, vicinities, buildings
- Arrest or detain people

**Prosecutors**

- File charges or petitions for judicial decision
- Seek indictments
- Drop cases
- Reduce charges
- Recommend sentences

**Judges or Magistrates**

- Set bail or conditions for release
- Accept pleas
- Determine delinquency
- Dismiss charges
- Impose sentences
- Revoke probation

**FIGURE 6.16** DISCRETIONARY POWERS IN LAW ENFORCEMENT

police departments because they constitute the vast majority of the law enforcement community.

Large urban police departments are made up of a chain of command (similar to the military), with ranks such as officer, sergeant, lieutenant, and captain, and each rank must follow specific rules and procedures. However, individual officers maintain a degree of discretion in the decisions they make as they respond to calls and try to apprehend fleeing or violent offenders. The problem of police discretion is most acute when decisions are made to use force (such as grabbing, pushing, or hitting a suspect) or deadly force (shooting and killing a suspect). Generally, deadly force is allowed only in situations in which a suspect is engaged in a felony, is fleeing the scene of a felony, or is resisting arrest and has endangered someone's life.

Although many police departments have worked to improve their public image in recent years, the practice of *racial profiling*—the use of ethnic or racial background as a means of identifying criminal suspects—remains a highly charged issue. Officers in some police departments have singled out for discriminatory treatment African Americans, Latinos/as, and other people of color, treating them more harshly than white (Euro-American) individuals. However, police department officials typically contend that race is only one factor in determining why individuals are questioned or detained as they go about everyday activities such as driving a car or walking down the street. By contrast, equal-justice advocacy groups argue that differential treatment of minority-group members amounts to a race-based double standard, which they believe exists not only in police work but also throughout the criminal justice system (e.g., Ridgeway, 2006; Smith and Hatter, 2011).

The belief that differential treatment takes place on the basis of race contributes to a negative image of police among many people of color who believe that they have been hassled by police officers, and this assumption is intensified by the fact that police departments have traditionally been made up of white male personnel at all levels. In recent years this situation has slowly begun to change. Slightly less than 25 percent of all *sworn officers*—those who have taken an oath and been given the powers to make arrests and use necessary force in accordance with their duties—are women and minorities. Women accounted for about 13 percent of all sworn officers in local police departments in 2013. The largest percentage of women and minority police officers are located in cities with a population of 250,000 or more. African Americans make up a larger percentage of the police department in cities with a larger proportion of African American residents (such as Detroit), but Latinos/as constitute a larger percentage in cities such as San Antonio and El Paso, Texas, where Latinos/as make up a larger proportion of the population. Women officers of all races are more likely to be employed in departments in cities of more than 250,000 as compared with smaller communities (cities of fewer than 50,000), where women officers typically constitute a small percentage of the force.

Police departments now place greater emphasis on *community-oriented policing*—an approach to law enforcement in which officers maintain a presence in the community, walking up and down the streets or riding bicycles, getting to know people, and holding public service meetings at schools, churches, and other neighborhood settings. Community-oriented policing is often limited by budget constraints and the lack of available personnel to conduct this type of "hands-on" community involvement. In many jurisdictions, police officers believe that they have only enough time to keep up with reports of serious crime and life-threatening occurrences; in this view, the level of available personnel and resources does not allow officers to take on a greatly expanded role in the community.

## The Courts

Criminal courts determine the guilt or innocence of those persons accused of committing a crime. In theory, justice is determined in an adversarial process in which the prosecutor (an attorney who represents the state) argues that the accused is guilty, and the defense attorney asserts that the accused is innocent. In reality, judges have considerable discretion but are still constrained in some cases by *structured sentencing* guidelines. Structured sentencing is also referred to as *determinate sentencing* or *mandatory sentencing*. A determinate sentence sets the term of imprisonment at a fixed period of time (such as three years) for a specific offense. Mandatory sentencing guidelines are established by law and require that a person convicted of a specific offense or series of offenses be given a penalty within a fixed range. Although these practices limit judicial discretion in sentencing, many critics are concerned about other negative effects that these sentencing approaches may have on the accused and the criminal justice system.

**criminal justice system** the local, state, and federal agencies that enforce laws, adjudicate crimes, and treat and rehabilitate criminals.

Prosecuting attorneys also have considerable leeway in deciding which cases to prosecute and when to negotiate a plea bargain with a defense attorney. As cases are sorted through the legal machinery, a steady attrition occurs. At each stage, various officials determine what alternatives will be available for those cases still remaining in the system.

About 90 percent of criminal cases are never tried in court; instead, they are resolved by plea bargaining, a process in which the prosecution negotiates a reduced sentence for the accused in exchange for a guilty plea (Siegel, 2012). Defendants (especially those who are poor and cannot afford to pay an attorney) may be urged to plead guilty to a lesser crime in return for not being tried for the more serious crime for which they were arrested. Prison sentences given in plea bargains vary widely from one region to another and even from judge to judge within one state.

Those who advocate the practice of plea bargaining believe that it allows for individualized justice for alleged offenders because judges, prosecutors, and defense attorneys can agree to a plea and to a punishment that best fits the offense and the offender. They also believe that this process helps reduce the backlog of criminal cases in the court system as well as the lengthy process often involved in a criminal trial. However, those who seek to abolish plea bargaining believe that this practice leads to innocent people pleading guilty to crimes they have not committed or pleading guilty to a crime other than the one they actually committed because they are offered a lesser sentence. More-serious crimes, such as murder, felonious assault, and rape, are more likely to proceed to trial than other forms of criminal conduct; however, many of these cases do not reach the trial stage.

One of the most important activities of the court system is establishing the sentence of the accused after he or she has been found guilty or has pleaded guilty. Typically, sentencing involves the following kinds of sentences or dispositions: fines, probation, alternative or intermediate sanctions (such as house arrest or electronic monitoring), incarceration, and capital punishment (Siegel, 2012). However, adult courts operate differently from those established for juvenile offenders.

**Juvenile Courts** Juvenile courts were established under a different premise from courts for adults. Under the doctrine of *parens patriae* (the state as parent), the official purpose of juvenile courts has been to care for, rather than punish, youthful offenders. In theory, less weight is given to offenses and more weight to the youth's physical, mental, or social condition. The juvenile court seeks to change or resocialize offenders through treatment or therapy, not to punish them. Consequently, judges in juvenile courts are given relatively wide latitude, or discretion, in the decisions they mete out regarding young offenders.

Unlike adult offenders, juveniles are not always represented by legal counsel. A juvenile hearing is not a trial but rather an informal private hearing before a judge or probation officer, with only the young person and a parent or guardian present. No jury is convened, and the juvenile offender does not cross-examine his or her accusers (Figure 6.17). In addition, the offender is not "sentenced"; rather, the case is "adjudicated" or "disposed of." Finally, the offender is not "punished" but instead may be "remanded to the custody" of a youth authority in order to receive training, treatment, or care.

Because of judicial discretion, courts may treat juveniles differently based on gender. Considerable disparity exists in the disposition of juvenile cases, with much of the variation thought to result from judges' beliefs rather than objective facts in the case. Female offenders are more likely than males to be institutionalized for committing status offenses such as truancy, running away from home, and other offenses that serve as "buffer charges" for suspected sexual misconduct.

Disparity also exists on the basis of race and class. Judges tend to see youths from white, middle- or upper-class families as being very much like their children and to believe that the families will take care of the problem on their own. They may view juveniles from lower-income families or from other

**FIGURE 6.17** Although juries play an important role in the criminal justice system, juries are not convened in cases involving juvenile offenders. Why might this be so?

© iStockphoto.com/Katherine Moffitt

racial–ethnic groups as delinquents in need of attention from authorities. Furthermore, some judges view gang members from impoverished central cities as "guilty by association" because of their companions.

The political climate may have an effect on how judges dispose of juvenile cases. In the process of dealing with the public perception that the juvenile justice system is too lenient, some judges may have inadvertently contributed to other problems. Many more youths have been remanded to overcrowded juvenile detention facilities that are unable to provide necessary educational, health, and social services. Based on a judge's discretion, many juvenile offenders are incarcerated under indeterminate sentences and placed in a detention facility that may serve merely as a school for adult criminality.

## Punishment and Corrections

Although the United States makes up less than 5 percent of the world's population, our nation accounts for almost 25 percent of the world's prison population. Some analysts suggest that our laws prescribe greater punishment for some offenses than those in other nations, resulting in Americans being locked up for crimes that would rarely result in prison sentences in other countries. In 2012 approximately 1.57 million inmates were incarcerated in state and federal prisons in the United States. When local jails are included, the number rises to about 2.3 million people who are incarcerated at any given time. It should be noted that jails differ from prisons. Most jails are run by local governments or a sheriff's department. They are designed to hold people before they make bail, when they are awaiting trial, or when they are serving short sentences for committing a misdemeanor. By contrast, prisons are operated by state governments and the Federal Bureau of Prisons, and are designed to hold individuals convicted of felonies. Some prisons are operated by private contractors that build and control the facilities while receiving public monies for their operation. Both jails and prisons are based on the assumption that punishment and/or corrections are necessary to protect the public good and to effectively deal with those who violate laws.

**Punishment is any action designed to deprive a person of things of value (including liberty) because of some offense the person is thought to have committed.** Historically, punishment has had four major goals:

1. *Retribution* is punishment that a person receives for infringing on the rights of others. Retribution imposes a penalty on the offender and is based on the premise that the punishment should fit the crime: The greater the degree of social harm, the more the offender should be punished. For example, an individual who murders should be punished more severely than one who shoplifts.

2. *General deterrence* seeks to reduce criminal activity by instilling a fear of punishment in the general public. However, we most often focus on *specific deterrence,* which inflicts punishment on specific criminals to discourage them from committing future crimes. Recently, criminologists have debated whether imprisonment has a deterrent effect, given the fact that high rates (30–50 percent) of those who are released from prison become recidivists (previous offenders who commit new crimes).

3. *Incapacitation* is based on the assumption that offenders who are detained in prison or are executed will be unable to commit additional crimes. This approach is often expressed as "lock 'em up and throw away the key!" In recent years, more emphasis has been placed on *selective incapacitation,* which means that offenders who repeat certain kinds of crimes are sentenced to long prison terms.

4. *Rehabilitation* seeks to return offenders to the community as law-abiding citizens by providing therapy or vocational or educational training. Based on this approach, offenders are treated, not punished, so that they will not continue their criminal activity. However, many correctional facilities are seriously understaffed and underfunded in the rehabilitation programs that exist. The job skills (such as agricultural work) that many offenders learn in prison do not transfer to the outside world, nor are offenders given any assistance in finding work that fits their skills once they are released.

Other approaches have also been advocated for dealing with criminal behavior. Key among these is the idea of *restoration,* which is designed to repair the damage done to the victim and the community by an offender's criminal act. This approach is based on the *restorative justice perspective,* which states that the criminal justice system should promote a peaceful and just society; therefore, the system should focus on peacemaking rather than on punishing offenders. Advocates of this approach believe that punishment of offenders actually encourages crime rather than deterring it and are in favor of approaches such as probation with treatment. Opponents of this approach suggest that increased punishment of offenders leads to lower crime rates

**punishment** any action designed to deprive a person of things of value (including liberty) because of some offense the person is thought to have committed.

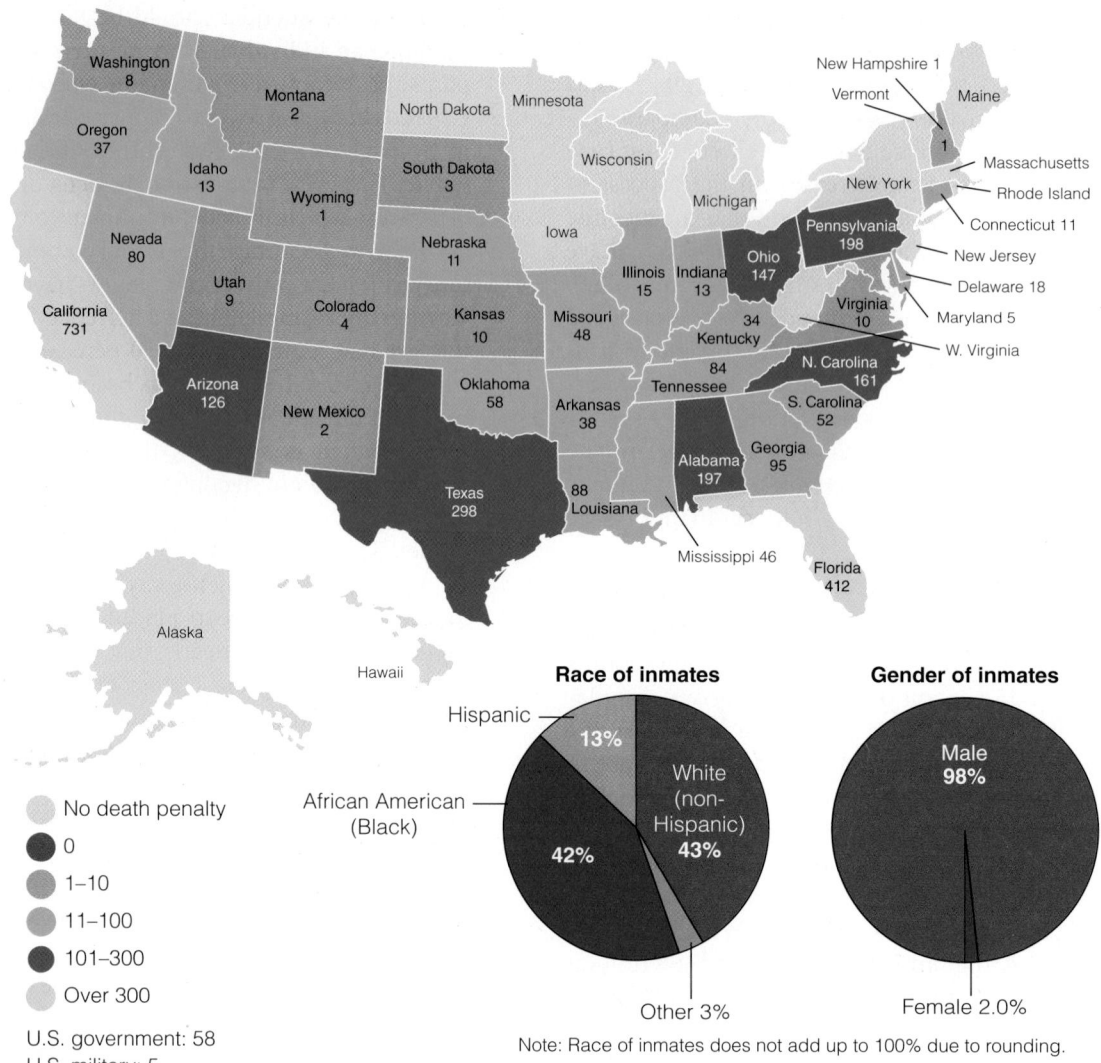

**FIGURE 6.18** DEATH ROW CENSUS, APRIL 1, 2013
Death row inmates are heavily concentrated in certain states.

Source: Death Penalty Information Center, 2013. © Cengage Learning

and that the restorative justice approach amounts to "coddling criminals." However, numerous restorative justice programs are now in operation, and many have been found to reduce recidivism rates below the ones associated with more-conventional criminal justice sanctions (Barkan, 2012).

Instead of the term *punishment,* the term *corrections* is often used. It refers not only to prisons and jails but also to a number of programs and organizations that manage individuals who have been either accused or convicted of crimes. Included in the field of corrections are halfway houses, probation, work release and education programs, parole supervision, counseling, and community service.

## The Death Penalty

Historically, removal from the group has been considered one of the ultimate forms of punishment. For many years, capital punishment, or the death penalty, has been used in the United States as an appropriate and justifiable response to very serious crimes. In 2013, 39 inmates were executed (as contrasted with 98 in 1999), and 3,108 people awaited execution, having received the death penalty under federal law or the laws of one of the states that have the death penalty (Death Penalty Information Center, 2013). By far, the largest numbers of death row inmates are in states such as California, Florida, Texas, Pennsylvania, and Alabama (see Figure 6.18).

Because of the finality of the death penalty, it has been a subject of much controversy and numerous U.S. Supreme Court debates about the decision-making process involved in capital cases. In 1972 the Supreme Court ruled (in *Furman v. Georgia*) that *arbitrary* application of the death penalty violates the Eighth Amendment to the Constitution but that the death penalty itself is not unconstitutional. In other words, capital punishment is legal if it is fairly imposed. Although there have been a number of cases involving death penalty issues before the Supreme Court since that time, the Court typically has upheld the constitutionality of this practice. Yet the fact remains that racial disparities are highly evident in the death row census and executions. Over 75 percent of the murder victims in cases that resulted in an execution were white despite the fact that nationally only about 50 percent of murder victims are white. In Louisiana, for example, the odds of a death sentence were 97 percent higher for perpetrators whose victim was white than for those whose victim was black (Death Penalty Information Center, 2013).

People who have lost relatives and friends as a result of criminal activity often see the death penalty as justified. However, capital punishment raises many doubts for those who fear that innocent individuals may be executed for crimes they did not commit. According to the Death Penalty Information Center (2013), more than 140 individuals have been released from death row since 1973 based on evidence that was later presented to demonstrate their innocence. For still others, the problem of racial discrimination in the sentencing process poses troubling questions. Other questions involve the execution of those who are believed to be insane and of those defendants who did not have effective legal counsel during their trial. In 2002, for example, the Supreme Court ruled (in *Atkins v. Virginia*) that executing the mentally retarded is unconstitutional. In another landmark case (*Ring v. Arizona*), the Court ruled that juries, not judges, must decide whether a convicted murderer should receive the death penalty.

Executions resumed in 2008 after a *de facto* moratorium was lifted by the Supreme Court when it decided to uphold lethal injection as not constituting "cruel and unusual punishment" despite the fact that some scientists disagree. Although only 78 new death sentences were handed down in 2012 (after a high of 328 in 1994), the issue of the death penalty is far from resolved; the debate, which has taken place for more than two centuries, will no doubt continue for many years in the future.

# U.S. Deviance and Crime in the Future

Three pressing questions pertaining to deviance and crime in the United States will continue to face us in the future: Is the solution to our "crime problem" more law and order? Is equal justice under the law possible? Is more-stringent gun control the way to reduce violent crime in the United States? (See "Sociology and Social Policy" for this discussion.)

Although many people in the United States agree that crime is one of the most important problems in this country, they are divided over what to do about it.

Some of the frustration about crime might be based on unfounded fears; studies show that the *overall* crime rate has been decreasing slightly in recent years.

One thing is clear: The existing criminal justice system cannot solve the "crime problem." If only a small percentage of all crimes result in arrest, and even a smaller percentage of those lead to a conviction in serious cases, and even less than 5 percent of those result in a jail term, the "lock 'em up and throw away the key" approach has little chance of succeeding. Nor does the high rate of recidivism among those who have been incarcerated speak well for the rehabilitative efforts of our existing correctional facilities. Reducing street crime may hinge on finding ways to short-circuit criminal behavior. Likewise, if corporate elites and wealthy offenders continue to get away with crimes that leave their employees and millions of others without their homes, pensions, and life savings, not only individuals and families but also the entire nation will continue to suffer from the consequences of misplaced priorities and a system that benefits the few at the expense of the law-abiding many who attempt to follow the rules and adhere to the system that they have been socialized to accept.

One of the greatest challenges is juvenile offenders, who may become the adult criminals of tomorrow. However, instead of military-style boot camps or other stopgap measures, *structural solutions*—such as more and better education and jobs, affordable housing, more equality and less discrimination, and socially productive activities—are needed to reduce street crime. In the past, structural solutions such as these have made it possible for immigrants who initially committed street crimes to leave the streets, get jobs, and lead productive lives. Ultimately, the best approach for reducing delinquency and crime would be prevention: to work with young people *before* they become juvenile offenders to help them establish family relationships, build self-esteem, choose a career, and get an education that will help them pursue that career. Criminologist Steven E. Barkan (2012) proposes seven crime-reduction strategies that are more structural in nature:

1. Create decent jobs that pay a living wage.
2. Provide economic aid for people who are unemployed or are barely making it.
3. End racial segregation in housing.
4. Strengthen social integration and social institutions in urban neighborhoods.
5. Reduce housing and population density.
6. Change male socialization.
7. Reduce economic inequality.

Many people still ask if equal justice under the law is possible. As long as racism, sexism, classism, and ageism exist in our society, many individuals will see deviant and criminal behavior through a selective lens. To solve the problems addressed in this chapter, we must ask ourselves what we can do to ensure the rights of everyone, including the poor, people of color, and women and men alike. Many of us can counter classism, racism, sexism, and ageism where they occur. Perhaps the only way that the United States can have equal justice under the law in the future (and, perhaps, less crime as a result) is to promote social justice for individuals regardless of their race, class, gender, or age.

## Transnational Crime and the Global Criminal Economy

Transnational crime occurs across multiple national borders. This type of crime not only involves crossing borders between countries but also crimes in which crossing national borders is essential to the criminal activity. Much of transnational crime is conducted by organized criminal groups that use systematic violence and corruption to achieve their goals. Often these criminal networks are able to prey on less-powerful governments that do not have the resources to oppose them. According to the National Institute of Justice (2011), transnational crime should be of concern to people living in the United States because this type of criminal activity has a detrimental effect on everyone, not just the nations that are destabilized by these activities, often through the use of bribery, violence, or terror.

Transnational crime is fueled by globalization, which has brought increased travel, expanded international trade, and advances in telecommunications and computer technology. This type of criminal activity cannot be controlled by one nation alone. How much money and other resources change hands in the global criminal economy? Although the exact amount of profits and financial flows originating in the global criminal economy is impossible to determine, the United Nations Conference on Global Organized Crime estimated that more than $600 billion (in U.S. currency) per year is accrued in the global trade in drugs alone. Profits from all kinds of global criminal activities are estimated to be as high as $5 trillion per year (United Nations Development Programme, 2011). Some analysts believe that even this figure may underestimate the true nature and extent of the global criminal economy. The highest-income-producing activities of global criminal organizations include trafficking in drugs, weapons, and nuclear material; smuggling of things and people

(including migrants); trafficking in women and children for the sex industry; and trafficking in body parts such as corneas and major organs for the medical industry. Undergirding the entire criminal system are money laundering and various complex financial schemes and international trade networks that make it possible for people to use the resources they obtain through illegal activity for the purposes of consumption and investment in the formal ("legitimate") economy. Theft of critical U.S. intellectual property, including intrusions into corporate computer networks, makes this country more vulnerable to significant business losses (Figure 6.19). Cybercrime poses a threat to banking, stock markets, and credit card services, just as penetration of intelligence services makes nations more vulnerable to terrorism.

Can anything be done about transnational organized crime? The United Nations Office of Drugs and Crime (2014) suggests that specific steps must be taken: (1) coordination at the international level for identifying, investigating, and prosecuting the people and groups behind the crimes; (2) education and awareness raising about these kinds of crime and how they affect individuals' everyday lives; (3) intelligence and technology to help law enforcement officials combat powerful criminal networks; and (4) assistance for developing countries to help them build their capacity for countering terrorist threats.

In the United States, the White House has also developed a strategy to combat transnational organized crime that focuses on stemming the southward flow of guns and money that contribute to an increase in drug violence. From this perspective, the demand for illegal drugs in the United States fuels the global drug trade, which is a key source of funding for transnational organized crime. If that source can be reduced or cut off, this will be a beginning toward

**FIGURE 6.19** How might international software piracy affect the U.S. economy?

# SOCIOLOGY and Social Policy

## The Long War Over Gun Control

The good news is there's already a growing consensus for us to build from. A majority of Americans support banning the sale of military-style assault weapons. A majority of Americans support banning the sale of high-capacity ammunition clips. A majority of Americans support laws requiring background checks before all gun purchases, so that criminals can't take advantage of legal loopholes to buy a gun from somebody who won't take the responsibility of doing a background check at all.

> —President Barack Obama's comments on the urgency of enacting social policies that will deter mass shootings and other gun-related violence (whitehouse.gov, 2012)

The only thing that stops a bad guy with a gun is a good guy with a gun.

> —National Rifle Association Chief Executive Wayne LaPierre, responding to demands for more-effective gun control in the aftermath of the Sandy Hook Elementary School shootings in Newtown, Connecticut (qtd. in Gold and Mason, 2012)

Pro-gun-control advocates believe that we need social policies that regulate the gun industry and gun ownership. However, opponents of gun control measures argue that regulation will not curb random violence perpetrated by a few disturbed or frustrated individuals and that more, rather than fewer, people need to be armed to protect others against future mass violence.

Some people view social policy as a means of "declaring war" on a problem such as school violence (Best, 1999). Social policy discussions on gun-related violence typically focus on how to win the "war" on guns and how to keep violent incidents from happening again. However, a lack of consensus exists on both the causes of gun violence and what should be done about it. Some favor regulation of the gun industry and gun ownership; others believe that additional regulations will not curb random violence perpetrated by individuals.

Underlying the arguments for and against gun control are these words from the Second Amendment to the U.S. Constitution: "A well regulated Militia, being necessary to the security of a free State, the right of the people to keep and bear Arms, shall not be infringed." Those in favor of legislation to regulate the gun industry and gun ownership argue that the Second Amendment does not guarantee an individual's right to own guns: The right to keep and bear arms applies only to those citizens who do so as part of an official state militia. However, in 2008 the U.S. Supreme Court ruled in *District of Columbia v. Heller* that the Second Amendment protects an individual's right to own a gun for personal use. From this perspective, individuals possess a constitutionally protected right to have a loaded gun at home for self-defense. This argument has been made for many years by the National Rifle Association (NRA), one of the most powerful interest groups in Washington and in legislatures across the nation. The NRA has stated that gun control regulations violate the individual's constitutional right to own a gun. The NRA's response to recent school violence has been to call for armed guards in all schools.

Those who call for more-stringent gun control laws in the United States point out, for comparison, that in the horrific attack on 22 children and one adult that occurred in the Henan Province of China near the date of the Newtown mass killings in 2012, everyone under attack survived the violence because the perpetrator wielded only a knife, not a semi-automatic assault rifle (Associated Press, 2012; CNN.com, 2012b). The fast and extremely lethal nature of gun attacks became all the more poignant in the contrast between these two events.

What solutions exist for the quandary over gun regulations? Declaring war on gun-related violence is difficult because it is virtually impossible to rally our political leaders and the general public behind a single policy. Instead, time is spent arguing over how to proceed and how to identify the real enemy, and gun violence remains a chronic concern in the United States. Clearly, this is an issue in which we have yet to successfully address a pressing problem that is continuing to take precious lives in this nation.

### reflect & analyze

**Do you believe that arming more people would reduce mass shootings in the United States? Should armed guards be placed in all schools? Should college students be allowed to carry guns on campus for their own safety? What do you think?**

reducing problems of terrorism and insurgent networks. Among the policy objectives set forth in the U.S. National Security Strategy are the following:

Protect Americans and our partners from the harm, violence, and exploitation of transnational criminal networks. . . . We will target the networks that pose the gravest threat to citizen safety and security, including those that traffic illicit drugs, arms, and people—especially women and children; sell and distribute substandard, tainted, and counterfeit goods; rob Americans of their prosperity; carry out kidnappings for ransom and extortion; and seek to terrorize and intimidate through acts of torture and murder. (whitehouse.gov, 2011)

Will the United States and other nations be able to curb transnational organized crime? This remains to be seen; however, public safety, public health, democratic institutions, and economic stability rely on the ability of the U.S. government and others to combat networks that pose a strategic threat to Americans and U.S. interests. Crime around the world, as well as at home, can be a destabilizing influence on the social order.

# chapter REVIEW

## LO1 What is deviance, and when is deviant behavior considered a crime?

Deviance is any behavior, belief, or condition that violates significant social norms in the society or group in which it occurs. Some forms of deviant behavior violate the criminal law, which defines the behaviors that society labels as criminal. Sociologists are interested in what types of behavior are defined by societies as "deviant," who does that defining, how individuals become deviant, and how those individuals are dealt with by society.

## LO2 What are the key functionalist perspectives on deviance?

Functionalist perspectives on deviance include strain theory and opportunity theory. Strain theory focuses on the idea that when people are denied legitimate access to cultural goals, such as a good job or a nice home, they may engage in illegal behavior to obtain them. Opportunity theory suggests that for deviance to occur, people must have access to illegitimate means to acquire what they want but cannot obtain through legitimate means.

## LO3 What are the key ideas of conflict explanations of deviance and crime that focus on power relations, capitalism, feminism, and the intersection of race, class, and gender?

Conflict theorists who focus on power relations in society suggest that the lifestyles considered deviant by political and economic elites are often defined as illegal. Marxist conflict theorists link deviance and crime to the capitalist society, which divides people into haves and have-nots, leaving crime as the only source of support for those at the bottom of the economic ladder. Feminist approaches to deviance focus on the relationship between gender and deviance. Multiracial feminist approaches have examined how the intersecting systems of race, class, and gender act as "structuring forces" that affect how people act, what opportunities they have available, and how their behavior is socially defined.

## LO4 What are some key symbolic interactionist perspectives on deviance, including differential association theory, social bond theory, and labeling theory?

According to symbolic interactionists, deviance is learned through interaction with others. Differential association theory states that individuals have a greater tendency to deviate from societal norms when they frequently associate with persons who tend toward deviance instead of conformity. According to social control theories, everyone is capable of committing crimes, but social bonding (attachments to family and to other social institutions) keeps many from doing so. According to labeling theory, deviant behavior is that which is labeled deviant by those in powerful positions.

## LO5 How do postmodern perspectives on deviance differ from other theoretical approaches?

Postmodernist views on deviance focus on how the powerful control others through discipline and surveillance. This control may be maintained through largely invisible forces such as the Panoptican, as described by Michel Foucault, or by newer technologies that place everyone—not just "deviants"—under constant surveillance by authorities, who use their knowledge as power over others.

## LO6 How do sociologists define the following types of crime: violent crime, property crime, public order crime, occupational crime, corporate crime, organized crime, and political crime?

Violent crime consists of actions involving force or the threat of force against others. Property crimes include burglary, motor vehicle theft, larceny-theft, and arson. Public order crimes involve an illegal action voluntarily engaged in by the participants, such as prostitution. Occupational crime comprises illegal activities committed by people in the course of their employment or financial affairs. Corporate crime consists of illegal acts committed by corporate employees on behalf of the corporation and with its support. Organized crime is a business operation that supplies illegal goods and services for profit. Political crime refers to illegal or unethical acts involving the usurpation of power by government officials or illegal/unethical acts perpetrated against the government by outsiders seeking to make a political statement, undermine the government, or overthrow it.

## LO7 Why may official statistics not be a good indicator of how many crimes are committed, particularly in regard to factors such as age, race, gender, and class?

Official crime statistics are taken from the Uniform Crime Report, which lists crimes reported to the police, and the National Crime Victimization Survey, which interviews households to determine the incidence of crimes, including those not reported to police. Studies show that many more crimes are committed than are officially reported. Age is the key factor in crime statistics. Younger people are more likely to have the highest criminal victimization rates. The elderly tend to be fearful of crime but are the least likely to be victimized. Males are arrested for significantly more crimes than females. Persons from lower socioeconomic backgrounds are more likely to be arrested for violent and property crimes; white-collar crime is more likely to occur among the upper socioeconomic classes.

## LO8 What are the components of the criminal justice system and the goals of punishment?

The criminal justice system refers to the local, state, and federal agencies that enforce laws, adjudicate crimes, and treat and rehabilitate criminals. The system includes the police, the courts, the correctional facilities, and the people in myriads of police agencies, courts, prosecutorial agencies, correctional institutions, and probation and parole departments. Historically, punishment has had four major goals: retribution, general deterrence, incapacitation, and rehabilitation.

## key TERMS

corporate crime  181

crime  166

criminal justice system  188

criminology  167

deviance  164

differential association theory  173

illegitimate opportunity structures  170

Internet crime  181

juvenile delinquency  166

labeling theory  175

occupational (white-collar) crime  179

organized crime  182

political crime  183

primary deviance  176

property crimes  179

punishment  191

rational choice theory of deviance  174

secondary deviance  176

social bond theory  175

social control  166

strain theory  168

terrorism  184

tertiary deviance  176

victimless crimes  179

violent crime  179

## questions for CRITICAL THINKING

1. Why does the crime rate fluctuate in the United States? How might we use sociology to explain the fact that the crime rate has not increased dramatically even in the recent recession and the slow economic recovery?

2. Should so-called victimless crimes, such as prostitution and recreational drug use, be decriminalized? Do these crimes harm society?

3. As a sociologist armed with a sociological imagination, how would you propose to deal with the problem of crime in the United States and around the world?

# 7 CLASS AND STRATIFICATION IN THE UNITED STATES

## LEARNING OBJECTIVES

**LO1** Define *social stratification* and distinguish among three major systems of stratification.

**LO2** Explain Karl Marx's views on social class and stratification.

**LO3** Discuss Max Weber's multidimensional approach to social stratification.

**LO4** List and note the key characteristics of U.S. social classes.

**LO5** Distinguish between income inequality and wealth inequality.

**LO6** Describe three important consequences of inequality in the United States.

**LO7** Identify the characteristics of the U.S. poor based on age, gender, and race/ethnicity.

**LO8** Compare and contrast functionalist, conflict, and symbolic interactionist perspectives on social inequality.

# Sociology and Everyday Life
## The Power of Class

When they were ten years old, the middle-class youth [in my study] seemed worldly, blasé, and hard to impress. For them, pizza parties were very common and thus no special treat. Spring concerts drew shrugs. . . . Although the working-class and poor children were the same age as the middle-class children, they seemed younger, bouncier, and more childlike. They smiled broadly while on stage for the spring concert, were ecstatic over a pizza party, and entertained themselves for hours on weekends and evenings. Ten years later, the pattern had reversed: it was the middle-class youth who seemed younger. Now college students, they were excited about the way the world was opening up for them. . . . By contrast, the working-class and poor youth were generally working full-time in jobs they did not like, and they had various pressing responsibilities such as raising children, paying for food and board, and making monthly car payments. . . . [A]s the children moved from fourth grade into adulthood, the power of class pushed their lives in such different directions that I could not pose the

How does class affect young people?

Many people living in the United States want to achieve the American Dream. Throughout the history of this country, people have aspired to have more than their parents and grandparents. Young people from low-income and poverty-level families are often no exception. Upward mobility in the U.S. class system has been linked to opportunities in education and employment; however, recent national and global economic crises have made it more difficult for many people to achieve their dream. Some sociological studies, such as the one conducted by Annette Lareau (2011), also emphasize that the social-class origins of children have powerful and long-term effects on young people because more-affluent parents are typically better able to orchestrate their children's lives in institutional settings such as schools.

From a sociological perspective, this issue of class-based inequalities experienced by individuals brings us to a larger, macrolevel question in studying class and stratification in the United States: What is the American Dream? Simply stated, the American Dream is the belief that if people work

hard and play by the rules, they will have a chance to get ahead. Moreover, each generation will be able to have a higher standard of living than that of its parents. The American Dream is based on the assumption that people in the United States have equality of opportunity regardless of their race, creed, color, national origin, gender, or religion.

For some people the American Dream means that each subsequent generation will be able to acquire more material possessions and wealth than family members in the preceding generation. For other people the American Dream means having an opportunity to secure a job, own a home, and get a good education for their children. For still others it is the chance to rise from poverty by working hard enough.

When we talk about the American Dream, it is important to realize that not all people will achieve success. The way a society is stratified (arranged from top to bottom) has a major influence on a person's position in the class structure. In this chapter we examine systems of social stratification and describe how the U.S. class system may make it easier

PART 3 | SOCIAL INEQUALITY

200

same interview questions to the group as a whole. Middle-class youths' interviews were filled with questions about their college preparation classes, college searches, college choice, and college adjustment. . . . Working-class and poor youths' interviews were filled with discussions of their difficulties in high school, challenges and work, and uncertain future goals. . . . The follow-up study suggests that over time the gap that existed between the families when the children were ten widened rather than narrowed. . . . [I]t is important to recognize that in American society, people who are blessed with class advantages tend to be unaware of these benefits and privileges. . . . They downplay, or do not even notice, the social class benefits bestowed upon them.

—sociologist Annette Lareau (2011: 309–310) describing how class influences white and African American children in poor-, working-, and middle-class families as she revisited young people from her earlier study who had moved from fourth grade to adulthood

## How Much Do You Know About Wealth, Poverty, and the American Dream?

| True | False | |
|------|-------|---|
| T | F | 1. Most U.S. adults say that they have achieved the American Dream. |
| T | F | 2. Individuals over age 65 have the highest rate of poverty. |
| T | F | 3. Compared with other developed nations, the United States has one of the highest rates of childhood poverty. |
| T | F | 4. In the United States a family of four is considered to be "poor" if the household earns less than $40,000. |

| True | False | |
|------|-------|---|
| T | F | 5. The wealthiest person in the world is from the United States. |
| T | F | 6. Slavery still exists in the United States. |
| T | F | 7. Some states have a higher minimum wage than the federal minimum wage. |
| T | F | 8. Since the 1970s, the gap between the rich and the poor in the United States has decreased significantly. |

Answers on page 202.

for some individuals to attain (or maintain) top positions in society while others have great difficulty moving up from poverty or low-income origins. But before we explore class and stratification, test your knowledge of wealth, poverty, and the American Dream by taking the "Sociology and Everyday Life" quiz.

## What Is Social Stratification?

**LO1** *Social stratification* **is the hierarchical arrangement of large social groups based on their control over basic resources.** Stratification involves patterns of structural inequality that are associated with membership in each of these groups, as well as the ideologies that support inequality. Sociologists examine the social groups that make up the hierarchy in a society and seek to determine how inequalities are structured and persist over time.

Max Weber's term *life chances* **refers to the extent to which individuals have access to important societal resources such as food, clothing, shelter, education, and health care.** According to sociologists, more-affluent people typically have better life chances than the less affluent; they have greater access to quality education, safe neighborhoods, high-quality nutrition and health care, police and private security protection, and an extensive array of other goods and services. In contrast, persons with low and poverty-level incomes tend to have limited access to these resources. *Resources* are anything valued in a society, ranging from money and property to medical care and education; they are

**social stratification** the hierarchical arrangement of large social groups based on their control over basic resources.

**life chances** Max Weber's term for the extent to which individuals have access to important societal resources such as food, clothing, shelter, education, and health care.

# ANSWERS to the Sociology Quiz on Wealth, Poverty, and the American Dream

**1. False.** In a national poll, less than a third (31 percent) of U.S. adults said that they had achieved the American Dream.

**2. False.** The age group that includes people ages 65 and over actually has the lowest rate of poverty of any age group in the United States. Several decades ago the poverty rate of the elderly was much higher than it is today, but largely because of Social Security, rates of poverty among the elderly have dropped significantly.

**3. True.** The United States has the second-highest rate of childhood poverty in developed nations. Approximately 23 percent of children live in poverty in this nation.

**4. False.** In 2013 a family of four in the 48 contiguous U.S. states would need to earn less than $23,550 to be considered poor. By contrast, the poverty guideline for a family of four was $29,440 in Alaska and $27,090 in Hawaii.

**5. False.** *Forbes* magazine's 2013 list of the richest people in the world identified Mexican telecom entrepreneur Carlos Slim Helu and his family as the wealthiest persons in the world, with a net worth of $73 billion.

**6. True.** The U.S. State Department estimates that between 17,000 and 20,000 people are brought to the United States each year and are enslaved in such jobs as farm work, domestic work, and sex work.

**7. True.** Several states have passed laws that establish a higher minimum wage than the federal minimum wage of $7.25 per hour in 2014. For example, in 2014 the minimum wage in Washington State was $9.32, and states such as Massachusetts were poised to raise their minimum wage to more than $11 per hour.

**8. False.** The gap between the rich and the poor has widened significantly. In 1979 the richest 1 percent of Americans took in less than 9 percent of the country's total income. By 2013, the richest 1 percent took in more than 25 percent of the nation's total income and owned about 43 percent of the nation's wealth.

Sources: Based on DeNavas-Walt, Proctor, and Smith, 2012; *Federal Register*, 2013; and *Forbes*, 2012, 2013. © Cengage Learning.

---

considered to be scarce because of their unequal distribution among social categories. If we think about the valued resources available in the United States, for example, the differences in life chances are readily apparent. As one analyst suggested, "Poverty narrows and closes life chances. The victims of poverty experience a kind of arteriosclerosis of opportunity. Being poor not only means economic insecurity, it also wreaks havoc on one's mental and physical health" (Ropers, 1991: 25). Our life chances are intertwined with our class, race, gender, and age.

All societies distinguish among people by age. Young children typically have less authority and responsibility than older persons. Older persons, especially those without wealth or power, may find themselves at the bottom of the social hierarchy. Similarly, all societies differentiate between females and males: Women are often treated as subordinate to men. From society to society, people are also treated differently as a result of their religion, race/ethnicity, appearance, physical strength, disabilities, or other distinguishing characteristics. All of these differentiations result in inequality. However, systems of stratification are also linked to the specific economic and social structure of a society and to a nation's position in the system of *global* stratification, which is so significant for understanding social inequality that we will devote the next chapter to this topic.

## Systems of Stratification

Around the globe, one of the most important characteristics of systems of stratification is their degree of flexibility. Sociologists distinguish among such systems based on the extent to which they are open or closed. In an *open system*, the boundaries between levels in the hierarchies are more flexible and may be influenced (positively or negatively) by people's achieved statuses. Open systems are assumed to have some degree of social mobility. *So-cial mobility* **is the movement of individuals or groups from one level in a stratification system to another.** This movement can be either upward or downward. *Intergenerational mobility* **is the social movement experienced by family members from one generation to the next.** For example,

Sarah's father is a carpenter who makes good wages in good economic times but is often unemployed when the construction industry slows to a standstill. Sarah becomes a neurologist, earning $350,000 a year, and moves from the working class to the upper-middle class. Between her father's generation and her own, Sarah has experienced upward social mobility.

By contrast, *intergenerational mobility* **is the social movement of individuals within their own lifetime.** Consider, for example, RaShandra, who began her career as a high-tech factory worker and through increased experience and taking specialized courses in her field became an entrepreneur, starting her own highly successful Internet-based business. RaShandra's advancement is an example of upward intragenerational social mobility. However, note that both intragenerational mobility and intergenerational mobility may be downward as well as upward.

In a *closed system*, the boundaries between levels in the hierarchies of social stratification are rigid, and people's positions are set by ascribed status (Figure 7.1). Open and closed systems are ideal-type constructs; no actual stratification system is completely open or closed. The systems of stratification that we will examine—slavery, caste, and class—are characterized by different hierarchical structures and varying degrees of mobility. Let's examine these three systems of stratification to determine how people acquire their positions in each and what potential for social movement they have.

## Slavery

*Slavery* **is an extreme form of stratification in which some people are owned or controlled by others for the purpose of economic or sexual exploitation.** It is a closed system in which people designated as "slaves" are treated as property and have

James Brunker/Alamy

**FIGURE 7.1** Max Weber's term *life chances* refers to the extent to which people have access to resources such as food, clothing, shelter, education, and health care. How might the life chances of the people living in these buildings be different?

little or no control over their lives. Those of us living in the United States are aware of the legacy of slavery in our own country, beginning in the 1600s, when slaves were forcibly imported to the United States as a source of cheap labor. Slavery was defined in law and custom by the 1750s, making it possible for one person to own another person. In fact, early U.S. presidents including George Washington, James Madison, and Thomas Jefferson owned slaves.

As practiced in the United States, slavery had four primary characteristics: (1) it was lifelong and was inherited (children of slaves were considered to be slaves); (2) slaves were considered property, not human beings; (3) slaves were denied rights; and (4) coercion was used to keep slaves "in their place." Although most slaves were powerless to bring about change, some were able to challenge slavery—or at least their position in the system—by engaging in activities such as sabotage, intentional carelessness, work slowdowns, or running away from owners and working for the abolition of slavery. Despite the fact that slavery in this country officially ended many years ago, many sociologists believe that its legacy is deeply embedded in current patterns of prejudice and discrimination against African Americans.

**social mobility** the movement of individuals or groups from one level in a stratification system to another.

**intergenerational mobility** the social movement experienced by family members from one generation to the next.

**intragenerational mobility** the social movement of individuals within their own lifetime.

**slavery** an extreme form of stratification in which some people are owned or controlled by others for the purpose of economic or sexual exploitation.

# SOCIOLOGY *in Global Perspective*

## A Day in Your Life: How Are You Touched by Modern Slavery?

According to the U.S. Department of State (2012), many of the things that we enjoy on a daily basis may have been produced or touched by those held in involuntary servitude. Consider this typical day in the life of a person:

### 6:00 A.M.: Wake Up and Get Ready for the Day

The clothes on your back could have been produced by a man, woman, or child in a garment factory in Asia, the Middle East, or Latin America who is subjected to forced labor, including withholding of passports, no pay, long working hours to meet quotas, and physical and sexual abuse. The jewelry you wear may include gold mined by trafficked children in Africa, Asia, and Latin America.

### 8:00 A.M.: Sit Down at Your Desk

The electronics you use may be dependent on minerals that are produced in conflict-affected areas of Africa where children and adults are forced to work in mines under conditions of forced labor and sexual servitude.

### 10:00 A.M.: Take a Caffeine Break

The coffee you drink may have been touched by modern slaves who work under conditions of forced labor on coffee plantations in Latin America and Africa. The sugar in your coffee may have come from plantations where children and men in Latin America, Asia, and Africa are subjected to conditions of forced labor and debt bondage.

### 12:00 P.M.: Eat Lunch

The fish you eat may have been caught by men in Southeast Asia and by children as young as four years old in West Africa, who are subjected to conditions of forced labor where they have been deprived of wages, food, water, and shelter while working long hours and suffering abuse.

---

Slavery is not simply an unfortunate historical legacy in American society. The U.S. State Department estimates that as many as 17,500 foreigners are brought to the United States every year and enslaved. The three main areas of "trafficking" people into the United States for the purpose of enslavement are agricultural slavery, domestic servitude, and sexual exploitation. The problem of global slavery and human trafficking is so prevalent in the second decade of the twenty-first century that the U.S. president continues to issue an annual proclamation setting forth a prevention month to call attention to the problem. "Sociology in Global Perspective" shows how modern slavery may affect your daily life without your ever knowing it.

## The Caste System

Like slavery, caste is a closed system of social stratification. A *caste system* is a system of social inequality in which people's status is permanently determined at birth based on their parents' ascribed characteristics. Vestiges of caste systems exist in contemporary India and South Africa.

In India, caste is based in part on occupation; thus, families typically perform the same type of work from generation to generation (Figure 7.2). By contrast, the caste system of South Africa was based on racial classifications and the belief of white South Africans (Afrikaners) that they were morally superior to the black majority. Until the 1990s, the Afrikaners controlled the government, the police, and the military by enforcing *apartheid*—the separation of the races. Blacks were denied full citizenship and restricted to segregated hospitals, schools, residential neighborhoods, and other facilities. Whites held almost all of the desirable jobs; blacks worked as manual laborers and servants.

In a caste system, marriage is endogamous, meaning that people are allowed to marry only within their own group. In India, parents traditionally have selected marriage partners for their children. In South Africa, interracial marriage was illegal until 1985.

Cultural beliefs and values sustain caste systems. Hinduism, the primary religion of India, reinforced the caste system by teaching that people should accept their fate in life and work hard as a moral duty. Caste systems grow weaker as societies industrialize; the values reinforcing the system break down, and people start to focus on the types of skills needed for industrialization.

As we have seen, in closed systems of stratification, group membership is hereditary, and it is almost impossible to move up within the structure. Custom and law frequently perpetuate privilege and

**2:00 P.M.: Afternoon Snack**

The chocolate snack that you eat may have come from plantations in Africa where the children who produce the cocoa—the key ingredient in chocolate—are subjected to conditions of forced labor. As many as 300,000 children may work in cocoa production worldwide.

**4:00 P.M.: Drive to Meeting**

The tires on your car are made of rubber, which may have been produced on rubber plantations in Asia and Africa where entire families are forced to work excessive hours on the plantations for little or no pay and in hazardous working conditions.

**6:00 P.M.: Arrive at Home**

Bricks in the walls of your house may have been produced in brick kilns in Asia or Latin America by men, women, and children who are forced to work in hazardous conditions as victims of bonded labor.

**8:00 P.M.: Enjoy Dinner**

The food you eat for dinner may have been touched by men and children subjected to forced labor—with little or no pay and extensive physical and emotional abuse—on cattle ranches and farms in the United States, Latin America, and Africa.

**11:00 P.M.: Go to Bed**

The cotton in your sheets may have been picked by men, women, and children—some as young as three years old—in cotton fields in Central Asia and Africa.

According to the U.S. Department of State, this scenario accounts for only one day. If you want to see how modern slavery may touch the rest of your life, check out the Slavery Footprint website.

## reflect & analyze

**Why is it important for us to be aware of human slavery? Is this concern relevant to our daily lives? Is there any relationship between the American Dream and the exploitation of people on a global basis?**

Source: U.S. Department of State, 2012.

---

ensure that higher-level positions are reserved for the children of the advantaged.

## The Class System

The *class system* **is a type of stratification based on the ownership and control of resources and on the type of work that people do.** At least theoretically, a class system is more open than a caste system because the boundaries between classes are less distinct than the boundaries between castes. In a class system, status comes at least partly through achievement rather than entirely by ascription.

In class systems, people may become members of a class other than that of their parents through both intergenerational and intragenerational mobility, either upward or downward. *Horizontal mobility* occurs when people experience a gain or loss in position and/or income that does not produce a change in their place in the class structure. For example, a person may get a pay increase and a more prestigious title but still not move from one class to another.

By contrast, movement up or down the class structure is *vertical mobility*. For example, Bruce is a physician, but he was the first person in his family to attend college, much less to graduate from medical school. His father was a day laborer picked up each day by contractors outside the local lumberyard to work on various building sites. His mother was a stay-at-home-mom who took care of Bruce and his four younger siblings. Bruce's parents did not complete high school and had little support from home because their families had limited economic means. In high school, Bruce became involved in the Upward Bound program, which encourages young people to remain in school and provides mentoring, tutoring, and enrichment activities that ultimately helped him achieve his dream.

Bruce's situation reflects upward mobility; however, people may also experience downward mobility, caused by any number of reasons, including a lack of jobs, low wages and employment instability, marriage to someone with fewer resources and less power than oneself, and changing social conditions.

**caste system** a system of social inequality in which people's status is permanently determined at birth based on their parents' ascribed characteristics.

**class system** a type of stratification based on the ownership and control of resources and on the type of work that people do.

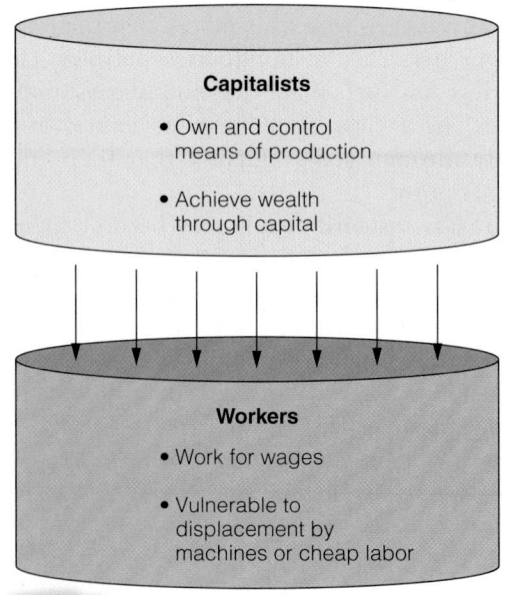

**FIGURE 7.2** Systems of stratification include slavery, caste, and class. As shown here, the life chances of people living in each of these systems differ widely.

# Classical Perspectives on Social Class

Early sociologists grappled with the definition of class and the criteria for determining people's location in the class structure. Both Karl Marx and Max Weber viewed class as an important determinant of social inequality and social change, and their works have had a profound influence on how we view the U.S. class system today.

## Karl Marx: Relationship to the Means of Production

**LO2** According to Karl Marx, class position and the extent of our income and wealth are determined by our work situation, or our relationship to the means of production. As we have previously seen, Marx stated that capitalistic societies consist of two classes—the capitalists and the workers. The **capitalist class (bourgeoisie) consists of those who own the means of production**—the land and capital necessary for factories and mines, for example. The **working class (proletariat) consists of those who must sell their labor to the owners in order to earn enough money to survive** (see Figure 7.3).

According to Marx, class relationships involve inequality and exploitation. Capitalists maximize their profits by exploiting workers, paying them less than the resale value of what they produce but do not own. This exploitation results in workers' *alienation*—**a feeling of powerlessness and estrangement from other people and from oneself**. In Marx's view,

**Capitalists**

- Own and control means of production

- Achieve wealth through capital

**Workers**

- Work for wages

- Vulnerable to displacement by machines or cheap labor

**FIGURE 7.3** MARX'S VIEW OF STRATIFICATION
© Cengage Learning

alienation develops as workers manufacture goods that embody their creative talents but the goods do not belong to them. Workers are also alienated from the work itself because they are forced to perform it in order to live. Because the workers' activities are not their own, they feel self-estrangement. Moreover, workers are separated from others in the factory because they individually sell their labor power to the capitalists as a commodity.

In Marx's view, the capitalist class maintains its position at the top of the class structure by control of the society's *superstructure*, which is composed of the government, schools, churches, and other social institutions that produce and disseminate ideas perpetuating the existing system of exploitation (Figure 7.4). Marx predicted that the exploitation of workers by the capitalist class would ultimately lead to **class conflict—the struggle between the capitalist class and the working class.** According to Marx, when the workers realized that capitalists were the source of their oppression, they would overthrow the capitalists and their agents of social control, leading to the end of capitalism. The workers would then take over the government and create a more egalitarian society.

Why has no workers' revolution occurred? Capitalism may have persisted because it has changed significantly since Marx's time. Individual capitalists no longer own and control factories and other means of production; today, ownership and control have largely been separated. For example, contemporary transnational corporations are owned by a multitude of stockholders but run by paid officers and managers. Similarly, many (but by no means all) workers have experienced a rising standard of living, which may have contributed to a feeling of complacency. During the twentieth century, workers pressed for salary increases and improvements in the workplace through their activism and labor union membership. They also gained more legal protection in the form of workers' rights and benefits such as workers' compensation insurance for job-related injuries and disabilities. For these reasons, and because of a myriad of other complex factors, the workers' revolution predicted by Marx has not come to pass. However, the failure of his prediction does not mean that his analysis of capitalism and his theoretical contributions to sociology are without validity.

Patrick T. Fallon/Bloomberg/Getty Images

**FIGURE 7.4** Karl Marx believed that class relationships involve inequality and exploitation. How is his idea represented in this photo?

Marx had a number of important insights into capitalist societies. First, he recognized the economic basis of class systems (Gilbert, 2011). Second, he noted the relationship between people's social location in the class structure and their values, beliefs, and behavior. Finally, he acknowledged that classes may have opposing (rather than complementary) interests. For example, capitalists' best interests are served by a decrease in labor costs and other expenses and a corresponding increase in profits; workers' best interests are served by well-paid jobs, safe working conditions, and job security.

## Max Weber: Wealth, Prestige, and Power

**LO3** Max Weber's analysis of class builds upon earlier theories of capitalism (particularly those by Marx) and of money (particularly those by Georg Simmel, as discussed in Chapter 1). Living in the late nineteenth and early twentieth

**capitalist class (or bourgeoisie)** Karl Marx's term for those who own the means of production.

**working class (or proletariat)** Karl Marx's term for those who must sell their labor to the owners in order to earn enough money to survive.

**alienation** a feeling of powerlessness and estrangement from other people and from oneself.

**class conflict** Karl Marx's term for the struggle between the capitalist class and the working class.

**FIGURE 7.5** What level of wealth, power, and prestige do you believe these women have?

centuries, Weber was in a unique position to see the transformation that occurred as individual, competitive, entrepreneurial capitalism went through the process of shifting to bureaucratic, industrial, corporate capitalism. As a result, Weber had more opportunity than Marx to see how capitalism changed over time.

Weber agreed with Marx's assertion that economic factors are important in understanding individual and group behavior. However, Weber emphasized that no single factor (such as economic divisions between capitalists and workers) was sufficient for defining the location of categories of people within the class structure. According to Weber, the access that people have to important societal resources (such as economic, social, and political power) is crucial in determining people's life chances. To highlight the importance of life chances for categories of people, Weber developed a multidimensional approach to social stratification that reflects the interplay among wealth, prestige, and power. In his analysis of these dimensions of class structure, Weber viewed the concept of "class" as an *ideal type* (that can be used to compare and contrast various societies) rather than as a specific social category of "real" people (Bourdieu, 1984).

**Wealth is the value of all of a person's or family's economic assets, including income, personal property, and income-producing property.** Weber placed categories of people who have a similar level of wealth and income in the same class (Figure 7.5). For example, he identified a privileged commercial class of *entrepreneurs*—wealthy bankers, ship owners, professionals, and merchants who possess similar financial resources. He also described a class of *rentiers*—wealthy individuals who live off their investments and do not have to work. According to Weber, entrepreneurs and rentiers have much in common. Both are able to purchase expensive consumer goods, control other people's opportunities to acquire wealth and property, and monopolize costly status privileges (such as education) that provide contacts and skills for their children.

Weber divided those who work for wages into two classes: the middle class and the working class. The middle class consists of white-collar workers, public officials, managers, and professionals. The working class consists of skilled, semiskilled, and unskilled workers.

The second dimension of Weber's system of stratification is *prestige*—**the respect or regard that a person or status position is given by others.** Fame, respect, honor, and esteem are the most common forms of prestige. A person who has a high level of prestige is assumed to receive deferential and respectful treatment from others. Weber suggested that individuals who share a common level of social prestige belong to the same status group regardless

**208**

**FIGURE 7.6** WEBER'S MULTIDIMENSIONAL APPROACH TO SOCIAL STRATIFICATION
According to Max Weber, wealth, power, and prestige are separate continuums. Individuals may rank high in one dimension and low in another, or they may rank high or low in more than one dimension. Also, individuals may use their high rank in one dimension to achieve a comparable rank in another. How does Weber's model compare with Marx's approach, as shown in Figure 7.3?

© Cengage Learning

of their level of wealth. They tend to socialize with one another, marry within their own group of social equals, spend their leisure time together, and safeguard their status by restricting outsiders' opportunities to join their ranks.

The other dimension of Weber's system is *power*—**the ability of people or groups to achieve their goals despite opposition from others.** The powerful can shape society in accordance with their own interests and direct the actions of others. According to Weber, bureaucracies hold social power in modern societies; individual power depends on a person's position within the bureaucracy. Weber suggested that the power of modern bureaucracies was so strong that even a workers' revolution (as predicted by Marx) would not lessen social inequality.

Weber stated that wealth, prestige, and power are separate continuums on which people can be ranked from high to low, as shown in Figure 7.6. Individuals may be high on one dimension while being low on another. For example, people may be very wealthy but have little political power (for example, a recluse who has inherited a large sum of money). They may also have prestige but not wealth (for instance, a college professor who receives teaching excellence awards but lives on a relatively low income). In Weber's multidimensional approach, people are ranked on all three dimensions. Sociologists often use the term *socioeconomic status (SES)* **to refer to a combined measure that attempts to classify individuals, families, or households in**

**terms of factors such as income, occupation, and education to determine class location.**

What important contribution does Weber make to our understanding of social stratification and class? Weber emphasized that people behave according to both their economic interests and their values. He also added to Marx's insights by developing a multidimensional explanation of the class structure and by identifying additional classes. Both Marx and Weber emphasized that capitalists and workers are the primary players in a class society, and both noted the importance of class to people's life chances. However, they saw different futures for capitalism and the social system. Marx saw these structures being overthrown; Weber saw the increasing bureaucratization of life even without capitalism.

**wealth** the value of all of a person's or family's economic assets, including income, personal property, and income-producing property.

**prestige** the respect or regard that a person or status position is given by others.

**power** the ability of people or groups to achieve their goals despite opposition from others.

**socioeconomic status (SES)** a combined measure that attempts to classify individuals, families, or households in terms of factors such as income, occupation, and education to determine class location.

# Contemporary Sociological Models of the U.S. Class Structure

**LO4** How many social classes exist in the United States? What criteria are used for determining class membership? No broad consensus exists about how to characterize the class structure in this country. In fact, many people deny that class distinctions exist. Most people like to think of themselves as middle class; it puts them in a comfortable middle position—neither rich nor poor. Sociologists have developed two models of the class structure: One is based on a Weberian approach, the other on a Marxian approach. We will examine both models briefly.

## The Weberian Model of the U.S. Class Structure

Expanding on Weber's analysis of class structure, sociologist Dennis Gilbert (2011) uses a model of social classes based on three elements: (1) education, (2) occupation of family head, and (3) family income (see Figure 7.7).

**The Upper (Capitalist) Class** The upper class is the wealthiest and most powerful class in the United States. About 1 percent of the population is included in this class, whose members own substantial income-producing assets and operate on both the national and international levels. According to Gilbert (2011), people in this class have an

influence on the economy and society far beyond their numbers.

Some models further divide the upper class into upper-upper ("old money") and lower-upper ("new money") categories (Warner and Lunt, 1941; Coleman and Rainwater, 1978; Kendall, 2002). Members of the upper-upper class come from prominent families, which possess great wealth that they have held for several generations. In the past, family names—such as Rockefeller, Mellon, Du Pont, and Kennedy—were well-known and often held in high esteem. Today, some upper-class family names are well-known, but many of the individuals and families involved have often made their fortunes much more recently. Persons in the upper-upper class tend to have strong feelings of in-group solidarity. They belong to the same exclusive clubs and support high culture (such as the opera, symphony orchestras, ballet, and art museums). Children are educated in prestigious private schools and Ivy League universities; many acquire strong feelings of privilege from birth, as upper-class author Lewis H. Lapham (1988: 14) states:

> Together with my classmates and peers, I was given to understand that it was sufficient accomplishment merely to have been born. Not that anybody ever said precisely that in so many words, but the assumption was plain enough, and I could confirm it by observing the mechanics of the local society. A man might become a drunkard, a concert pianist or an owner of companies, but none of these occupations would have an important bearing on his social rank.

Children of the upper class are socialized to view themselves as different from others; they also learn that they are expected to marry within their own class (Warner and Lunt, 1941; Mills, 1959a; Kendall, 2002).

Members of the lower-upper class may be extremely wealthy but not have attained as much prestige as members of the upper-upper class. The "new rich" have earned most of their money in their own lifetime as entrepreneurs, presidents of major corporations, sports or entertainment celebrities, or top-level professionals. For some members of the lower-upper class, the American Dream has become a reality. Others still desire the respect of members of the upper-upper class.

**The Upper-Middle Class** Persons in the upper-middle class are often highly educated professionals who have built careers as physicians, attorneys, stockbrokers, or corporate managers. Others derive their income from family-owned businesses. According to Gilbert (2011), about 14 percent of the

**FIGURE 7.7** STRATIFICATION BASED ON EDUCATION, OCCUPATION, AND INCOME

© Cengage Learning

U.S. population is in this category. A combination of three factors qualifies people for the upper-middle class: university degrees, authority and independence on the job, and high income. Of all the class categories, the upper-middle class is the one that is most shaped by formal education.

Over the past fifty years, Asian Americans, Latinos/as, and African Americans have placed great importance on education as a means of attaining the American Dream. Many people of color have moved into the upper-middle class by acquiring higher levels of education. However, higher levels of educational attainment are not always a guarantee of wealth and success for persons in any racial–ethnic category.

**The Middle Class** In past decades a high school diploma was necessary to qualify for most middle-class jobs. Today, two-year or four-year college degrees have replaced the high school diploma as an entry-level requirement for employment in many middle-class occupations, including medical technicians, nurses, legal and medical assistants, lower-level managers, semiprofessionals, and nonretail sales workers. An estimated 30 percent of the U.S. population is in the middle class, even though most people in this country think of themselves as middle class. Nowhere is this myth of the vast middle class more prevalent than in television shows, which for decades have focused on an idealized notion of the middle class or the debunking of that myth.

Traditionally, most middle-class occupations have been relatively secure and have provided more opportunities for advancement (especially with increasing levels of education and experience) than working-class positions. Recently, however, several factors have eroded the American Dream for this class: (1) the housing crisis, beginning with an overpriced housing market and too-easy credit, followed by high rates of foreclosures and a banking meltdown; (2) a high rate of job loss and instability, with long-term unemployment and blocked mobility on the job; and (3) national economic sluggishness, starting with the 2007 U.S. recession and then a promised economic recovery that still had not fully occurred as of 2014, the time of this writing.

**The Working Class** An estimated 30 percent of the U.S. population is in the working class. The core of this class is made up of semiskilled machine operators who work in factories and elsewhere. Members of the working class also include some workers in the service sector, as well as clerks and salespeople whose job responsibilities involve routine, mechanized tasks requiring little skill beyond basic literacy and a brief period of on-the-job training (Gilbert, 2011). Some people in the working class are employed in *pink-collar occupations*—**relatively low-paying, nonmanual, semiskilled positions primarily held by women,** such as day-care workers, checkout clerks, cashiers, and restaurant servers.

How does life in working-class families compare with that of individuals in middle-class families? According to sociologists, working-class families not only earn less than middle-class families, but they also have less financial security, particularly because of high rates of layoffs and plant closings in some regions of the country. Few people in the working class have more than a high school diploma, and many have less, which makes job opportunities scarce for them in a "high-tech" society (Gilbert, 2011). Others find themselves in low-paying jobs in the service sector of the economy, particularly fast-food restaurants, a condition that often places them among the working poor.

**The Working Poor** The working poor account for about 20 percent of the U.S. population. Members of the working-poor class live from just above to just below the poverty line; they typically hold unskilled jobs, seasonal migrant jobs in agriculture, lower-paid factory jobs, and service jobs (such as counter help at restaurants). Employed single mothers often belong to this class; consequently, children are overrepresented in this category. African Americans and other people of color are also overrepresented among the working poor. To cite only one example, in the United States today there are two white hospital orderlies to every one white physician, whereas there are twenty-five African American orderlies to every one African American physician (Gilbert, 2011). For the working poor, living from paycheck to paycheck makes it impossible to save money for emergencies such as periodic or seasonal unemployment, which is a constant threat to any economic stability they may have (Figure 7.8).

Over a decade ago, social critic and journalist Barbara Ehrenreich (2001, 2011) left her upper-middle-class lifestyle for a period of time to see if it was possible for the working poor to live on the wages that they were being paid as restaurant servers, salesclerks at discount department stores, aides at nursing homes, housecleaners for franchise maid services, or similar jobs. She conducted her research by actually holding those jobs for periods of time and seeing if she could live on the wages that she received. Through her research, Ehrenreich

**pink-collar occupations** relatively low-paying, nonmanual, semiskilled positions primarily held by women.

**FIGURE 7.8** In which segment of the class structure would sociologists place this worker? What are the key elements of that social class?

persuasively demonstrated that people who work full time, year round, for poverty-level wages must develop survival strategies that include such things as help from relatives or constantly moving from one residence to another in order to have a place to live. Like many other researchers, Ehrenreich found that minimum-wage jobs cannot cover the full cost of living, such as rent, food, and the rest of an adult's monthly needs, even without taking into consideration the needs of children or other family members. If Ehrenreich's study were to be replicated, we would probably find that the tasks she sought to accomplish are even more difficult today because jobs have become more scarce, more people are working part time who desire full-time employment, and the cost of living continues to increase in many regions.

At some point in our lives, most of us have held a job paying the minimum wage, and we know the limitations of trying to survive on such low earnings. The federal *minimum wage* is the hourly rate that (with certain exceptions) is the lowest amount an employer can legally pay its employees (each state may adopt a higher minimum wage but not a lower one). In 2009 the federal minimum wage was raised to $7.25 per hour, where it remains today. A person earning minimum wage and working forty hours every week, fifty-two weeks per year (in other words, no time off, no vacation), would still earn an amount slightly above the official poverty line (and slightly below that line for a person with two children). Several states have a minimum wage above the federal requirement: Washington State's minimum wage is $9.32 per hour—the highest minimum wage of all the states. Increasing social and economic inequality in the United States has been partly attributed to the vast divide between the low wages paid to workers, based on a low federal minimum wage, and the astronomically high salaries and compensation packages given to some major corporate CEOs.

**The Underclass** According to Gilbert (2011), people in the *underclass* **are poor, seldom employed, and caught in long-term deprivation that results from low levels of education and income and high rates of unemployment.** Some are unable to work because of age or disability; others experience discrimination based on race/ethnicity. Single mothers are overrepresented in this class because of the lack of jobs, lack of affordable child care, and many other impediments to the mother's future and that of her children. People without a "living wage" often must rely on public or private assistance programs for their survival. About 5 percent of the U.S. population is in this category, and the chances of their children moving out of poverty are about fifty-fifty (Gilbert, 2011).

Gaining work-related skills and having employment opportunities are two critical issues for people on the lowest rungs of the class ladder. Many of the jobs that exist today require specialized knowledge or skills that are inaccessible to people in the underclass. Skills and jobs are essential for people to have the opportunity to earn a decent wage; have medical coverage; live meaningful, productive lives; and raise their children in a safe environment. These issues are closely tied to the American Dream that we have been discussing in this chapter.

## The Marxian Model of the U.S. Class Structure

The earliest Marxian model of class structure identified ownership or nonownership of the means of production as the distinguishing feature of classes. From this perspective, classes are social groups organized around property ownership, and social stratification is created and maintained by one group in order to protect and enhance its own economic interests. Moreover, societies are organized around classes in conflict over scarce resources. Inequality results from the more powerful exploiting the less powerful.

Contemporary Marxian (or conflict) models examine class in terms of people's relationship to others in the production process. For example, conflict theorists attempt to determine the degree of control that workers have over the decision-making process

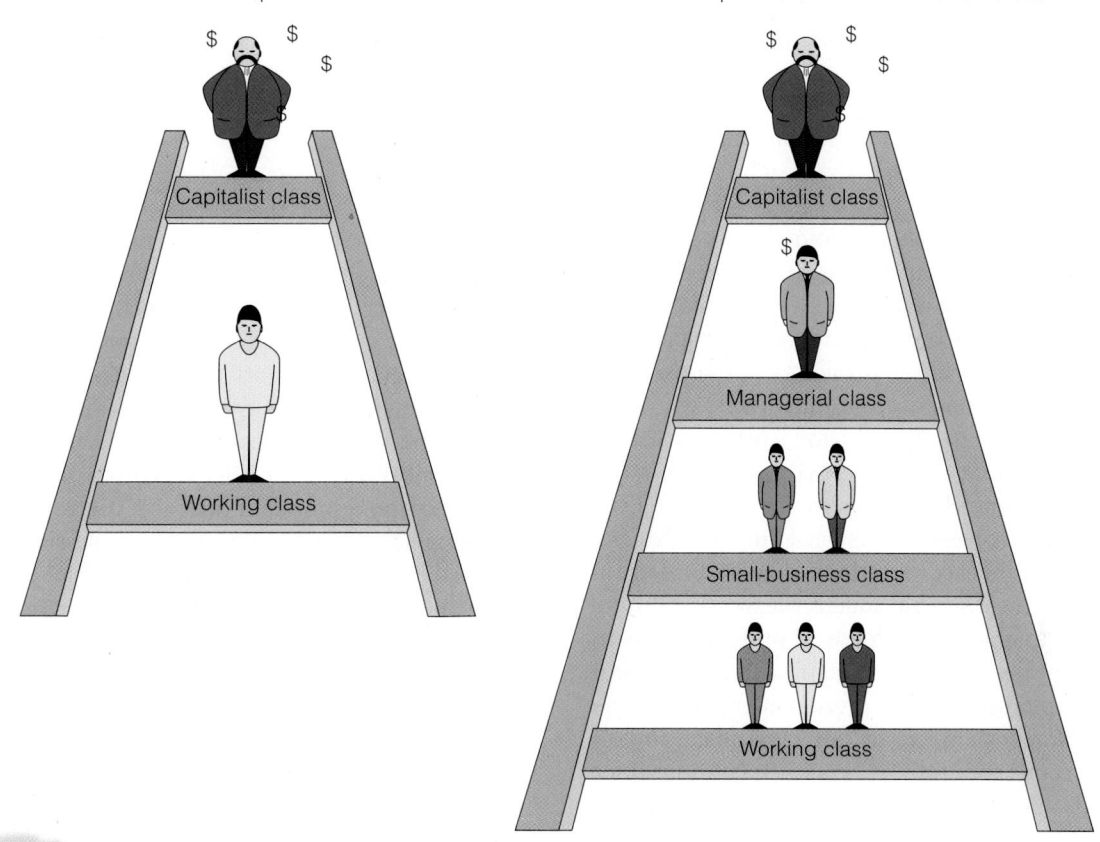

| Marx's Model | Wright's Model |
|---|---|
| Based on relationship to the means of production: | Takes into account both ownership and control of the means of production and the labor of others: |

**FIGURE 7.9** COMPARISON OF MARX'S AND WRIGHT'S MODELS OF CLASS STRUCTURE

© Cengage Learning

and the extent to which they are able to plan and implement their own work. They also analyze the type of supervisory authority, if any, that a worker has over other workers. According to this approach, most employees are a part of the working class because they do not control either their own labor or that of others.

Erik Olin Wright (1979, 1985, 1997, 2010), one of the leading stratification theorists to examine social class from a Marxian perspective, has concluded that Marx's definition of "workers" does not fit the occupations found in advanced capitalist societies. For example, many top executives, managers, and supervisors who do not own the means of production (and thus would be "workers" in Marx's model) act like capitalists in their zeal to control workers and maximize profits. Likewise, some experts hold positions in which they have control over money and the use of their own time even though they are not owners. Wright views Marx's category of "capitalist" as being too broad as well. For instance, small-business owners might be viewed as capitalists because they own their own tools and have a

few people working for them, but they have little in common with large-scale capitalists and do not share the interests of factory workers. Figure 7.9 compares Marx's and Wright's models.

Wright (1979) also argues that classes in modern capitalism cannot be defined simply in terms of different levels of wealth, power, and prestige, as in the Weberian model. Consequently, he outlines four criteria for placement in the class structure: (1) ownership of the means of production, (2) purchase of the labor of others (employing others), (3) control of the labor of others (supervising others on the job), and (4) sale of one's own labor (being employed by someone else). Wright (1978) assumes that these criteria can be used to determine the class placement of

**underclass** those who are poor, seldom employed, and caught in long-term deprivation that results from low levels of education and income and high rates of unemployment.

213

all workers, regardless of race/ethnicity, in a capitalist society. Let's take a brief look at Wright's (1979, 1985) four classes—(1) the capitalist class, (2) the managerial class, (3) the small-business class, and (4) the working class—so that you can compare them to those found in the Weberian model.

**The Capitalist Class** According to Wright, the capitalist class holds most of the wealth and power in society through ownership of capital—for example, banks, corporations, factories, mines, news and entertainment industries, and agribusiness firms. The "ruling elites," or "ruling class," within the capitalist class hold political power and are often elected or appointed to influential political and regulatory positions.

This class is composed of individuals who have inherited fortunes, own major corporations, or are top corporate executives with extensive stock holdings or control of company investments. Even though many top executives have only limited *legal ownership* of their corporations, they have substantial economic ownership and exert extensive control over investments, distribution of profits, and management of resources. The major sources of income for the capitalist class are profits, interest, and very high salaries. Members of this class make important decisions about the workplace, including which products and services to make available to consumers and how many workers to hire or fire.

*Forbes* magazine's 2013 list of the richest people in the world identified Mexican telecom entrepreneur Carlos Slim Helu as the wealthiest person in the world, with a net worth of $73 billion. Bill Gates (cofounder of Microsoft Corporation) was the second-wealthiest capitalist, with a net worth of $67 billion (*Forbes*, 2013). Although men who made the *Forbes* list of the wealthiest people gained their fortunes through entrepreneurship or being CEOs of large corporations, many women who made the list acquired their wealth typically through inheritance, marriage, or both. Of the top twenty richest people in the United States, only three are women: Christy and Alice Walton—members of the family of Sam Walton, founder of Walmart—and Jacqueline Mars of the Mars Candy family.

**The Managerial Class** People in the managerial class have substantial control over the means of production and over workers. However, these upper-level managers, supervisors, and professionals usually do not participate in key corporate decisions such as how to invest profits. Lower-level managers may have some control over employment practices, including the hiring and firing of some workers.

Top professionals such as physicians, attorneys, accountants, and engineers may control the structure of their own work; however, they typically do not own the means of production and may not have supervisory authority over more than a few people. Even so, they may influence the organization of work and the treatment of other workers. Members of the capitalist class often depend on these professionals for their specialized knowledge.

**The Small-Business Class** This class consists of people who are self-employed—small-business owners and craftspeople—who may hire a small number of employees but largely do their own work. Some members own businesses such as "mom-and-pop" grocery stores, retail clothing stores, and jewelry stores. Others are doctors and lawyers who receive relatively high incomes from selling their own services. Some of these professionals now share attributes with members of the capitalist class because they have formed corporations that hire and control the employees who produce profits for the professionals.

It is in the small-business class that we find many people's hopes of achieving the American Dream (Figure 7.10). Recent economic trends, including corporate downsizing, telecommuting, and the movement of jobs to other countries, have encouraged more people to think about being self-employed. About one in nine U.S. workers is self-employed (Hipple, 2010). Self-employment is more common among men, whites, Asians, and older workers who retire from wage and salary jobs and may become self-employed to supplement their retirement income.

**The Working Class** The working class is made up of a number of subgroups, one of which is blue-collar workers, some of whom are highly skilled and well paid and others of whom are unskilled and poorly paid. Skilled blue-collar workers include electricians, plumbers, and carpenters; unskilled blue-collar workers include janitors and gardeners.

White-collar workers are another subgroup of the working class. Referred to by some as a "new middle class," these workers are actually members of the working class because they do not own the means of production, do not control the work of others, and are relatively powerless in the workplace. Secretaries, administrative assistants and other clerical workers, and sales workers are members of the white-collar faction of the working class. They take orders from others and tend to work under constant supervision. Thus, these workers are at the bottom of the class structure in terms of domination and control in the workplace. The working class contains about half of all employees in the United States.

Although Marxian and Weberian models of the U.S. class structure show differences in people's

**FIGURE 7.10** Many immigrants believe they can achieve the American Dream by starting a small business, such as a store, restaurant or newsstand, as seen here in a New York City subway. Skilled laborers and tradespeople sometimes take another approach to small business by offering services, such as landscaping, plumbing, carpentry, and dog grooming.

occupations and access to valued resources, neither fully reflects the nature and extent of inequality in the United States. In the next section we will take a closer look at the unequal distribution of income and wealth in the United States and the effects of inequality on people's opportunities and life chances.

# Inequality in the United States

Throughout human history, people have argued about the distribution of scarce resources in society. Disagreements often center on whether the share we get is a fair reward for our effort and hard work. Recently, social analysts have pointed out that (except during temporary economic downturns) the old maxim "the rich get richer" continues to be valid in the United States. To understand how this happens, we must take a closer look at the distribution of income and wealth in this country.

## Distribution of Income and Wealth

**LO5** Money is essential for acquiring goods and services. People without money cannot purchase food, shelter, clothing, medical care, legal aid, education, and the other things they

need or desire. Money—in the form of both income and wealth—is very unevenly distributed in the United States. Among high-income nations, the United States remains number one in inequality of income distribution.

**Income Inequality** *Income* **is the economic gain derived from wages, salaries, income transfers (governmental aid), and ownership of property.** Or, to put it another way, "income refers to money, wages, and payments that periodically are received as returns for an occupation or investment" (Kerbo, 2000: 19). Data from the U.S. Census Bureau typically provide income estimates that are based solely on money income before taxes and do not include the value of noncash benefits such as health care coverage or retirement benefits.

National distribution of income can be compared to a pizza that has been sliced into portions, ranging in size from very small to generous, for distribution among segments of the population. Figure 7.11 shows

**income** the economic gain derived from wages, salaries, income transfers (governmental aid), and ownership of property.

Bo Zaunders/Corbis

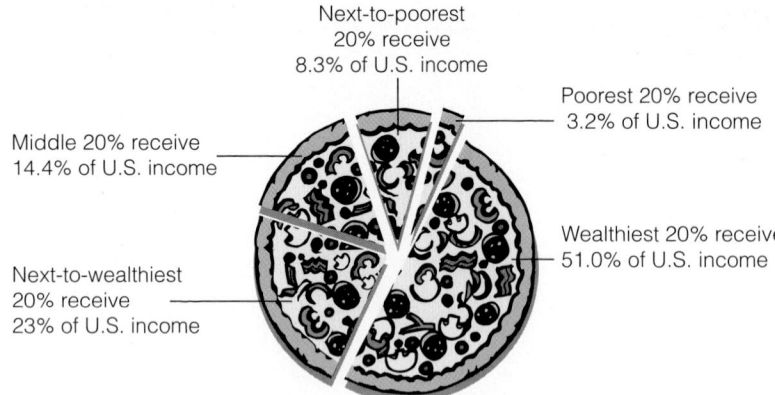

Next-to-poorest
20% receive
8.3% of U.S. income

Poorest 20% receive
3.2% of U.S. income

Middle 20% receive
14.4% of U.S. income

Wealthiest 20% receive
51.0% of U.S. income

Next-to-wealthiest
20% receive
23% of U.S. income

**FIGURE 7.11** DISTRIBUTION OF PRETAX INCOME IN THE UNITED STATES, 2011
Thinking of personal income in the United States (before taxes) as a large pizza helps us to see which segments of the population receive the largest and smallest portions.

Source: DeNavas-Walt, Proctor, and Smith, 2012.

that in 2012 the wealthiest 20 percent of U.S. households received more than half of the total income "pie," while the poorest 20 percent of households received slightly more than 3 percent of all income. Although not shown on the pie chart, the top 5 percent *alone* received slightly more than 22 percent of all income—an amount greater than that received by the bottom 40 percent of all households.

The gap between the rich and the poor has widened significantly in recent decades. Between 1981 and 2012, the median income of the top one-fifth of U.S. households increased by more than four times—about twice the rate of increase of the bottom one-fifth of families (U.S. Census Bureau Historical Income Tables, 2012; DeNavas-Walt, Proctor, and Smith, 2013) (see Figure 7.12). Although the recent recession may have reduced the annual income of some rich Americans, this does not mean that they

have become poor. Their income may have dropped by several million dollars, causing them to slightly modify their affluent lifestyles, but their overall standard of living is protected by other wealth they possess. For example, the net worth of top U.S. billionaires such as Bill Gates and Warren Buffett dropped several billion dollars, but they are far from living in poverty. However, the picture is different for people situated in the middle and bottom sectors of the income pie, who are faced with high mortgage rates or rent payments and increasing costs for food, fuel, transportation, and other necessities.

Income distribution varies by race/ethnicity as well as class. Figure 7.13 compares median household income by race/ethnicity, showing not only the disparity among groups but also the consistency of that disparity over almost two decades. Although households across all racial–ethnic categories have

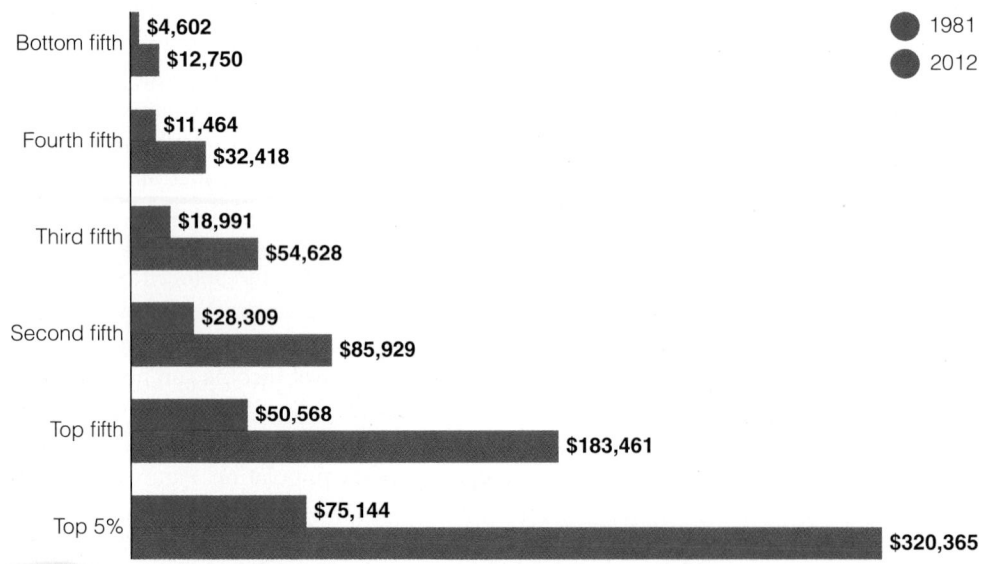

● 1981
● 2012

Bottom fifth
$4,602
$12,750

Fourth fifth
$11,464
$32,418

Third fifth
$18,991
$54,628

Second fifth
$28,309
$85,929

Top fifth
$50,568
$183,461

Top 5%
$75,144
$320,365

**FIGURE 7.12** MEDIAN HOUSEHOLD INCOME IN THE UNITED STATES
This chart shows the distribution of before-tax income in the United States. Notice the dramatic increase in income for the top 5 percent of households. During the past three decades the difference in income between the richest and poorest has become much more pronounced.

Source: U.S. Census Bureau Historical Income Tables, 2012.

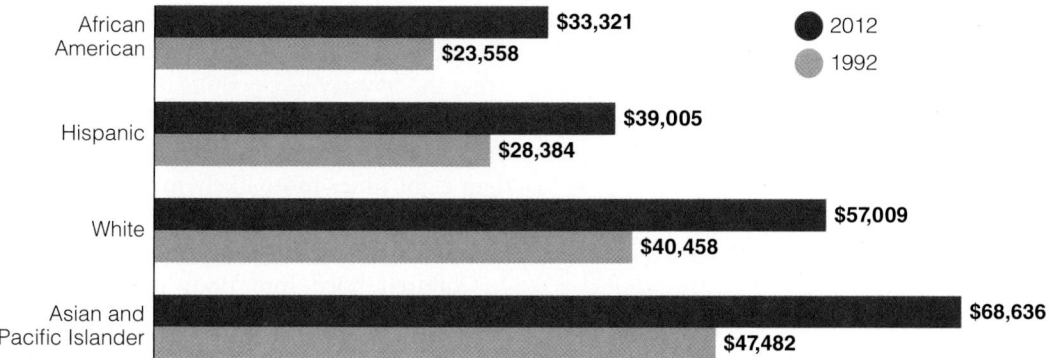

**FIGURE 7.13** MEDIAN HOUSEHOLD INCOME BY RACE/ETHNICITY IN THE UNITED STATES

Sources: DeNavas-Walt, Cleveland, and Webster, 2003; DeNavas-Walt, Proctor, and Smith, 2012. © Cengage Learning

experienced some decline in real annual median income, the income gap between African American (black) households and the households of whites (non-Hispanic) and Asians or Pacific Islanders is striking. In 2012 African American households had the lowest median income, $33,321, as compared with Asian households, which had the highest median, $68,636. Non-Hispanic white households had a median income of $57,009, as compared with Hispanic households, which averaged $39,005. Although not shown in Figure 7.13, the 2012 median household income for all races was $51,017.

**Wealth Inequality** Income is only one aspect of wealth. Wealth includes property such as buildings, land, farms, houses, factories, and cars, as well as other assets such as bank accounts, corporate stocks, bonds, and insurance policies. Wealth is computed by subtracting all debt obligations and converting the remaining assets into cash. Wealth is also referred to as "net worth," and it is a very important component in determining the standard of living for families and households. For most people in the United States, wealth is invested primarily in property that generates no income, such as a home or car. By contrast, the wealth of elites is often in the form of income-producing property or investments that make money for them.

Wealth is even more unevenly distributed than income. In 2011 (the latest year for which data are available), the top 1 percent of U.S. households owned more than 40 percent of the nation's approximately $54 trillion in wealth while earning about 19 percent of all income. The net worth of the wealthiest 1 percent was 290 times greater than the net worth of the median U.S. household—the highest ratio on record. (The "median U.S. household" refers to families that are in the exact middle of the distribution for their group. Half of the people in the group will have more than the median, and half will have less.) Many people in less-affluent income

categories live on wages from their jobs; however, the truly wealthy become richer primarily through investments in stocks, bonds, real estate, and other financial endeavors, not from wages.

As the United States has slowly tried to move beyond the Great Recession of 2007–2009, the wealthiest households have continued to outgain everyone else in terms of wealth accumulation and net worth. *Net worth* is the total household wealth after secured debts (such as home mortgages and vehicle loans) and unsecured debts (such as credit card debt and some bank loans) have been subtracted. Between 2009 and 2011, the mean net worth of households in the upper 7 percent of wealth distribution rose by about 28 percent while the mean net worth in the lower 93 percent dropped by 4 percent (Fry and Taylor, 2013). *Mean wealth* is the average wealth for each household in the category. For example, in 2011 the mean wealth of the wealthiest 7 percent in the United States was $3,173,895. By contrast, the mean wealth of the lower 93 percent was about $133,817. As previously stated, most of this wealth was tied up in the value of homes (Fry and Taylor, 2013).

The racial wealth gap continues to widen in terms of both wealth creation and maintenance. Factors closely associated with the growing racial wealth divide are as follows: (1) the number of years of homeownership; (2), household income; (3) unemployment, which is more prevalent among African American families; (4) possession of a college education; and (5) inheritance, financial support from family or friends, and preexisting family wealth (Shapiro, Meschede, and Osoro, 2013). Residential segregation artificially lowers the cost of housing and the building of home equity for persons of color who own residences in nonwhite neighborhoods. Moreover, for many years, white Americans have been more likely to inherit money from their families than are persons of color. Typically, white Americans have also received more parental assistance in making down payments on home purchases because

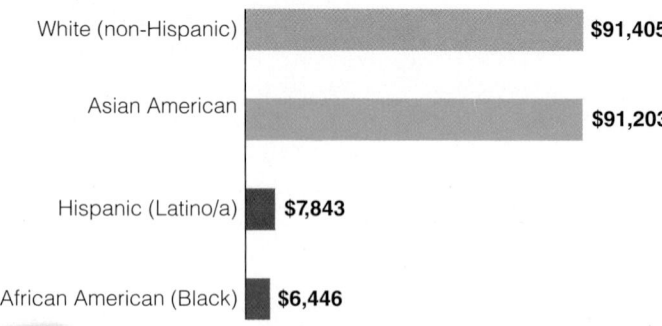

**FIGURE 7.14** RACIAL DIVIDE IN NET WORTH, SHOWN IN 2012 DOLLARS

their families have had greater financial reserves. White Americans have also had easier access to credit or have been able to acquire credit with lower interest rates and lending costs (Shapiro, Meschede, and Osoro, 2013).

Given these factors, the median wealth of white households is about twenty times higher than that of black (African American) households and eighteen times that of Hispanic (Latino/a) households. During the U.S. economic crisis of 2007–2009, household median wealth plummeted for many persons of color because they lacked homeownership or experienced foreclosures that disproportionately affected African American and Latino/a households. Figure 7.14 shows the racial divide in net worth in 2012 dollars. White (non-Hispanic) Americans, at $91,405, and Asian Americans, at $91,203, have much higher net worth than Hispanic Americans (Latinos/as), at $7,843, and African Americans, at $6,446 (Pew Research Social & Demographic Trends, 2013b).

## Consequences of Inequality

**LO6** Income and wealth are not simply statistics; they are intricately related to the American Dream and our individual life chances. Persons with a high income or substantial wealth have more control over their own lives. They have greater access to goods and services; they can afford better housing, more education, and a wider range of medical services. Persons with less income, especially those living in poverty, must spend their limited resources to acquire the basic necessities of life.

**Physical Health, Mental Health, and Nutrition** People who are wealthy, are well educated, and have high-paying jobs are much more likely to be healthy than are poor people. As people's economic status increases, so does their health status. The poor have shorter life expectancies and are at greater risk for chronic illnesses such as type 2 diabetes, heart disease, and cancer, as well as infectious diseases

such as tuberculosis. Compared with adults living in households with incomes at least four times the poverty level, adults living below the poverty line are five times more likely to report their health as being fair or poor and more than eight times more likely to report serious psychological distress (National Center for Health Statistics, 2011).

Children born into poor families are at much greater risk of dying during their first year of life. Some die from disease, accidents, or violence. Others are unable to survive because they are born with low birth weight, a condition linked to birth defects and increased probability of infant mortality. Low birth weight in infants is attributed, at least in part, to the inadequate nutrition received by many low-income pregnant women. Most of the poor do not receive preventive medical and dental checkups; many do not receive adequate medical care after they experience illness or injury (Children's Defense Fund, 2012b).

Many high-poverty areas lack an adequate supply of doctors and medical facilities. Even in areas where such services are available, the inability to pay often prevents people from seeking medical care when it is needed. Some "charity" clinics and hospitals may provide indigent patients (those who cannot pay) with minimal emergency care but make them feel stigmatized in the process. For some poor individuals living in states that did not expand Medicaid coverage under the Affordable Care Act, the lack of adequate health insurance remains a pressing problem. The Census Bureau classifies health insurance coverage as private coverage or government coverage. Private health insurance is a plan provided through an employer or a union, or is a plan purchased by an individual from a private company. By contrast, government health insurance includes such programs as Medicare, Medicaid, military health care, the Children's Health Insurance Program (CHIP), and individual state health plans.

As provisions of the Affordable Care Act were being implemented in 2013–2014, approximately 48 million adults in the United States—15.7 percent of the U.S. population—had no health insurance coverage (DeNavas-Walt, Proctor, and Smith, 2013). Figure 7.15 shows that the rate of uninsured people increases as household income decreases. Nearly one-fourth of all households with income less than $25,000 were uninsured in 2012, as compared with only 7.9 percent of households uninsured when income was $75,000 and over (DeNavas-Walt, Proctor, and Smith, 2013). Many people rely on their employers for health coverage; however, some employers are cutting back on health coverage, particularly for employees' family members. Overall, the number

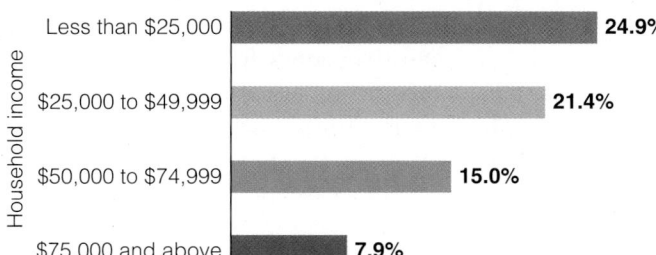

**FIGURE 7.15** RATE OF UNINSURANCE BY HOUSEHOLD INCOME, 2012

Source: DeNavas-Walt, Proctor, and Smith, 2013. © Cengage Learning

and percentage of people without health insurance decreased slightly between 2011 and 2013, partly as a result of an increase in the percentage and number of people who were covered by Medicaid or Medicare, and the initial enrollment of individuals and households in plans established by the Affordable Care Act.

In 2012 the percentage of people covered by government health programs increased slightly, to 32.6 percent from 32.2 percent in 2011. The number of people covered by government health programs grew to 101.5 million in 2012, as compared to 99.5 million in 2011. Medicaid accounted for 50.9 million people (16.4 percent) in 2012. As discussed in more detail in Chapter 14, Medicaid is government-funded health insurance coverage for persons who meet specific income guidelines and other criteria. In addition, Medicare coverage (for older individuals and those with certain disabilities) also increased in 2012, to 15.7 percent and 48.9 million, as compared to 15.2 percent and 46.9 million in 2011 (DeNavas-Walt, Proctor, and Smith, 2013).

When the effects of the Affordable Care Act are seen in the future, further reductions in the number and percentage of uninsured persons will probably occur because the law includes a requirement that individuals must purchase coverage under various circumstances. Those who do not enroll in a health care plan will be assessed increasing penalties (estimated to range from $95 to $695 per adult and $47.50 to $347.50 per child up to a specified dollar figure) in the years 2014–2016 and beyond. (This topic is further discussed in Chapter 14, "Health, Health Care, and Disability.")

Although the precise relationship between class and health is not known, social analysts and medical practitioners suggest that people without insurance are less likely than those with insurance to receive preventive care and services for major health conditions and chronic diseases. Individuals with higher income and wealth tend to smoke less, exercise more, maintain a healthy body weight, and eat nutritious meals. In addition, many lower-paying jobs are often the most dangerous and have the greatest health hazards. Black lung disease, cancers caused by asbestos, and other environmental hazards found in the workplace are more likely to affect manual laborers and low-income workers, as are job-related accidents.

Good health is basic to good life chances; in turn, adequate nutrition is essential for good health. Hunger is related to class position and income inequality. Studying the problem of hunger has become more complex in recent years because the U.S. Department of Agriculture stopped using the word *hunger* in its reports in 2006. *Food insecure* is now used to identify people whose access to adequate food is limited by lack of money and other resources (Coleman-Jensen, Nord, and Singh, 2013). People living with food insecurity include those who are unable to afford the basics, those who are unable to get to the grocery store, and those who are unable to find fresh, nutritious produce and other foods to eat because they are surrounded with fast-food stores that do not provide foods with proper nutrition (Dolnick, 2010). In 2012, 49 million Americans lived in households that were food insecure at some time during the year. Of the 14.5 percent of the U.S. population that reported that they were in food-insecure households, 8.8 percent were in households with "low food security" and 5.7 percent were in those with "very low food security" (Coleman-Jensen, Nord, and Singh, 2013). Indicators of "low food security" are factors such as "worrying food would run out," "food bought did not last," and "could not afford balanced meals." "Very low food security" is characterized by these factors, plus "being hungry but not eating," "losing weight," "did not eat whole day," and "did not eat whole day for three months or longer." Rates of food insecurity are highest in households with incomes near or below the poverty line, households with children headed by single women or single men, and black and Hispanic households.

The number of people who receive benefits from the Supplemental Nutrition Assistance Program (SNAP)—formerly called "food stamps"—remains at about 47 million in 2013–2014. The benefits average about $133 per person per month, or the equivalent of less than $1.40 per person per meal in 2014. However, it is estimated that SNAP has helped keep nearly 4.9 million people from being food insecure, including 2.2 million children (Rosenbaum, 2013). Other food assistance comes from the National School Lunch Program, which provides an average of 31.6 million children with food in more than 100,000 public and private schools and residential child-care facilities. Approximately 58 percent of all meals in the National School Lunch Program are served free,

**FIGURE 7.16** Conflict theorists see schools as agents of the capitalist class system that perpetuate social inequality: Upper-class students are educated in well-appointed environments such as the one shown here, whereas children of the poor tend to go to antiquated schools with limited facilities.

*(Image Source/Getty Images)*

another 8 percent are provided at reduced prices, and the remaining one-third are paid for by either students or schools (Coleman-Jensen, Nord, and Singh, 2013).

**Housing** As discussed in Chapter 4, homelessness is a major problem in the United States. It is estimated that about 1.6 million (or 1 in 45 children) are homeless each year and that 40 percent of all homeless children are 5 years of age or younger (Children's Defense Fund, 2012b). The lack of affordable housing is a pressing concern for many low-income individuals and families. Housing is considered affordable when a household spends no more than 30 percent of its income on rent or mortgage. But about 12 million renter and homeowner households pay more than half of their annual income for housing (U.S. Department of Housing and Urban Development, 2011).

Lack of *affordable* housing is one central problem brought about by economic inequality. Another concern is *substandard* housing, which refers to facilities that have inadequate heating, air conditioning, plumbing, electricity, or structural durability. Structural problems—caused by faulty construction or lack of adequate maintenance—exacerbate the potential for other problems such as damage from fire, falling objects, or floors and stairways collapsing. A final issue is the extent to which a family lives in a neighborhood with a high poverty rate. Studies have shown that high school graduation rates are as much as 20 percentage points lower

in neighborhoods with a high poverty rate than graduation rates in higher-income areas (Children's Defense Fund, 2012b).

**Education** Educational opportunities and life chances are directly linked (Figure 7.16). Some functionalist theorists view education as the "elevator" to social mobility. Improvements in the educational achievement levels (measured in number of years of schooling completed) of the poor, people of color, and white women have been cited as evidence that students' abilities are now more important than their class, race, or gender. From this perspective, inequality in education is declining, and students have an opportunity to achieve upward mobility through achievements at school. Functionalists generally see the education system as flexible, allowing most students the opportunity to attend college if they apply themselves (Ballantine and Hammack, 2012).

In contrast, most conflict theorists stress that schools are agencies for reproducing the capitalist class system and perpetuating inequality in society. From this perspective, education perpetuates poverty. Parents with low educational attainment and limited income are often not able to provide the same educational opportunities for their children as families where at least one parent has more formal education.

Today, great disparities exist in the distribution of educational resources. Because funding for education comes primarily from local property taxes, school districts in wealthy suburban areas generally pay higher teachers' salaries, have newer buildings, and provide state-of-the-art equipment. By contrast, schools in poorer areas have a limited funding base. Students in central-city schools and poverty-stricken rural areas often attend dilapidated schools that lack essential equipment and teaching materials. As far back as the early 1990s, author Jonathan Kozol (1991, qtd. in Feagin and Feagin, 1994: 191) documented the effect of a two-tiered system on students, which remains in many of today's schools:

> Kindergartners are so full of hope, cheerfulness, high expectations. By the time they get into fourth grade, many begin to lose heart. They see the score, understanding they're not getting what others are getting. . . . They see suburban schools on television. . . . They begin to get the point that they are not valued much in our

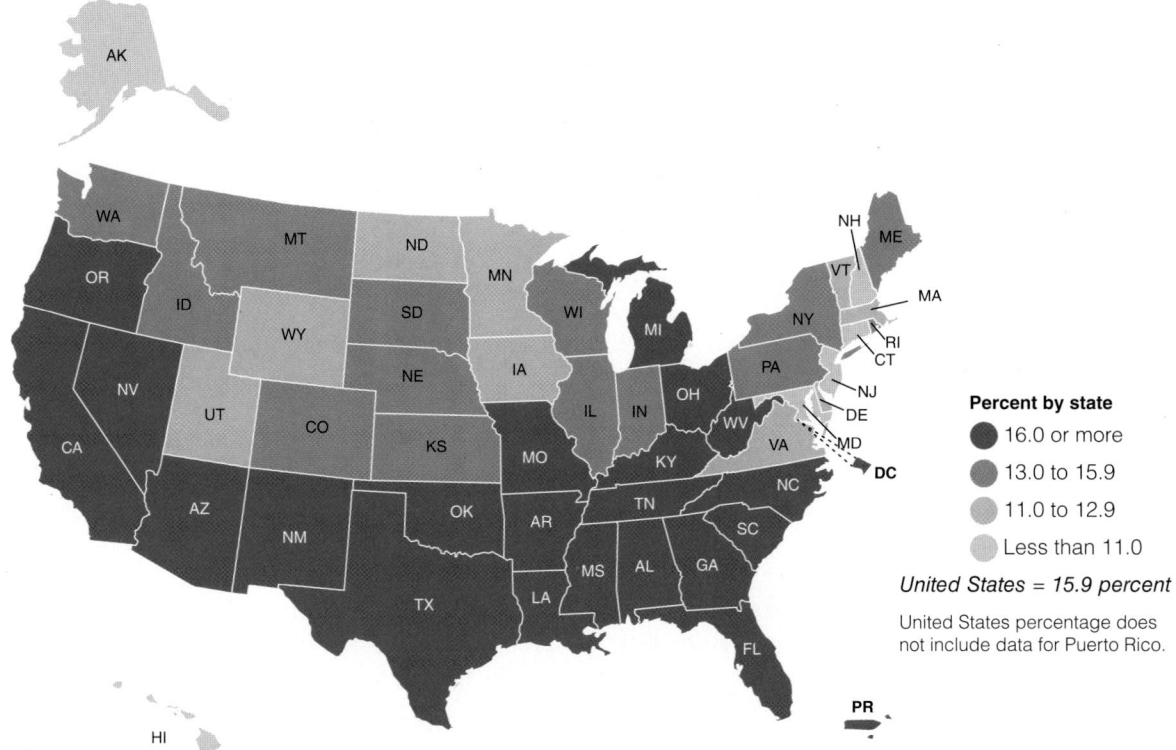

**FIGURE 7.17** PERCENTAGE OF PEOPLE IN POVERTY IN THE PAST 12 MONTHS BY STATE, 2012

Source: DeNavas-Walt, Proctor, and Smith, 2013.

society. By the time they are in junior high, they understand it. "We have eyes and we can see; we have hearts and we can feel. . . . We know the difference."

As previously noted, poverty extracts such a toll that many young people will not have the opportunity to finish high school, much less enter college.

## Poverty in the United States

**LO7** So far, we have examined various forms of inequality in the United States and their effects. Let's now focus more closely on the problem of poverty in this country.

The United States has the highest rate of poverty among wealthy countries. When many people think about poverty, they think of people who are unemployed or on welfare. However, many hardworking people with full-time jobs live in poverty. The U.S. Social Security Administration has established an *official poverty line,* **which is based on what the federal government considers to be the minimum amount of money required for living at a subsistence level.** The poverty level (or poverty line) is computed by determining the cost of a minimally nutritious diet (a low-cost food budget on which a family could survive nutritionally on a short-term, emergency basis) and multiplying this figure by

three to allow for nonfood costs. In 2012 about 15 percent of the U.S. population had income below the poverty level, which means that approximately 46.5 million people lived below the official government poverty level of $23,283 for a family of four with two adults and two children under the age of eighteen (DeNavas-Walt, Proctor, and Smith, 2013). In 2013 the federal government increased the official poverty level for a family of four to $23,550. Although data are not currently available for the total number and percentage of all families living in poverty in 2013–2014, some analysts believe that the poverty rate will be relatively unchanged because no major changes have occurred in recent years in the nature and extent of this problem.

Poverty rates vary widely across the United States. The percentage of people living below the poverty line is higher in some states and regions than in others (see Figure 7.17). In 2012 the highest poverty rates by state were in Mississippi (24.2 percent) and New Mexico (20.8 percent). The lowest

**official poverty line** the income standard that is based on what the federal government considers to be the minimum amount of money required for living at a subsistence level.

poverty rates were in New Hampshire (10.0 percent), Alaska (10.1 percent), Maryland (10.3 percent), Connecticut (10.7 percent), and New Jersey (10.8 percent). Patterns in poverty are consistent: Between 2011 and 2012, poverty rates did not change for 45 states and the District of Columbia. However, the number and percentage of people in poverty increased in the states of California, Mississippi, and New Hampshire during that same period (Bishaw, 2013).

When sociologists define poverty, they distinguish between absolute and relative poverty. *Ab-**solute poverty** exists when people do not have the means to secure the most basic necessities of life.* This definition comes closest to that used by the federal government. Absolute poverty often has life-threatening consequences, such as when a homeless person freezes to death on a park bench. By comparison, *relative poverty **exists when people may be able to afford basic necessities but are still unable to maintain an average standard of living.*** A family must have income substantially above the official poverty line in order to afford the basic necessities, even when these are purchased at the lowest possible cost. But many families do not earn enough money to afford living comfortably and must survive on an economy budget, as described below:

> Members of families existing on the economy budget never go out to eat, for it is not included in the food budget; they never go out to a movie, concert, or ball game or indeed to any public or private establishment that charges admission, for there is no entertainment budget; they have no cable television, for the same reason; they never purchase alcohol or cigarettes; never take a vacation or holiday that involves any motel or hotel or, again, any meals out; never hire a baby-sitter or have any other paid child care; never give an allowance or other spending money to the children; never purchase any lessons or home-learning tools for the children; never buy books or records for the adults or children, or any toys, except in the small amounts available for birthday or Christmas presents ($50 per person over the year); never pay for a haircut; never buy a magazine; have no money for the feeding or veterinary care of any pets; and, never spend any money for preschool for the children, or educational trips for them away from home, or any summer camp or other activity with a fee.

Take a guess: When do you think this statement was written? Recently? No, this statement was written more than twenty years ago (1992) by social scientists John Schwarz and Thomas Volgy to describe the limited resources of people at or on the edge of poverty. Today, many people live in similar or worse conditions. Some participate in the occasional protest, such as the brief Occupy movement; others hope that they will not sink to the bottom rungs of poverty.

## Who Are the Poor?

Poverty in the United States is not randomly distributed, but rather is highly concentrated according to age, gender, and race/ethnicity.

**Age** In the past, persons over age 65 were at the greatest risk of being poor; however, older individuals today have the lowest poverty rate of all age categories, largely because of Social Security. Social Security, actually titled "Old Age, Survivors, Disability, and Health Insurance," is a federal insurance program established in 1935 that protects against loss of income caused by retirement, disability, or death. When Social Security was established, it was intended to supplement other savings and assets. But for more than half of Americans over age 65, Social Security provides more than fifty percent of their income, and without Social Security income, nearly half of all U.S. seniors would be living in poverty (Edwards, Turner, and Hertel-Fernandez, 2011). Instead, fewer than one in ten senior citizens lives below the poverty line, making Social Security the most successful antipoverty program in the United States. However, because Social Security benefits are based on the number of years of paid employment and preretirement earnings, women and minorities, who often earn less during their employment years, receive less in Social Security benefits and are less protected against poverty in old age (AARP.org, 2012).

The child poverty rate in the United States is higher than in any other industrialized country. In the United States today, children under age 18 have the highest rate of poverty, followed closely by young adults ages 18 to 24 (see Figure 7.18). In the latest available statistics, there has been no significant change in the poverty rate for U.S. children under 18 (16.1 million in 2012, which is essentially unchanged from 2011). More than one in five children in America are considered to be poor, and their poverty rate is 21 percent higher than before the Great Recession. In 2012 there were 2.75 million more children living in poverty than there were in 2007 (Children's Defense Fund, 2013). And it must be remembered that these figures do not take into account the large number of children whose families hover just above the official poverty line. The precarious position of African American

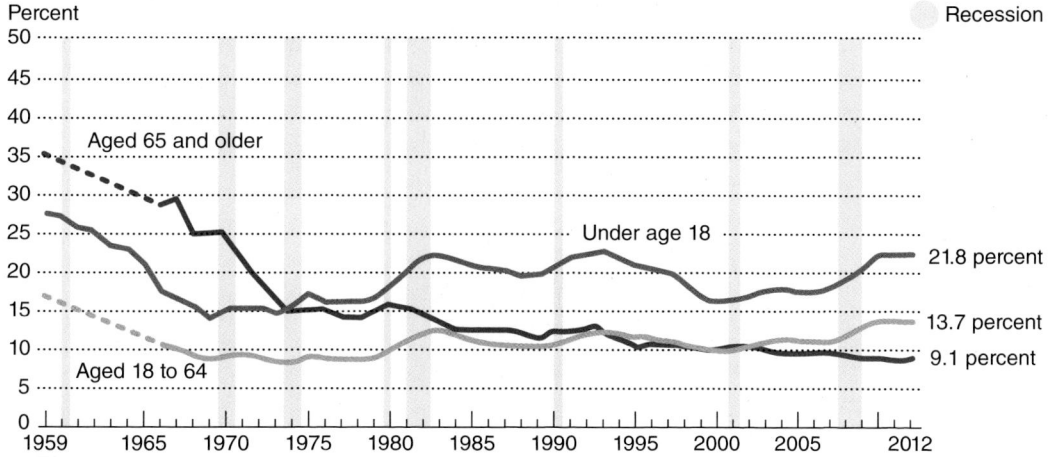

**FIGURE 7.18** U.S. POVERTY RATES BY AGE, 1959–2012

Note: The data points are placed at the midpoints of the respective years. Data for people ages 18 to 64 and 65 and older are not available from 1960 to 1965.

Source: U.S. Census Bureau. Current Population Survey, 1960 to 2012 Annual Social and Economic Supplements.

and Latino/a children is even more striking. For people under age 18 in 2012, about 37.9 percent of African Americans and 33.8 percent of Hispanics (Latinos/as) were living in poverty, as compared with 12.3 percent of white (non-Hispanic) and 13.8 percent of Asian American children (Children's Defense Fund, 2013).

**Gender** In 2012 the poverty rate was 13.6 percent for males of all ages, as compared with 16.3 percent for females of all ages. However, these figures do not tell the entire story. Gender differences in poverty rates are more pronounced for people ages 65 and older and among younger women who head single-parent families. The poverty rate for women ages 65 and older was 11 percent, while it was 6.6 percent for men (DeNavas-Walt, Proctor, and Smith, 2013). In her now-classic study the sociologist Diana Pearce (1978) coined a term to describe this problem of gender-specific poverty: The *feminization of poverty* **refers to the trend in which women are disproportionately represented among individuals living in poverty.** Over the past three decades since Pearce's study, women have continued to face a higher risk of being poor because they bear the major economic and emotional burdens of raising children when they are single heads of households. This problem is compounded by the fact that the female-to-male earnings ratio was 84 percent in 2012, which means that among all working men and women ages 16 and older, women's hourly wages were 84 percent of those of men (Pew Research Social & Demographic Trends, 2013a). This gap has narrowed somewhat in recent years, attributed primarily to women's rising earnings because they have more education, greater labor force participation, and an increasing

presence in higher-paying occupations than in the past. This change does little to help low-income women living in poverty who have limited educational and workplace opportunities and are caught in dead-end jobs.

Does the feminization of poverty explain poverty in the United States today? Clearly, this thesis highlights a genuine problem—the link between gender and poverty (Figure 7.19). However, all women are not equally vulnerable to poverty: Many in the upper and upper-middle classes have the financial resources, education, and skills to support themselves regardless of the presence of a man in the household. Moreover, poverty is everyone's problem, not just women's. When women are impoverished, so are their children. Likewise, many of the poor in our society are men, especially men who are chronically unemployed, older men, homeless men, men with disabilities, and men of color.

**Race/Ethnicity** In 2012 whites (non-Hispanic) accounted for slightly less than 63 percent of the

**absolute poverty** a level of economic deprivation that exists when people do not have the means to secure the most basic necessities of life.

**relative poverty** a level of economic deprivation that exists when people may be able to afford basic necessities but are still unable to maintain an average standard of living.

**feminization of poverty** the trend in which women are disproportionately represented among individuals living in poverty.

total U.S. population but only 40.7 percent of people in poverty. Whites (non-Hispanic) had the lowest rate of poverty—9.7 percent—of any racial–ethnic group. The highest rate of poverty was among Native Americans and Alaska Natives at 29.1 percent, followed by African Americans (blacks) at 27.2 percent. Next in line were Hispanics (25.6 percent) and finally Asian Americans (11.7 percent) (DeNavas-Walt, Proctor, and Smith, 2013; U.S. Census Bureau News, 2013). Some demographic factors contribute to slight fluctuations in poverty rates by race/ethnicity, but overall rates remain persistently stubborn to change.

### Economic and Structural Sources of Poverty

Social inequality and poverty have both economic and structural sources. Unemployment is a major cause of contemporary poverty that involves both economic and structural components. Tough economic times provide fewer opportunities for individuals to get a position that may help them gain a toehold in U.S. society. In December 2013, 10.4 million persons were unemployed in the United States, making the unemployment rate 6.7 percent. Although the unemployment rate for adult women was slightly lower (at 6.0 percent) than the average, much higher rates were reported for teenagers (20.2 percent) and African Americans (11.9 percent). Nearly 38 percent of the unemployed (3.9 million) had been jobless for 27 weeks or more in 2013 (U.S. Bureau of Labor Statistics, 2014c). At the beginning of 2014, extended unemployment benefits were cut off to about 1.4 million people who had been jobless for more than 27 weeks and who had exhausted their state unemployment compensation. Almost 3.5 million were out of work for six months or longer and were not receiving unemployment payments in 2014 because Congress did not pass a bill to reinstate such long-term benefits. In 2013 unemployment rates declined slightly from previous years; however, more people did not necessarily find work. For example, individuals are no longer counted as "unemployed" if they drop out of the labor market altogether by deciding to retire or are no longer actively seeking work.

In addition to unemployment, low wages paid for many jobs are another major cause of poverty: Extensive debate has taken place recently about the inability of individuals and families to live on the federal minimum wage of $7.25 per hour. Although most working families are not officially poor, many are "near poor" or "low income," struggling to pay for basic needs such as housing, health care, food, child care, and transportation.

Structural problems contribute to both unemployment and underemployment. Corporations have been disinvesting in the United States, displacing millions of people from their jobs. This displacement is referred to as the *deindustrialization* of America. Even as they have closed their U.S. factories and plants, such as portions of the automobile industry, many corporations have opened new facilities in other countries where lower-wage labor exists because people will, of necessity, work for lower wages. Many analysts have demonstrated how the relocation of domestic manufacturing offshore has drained millions of manufacturing jobs from the U.S. economy.

*Job deskilling*—**a reduction in the proficiency needed to perform a specific job**—leads to a corresponding reduction in the wages for that job or in the use of nonhuman technologies to perform the work. This kind of deskilling has resulted from the computerization and automation of the workplace. A significant step in job deskilling occurs when the primary responsibility of human operators is to monitor automated systems. The eventual outcome of such deskilling is that human operators either do not know what to do when the system fails or they are replaced entirely by automated technology. Other structural problems include the overall shift from manufacturing to service occupations in the United States, which has resulted in the loss of higher-paying positions and their replacement with

**FIGURE 7.19** Many women are among the "working poor," who, although employed full time, have jobs in service occupations that are typically lower paying and less secure than jobs in other sections of the labor market. Does the nature of women's work contribute to the feminization of poverty in the United States?

Sonda Dawes/The Image Works

# Sociology WORKS!

## Can Compassion Rebuild a Nation?

Our society recognizes a moral obligation to provide a helping hand to those in need, but those in poverty have been getting only the back of the hand. They receive little or no public assistance. Instead, they are scolded and told that they have caused their own misfortunes. This is our "compassion gap"—a deep divide between our moral commitments and how we actually treat those in poverty.

In this statement the sociologists Fred Block, Anna C. Korteweg, and Kerry Woodward (2008: 166) describe the contradiction between our nation's alleged moral commitment to alleviating poverty and how we actually treat people who live in poverty. Children, single mothers with children, and people of color (particularly African Americans and Latinos/as) make up a disproportionate segment of the nation's poorest groups, and individuals in these categories are the persons most disadvantaged by arguments asserting that the poor have no one but themselves to blame for their poverty.

Numerous sociological studies regarding wealth and poverty demonstrate how structural factors contribute to the ability of some individuals to achieve the American Dream whereas others are hampered in achieving that goal by factors that are beyond their control. Yet, according to the Economic Mobility Project, policy makers do little to alleviate poverty in the United States because of the widely held belief that the American Dream should provide everyone with equality of *opportunity* but not necessarily equality of *outcome*. Indeed, research continues to reveal that structural factors beyond the control of individuals are important in determining where a person's place will be in the U.S. class system.

Can we keep the American Dream alive for all? According to Block, Korteweg, and Woodward (2008), we must take a number of specific steps to revitalize the American Dream for more people in this country and to reverse the compassion gap. For example, these sociologists argue that we must make people aware of how far that social reality has departed from the ideals of the American Dream. As a nation, they argue, we must also take action to deal with the costs of four critical services that have risen much more rapidly than people's wages and the rate of inflation: health care, higher education, high-quality child care, and housing.

As sociologists and other social analysts have suggested, if we as a nation are to claim that we have a commitment to compassion, we must make it our collective responsibility to help remove the structural barriers that currently reduce opportunities, mobility, and a chance for a better way of life for millions of Americans: "True compassion requires that we build a society in which every person has a first chance, a second chance, and, if needed, a third and fourth chance, to achieve the American Dream" (Block, Korteweg, and Woodward, 2008: 175).

### reflect & analyze

**Consider the area where you live. Do you know of people there whose situations could be improved if they were given a greater opportunity to take charge of their lives? How could such a change happen?**

---

lower-paying and less-secure positions that do not offer the wages, job stability, or advancement potential of the disappearing jobs.

Technological advances and changing patterns of consumerism have also contributed to unemployment in the United States and other high-income nations. For example, the introduction of smartphones, tablets, and other electronic devices means that fewer people now own a watch, camera, calculator, printed calendar, or numerous other features (or apps) found on the typical smartphone or tablet. As a result, fewer people are needed to design, make, and repair numerous items that created jobs for millions in the past. The Internet has also contributed to a decline in hundreds of thousands of jobs in the postal, publishing, and printing industries as people now e-mail or text one another or self-publish their ideas rather than going the traditional route. For example, thousands of postal jobs have disappeared because of a dramatic reduction in the number of items to be delivered, and automated systems have reduced the number of workers employed at the post office itself.

### Solving the Poverty Problem

The United States has attempted to solve its poverty problem in several ways. One of the most enduring is referred to as social welfare. When most people think of "welfare," they think of food stamps (currently called SNAP) and programs such as Temporary Assistance for Needy Families (TANF) or the earlier program it replaced, Aid to Families with Dependent Children (AFDC). Some who receive benefits from welfare programs tend to be stigmatized, even when our nation describes itself as having compassion for the less fortunate (see "Sociology Works!").

**job deskilling** a reduction in the proficiency needed to perform a specific job.

FIGURE 7.20 This California electronic benefit transfer (EBT) card represents a modern approach to helping people of limited income purchase groceries. Data-encoded cards such as this one were developed to prevent the trading or selling of traditional food stamps. However, one drawback of this technology is that many of California's popular farmers' markets are not able to process EBT cards.

## Sociological Explanations of Social Inequality in the United States

**LO8** Obviously, some people are disadvantaged as a result of social inequality (Figure 7.20). In this section we examine some sociological explanations of this issue. In doing so we will see how different sociologists answer this question: Is inequality always harmful to society?

### Functionalist Perspectives

According to the well-known classical sociologists Kingsley Davis and Wilbert Moore (1945), inequality is not only inevitable but also necessary for the smooth functioning of society. The Davis–Moore thesis, which has become the definitive functionalist explanation for social inequality, can be summarized as follows:

1. All societies have important tasks that must be accomplished and certain positions that must be filled.
2. Some positions are more important for the survival of society than others.
3. The most important positions must be filled by the most qualified people (see Figure 7.21).
4. The positions that are the most important for society and that require scarce talent, extensive training, or both must be the most highly rewarded.
5. The most highly rewarded positions should be those that are functionally unique (no other position can perform the same function) and on which other positions rely for expertise, direction, or financing.

Davis and Moore use the physician as an example of a functionally unique position. Doctors are very important to society and require extensive training, but individuals would not be motivated to go through years of costly and stressful medical training without incentives to do so. The Davis–Moore thesis assumes that social stratification results in *meritocracy*—**a hierarchy in which all positions are rewarded based on people's ability and credentials.**

A key problem with the Davis–Moore thesis is that it ignores inequalities based on inherited wealth and intergenerational family status. The thesis assumes that economic rewards and prestige are the only effective motivators for people and fails to take into account other intrinsic aspects of work, such as self-fulfillment. It also does not adequately explain how such a reward system guarantees that the most-qualified people will gain access to the most highly rewarded positions.

### Conflict Perspectives

From a conflict perspective, people with economic and political power are able to shape and distribute the rewards, resources, privileges, and opportunities in society for their own benefit. Conflict theorists do not believe that inequality serves as a motivating force for people; they argue that powerful individuals and groups use ideology to maintain their favored positions at the expense of others. Core values in the United States emphasize the importance of material possessions, hard work, individual initiative to get ahead, and behavior that supports the existing social structure. These same values support the prevailing resource-distribution system and contribute to social inequality.

Are wealthy people smarter than others? According to conflict theorists, certain stereotypes suggest that this is the case; however, the wealthy may actually be "smarter" than others only in the sense of having "chosen" to be born to wealthy parents from whom they could inherit assets. Conflict theorists also note that laws and informal social norms support inequality in the United States. For the first half of the twentieth century, both legalized and

AP Images/Rich Pedroncelli

**FIGURE 7.21** According to a functionalist perspective, people such as these Harvard Law School graduates attain high positions in society because they are the most qualified and they work the hardest. Is our society a meritocracy? How would conflict theorists answer this question?

institutionalized segregation and discrimination reinforced employment discrimination and produced higher levels of economic inequality. Although laws have been passed to make these overt acts of discrimination illegal, many forms of discrimination still exist in educational and employment opportunities.

## Symbolic Interactionist Perspectives

Symbolic interactionists focus on microlevel concerns and usually do not analyze larger structural factors that contribute to inequality and poverty. However, many significant insights on the effects of wealth and poverty on people's lives and social interactions can be derived from applying a symbolic interactionist approach. Using qualitative research methods and influenced by a symbolic interactionist approach, researchers have collected the personal narratives of people across all social classes, ranging from the wealthiest to the poorest people in the United States.

A few studies provide rare insights into the social interactions between people from vastly divergent class locations. In sociologist Judith Rollins's (1985) classic study of the relationship between household workers and their employers, she examined rituals of deference that were often demanded by elite white women of their domestic workers, who were frequently women of color. According to the sociologist Erving Goffman (1967), *deference* is a type of ceremonial activity that functions as a symbolic means whereby appreciation is regularly conveyed to a recipient. In fact, deferential behavior between nonequals (such as employers and employees) confirms the inequality of the relationship and each party's position in the relationship relative to the other. Rollins identified three types of linguistic deference between domestic workers and their employers: use of the first names of the workers, contrasted with titles and last names (Mrs. Adams, for example) of the employers; use of the term *girls* to refer to female household workers regardless of their

**meritocracy** a hierarchy in which all positions are rewarded based on people's ability and credentials.

## Sociological Explanations of Social Inequality in the United States

| | |
|---|---|
| Functionalist Perspectives | Some degree of social inequality is necessary for the smooth functioning of society (in order to fill the most important positions) and thus is inevitable. |
| Conflict Perspectives | Powerful individuals and groups use ideology to maintain their favored positions in society at the expense of others. |
| Symbolic Interactionist Perspectives | The beliefs and actions of people reflect their class location in society. |

age; and deferential references to employers, such as "Yes, ma'am." Spatial demeanor, including touching and how close one person stands to another, is an additional factor in deference rituals across class lines. Rollins (1985: 232) concludes that

> The employer, in her more powerful position, sets the essential tone of the relationship; and that tone . . . is one that functions to reinforce the inequality of the relationship, to strengthen the employer's belief in the rightness of her advantaged class and racial position, and to provide her with justification for the inegalitarian social system.

Many concepts introduced by the sociologist Erving Goffman (1959, 1967) could be used as springboards for examining microlevel relationships between inequality and people's everyday interactions. What could you learn about class-based inequality in the United States by using a symbolic interactionist approach to examine a setting with which you are familiar?

The Concept Quick Review summarizes the three major perspectives on social inequality in the United States.

## U.S. Stratification in the Future

The United States continues to face one of the greatest economic challenges it has experienced since the Great Depression of the 1930s. Although we have strong hopes that the American Dream will remain alive and well, many people are concerned about the lack of upward mobility for many Americans and a decline for others. The nationwide slump in housing and jobs has distressed people across all income levels, and continued high rates of unemployment and a shifting stock market bring about weekly predictions that things are either getting only slightly better or are becoming worse. Perhaps one of the most critical factors contributing to a lack of optimism about future mobility in the United States is the vast wealth gap between the rich and everyone else, and the depth of poverty in this country.

Given the current economic situation, it is difficult to predict the future of the U.S. system of stratification. What will happen with the great economic imbalance in the United States? Economist and former Secretary of Labor Robert Reich (2010: 146) has summed up the problem as follows:

> None of us can thrive in a nation divided between a small number of people receiving an ever larger share of the nation's income and wealth, and everyone else receiving a declining share. The lopsidedness not only diminishes economic growth but also tears at the fabric of our society. America cannot succeed if the basic bargain at the heart of our economy remains broken. The most fortunate among us who have reached the pinnacles of power and success depend on a stable economic and political system. That stability rests on the public's trust that the system operates in the interest of us all. Any loss of such trust threatens the well-being of everyone.

Given this assessment, politicians, business leaders, and ordinary people must do all they can to reinvigorate the American Dream, or everyone's future—young and old alike—will look much dimmer as we continue through the twenty-first century.

Are we sabotaging our future if we do not work constructively to eliminate poverty? It has been said that a chain is no stronger than its weakest link. If we apply this idea to the problem of poverty, then it is to our advantage to see that those who cannot find work or do not have a job that provides a living wage receive adequate training and employment. Innovative programs can combine job training with producing something useful to meet the immediate needs of people living in poverty. Children of today—the adults of tomorrow—need nutrition, education, health care, and safety as they grow up.

# chapter REVIEW

## LO1 What is social stratification, and how do the three major systems of stratification compare?

Social stratification is the hierarchical arrangement of large social groups based on their control over basic resources. People are treated differently based on where they are positioned within the social hierarchies of class, race, gender, and age. Stratification systems include slavery, caste, and class. Slavery, an extreme form of stratification in which people are owned or controlled by others, is a closed system. The caste system is also a closed one in which people's status is determined at birth based on their parents' position in society. The class system, which exists in the United States, is a type of stratification based on ownership of resources and on the type of work that people do.

## LO2 How did Karl Marx view social class and stratification?

Marx viewed social class as a key determinant of social inequality and social change. For Marx, class position and the extent of our income and wealth are determined by our work situation, or our relationship to the means of production. Marx stated that capitalistic societies consist of two classes—the capitalists and the workers—and that class relationships involve inequality and exploitation.

## LO3 What is Max Weber's multidimensional approach to social stratification?

Weber emphasized that no single factor (such as economic divisions between capitalists and workers) is sufficient for defining the location of categories of people within the class structure. Weber developed a multidimensional concept of stratification that focuses on the interplay of wealth, prestige, and power.

## LO4 What are the key characteristics of social classes in the United States?

No broad consensus exists about how to characterize the class structure in this country. Sociologists have developed two models of the class structure: One is based on a Weberian approach, the other on a Marxian approach. In the Weberian-based approach, social classes are based on three elements: (1) education, (2) occupation of family head, and (3) family income. This approach to class structure consists of the upper class, the upper-middle class, the middle class, the working class, the working poor, and the underclass. Contemporary Marxian models examine class in terms of people's relationship to others in the production process.

## LO5 What is the difference between income inequality and wealth inequality?

Income is the economic gain derived from wages, salaries, income transfers (governmental aid), and ownership of property. In 2011 the wealthiest 20 percent of U.S. households received more than half of the total income "pie," while the poorest 20 percent of households received slightly more than 3 percent of all income. Wealth includes property such as buildings, land, farms, houses, factories, and cars, as well as other assets such as bank accounts, corporate stocks, bonds, and insurance policies. Wealth is even more unevenly distributed than income.

## LO6 What are three important consequences of inequality in the United States?

The stratification of society into different social groups results in wide discrepancies in income and wealth and in variable access to available goods and services. People with high income or wealth have greater opportunity to control their own lives. They can afford better housing, more education, and a wider range of medical services. People with less income have fewer life chances and must spend their limited resources to acquire basic necessities.

## LO7 What are the characteristics of the U.S. poor based on age, gender, and race/ethnicity?

Age, gender, and race tend to be factors in poverty. Children have a greater risk of being poor than do the elderly, and women have a higher rate of poverty than do men. Although whites account for approximately two-thirds of those below the poverty line, people of color account for a disproportionate share of the impoverished in the United States.

## LO8 How do functionalist, conflict, and symbolic interactionist perspectives on social inequality compare?

Functionalist perspectives view classes as broad groupings of people who share similar levels of

**229**

privilege on the basis of their roles in the occupational structure. According to the Davis–Moore thesis, stratification exists in all societies, and some inequality is not only inevitable but also necessary for the ongoing functioning of society. The positions that are most important within society and that require the most talent and training must be highly rewarded. Conflict perspectives on class are based on the assumption that social stratification is created and maintained by one group (typically the capitalist class) in order to enhance and protect its own economic interests. Conflict theorists measure class according to people's relationships with others in the production process. Unlike functionalist and conflict perspectives, which focus on macrolevel inequalities in societies, symbolic interactionist views focus on microlevel inequalities such as how class location may positively or negatively influence one's identity and everyday social interactions. Symbolic interactionists use terms such as *deference* to explain how class binds some individuals together while categorically separating out others.

# key TERMS

absolute poverty  222

alienation  206

capitalist class (bourgeoisie)  206

caste system  204

class conflict  207

class system  205

feminization of poverty  223

income  215

intergenerational mobility  202

intragenerational mobility  203

job deskilling  224

life chances  201

meritocracy  226

official poverty line  221

pink-collar occupations  211

power  209

prestige  208

relative poverty  222

slavery  203

social mobility  202

social stratification  201

socioeconomic status (SES)  209

underclass  212

wealth  208

working class (proletariat)  206

# questions for CRITICAL THINKING

1. Based on Max Weber's multidimensional class model of wealth, power, and prestige, how do entertainers and reality TV personalities, such as Lady Gaga and the Kardashians, become wealthy and well-known worldwide? Where do individuals such as these fit in Weber's system of stratification?

2. Should employment be based on meritocracy, need, or affirmative-action policies?

3. What might happen in the United States if the gap between rich and poor continues to widen?

# 8 GLOBAL STRATIFICATION

## LEARNING OBJECTIVES

**LO1** Define *global stratification* and explain how it contributes to economic inequality within and between nations.

**LO2** Describe the levels of development approach for studying global inequality.

**LO3** Identify the World Bank classifications of nations into four economic categories and explain why measuring wealth and poverty on a global basis is a difficult process.

**LO4** Discuss the relationship between global poverty and key human development issues such as life expectancy, health, education, and literacy.

**LO5** Define modernization theory and list the four stages of economic development identified by Walt Rostow.

**LO6** Define dependency theory and explain why this theory is often applied to newly industrializing countries.

**LO7** Describe the key components of world systems theory and identify the three major types of nations set forth in this theory.

**LO8** Discuss the new international division of labor theory and explain how it might be useful in the twenty-first century.

Viviane Moos/Corbis News/Corbis

233

# Leaving the Snare of Poverty

**Angeline Mugwendere's parents were impoverished farmers in Zimbabwe,** and she was mocked by classmates when she went to school barefoot and

How does poverty affect the educational opportunities of children in poor countries such as Zimbabwe?

in a torn dress with nothing underneath. Teachers would sternly send her home to collect school fees that were overdue, even though everyone knew there was no way her family could pay them. Yet Angeline suffered the humiliations and teasing and pleaded to be allowed to remain in school. Unable to buy school supplies, she cadged [asked for] what she could.

"At break time, I would go to a teacher's house and say, 'Can I wash your dishes?'" she remembers. "And in return, they would sometimes give me a pen."

—In their best-selling book *Half the Sky*, the journalists Nicholas D. Kristof and Sheryl WuDunn (2009: 179) describe how some impoverished girls have benefited from efforts to eradicate global poverty. Through help from organizations such as Campaign for Female Education (Camfed), Angeline acquired an education, moved out of poverty, and eventually became Camfed's executive director.

*David Snyder/Dreamstime.com*

lobal poverty affects people in a variety of ways: For some, it means absolute poverty because they do not have even the basic necessities to survive. For others, poverty is relative: Their standard of living remains below that of other people who reside in their nation and around the world. Without the intervention of individuals or organizations such as the one that helped Angeline attain more education and find productive employment, young people may be mired in hardscrabble poverty from which there is little possibility of escape. Consequently, their children are also born into poverty, and the cycle continues across generations.

Regardless of where people live in the world, social and economic inequalities are pressing daily concerns. Poverty and inequality know no political boundaries or national borders. Even within countries that are designated as "high income," many people live in poverty. Likewise, some wealthy people live in "low-income nations." Disparities between the rich and the poor within one high-income country may be greater than inequalities based on wealth and income that exist among some people who live in nations identified as middle income or low income. In this chapter we examine global stratification and inequality and discuss sociological perspectives that have been developed to explain the nature and extent of this problem. Before reading on, test your knowledge of global wealth and poverty by taking the "Sociology and Everyday Life" quiz.

## Wealth and Poverty in Global Perspective

**LO1** What do we mean by global stratification? *Global stratification* **refers to the unequal distribution of wealth, power, and prestige on a global basis, resulting in people having vastly different lifestyles and life chances both within and among the nations of the world.** Just as the United States is divided into classes, the world is divided into unequal segments characterized by extreme differences in wealth and poverty.

# How Much Do You Know About Global Wealth and Poverty?

**True  False**

| | | |
|---|---|---|
| T | F | 1. Poverty has been increasing in the United States but decreasing in other nations because of globalization. |
| T | F | 2. The assets of the world's 200 richest people are more than the combined income of over 50 percent of the world's population. |
| T | F | 3. More than one billion people worldwide live below the international poverty line, earning less than $1.25 each day. |
| T | F | 4. Although poverty is a problem in most areas of the world, relatively few people die of causes arising from poverty. |
| T | F | 5. In low-income countries, the problem of poverty is unequally shared by men and women. |

**True  False**

| | | |
|---|---|---|
| T | F | 6. The majority of people with incomes below the poverty line live in urban areas of the world. |
| T | F | 7. Two-thirds of adults (15 years and older) worldwide who are not able to read and write are men. |
| T | F | 8. Poor people in low-income countries meet most of their energy needs by burning wood, dung, and agricultural wastes, which increases health hazards and environmental degradation. |

Answers on page 237.

For example, the income gap between the richest and the poorest people of the world population continues to widen (see Figure 8.1).

As previously defined, *high-income countries* have highly industrialized economies; technologically advanced industrial, administrative, and service occupations; and relatively high levels of national and per capita (per person) income. In contrast, *middle-income countries* have industrializing economies, particularly in urban areas, and moderate levels of national and personal income. *Low-income countries* have little industrialization and low levels of national and personal income.

Although some progress has been made in reducing extreme poverty and child mortality rates while improving health and literacy rates in some lower-income countries, the overall picture remains bleak. Many people have sought to address the issue of world poverty and to determine ways in which resources can be used to meet the urgent challenge of poverty. However, not much progress has been made on this front despite a great deal of talk and billions of dollars in "foreign aid" flowing from high-income nations to low-income nations. The idea of "development" has become one of the primary means used in attempts to reduce social and economic inequalities and alleviate the worst effects of poverty in the less industrialized nations of the world.

As we take a closer look at global stratification, there are a number of problems inherent in studying this issue, one of which is what terminology should be used to describe various nations. As we shall now see, a lack of consensus exists among political, economic, and social leaders on this topic.

**global stratification** the unequal distribution of wealth, power, and prestige on a global basis, resulting in people having vastly different lifestyles and life chances both within and among the nations of the world.

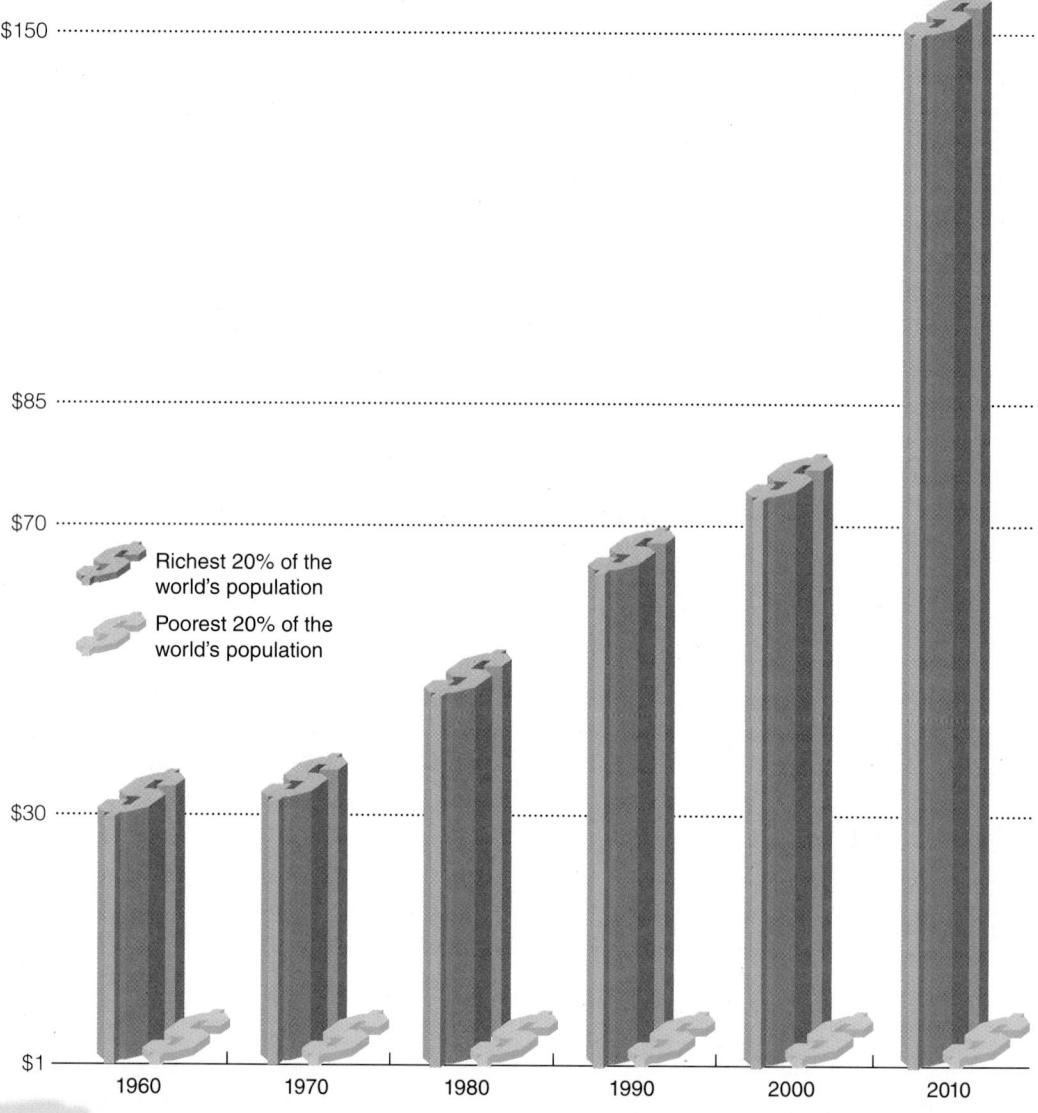

**FIGURE 8.1** INCOME GAP BETWEEN THE WORLD'S RICHEST AND POOREST PEOPLE

The income gap between the richest and poorest people in the world continued to grow between 1960 and 2010 (the last year for which information is available). As this figure shows, in 1960 the highest-income percent of the world's population received $30 for each dollar received by the lowest-income percent. By 2010, the disparity had increased: $150 to $1.

# Problems in Studying Global Inequality

One of the primary problems encountered by social scientists studying global stratification and social and economic inequality is what terminology should be used to refer to the distribution of resources in various nations. During the past sixty years, major changes have occurred in the way that inequality is addressed by organizations such as the United Nations and the World Bank. Most definitions of inequality are based on comparisons of levels of income or economic development, whereby countries are identified in terms of the "three worlds" or upon their levels of economic development.

# The "Three Worlds" Approach

After World War II, the terms "First World," "Second World," and "Third World" were introduced by social analysts to distinguish among nations on the basis of their levels of economic development and the standard of living of their citizens. *First World* nations were said to consist of the rich, industrialized nations that primarily had capitalist economic systems and democratic political systems. The most frequently noted First World nations were the United States, Canada, Japan, Great Britain, Australia, and New Zealand. *Second World* nations were said to be countries with at least a moderate level of economic development and a

## ANSWERS to the Sociology Quiz on Global Wealth and Poverty

**1. False.** In the twenty-first century, the income gap between the richest and poorest has increased on a global basis as well as in the United States.

**2. True.** Assets of the world's 200 richest people are more than the combined income of 50 percent of the world's population (United Nations Development Programme, 2010).

**3. True.** The World Bank estimates that nearly 1.3 billion people (or 22 percent of the developing world's population) live below the international poverty line, which is defined as earning less than $1.25 each day (Gill, 2012).

**4. False.** One of the consequences of extreme poverty is hunger, and millions of people—including six million children under the age of five—die of hunger-related diseases or chronic malnutrition each year.

**5. True.** In almost all low-income countries (as well as middle- and high-income countries), poverty is a more chronic problem for women because of sexual discrimination, resulting in a lack of educational and employment opportunities.

**6. False.** Although the number of poor people residing in urban areas is growing rapidly, the majority of people with incomes below the poverty line live in rural areas of the world.

**7. True.** Two-thirds of adults worldwide who are unable to read and write are women. Gender disparity is greatest in southern Asia, where slightly more than half of all women are able to read and write.

**8. True.** Although these fuels are inefficient and harmful to health, many low-income people cannot afford appliances, connection charges, and so forth. In some areas, electric hookups are not available.

moderate standard of living. These nations included China, North Korea, Vietnam, Cuba, and portions of the former Soviet Union. According to social analysts, although the quality of life in Second World nations was not comparable to that of life in the First World, it was far greater than that of people living in the *Third World*—the poorest countries, with little or no industrialization and the lowest standards of living, shortest life expectancies, and highest rates of mortality. Examples of these nations were the Democratic Republic of the Congo, Liberia, Niger, and Sierra Leone.

### The Levels of Development Approach

**LO2** Among the most controversial terminology used for describing world poverty and global stratification has been the language of development. Terminology based on levels of development includes concepts such as developed nations, developing nations, less-developed nations, and underdevelopment. Let's look first at the contemporary origins of the idea of "underdevelopment" and "underdeveloped nations."

Following World War II, the concepts of *underdevelopment* and *underdeveloped nations* emerged out of the Marshall Plan (named after U.S. Secretary of State George C. Marshall), which provided massive sums of money in direct aid and loans to rebuild the European economic base destroyed during World War II. Given the Marshall Plan's success in rebuilding much of Europe, U.S. political leaders decided that the Southern Hemisphere nations that had recently been released from European colonialism could also benefit from a massive financial infusion and rapid economic development. Leaders of the developed nations argued that urgent problems such as poverty, disease, and famine could be reduced through the transfer of finance, technology, and experience from the developed nations to lesser-developed countries. From this viewpoint, economic development is the primary way to solve the poverty problem: Hadn't economic growth brought the developed nations to their own high standard of living?

Ideas regarding *underdevelopment* were popularized by President Harry S. Truman in his 1949 inaugural address. According to Truman, the nations in the Southern Hemisphere were "underdeveloped areas" because of their low gross national product, which today is referred to as *gross national income* (GNI)—a term that refers to all the goods and services produced in a country in a given year, plus the net income earned outside the country by individuals or corporations. If nations could increase their GNI, then social and economic inequality among the citizens within the country could also be reduced. Accordingly, Truman believed that it was necessary to assist the people of economically

**FIGURE 8.2** Some analysts believe that growing global social and economic inequality is related to high rates of population growth taking place in underdeveloped nations. Why might this be so?

underdeveloped areas to raise their *standard of living,* by which he meant material well-being that can be measured by the quality of goods and services that may be purchased by the per capita national income. Thus, an increase in the standard of living meant that a nation was moving toward economic development, which typically included the exploitation of natural resources by industrial development.

What has happened to the issue of development since the post–World War II era? After several decades of economic development fostered by organizations such as the United Nations and the World Bank, it became apparent by the 1970s that improving a country's GNI did not tend to reduce the poverty of the poorest people in that country. In fact, global poverty and inequality were increasing, and the initial optimism of a speedy end to underdevelopment faded.

Why did inequality increase even with greater economic development? Some analysts in the developed nations began to link growing social and economic inequality on a global basis to relatively high rates of population growth taking place in the underdeveloped nations (Figure 8.2). Organizations such as the United Nations and the World Health Organization stepped up their efforts to provide family planning services to the populations so that they could control their own fertility. However, population researchers are now aware that issues such as population growth, economic development, and environmental problems must be seen as interdependent concerns. After the U.N. Conference on Environment and Development in Rio de Janeiro, Brazil (the "Earth Summit"), in 1992, terms such as *underdevelopment* were dropped by many analysts.

## Classification of Economies by Income

**LO3** The World Bank classifies nations into four economic categories and establishes the upper and lower limits for the gross national income (GNI) in each category. *Low-income economies* had a GNI per capita of less than $1,035 in 2013, *lower-middle-income economies* had a GNI per capita between $1,036 and $4,085, *upper-middle-income economies* had a GNI per capita between $4,086 and $12,615, and *high-income economies* had a GNI per capita of $12,616 or more (World Bank, 2013).

### Low-Income Economies

Currently, fewer than twenty nations are classified by the World Bank (2013) as low-income economies. In low-income economies, many people engage in

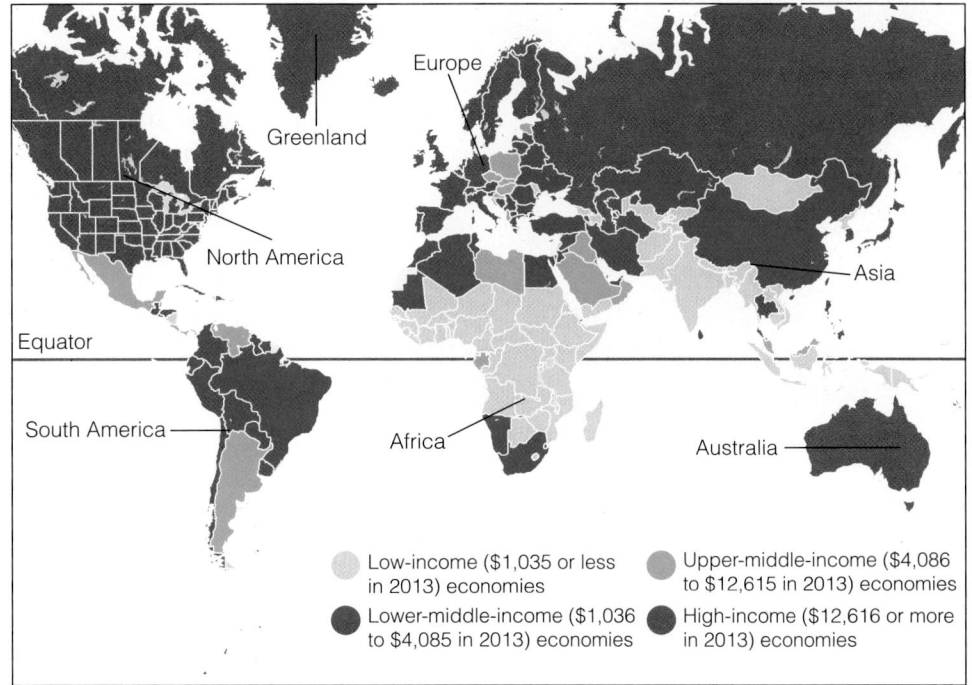

**FIGURE 8.3** HIGH-, MIDDLE-, AND LOW-INCOME ECONOMIES IN GLOBAL PERSPECTIVE
How does the United States compare with other nations with regard to income?

Source: World Bank, 2012. © Cengage Learning

agricultural pursuits, reside in nonurban areas, and are impoverished. As shown in Figure 8.3, low-income economies are primarily found in countries in Asia and Africa, where half of the world's population resides. Between 1990 and 2012, income per capita rose in low-income nations and impoverished regions such as sub-Saharan Africa.

Among those most affected by poverty in low-income economies are women and children. Why is this true? Fertility rates remain high in low-income economies. In all nations the poor have higher fertility rates than the wealthy residing within the same country. Other factors that contribute to the poverty of women and children are lack of educational opportunity, disadvantage in control over resources and assets in the household, gender disparities in work, and lower overall pay than men. To learn more about gender inequality worldwide, go online to the World Bank's *World Development Report.*

## Middle-Income Economies

About one-third of the world's population resides in a middle-income economy. As previously stated, the World Bank subdivides the middle-income economies into two categories—the lower middle income ($1,036 to $4,085 in 2013) and the upper middle income ($4,086 to $12,615). Countries classified as lower middle income include Haiti, Ghana, India, and Timor-Leste in Southeast Asia. In recent years,

millions of people have migrated from the world's poorest nations in hopes of finding better economic conditions elsewhere.

As compared with lower-middle-income economies, nations having upper-middle-income economies typically have a somewhat higher standard of living. Nations with upper-middle-income economies include Angola, Bhutan, China, Iran, Costa Rica, and Gabon. Some of these nations export a diverse variety of goods and services, ranging from manufactured goods to raw materials and fuels. For example, Kazakhstan, in Central Asia, is well-known for tobacco growing and harvesting; however, this country has been accused of hazardous child labor practices by the United Nations and the international media (see "Child Labor Issues in the Media").

## High-Income Economies

High-income economies (a gross national income per capita of $12,616 or more in 2013) are found in nations such as the United States, Canada, Japan, Australia, Portugal, Ireland, Israel, Italy, Norway, and Germany. In 2013 nations such as Antigua and Barbuda, Chile, Latvia, Lithuania, Russian Federation, and Uruguay moved into the high-income category based on gross national income per capita for the previous year. According to the World Bank, people in high-income economies typically have a higher standard of living than those in low- and

**239**

the movement of work to other regions of the country or to other nations has contributed to shifts in the global marketplace. Meanwhile, the 2008 U.S. financial crisis led to a global economic recession that has produced higher rates of unemployment. It is estimated that more than 200 million people worldwide are unemployed as job opportunities have failed to grow as rapidly as the global workforce (International Labor Organization, 2013). (The Concept Quick Review describes economies classified by income.)

**FIGURE 8.4** Vehicle manufacturing is an example of the global assembly line, where jobs and economic resources move from one nation to another. This form of capital flight has contributed to the economic woes of millions of U.S. workers who have been unable to find similar employment after their jobs disappeared.

middle-income economies, but income is only one indicator of overall human development.

Nations with high-income economies continue to dominate the world economy, despite the fact that shifts in the global marketplace have affected workers who have a mismatch between their schooling and workplace skills and the availability of job opportunities. Another problem is *capital flight*—the movement of jobs and economic resources from one nation to another—because personnel in transnational corporations have found a ready-and-willing pool of workers worldwide to perform jobs for lower wages (Figure 8.4). In the United States and other industrialized nations, the process of *deindustrialization*—the closing of plants and factories—because of their obsolescence or

## Measuring Global Wealth and Poverty

On a global basis, measuring wealth and poverty is a difficult task because of conceptual problems and problems in acquiring comparable data from various nations. As well, over time some indicators, such as the literacy rate, become less useful in helping analysts determine what progress is being made in reducing poverty.

### Absolute, Relative, and Subjective Poverty

How is poverty defined on a global basis? Isn't it more a matter of comparison than an absolute standard? According to social scientists, defining poverty involves more than comparisons of personal or household income; it also involves social judgments made by researchers. From this point of view, *absolute poverty*—previously defined as a condition in which people do not have the means to secure the most basic necessities of life—would be measured by comparing personal

**[ concept quick review ]**

### Classification of Economies by Income

|  | Low-Income Economies | Middle-Income Economies | High-Income Economies |
|---|---|---|---|
| **Previous categorization** | Third World, underdeveloped | Second World, developing | First World, developed |
| **Per capita income (GNI)** | $1,035 or less | Lower middle: $1,036 to $4,085<br><br>Upper middle: $4,086 to $12,615 | $12,616 or more |
| **Type of economy** | Largely agricultural | Diverse, from agricultural to manufacturing | Information-based and postindustrial |

Source: World Bank, 2013.

# CHILD LABOR ISSUES IN THE MEDIA

## Does the News Affect Your Views on Child Labor?

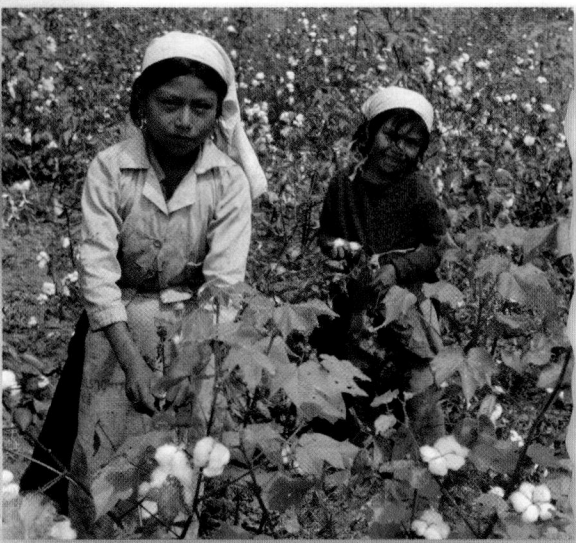

How do pictures like this as well as news stories affect your views on child labor?

**News Article 1:**

One woman said children as young as 10 working in the fields developed red rashes on their stomachs and necks as they harvested tobacco for use in cigarettes made by Philip Morris. . . . Human Rights Watch plans to release a report on Wednesday showing that child and forced labor is widespread on farms that supply a cigarette factory owned by Philip Morris International in Kazakhstan, in Central Asia. (Kramer, 2010)

**News Article 2:**

Philip Morris International Inc. said it would change its policies after a report said it bought tobacco from Eurasian farms that used forced and child labor. (UPI.com, 2010)

News articles such as these call our attention to the fact that an estimated 158 million children ages five to fourteen are engaged in child labor—one in six children in the world—and that many of these children are engaged in hazardous situations or conditions (unicef.org, 2011). Tobacco growing and harvesting are particularly dangerous because of factors such as the physical difficulty of the work; long hours; exposure to high heat, the sun, pesticides, and fumes from tobacco plants; and other health risks associated with the handling of tobacco plants.

Media articles about child labor are often written in such a manner that corporate spokespersons have a distinct opportunity to frame how the story is told. In the case of Philip Morris, a spokesperson stated that the company was unaware that child labor was involved in the growing and harvesting of the tobacco. The individual went on to say that no tobacco raised on the farms that employ child labor went into cigarettes sold outside of the former Soviet countries (UPI.com, 2010).

Should the United States become involved in child labor issues? In the Philip Morris case, some analysts stated that the United States should not get involved with child labor practices in other nations because the cigarettes were not sold in the United States and many of the children's parents were migrant laborers from neighboring Central Asian countries, particularly impoverished Kyrgyzstan (officially the Kyrgyz Republic), who relied on the money to keep their families fed.

However, in the United States, as in most other high-income countries, most people believe that young children should not be toiling in fields and factories—whether in this country or in another—and that they should not be working long hours, be exposed to high levels of nicotine, or drink from irrigation channels contaminated with pesticides, as was reported in the Philip Morris case. Rather, we believe that children should be in school, obtaining an education that will allow them to better themselves, and that they should have plenty of spare time to be, well, just to be *kids*.

Our nation's laws restrict the number of hours that people under age sixteen can work and also list certain occupations that are deemed to be too hazardous for young workers to perform. But what about other countries? Should we have any say about their social policies in regard to child labor practices? Are we engaging in ethnocentrism if we try to impose our laws on people in other nations? In some nations, child labor in agricultural fields or factories is viewed as a necessity for the family's and the country's economic survival. Parents can earn more money if they bring their children to help with the harvest, and, even then, many families make only a few hundred dollars, after paying for travel and lodging, for a hard half-year's work.

Some analysts believe that the best way to end abusive child labor is public awareness of the problem through informed media coverage that explains all sides of the issue, putting pressure on political officials and business elites around the globe, and demanding better educational opportunities for children in low-income countries. This is where conventional media and social networking both play important roles in informing people worldwide about child labor abuses and about the kind of products that are produced by global sweatshops in which children are employed. To reduce child labor, however, we must also be willing to spend more for the goods we buy (such as sporting goods celebrating our favorite teams) that are manufactured in lower-income nations. Are we willing to pay that price?

## reflect & analyze

**To what extent do you think the media inform us about harmful labor practices around the world? Can you think of examples that show how our purchasing habits affect people worldwide?**

or household income or expenses with the cost of buying a given quantity of goods and services. The World Bank has defined absolute poverty as living on less than $1.25 per day. Similarly, *relative poverty*—which exists when people may be able to afford basic necessities but are still unable to maintain an average standard of living—would be measured by comparing one person's income with the incomes of others. Finally, *subjective poverty* would be measured by comparing the actual income against the income earner's expectations and perceptions.

However, for low-income nations in a state of economic transition, data on income and levels of consumption are typically difficult to obtain and are often ambiguous when they are available. Defining levels of poverty involves several dimensions: (1) how many people are poor, (2) how far below the poverty line people's incomes fall, and (3) how long that people have been poor (is the poverty temporary or long term?). Figure 8.5 provides a unique portrayal of human poverty in which the territory size shows the proportion of the world population living in poverty in each region.

### The Gini Coefficient and Global Quality-of-Life Issues

One measure of income inequality is the *Gini coefficient,* which measures the degree of inequality in the distribution of family income in a country. In technical terms the *Gini Index* measures the extent to which the distribution of income (or consumption expenditures) among individuals or households within an economy deviates from a perfectly equal distribution. The lower a country's score on the Gini coefficient, the more equal the income distribution. The index ranges from zero (meaning that everyone has the same income) to 100 (one person receives all the income). Based on this measure, income inequality is highest in countries such as Seychelles (65.8) and South Africa (63.1) and lowest in nations such as Norway (25.8) and Sweden (25.0) (World Bank, 2012). By comparison, the United States has a Gini coefficient of 40.8. Countries such as Brazil and South Africa have remained high on this scale (indicating higher levels of inequality), while many European countries have maintained lower levels of inequality. But the United States has become noticeably more unequal, passing European countries such as Italy and France, and moving closer to nations such as China (42.5) in unequal distribution of income. However, as the World Bank (2012) notes, these data are not always strictly comparable because of differing methods and types of data collection in various countries.

## Global Poverty and Human Development Issues

**LO4** Income disparities are not the only factor that defines poverty and its effect on people. Although the average income per person in lower-income countries has doubled in the past thirty years and for many years economic

**FIGURE 8.5** PROPORTION OF WORLD'S POPULATION LIVING IN POVERTY (BY REGION)

growth has been seen as the primary way to achieve development in low-income economies, the United Nations since the 1970s has more actively focused on human development as a crucial factor in fighting poverty. In 1990 the United Nations Development Program introduced the *Human Development Index (HDI)*, establishing three new criteria—in addition to GNI—for measuring the level of development in a country: life expectancy, education, and living standards. According to the United Nations, human development is the process of increasing the number of choices that people have so that they can lead life to its fullest and be able to take action for themselves to improve their lives (United Nations Development Programme, 2013). (Figure 8.6 compares indicators such as life expectancy and per capita gross national income of various regions around the world.) The United Nations continues to monitor the progress of nations in regard to life expectancy, educational attainment, and other factors that are related to length and quality of life, as discussed in "Sociology and Social Policy."

The top level of development category used by the United Nations is "Very High Human Development." According to the *Human Development Report,* on average, people who live in countries in the highest human development categories are better educated, will live longer, and will earn more. In a nation with very high human development, for example, the per capita income ranges on average to more than $33,391 per year, as compared to about $2,010 in countries with a low high-development index. The top three countries in the HDI are Norway, Australia, and the United States. By contrast, Mozambique, the Democratic Republic of the Congo, and Niger are the three bottom countries in the 2013 HDI. In terms of human development, a child born in the Democratic Republic of the Congo in 2012, for example, has a life expectancy of 48.7 years—32 years less than a child born in the same year in Norway, who will have a life expectancy of 81.3 years (United Nations Development Programme, 2013).

## Life Expectancy

Although some advances have been made in middle- and low-HDI countries regarding increasing life expectancy, major problems still exist. The average life expectancy at birth of people in medium- or middle-HDI countries remains about 10 years less (69.9 years) than that of people in very-high-HDI countries (80.1 years). Moreover, as the above example of the Democratic Republic of the Congo and Norway illustrates, the life expectancy of people in low-HDI nations is about 20 years lower than that of people in very-high-HDI nations. Consider these figures: A child born in a low-HDI country has a life expectancy of just 59.1 years, while the average in medium-HDI countries is 69.9 years. Regions also vary on the basis of life expectancies: An infant born in 2012 in Latin America or the Caribbean had an average life expectancy at birth of 74.7 years, while a child born in sub-Saharan Africa in the same year had a life expectancy at birth of 54.9 years (United Nations Development Programme, 2013).

One major cause of shorter life expectancy in low-income nations is the high rate of infant

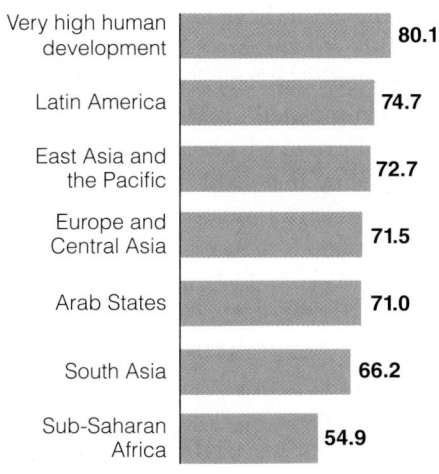

**a. Life expectancy in years for persons born in 2012**

| | |
|---|---|
| Very high human development | 80.1 |
| Latin America | 74.7 |
| East Asia and the Pacific | 72.7 |
| Europe and Central Asia | 71.5 |
| Arab States | 71.0 |
| South Asia | 66.2 |
| Sub-Saharan Africa | 54.9 |

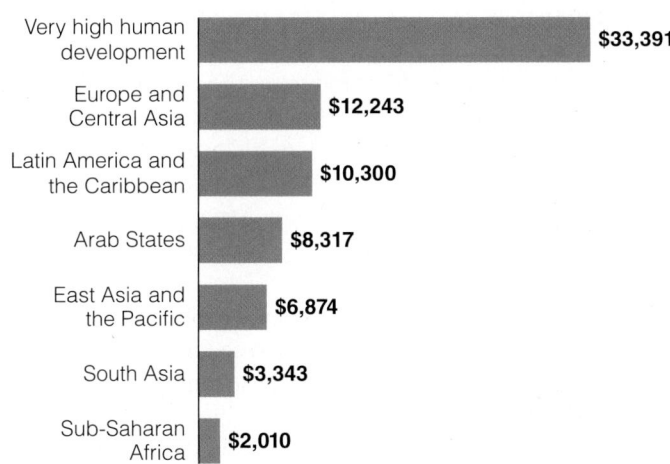

**b. Per capita gross national income in U.S. dollars (no date)**

| | |
|---|---|
| Very high human development | $33,391 |
| Europe and Central Asia | $12,243 |
| Latin America and the Caribbean | $10,300 |
| Arab States | $8,317 |
| East Asia and the Pacific | $6,874 |
| South Asia | $3,343 |
| Sub-Saharan Africa | $2,010 |

**FIGURE 8.6** INDICATORS OF HUMAN DEVELOPMENT

Source: United Nations Development Programme, 2013. © Cengage Learning

**FIGURE 8.7** In an effort to reduce poverty, some nations have developed adult literacy programs so that people can gain an education that will help lift them out of poverty. In regions such as Solomuna, Eritrea, women's literacy is a particularly crucial issue.

mortality. The *infant mortality rate* refers to the number of deaths per thousand live births in a calendar year. Low-income countries typically have higher rates of illness and disease, and they do not have adequate health care facilities. Malnutrition is a common problem among children, many of whom are underweight, stunted, and anemic—a nutritional deficiency with serious consequences for child mortality.

Among adults and children alike, life expectancies are strongly affected by hunger and malnutrition. It is estimated that people in the United States spend $60–$66 billion each year on weight-loss products and endeavors such as joining a gym, whereas the world's poorest people suffer from chronic malnutrition, and many die from hunger-related diseases. Inadequate nutrition affects people's ability to work and to earn the income necessary for a minimum standard of living. Although some gains have been made in reducing the rate of malnourishment in some lower-income nations, about one billion people around the world are malnourished, and 63 percent of these are in Asia and the Pacific, 26 percent in sub-Saharan Africa, and 1 percent in developed countries (United Nations Development Programme, 2013).

On the plus side of the life expectancy problem, some nations have made positive gains, seeing average life expectancy increase in the past four decades. For example, life expectancy has risen by 5.5 years between 2000 and 2012 in sub-Saharan Africa (United Nations Development Programme, 2013). Of course, problems of illness and mortality remain grave in regions such as sub-Saharan Africa, where overall longevity and quality of life are highly problematic.

## Health

*Health* refers to a condition of physical, mental, and social well-being. In other words, it is more than the absence of illness or disease. Many people in low-income nations are far from having physical, mental, and social well-being. In fact, more than 25 million people die each year from AIDS, malaria, tuberculosis, pneumonia, diarrheal diseases, measles, and other infectious and parasitic illnesses. According to the World Health Organization, infectious diseases are far from under control in many nations: Infectious and parasitic diseases are the leading killers of children and young adults, and these diseases have a direct link to environmental conditions and poverty, especially in terms of unsanitary and overcrowded living conditions.

Some middle-income countries are experiencing rapid growth in degenerative diseases such as cancer and coronary heart disease, and many more deaths are expected from smoking-related illnesses. Despite the decrease in tobacco smoking in high-income countries, globally there has been an increase in per capita consumption of tobacco products. Today, nearly 80 percent of the world's one billion smokers live in low- or middle-income nations (World Health Organization, 2013).

## Education and Literacy

According to the *Human Development Report* (United Nations Development Programme, 2013), education is fundamental to improving life chances and reducing both individual and national poverty. People with more years of formal education tend to earn higher wages and have better jobs. Progress in education has been made in many nations, and people around the world have higher levels of education than in the past (Figure 8.7). For this reason the *Human Development Report* uses "mean years of schooling"—completed years of educational attainment—and "expected years of schooling"—the years of schooling that a child can expect to receive given current enrollment rates—to measure progress in education.

In nations with very high human development, the mean years of schooling received by people ages 25 and older is 11.5 years; in low-HDI nations, the average number is only 4.2 years. Medium-HDI countries have 6.3 mean years of schooling, compared with 8.8 years for high-HDI countries. However, the United Nations also calculates expected years of schooling because the agency believes that these figures better reflect changing education opportunities in developing countries. For example, the average incoming elementary school student

# SOCIOLOGY and Social Policy

## Fighting Extreme Poverty One Social Policy at a Time

The primary aim of the United Nations Millennium Development Goals "should be to eradicate extreme poverty. That should be our rallying cry. We can do it in our generation."

—At a UN conference, UK Prime Minister David Cameron, chairman of the G8 organization (Canada, France, Germany, Italy, Japan, Russia, the United Kingdom, and the United States) made this call for a rapid reduction in extreme poverty worldwide (qtd. in Wintour, 2013).

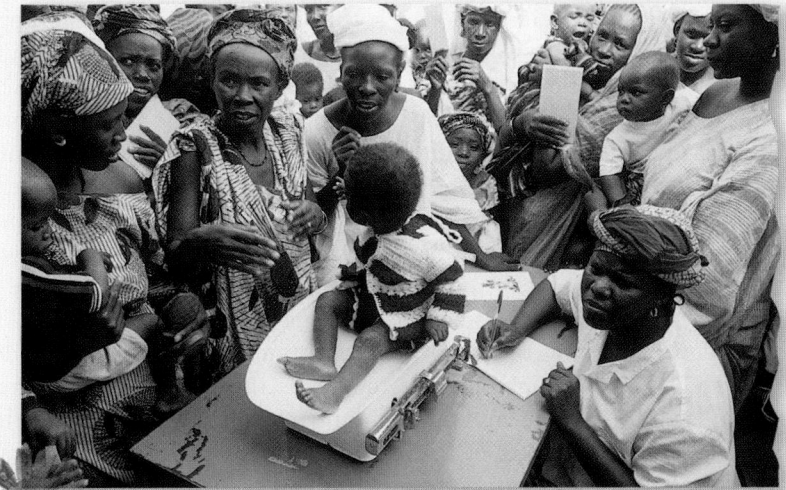

Malnutrition is a widespread problem in many low-income nations. What other kinds of health problems are related to global poverty?

Organizations such as the United Nations establish social policy initiatives to reduce extreme poverty (defined as living on under $1.25 per day) and to promote education, health, and well-being. Developing effective policies for dealing with global inequality and extreme poverty is crucial for bringing about social change. Social policy is an area of research and action that looks at the social relations necessary for people's well-being and works to build systems that promote well-being. The United Nations established eight Millennium Development Goals to be achieved and continues to monitor efforts in each of these areas: end poverty and hunger; ensure universal education, gender equality, child health, and maternal health; combat HIV/AIDS; work toward environmental sustainability; and establish global partnerships (United Nations, 2013).

All G8 leaders signed on to help implement these policies, and in 2012 the World Bank announced that the first goal, cutting extreme poverty in half, was ahead of the 2015 deadline. Although it is an outstanding accomplishment that an estimated 663 million people moved above the extreme poverty line (of $1.25 per day), it is important to recall that they still remain very poor, even at $2 per day, based on the standards of middle- and higher-income nations. Looking at differences in urban and rural areas, the bad news is that extreme poverty engulfs almost 29 percent of the urban population and 34 percent of the rural population. However, the good news is that urban poverty is down about 7 percent and rural poverty has decreased by more than 9 percent since 2005.

Will it be possible to eliminate poverty through social policy? According to David Cameron, it will be possible to eradicate poverty only if people see the issue as "not only about money" (qtd. in Gill, 2012). He suggests that these goals will be impossible to attain if there is widespread corruption, lack of justice, and limited access to law. As a result, Cameron believes that people must stand up for these governance issues in order to eliminate extreme poverty and help reduce overall inequality.

### reflect & analyze

Do you believe that social policy is an effective tool for reducing poverty? Are global poverty-reduction policies established by international organizations such as the United Nations effective in reducing poverty? Should each nation have its own policies to reduce inequality?

in a low-HDI country is expected to complete 8.5 years of schooling, a sum that is about equal to the 8.8 years of schooling currently possessed by adults 25 years of age and older in high-HDI countries. As the most recent report of the United Nations Development Programme (2013: 23) stated, "Overall, most low HDI countries have achieved or are advancing towards full enrollment in elementary school and more than 50 percent enrollment in secondary school."

What is the relationship between education and literacy? The United Nations Educational, Scientific and Cultural Organization (UNESCO) defines a literate person as "someone who can, with understanding, both read and write a short, simple statement on their everyday life" (United Nations Regional Information Center for Western Europe, 2014). Based on this definition, people who can write only with figures, their name, or a memorized phrase are not considered literate. Literacy was previously used by the United Nations as a measure to determine education in relation to levels of human development. As improvements occurred in global literacy rates, researchers found that other measures of educational attainment were more useful in assessing the knowledge dimension of human development.

Literacy rates continue to rise among adults and youths, and gender gaps are narrowing. It is estimated that 84 percent of the global adult population (ages 15 and above) is able to read and write. Although 89 percent of young people around the world have basic literacy and numeracy skills, there are still more than 123 million young people who are unable to read or write (United Nations, 2013). The largest increases in youth literacy rates were identified in Northern African and Southern Asia. The literacy rate among young women is growing faster than that of young men, particularly in these areas of the world. Although the literacy rate among adult women has continued to rise, women still represent two-thirds of those who are illiterate worldwide (United Nations, 2013). Literacy is crucial for everyone, but it is especially important for women because it has been closely linked to decreases in fertility, improved child health, and increased earnings potential.

## Persistent Gaps in Human Development

Some middle- and lower-income countries have made progress in certain indicators of human development. The gap between some richer and middle- or lower-income nations has narrowed significantly for life expectancy, health, education, and income. For example, Niger and the Democratic Republic of the Congo have experienced many challenges but have been able to make strides in HDI improvements. Some of the countries in Africa that have seen notable improvements in school attendance, life expectancy, and per capita income growth have recently emerged from lengthy periods of armed conflict within their borders. However, the overall picture for the world's poorest people remains dismal. The gap between the richest and poorest countries has widened to a gulf. As previously stated, the countries with the highest percentages of "poor" on the Multidimensional Poverty Index are all in Africa: Ethiopia

(87 percent), Liberia (84 percent), and Mozambique (79 percent). Overall, however, South Asia has the largest absolute number of multidimensionally poor people, and about 612 million of them live in India alone (United Nations Development Programme, 2013).

Poverty, food shortages, hunger, and rapidly growing populations are pressing problems for at least two billion people, most of them women and children living in a state of absolute poverty. Although more women around the globe have paid employment than in the past, women are still finding themselves in poverty because of increases in single-person and single-parent households headed by women and the fact that low-wage work is often the only source of livelihood available to them.

Human development research has reached a surprising conclusion: Economic growth and higher incomes in low- and medium-development nations are not always necessary to bring about improvements in health and education. According to the *Human Development Report,* technological improvements and changes in societal structure allow even poorer countries to bring about significant changes in health and education even without significant gains in income.

## Theories of Global Inequality

**LO5** Social scientists have developed a variety of theories that view the causes and consequences of global inequality. We will examine the development approach and modernization theory, dependency theory, world systems theory, and the new international division of labor theory.

## Development and Modernization Theory

According to some social scientists, global wealth and poverty are linked to the level of industrialization and economic development in a given society (Figure 8.8). Although the process by which a nation industrializes may vary somewhat, industrialization almost inevitably brings with it a higher standard of living and some degree of social mobility for individual participants in the society. Specifically, the traditional caste system becomes obsolete as industrialization progresses. Family status, race/ethnicity, and gender are said to become less significant in industrialized nations than in agrarian-based societies. As societies industrialize, they also urbanize as workers locate their residences near factories, offices, and other places of work. Consequently, urban values and folkways overshadow the beliefs and practices of the rural areas. Analysts using a development framework typically view industrialization and economic development as essential steps that nations must go through in order to reduce poverty and increase life chances for their citizens.

The most widely known development theory is **modernization theory—a perspective that links global inequality to different levels of economic development and suggests that low-income economies can move to middle- and high-income economies by achieving self-sustained economic growth.** According to modernization theory, the low-income, less-developed nations can improve their standard of living only with a period of intensive economic growth and accompanying changes in people's beliefs, values, and attitudes toward work. As a result of modernization, the values of people in developing countries supposedly become more similar to those of people in high-income nations. The number of hours that people work at their jobs each week is one measure of the extent to which individuals subscribe to the *work ethic,* a core value widely believed to be of great significance in the modernization process. Of course, this assumption may be false because much research indicates that people in developing nations typically work longer hours per day than individuals in industrialized nations.

Perhaps the best-known modernization theory is that of Walt W. Rostow (1971, 1978), who, as an economic advisor to President John F. Kennedy, was highly instrumental in shaping U.S. foreign policy toward Latin America in the 1960s. Rostow suggested that all countries go through four stages of economic development, with identical content, regardless of when these nations started the process of industrialization. He compared the stages of economic development to an airplane ride. The first stage is the *traditional stage*, in which very little social change takes place, and people do not think much about changing their current circumstances. According to Rostow, societies in this stage are slow to change because the people hold a fatalistic value system, do not subscribe to the work ethic, and save very little money. The second stage is the *take-off stage*—a period of economic growth accompanied by a growing belief in individualism, competition, and achievement. During this stage people start to look toward the future, to save and invest money, and to discard traditional values. According to Rostow's modernization theory, the development of capitalism is essential for the transformation from a traditional, simple society to a modern, complex

Chris Hondros/Getty Images News/Getty Images

**FIGURE 8.8** Poverty and war continue to devastate low-income countries such as Afghanistan. Many low-income countries receive aid from industrialized nations through initiatives such as the World Food Program. Modernization theory links global inequality to levels of economic development, but factors such as war and internal conflict also greatly contribute to patterns of global inequality.

one. With the financial help and advice of the high-income countries, low-income countries will eventually be able to "fly" and enter the third stage of economic development. In the third stage the country moves toward *technological maturity*. At this point, the country will improve its technology, reinvest in new industries, and embrace the beliefs, values, and social institutions of the high-income, developed nations. In the fourth and final stage the country reaches the phase of *high mass consumption* and a correspondingly high standard of living.

Modernization theory has had both its advocates and its critics. According to proponents of this approach, studies have supported the assertion that economic development occurs more rapidly in a capitalist economy. In fact, the countries that have been most successful in moving from low- to middle-income status typically have been those that are most centrally involved in the global capitalist economy. For example, the nations of East Asia have successfully made the transition from low-income to higher-income economies through factors such

**modernization theory** a perspective that links global inequality to different levels of economic development and suggests that low-income economies can move to middle- and high-income economies by achieving self-sustained economic growth.

247

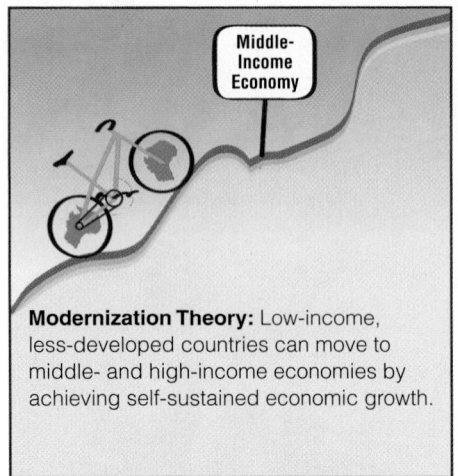

**Modernization Theory:** Low-income, less-developed countries can move to middle- and high-income economies by achieving self-sustained economic growth.

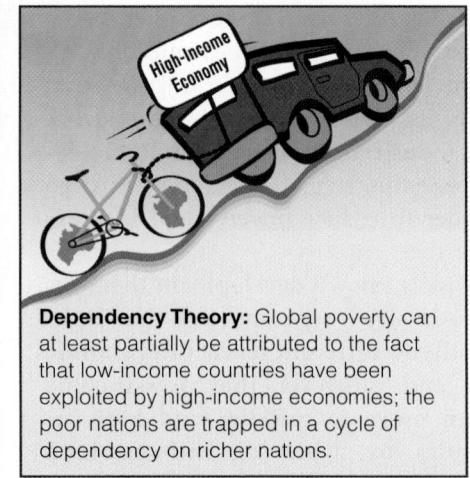

**Dependency Theory:** Global poverty can at least partially be attributed to the fact that low-income countries have been exploited by high-income economies; the poor nations are trapped in a cycle of dependency on richer nations.

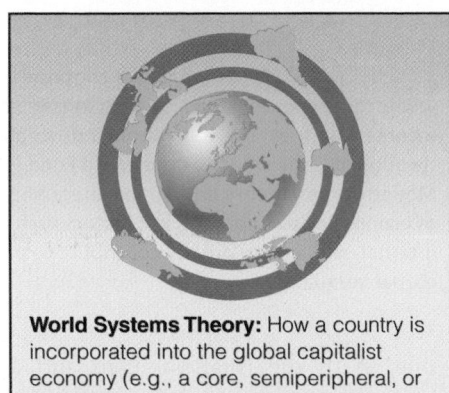

**World Systems Theory:** How a country is incorporated into the global capitalist economy (e.g., a core, semiperipheral, or peripheral nation) is the key feature in determining how economic development takes place in that nation.

**The New International Division of Labor Theory:** Commodity production is split into fragments, each of which can be moved (e.g., by a transnational corporation) to whichever part of the world can provide the best combination of capital and labor.

**FIGURE 8.9** APPROACHES TO STUDYING GLOBAL INEQUALITY
What causes global inequality? Social scientists have developed a variety of explanations, including the four theories shown here.
© Cengage Learning

as a high rate of savings and the fostering of a market economy.

Critics of modernization theory point out that it tends to be Eurocentric in its analysis of low-income countries, which it implicitly labels as backward. In particular, modernization theory does not take into account the possibility that all nations do not industrialize in the same manner. In contrast, some analysts have suggested that modernization of low-income nations today will require novel policies, sequences, and ideologies that are not accounted for by Rostow's approach.

Which sociological perspective is most closely associated with the development approach? Modernization theory is based on a market-oriented perspective which assumes that "pure" capitalism is good and that the best economic outcomes occur when governments follow the policy of laissez-faire (or hands-off) regarding business, giving capitalists the opportunity to make the "best" economic decisions, unfettered by government restraints

or cumbersome rules and regulations. In today's global economy, however, many analysts believe that national governments are no longer central corporate decision makers and that transnational corporations determine global economic expansion and contraction. Therefore, corporate decisions to relocate manufacturing processes around the world make the rules and regulations of any one nation irrelevant and national boundaries obsolete. Just as modernization theory most closely approximates a functionalist approach to explaining inequality, dependency theory, world systems theory, and the new international division of labor theory are perspectives rooted in the conflict approach. All four of these approaches are depicted in Figure 8.9.

## Dependency Theory

**L06** *Dependency theory* states that global poverty can at least partially be attributed to the fact that the low-income countries

**have been exploited by the high-income countries.** Analyzing events as part of a particular historical process—the expansion of global capitalism—dependency theorists see the greed of the rich countries as a source of increasing impoverishment of the poorer nations and their people. Dependency theory disputes the notion of the development approach, and modernization theory specifically, that economic growth is the key to meeting important human needs in societies. In contrast, the poorer nations are trapped in a cycle of structural dependency on the richer nations because of their need for infusions of foreign capital and external markets for their raw materials, making it impossible for the poorer nations to pursue their own economic and human development agendas.

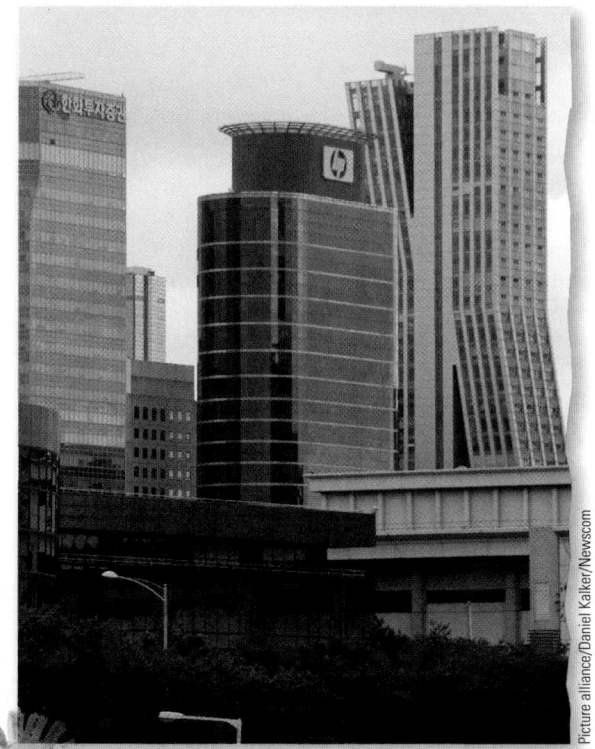

Picture alliance/Daniel Kalker/Newscom

**FIGURE 8.10** A variety of factors—such as foreign investment and the presence of transnational corporations—have contributed to the economic growth of nations such as South Korea.

Dependency theory has been most often applied to the newly industrializing countries (NICs) of Latin America, whereas scholars examining the NICs of East Asia found that dependency theory had little or no relevance to economic growth and development in that part of the world. Therefore, dependency theory had to be expanded to encompass transnational economic linkages that affect developing countries, including foreign aid, foreign trade, foreign direct investment, and foreign loans. On the one hand, in Latin America and sub-Saharan Africa, transnational linkages such as foreign aid, investments by transnational corporations, foreign debt, and export trade have been significant impediments to development within a country. On the other hand, East Asian countries such as Taiwan, South Korea, and Singapore have historically also had high rates of dependency on foreign aid, foreign trade, and interdependence with transnational corporations but have still experienced high rates of economic growth despite this dependency (Figure 8.10).

Dependency theory makes a positive contribution to our understanding of global poverty by noting that "underdevelopment" is not necessarily the cause of inequality. Rather, it points out that exploitation not only of one country by another but of countries by transnational corporations may limit or retard economic growth and human development in some nations.

## World Systems Theory

**L07** *World systems theory* **is a perspective that examines the role of capitalism, and particularly the transnational division of labor, in a truly global system held together by economic ties.** From this approach, global inequality does not emerge solely as a result of the exploitation of one country by another. Instead, economic domination involves a complex world system in which the industrialized, high-income nations benefit from other nations and exploit their citizens. This approach is most closely associated with the sociologist Immanuel Wallerstein (1979, 1984, 2011), who believed that a country's mode of incorporation into the capitalist work economy is the key feature in determining how economic development takes place in that nation.

According to world systems theory, the capitalist world economy is a global system divided into a hierarchy of three major types of nations—core, semiperipheral, and peripheral—in which

**dependency theory** the belief that global poverty can at least partially be attributed to the fact that the low-income countries have been exploited by the high-income countries.

**world systems theory** a perspective that examines the role of capitalism, and particularly the transnational division of labor, in a truly global system held together by economic ties.

upward or downward mobility is conditioned by the resources and obstacles that characterize the international system. **Core nations are dominant capitalist centers characterized by high levels of industrialization and urbanization.** Core nations such as the United States, Japan, and Germany possess most of the world's capital and technology. Even more importantly for their position of domination, they exert massive control over world trade and economic agreements across national boundaries. Some cities in core nations are referred to as *global cities* because they serve as international centers for political, economic, and cultural concerns. New York, Tokyo, and London are the largest global cities, and they are often referred to as the "command posts" of the world economy.

*Semiperipheral nations* **are more developed than peripheral nations but less developed than core nations.** Nations in this category typically provide labor and raw materials to core nations within the world system. These nations constitute a midpoint between the core and peripheral nations that promotes the stability and legitimacy of the three-tiered world economy. These nations include South Korea and Taiwan in East Asia, Mexico and Brazil in Latin America, India in South Asia, and Nigeria and South Africa in Africa. Only two global cities are located in semiperipheral nations: São Paulo, Brazil, which is the center of the Brazilian economy, and Singapore, which is the economic center of a multicountry region in Southeast Asia. According to Wallerstein, semiperipheral nations exploit peripheral nations, just as the core nations exploit both the semiperipheral and the peripheral nations.

Most low-income countries in Africa, South America, and the Caribbean are *peripheral nations*—**nations that are dependent on core nations for capital, have little or no industrialization (other than what may be brought in by core nations), and have uneven patterns of urbanization.** According to Wallerstein (1979, 1984, 2011), the wealthy in peripheral nations benefit from the labor of poor workers and from their own economic relations with core-nation capitalists, whom they uphold in order to maintain their own wealth and position. At a global level, uneven economic growth results from capital investment by core nations; disparity between the rich and the poor within the major cities in these nations is increased in the process. The U.S./Mexico border is an example of disparity and urban growth: Transnational corporations have built *maquiladora* plants so that goods can be assembled by low-wage workers to keep production costs down. As compared to a federal minimum wage of $7.25 per hour

in the United States, the minimum wage in Mexico is about $5.10 per *day*. Many people are paid more than this wage, but often not enough to be a living wage. For example, a person working in a factory that builds pickups and tractor-trailer cabs might earn $7.50 per day. Figure 8.11 describes this process. In 2013 about 5,000 *maquiladora* plants were located in Mexico, employing nearly 2 million people in factories manufacturing apparel; electronic accessories such as computers, televisions, and small appliances; and motor vehicles and parts.

As Wallerstein's world systems theory would suggest, one of the threats to the Mexican *maquiladora* plants is the movement of work by some transnational corporations to China and some of the countries in Central America, where wages are even lower.

Not all social analysts agree with Wallerstein's perspective on the hierarchical position of nations in the global economy. However, nations throughout the world are influenced by a relatively small number of cities and transnational corporations that have prompted a shift from an international to a more global economy. Wallerstein (1991, 2011) acknowledged that world systems theory is an "incomplete, unfinished critique" of long-term, large-scale social change that influences global inequality.

## The New International Division of Labor Theory

**LO8** Although the term *world trade* has long implied that there is a division of labor between societies, the nature and extent of this division have recently been reassessed based on the changing nature of the world economy. According to the *new international division of labor theory,* **commodity production is being split into fragments that can be assigned to whichever part of the world can provide the most profitable combination of capital and labor.** Consequently, the new international division of labor has changed the pattern of geographic specialization between countries, whereby high-income countries have now become dependent on low-income countries for labor. The low-income countries provide transnational corporations with a situation in which they can pay lower wages and taxes and face fewer regulations regarding workplace conditions and environmental protection. Overall, a global manufacturing system that includes the offshoring of jobs has emerged in which transnational corporations establish labor-intensive, assembly-oriented export production, ranging from textiles and clothing to technologically sophisticated

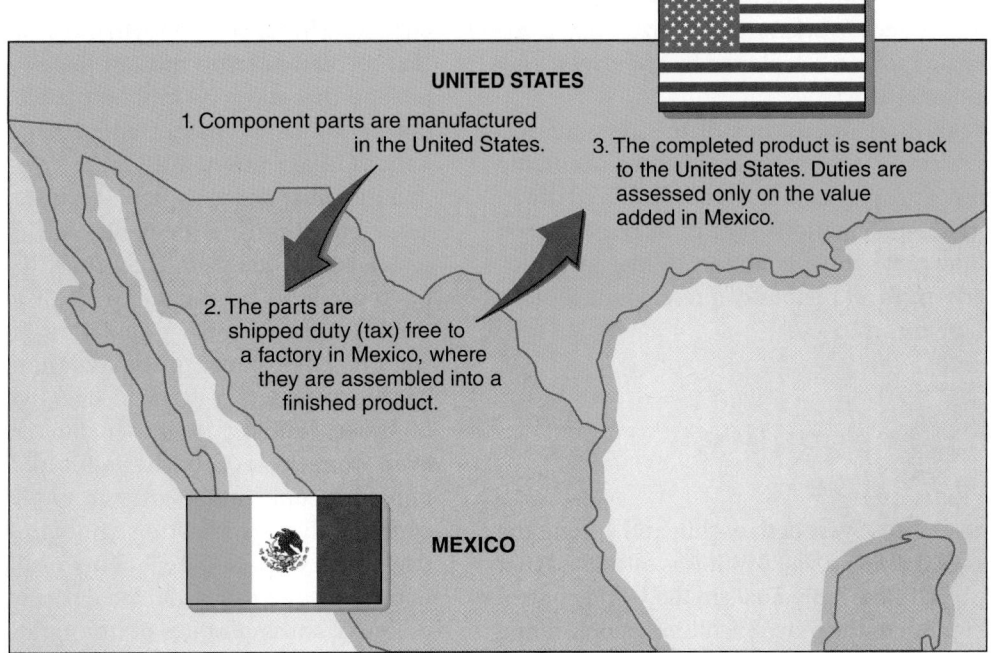

**FIGURE 8.11** *MAQUILADORA* PLANTS
Here is the process by which transnational corporations establish plants in Mexico so that profits can be increased by using low-wage workers there to assemble products that are then brought into the United States for sale.

© Cengage Learning

exports such as computers, in middle- and lower-income nations.

At the same time, manufacturing technologies are shifting from the large-scale, mass-production assembly lines of the past toward a more flexible production process involving microelectronic technologies. Even service industries—such as processing insurance claims forms, reading MRI and CT scans, and preparing tax forms—that were formerly thought to be less mobile have become exportable through electronic transmission and the Internet. These activities have been referred to as *global commodity chains,* a complex pattern of international labor and production processes that results in a finished commodity ready for sale in the marketplace.

Some commodity chains are producer-driven, whereas others are buyer-driven. *Producer-driven commodity chains* is the term used to describe industries in which transnational corporations play a central part in controlling the production process. Industries that produce automobiles, computers, and other capital- and technology-intensive products are typically producer-driven. In contrast, *buyer-driven commodity chains* is the term used to refer to industries in which large retailers (such as Walmart–like corporations), brand-name merchandisers, and trading companies set up decentralized production networks in various middle- and low-income countries. This type of chain is most common in labor-intensive, consumer-goods industries such as toys, garments, and footwear. Athletic footwear companies such as Nike and Reebok are examples of the buyer-driven model. Because these products tend to be labor intensive at the manufacturing stage, the

**core nations** according to world systems theory, nations that are dominant capitalist centers characterized by high levels of industrialization and urbanization.

**semiperipheral nations** according to world systems theory, nations that are more developed than peripheral nations but less developed than core nations.

**peripheral nations** according to world systems theory, nations that are dependent on core nations for capital, have little or no industrialization (other than what may be brought in by core nations), and have uneven patterns of urbanization.

**new international division of labor theory** the perspective that views commodity production as being split into fragments that can be assigned to whichever part of the world can provide the most profitable combination of capital and labor.

typical factory system is very competitive and globally decentralized. Workers in buyer-driven commodity chains are often exploited by low wages, long hours, and poor working conditions.

Although most discussions of the new international division of labor focus on changes occurring in the lives of people residing in industrialized urban areas of developing nations, it must be remembered that millions of people continue to live in grinding poverty in rural regions of these countries (see "Sociology Works!").

## Global Inequality in the Future

Social inequality is vast both within and among the countries of the world. In 2014 it was estimated that 85 of the wealthiest individuals in the world control as much wealth as the other 3.5 billion people living on the planet! No, you did not read that incorrectly: It is 85 persons as compared with 3.5 billion. Even in high-income nations where wealth is highly concentrated, many poor people coexist with the affluent. In middle- and low-income countries, there are small pockets of wealth in the midst of poverty and despair. Although some political and business elites in local economies benefit greatly from partnerships with transnational corporations, everyday people residing in these nations have often continued to be exploited in both industrial and agricultural work. In China, for example, some people have accumulated vast wealth in urban areas while poverty has increased dramatically in some regions and particularly in rural areas.

What are the future prospects for greater equality across and within nations? Not all social scientists agree on the answer to this question. Depending on the theoretical framework that they apply in studying global inequality, social analysts may describe either an optimistic or a pessimistic scenario for the future. Moreover, some analysts highlight the human rights issues embedded in global inequality, whereas others focus primarily on an economic framework.

In some regions, high rates of unemployment and persistent poverty undermine human development and future possibilities for socioeconomic change. In the second decade of the twenty-first century, high unemployment rates in the United States and the Euro zone (nations in the European Union such as Italy, France, and Germany) continue to threaten global economic growth and limit opportunities for unemployed and underemployed workers and young people seeking to get into the labor market. Worldwide, millions of young people are unemployed, and many are destined to experience long-term unemployment. Estimates suggest that as many as 200 million people are without a job and that about 39 million people have dropped out of the labor market altogether (International Labor Organization, 2013).

Problems such as long-term unemployment contribute to gross inequality, which has high financial and quality-of-life costs to people, even for those who are not the poorest of the poor. In the future, continued population growth, urbanization, environmental degradation, and violent conflict threaten even the meager living conditions of those residing in low-income nations. However, from this approach the future looks dim not only for people in low-income and middle-income countries but also for those in high-income countries, who will see their quality of life diminish as natural resources are depleted, the environment is polluted, and high rates of immigration and global political unrest threaten the high standard of living that many people have come to enjoy. According to some social analysts, transnational corporations and financial institutions such as the World Bank and the International Monetary Fund will further solidify and control a globalized economy, which will transfer the power to make significant choices to these organizations and away from the people and their governments. Further loss of resources and means of livelihood will affect people and countries around the globe.

As a result of global corporate domination, there could be a leveling out of average income around the world, with wages falling in high-income countries and wages increasing significantly in low- and middle-income countries. If this pessimistic scenario occurs, there is likely to be greater polarization of the rich and the poor and more potential for ethnic and national conflicts over such issues as worsening environmental degradation and who has the right to natural resources. For example, pulp-and-paper companies in Indonesia, along with palm-oil plantation owners, have continued clearing land for crops by burning off vast tracts of jungle, producing high levels of air and water pollution across seven Southeast Asian nations and creating havoc for millions of people.

On the other hand, a more optimistic scenario is also possible. With modern technology and worldwide economic growth, it might be possible to reduce absolute poverty and to increase people's opportunities. Among the trends cited by the *Human Development Report* (United Nations Development Programme, 2013) that have the potential to bring about more-sustainable patterns

# Sociology WORKS!

## The Relationship Between Place and Global Poverty

How should I be happy? I'm poor, and my parents have illnesses.

> —Teng Ling Dang earns about $12 each day working at a brick factory and told journalists that he cannot afford to get married or leave his rural area because of his elderly parents (qtd. in Gross, 2011).

Compared to rich people, our life is not that good—but it is much better than before.

> —Xu Zuxian, a grandmother in a remote part of Guizhou Province, describes how rural poverty may be decreasing slightly since Communist Party leaders vowed to address the wealth gap between rural and urban areas in China (qtd. in Branigan, 2012).

Teng Ling Dang lives in Ta Ping, a village with 103 residents, which has largely been untouched by the economic boom in Xi'an and other rapidly growing cities in China. By contrast, Xu Zuxian hopes that Xi Jinping, China's new Communist Party leader, will continue efforts to uplift impoverished rural areas (Jacobs, 2013).

As more people have attained great wealth in China's urban centers in recent decades, poverty in the rural areas, mountainous regions, and deserts has remained severe. In rural areas, many farmers and their families have lived on what they could make from selling their crops and on money sent from relatives working in cities far away. In urban areas, vast skyscrapers and expensive designer boutiques have become visual representations of the affluence that has largely eluded the rural population.

For many years, sociologists studying poverty have focused on differences in rural and urban poverty. Throughout the world, they have found that *place* does matter when it comes to finding the deepest pockets of poverty (Rural Policy Research Institute, 2004). Where people live strongly influences how much money they will make, and income inequalities are important indicators of the life chances of entire families. In some developing countries, the rural poor rely primarily on agriculture, fishing, forestry, and sometimes small-scale industries and services for their livelihood. When they are unable to derive sufficient economic resources from these endeavors, little else is available for them. Some migrate to urban centers in hopes of finding new opportunities, but many remain behind, living in grinding poverty. When sociologists speak of "place," they are referring to such things as an area's natural environment, which includes its climate, natural resources, and degree of isolation (Rural Policy Research Institute, 2004). Place also involves the economic structure in the area, such as the extent to which adequate amounts of food can be raised to meet people's needs or whether an individual can earn sufficient money to purchase food.

Can rural poverty be reduced? According to some social policy analysts, broad economic stability, competitive markets, and public investment in *physical* and *social* infrastructure are important prerequisites for a reduction in rural poverty in developing nations. From this perspective, a major reduction in rural poverty in China will occur only if people have access to land and credit, education, health care, support services, and food through well-designed public works programs and other transfer mechanisms.

### reflect & analyze

**Whether changes that reduce poverty will occur in China's future remains to be seen, but the sociological premise that place matters in regard to poverty remains a valid assumption in helping us explain global poverty and inequality. Can you apply this idea to rural and urban areas with which you are familiar in the United States? Why are issues such as this important to each of us even if we do not live in a rural area and have no personal experience with poverty?**

of development are the various forms of socioeconomic progress made in many low- and middle-income countries over the past thirty years through technological, social, and environmental improvements. For example, technological innovation continues to improve living standards for some people. Fertility rates are declining in some regions (but remain high in others, where there remains grave cause for concern about the availability of adequate natural resources for the future). Finally, health and education may continue to improve in lower-income countries. According to the *Human Development Report*, healthy, educated populations are crucial for the future in order to reduce global poverty. The education of women is of primary importance in the future if global inequality is to be reduced. All aspects of schooling and training are crucial for the future, including agricultural extension services in rural areas to help women farmers in regions such as western Kenya produce more crops to feed their families (United Nations Development Programme, 2013). From this viewpoint, we can enjoy prosperity only by ensuring that other people have the opportunity to survive and thrive in their own surroundings. The problems associated with global poverty are therefore of interest to a wide-ranging set of countries and people.

# chapter REVIEW

## LO1 What is global stratification, and how does it contribute to economic inequality?

Global stratification refers to the unequal distribution of wealth, power, and prestige on a global basis, which results in people having vastly different lifestyles and life chances both within and among the nations of the world. Today, the income gap between the richest and the poorest percent of the world population continues to widen, and within some nations the poorest one-fifth of the population has an income that is only a slight fraction of the overall average per capita income for that country.

## LO2 What is the levels of development approach for studying global inequality?

One of the primary problems encountered by social scientists studying global stratification and social and economic inequality is what terminology should be used to refer to the distribution of resources in various nations. Most definitions of inequality are based on comparisons of levels of income or economic development, whereby countries are identified in terms of the "three worlds" or upon their levels of economic development. Terminology based on levels of development includes concepts such as developed nations, developing nations, less-developed nations, and underdevelopment.

## LO3 How does the World Bank classify nations into four economic categories, and why is measuring wealth and poverty on a global basis a difficult process?

The World Bank classifies nations into four economic categories and establishes the upper and lower limits for the GNI in each category. In 2013 low-income economies had a GNI per capita of less than $1,035, lower-middle-income economies had a GNI per capita between $1,036 and $4,085, upper-middle-income economies had a GNI per capita between $4,086 and $12,615, and high-income economies had a GNI per capita of $12,616 or more. Defining poverty is more than just personal and household income; it also involves social judgments made by researchers. Absolute poverty is a condition in which people do not have the means to secure the most basic necessities of life. It is measured by comparing personal or household income or expenses with the cost of buying a given quantity of goods and services. Relative poverty exists when

people may be able to afford basic necessities but are still unable to maintain an average standard of living. This is measured by comparing one person's income with the incomes of others.

## LO4 What is the relationship between global poverty and key human development issues such as life expectancy, health, education, and literacy?

Income disparities are not the only factor that defines poverty and its effect on people. The United Nations' Human Development Index measures the level of development in a country through indicators such as life expectancy, infant mortality rate, proportion of underweight children under age five, and adult literacy rate for low-income, middle-income, and high-income countries. People who live in countries in the highest human development categories are, on average, better educated, will live longer, and will earn more. The adult literacy rate in the low-income countries is significantly lower than that of high-income countries, and for women the rate is even lower.

## LO5 What is modernization theory, and what are the four stages of economic development identified by Walt Rostow?

Modernization theory is a perspective that links global inequality to different levels of economic development and suggests that low-income economies can move to middle- and high-income economies by achieving self-sustained economic growth. Walt Rostow suggested that all countries go through four stages of economic development, with identical content, regardless of when these nations started the process of industrialization. The stages of economic development are as follows: the traditional stage, in which very little social change takes place and people do not think much about changing their current circumstances. The second stage is the take-off stage—a period of economic growth accompanied by a growing belief in individualism, competition, and achievement. In the third stage the country moves toward technological maturity. In the fourth and final stage the country reaches the phase of high mass consumption and a correspondingly high standard of living.

## LO6 What is dependency theory, and why is this theory often applied to newly industrializing countries?

Dependency theory states that global poverty can at least partially be attributed to the fact that the low-income countries have been exploited by the high-income countries. Whereas modernization theory focuses on how societies can reduce inequality through industrialization and economic development, dependency theorists see the greed of the rich countries as a source of increasing impoverishment of the poorer nations and their people.

**LO7** What is world systems theory, and what are the three major types of nations set forth in this theory?

According to world systems theory, the capitalist world economy is a global system divided into a hierarchy of three major types of nations: Core nations are dominant capitalist centers characterized by high levels of industrialization and urbanization, semiperipheral nations are more developed than peripheral nations but less developed than core nations, and peripheral nations are those countries that are dependent on core nations for capital, have little or no industrialization (other than what may be brought in by core nations), and have uneven patterns of urbanization.

**LO8** What is the new international division of labor theory, and how might it be useful in the twenty-first century?

The new international division of labor theory is based on the assumption that commodity production is split into fragments that can be assigned to whichever part of the world can provide the most profitable combination of capital and labor. This division of labor has changed the pattern of geographic specialization between countries, whereby high-income countries have become dependent on low-income countries for labor. The low-income countries provide transnational corporations with a situation in which they can pay lower wages and taxes, and face fewer regulations regarding workplace conditions and environmental protection.

# key TERMS

core nations  250

dependency theory  248

global stratification  234

modernization theory  247

new international division
   of labor theory  250

peripheral nations  250

semiperipheral nations  250

world systems theory  249

# questions for CRITICAL THINKING

1. You have decided to study global wealth and poverty. How would you approach your study? What research methods would provide the best data for analysis? What might you find if you compared your research data with popular presentations—such as films and advertising—of everyday life in low- and middle-income countries?

2. How would you compare the lives of poor people living in the low-income nations of the world with those in central cities and rural areas of the United States? In what ways are their lives similar? In what ways are they different?

3. Should U.S. foreign policy include provisions for reducing poverty in other nations of the world? Should U.S. domestic policy include provisions for reducing poverty in the United States? How are these issues similar? How are they different?

4. Using the theories discussed in this chapter, devise a plan to alleviate global poverty. Assume that you have the necessary wealth, political power, and other resources necessary to reduce the problem. Share your plan with others in your class, and create a consolidated plan that represents the best ideas and suggestions presented.

# 9 RACE AND ETHNICITY

## LEARNING OBJECTIVES

**LO1** Distinguish between the terms *race* and *ethnicity*.

**LO2** Explain how racial and ethnic classifications continue to change in the United States.

**LO3** Define *prejudice, stereotypes, racism, scapegoat,* and *discrimination*.

**LO4** Compare the major sociological perspectives on race and ethnic relations.

**LO5** Discuss the unique historical experiences of Native Americans and WASPs in the United States.

**LO6** Describe how slavery, segregation, lynching, and persistent discrimination have uniquely affected the African American experience in this country.

**LO7** Identify the major categories of Asian Americans and describe their historical and contemporary experiences.

**LO8** Describe the unique experiences of Latinos/as (Hispanics) and Middle Eastern Americans in the United States.

Lisette Le Bon/SuperStock

257

# A New Multiracial Awareness

**My whole life, up until I was eight years old, I thought that I was white.** Looking back, I thought how did I think that, because all my brothers and sisters have blond hair and blue eyes, and I obviously don't. I don't consider myself to be half anything unless someone asks what I'm really composed of because I don't see myself as relating sometimes to this and sometimes to that—I'm just me!

—Laura Wood, a University of Maryland student, explaining how she perceives her mixed-race identity as the daughter of a black father and a white mother (qtd. in Orr and Saulny, 2011)

In the United States, having more than one racial background, formerly a rare situation, is becoming more and more common.

**The Federal Department of Education would categorize Michelle López-Mullins** a university student who is of Peruvian, Chinese, Irish, Shawnee, and Cherokee descent—as "Hispanic." But the National Center for Health Statistics, the government agency that tracks data on births and deaths, would see her as "Asian" racially and Hispanic ethnically. And what does Ms. López-Mullins's birth certificate from the state of Maryland say? It doesn't mention her race. Ms. López-Mullins, twenty, usually marks "other" on

Look around you. Today's college campus is a reflection of important changes that are occurring in the racial and ethnic composition of the United States, including a dramatic increase in the number of individuals who identify themselves as "mixed race." According to Susan Saulny (2011a: A1), "The crop of students moving through college right now includes the largest group of mixed-race people ever to come of age in the United States, and they are only the vanguard: the country is in the midst of a demographic shift driven by immigration and intermarriage."

Similar to the changing face of many college campuses, college sports and professional sports have a growing number of athletes who self-identify as "mixed race." Professional golfer Tiger Woods, who has received extensive publicity about his professional and private life, is one of the best-known "mixed-race" athletes because of media coverage regarding the fact that he is one-half African American, one-fourth Chinese, and one-fourth Native American (Mixedfolks.com, 2011).

For more than a century, most people in the United States thought of themselves in singular racial terms, and on official documents they were identified as being of one racial or ethnic category. However, this country is in the middle of a major demographic shift because multiracial and multiethnic Americans are rapidly growing in number, and the U.S. Census Bureau and other governmental agencies are increasingly grouping them together as "mixed race."

How has this happened? Today, one in seven new marriages is between spouses of different races or ethnicities, and immigration in the past two decades has brought large numbers of new residents, particularly from Asia and Mexico, to this country. In the late 1990s the U.S. Census Bureau abandoned the idea of allowing individuals to classify themselves simply as "multiracial" and instead gave people the option to check more than one racial or ethnic category on census and other government forms. As a result, many young people today no longer adhere to "old" color lines that have defined this country for generations: They favor a much more fluid sense of identity, as indicated by the situations of Laura Wood and Michelle López-Mullins.

In this chapter we examine prejudice, discrimination, sociological perspectives on race and ethnicity, surveys these days, but when she filled out a census form last year, she chose Asian, Hispanic, Native American, and white.

—Susan Saulny (2011b: A1), a *New York Times* journalist, describing the numerous ways that persons such as Michelle López-Mullins are classified in the United States, a rapidly changing, multiracial nation

## How Much Do You Know About Race, Ethnicity, and Sports?

| True | False | |
|------|-------|---|
| T | F | 1. Some deeply held ideas about race and racial differences are often expressed in our beliefs about sports and athletic ability. |
| T | F | 2. Most elite sprinters are of West African descent because they possess a greater percentage of fast-twitch muscles that can be attributed to a specific racial gene. |
| T | F | 3. The chances (statistical odds) of becoming a professional athlete are better than the chance of getting struck by lightning or writing a *New York Times* best seller. |
| T | F | 4. In the twenty-first century the number of African Americans in head football coaching positions in the 120 Football Bowl Subdivision (FBS), formerly known as the NCAA Division I-A, has increased significantly. |

| True | False | |
|------|-------|---|
| T | F | 5. Today, all university presidents and athletic directors at the 125 FBS (NCAA Division I) colleges are white. |
| T | F | 6. African Americans account for more than half of all FBS football student-athletes. |
| T | F | 7. All of the conference commissioners of the FBS are white men. |
| T | F | 8. Visible racial–ethnic diversity on college sports teams has little effect on how younger athletes perceive of their future educational and sporting opportunities in college and beyond. |

Answers on page 260.

and commonalities and differences in the experiences of racial and ethnic groups in the United States. In the process, sports is used as an example of the effects of race and ethnicity on people's lives because this area of social life shows us how, for more than a century, people who have been singled out for negative treatment on the basis of their perceived race or ethnicity have sought to overcome prejudice and discrimination through their determination in endeavors that can be judged based on objective standards ("What's the final score?") rather than subjective standards ("Do I like this person based on characteristics such as race or ethnicity?"). Before reading on, test your knowledge about race, ethnicity, and sports by taking the "Sociology and Everyday Life" quiz.

## Race and Ethnicity

What is race? Some people think it refers to skin color (the Caucasian "race"); others use it to refer to a religion (the Jewish "race"), nationality (the British "race"), or the entire human species (the human "race") (Marger, 2012). Popular usages of the word

have been based on the assumption that a race is a grouping or classification based on *genetic* variations in physical appearance, particularly skin color. However, social scientists, biologists, and genetic anthropologists dispute the idea that biological race is a meaningful concept. Researchers with the Human Genome Project, which was commissioned to map all of the genes on the 23 pairs of human chromosomes and to sequence the 3.1 billion DNA base pairs that make up the chromosomes, made this statement about genes and race:

DNA studies do not indicate that separate classifiable subspecies (races) exist within modern humans. While different genes for physical traits such as skin and hair color can be identified between individuals, no consistent patterns of genes across the human genome exist to distinguish one race from another. There also is no genetic basis for divisions of human ethnicity. People who have lived in the same geographic region for many generations may have some alleles in common [an allele is one member of a pair or series of genes that occupy a specific position on a specific

**259**

# ANSWERS to the Sociology Quiz on Race, Ethnicity, and Sports

**1. True.** Many people continue to believe that differences in athletic abilities can largely be attributed to racial physiology (such as in the term "natural athlete") rather than personal attributes such as "hard worker" or "good decision maker."

**2. False.** This is an example of the "white men can't jump" assumption that links racial differences to athletic ability; however, no "racial" gene has been found to account for differences in muscle fiber in black runners and white runners. (Human muscles contain a genetically determined mixture of both slow and fast fiber types. Fast-twitch fibers produce force at a higher rate for short bursts of speed and can be an asset to a short-distance sprinter. Slow-twitch fibers produce a lower rate of force that lasts longer and can be an asset to a distance runner.)

**3. False.** A person has a better chance of getting struck by lightning or writing a *New York Times* best seller (or marrying a millionaire) than becoming a professional athlete, where the current odds are about 24,550 to 1.

**4. True.** Some changes have occurred in the number of African Americans in football head coaching positions. In 2013 there were 13 African American head coaches out of 125 FBS institutions. This figure is down from 15 in 2012 and an all-time high of 19 in 2011. There was one Latino head coach and two Asian American/Pacific Islander head coaches.

**5. False.** In the 2013–2014 academic year, nearly 89 percent of all FBS presidents and 85 percent of athletic directors were white. At one time, all presidents and athletic directors of NCAA Division I schools were white (non-Hispanic).

**6. True.** In the 2013 season the percentage of African American football student-athletes was 54.4 percent, as compared to nearly 42 percent for white players, 2 percent each for Latino and Asian American, and 0.1 percent for Native American players.

**7. True.** All eleven FBS conference commissioners, who make major decisions about how their conferences are run, are white men, despite the relatively large number of student-athletes and coaches who are persons of color and/or women.

**8. False.** Many children and young adolescents look upon college athletes as role models. When younger people see players on the team who look like them, they may be more disposed to believe that they too could attend college and participate in sports.

Source: Based on Lapchick, 2014. © Cengage Learning

chromosome], but no allele will be found in all members of one population and in no members of any other. (genomics.energy.gov, 2007)

The idea of race has little meaning in a biological sense because of the enormous amount of interbreeding that has taken place within the human population. For these reasons, sociologists sometimes place "race" in quotation marks to show that categorizing individuals and population groups on biological characteristics is neither accurate nor based on valid distinctions between the genetic makeup of differently identified "races."

Today, sociologists emphasize that race is a *socially constructed reality,* not a biological one. Race as a *social construct* means that races as such do not actually exist but that some groups are still racially defined because the *idea* persists in many people's minds that races are distinct biological categories with physically distinguishable characteristics and a shared common cultural heritage. The process of creating a socially constructed reality involves three key activities: *collective agreement, imposition,* and *acceptance of a specific construction* (Lusca, 2008).

*Collective agreement* means that people jointly agree on the idea of race and that they accept that it exists as an important component in how we describe or explain an individual's experiences in everyday life. Examples of collective agreement include a widely held acceptance of the view that "racial differences affect people's athletic ability" or of the assumption that "physical differences based on race cause cultural differences among various distinct categories of people."

*Imposition* refers to the fact that throughout much of human history, the notion of race has been defined by members of dominant groups who have the power to establish a system that hierarchically organizes racial categories (as superior or inferior, for example) to establish and maintain permanent status differentials among individuals and groups. These differences are demonstrated by the level of access that dominant- and subordinate-group members have to necessary and desired goods and services, such as education, housing, employment, health care, and legal services.

Finally, *acceptance of a specific construction* means that ideas pertaining to race become so widely accepted that they are embedded in law and social customs in a society and become much more difficult to change or eliminate. When a significant *number* of people, or a number of *significant* people, accept a social construction as absolute and real, the prevailing group typically imposes its beliefs and practices upon others through tradition and law. Over time, ideas about race, inadequate or false though they may be, are passed on from generation to generation.

In sum, the social significance that people accord to race is more important than any biological differences that might exist among people who are placed in arbitrary categories. Although race does not exist in an objective way, it does have *real* consequences and effects in the social world.

## Comparing Race and Ethnicity

**LO1** A *race* is a category of people who have been singled out as inferior or superior, often on the basis of real or alleged physical characteristics such as skin color, hair texture, eye shape, or other subjectively selected attributes (Feagin and Feagin, 2012). Racial categories identified by the U.S. Census Bureau include white, black, Asian or Pacific Islander, and American Indian or Alaska Native.

As compared with race, *ethnicity* refers to one's cultural background or national origin. An **ethnic group is a collection of people distinguished, by others or by themselves, primarily on the basis of cultural or nationality characteristics** (Feagin and Feagin, 2012) (Figure 9.1). Ethnic groups share five main characteristics: (1) *unique cultural traits,* such as language, clothing, holidays, or religious practices; (2) *a sense of community;* (3) *a feeling of ethnocentrism;* (4) *ascribed membership from birth;* and (5) *territoriality,* or the tendency to occupy a distinct geographic area (such as Little Italy or Little Moscow) by choice and/or for self-protection. Examples of ethnic groups include Jewish Americans, Irish Americans, Italian Americans, and Russian Americans. Many people mistakenly believe that the classification "Hispanic" or "Latino/a" is a "race." According to the U.S. Census Bureau,

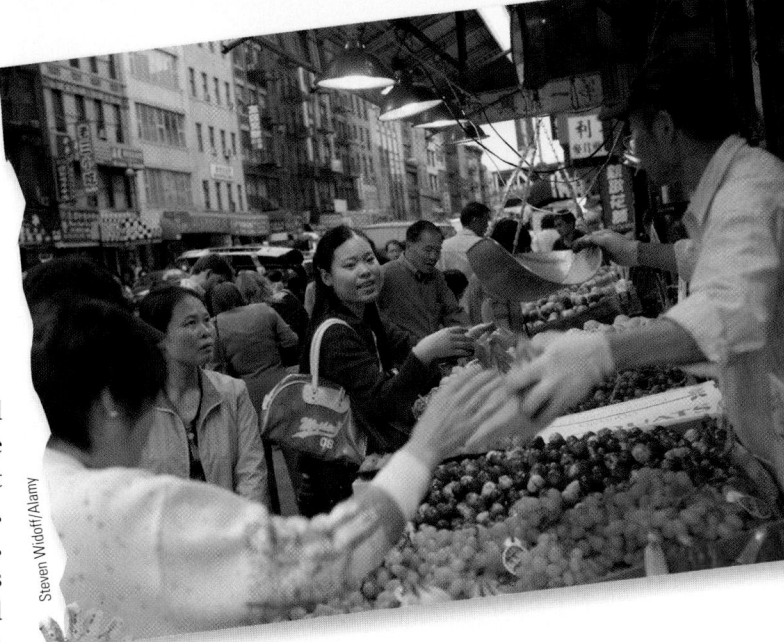

Steven Widoff/Alamy

**FIGURE 9.1** New York City's Chinatown is an ethnic enclave where people participate in social interaction with other individuals in their ethnic group and feel a sense of shared identity. Ethnic enclaves provide economic and psychological support for recent immigrants as well as for those who were born in the United States.

Hispanic or Latino/a is an ethnicity that refers to a person of Cuban, Mexican, Puerto Rican, South or Central American, or other Spanish culture or origin (Ennis, Rios-Vargas, and Albert, 2011). Persons who are Hispanic or Latino/a can be of any race.

Although some people do not identify with any ethnic group, others participate in social interaction with individuals in their ethnic group and feel a sense of common identity based on cultural characteristics such as language, religion, or politics. However, ethnic groups are not only influenced by their own history but also by patterns of ethnic domination and subordination in societies. It is important to note that terminology pertaining to racial–ethnic groups is continually in flux and that people within the category as well as outsiders often contest these changes. Examples include the use of *African American,* as compared to *black,* and *Hispanic,* as compared to *Latino/a.*

**race** a category of people who have been singled out as inferior or superior, often on the basis of real or alleged physical characteristics such as skin color, hair texture, eye shape, or other subjectively selected attributes.

**ethnic group** a collection of people distinguished, by others or by themselves, primarily on the basis of cultural or nationality characteristics.

## The Social Significance of Race and Ethnicity

Race and ethnicity take on great social significance because how people act in regard to these terms drastically affects other people's lives, including what opportunities they have, how they are treated, and even how long they live. According to the now-classic work on the effects of race by the sociologists Michael Omi and Howard Winant (1994: 158), race "permeates every institution, every relationship, and every individual" in the United States:

> As we . . . compare real estate prices in different neighborhoods, select a radio channel to enjoy while we drive to work, size up a potential client, customer, neighbor, or teacher, stand in line at the unemployment office, or carry out a thousand other normal tasks, we are compelled to think racially, to use the racial categories and meaning systems into which we have been socialized. (Omi and Winant, 1994: 158)

Historically, stratification based on race and ethnicity has pervaded all aspects of political, economic, and social life. Consider sports as an example. Throughout the early history of the game of baseball, many African Americans had outstanding skills as players but were categorically excluded from Major League teams because of their skin color. Even in 1947, after Jackie Robinson broke the "color line" to become the first African American in the Major Leagues, his experience was marred by racial slurs, hate letters, death threats against his infant son, and assaults on his wife (Ashe, 1988; Peterson, 1992/1970). With some professional athletes from diverse racial–ethnic categories having multimillion-dollar contracts and lucrative endorsement deals, it is easy to assume that racism in sports—as well as in the larger society—is a thing of the past. However, this *commercialization* of sports does not mean that racial prejudice and discrimination no longer exist (Coakley, 2009).

## Racial Classifications and the Meaning of Race

**LO2** If we examine racial classifications throughout history, we find that in ancient Greece and Rome, one's race was the group to which she or he belonged, associated with an ancestral place and culture. From the Middle Ages until about the eighteenth century, a person's race was based on family and ancestral ties, in the sense of a *line,* or ties to a national group. During the eighteenth century, physical differences such as the darker skin hues of Africans became associated with race, but racial divisions were typically based on differences in religion and cultural tradition rather than on human biology. With the intense (though misguided) efforts that surrounded the attempt to justify black slavery and white dominance in all areas of life during the second half of the nineteenth century, *races* came to be defined as distinct biological categories of people who were not all members of the same family but who shared inherited physical and cultural traits that were alleged to be different from those traits shared by people in other races. Hierarchies of races were established, placing the "white race" at the top, the "black race" at the bottom, and others in between.

However, racial classifications in the United States have changed over the past century. If we look at U.S. Census Bureau classifications, for example, we can see how the meaning of race continues to change. First, race is defined by perceived skin color: white or nonwhite. Whereas one category exists for "whites" (who vary considerably in actual skin color and physical appearance), all of the remaining categories are considered "nonwhite."

Second, categories of official racial classifications may (over time) create a sense of group membership or "consciousness of kind" for people within a somewhat arbitrary classification. When people of European descent were classified as "white," some began to see themselves as different from "nonwhite." Consequently, Jewish, Italian, and Irish immigrants may have felt more a part of the Northern European white mainstream in the late eighteenth and early nineteenth centuries. Whether Chinese Americans, Japanese Americans, Korean Americans, and Filipino Americans come to think of themselves collectively as "Asian Americans" because of official classifications remains to be seen.

Third, racial purity is assumed to exist. Prior to the 2000 census, for example, the true diversity of the U.S. population was not revealed in census data because multiracial individuals were forced to either select a single race as being their "race" or to select the vague category of "other." Census 2000 made it possible—for the first time—for individuals to classify themselves as being of more than one race (see "Census Profiles: Percentage Distribution of Persons Reporting Two or More Races"). In the 2010 census, nine million people in the United States—about 3 percent of the total population—identified themselves as multiracial (Humes, Jones, and Ramirez, 2011). Between 2000 and 2010, the percentage of Americans identifying as more than one race increased by 32 percent. Among U.S. children, the mixed-race population increased by nearly 50 percent, making mixed-race children the fastest-growing youth

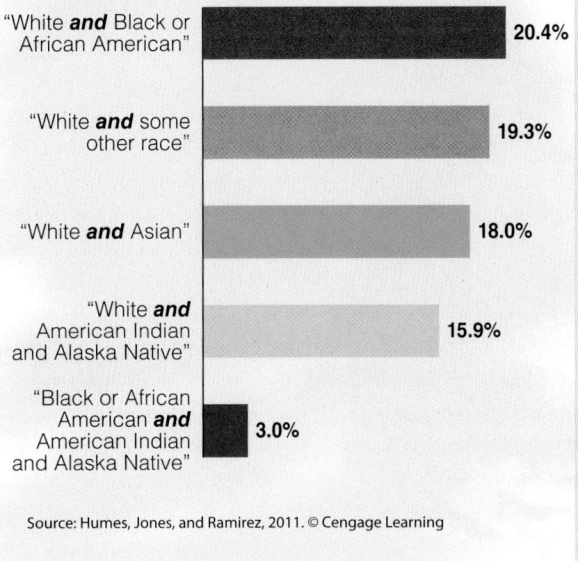
group in the United States (Saulny, 2011a). With one in seven new marriages in the United States involving spouses of different races or ethnicities, the multiracial population is likely to continue to increase (Passel, Wang, and Taylor, 2010).

Multiracial individuals do not always identify as such. Although President Barack Obama is the product of an interracial couple (mother is white, father black), for the 2010 Census Obama checked only one box: black (Saulny, 2011b). For many multiracial individuals, choosing a racial–ethnic identity is not simple. Consider Ms. López-Mullins (introduced at the beginning of the chapter), a mixed-race college student whose father is both Asian and Latino, and whose mother, with her long blonde hair, is mostly European in ancestry but is mixed with some Cherokee and Shawnee Native American. In grade school, Ms. López-Mullins was frequently asked "What are you?" and "Where are you from?" As she explains,

I hadn't even learned the word "Hispanic" until I came home from school one day and asked my dad what I should refer to him as, to express what I am. . . . Growing up with my parents, I never thought we were different from any other family. . . . I was always having to explain where my parents are from because just saying "I'm from Takoma Park, Maryland," was not enough. . . . Saying "I'm an American" wasn't enough. . . . Now when people ask what I am, I say, "How much time do you have?" . . . Race will not automatically tell you my story. (qtd. in Saulny, 2011b)

When asked which box she checks on forms such as the census, Ms. López-Mullins replied, "Hispanic, white, Asian American, Native American. . . . I'm pretty much checking everything" (qtd. in Saulny, 2011b).

As noted earlier, the way people are classified remains important because such classifications affect their access to employment, education, housing, social services, federal aid, and other public and private goods and services that might be available to them.

## Dominant and Subordinate Groups

The terms *majority group* and *minority group* are widely used, but their meanings are less clear as the composition of the U.S. population continues to change. Accordingly, many sociologists prefer the terms *dominant* and *subordinate* to identify power relationships that are based on perceived racial, ethnic, or other attributes and identities. To sociologists, a **dominant group** **is a racial or ethnic group that has the greatest power and resources in a society** (Feagin and Feagin, 2012). In the United States, whites with Northern European ancestry (often referred to as Euro-Americans, white Anglo-Saxon Protestants, or WASPs) have been considered to be the dominant group for many years. A **subordinate group** **is one whose members, because of physical or cultural characteristics, are disadvantaged and subjected to unequal treatment and discrimination by the dominant group.** Historically, African Americans and other persons of color have been considered to be subordinate-group members, particularly when they are from lower-income categories. Some individuals who are identified with subordinate groups

**dominant group** a racial or ethnic group that has the greatest power and resources in a society.

**subordinate group** a group whose members, because of physical or cultural characteristics, are disadvantaged and subjected to unequal treatment and discrimination by the dominant group.

**FIGURE 9.2** Contemporary prejudice and discrimination cannot be understood without taking into account the historical background. School integration in the 1950s was accomplished despite white resistance. Today, integration in education, housing, and many other areas of social life remains a pressing social issue.

gain support from each other by living in an ethnic enclave where they share a common identity.

It is important to note that, in the sociological sense, the word *group* as used in these two terms is misleading because people who merely share ascribed racial or ethnic characteristics do not constitute a group. However, the terms *dominant group* and *subordinate group* do give us a way to describe relationships of advantage/disadvantage and power/exploitation that exist in contemporary nations.

## Prejudice

**LO3** Although there are various meanings of the word, sociologists define **prejudice as a negative attitude based on faulty generalizations about members of specific racial, ethnic, or other groups**. The term *prejudice* is from the Latin words *prae* ("before") and *judicium* ("judgment"), which means that people may be biased either for or against members of other groups even before they have had any contact with them. Although prejudice can be either *positive* (bias in favor of a group—often our own) or *negative* (bias against a group—one we deem less worthy than our own), it most often refers to the negative attitudes that people may have

about members of other racial or ethnic groups (Figure 9.2).

### Stereotypes

Prejudice is rooted in ethnocentrism and stereotypes. When used in the context of racial and ethnic relations, *ethnocentrism* refers to the tendency to regard one's own culture and group as the standard—and thus superior—whereas all other groups are seen as inferior. Ethnocentrism is maintained and perpetuated by **stereotypes—overgeneralizations about the appearance, behavior, or other characteristics of members of particular categories.**

Although stereotypes can be either positive or negative, examples of negative stereotyping abound in sports. Think about the Native American names, images, and mascots used by sports teams such as the Atlanta Braves, Cleveland Indians, and Washington Redskins. Members of Native American groups have been actively working to eliminate the use of stereotypic mascots (with feathers, buckskins, beads, spears, and "warpaint"), "Indian chants," and gestures (such as the "tomahawk chop"), which they claim trivialize and exploit Native American culture. In 2010 Wisconsin passed a law that allows residents of a school district to lodge complaints regarding race-based names of school sports teams in an effort to eliminate any hostile environment that might be created by the use of such names and imagery (Keen, 2010). College and university sports teams with Native American names and logos also remain the subject of controversy in the twenty-first century. In 2005 the NCAA prohibited the use of Native American mascots and nicknames in bowl games and other post-season competition (Keen, 2010). According to sociologist Jay Coakley (2009), the use of stereotypes and words such as *redskin* symbolizes a lack of understanding of the culture and heritage of native peoples and is offensive to many Native Americans. Although some people see these names and activities as "innocent fun," others view them as a form of racism.

## Racism

What is racism? **Racism is a set of attitudes, beliefs, and practices that is used to justify the superior treatment of one racial or ethnic group and the inferior treatment of another racial or ethnic group.** The world has seen a long history of racism: It can be traced from the earliest civilizations. At various times throughout U.S. history, various categories of people, including Irish Americans, Italian Americans, Jewish Americans, African Americans, and Latinos/as, have been the objects of racist ideology.

Racism may be overt or subtle. *Overt racism* is blatant and may take the form of public statements about the "inferiority" of members of a racial or ethnic group. In sports, for example, calling a player of color a derogatory name, participating in racist chanting during a sporting event, and writing racist graffiti in a team's locker room are all forms of overt racism. These racist actions are blatant, but subtle forms of racism are often hidden from sight and more difficult to recognize. Examples of *subtle racism* in sports include those descriptions of African American athletes which suggest that they have "natural" abilities and are better suited for team positions requiring speed and agility. By contrast, whites are described as having the intelligence, dependability, leadership, and decision-making skills needed in positions requiring higher levels of responsibility and control.

Racism tends to intensify in times of economic uncertainty and high rates of immigration. Recently, relatively high rates of immigration in nations such as the United States, Canada, England, France, and Germany have been accompanied by an upsurge in racism and racial conflict (see "Sociology in Global Perspective").

Michael Greenlar/The Image Works

**FIGURE 9.3** According to the frustration–aggression hypothesis, members of white supremacy groups such as the Ku Klux Klan often use members of subordinate racial and ethnic groups as scapegoats for societal problems over which they have no control.

## Theories of Prejudice

Are some people more prejudiced than others? To answer this question, some theories focus on how individuals may transfer their internal psychological problem onto an external object or person. Others look at factors such as social learning and personality types.

The *frustration–aggression hypothesis* states that people who are frustrated in their efforts to achieve a highly desired goal will respond with a pattern of aggression toward others (Dollard et al., 1939). The object of their aggression becomes the **scapegoat—a person or group that is incapable of offering resistance to the hostility or aggression of others** (Marger, 2012) (Figure 9.3). Scapegoats are often used as

**prejudice** a negative attitude based on faulty generalizations about members of specific racial, ethnic, or other groups.

**stereotypes** overgeneralizations about the appearance, behavior, or other characteristics of members of particular categories.

**racism** a set of attitudes, beliefs, and practices that is used to justify the superior treatment of one racial or ethnic group and the inferior treatment of another racial or ethnic group.

**scapegoat** a person or group that is incapable of offering resistance to the hostility or aggression of others.

# SOCIOLOGY *in Global Perspective*

## Racism in European Football

**News Bulletin:** AC Milan player Kevin-Prince Boateng leads his team off the field to protest racial abuse at the hands of the crowd for the opposing team. Boateng kicked the ball into the crowd and walked off with his teammates following him, thus denying the crowd the European soccer (football) game they came to see (reported in *The Guardian*, January 4, 2013). Afterwards, Boateng tweeted: "Shame that these things still happen . . . #StopRacismforever" (CNN.com, 2013).

**B**ut the experience of Boateng's team was not an isolated incident: Racism in European football has existed for many years, as further evidenced by the following comment from Djibril Cisse, a French player born in Africa's Ivory Coast, describing the racism and abuse that he has endured (qtd. in farenet.org, 2011):

> I was their target. Some of them hit me and I retaliated to defend myself. It's a disgrace. I'm not going to let things lie. . . . This season the atmosphere in the stands has got worse. Twice

Mark Eite/Aflo Co. Ltd./Alamy

Sporting events, such as soccer in Europe and football in the United States, are sometimes tainted by the actions of fans who demonstrate racist behavior by their comments about or actions toward players, officials, or other fans. What can be done to reduce such harmful behavior?

---

substitutes for the actual source of the frustration. For example, members of subordinate racial and ethnic groups are often blamed for societal problems (such as unemployment or an economic recession) over which they have no control.

According to some symbolic interactionists, prejudice results from social learning; in other words, it is learned from observing and imitating significant others, such as parents and peers. Initially, children do not have a frame of reference from which to question the prejudices of their relatives and friends. When they are rewarded with smiles or laughs for telling derogatory jokes or making negative comments about outgroup members, children's prejudiced attitudes may be reinforced.

Psychologist Theodor W. Adorno and his colleagues (1950) concluded that highly prejudiced individuals tend to have an **authoritarian personality, which is characterized by excessive conformity, submissiveness to authority, intolerance, insecurity, a high level of superstition, and rigid, stereotypic thinking.** This type of personality is most likely to develop in a family environment in which dominating parents who are anxious about status use physical discipline but show very little love in raising their children (Adorno et al., 1950). Other scholars have linked prejudiced attitudes to traits such as submissiveness to authority, extreme anger toward outgroups, and conservative religious and political beliefs (Altemeyer, 1981, 1988; Weigel and Howes, 1985).

Whereas prejudice is an attitude, **discrimination involves actions or practices of dominant-group members (or their representatives) that have a harmful effect on members of a subordinate group.** Prejudiced attitudes do not always lead to discriminatory behavior. As shown in Figure 9.4, the sociologist Robert K. Merton (1949) identified four combinations of attitudes and responses. *Unprejudiced nondiscriminators* are not personally prejudiced and do not discriminate against others. For example, two players on a professional sports team may be best friends although they are of different races. *Unprejudiced discriminators* may have no personal prejudice but still engage in discriminatory behavior because of peer-group pressure or economic, political, or social interests. For example, in some sports a coach might feel no prejudice toward African American players but believe that white fans will accept only a certain percentage of people of color on the team. *Prejudiced nondiscriminators* hold personal already, I have been the victim of monkey chants, of racist insults. It was the same thing again on Saturday night. They managed to get inflatable bananas into the stadium that they were shaking around every time I touched the ball.

Soccer is Europe's equivalent of football in the United States. Sadly, UEFA (Union of European Football Association), which is European soccer's ruling body, has received numerous reports of racist behavior at soccer matches throughout Europe, including black footballers and fans being abused, harassed, and attacked. Accusations of such racist behavior toward players of other nationalities or ethnic groups have been lodged against sports fans and some team members at various athletic events throughout Europe. At some soccer matches, fans shout racist comments during the game; throw bottles, cans, and other missiles at players; and pass out derogatory literature about other ethnic groups or nationalities outside the stadium.

In an effort to eliminate such behavior, organizations such as UEFA and FARE (Football Against Racism in Europe) have developed educational programs for schools and are attempting to educate the general public about the destructive nature of racism. UEFA has developed a specific plan that calls for actions on the part of soccer clubs such as having statements in printed programs and during public address announcements that racism will not be tolerated and that specific actions will be taken against those who engage in racist chanting or similar behavior. In fact, in 2012 the UEFA opened disciplinary proceedings against Serbia and England following brawls and racist chants at various matches.

Will these steps help to reduce racist behavior at European sporting events? Although these actions are a step in the right direction, over the past decade such behavior has continued at some games. But it is important that UEFA officials and others have acknowledged that a problem does exist and that the problem is not limited to a small handful of fans and players. In the final analysis, whether in Europe or other regions of the world, consistent efforts are needed by everyone to reduce racist behavior at football games and throughout everyday life.

## reflect & analyze

**Have you witnessed truly obnoxious verbal behavior by other spectators at a sporting event? How did you react? Also, some people believe that such behavior is constitutionally protected—a matter of "free speech." Do you agree? Why or why not?**

| | Prejudiced attitude? | Discriminatory behavior? |
|---|---|---|
| Unprejudiced nondiscriminator | No | No |
| Unprejudiced discriminator | No | Yes |
| Prejudiced nondiscriminator | Yes | No |
| Prejudiced discriminator | Yes | Yes |

**FIGURE 9.4** MERTON'S TYPOLOGY OF PREJUDICE AND DISCRIMINATION
Merton's typology shows that some people may be prejudiced but not discriminate against others. Do you think that it is possible for a person to discriminate against some people without holding a prejudiced attitude toward them? Why or why not?

© Cengage Learning

prejudices but do not discriminate because of peer pressure, legal demands, or a desire for profits. For example, a coach with prejudiced beliefs may hire an African American player to enhance the team's ability to win. Finally, *prejudiced discriminators* hold personal prejudices and actively discriminate against others. For example, a baseball umpire who is personally prejudiced against African Americans may intentionally call a play incorrectly based on that prejudice.

Discriminatory actions vary in severity, from the use of derogatory labels to violence against individuals and groups. The ultimate form of discrimination

**authoritarian personality** a personality type characterized by excessive conformity, submissiveness to authority, intolerance, insecurity, a high level of superstition, and rigid, stereotypic thinking.

**discrimination** actions or practices of dominant-group members (or their representatives) that have a harmful effect on members of a subordinate group.

occurs when people are considered to be unworthy to live because of their race or ethnicity. **Genocide is the deliberate, systematic killing of an entire people or nation.** Examples of genocide include the killing of thousands of Native Americans by white settlers in North America and the extermination of six million European Jews by Nazi Germany. More recently, the term *ethnic cleansing* has been used to define a policy of "cleansing" geographic areas by forcing persons of other races or religions to flee—or die.

Discrimination also varies in how it is carried out. Individuals may act on their own, or they may operate within the context of large-scale organizations and institutions, such as schools, churches, corporations, and governmental agencies. How does individual discrimination differ from institutional discrimination? **Individual discrimination consists of one-on-one acts by members of the dominant group that harm members of the subordinate group or their property.** Individual discrimination is often considered to be based on the prejudicial beliefs of bigoted individuals who overtly express those beliefs through discriminatory actions. For example, a college student may write racist graffiti on the dorm door of another student because the perpetrator possesses bigoted attitudes about the superiority or inferiority of others based on their race or ethnicity.

However, sociologists emphasize that individual discrimination is not purely *individual*. As sociologists in the past moved beyond studying individual racial discrimination, they found that a close relationship exists between individual discrimination and institutional discrimination because they are two aspects of the same phenomenon. Simply stated, when individuals engage in racial discrimination, their actions are *shaped* by structural racial inequalities in the existing society or social system, and, in turn, their actions *reinforce* existing large-scale patterns of discrimination, which we refer to as institutional discrimination. **Institutional discrimination consists of the day-to-day practices of organizations and institutions that have a harmful effect on members of subordinate groups.** For example, a bank might consistently deny loans to people of a certain race or a university might not accept additional Asian American students in its first-year class or medical school because of an institutional assumption that persons in this racial–ethnic category are already overrepresented at the school. However, it is important to note that institutional discrimination is carried out by the *individuals* who implement the policies and procedures of organizations.

Sociologist Joe R. Feagin has identified four major types of discrimination:

1. *Isolate discrimination* is harmful action intentionally taken by a dominant-group member against a member of a subordinate group. This type of discrimination occurs without the support of other members of the dominant group in the immediate social or community context. For example, a prejudiced judge may give harsher sentences to African American defendants but may not be supported by the judicial system in that action.

2. *Small-group discrimination* is harmful action intentionally taken by a limited number of dominant-group members against members of subordinate groups. This type of discrimination is not supported by existing norms or other dominant-group members in the immediate social or community context. For example, a small group of white students may hang nooses (that signify the practice of racial lynching in the past) on the door of an African American professor's office without the support of other students or faculty members.

3. *Direct institutionalized discrimination* is organizationally prescribed or community-prescribed action that intentionally has a differential and negative impact on members of subordinate groups. These actions are routinely carried out by a number of dominant-group members based on the norms of the immediate organization or community (Feagin and Feagin, 2012). In the past, the intentional exclusion of people of color from public accommodations was an example of this type of discrimination.

4. *Indirect institutionalized discrimination* refers to practices that have a harmful effect on subordinate-group members even though the organizationally or community-prescribed norms or regulations guiding these actions were initially established with no intent to harm. For example, special education classes were originally intended to provide extra educational opportunities for children with various types of disabilities. However, critics claim that these programs have amounted to racial segregation in many school districts.

5. Various types of racial and ethnic discrimination call for divergent remedies if we are to reduce discriminatory actions and practices in contemporary social life. Since the 1950s and 1960s, many U.S. sociologists have analyzed the complex relationship between prejudice and discrimination. Some have reached the conclusion that prejudice is difficult, if not seemingly impossible, to eradicate because of the deeply held racist beliefs and attitudes that are often passed on from person to person and from one generation to the next. However, the persistence of prejudicial attitudes and beliefs does not mean that racial

# Sociology WORKS!

## A Starting Point for Social Change

**Question:** Do you think it is possible to reduce racial and ethnic prejudice in the United States by attacking discrimination at the societal level?

**Answer:** Well, I think since it's individuals who hate each other and do mean stuff, they should work it out for themselves. We don't need the government telling us what to do or how to behave. Like, my dad runs a company, and he says that the government should "butt out" of our business.
   —"Brian," an introductory sociology student, stating why he believes that the government has "no business" trying to reduce discrimination (author's notes)

**B**rian's comment is typical of how many people feel about court rulings and government initiatives over the past 60 years that have sought to reduce the corrosive effects of racial and ethnic discrimination on American life. Based on the widely held axiom that "prejudice causes discrimination," many people argue that discrimination can be alleviated only through changing the attitudes of individuals. From this perspective, discrimination will go away over time if people are encouraged to shed their negative attitudes about members of other racial or ethnic groups. Discarding negative stereotypes about other groups and bringing to light the truth about popular myths regarding the superiority of one's own race, ethnic group, or nationality are widely seen as the best ways of bringing about positive social change. Diversity training sessions held in schools and at the workplace are a classic example of this approach.

Since the civil rights era in the 1960s, however, sociologists have demonstrated that discrimination can be tackled up front

and now—rather than waiting for prejudice to diminish—so that people in subordinate racial–ethnic categories can gain a greater measure of human dignity and a variety of opportunities that they otherwise would not have. For example, establishing social policies and laws to eliminate specific practices of segregation and discrimination in education, employment, housing, health care, law enforcement, and other areas of public life has served as a significant starting point for social change. By enacting legislation that prohibits discrimination based on race, ethnic origin, and color, our nation has sought to provide people with greater equality. These changes have, over time, reduced the prejudiced attitudes of many people on a wide variety of issues.

Although progress has been made in reducing some aspects of overt prejudice and institutional discrimination, racism is clearly not a thing of the past. However, if previous sociological research tells us anything about the future, it is that we must continue to tackle not only individual prejudices and discriminatory conduct but also the larger, societal patterns of discrimination—embedded in the organizations and institutions of which we are a part—that restrict freedom, opportunities, and quality of life for all people.

### reflect & analyze

**Using sports as an example, let's think about these questions: How might college sporting events serve to reduce prejudice and discrimination on campus and beyond? How might these same events serve to perpetuate negative stereotypes and popular myths about racial and ethnic "differences"? What do you think?**

and ethnic discrimination should be allowed to flourish until such time as prejudice is effectively eliminated. From this approach, discrimination must be aggressively tackled through demands for change and through policies that specifically target patterns of discrimination (see "Sociology Works!").

## Sociological Perspectives on Race and Ethnic Relations

**LO4** Symbolic interactionist, functionalist, and conflict analysts examine race and ethnic relations in different ways. Functionalists focus on the macrolevel intergroup processes that occur between members of dominant and subordinate groups in society. Conflict theorists analyze

power and economic differentials between the dominant group and subordinate groups. Symbolic interactionists examine how microlevel contacts between people may produce either greater racial tolerance or increased levels of hostility.

**genocide** the deliberate, systematic killing of an entire people or nation.

**individual discrimination** behavior consisting of one-on-one acts by members of the dominant group that harm members of the subordinate group or their property.

**institutional discrimination** the day-to-day practices of organizations and institutions that have a harmful effect on members of subordinate groups.

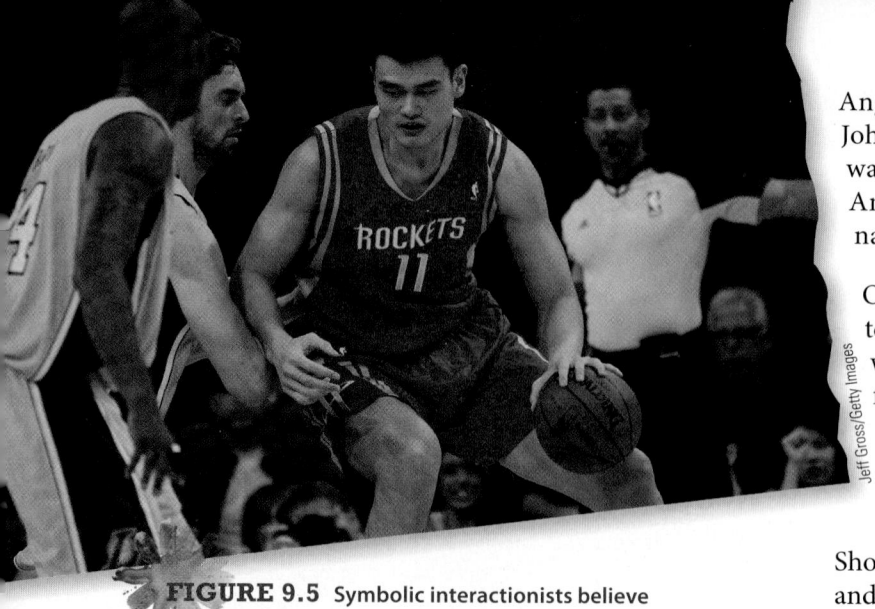

**FIGURE 9.5** Symbolic interactionists believe that intergroup contact can reduce stereotyping and prejudice if group members have equal status, pursue the same goals and cooperate to achieve them, and receive positive feedback when they interact with one another in positive ways. How do sports teams enable such interaction?

## Symbolic Interactionist Perspectives

What happens when people from different racial and ethnic groups come into contact with one another? Symbolic interactionists claim that intergroup contact may either intensify or reduce racial and ethnic stereotyping and prejudice, depending on the context. In the *contact hypothesis,* symbolic interactionists point out that contact between people from divergent groups should lead to favorable attitudes and behavior *when certain factors are present:* Members of each group must (1) have equal status, (2) pursue the same goals, (3) cooperate with one another to achieve their goals, and (4) receive positive feedback when they interact with one another in positive, nondiscriminatory ways (Figure 9.5). However, if these factors are not present, intergroup contact may lead to increased stereotyping and prejudice.

Of course, intergroup contact often does not include the four factors described above. What then happens when individuals meet someone who does not conform to their existing stereotype? According to symbolic interactionists, they frequently ignore anything that contradicts the stereotype, or they interpret the situation to support their prejudices. For example, a person who does not fit the stereotype may be seen as an exception—"You're not like other [persons of a particular race]." Conversely, when a person is seen as conforming to a stereotype, he or she may be treated simply as one of "you people." Former Los Angeles Lakers basketball star Earvin "Magic" Johnson (1992: 31–32) described how he was categorized along with all other African Americans when he was bused to a predominantly white school:

On the first day of [basketball] practice, my teammates froze me out. Time after time I was wide open, but nobody threw me the ball. At first I thought they just didn't see me. But I woke up after a kid named Danny Parks looked right at me and then took a long jumper. Which he missed.

I was furious, but I didn't say a word. Shortly after that, I grabbed a defensive rebound and took the ball all the way down for a basket. I did it again and a third time, too.

Finally Parks got angry and said, "Hey, pass the [bleeping] ball."

That did it. I slammed down the ball and glared at him. Then I exploded. "I *knew* this would happen!" I said. "That's why I didn't want to come to this [bleeping] school in the first place!"

"Oh, yeah? Well, you people are all the same," he said. "You think you're gonna come in here and do whatever you want? Look, hotshot, your job is to get the rebound. Let us do the shooting."

Eventually, Johnson and Parks were able to work out most of their differences. "There's nothing like winning to help people get along," Johnson explained (1992: 32). Although we might hope that nothing like this happens today, there is much evidence that discrimination occurs in many sports and social settings. Symbolic interactionist perspectives make us aware of the importance of intergroup contact and the fact that it may either intensify or reduce racial and ethnic stereotyping and prejudice.

## Functionalist Perspectives

How do members of subordinate racial and ethnic groups become a part of the dominant group? To answer this question, early functionalists studied immigration and patterns of dominant- and subordinate-group interactions.

**Assimilation** *Assimilation* **is a process by which members of subordinate racial and ethnic groups become absorbed into the dominant culture.** To some analysts, assimilation is functional because it contributes to the stability of society by minimizing group differences that might otherwise result in hostility and violence.

Assimilation occurs at several distinct levels, including the cultural, structural, biological, and psychological stages. *Cultural assimilation,* or *acculturation,* occurs when members of an ethnic group adopt dominant-group traits, such as language, dress, values, religion, and food preferences. Cultural assimilation in this country initially followed an "Anglo conformity" model; members of subordinate ethnic groups were expected to conform to the culture of the dominant white Anglo-Saxon population (Gordon, 1964). However, members of some groups refused to be assimilated and sought to maintain their unique cultural identity.

*Structural assimilation,* or *integration,* occurs when members of subordinate racial or ethnic groups gain acceptance in everyday social interaction with members of the dominant group. This type of assimilation typically starts in large, impersonal settings such as schools and workplaces, and only later (if at all) results in close friendships and intermarriage. *Biological assimilation,* or *amalgamation,* occurs when members of one group marry those of other social or ethnic groups. Biological assimilation has been more complete in some other countries, such as Mexico and Brazil, than in the United States.

*Psychological assimilation* involves a change in racial or ethnic self-identification on the part of an individual. Rejection by the dominant group may prevent psychological assimilation by members of some subordinate racial and ethnic groups, especially those with visible characteristics such as skin color or facial features that differ from those of the dominant group.

**Ethnic Pluralism** Instead of complete assimilation, many groups share elements of the mainstream culture while remaining culturally distinct from both the dominant group and other social and ethnic groups. **Ethnic pluralism is the coexistence of a variety of distinct racial and ethnic groups within one society.**

*Equalitarian pluralism,* or *accommodation,* is a situation in which ethnic groups coexist in equality with one another. Switzerland has been described as a model of equalitarian pluralism; more than six million people with French, German, and Italian cultural heritages peacefully coexist there. *Inequalitarian pluralism,* or *segregation,* exists when specific ethnic groups are set apart from the dominant group and have unequal access to power and privilege. **Segregation is the spatial and social separation of categories of people by race, ethnicity, class, gender, and/or religion.** Segregation may be enforced by law. *De jure segregation* refers

**FIGURE 9.6** Segregation exists when specific ethnic groups are set apart from the dominant group and have unequal access to power and privilege. What examples of segregation do you see today?

to laws that systematically enforced the physical and social separation of African Americans in all areas of public life. For example, Jim Crow laws legalized the separation of the races in public accommodations (such as hotels, restaurants, transportation, hospitals, jails, schools, churches, and cemeteries) in the southern United States after the Civil War (Feagin and Feagin, 2012) (Figure 9.6).

Segregation may also be enforced by custom. *De facto segregation*—racial separation and inequality enforced by custom—is more difficult to document than de jure segregation. For example, residential segregation is still prevalent in many U.S. cities; owners, landlords, real estate agents, and apartment managers often use informal mechanisms to maintain their properties for "whites only." Even middle-class people of color find that racial polarization is fundamental to the residential layout of many cities.

Although functionalist explanations provide a description of how some early white ethnic immigrants assimilated into the cultural mainstream,

**assimilation** a process by which members of subordinate racial and ethnic groups become absorbed into the dominant culture.

**ethnic pluralism** the coexistence of a variety of distinct racial and ethnic groups within one society.

**segregation** the spatial and social separation of categories of people by race, ethnicity, class, gender, and/or religion.

they do not adequately account for the persistent racial segregation and economic inequality experienced by people of color.

## Conflict Perspectives

Conflict theorists focus on economic stratification and access to power in their analyses of race and ethnic relations. Some emphasize the caste-like nature of racial stratification, others analyze class-based discrimination, and still others examine internal colonialism and gendered racism.

**The Caste Perspective** The caste perspective views racial and ethnic inequality as a permanent feature of U.S. society. According to this approach, the African American experience must be viewed as different from that of other racial or ethnic groups. African Americans were the only group to be subjected to slavery; when slavery was abolished, a caste system was instituted to maintain economic and social inequality between whites and African Americans (Feagin and Feagin, 2012).

The caste system was strengthened by *antimiscegenation laws,* which prohibited sexual intercourse or marriage between persons of different races. Most states had such laws, which were later expanded to include relationships between whites and Chinese, Japanese, and Filipinos. Although these laws were declared unconstitutional in 1967, roughly one in ten (11 percent) of U.S. adults say they disapprove of marriage between blacks and whites (Gallup Organization, 2011b).

**Class Perspectives** Although the caste perspective points out that racial stratification may be permanent because of structural elements such as the law, it has been criticized for not examining the role of class in perpetuating racial inequality. Class perspectives emphasize the role of the capitalist class in racial exploitation. Based on early theories of race relations by the African American scholar W. E. B. Du Bois, the sociologist Oliver C. Cox (1948) suggested that African Americans were enslaved because they were the cheapest and best workers the owners could find for heavy labor in mines and on plantations. Thus, the profit motive of capitalists, not skin color or racial prejudice, accounts for slavery.

Sociologists have also debated the relative importance of class and race in explaining the unequal life chances of African Americans. Sociologists William Julius Wilson and Richard P. Taub (2007) have suggested that race, cultural factors, social psychological variables, and social class must *all* be taken into account in examining the life chances of

"inner-city residents." Their analysis focuses on how race, ethnicity, and class tensions are all important in assessing how residents live their lives in four low-income Chicago neighborhoods and why this finding is important for the rest of America as well.

How do conflict theorists view the relationship among race, class, and sports? Simply stated, sports reflects the interests of the wealthy and powerful. At all levels, sports exploits athletes (even highly paid ones) in order to gain high levels of profit and prestige for coaches, managers, and owners. In particular, African American athletes and central-city youths are exploited by the message of rampant consumerism. Many are given the unrealistic expectation that sports can be a ticket out of the ghetto or barrio. If they try hard enough (and wear the right athletic gear), they too can become wealthy and famous.

**Internal Colonialism** Why do some racial and ethnic groups continue to experience subjugation after many years? According to the sociologist Robert Blauner (1972), groups that have been subjected to internal colonialism remain in subordinate positions longer than groups that voluntarily migrated to the United States. ***Internal colonialism occurs when members of a racial or ethnic group are conquered or colonized and forcibly placed under the economic and political control of the dominant group.*** This idea has been so widely received that it is often referred to as the "Blauner hypothesis" and is still used in research.

In the United States, indigenous populations (including groups known today as Native Americans and Mexican Americans) were colonized by Euro-Americans and others who invaded their lands and conquered them. In the process, indigenous groups lost property, political rights, aspects of their culture, and often their lives. The capitalist class acquired cheap labor and land through this government-sanctioned racial exploitation. The effects of past internal colonialism are reflected today in the number of Native Americans who live on government reservations and in the poverty of Mexican Americans who lost their land and had no right to vote (Figure 9.7).

The internal colonialism perspective is rooted in the historical foundations of racial and ethnic inequality in the United States. However, it tends to view all voluntary immigrants as having many more opportunities than do members of colonized groups. Thus, this model does not explain the continued exploitation of some immigrant groups, such as the Chinese, Filipinos, Cubans, Vietnamese, and Haitians, and the greater acceptance of others, primarily those from Northern Europe.

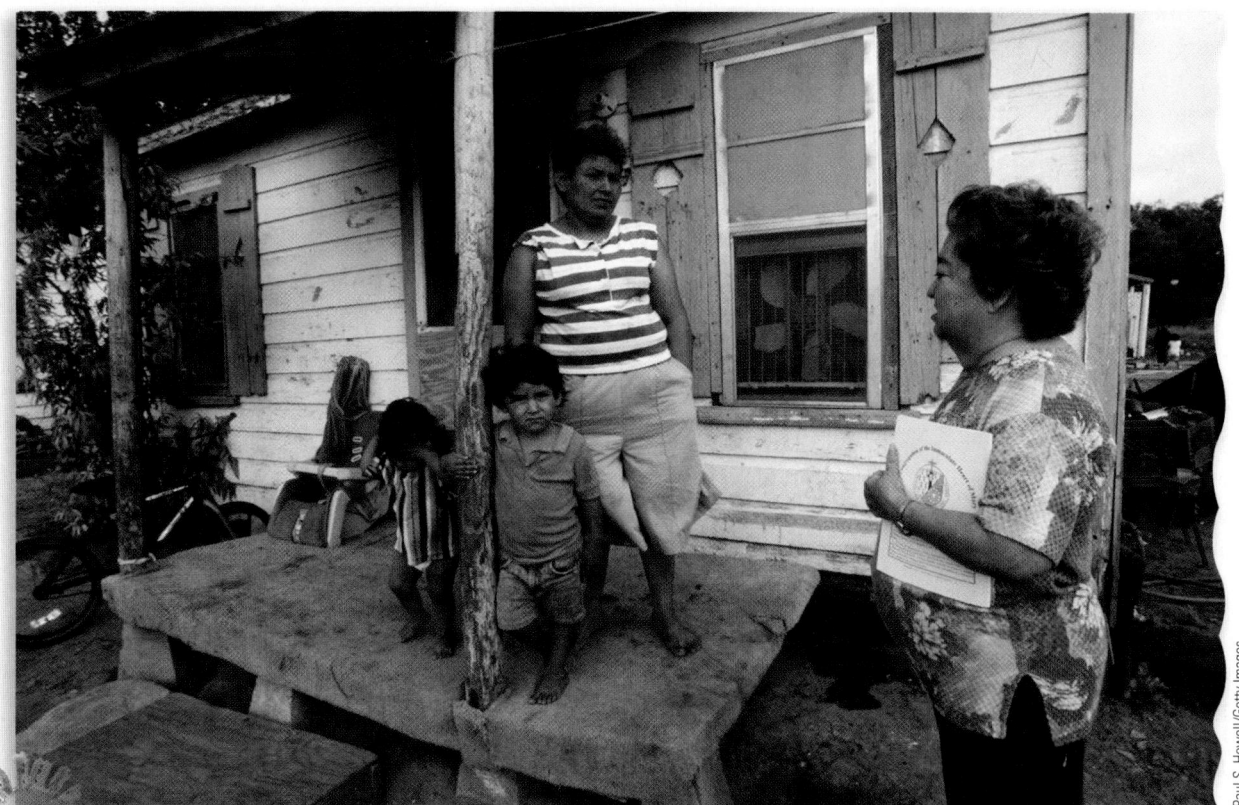

**FIGURE 9.7** Grinding poverty is a pressing problem for families living along the border between the United States and Mexico. Economic development has been limited in areas where *colonias* such as this one are located, and the wealthy have derived far more benefit than others from recent changes in the global economy. How might the concept of internal colonialism be used to explain the impoverished conditions shown in this photo?

**The Split-Labor-Market Theory** Who benefits from the exploitation of people of color? Dual- or split-labor-market theory states that white workers and members of the capitalist class both benefit from the exploitation of people of color. *Split labor market* **refers to the division of the economy into two areas of employment, a primary sector or upper tier composed of higher-paid (usually dominant-group) workers in more-secure jobs, and a secondary sector or lower tier composed of lower-paid (often subordinate-group) workers in jobs with little security and hazardous working conditions** (Bonacich, 1972, 1976). According to this perspective, white workers in the upper tier may use racial discrimination against nonwhites to protect their positions. These actions most often occur when upper-tier workers feel threatened by lower-tier workers hired by capitalists to reduce labor costs and maximize corporate profits. In the past, immigrants were a source of cheap labor that employers could use to break strikes and keep wages down. Throughout U.S. history, higher-paid workers have responded with racial hostility and joined movements to curtail immigration and thus do away with the source of cheap labor (Marger, 2012).

Proponents of the split-labor-market theory suggest that white workers benefit from racial and ethnic antagonisms. However, these analysts typically do not examine the interactive effects of race, class, and gender in the workplace.

**Perspectives on Race and Gender** The term *gendered racism* refers to the interactive effect of racism and sexism on the exploitation of women

**internal colonialism** according to conflict theorists, a practice that occurs when members of a racial or ethnic group are conquered or colonized and forcibly placed under the economic and political control of the dominant group.

**split labor market** the division of the economy into two areas of employment, a primary sector or upper tier composed of higher-paid (usually dominant-group) workers in more-secure jobs, and a secondary sector or lower tier composed of lower-paid (often subordinate-group) workers in jobs with little security and hazardous working conditions.

**gendered racism** the interactive effect of racism and sexism on the exploitation of women of color.

of color. According to the social psychologist Philomena Essed (1991), women's particular position must be explored within each racial or ethnic group because their experiences will not have been the same as men's in each grouping.

Capitalists do not equally exploit all workers. Gender and race or ethnicity are important in this exploitation. Historically, white men have monopolized the high-paying primary labor market. Many people of color and white women hold lower-tier jobs. Below that tier is the underground sector of the economy, characterized by illegal or quasi-legal activities such as drug trafficking, prostitution, and working in sweatshops that do not meet minimum wage and safety standards. Many undocumented workers and some white women and people of color attempt to earn a living in this sector.

**Racial Formation** The *theory of racial formation* **states that actions of the government substantially define racial and ethnic relations in the United States.** Government actions range from race-related legislation to imprisonment of members of groups believed to be a threat to society. Sociologists Michael Omi and Howard Winant (2013) suggest that the U.S. government has shaped the politics of race through actions and policies that cause people to be treated differently because of their race. For example, immigration legislation reflects racial biases. The Naturalization Law of 1790 permitted only white immigrants to qualify for naturalization; the Immigration Act of 1924 favored Northern Europeans and excluded Asians and Southern and Eastern Europeans.

Social protest movements of various racial and ethnic groups periodically challenge the government's definition of racial realities. When this social rearticulation occurs, people's understanding about race may be somewhat restructured. For example, the African American protest movements of the 1950s and 1960s helped redefine the rights of people of color in the United States.

## An Alternative Perspective: Critical Race Theory

Emerging out of scholarly law studies on racial and ethnic inequality, *critical race theory* derives its foundation from the U.S. civil rights tradition.

## [ concept quick review ]

### Sociological Perspectives on Race and Ethnic Relations

| | Focus | Theory/Hypothesis |
|---|---|---|
| Symbolic Interactionist | Microlevel contacts between individuals | Contact hypothesis |
| Functionalist | Macrolevel intergroup processes | 1. Assimilation<br>  a. cultural<br>  b. biological<br>  c. structural<br>  d. psychological<br>2. Ethnic pluralism<br>  a. equalitarian pluralism<br>  b. inequalitarian pluralism (segregation) |
| Conflict | Power/economic differentials between dominant and subordinate groups | 1. Caste perspective<br>2. Class perspective<br>3. Internal colonialism<br>4. Split labor market<br>5. Gendered racism<br>6. Racial formation |
| Critical Race Theory | Racism as an ingrained feature of society that affects everyone's daily life | Laws may remedy overt discrimination but have little effect on subtle racism. Interest convergence is required for social change. |

# SPORTS IN THE MEDIA

## Media Coverage of Minorities in Sports

The most popular sports in the United States are the NFL, MLB, and NBA. The majority of these players are minorities . . . and just like in society when minorities are dominating something there will always be a racial element to it. . . . I don't believe the media are racists. What I do believe is the media have a bias against athletes who don't fit into their image of a *"classic athlete."* I call it *"Michael Jordan Syndrome."* . . . The media expects every athlete to act, speak and be like Mike and if they aren't they are treated and covered differently.

—Robert Littal (2010), the African American creator and CEO of BlackSportsOnline, discussing his ideas about race and sports reporting, in an era when middle-aged white men have dominated sports journalism. (Reprinted by permission of Robert Littal [www.blacksportsonline.com].)

How do journalists typically frame sports stories that involve racially related issues? Although some sports journalists make an effort to discuss relevant race-related issues as they occur and to offer balanced coverage of all participants' viewpoints, two major negative framing devices are prevalent among today's sports journalists. The first is the *color-blind approach,* which denies the racial underpinnings of an incident where discrimination was evident, such as when a coach, player, fan, or reporter refers to another participant by an offensive, racially derogatory name or description.

From this perspective, discriminatory language and actions are not about institutional racism but about isolated individuals who exhibit bad sports behavior and should receive the equivalent of a "foul" call in a game. In contrast to the color-blind approach, the *new racism approach* demonizes black athletes and demands that their attitudes, appearance, and actions be patterned after those of white athletes or of respected African American sports figures.

According to researchers at Penn State University's John Curley Center for Sports Journalism (2007), sports reporting often emphasizes individualism while ignoring societal problems: "We see it in sports with the general demonizing of black athletes coupled with a failure to recognize institutional racism in sports." Some white athletes are able to maintain their popularity and "good-guy" reputations despite their well-documented off-the-field bad behavior, even as some black athletes with similar behavior become synonymous with "bad boys" and all that's wrong with college or professional sports (Rogers, 2011).

Is there a way to improve reporting so that it is a more accurate reflection of what is going on in the world of sports, particularly situations involving race and possible racism? Richard Lapchick (2010), director of the Institute for Diversity and Ethics in Sports, believes that achieving greater diversity in the media itself—currently dominated by white reporters and editors—will enhance the overall quality of sports reporting. In the words of sports journalist Thabiti Lewis (2010: xiv), "As a writer, I deem it my duty to utilize sport as a vehicle for discussing and correcting the subtle nuances of the complex images and realities of race. . . . But the bitter truth is that in American sports culture—like American society—the treatment and depiction of athletes of color is plagued by race, which influences White self-perception, while attacking Black self-reliance, self-culture, and individual expression."

## reflect & analyze

**How do you think sports journalism might be improved to better reflect the accomplishments and concerns of athletes across lines of race and gender?**

---

Critical race theory has several major premises, including the belief that racism is such an ingrained feature of U.S. society that it appears to be ordinary and natural to many people (Delgado, 1995). As a result, civil rights legislation and affirmative action laws (formal equality) may remedy some of the more overt, blatant forms of racial injustice but have little effect on subtle, business-as-usual forms of racism that people of color experience as they go about their everyday lives. Although many minority-group members participate in collegiate and professional sports, studies of sports and media show that overt and covert forms of racism persist in the twenty-first century (see "Sports in the Media").

According to this approach, the best way to document racism and ongoing inequality in society is to listen to the lived experiences of people who have encountered such discrimination. In this way we can learn what actually happens in regard to racial oppression and the many effects it has on people, including alienation, depression, and certain physical illnesses. Central to this argument is

**theory of racial formation** the idea that actions of the government substantially define racial and ethnic relations in the United States.

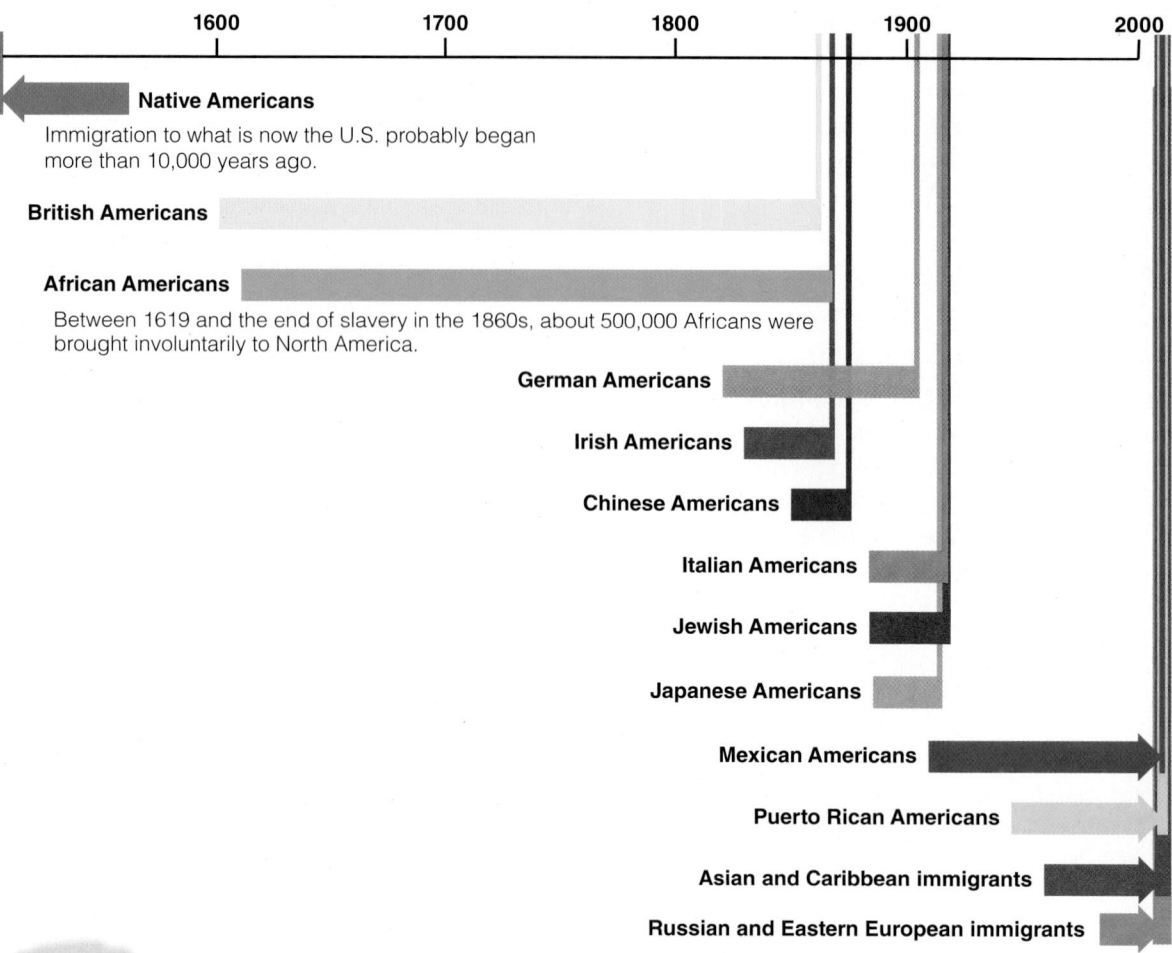

**FIGURE 9.8** TIME LINE OF RACIAL AND ETHNIC GROUPS IN THE UNITED STATES
© Cengage Learning

the belief that *interest convergence* is a crucial factor in bringing about social change. According to the legal scholar Derrick Bell, white elites tolerate or encourage racial advances for people of color *only* if the dominant-group members believe that their own self-interest will be served in so doing (cited in Delgado, 1995). From this approach, civil rights laws have typically benefited white Americans as much (or more) as people of color because these laws have been used as mechanisms to ensure that "racial progress occurs at just the right pace: change that is too rapid would be unsettling to society at large; change that is too slow could prove destabilizing" (Delgado, 1995: xiv). The Concept Quick Review outlines the key aspects of each sociological perspective on race and ethnic relations.

# Racial and Ethnic Groups in the United States

How do racial and ethnic groups come into contact with one another? How do they adjust to one another and to the dominant group over time? Sociologists have explored these questions extensively;

however, a detailed historical account of the unique experiences of each group is beyond the scope of this chapter. Instead, we will look briefly at intergroup contacts. In the process, sports will be used as an example of how members of some groups have attempted to gain upward mobility and become integrated into society.

## Native Americans and Alaska Natives

**LO5** Native Americans and Alaska Natives are believed to have migrated to North America from Asia thousands of years ago, as shown on the time line in Figure 9.8. One of the most widely accepted beliefs about this migration is that the first groups of Mongolians made their way across a natural bridge of land called Beringia into present-day Alaska. From there, they moved to what is now Canada and the northern United States, eventually making their way as far south as the tip of South America.

As schoolchildren are taught, Spanish explorer Christopher Columbus first encountered the native inhabitants in 1492 and referred to them as "Indians." When European settlers (or invaders) arrived on this continent, the native inhabitants' way of life was changed forever. Experts estimate that approximately two million native inhabitants lived in North America at that time; however, their numbers had been reduced to fewer than 240,000 by 1900.

**Genocide, Forced Migration, and Forced Assimilation** Native Americans have been the victims of genocide and forced migration. Although the United States never had an official policy that set in motion a pattern of deliberate extermination, many Native Americans were either massacred or died from European diseases (such as typhoid, smallpox, and measles) and starvation. In battle, Native Americans were often no match for the Europeans, who had "modern" weaponry. Europeans justified their aggression by stereotyping the Native Americans as "savages" and "heathens."

After the Revolutionary War, the federal government offered treaties to the Native Americans so that more of their land could be acquired for the growing white population. Scholars note that the government broke treaty after treaty as it engaged in a policy of wholesale removal of indigenous nations in order to clear the land for settlement by Anglo-Saxon "pioneers." Entire nations were forced to move in order to accommodate the white settlers. The "Trail of Tears" was one of the most disastrous of the forced migrations. In the coldest part of the winter of 1832, over half of the members of the Cherokee Nation died during or as a result of their forced relocation from the southeastern United States to the Indian Territory in Oklahoma.

Native Americans were subjected to forced assimilation on the reservations after 1871. Native American children were placed in boarding schools operated by the Bureau of Indian Affairs to hasten their assimilation into the dominant culture. About 98 percent of native lands had been expropriated by 1920. This process was aided by the Dawes Act (1877), which allowed the federal government to usurp Native American lands for the benefit of corporations and other nonnative settlers who sought to turn a profit from oil and gas exploration and grazing.

**Native Americans and Alaska Natives Today** Currently, about 5.2 million Native Americans and Alaska Natives, including those of more than one race, live in the United States, including Aleuts, Inuit (Eskimos), Cherokee, Navajo, Choctaw, Chippewa, Sioux, and more than 500 other nations of varying sizes and different locales. Native Americans and Alaska Natives made up about 2 percent of the total U.S. population in 2012. There

is a wide diversity among people in this category: Each nation has its own culture, history, and unique identity, and more than 250 Native American languages are spoken today. Slightly more than 20 percent of Native Americans and Alaska Natives ages five and older reported that they spoke a language other than English at home.

Although Native Americans live in a number of states, they are concentrated in specific regions of the country. About 22 percent of American Indians and Alaska Natives reside in federal American Indian reservations and/or off-reservation trust lands, or in other tribal-designated areas. There are 325 federally recognized American Indian reservations in this country and a total of 618 American Indian and Alaska Native legal and statistical areas included in the U.S. Census Bureau Statistics ("U.S. Census Bureau Facts on American Indian and Alaska Native Heritage Month," 2013).

Data continue to show that Native Americans are the most disadvantaged racial or ethnic group in the United States in terms of income, employment, housing, nutrition, and health. In 2012 almost 30 percent of American Indians and Alaska Natives were in poverty at a time when the poverty rate for the nation as a whole was under 16 percent. The life chances of Native Americans who live on reservations are especially limited. They have the highest rates of infant mortality and death by exposure and malnutrition. They also have high rates of suicide, substance abuse, and school violence (*CDC Morbidity and Mortality Weekly Report, 2013*).

Historically, Native Americans have had very limited educational opportunities and a very high rate of unemployment. Since the introduction of six tribally controlled community colleges in the 1970s, a growing network of 37 tribal colleges and universities now serves over 30,000 students from more than 250 tribal nations (see Figure 9.9). This network has been successful in providing some Native Americans with the education they need to move into the ranks of the skilled working class and beyond (Figure 9.10). Across the nation, Native Americans own and operate many types of enterprises, such as construction companies, computer-graphic-design firms, grocery stores, and management consulting businesses. Casino gambling operations and cigarette shops on Native American reservations—resulting from a reinterpretation of federal law in the 1990s—have brought more income to some of the nations, but this change has not been without its critics, many of whom believe that these businesses result in new problems for Native Americans.

In 2009 Native Americans received a $3.4 billion settlement from the federal government after the

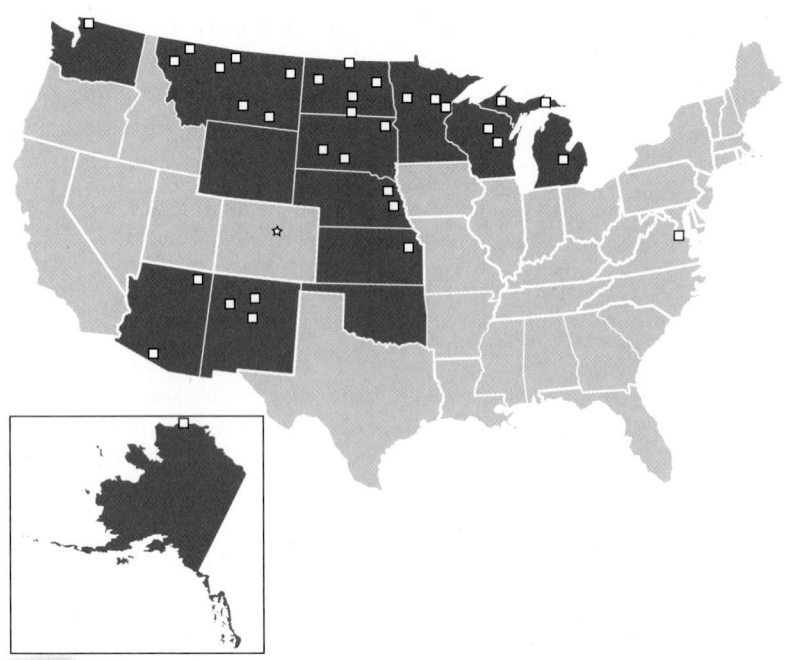

**FIGURE** 9.9 U.S. TRIBAL COLLEGES AND UNIVERSITIES

Source: American Indian College Fund, 2011. © Cengage Learning

conclusion of *Cobell v. Salazar,* a thirteen-year-old lawsuit that accused the government of mishandling revenues generated by the extraction of natural resources from American Indian land trusts as a result of the Dawes Act. Although the federal government was responsible for leasing tribal lands for use by mining, lumber, oil, and gas industries and passing on royalty payments to the Native Americans to whom the lands belonged, Native Americans derived little benefit because of the government's massive abuse of the trust funds.

Native Americans are currently in a transition from a history marked by prejudice and discrimination to a contemporary life in which they may find new opportunities. Many see the challenge for Native Americans today as erasing negative stereotypes while maintaining their heritage and obtaining recognition for their contributions to this nation's development and growth.

**Native Americans and Sports** Early in the twentieth century, Native Americans such as Jim Thorpe gained national visibility as athletes in football, baseball, and track and field. Teams at boarding schools such as the Carlisle Indian Industrial School in

Pennsylvania and the Haskell Institute in Kansas were well known. However, after the first three decades of the twentieth century, Native Americans became less prominent in sports. Native American scholar Joseph B. Oxendine (2003) attributes the lack of athletic participation to these factors: (1) a reduction in opportunities for developing sports skills, (2) restricted opportunities for participation, and (3) a lessening of Native Americans' interest in competing with and against non–Native Americans. However, in the twenty-first century, Native Americans

**FIGURE 9.10** Historically, Native Americans have had a low rate of college attendance. However, the development of a network of tribal colleges has provided them with a local source for upward mobility.

PART 3 | SOCIAL INEQUALITY

slowly began making their mark in a few professional sports, such as golf (Notah Begay III), lacrosse (Brett Bucktooth), bowling (Mike Edwards), rodeo (Clint Harry), and baseball (Kyle Lohse).

## White Anglo-Saxon Protestants (British Americans)

Whereas Native Americans have been among the most disadvantaged peoples, white Anglo-Saxon Protestants (WASPs) have been the most privileged group in this country. Although many English settlers initially came to North America as indentured servants or as prisoners, they quickly emerged as the dominant group, creating a core culture (including language, laws, and holidays) to which all other groups were expected to adapt. Most of the WASP immigrants arriving from Northern Europe were advantaged over later immigrants because they were highly skilled and did not experience high levels of prejudice and discrimination.

**Class, Gender, and WASPs** Like members of other racial and ethnic groups, not all WASPs are alike. Social class and gender affect their life chances and opportunities. For example, members of the working class and the poor do not have political and economic power; men in the capitalist class do. WASPs constitute the majority of the upper class and maintain cohesion through listings such as the *Social Register* and interactions with one another in elite settings such as private schools and country clubs (Kendall, 2002).

**WASPs and Sports** Family background, social class, and gender play an important role in the sports participation of WASPs. Contemporary North American football was invented at the Ivy League colleges and was dominated by young, affluent WASPs who had the time and money to attend college and participate in sports activities. Today, whites are more likely than any other racial or ethnic group to become professional athletes in all sports except football and basketball. Although current data are not available to document differences among racial and ethnic categories by types of sports, we know that the probability of competing in athletics beyond the high school interscholastic level is extremely low. For example, only .03 percent of high school men's basketball players will become professional athletes, as will only .02 percent of women's basketball players. For football, the percentage of high school players who will become professional athletes is .08 percent; for baseball, .4 percent; for men's ice hockey, .4 percent; and for men's soccer, .08 percent. Even the odds of advancing from high school athletics to NCAA college sports remain low: 3.2 percent for men's basketball, 3.6 percent for women's basketball, 6.1 percent for football, 6.6 percent for baseball, 10.7 percent for men's ice hockey, and 5.7 percent for men's soccer (National Collegiate Athletic Association, 2012).

Affluent WASP women participated in intercollegiate women's basketball in the late 1800s, and various other sporting events were used as a means to break free of restrictive codes of femininity. Until fairly recently, however, most women have had little chance for any involvement in college and professional sports.

## African Americans

**LO6** The African American (black) experience has been one uniquely marked by slavery, segregation, and persistent discrimination. There is a lack of consensus about whether *African American* or *black* is the most appropriate term to refer to the 38.9 million Americans of African descent who live in the United States today. Those who prefer the term *black* point out that it incorporates many African-descent groups living in this country that do not use *African American* as a racial or ethnic self-description. For example, people who trace their origins to Haiti, Puerto Rico, or Jamaica typically identify themselves as "black" but not as "African American." Although African Americans reside throughout the United States, eighteen states have an estimated black population of at least one million. About 3.5 million African Americans lived in New York State in 2011, but Mississippi had the largest share of blacks (38 percent) in its total population (see Figure 9.11).

Although the earliest African Americans probably arrived in North America with the Spanish conquerors in the fifteenth century, most historians trace their arrival to about 1619, when the first groups of indentured servants were brought to the colony of Virginia. However, by the 1660s, indentured servitude had turned into full-fledged slavery because of the enactment of laws that sanctioned the enslavement of African Americans. Although the initial status of persons of African descent in this country may not have been too different from that of the English indentured servants, all of that changed with the passage of laws turning human beings into property and making slavery a status from which neither individuals nor their children could escape (Franklin, 1980).

Between 1619 and the 1860s, about 500,000 Africans were forcibly brought to North America, primarily to work on southern plantations, and these actions were justified by the devaluation and

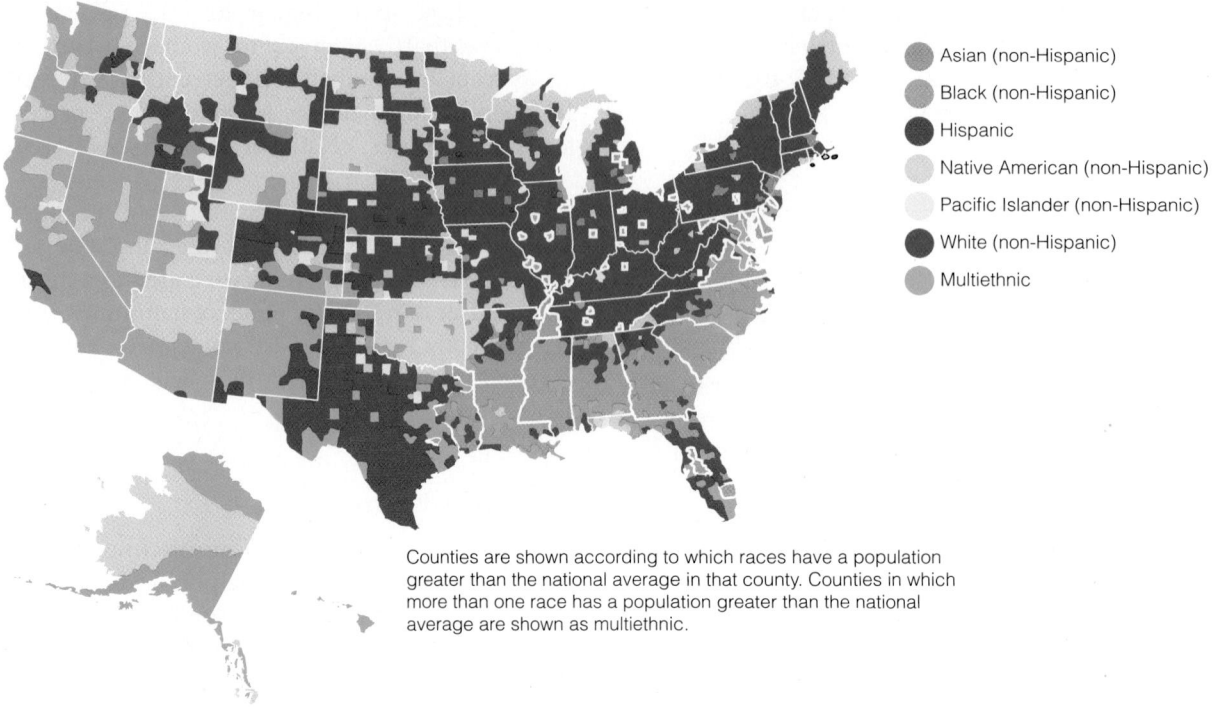

Counties are shown according to which races have a population greater than the national average in that county. Counties in which more than one race has a population greater than the national average are shown as multiethnic.

**FIGURE 9.11** U.S. RACIAL AND ETHNIC DISTRIBUTION

While minority populations do continue to grow, regional differences in racial makeup are still quite pronounced, as this map shows.

Source: U.S. Census Bureau, 2011f.

stereotyping of African Americans. Some analysts believe that the central factor associated with the development of slavery in this country was the plantation system, which was heavily reliant on cheap and dependable manual labor. Slavery was primarily beneficial to the wealthy southern plantation owners, but many of the stereotypes used to justify slavery were eventually institutionalized in southern custom and practice (Wilson, 1978). However, some slaves and whites engaged in active resistance against slavery and its barbaric practices, eventually resulting in slavery being outlawed in the northern states by the late 1700s. Slavery continued in the South until 1863, when it was abolished by the Emancipation Proclamation (Takaki, 1993).

**Segregation and Lynching** Gaining freedom did not give African Americans equality with whites. African Americans were subjected to many indignities because of race. Through informal practices in the North and *Jim Crow laws* in the South, African Americans experienced segregation in housing, employment, education, and all public accommodations. African Americans who did not stay in their "place" were often the victims of violent attacks and lynch mobs (Franklin, 1980). *Lynching* is a killing carried out by a group of vigilantes seeking revenge for an actual or imagined crime by the victim. The

practice of lynching was used by whites to intimidate African Americans into staying "in their place." It is estimated that as many as 6,000 lynchings occurred from the end of the Civil War to the present, at least half of which have gone unrecorded (Feagin and Feagin, 2012). In spite of all odds, many African American women and men resisted oppression and did not give up in their struggle for equality.

**Discrimination** In the twentieth century the lives of many African Americans were changed by industrialization and two world wars. When factories were built in the northern United States, many African American families left the rural South in hopes of finding jobs and a better life.

During World Wars I and II, African Americans were a vital source of labor in war production industries; however, racial discrimination continued both on and off the job. In World War II, many African Americans fought for their country in segregated units in the military; after the war, they sought—and were denied—equal opportunities in the country for which they had risked their lives.

African Americans began to demand sweeping societal changes in the 1950s. Initially, the Reverend Dr. Martin Luther King Jr., and the civil rights movement used *civil disobedience*—nonviolent action seeking to change a policy or law by refusing to comply with it—to call attention to racial inequality and to demand greater inclusion of African Americans in all areas of public life. Subsequently, leaders of the Black Power movement, including Malcolm X and Marcus Garvey, advocated black pride and racial awareness among African Americans. Gradually, racial segregation was outlawed by the courts and the federal government. For example, the Civil Rights Acts of 1964 and 1965 sought to do away with discrimination in education, housing, employment, and health care.

Affirmative action programs were instituted in both public-sector and private-sector organizations in an effort to bring about greater opportunities for African Americans and other previously excluded groups. *Affirmative action* refers to policies or procedures that are intended to promote equal opportunity for categories of people deemed to have been previously excluded from equality in education, employment, and other fields on the basis of characteristics such as race or ethnicity. Critics of affirmative action often assert that these policies amount to *reverse discrimination*—a person who is better qualified being denied a position because another person received preferential treatment as a result of affirmative action.

**African Americans Today** According to the 2010 U.S. Census, blacks or African Americans make up slightly less than 13 percent of the U.S. population. Some are descendants of families that have been in this country for many generations; others are recent immigrants from Africa and the Caribbean. Black Haitians make up the largest group of recent Caribbean immigrants; others come from Jamaica and Trinidad and Tobago. Recent African immigrants are primarily from Nigeria, Ethiopia, Ghana, and Kenya. They have been simultaneously "pushed" out of their countries of origin by severe economic and political turmoil and "pulled" by perceived opportunities for a better life in the United States. Recent immigrants are often victimized by the same racism that has plagued African Americans as a people for centuries.

Since the 1960s, many African Americans have made significant gains in politics, education, employment, and income. Between 1964 and 2010, the number of African Americans elected to political office (on local, state, and federal levels) increased from about 100 to more than 10,000 nationwide (Hayes, 2011). At the local level, African Americans have won mayoral elections in many major cities that have large African American populations, such as Atlanta, Houston, New Orleans, Philadelphia, and Washington, D.C. At the national level, the top

Scott Andrews—Pool via CNP/Newscom

**FIGURE 9.12** In August 2008 Barack Obama made history by becoming the first African American to receive the presidential nomination of a major political party, and on Election Day he was voted in as the first African American president of the United States. In 2012 Obama was reelected to his second term and is shown here taking the oath of office from Chief Justice John Roberts on January 21, 2013.

accomplishment of an African American in politics was the election of Barack Obama to the presidency in 2008 and his reelection to a second term of office in 2012 (Figure 9.12).

Despite these political gains, however, African Americans represented less than 8 percent of the members of the 112th U.S. Congress. In 2012 there were 43 African Americans in Congress. However, all were in the House of Representatives: None were in the U.S. Senate. Over the past five decades, some African Americans have made impressive occupational gains and joined the ranks of professionals in the upper-middle class. Others have achieved great wealth and fame as entertainers, professional athletes, and entrepreneurs. However, even those who make millions of dollars per year and live in affluent neighborhoods are not always exempt from racial prejudice and discrimination. And although some African Americans have made substantial occupational and educational gains, many more have not. The African American unemployment rate remains twice as high as that of whites.

**African Americans and Sports** In recent decades, many African Americans have seen sports as a possible source of upward mobility because other means have been unavailable. However, their achievements in sports have often been attributed to "natural ability" and not determination and hard work. Sociologists have rejected such biological explanations for African Americans' success in sports and have focused instead on explanations rooted in the structure of society.

During the slavery era a few African Americans gained better treatment and, occasionally, freedom by winning boxing matches on which their owners had bet large sums of money (McPherson, Curtis, and Loy, 1989). After emancipation, some African Americans found jobs in horse racing and baseball. For example, fourteen of the fifteen jockeys in the first Kentucky Derby (in 1875) were African Americans. A number of African Americans played on baseball teams; a few played in the Major Leagues until the Jim Crow laws forced them out. Then they formed their own "Negro" baseball and basketball leagues (Peterson, 1992/1970).

Since Jackie Robinson broke baseball's "color line," in 1947, many African American athletes have played collegiate and professional sports. Even now, however, persistent class inequalities between whites and African Americans are reflected in the fact that until recently, African Americans have primarily excelled in sports (such as basketball or football) that do not require much expensive equipment and specialized facilities in order to develop athletic skills. According to one sports analyst, African Americans typically participate in certain sports and not others because of the *sports opportunity structure*—the availability of facilities, coaching, and competition in the schools and community recreation programs in their area (Phillips, 1993).

Regardless of the sport in which they participate, African American men athletes continue to experience inequalities in assignment of coaching, management, and ownership opportunities in professional sports. In recent years, only six of the thirty-two National Football League head coaches and twenty-eight Division I NCAA head coaches in college football were African American. However, in the elite "Football Bowl Subdivision"—postseason (bowl-eligible) competitors—only nineteen schools (or 16 percent) have African American head coaches. By contrast, nearly half (46 percent) of all college football players are African American (Solomon, 2012). Today, African Americans remain significantly underrepresented in other sports, including hockey, skiing, figure skating, golf, volleyball, softball, swimming, gymnastics, sailing, soccer, bowling, cycling, and tennis.

## White Ethnic Americans

The term *white ethnic Americans* is applied to a wide diversity of immigrants who trace their origins to Ireland and to Eastern and Southern European countries such as Poland, Italy, Greece, Germany, Yugoslavia, and Russia and other former Soviet republics. Unlike WASPs, who migrated primarily from Northern Europe and assumed a dominant cultural position in society, white ethnic Americans arrived late in the nineteenth century and early in the twentieth century to find relatively high levels of prejudice and discrimination directed at them by nativist organizations that hoped to curb the entry of non-WASP European immigrants. Because many of the people in white ethnic American categories were not Protestant, they experienced discrimination because they were Catholic, Jewish, or members of other religious bodies, such as the Eastern Orthodox churches.

**Discrimination Against White Ethnics**
Many white ethnic immigrants entered the United States between 1830 and 1924. Irish Catholics were among the first to arrive, driven out of Ireland by English oppression and famine and seeking jobs in the United States (Feagin and Feagin, 2012). When they arrived, they found that British Americans controlled the major institutions of society. The next arrivals were Italians, who had been recruited for low-wage industrial and construction jobs. British Americans viewed Irish and Italian immigrants as "foreigners": The Irish were stereotyped as ape-like, filthy, bad-tempered, and heavy drinkers; the Italians were depicted as lawless, knife-wielding thugs looking for a fight, "dagos," and "wops" (short for "without papers") (Feagin and Feagin, 2012).

Both Irish Americans and Italian Americans were subjected to institutionalized discrimination in employment. Employment ads read "Help Wanted—No Irish Need Apply" and listed daily wages at $1.30–$1.50 for "whites" and $1.15–$1.25 for "Italians" (Gambino, 1975: 77). In spite of discrimination, white ethnics worked hard to establish themselves in the United States, often founding mutual self-help organizations and becoming politically active (Mangione and Morreale, 1992).

Between 1880 and 1920, a wave of Eastern European Jewish immigrants arrived in the United States and settled in the Northeast. Jewish Americans differ from other white ethnic groups in that some focus their identity primarily on their religion whereas others define their Jewishness in terms of ethnic group membership (Feagin and Feagin, 2012). In any case, Jews continued to be the victims of *anti-Semitism*—prejudice, hostile attitudes, and discriminatory behavior targeted at Jews. For example, signs in hotels read "No Jews Allowed," and some "help wanted" ads stated "Christians Only" (Levine, 1992: 55). In spite of persistent discrimination,

Jewish Americans achieved substantial success in many areas, including business, education, the arts and sciences, law, and medicine.

**White Ethnics and Sports** Sports provided a pathway to assimilation for many white ethnics. The earliest collegiate football players who were not white Anglo-Saxon Protestants were of Irish, Italian, and Jewish ancestry. Sports participation provided educational opportunities that some white ethnics would not have had otherwise.

Boxing became a way to make a living for white ethnics who did not participate in collegiate sports (Figure 9.13). Boxing promoters encouraged ethnic rivalries to increase their profits, pitting Italians against Irish or Jews, and whites against African Americans (Levine, 1992; Mangione and Morreale, 1992). Eventually, Italian Americans graduated from boxing into baseball and football. Jewish Americans found that sports lessened the shock of assimilation and gave them an opportunity to refute stereotypes about their physical weaknesses and to counter anti-Semitic charges that they were "unfit to become Americans" (Levine, 1992: 272). Today, assimilation is so complete that little attention is paid to the origins of white ethnic athletes.

## Asian Americans

**LO7** The U.S. Census Bureau uses the term *Asian American*s to designate the many diverse groups with roots in Asia. Chinese and Japanese immigrants were among the earliest Asian Americans. Many Filipinos, Asian Indians, Koreans, Vietnamese, Cambodians, Pakistani, and Indonesians have arrived more recently. Although Asians constitute less than 6 percent of the U.S. population, they are the fastest-growing minority group. From 2000 to 2010 the U.S. Asian population increased 46 percent, and in 2011 about 18.2 million people in the United States identified themselves as residents of Asian descent or Asian in combination with one or more other races. The largest Asian American population is in California, where more than 5.8 million reside, followed by New York with 1.7 million. In Hawaii, persons in the Asian American-alone and the Asian American-combination population account for 57 percent of the state's total population ("U.S. Census Bureau Facts on Asian/Pacific American Heritage Month," 2013).

**Chinese Americans** Chinese Americans are the largest Asian American group (U.S. Census Bureau, 2011c). The initial wave of Chinese immigration

Bettmann/Corbis

**FIGURE 9.13** Early-twentieth-century Jewish American and Italian American boxers not only produced intragroup ethnic pride but also earned a livelihood through boxing matches.

occurred between 1850 and 1880, when more than 200,000 Chinese men were "pushed" from China by harsh economic conditions and "pulled" to the United States by the promise of gold in California and employment opportunities in the construction of transcontinental railroads. Far fewer Chinese women immigrated; however, many of them were brought to the United States against their will and forced into prostitution, where they were treated like slaves (Takaki, 1993).

Chinese Americans were subjected to extreme prejudice and stereotyped as "coolies," "heathens," and "Chinks." Some Asians were attacked and even lynched by working-class whites who feared that they were losing their jobs to these immigrants. Passage of the Chinese Exclusion Act of 1882 brought Chinese immigration to a halt. The Exclusion Act was not repealed until World War II, when Chinese Americans who were contributing to the war effort by working in defense plants pushed for its repeal (Takaki, 1993). After immigration laws were further relaxed in the 1960s, the second and largest wave of Chinese immigration occurred, with immigrants coming primarily from Hong Kong and Taiwan. These recent immigrants have had more education and workplace skills than earlier arrivals, and they brought families and capital with them to pursue the American Dream.

Today, about four million Asians of Chinese descent live in the United States. Many reside in large urban enclaves in California, Texas, New York, and Hawaii. As a group, Asian Americans have enjoyed considerable upward mobility: The median household income for U.S. Asian

Americans (Chinese and others) is $67,885. Some Chinese Americans have become highly successful professionals and business entrepreneurs. However, many Chinese Americans remain in the lower tier of the working class—providing low-wage labor in personal services, repair, and maintenance (Pew Research Center, 2012c).

**Japanese Americans** Most of the early Japanese immigrants were men who worked on sugar plantations in the Hawaiian Islands in the 1860s. Like Chinese immigrants, the Japanese American workers were viewed as a threat by white workers, and immigration of Japanese men was curbed in 1908. However, Japanese women were permitted to enter the United States for several years thereafter because of the shortage of women on the West Coast. Although some Japanese women married white men, laws prohibiting interracial marriage stopped this practice.

With the exception of the forced migration and genocide experienced by Native Americans and the enslavement of African Americans, Japanese Americans experienced one of the most vicious forms of discrimination ever sanctioned by U.S. laws. During World War II, when the United States was at war with Japan, nearly 120,000 Japanese Americans were placed in internment camps, where they remained for more than two years despite the total lack of evidence that they posed a security threat to this country (Takaki, 1993) (Figure 9.14). This action was a direct violation of the citizenship rights of many *Nisei* (second-generation Japanese Americans), who were born in the United States. Only Japanese Americans were singled out for such harsh treatment; German Americans avoided this fate even though the United States was also at war with Germany. Four decades later, the U.S. government issued an

**FIGURE 9.14** During World War II, nearly 120,000 Japanese Americans—some of whom are still alive today—were interned in camps such as the Manzanar Relocation Center in California, where this statue memorializes their ordeal.

apology for its actions and eventually paid $20,000 each to some of those who had been placed in internment camps (Daniels, 1993; Takaki, 1993).

Since World War II, many Japanese Americans have been very successful. The household annual income of Japanese Americans is $65,390, which is slightly lower than for all Asian American households but higher than for all U.S. households ($49,800). However, many Japanese Americans (and other Asian Americans) live in states that not only have higher incomes but also higher costs of living than the national average (Le, 2013).

**Korean Americans** The first wave of Korean immigrants were male workers who arrived in Hawaii between 1903 and 1910. The second wave came to the U.S. mainland following the Korean War in 1954 and was made up primarily of the wives of servicemen and Korean children who had lost their parents in the war. The third wave arrived after the Immigration Act of 1965 permitted well-educated professionals to migrate to the United States. Korean Americans have helped one another open small businesses by pooling money through the *kye*—an association that grants members money on a rotating basis to gain access to more capital.

Today, an estimated 1.3 million Korean Americans reside in the United States, making up the fifth-largest group among Asian Americans and about 10 percent of the adult Asian population in the nation. Slightly more than half of Korean Americans speak English proficiently, as compared with 63 percent of all Asian Americans and 90 percent of the U.S. population overall. Many Korean Americans live in California and New York, where there is a concentration of Korean-owned businesses. The median annual household income for Korean Americans is $50,000, which is lower than for all Asians ($66,000) but slightly higher than for the U.S. population as a whole ($49,800) (Pew Research Center, 2012c).

**Filipino Americans** Today, Filipino Americans constitute the second-largest category of Asian Americans, with about 3.4 million U.S. residents reporting that they are Filipino alone or in combination with one or more additional racial–ethnic categories (Pew Research Center, 2012c). To understand the status of Filipino Americans, it is important to look at the complex relationship between the Philippine Islands and the U.S. government. After Spain lost the Spanish-American War, the United States established colonial rule over the islands, a rule that lasted from 1898 to 1946. Despite control by the United States, Filipinos were not granted U.S. citizenship, but male Filipinos were allowed to migrate to Hawaii and the U.S. mainland to work in agriculture and in fish canneries in Seattle and Alaska. Like other Asian Americans, Filipino Americans were accused of taking jobs away from white workers and suppressing wages, and Congress restricted Filipino immigration to fifty people per year between the Great Depression and the aftermath of World War II.

The second wave of Filipino immigrants came following the Immigration Act of 1965, when large numbers of physicians, nurses, technical workers, and other professionals moved to the U.S. mainland. Most Filipinos have not had the start-up capital necessary to open their own businesses, and many have been employed in the low-wage sector of the service economy. However, the average household income of Filipino American families is relatively high at $75,000 because, among other reasons, Filipinos have one of the highest levels of educational attainment among Asian Americans (Pew Research Center, 2012c).

**Indochinese Americans** Indochinese Americans include people from Vietnam, Cambodia, Thailand, and Laos. Vietnamese refugees who had the resources to flee at the beginning of the Vietnam War were the first to arrive. Next came Cambodians and lowland Laotians, referred to as "boat people" by the media. Many who tried to immigrate did not survive at sea; others were turned back when they reached this country or were kept in refugee camps for long periods of time. When they arrived in the United States, inflation was high, the country was in a recession, and many native-born citizens feared that they would lose their jobs to these new refugees, who were willing to work very hard for low wages.

According to the 2010 U.S. Census, about 1.31 million adult Vietnamese Americans reside in the United States, constituting the fourth-largest group of Asian Americans. About 84 percent of Vietnamese Americans are foreign born, and less than half (41 percent) speak English proficiently. However, nearly 80 percent of Vietnamese Americans possess U.S. citizenship. The median age of Vietnamese Americans is 41, which is slightly less than the median age of 45 for the entire U.S. population. Their median annual household income of $53,400 is higher than that of the U.S. population as a whole but lower than the $66,000 for all U.S. Asian Americans (Pew Research Center, 2012c).

Like Vietnamese Americans, other Indochinese Americans from Cambodia, Thailand, and Laos are often first- or second-generation residents of the United States; about half live in the western states, especially California. Even though most first-generation Indochinese immigrants spoke no English when they arrived in this country, their children and grandchildren have done very well in school and have been stereotyped as "brains."

**Asian Americans and Sports** Asian American athletes have begun to receive recognition in a variety of sports, winning acclaim in the Olympics and in other major sports: Kyla Ross (gymnastics), Nathan Adrian (swimming), Jeremy Lin (basketball), Nonito Donaire (boxing), Julie Chu (ice hockey), Ed Wang (football), and Ichiro Suzuki and Tim Lincecum (baseball). These and a number of other Asian Americans have been recognized as top athletes. Sports analysts have pointed out the importance of having outstanding Asian American athletes because they provide role models for all young people, but especially for their own communities, exemplifying the integrity, discipline, and hard work that are necessary to become a success in sports and in life.

## Latinos/as (Hispanic Americans)

**LO8** The terms *Latino* (for males), *Latina* (for females), and *Hispanic* are used interchangeably to refer to people who trace their origins to Spanish-speaking Latin America and the Iberian peninsula. However, as racial–ethnic scholars have pointed out, the label *Hispanic* was first used by the U.S. government to designate people of Latin American and Spanish descent living in the United States, and it has not been fully accepted as a source of identity by the more than 53 million Latinos/as who live in the United States today and constitute 17 percent of the nation's total population ("U.S. Census Bureau Facts on Hispanic Heritage Month," 2013). Instead, many of the people who trace their roots to Spanish-speaking countries think of themselves as Mexican Americans, Chicanos/as, Puerto Ricans, Cuban Americans, Salvadorans, Guatemalans, Nicaraguans, Costa Ricans, Argentines, Hondurans, Dominicans, or members of other categories. Many also think of themselves as having a combination of Spanish, African, and Native American ancestry.

**Mexican Americans or Chicanos/as** Mexican Americans—including both native- and foreign-born people of Mexican origin—are the largest segment (approximately two-thirds) of the Latino/a population in the United States and total about 32.9 million. Most Mexican Americans live in the southwestern region of the United States, although more have moved throughout the United States in recent years.

Immigration from Mexico is the primary vehicle by which the Mexican American population grew in this country. Initially, Mexican-origin workers came to work in agriculture, where they were viewed as a readily available cheap and seasonal labor force. Many initially entered the United States as undocumented workers ("illegal aliens"); however, they were more vulnerable to deportation than other illegal immigrants because of their visibility and the proximity of their country of origin. For more than a century, there has been a "revolving door" between the United States and Mexico that has been open when workers were needed and closed during periods of economic recession and high rates of U.S. unemployment.

Mexican Americans have long been seen as a source of cheap labor, while at the same time they have been stereotyped as lazy and unwilling to work. As has been true of other groups, when white workers viewed Mexican Americans as a threat to their jobs, they demanded that the "illegal aliens" be sent back to Mexico. Consequently, U.S. citizens who happen to be Mexican American have been asked for proof of their citizenship, especially when anti-immigration sentiments are running high. Many Mexican American families have lived in the United States for five or six generations—they have fought in wars, made educational and political gains, and consider themselves to be solid U.S. citizens. Thus, it is a great source of frustration for them to be viewed as illegal immigrants or to be asked "How long have you been in this country?"

The U.S. recession that began in 2007 and the gradual economic recovery of the second decade of the twenty-first century considerably reduced the flow of immigration from Mexico to the United States. The collapse of the U.S. housing market reduced the number of jobs in the construction industry, and other employment opportunities were also lost as the financial crisis took away positions in manufacturing, personal service, leisure, and other sectors. However, it is clear that Mexican Americans will continue to make a major contribution to the U.S. population because the

Mexican-origin population increased by more than 50 percent (from 20.6 million to 32.9 million)—the largest numerical increase of any racial or ethnic category—between 2000 and 2010 (Pew Research Center, 2012a). An additional 1.1 million Hispanics were added to the nation's population between July 1, 2011, and July 1, 2012, accounting for nearly half of the approximately 2.3 million people who were added to the nation's population during this period ("U.S. Census Bureau Facts on Hispanic Heritage Month," 2013).

**Puerto Ricans** Today, the 4.7 million Puerto Rican Americans residing in the United States make up 9.2 percent of Hispanic-origin people in this country. When Puerto Rico became a territory of the United States, in 1917, Puerto Ricans acquired U.S. citizenship and the right to move freely to and from the mainland. In the 1950s, many migrated to the mainland when the Puerto Rican sugar industry collapsed, settling primarily in New York and New Jersey. Today, more than half of all Puerto Rican Americans reside in the Northeast, followed by the South, primarily Florida (Pew Research Center, 2012a).

Although living conditions have improved substantially for some Puerto Ricans, life has been difficult for the many living in poverty in Spanish Harlem and other barrios. Nevertheless, in recent years Puerto Ricans have made dramatic advances in education, the arts, and politics. Puerto Rican Americans have higher levels of educational attainment than the Hispanic population overall: Among Puerto Ricans ages 25 and older, 16 percent have obtained at least a bachelor's degree, as compared to 13 percent of all U.S. Hispanics (Pew Research Center, 2012a). However, the annual median household income ($36,000) of Puerto Rican Americans is considerably less than that of the U.S. population as a whole.

**Cuban Americans** Cuban Americans (totaling about 2.0 million) live primarily in the Southeast, especially Florida. They are the fourth-largest population of Hispanic origin living in the United States, constituting about 3.5 percent of the total U.S. Hispanic population (Pew Research Hispanic Trends Project, 2013). As a group, they have fared somewhat better than other Latinos/as because many Cuban immigrants were affluent professionals and businesspeople who fled Cuba after Fidel Castro's 1959 Marxist revolution. This early wave of Cuban immigrants has median incomes well above those of other Latinos/as; however, this group is still below the national average. The second wave of Cuban Americans, arriving in the 1970s, has fared worse. Many had been released from prisons and mental hospitals in Cuba, and their arrival fueled an upsurge in prejudice against all Cuban Americans. The more-recent arrivals have developed their own ethnic and economic enclaves in Miami's Little Havana, and many of the earlier immigrants have become mainstream professionals and entrepreneurs.

As of 2014, the United States accepts 20,000 Cubans each year via legal immigration and also allows legal residency to those individuals and families that manage to reach U.S. shores. If Cubans are picked up at sea as they attempt to cross the Florida Straits to enter the United States, they are turned back under what is known as the "wet foot, dry foot policy" (Frank, 2014). In 2013 Cuba relaxed long-standing restrictions on travel by its residents to the United States, making it possible for more of them to travel legally and for some to gain permanent residence in this country. It remains to be seen whether these gradual changes in immigration policy between the two countries will increase the overall population of Cubans as permanent residents or U.S. citizens (Frank, 2014).

**Latinos/as and Sports** For more than a century, Latinos have played Major League Baseball in the United States (Figure 9.15). Originally, Cubans, Puerto Ricans, and Venezuelans were selected for their light skin as well as for their skill as players. Baseball had become a major means of assimilation for earlier Latinos in the United States. By 2012, Latinos represented more than 25 percent of Major League Baseball players, and growing numbers also participated in football, hockey, and basketball at all levels of competition. In women's sports, golf, soccer, and basketball have also had rising numbers of Latina athletes.

In addition to baseball, Latinos have made impressive gains in Major League Soccer in the United States and Canada; however, many people in the United States are less enthusiastic fans of soccer as compared to football. Consequently, salaries of professional soccer players are lower than in the high-profile sports, fan bases are smaller, and revenues from sales of team clothing are not as lucrative. However, the sixteen U.S. and three Canadian soccer teams provide opportunities for Latino athletes to be highly visible in professional sports in regions with a large Latino/a population.

JORGE LEMUS/Ai Wire/Landov

**FIGURE 9.15** Professional sports, particularly baseball, increasingly reflects the growing racial–ethnic and national diversity of the U.S. population.

## Middle Eastern Americans

Since 1970, many immigrants have arrived in the United States from countries located in the "Middle East," which is the geographic region from Afghanistan to Libya and including Arabia, Cyprus, and Asiatic Turkey. Placing people in the "Middle Eastern" American category is somewhat like placing wide diversities of people in the categories of Asian American or Latino/a; some U.S. residents trace their origins to countries such as Algeria, Bahrain, Egypt, Iran, Iraq, Jordan, Kuwait, Lebanon, Libya, Morocco, Oman, Palestine, Qatar, Saudi Arabia, Syria, Tunisia, UAE (United Arab Emirates), and Yemen (Asi and Beaulieu, 2013).

Middle Eastern Americans speak a variety of languages and have diverse religious backgrounds: Some are Muslim, some are Coptic Christian, and others are Melkite Catholic. Although some are from working-class families, Lebanese Americans, Syrian Americans, Iranian Americans, and Kuwaiti Americans primarily come from middle- and upper-income family backgrounds. Between 2006 and 2010, for example, the median household income for Lebanese Americans was $67,264, as contrasted with $32,075 for Iraqi Americans (Asi and Beaulieu, 2013).

**Arab Americans** In the twenty-first century, about 3.5 million people in the United States identify their family's country of origin as being an Arab country. The primary countries of origin are Lebanon, Syria, Palestine, Egypt, and Iraq. Although Arab Americans live throughout the United States, nearly half live in California, Michigan, New York, Florida, and New Jersey. One-third of all Arab Americans reside in one of three major metropolitan areas: Detroit, Los Angeles, and New York. Most Arab Americans were born in the United States, and over 80 percent are U.S. citizens.

Since the 2010 U.S. Census, the population of Arab Americans in the United States has become better known because of campaigns promoting the slogan "check it right, you ain't White" that were launched by various Arab American groups to encourage Arab Americans to check the "Other" box when they filled out the 2010 Census form and then to identify themselves as "Arab" or to indicate their specific country of origin.

## Iranian (Persian) Americans

About 1.5 million Iranian Americans live in the United States today. However, no official statistics are available because these data are not collected by the Census Bureau. Instead, the annual American Community Survey, a sample survey, asks questions of ancestry that provide this information.

The terms *Iranian American* and *Persian American* are used interchangeably because Iran was called Persia prior to 1935. Many Iranian Americans refer to themselves as "Persian" rather than "Iranian" because of the perceived negativity associated with the political history of the country of Iran and its relationship to the United States. It should be noted that Persian Americans are not considered to be Arab because they speak Farsi and have a different culture.

The most extensive immigration of Iranians to the United States began in the late 1970s and early 1980s, when early immigrants, particularly college students, left their homeland as the Iranian revolution was taking place. When the Islamic Republic was established after the revolution, many Iranian students decided to remain in the United States, and other Iranians also left their country and established a new life in this nation.

Today, the United States has the highest number of Iranian residents outside of Iran. More than 80 percent of Iranian Americans are U.S. citizens. Many Iranian Americans have high levels of educational attainment and are employed in professional positions in business, academia, and science.

**Discrimination** Despite such high level of achievement, Iranian Americans, like Arab Americans, have experienced persistent discrimination, particularly if they are Muslim (Figure 9.16) or if there has been a recent threat of a terrorist attack in the United States. Following the September 11, 2001, attacks on the United States by terrorists whose origins were traced to the Middle East, there was an escalation in the number of hate crimes and other types of discrimination against persons assumed to be Arabs, Arab Americans, Iranian Americans, or Muslims. In the aftermath of this terrorist attack,

**FIGURE 9.16** Muslims in the United States who wear traditional attire may face prejudice and/or discrimination as they go about their daily lives.

Jim West/The Image Works

the U.S. Patriot Act was passed. This law gave the federal government greater authority to engage in searches and surveillance of persons suspected of terrorist activity than in the past. The Patriot Act caused heightened concern among many individuals and groups because it was believed that this law might be used to target individuals who appear to be of Middle Eastern origins. In 2011 President Barack Obama signed the Patriot Sunsets Extension Act, which provided for a four-year extension of three key provisions in the original act, allowing for roving wiretaps, searches of business records, and continued surveillance of individuals suspected of terrorist-related activities not linked to terrorist groups.

What about the Muslim experience in the United States? In cities across this country, Muslims have established social, economic, and ethnic enclaves for social stability and personal safety. Islamic schools and centers often bring together people from a diversity of countries such as Egypt and Pakistan. Many Muslim leaders and parents focus on how to raise children to be good Muslims and good U.S. citizens. In the middle of the second decade of the twenty-first century, some Middle Eastern Americans experience discrimination based on their speech patterns, appearance, and clothing (such as the *hijabs,* or "head-to-toe covering" that leaves only the face exposed, which many girls and women wear). According to studies by the Pew Research Center, 20 percent of U.S. Muslims state that they have experienced discrimination, prejudice, and unfair treatment. Nearly 30 percent indicate that they are aware of negative views about Muslims, 15 percent mention ignorance about Islam, and 7 percent indicate that there are religious and cultural problems between Muslims and non-Muslims (Lipka, 2013). The idea that Middle Easterners, particularly those identified as Muslim, are somehow associated with terrorism has been difficult to remove from media representations and some people's thinking, producing ongoing hardship for many upstanding citizens of this nation.

**Middle Eastern Americans and Sports**
Although an increasing number of Islamic schools now focus on sports for teenage boys, overall there has been less emphasis placed on competitive athletics among Middle Eastern Americans when compared to other groups. Because of sports traditions in their countries of origin, some Middle Eastern Americans play golf or soccer. As well, some Iranian Americans follow the soccer careers of professional players from Iran who now play for German, Austrian, Belgian, and Greek clubs. Over time, sports participation will probably continue to increase among Middle Eastern American males, particularly in soccer and golf; however, girls and women in more-traditional Muslim families typically have not participated in athletic activities unless such events are conducted privately.

# Global Racial and Ethnic Inequality in the Future

Throughout the world, many racial and ethnic groups seek *self-determination*—the right to choose their own way of life. As many nations are currently structured, however, self-determination is impossible.

## Worldwide Racial and Ethnic Struggles

The cost of self-determination is the loss of life and property in ethnic warfare. Ethnic violence has persisted in Mali, Myanmar, Bangladesh, India, China, South Sudan, and many other regions where hundreds of thousands have died from warfare, disease, and refugee migration. Ethnic wars have a high price even for survivors, whose life chances can become bleaker even after the violence subsides.

In the twenty-first century the struggle between the Israeli government and various Palestinian factions over the future and borders of Palestine continues to make headlines. Discord in this region has heightened tensions among people not only in Israel and Palestine but also in the United States and around the world as deadly clashes continue and political leaders are apparently unable to reach a lasting solution to the decades-long strife.

## Growing Racial and Ethnic Diversity in the United States

Racial and ethnic diversity is increasing in the United States. African Americans, Latinos/as, Asian Americans, Native Americans, and mixed-race individuals constitute more than a third (36.3 percent) of the U.S. population—up from 30.9 percent in 2000 (Humes, Jones, and Ramirez, 2011). As shown in Figure 9.17, states vary in their percentage of population that are minorities. Today, non-Hispanic white Americans make up 63.3 percent of the population, in contrast to 80 percent in 1980. It is predicted that by 2056, the roots of the average U.S. resident will be in Africa, Asia, Hispanic countries, Pacific islands, and the Middle East—not white Europe.

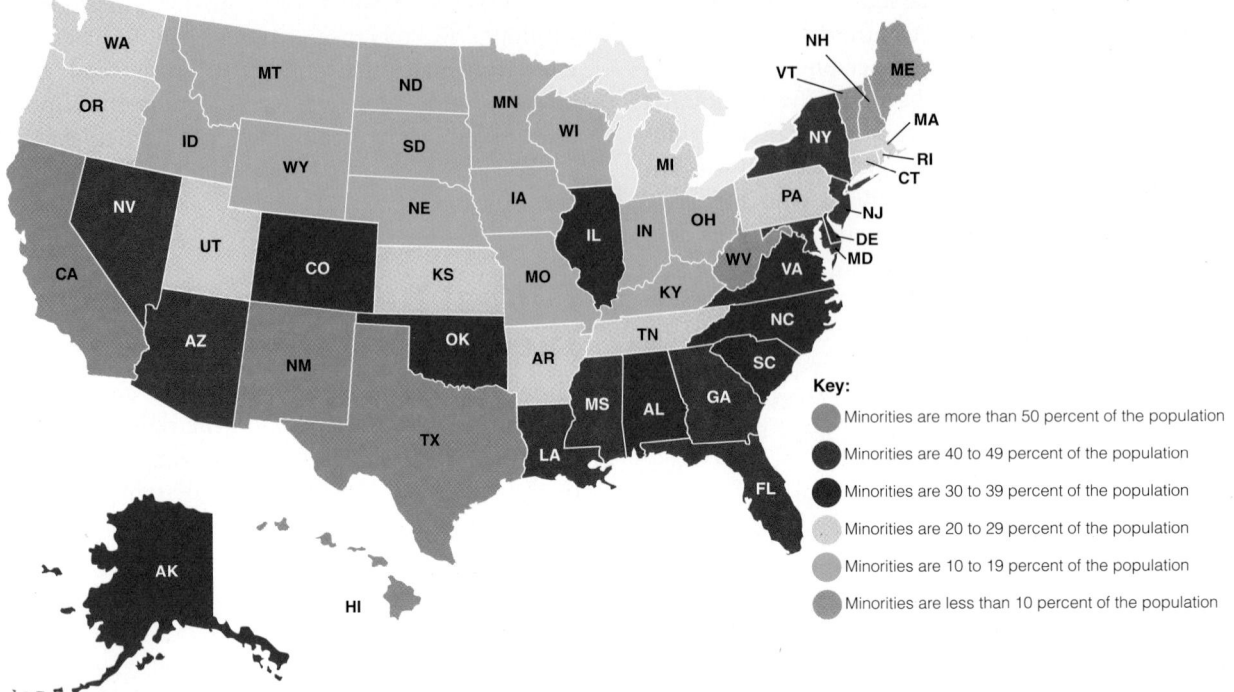

**FIGURE 9.17** MINORITY POPULATIONS AND PERCENTAGES* BY STATE, 2010
*Percentages are rounded to nearest whole number.

Source: Humes, Jones, and Ramirez, 2011. © Cengage Learning

Key:
- Minorities are more than 50 percent of the population
- Minorities are 40 to 49 percent of the population
- Minorities are 30 to 39 percent of the population
- Minorities are 20 to 29 percent of the population
- Minorities are 10 to 19 percent of the population
- Minorities are less than 10 percent of the population

What effect will these changes have on racial and ethnic relations? Several possibilities exist. On the one hand, conflicts may become more overt and confrontational as people continue to use *sincere fictions*—personal beliefs that reflect larger societal mythologies, such as "I am not a racist" or "I have never discriminated against anyone"—even when these are inaccurate perceptions (Feagin and Vera, 1995). Although the term *sincere fictions* was coined two decades ago, we face the real possibility in the future that interethnic tensions, as well as many other forms, may increase as competition for scarce resources such as education, jobs, and valued goods in society continues to grow and the U.S. population continues to age.

On the other hand, there is reason for cautious optimism. Throughout U.S. history, members of diverse racial and ethnic groups have struggled to gain the freedom and rights that were previously withheld from them. Today, minority grassroots organizations are pressing for affordable housing, job training, and educational opportunities. Movements composed of both whites and people of color continue to oppose racism in everyday life, to seek to heal divisions among racial groups, and to teach children about racial tolerance. Many groups hope not only to affect their own microcosm but also to contribute to worldwide efforts to end racism.

To eliminate racial discrimination, it will be necessary to equalize opportunities in schools and workplaces. According to Michael Omi and Howard Winant (2013), it is important for us to be aware of race, rather than ignoring it, if we wish to challenge the problem of racism. If we are aware that race as a social construction exists and has meaning in everyday life, we will gain the political insights necessary to mobilize ourselves and others against injustice and inequality in our society.

# chapter REVIEW

## LO1 How do race and ethnicity differ?

A race is a category of people who have been singled out as inferior or superior, often on the basis of physical characteristics such as skin color, hair texture, or eye shape. An ethnic group is a collection of people distinguished primarily by cultural or national characteristics, including unique cultural traits, a sense of community, a feeling of ethnocentrism, ascribed membership, and territoriality.

## LO2 How do racial and ethnic classifications continue to change in the United States?

Racial classifications in the United States have changed over the past century. If we look at U.S. Census Bureau classifications, for example, we can see how the meaning of race continues to change. First, race is defined by perceived skin color: white or nonwhite. Census 2000 made it possible—for the first time—for individuals to classify themselves as being of more than one race.

## LO3 How are *prejudice, stereotypes, racism, scapegoat,* and *discrimination* defined?

Prejudice is a negative attitude often based on stereotypes, which are overgeneralizations about the appearance, behavior, or other characteristics of all members of a group. Stereotypes are overgeneralizations about the appearance, behavior, or other characteristics of members of particular categories. Racism is a set of attitudes, beliefs, and practices that is used to justify the superior treatment of one racial or ethnic group and the inferior treatment of another racial or ethnic group. A scapegoat is a person or group that is incapable of offering resistance to the hostility or aggression of others. Discrimination involves actions or practices of dominant-group members that have a harmful effect on members of a subordinate group.

## LO4 How do sociologists view racial and ethnic group relations?

Symbolic interactionists claim that intergroup contact may either intensify or reduce racial and ethnic stereotyping and prejudice, depending on the context. In the *contact hypothesis,* symbolic interactionists point out that contact between people from divergent groups should lead to favorable attitudes and behavior *when certain factors are present.* Functionalists stress that members of subordinate groups become a part of the mainstream through assimilation, the process by which members of subordinate groups become absorbed into the dominant culture. Conflict theorists focus on economic stratification and access to power in race and ethnic relations. The caste perspective views inequality as a permanent feature of society, whereas class perspectives focus on the link between capitalism and racial exploitation. According to racial formation theory, the actions of the U.S. government substantially define racial and ethnic relations.

## LO5 What are the historical experiences of Native Americans and WASPs in the United States?

Experts estimate that approximately two million native inhabitants lived in North America in 1492; their numbers had been reduced to fewer than 240,000 by 1900. Native Americans have been the victims of genocide and forced migration. After the Revolutionary War, the federal government broke treaty after treaty as it engaged in a policy of wholesale removal of indigenous nations in order to clear the land for settlement by Anglo-Saxon "pioneers." Data continue to show that Native Americans are the most disadvantaged racial or ethnic group in the United States in terms of income, employment, housing, nutrition, and health. Whereas Native Americans have been among the most disadvantaged peoples, white Anglo-Saxon Protestants (WASPs) have been the most privileged group in this country. Although many English settlers initially came to North America as indentured servants or as prisoners, they quickly emerged as the dominant group, creating a core culture (including language, laws, and holidays) to which all other groups were expected to adapt.

## LO6 How have slavery, segregation, lynching, and persistent discrimination uniquely affected the African American experience in this country?

The African American (black) experience has been one uniquely marked by slavery, segregation, and persistent discrimination. Between 1619 and the 1860s, about 500,000 Africans were forcibly brought to North America, primarily to work on southern plantations, and these actions were justified by the devaluation and stereotyping of African Americans. Following the abolishment of slavery in 1863, African Americans were still subjected to segregation, discrimination, and lynchings. Despite civil rights legislation and economic and political gains by many African Americans, racial prejudice and discrimination continue to exist.

## LO7 What are the major categories of Asian Americans, and what are their historical and contemporary experiences?

The term *Asian Americans* designates the many diverse groups with roots in Asia. Chinese and Japanese immigrants were among the earliest Asian Americans. Many Filipinos, Asian Indians, Koreans, Vietnamese, Cambodians, Pakistani, and Indonesians have arrived more recently. The subgroups are listed as Chinese Americans (the largest Asian American group), Japanese Americans, Korean Americans, Filipino Americans (the second-largest category of Asian Americans), and Indochinese Americans (which include people from Vietnam, Cambodia, Thailand, and Laos). Asian American immigrants as a group have enjoyed considerable upward mobility in U.S. society in recent decades, but many Asian Americans still struggle to survive by working at low-paying jobs and living in urban ethnic enclaves.

## LO8 What have been the unique experiences of Latinos/as (Hispanics) and Middle Eastern Americans in the United States?

Mexican Americans—including both native- and foreign-born people of Mexican origin—are the largest segment (approximately two-thirds) of the Latino/a population in the United States. Today, Puerto Rican Americans make up 9 percent of Hispanic-origin people in the United States. Although some Latinos/as have made substantial political, economic, and professional gains in U.S. society, as a group they are nevertheless subjected to anti-immigration sentiments. Since 1970, many immigrants have arrived in the United States from countries located in the "Middle East," which is the geographic region from Afghanistan to Libya and including Arabia, Cyprus, and Asiatic Turkey. Middle Eastern immigrants to the United States speak a variety of languages and have diverse religious backgrounds. Because they generally come from middle-class backgrounds, they have made inroads into mainstream U.S. society. However, some Middle Eastern Americans experience discrimination based on their speech patterns, appearance, and clothing. The idea that Middle Easterners are somehow associated with terrorism has also been difficult to remove from media representations and some people's thinking, producing ongoing hardship for many upstanding citizens of this nation.

## key TERMS

assimilation 270

authoritarian personality 266

discrimination 266

dominant group 263

ethnic group 261

ethnic pluralism 271

gendered racism 273

genocide 268

individual discrimination 268

institutional discrimination 268

internal colonialism 272

prejudice 264

race 261

racism 265

scapegoat 265

segregation 271

split labor market 273

stereotypes 264

subordinate group 263

theory of racial formation 274

## questions for CRITICAL THINKING

1. Do you consider yourself defined more strongly by your race or by your ethnicity? How so?
2. Given that subordinate groups have some common experiences, why is there such deep conflict between some of these groups?

3. What would need to happen in the United States, both individually and institutionally, for a positive form of ethnic pluralism to flourish in the twenty-first century?

# 10 SEX, GENDER, AND SEXUALITY

SelectStock/Vetta/Getty Images

## LEARNING OBJECTIVES

**LO1** Distinguish between sex and gender.

**LO2** Discuss discrimination based on sexual orientation.

**LO3** Define *gender role, gender identity, body consciousness,* and *sexism.*

**LO4** Describe how the division of labor between women and men differs in various kinds of societies.

**LO5** Identify the primary agents of gender socialization and note their role in socializing people throughout life.

**LO6** Discuss ways in which the contemporary workplace reflects gender stratification.

**LO7** Compare functionalist and conflict perspectives on gender inequality.

**LO8** Describe the feminist perspectives on gender inequality.

# Sociology and Everyday Life

# Child Beauty Pageants and the Sexualization of Girls

**Paisley would not be as well-known if it were** not for that outfit [a pint-sized costume

What sociological concerns do child beauty competitions raise?

*Evan Hurd/Sygma/Corbis*

resembling the clothing of a prostitute portrayed by Julia Roberts in *Pretty Woman*]. . . . Well, at that pageant there was an option to do celebrity wear, and we thought about what we could wear with her being a brunette and Julia Roberts is my favorite actress of all time. I thought it was real cute to do Julia. She's 3, if she was 10 I never would have considered this. But as young as she is I thought it was very comical. . . .

[Turning to Paisley, who is getting dressed for another pageant.] . . . Can you show me how you're going to take your skirt off? [Paisley seductively removes her wraparound skirt to reveal fishnet stockings and a garter belt for her Lady Gaga routine.] . . . Hopefully people will not be calling her a prostitot again.

—Wendy Dickey, mother of a three-year-old child beauty pageant contestant, justifying the use of sexy, often age-inappropriate clothing to get judges' votes so that her daughter can win the pageant's top, Ultimate Grand Supreme, title (qtd. in NowPublic.com, 2011)

**S**hould children and adolescents be allowed to be in beauty pageants? The reality series *Toddlers & Tiaras* opens up the world of child beauty pageants to people who are largely unaware of their existence. Episodes feature numerous young contestants in glitzy outfits, including skimpy theme costumes, gowns, and swimwear. Pageant contenders spend many hours being prepared for their time on stage: They are often worked on by several adults at the same time, each with a different task, such as dressing the contestant, applying extensive makeup, putting in false teeth (flippers), gluing on fake eyelashes, and adding hairpieces to get the "big hair" look. Constant coaching, before and during the pageant, by parents and other "handlers" reminds contestants that they need to flirt with the judges and audience members using seductive gestures (such as blowing kisses) and facial expressions (such as fluttering false eyelashes while looking over their shoulder), and even just in how they walk. For

their efforts, child pageant contestants and their parents are awarded with money, sparkly crowns, and tall trophies. However, unsuccessful contestants may leave feeling that they are "losers" because they were not pretty enough or talented enough to be named the winner. When shows such as *Toddlers & Tiaras* move on to DVD and streaming video, other shows, such as *Kim of Queens* (featuring a former Miss Georgia who helps adolescent girls become future beauty queens), take their place, apparently because of viewer demands for the sexualization of young females.

Child and adolescent beauty competitions raise important sociological concerns about whether activities such as these contribute to young girls becoming overly conscious at an early age about their physical appearance, sexual appeal, body image, weight, and how others evaluate them. Advocates of youth beauty pageants believe that these competitions promote individual achievement and build

PART 3 | SOCIAL INEQUALITY

**296**

When Marleigh was born, the first thing I told my husband was, "Yeah! We've got us a pageant girl." I loved competing in pageants myself because I loved the talent portion where I could actually entertain the crowd that was in front of me.... Marleigh absolutely loves [pageants].... I am definitely a proud mother of my child. I love it when she's on stage. Hopefully, Marleigh will get [crowns along the way] and hopefully she will become Miss America 2028. So tune in!

—Amy, another pageant mom, describing how her daughter loves child beauty competitions despite the fact that the child routinely screams, cries, and hits her mother while on camera (*Toddlers & Tiaras*, 2009)

## How Much Do You Know About Gender, Body Image, and Sexualization?

**True  False**

T  F  1. Girls under 12 years old increasingly wear teen and adult clothing that is designed to accent specific features of the female body and highlight female sexuality.

T  F  2. Preteenage girls place greater emphasis on their body's competence to perform specific tasks (such as making good grades or successfully participating in sports activities) than on their physical appearance.

T  F  3. Many young girls and women believe that being even slightly "overweight" makes them less "feminine."

T  F  4. Physical attractiveness is a more central part of self-concept for women than for men.

**True  False**

T  F  5. Constant exposure to sexualized images of girls in media and society may "train" some men and women to perceive and label sexualized girls as "seductive" and in search of attention that is sexual in nature.

T  F  6. Thinness has always been the "ideal" body image for women.

T  F  7. Women bodybuilders have gained full acceptance in society.

T  F  8. The media play a significant role in shaping societal perceptions about the ideal female body.

Answers on page 299.

self-confidence. Critics of this multimillion-dollar industry claim that the pageants cater to parental insecurities, excessive competitiveness, and the exploitation of young girls, specifically for the financial gain and gratification of parents and the voyeurism of adult audiences (Brooks, 2011). From this perspective, pageants cause, or at least contribute to, low self-esteem, the sexual exploitation of children, and expectations of an unrealistic beauty ideal. In sum, critics argue that beauty pageants *sexualize* and *objectify* girls and women.

The American Psychological Association (2010) defines **sexualization as the acts or processes whereby an individual or group is seen as sexual in nature or persons become aware of their sexuality.** As such, sexualization can have both cognitive and emotional consequences that occur when persons are sexualized: (1) a person's value comes only from his or her sexual appeal or sexual behavior, to the exclusion of other characteristics; (2) a person is held to a standard that equates physical attractiveness (narrowly defined) with being sexy; (3) a person is sexually objectified; and/or (4) sexuality is inappropriately imposed upon a person.

What is objectification? *Objectification* is the process whereby some people treat other individuals as if they were objects or things, not human beings. For example, we objectify women—or men—when we judge them strictly on the basis of their physical appearance, rather than on their individual qualities, attributes, or actions (Hatton and Trautner, 2011). Although men may be objectified in some societies, the objectification of

**sexualization** the acts or processes whereby an individual or group is seen as sexual in nature or persons become aware of their sexuality.

## table 10.1

### The Objectification of Women

| General Aspects of Objectification | Objectification Based on Cultural Preoccupation with "Looks" |
|---|---|
| Women are responded to primarily as "females," whereas their personal qualities and accomplishments are of secondary importance. | Women are often seen as the objects of sexual attraction, not full human beings—for example, when they are stared at because of their physical characteristics. |
| Women are seen as "all alike." | Women are seen by some as depersonalized body parts—for example, "a piece of ass." |
| Women are seen as being subordinate and passive, so things can easily be "done to a woman"—for example, discrimination, harassment, and violence. | Depersonalized female sexuality is used for cultural and economic purposes—such as in the media, advertising, the fashion and cosmetics industries, and pornography. |
| Women are seen as easily ignored or trivialized. | Women are seen as being "decorative" and status-conferring objects to be bought (sometimes collected) and displayed by men and sometimes by other women. |
| Women are judged on appearance more than men are. | Women are evaluated according to prevailing, narrow "beauty" standards and often feel pressure to conform to appearance norms. |

Source: Adapted from Schur, 1983. © Cengage Learning

girls and women is widespread and particularly common in the United States and many other nations (see Table 10.1). In regard to child beauty pageants, organizers seek to deflect this criticism by providing contestants with an opportunity to have fun showing off their beauty and talent.

The way that people think about beauty pageants, particularly those for young children, provides important information about a society's cultural norms, expectations, and values in regard to women and men. This social construction of reality involves what we consider to be appropriate, or inappropriate, behavior for males and females at different ages. Obviously, some differences between men and women are biological in nature. However, many differences between the sexes are socially constructed. Studying sociology makes us aware of differences that relate to gender (a social concept) as well as differences that are based on a person's biological makeup, or sex. In this chapter we examine the issue of gender: what it is and how it affects us. Before reading on, test your knowledge about body image and gender by taking the "Sociology and Everyday Life" quiz.

## Sex: The Biological Dimension

**LO1** Whereas the word *gender* is often used to refer to the distinctive qualities of men and women (masculinity and femininity) that are *culturally* created, *sex* **refers to the biological and anatomical differences between females and males.** At the core of these biological and anatomical differences is the chromosomal

information transmitted at the moment a child is conceived. The mother contributes an X chromosome and the father either an X (which produces a female embryo) or a Y (which produces a male embryo). At birth, male and female infants are distinguished by *primary sex characteristics:* **the genitalia used in the reproductive process.** At puberty, an increased production of hormones results in the development of *secondary sex characteristics:* **the physical traits (other than reproductive organs) that identify an individual's sex.** For women, these include larger breasts, wider hips, and narrower shoulders; a layer of fatty tissue throughout the body; and menstruation. For men, they include development of enlarged genitals, a deeper voice, greater height, a more muscular build, and more body and facial hair.

### Intersexed and Transgendered Persons

Sex is not always clear-cut. An *intersexed person* **is an individual who is born with a reproductive or sexual anatomy that does not correspond to typical definitions of male or female; in other words, the person's sexual differentiation is ambiguous.** Formerly referred to as *hermaphrodites* by some in the medical community, intersexed persons may appear to be female on the outside at birth but have mostly male-type anatomy on the inside, or they may be born with genitals that appear to be in between the usual male and female types. For example, a chromosomally normal (XY) male may be born with a penis just one centimeter long and a urinary opening similar to

## ANSWERS to the Sociology Quiz on Gender, Body Image, and Sexualization

**1. True.** Clothing designers, marketers, and retail stores have worked to sell a unique identity to girls by way of clothing that highlights female sexuality (such as thongs) and having an overall adult appearance at an early age.

**2. False.** Research has shown that girls as young as 12 placed greater emphasis on their body's appearance than on their competence to perform specific activities and tasks.

**3. True.** More than half of all adult women in the United States are currently dieting, and over three-fourths of normal-weight women think they are "too fat." Recently, very young girls have developed similar concerns.

**4. True.** Women have been socialized to believe that being physically attractive is very important. Studies have found that weight and body shape are the central determinants of women's perception of their physical attractiveness.

**5. True.** Studies continue to document the fact that constant exposure to sexualized images of girls in media and society does encourage some people to view sexualized girls as "seductive" and to sometimes produce unwanted sexual advances.

**6. False.** The "ideal" body image for women has changed a number of times. A positive view of body fat has prevailed for most of human history; however, in the twentieth century in the United States, this view gave way to "fat aversion."

**7. False.** Although bodybuilding among women has gained some degree of acceptance, women bodybuilders are still expected to be very "feminine" and not to overdevelop themselves.

**8. True.** Women in the United States are bombarded by advertising, television programs, and films containing images of women that typically represent an ideal that most real women cannot attain.

Sources: Based on American Psychological Association, 2010; Grogan, 2007; Sciencedaily.com, 2008; and Zurbriggen and Roberts, 2012. © Cengage Learning

---

that of a female. However, although intersexuality is considered to be an inborn condition, intersexed anatomy is not always known or visible at birth. In fact, intersexual anatomy sometimes does not become apparent until puberty, or when an adult is found to be infertile, or when an autopsy is performed after death. It is possible for some intersexual people to live and die with intersexed anatomy but never know that the condition exists. According to the Intersex Society of North America (2011),

> Intersex is a socially constructed category that reflects real biological variation. Nature presents us with sex anatomy spectrums [, but] nature doesn't decide where the category of "male" ends and the category of "intersex" begins, or where the category of "intersex" ends and the category of "female" begins. *Humans decide.* Humans (today, typically doctors) decide how small a penis has to be, or how unusual a combination of parts has to be, before it counts as intersex. Humans decide whether a person with XXY chromosomes and XY chromosomes and androgen insensitivity will count as intersex.

Some people may be genetically of one sex but have a gender identity of the other. That is true for a *transgendered person*—an individual whose gender identity (self-identification as woman, man, neither, or both) does not match the person's assigned sex (identification by others as male, female, or intersex based on physical/genetic sex). Consequently, transgendered persons may believe that they have the opposite gender identity from that

**sex** the biological and anatomical differences between females and males.

**primary sex characteristics** the genitalia used in the reproductive process.

**secondary sex characteristics** the physical traits (other than reproductive organs) that identify an individual's sex.

**intersexed person** an individual who is born with a reproductive or sexual anatomy that does not correspond to typical definitions of male or female; in other words, the person's sexual differentiation is ambiguous.

**transgendered person** an individual whose gender identity (self-identification as woman, man, neither, or both) does not match the person's assigned sex (identification by others as male, female, or intersex based on physical/genetic sex).

**299**

**FIGURE 10.1** Chaz Bono (formerly known as Chastity, the only daughter of entertainers Sonny and Cher) underwent a physical and social gender transition from female to male. In the documentary film, *Becoming Chaz*, Bono describes his gender transformation and social activism for lesbian, gay, bisexual, and transgender rights. What influence do you think high-profile people like Chaz Bono have on the attitudes and actions of other people in regard to the LGBT community?

of their sex organs and may be aware of this conflict between gender identity and physical sex as early as the preschool years. Some transgendered individuals choose to take hormone treatments or have a sex-change operation to alter their genitalia so that they can have a body congruent with their sense of gender identity (Figure 10.1). Many then go on to lead lives that they view as being compatible with their true gender identity. But the issue of hormonal and surgical sex reassignment remains highly politicized. The "Standards of Care," a set of guidelines set up by the Harry Benjamin International Gender Dysphoria Association, establishes standards by which transgendered persons may obtain hormonal and surgical sex reassignment to help ensure that people choosing such options are informed about what is involved in a gender transition.

Western societies acknowledge the existence of only two sexes; some other societies recognize three—men, women, and *berdaches* (or *hijras* or *xaniths*): biological males who behave, dress, work, and are treated in most respects as women. The closest approximation of a third sex in Western societies is a ***cross-dresser*** (formerly known as a *transvestite*), **a male who dresses as a woman or a female who dresses as a man but does not alter his or her genitalia.** Although cross-dressers are not treated as a third sex, they

often "pass" for members of that sex because their appearance and mannerisms fall within the range of what is expected from members of the other sex. Most cross-dressers are heterosexual men, many of whom are married, but gay men, lesbians, and straight women may also be cross-dressers. Cross-dressing can occur in conjunction with homosexuality, but this is frequently not the case. Researchers and analysts continue to engage in dialogue about the correct terminology to use when referring to persons in the diverse groups that now make up this segment of the population.

## Sexual Orientation

***Sexual orientation*** **refers to an individual's preference for emotional–sexual relationships with members of the different sex (heterosexuality), the same sex (homosexuality), or both (bisexuality).** In referring to homosexuality, many organizations representing lesbian, gay, bisexual, and transgendered persons have adopted the acronym *LGBT* (or *GLBT*). The term *lesbian* refers to females who prefer same-sex relationships; *gay* refers to males who prefer same-sex relationships. As noted above, *bisexual* is the term used to describe a person's physical or romantic attraction to both males and females, whereas *transgendered* is a term applied to persons whose appearance, behavior, and/or gender identity does not match that individual's assigned sex.

What criteria have social scientists used to study sexual orientation? A definitive study of sexuality conducted by researchers at the University of Chicago established three criteria for identifying people as homosexual or bisexual: (1) *sexual attraction* to persons of one's own gender, (2) *sexual involvement* with one or more persons of one's own gender, and (3) *self-identification* as a gay, lesbian, or bisexual (Michael et al., 1994). According to these criteria, then, having engaged in a homosexual act does not necessarily classify a person as homosexual. In fact, many respondents in the University of Chicago study indicated that although they had at least one homosexual encounter when they were younger, they were no longer involved in homosexual conduct and never identified themselves as gay, lesbian, or bisexual.

**Measuring Sexual Orientation** It is difficult to determine how many people identify as LGBT because of a lack of official statistics. In 2012, for the first time the Gallup survey asked this question: "Do you, personally, identify as lesbian, gay, bisexual, or transgender?" More than 120,000 people

PART 3 | SOCIAL INEQUALITY

responded to this survey, making it the largest study of its kind to date, and about 3.4 percent of U.S. adults answered "yes" to the question, thereby self-identifying as lesbian, gay, bisexual, or transgendered (Gates and Newport, 2012). Unfortunately, it is not possible to separately consider differences among lesbians, gay men, bisexuals, or transgendered individuals because of the way the data were collected.

Gallup researchers emphasize that measuring sexual orientation and gender identity is "challenging since these concepts involve complex social and cultural patterns" (Gates and Newport, 2012: 2). Because of a lingering social stigma attached to the LGBT identity, people are not always forthcoming about their identity when asked to respond to a survey. As a result, an unknown number of individuals remain in what is often referred to as "the closet" and are not included in estimates of the LGBT population.

**LGBT Population Estimates** What does the Gallup study show about the LGBT community? Although the typical media portrayal of the LGBT community is of persons who are white (non-Hispanic), highly educated, and affluent, Gallup found that nonwhites are more likely than whites to identify as LGBT. For example, 4.6 percent of African Americans, 4.3 percent of Asian Americans, and 4.0 percent of Hispanics identified as LGBT in the survey. This means that about one-third (33 percent) of LGBT identifiers in the Gallup study are nonwhite, as compared with 27 percent of non-LGBT individuals. Persons with lower levels of education were also more likely to identify as LGBT: 3.5 percent of those who identified as LGBT had a high school education or less, 4.0 percent had some college education but not a college degree, 2.8 percent had a college degree, and 3.2 percent had postgraduate education (Gates and Newport, 2012). A higher proportion of people with lower incomes identify as LGBT: More than 5 percent of those with incomes of less than $24,000 per year self-identified, as compared with 2.8 percent of those making $60,000 per year or more. However, among Gallup participants who reported their income, about 16 percent of LGBT-identified individuals had incomes above $90,000 per year (Gates and Newport, 2012: 5).

Women accounted for more than 53 percent of LGBT individuals who self-identified in the Gallup study, and young adults between the ages of 18 and 29 were more than three times as likely as persons over the age of 65 to identify as LGBT. Younger women between the ages of 18 and 29 were more likely to self-identify (8.3 percent) than men the same age (4.6 percent). These figures may reflect continuing societal opposition among some political leaders and persons in the general public to

equal rights and opportunities for persons in the LGBT community. (To see the full report, go to the Gallup website and search for "Special Report: 3.4% of U.S. Adults Identify as LGBT.")

How valid are these estimates of the LGBT population in the United States? The figures from Gallup are relatively consistent with a previous study by the Williams Institute at the University of California at Los Angeles School of Law, in which researchers estimated that approximately nine million people (about 3.8 percent of all Americans) identify as gay, lesbian, bisexual, or transgendered. According to this report, bisexuals make up 1.8 percent of the U.S. population, with more women than men typically identifying as bisexual. Only a slightly smaller proportion (1.7 percent) identify as being gay or lesbian. Transgendered adults make up 0.3 percent of the population (Gates, 2011).

Gallup and other organizations will no doubt continue to engage in research on the LGBT population because many pressing social issues, including same-sex marriage and LGBT child-rearing rights, may be informed by additional information on the lives of LGBT persons and their families (Gates and Newport, 2012).

## Discrimination Based on Sexual Orientation

**LO2** The United States has numerous forms of discrimination based on sexual orientation. In most states, gay and lesbian couples cannot enter into legally recognized marital relationships because the states have passed constitutional amendments that limit marriage to a union between a man and a woman or because legislators have passed statutes with similar language. As well, lesbian and gay couples continue to fight for parental rights in a number of states because partners who want to adopt a child or are raising children together (typically from a previous heterosexual marriage) learn that only one partner is legally recognized as the child's parent or guardian.

In the past decade, many people have come forward to declare their support for the LGBT cause,

**cross-dresser** a male who dresses as a woman or a female who dresses as a man but does not alter his or her genitalia.

**sexual orientation** an individual's preference for emotional–sexual relationships with members of the different sex (heterosexuality), the same sex (homosexuality), or both (bisexuality).

demanding equal rights for all people to marry, to obtain custody of children, to adopt, and to have their property pass to one another at death—to do all the things that people in heterosexual marriages are permitted to do. As of June 2014, same-sex marriage was legal in eighteen states: California, Connecticut, Delaware, Hawaii, Illinois, Iowa, Maine, Maryland, Massachusetts, Minnesota, New Hampshire, New Mexico, New Jersey, New York, Pennsylvania, Rhode Island, Washington, and Vermont, along with the District of Columbia (Washington, D.C.). At that time, 32 states had gay marriage bans, through laws, constitutional amendments, or both. A higher percentage of the overall U.S. population now lives in states where same-sex marriage has been legalized either by statute or constitutional amendment. More states are likely to follow this path in the aftermath of two significant U.S. Supreme Court decisions in 2013. One case lifted bans on federal benefits for same-sex spouses who resided in states where gay couples are able to legally marry under state law. Striking down the 1996 Defense of Marriage Act, which denied federal benefits to gay couples even in those states where same-sex marriage is legal, the Supreme Court opened the way for more states to extend recognition for gay marriage. In another 2013 decision the Supreme Court ruled that California's Proposition 8, a 2008 voter initiative that ended same-sex marriages in that state, is unconstitutional and opened the way for marriage licenses once again to be issued to same-sex couples residing in California. However, neither of these rulings established that there is a federal constitutional right to same-sex marriage. Battles over this issue are likely to continue in states where same-sex marriage is barred as opponents of gay marriage continue to fight against legalized same-sex marriage on a state-by-state basis. As a result, some same-sex couples in the United States may choose to marry in other countries that provide legal recognition for same-sex unions, such as Canada, Belgium, Spain, and the Netherlands. Others in the LGBT population live in domestic partnerships, but many find this arrangement does not provide the legal protections afforded by state-sanctioned marriage.

The LGBT community has struggled to gain the same parental rights in regard to legal and physical custody of children as heterosexual couples, including the right to physical access or visitation with a child, and various other rights pertaining to the property and well-being of a child. If gay and lesbian couples are denied parental rights by law and in the courts of the land, they have little or no legal recourse and are unable to exert authority over their children's lives, health care, or property.

Housing discrimination is a problem in the LGBT community because the Fair Housing Act, which affords some redress for some other minority groups, does not apply. For example, gays and lesbians have been discriminated against by real estate agents who refuse to show them houses in "family-oriented" apartments, condo buildings, or neighborhoods. Some finance and insurance companies have treated same-sex couples differently from other prospective home buyers or lessees. Transgendered people have been particularly harmed by discriminatory practices in housing. One study found that transgendered respondents were four times more likely to live in extreme poverty, and one in five respondents stated that they had experienced homelessness at some time in the past because of their gender identity (thetaskforce.org, 2011).

Health care is another area of discrimination based on sexual orientation. Many LGBT people are turned away from health care facilities or face overt discrimination when they are sick and seeking medical care. Large numbers of LGBT people are unable to afford the high cost of health insurance coverage, and they are unable to acquire employer-provided health insurance because they are often not allowed by their partner's employer to be counted as a dependent under the partner's insurance plan. Still other LGBT individuals are denied insurance on the basis of preexisting conditions such as HIV/AIDS. The problem is especially pronounced among transgendered people, some of whom report that they have been refused care because of bias. Transgendered respondents also have over four times the national average of HIV infection, which contributes to some health care professionals' lack of desire to provide medical treatment. Further confounding the problem of discrimination is the race, ethnicity, and/or class of LGBT persons.

Occupational discrimination remains a pressing problem for people in the LGBT community. Despite laws in many states prohibiting discrimination in employment on the basis of sexual orientation, openly gay and lesbian people often experience bias in hiring, retention, and promotion in public-sector and private-sector employment.

One of the most widely publicized forms of discrimination against gays and lesbians was in the military, where the 1993 "Don't Ask, Don't Tell" policy implemented during the Clinton administration required that commanders not ask a service person about his or her sexual orientation. Gays and lesbians were allowed to serve in the military as long as they did not reveal their orientation. However, various studies showed that this policy led to differential treatment of many gays and lesbians in the military. As many as 13,000 military personnel may have been discharged under this law, and gay rights organizations advocated for its repeal, arguing that the rules were discriminatory and that they kept gay troops from seeking medical care or reporting domestic abuse for fear of being exposed and expelled from their military branch. In 2010 President Barack Obama signed the repeal of the policy, thus allowing gay and lesbian Americans to serve openly in the armed forces.

Various organizations of gays, lesbians, and transgendered persons have been unified in their desire to reduce discrimination and other forms of *homophobia*—**extreme prejudice and sometimes discriminatory actions directed at gays, lesbians, bisexuals, and others who are perceived as not being heterosexual.** Homophobia involves an aversion to LGBT people or their lifestyle or culture, and it sometimes includes behavior or an act, such as a hate crime, based on this aversion. Because of violence against LGBT individuals in the past, laws have been passed such as the Matthew Shepard and James Byrd Jr. Hate Crimes Prevention Act that attempt to prevent these crimes or to bring to justice individuals who perpetrate them.

**FIGURE 10.2** In 2010 Constance McMillen, a high school student in Aberdeen, Mississippi, wanted to attend the senior prom with another young woman as her date. The county school board refused. After a court blocked the school board's decision, McMillen and her friend attended the prom to find only five other students there. The "real" prom was apparently held in a secret location.

Some analysts use the term *heterosexism* to describe an ideological system that denies, denigrates, and stigmatizes any nonheterosexual form of behavior, identity, relationship, or community (Figure 10.2). This term is used as a parallel to other forms of prejudice and discrimination, including racism, sexism, ageism, and anti-Semitism. Clearly, from this perspective, issues pertaining to homosexuality and heterosexism are not just biological issues but also social constructions that involve societal customs and institutions. Let's turn to the cultural dimension of gender to see how socially constructed differences between females and males are crucial in determining how we identify ourselves as girls or boys, women or men.

## Gender: The Cultural Dimension

**LO3** *Gender* refers to the culturally and socially constructed differences between females and males found in the meanings, beliefs, and practices associated with "femininity" and "masculinity."** Although biological differences between women and men are very important, in reality most "sex differences" are socially constructed "gender differences." According to sociologists, social and cultural processes, not biological "givens," are most important in defining what females and males are, what they should do, and what sorts of relations do or should exist between them. In a now-classic statement, the sociologist Judith Lorber (1994: 6) summarized the importance of gender:

> Gender is a human invention, like language, kinship, religion, and technology; like them, gender organizes human social life in culturally patterned ways. Gender organizes social relations in everyday life as well as in the major social structures, such as social class and the hierarchies of bureaucratic organizations.

Virtually everything social in our lives is *gendered:* People continually distinguish between males and

**homophobia** extreme prejudice and sometimes discriminatory actions directed at gays, lesbians, bisexuals, and others who are perceived as not being heterosexual.

**gender** the culturally and socially constructed differences between females and males found in the meanings, beliefs, and practices associated with "femininity" and "masculinity."

**FIGURE 10.3** In what ways do these two figures—a man engaging in housework and a woman serving in the military—contradict traditional gender expectations? Do you see this trend as a healthy one?

females, and evaluate them differentially. Gender is an integral part of the daily experiences of both women and men (Kimmell and Messner, 2012).

A *microlevel* analysis of gender focuses on how individuals learn gender roles and acquire a gender identity. **Gender role refers to the attitudes, behavior, and activities that are socially defined as appropriate for each sex and are learned through the socialization process.** For example, in U.S. society males are traditionally expected to demonstrate aggressiveness and toughness, whereas females are expected to be passive and nurturing. As noted earlier, **gender identity is a person's perception of the self as female or male.** Typically established between eighteen months and three years of age, gender identity is a powerful aspect of our self-concept. Although this identity is an individual perception, it is developed through interaction with others. As a result, most people form a gender identity that matches their biological sex: Most biological females think of themselves as female, and most biological males think of themselves as male (see Figure 10.3). However, some people think of gender as a *continuum* (a continuous succession or whole) in which biological females perceive of themselves as more female than male, and biological males perceive of themselves as more male than female. Of course, this is a matter for individual consideration, as is the issue of body consciousness, which is also a part of gender identity. **Body consciousness is how a person perceives and feels about his or her body;** it also includes an awareness of social conditions in society that contribute to this self-knowledge. As we grow up, we become aware that the physical shape of our bodies subjects us to the approval or disapproval of others.

Being small and weak may be considered positive attributes for women, but they are considered negative characteristics for "true men."

A *macrolevel* analysis of gender examines structural features, external to the individual, that perpetuate gender inequality. Gender is embedded in the images, ideas, and language of a society and is used as a means to divide up work, allocate resources, and distribute power. For example, every society uses gender to assign certain tasks—ranging from child rearing to warfare—to females and to males, and differentially rewards those who perform these duties. These structures have been referred to as *gendered institutions,* meaning that gender is one of the major ways by which social life is organized in all sectors of society.

These institutions are reinforced by a *gender belief system,* which includes all the ideas regarding masculine and feminine attributes that are held to be valid in a society. This belief system is legitimated by religion, science, law, and other societal values. For example, gender belief systems may change over time as gender roles change. Many fathers take care of young children today while women are the primary income earners in the family, and there is a much greater acceptance of this change in roles by both partners. However, popular stereotypes about men and women, as well as cultural norms about gender-appropriate appearance and behavior, still linger and sometimes reinforce gendered institutions in society.

## The Social Significance of Gender

Gender is a social construction with important consequences in everyday life. Just as stereotypes regarding race/ethnicity have built-in notions of

**FIGURE 10.4** Not all anorexics are women, and not all bodybuilders are men. However, some analysts suggest that these two issues are manifestations of the same desire: to avoid having soft, flabby flesh.

superiority and inferiority, gender stereotypes hold that men and women are inherently different in attributes, behavior, and aspirations. Stereotypes define men as strong, rational, dominant, independent, and less concerned with their appearance. Women are stereotyped as more emotional, nurturing, dependent, and anxious about their appearance.

The social significance of gender stereotypes is illustrated by eating disorders. The three most common eating problems are anorexia, bulimia, and obesity. Some studies estimate that as many as 65 percent of women between the ages of 25 and 45 have disordered eating behaviors (sciencedaily.com, 2008). With *anorexia,* a person has an overriding obsession with food and thinness that constantly controls his or her activities and eating patterns, resulting in a body weight of less than 85 percent of the average weight for a person of that individual's age and height group. With *bulimia,* a person binges by consuming large quantities of food and then purges the food by induced vomiting, excessive exercise, laxatives, or subsequent fasting. In the past, *obesity* was defined as being 20 percent or more above a person's desirable weight, as established by the medical profession. Today, however, medical professionals use the BMI (body mass index) to define obesity. To determine this index, a person's weight in kilograms is divided by his or her height in meters and squared to yield the BMI. Obesity is defined as a BMI of 30 and above (about 30 pounds overweight for the average person).

In the past it was assumed that the individuals most likely to have eating disorders were white, middle-class, heterosexual women; however, such problems also exist among women of color, working-class women, lesbians, and some men. Explanations regarding the relationship between gender and eating disorders must take into account a complex array of social factors, including gender socialization and women's responses to problems such as racism and emotional, physical, and sexual abuse.

Bodybuilding is another gendered experience (Figure 10.4). *Bodybuilding* is the process of deliberately cultivating an increase in the mass and

**gender role** the attitudes, behavior, and activities that are socially defined as appropriate for each sex and are learned through the socialization process.

**gender identity** a person's perception of the self as female or male.

**body consciousness** how a person perceives and feels about his or her body.

**305**

# Sociology WORKS!

## Institutional Discrimination: Women in a Locker-Room Culture

**News Item: "SunTrust Banks Hit with Sexual Harassment Lawsuit by EEOC."**

Women employees of one of the nation's largest financial services providers claim that they worked in an environment similar to the 1960s "Mad Men" world, as portrayed on the popular AMC TV series, in which women are sexual objects to be groped, continually exposed to vulgar and demeaning language, and fearful of losing their job if they reject unwanted caresses and touching/grabbing by an all-powerful supervisor. Even when sexual harassment is reported to supervisors, the problem is often ignored. (U.S. Equal Employment Opportunity Commission, 2012)

Cases such as the one against SunTrust Banks, filed by the U.S. Equal Employment Opportunity Commission, focus on harmful behavior directed toward women in the workplace. Many of these actions affect women's opportunities for training, gaining new clients, promotion, and pay and bonus equity—all factors related to institutional discrimination, defined in Chapter 9 as the day-to-day practices of organizations and institutions that have a harmful impact on members of subordinate groups. According to many sex-discrimination lawsuits, a "locker room culture" or an "outdated corporate culture" prevails in many prestigious organizations where elite men make everyday decisions with little or no consideration for women who hold positions as managing directors, vice presidents, and associate employees.

Sociological theorizing and research have increased public awareness that the day-to-day practices of organizations and institutions may have a negative and differential effect on individuals who have historically been excluded from workplace settings such as financial-services providers, prestigious banks, and Wall Street firms, which traditionally have been dominated by white males. Although many gains have been made through legislation and litigation to reduce institutional discrimination, recent lawsuits demonstrate that much remains to be done before women truly have equal opportunities in the workplace.

### reflect & analyze

Why is it important for all employees to feel that they are being treated fairly at work? Are some employment settings more resistant to change than others? What do you think?

---

strength of the skeletal muscles by means of lifting and pushing weights. In the past, bodybuilding was predominantly a male activity; musculature connoted power, domination, and virility. Today, however, an increasing number of women engage in this activity. As gendered experiences, eating problems and bodybuilding have more in common than we might think. As some women's studies scholars have pointed out, the anorexic body and the muscled body are not opposites: Both are united against the common enemy of soft, flabby flesh. In other words, the *body* may be objectified through both compulsive dieting and compulsive bodybuilding.

### Sexism

*Sexism* **is the subordination of one sex, usually female, based on the assumed superiority of the other sex.** Sexism directed at women has three components: (1) negative attitudes toward women; (2) stereotypical beliefs that reinforce, complement, or justify the prejudice; and (3) discrimination—acts that exclude, distance, or keep women separate.

Can men be victims of sexism? Although women are more often the target of sexist remarks and practices, men can be victims of sexist assumptions. Examples of sexism directed against men are the assumption that men should not be employed in certain female-dominated occupations, such as nurse or elementary school teacher, and the belief that it is somehow more harmful for families when female soldiers are killed in battle than male soldiers.

Like racism, sexism is used to justify discriminatory treatment. Obvious manifestations of sexism are found in the undervaluing of women's work and in hiring and promotion practices that effectively exclude women from an organization or confine them to the bottom of the organizational hierarchy. Even today, some women who enter nontraditional occupations (such as firefighting and welding) or professions (such as dentistry, architecture, or investment banking) encounter hurdles that men do not face (see "Sociology Works!").

Sexism is interwoven with *patriarchy*—**a hierarchical system of social organization in which cultural, political, and economic structures are**

**table 10.2**

Technoeconomic Bases of Society

| | Hunting and Gathering | Horticultural and Pastoral | Agrarian | Industrial | Postindustrial |
|---|---|---|---|---|---|
| Change from Prior Society | — | Use of hand tools, such as digging stick and hoe | Use of animal-drawn plows and equipment | Invention of steam engine | Invention of computer and development of "high-tech" society` |
| Economic Characteristics | Hunting game, gathering roots and berries | Planting crops, domestication of animals for food | Labor-intensive farming | Mechanized production of goods | Information and service economy |
| Control of Surplus | None | Men begin to control societies | Men who own land or herds | Men who own means of production | Corporate shareholders and high-tech entrepreneurs |
| Women's Status | Relative equality | Decreasing in move to pastoralism | Low | Low | Varies by class, race, and age |

Source: Adapted from Lorber, 1994: 140. © Cengage Learning

controlled by men. By contrast, *matriarchy* is a hierarchical system of social organization in which cultural, political, and economic structures are controlled by women; however, few (if any) societies have been organized in this manner. Patriarchy is reflected in the way that men may think of their position as men as a given, whereas women may deliberate on what their position in society should be. As the sociologist Virginia Cyrus (1993: 6) explains, "Under patriarchy, men are seen as 'natural' heads of households, Presidential candidates, corporate executives, college presidents, etc. Women, on the other hand, are men's subordinates, playing such supportive roles as housewife, mother, nurse, and secretary." Gender inequality and a division of labor based on male dominance are nearly universal, as we will see in the following discussion on the origins of gender-based stratification.

## Gender Stratification in Historical and Contemporary Perspective

**LO4** How do tasks in a society come to be defined as "men's work" or "women's work"? Three factors are important in determining the gendered division of labor in a society: (1) the type of subsistence base, (2) the supply of and demand for labor, and (3) the extent to which women's child-rearing activities are compatible with certain types of work. *Subsistence* refers to the means by which a society gains the basic necessities of life, including food, shelter, and clothing. The three factors vary according to a society's *technoeconomic base*—the level of technology and the organization of the economy in a given society. Five such bases have been identified: hunting and gathering societies, horticultural and pastoral societies, agrarian societies, industrial societies, and postindustrial societies, as shown in Table 10.2.

### Hunting and Gathering Societies

The earliest-known division of labor between women and men is in hunting and gathering societies. While the men hunt for wild game, women gather roots and berries. A relatively equitable relationship exists because neither sex has the ability to provide all the food necessary for survival. When wild game is nearby, both men and women may hunt. When it is far away, hunting becomes incompatible with child rearing (which women tend to do because they breast-feed their young),

**sexism** the subordination of one sex, usually female, based on the assumed superiority of the other sex.

**patriarchy** a hierarchical system of social organization in which cultural, political, and economic structures are controlled by men.

**matriarchy** a hierarchical system of social organization in which cultural, political, and economic structures are controlled by women.

and women are placed at a disadvantage in terms of contributing to the food supply (Lorber, 1994). In most hunting and gathering societies, women are full economic partners with men; relations between them tend to be cooperative and relatively egalitarian (Bonvillain, 2001). Little social stratification of any kind is found because people do not acquire a food surplus.

## Horticultural and Pastoral Societies

In horticultural societies, which first developed ten to twelve thousand years ago, a steady source of food becomes available. People are able to grow their own food because of hand tools, such as the hoe. Women make an important contribution to food production because hoe cultivation is compatible with child care. A fairly high degree of gender equality exists because neither sex controls the food supply.

When inadequate moisture in an area makes planting crops impossible, *pastoralism*—the domestication of large animals to provide food—develops. Herding is primarily done by men, and women contribute relatively little to subsistence production in such societies. In some herding societies, women have relatively low status; their primary value is their ability to produce male offspring so that the family lineage can be preserved and enough males will exist to protect the group against attack.

In contemporary horticultural societies, women do most of the farming while men hunt game, clear land, work with arts and crafts, make tools, participate in religious and ceremonial activities, and engage in war. A combination of horticultural and pastoral activities is found in some contemporary societies in Asia, Africa, the Middle East, and South America. These societies are characterized by more gender inequality than in hunting and gathering societies but less than in agrarian societies (Bonvillain, 2001).

## Agrarian Societies

In agrarian societies, which first developed about eight to ten thousand years ago, gender inequality and male dominance become institutionalized. The most extreme form of gender inequality developed about five thousand years ago in societies in the fertile crescent around the Mediterranean Sea. Agrarian societies rely on agriculture—farming done by animal-drawn or mechanically powered plows and equipment. Because agrarian tasks require more labor and greater physical strength than horticultural ones, men become more involved in food production. It has been suggested that women are excluded from these tasks because they are viewed as too weak for the work and because child-care responsibilities are considered incompatible with the full-time labor that the tasks require.

Why does gender inequality increase in agrarian societies? Scholars cannot agree on an answer; some suggest that it results from private ownership of property. When people no longer have to move continually in search of food, they can acquire a surplus. Men gain control over the disposition of the surplus and the kinship system, and this control serves men's interests. The importance of producing "legitimate" heirs to inherit the surplus increases significantly, and women's lives become more secluded and restricted as men attempt to ensure the legitimacy of their children. Premarital virginity and marital fidelity are required; indiscretions are punished. However, some scholars argue that male dominance existed before the private ownership of property (Firestone, 1970; Lerner, 1986).

Male dominance is very strong in agrarian societies. Women are secluded, subordinated, and mutilated as a means of regulating their sexuality and protecting paternity. Most of the world's population currently lives in agrarian societies in various stages of industrialization.

## Industrial Societies

An *industrial society* is one in which factory or mechanized production has replaced agriculture as the major form of economic activity. As societies industrialize, the status of women tends to decline further. Industrialization in the United States created a gap between the nonpaid work performed by middle- and upper-class women at home and the paid work that was increasingly performed by men and unmarried girls. Husbands were responsible for being "breadwinners"; wives were seen as "homemakers."

This gendered division of labor increased the economic and political subordination of women. It also became a source of discrimination against women of color based on both their race and the fact that many of them had to work in order to survive. In the late 1800s and into the 1900s, many African American women were employed as domestic servants in affluent white households.

As people moved from a rural, agricultural lifestyle to an urban existence, body consciousness increased. People who worked in offices often became sedentary and exhibited physical deterioration

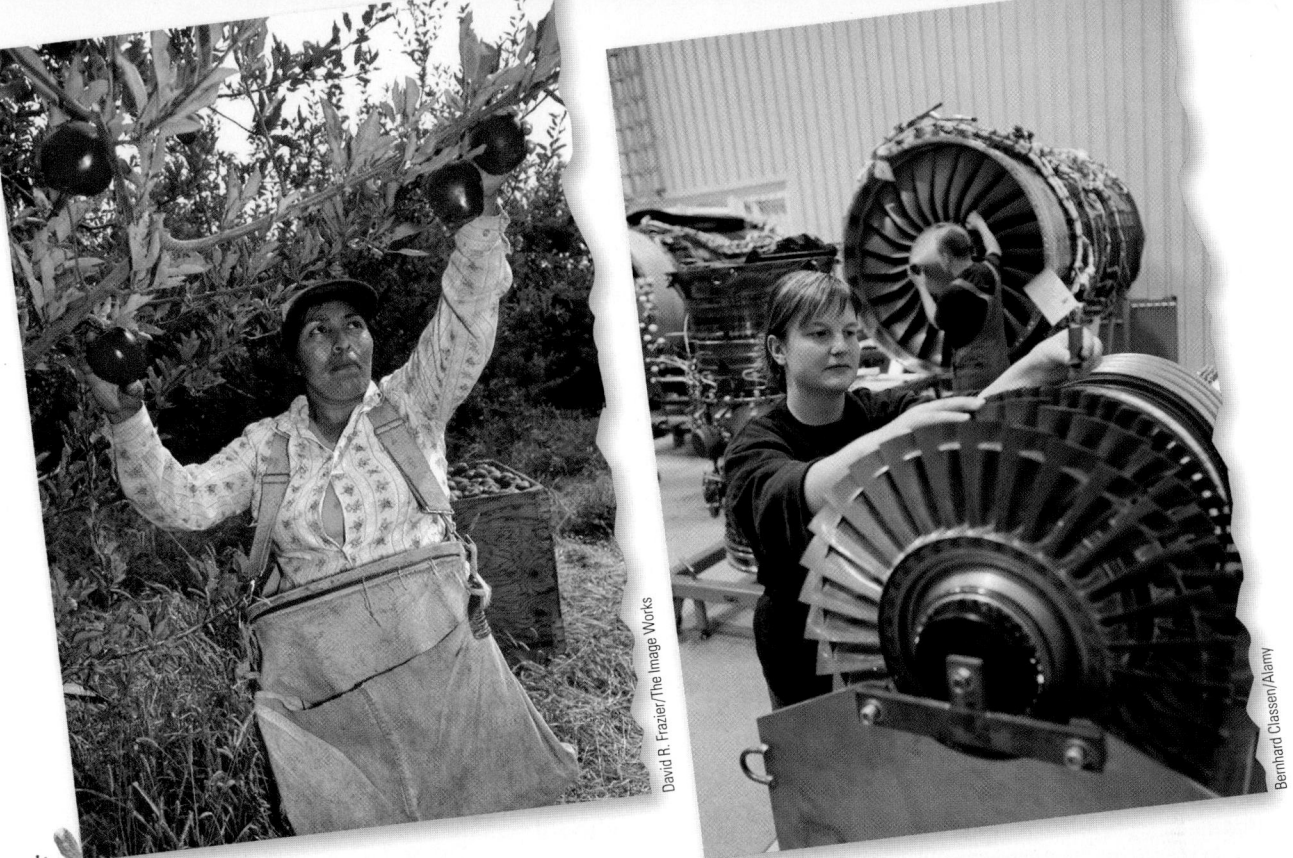

**FIGURE 10.5** In contemporary societies, women do a wide variety of work and are responsible for many diverse tasks. The women shown here are employed in the agricultural and the postindustrial sectors of the U.S. economy. How might issues of gender inequality differ for these two women? What issues might be the same for both of them?

David R. Frazier/The Image Works

Bernhard Classen/Alamy

from their lack of activity. As gymnasiums were built to fight this lack of physical fitness, images of masculinity shifted from the physique of the farmer or factory workman to the middle-class office man who exercised and lifted weights. As industrialization progressed and food became more plentiful, the social symbolism of women's body weight and size also changed, and middle-class women became more preoccupied with body fitness.

## Postindustrial Societies

As previously defined, *postindustrial societies* are ones in which technology supports a service- and information-based economy. In such societies the division of labor in paid employment is increasingly based on whether people provide or apply information or are employed in service jobs such as fast-food restaurant counter help or health care workers. For both women and men in the labor force, formal education is increasingly crucial for economic and social success. However, although some women have moved into entrepreneurial, managerial, and professional occupations, many others have remained in the low-paying service sector, which affords few opportunities for upward advancement (Figure 10.5).

How do new technologies influence gender relations in the workplace? Although some analysts presumed that technological developments would reduce the boundaries between women's and men's work, researchers have found that the gender stereotyping associated with specific jobs has remained remarkably stable even when the nature of work and the skills required to perform it have been radically transformed. Today, men and women continue to be segregated into different occupations, and this segregation is particularly visible within individual workplaces (as discussed later in the chapter).

How does the division of labor change in families in postindustrial societies? For a variety of reasons, more households are headed by women with no adult male present (see "Census Profiles: Single Mothers with Children Under 18"). In 2012 more than 10 million U.S. children lived with their mother only (as contrasted with just 1.9 million who resided with their father only). Among African American children, 50 percent lived with their mother only (Child Statistics, 2012). This means that women in these households truly have a double burden, both from family responsibilities and from the necessity of holding gainful employment in the labor force.

## Single Mothers with Children Under 18

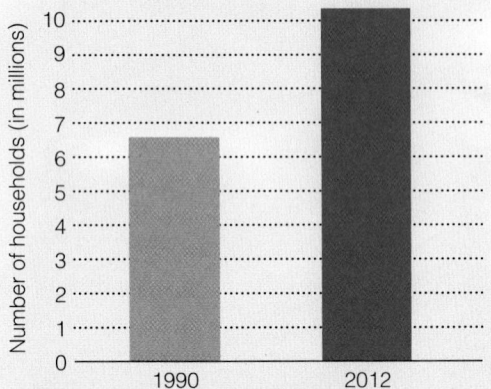

As shown above, the number of U.S. families headed by single mothers increased from almost 6,600,000 in 1990 to an estimated 10,323,000 in 2012. (For census purposes, a single mother is identified as a woman who is widowed, divorced, separated, or never married and who has children under 18 living at home.) In contrast, it is estimated that only 1,956,000 single male householders had children under age 18 living at home. In your opinion, what implications do figures such as these have for gender roles and family relationships in the future?

Source: Vespa, Lewis, and Kreider, 2013.

In postindustrial societies such as the United States, approximately 60 percent of adult women are in the labor force, meaning that finding time to care for children, help aging parents, and meet the demands of the workplace will continue to place a heavy burden on women, despite living in an information- and service-oriented economy.

How people accept new technologies and the effect that these technologies have on gender stratification are related to how people are socialized into gender roles. However, gender-based stratification remains rooted in the larger social structures of society, which individuals have little ability to control.

## Gender and Socialization

**L05** We learn gender-appropriate behavior through the socialization process. Our parents, teachers, friends, and the media all serve as gendered institutions that communicate to us our earliest, and often most lasting, beliefs about the social meanings of being male or female and thinking and behaving in masculine or feminine ways. Some gender roles have changed dramatically in recent years; others have remained largely unchanged over time.

Some parents prefer boys to girls because of stereotypical ideas about the relative importance of males and females to the future of the family and society. Research suggests that social expectations play a major role in this preference. We are socialized to believe that it is important to have a son, especially in the case of a first or only child. For many years it was assumed that only a male child could support his parents in their later years and carry on the family name.

Across cultures, boys are preferred to girls, especially when the number of children that parents can have is limited by law or economic conditions. In China and India, fewer girls are born each year than boys because a disproportionate number of female fetuses are aborted. Starting in the 1970s, China had a one-child-per-family law that favored males over females. However, in 2013 the policy was revised so that couples would be allowed to have two children if one parent was an only child. What effect this will have on the birth of female children remains to be seen. In India a strong cultural belief exists that a boy is an asset to his family while a girl is liability. Beliefs such as this contribute to the selective abortion of female fetuses. As a result of these past practices, nations such as China and India are faced with a shortage of marriageable young women and many other problems that result from an imbalance in the sex ratio. Perhaps seeing the consequences of favoring one sex over the other will produce new ideas among parents regarding sex and gender socialization.

## Parents and Gender Socialization

From birth, parents act differently toward children on the basis of the child's sex. Baby boys are perceived to be less fragile than girls and tend to be treated more roughly by their parents. Girl babies are thought to be "cute, sweet, and cuddly," and receive more-gentle treatment. Parents strongly influence the gender-role development of children by passing on—both overtly and covertly—their own beliefs about gender. Although contemporary parents tend to play more similarly with their male and female children than their own parents or grandparents might have played with them as they were growing up, there remains a difference in how they

**FIGURE 10.6** Are children's toys a reflection of their own preferences and choices? How do toys reflect gender socialization by parents and other adults?

respond toward their children based on gender even when "roughhousing" with them or engaging in sports events or other activities.

Children's toys reflect their parents' gender expectations (Figure 10.6). Gender-appropriate toys for boys include video games, trucks and other vehicles, sports equipment, and war toys such as guns and soldiers. Girls' toys include stuffed animals and dolls, makeup and dress-up clothing, and home-making items. Ads for children's toys appeal to boys and girls differently. Most girl and boy characters are shown in gender-specific toy commercials that target either females or males. These commercials typically show boys playing outdoors and engaging in competitive activities. Girls are more often engaged in cooperative play in the ads, and this is in keeping with gender expectations about their behavior (Kaklenberg and Hein, 2010).

When children are old enough to help with household chores, they are often assigned different tasks. Girls frequently spend more time doing housework than boys (Belkin, 2009). Boys are more likely to be paid for doing chores at home than girls (University of Michigan, 2007). Parents are more likely to assign maintenance chores (such as mowing the lawn) to boys, whereas domestic chores (such as shopping, cooking, clearing the table, and taking care of young siblings) are assigned to girls.

In the past, most studies of gender socialization focused on white, middle-class families and paid little attention to ethnic differences. According to earlier studies, children from middle- and upper-income families are less likely to be assigned gender-linked chores than children from lower-income backgrounds. In addition, gender-linked chore assignments occur less frequently in African American families, where both sons and daughters tend to be socialized toward independence, employment, and child care (McHale et al., 2006). In contrast, gender socialization in Hispanic (Latino/a) families suggests that adolescent females often receive different gender socialization from their parents than do their male siblings. Many Latinas are allowed less interaction with members of the opposite sex than are the adolescent males in their families. Rules for dating, school activities, and part-time jobs are more stringent for girls because many parents want to protect their daughters and keep them closer to home. Moreover, studies continue to show that many Latinas are primarily socialized by their families to become wives and mothers, while less emphasis is placed on educational attainment and careers (Landale and Oropesa, 2007). Some contemporary Latinas find they must struggle with both cultural and structural barriers to achieving their academic and professional goals.

Across classes and racial–ethnic categories, mothers typically play a stronger role in gender socialization of daughters, whereas fathers do more to socialize sons than daughters, particularly when it comes to racial and gender socialization (McHale et al., 2006). However, many parents are aware of the effect that gender socialization has on their children and make a conscientious effort to provide gender-neutral experiences for them.

## Peers and Gender Socialization

Peers help children learn prevailing gender-role stereotypes, as well as gender-appropriate and gender-inappropriate behavior. During the preschool years, same-sex peers have a powerful effect on how children see their gender roles. Children are more socially acceptable to their peers when they conform to implicit societal norms governing the "appropriate" ways that girls and boys should act in social situations and what prohibitions exist in such cases.

Male peer groups place more pressure on boys to do "masculine" things than female peer groups place on girls to do "feminine" things. For example, girls wear jeans and other "boy" clothes, play soccer and softball, and engage in other activities traditionally associated with males. By contrast, if a boy wears a dress, plays hopscotch with girls, and engages in other activities associated with being female, he will be ridiculed by his peers. This distinction between the relative value of boys' and girls' behaviors strengthens the cultural message that masculine activities and behavior are more important and more acceptable.

During adolescence, peers are often stronger and more effective agents of gender socialization than adults. Peers are thought to be especially important in boys' development of gender identity. Male bonding that occurs during adolescence is believed to reinforce masculine identity and to encourage gender stereotypical attitudes and behavior. For example, male peers have a tendency to ridicule and bully others about their appearance, size, and weight. Because peer acceptance is so important, such actions can have very harmful consequences.

As young adults, men and women still receive many gender-related messages from peers. Among college students, for example, peer groups are organized largely around gender relations and play an important role in career choices and the establishment of long-term intimate relationships. In a now-classical study of women college students at two universities (one primarily white, the other predominantly African American), anthropologists Dorothy C. Holland and Margaret A. Eisenhart (1990) found that the peer system propelled women into a world of romance in which their attractiveness to men counted most. Although peers initially did not influence the women's choices of majors and careers, they did influence whether the women continued to pursue their original goals, changed their course of action, or were "derailed." Subsequent research has also found that some African American women, as well as women from other racial–ethnic categories, may change their occupational aspirations partly based on peer-group influence and their social environment (Frome et al., 2006).

## Teachers, Schools, and Gender Socialization

From kindergarten through college, schools operate as a gendered institution. Teachers provide important messages about gender through both the formal content of classroom assignments and informal interactions with students. Sometimes, gender-related messages from teachers and other students reinforce gender roles that have been taught at home; however, teachers may also contradict parental socialization. During the early years of a child's schooling, teachers' influence is very powerful; many children spend more hours per day with their teachers than they do with their own parents.

According to some researchers, the quantity and quality of teacher–student interactions often vary between the education of girls and that of boys (Sadker and Zittleman, 2009). One of the messages that teachers may communicate to students is that boys are more important than girls. Research spanning the past thirty years shows that unintentional gender bias occurs in virtually all educational settings. **Gender bias consists of showing favoritism toward one gender over the other.** Researchers consistently find that teachers devote more time, effort, and attention to boys than to girls (Sadker and Zittleman, 2009). Males receive more praise for their contributions and are called on more frequently in class, even when they do not volunteer.

Teacher–student interactions influence not only students' learning but also their self-esteem (Sadker and Zittleman, 2009). A comprehensive study of gender bias in schools suggested that girls' self-esteem is undermined in school through such experiences as (1) a relative lack of attention from teachers; (2) sexual harassment by male peers; (3) the stereotyping and invisibility of females in textbooks, especially in science and math texts; and (4) test bias based on assumptions about the relative importance of quantitative and visual–spatial ability, as compared with verbal ability, that restricts some girls' chances of being admitted to the most-prestigious colleges and being awarded scholarships.

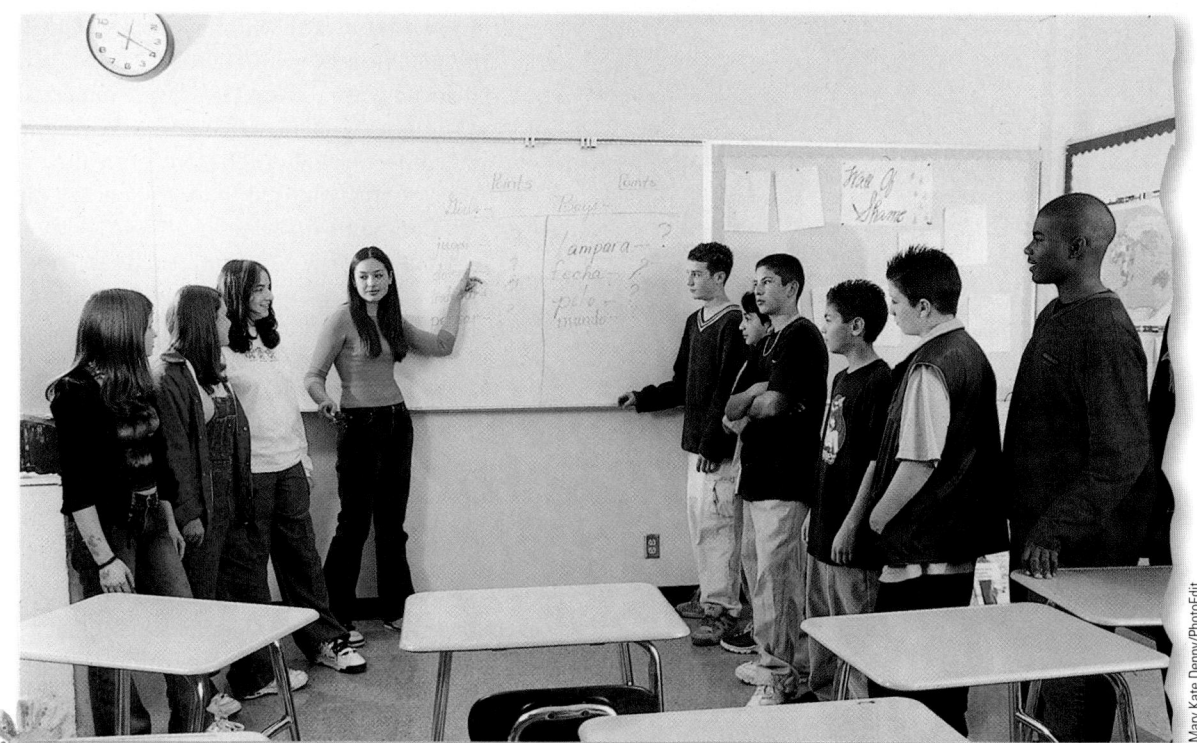

**FIGURE 10.7** Teachers often use competition between boys and girls because they hope to make a learning activity more interesting. Here, a middle-school girl leads other girls against boys in a Spanish translation contest. What are the advantages and disadvantages of gender-based competition in classroom settings?

Teachers also influence how students treat one another during school hours. Many teachers use sex segregation as a way to organize students, resulting in unnecessary competition between females and males (Figure 10.7). In addition, teachers may take a "boys will be boys" attitude when girls complain of sexual harassment. Even though sexual harassment is prohibited by law and teachers and administrators are obligated to investigate such incidents, the complaints may be dealt with superficially. If that happens, the school setting can become a hostile environment rather than a site for learning.

## Sports and Gender Socialization

Children spend more than half of their nonschool time in play and games, but the type of games played differs with the child's sex. Studies indicate that boys are socialized to participate in highly competitive, rule-oriented games with a larger number of participants than games played by girls. Girls have been socialized to play exclusively with others of their own age, in groups of two or three, in activities such as hopscotch and jump rope that involve a minimum of competitiveness. Other research shows that boys express much more favorable attitudes toward physical exertion and exercise than girls do. Some analysts believe this difference in attitude is linked to ideas about what is gender-appropriate behavior

for boys and girls. For males, competitive sport becomes a means of "constructing a masculine identity, a legitimated outlet for violence and aggression, and an avenue for upward mobility" (Lorber, 1994: 43). Recently, more girls have started to play soccer and softball and to participate in sports formerly regarded as exclusively "male" activities (Figure 10.8). Girls who go against the grain and participate in masculine play as children are more likely to participate in sports as young women and adults (Giuliano, Popp, and Knight, 2000).

Many women athletes believe that they have to manage the contradictory statuses of being both "women" and "athletes." An earlier study found that women college basketball players dealt with this contradiction by dividing their lives into segments. On the basketball court, the women "did athlete": They pushed, shoved, fouled, ran hard, sweated, and cursed. Off the court, they "did woman": After the game, they showered, dressed, applied makeup, and styled their hair, even if they were only getting in a van for a long ride home (Watson, 1987). A more recent study found that female athletes who played softball, soccer, or

**gender bias** behavior that shows favoritism toward one gender over the other.

**313**

**FIGURE 10.8** NCAA women's sports such as basketball are popularizing athletics for young women and making it easier for girls to become actively involved in sports at a young age.

basketball engaged in "apologetic behavior" after the game through their efforts to look feminine, their apologies for their aggression during the game, and the way in which they marked themselves as heterosexual (Davis-Delano, Pollock, and Vose, 2009).

## Mass Media and Gender Socialization

The media, including newspapers, magazines, television, and movies, are powerful sources of gender stereotyping. Although some critics argue that the media simply reflect existing gender roles in society, others point out that the media have a unique ability to shape ideas. Think of the impact that television might have on children if they spend one-third of their waking time watching it, as has been estimated. From children's cartoons to adult shows, television programs are sex typed, and many are male oriented. More male than female roles are shown, and male characters act strikingly different from female ones. Typically, males are more aggressive, constructive, and direct, and are rewarded for their actions. By contrast, females are depicted as acting deferential toward other people or as manipulating them through helplessness or seductiveness to get their way.

In prime-time television, a number of significant changes in the past three decades have reduced gender stereotyping; however, men still outnumber women as leading characters, and they are often "in charge" in any setting where both men's and women's roles are portrayed. Recently, retro-series on network and cable television have brought back an earlier era when men were dominant in public and family life and women played a subordinate role to them. For example, the award-winning series *Mad Men* (on AMC) is set in a 1960s New York advertising agency, where secretaries were expected to wear tight sweaters and skirts and bring men hot coffee throughout the day, while the men's wives were supposed to be the perfect companions and hostesses at home. Although many other TV series, such as *Modern Family,* have changed traditional norms, offering a wide diversity of families, including gay dads with a child, the shift to retro-gender roles in some television programming and films in the second decade of the twenty-first century has raised questions about the extent to which change actually occurs in the portrayal of women and men in the media.

Advertising—whether on television and billboards or in magazines and newspapers—can be very persuasive. The intended message is clear to many people: If they embrace traditional notions of masculinity and femininity, their personal and social success is assured; if they purchase the right products and services, they can enhance their appearance and gain power over other people. A study by the sociologist Anthony J. Cortese (2004) found that women—regardless of what they were doing in a particular ad—were frequently shown in advertising as being young, beautiful, and seductive. Although such depictions may sell products, they may also have the effect of influencing how we perceive ourselves and others with regard to issues of power and subordination (see "Gender in the Media").

## Adult Gender Socialization

Gender socialization continues as women and men complete their training or education and join the workforce. Men and women are taught the "appropriate" type of conduct for persons of their sex in a particular job or occupation—both by their employers and by coworkers. However, men's socialization usually does not include a measure of whether their work can be successfully combined with having a family; it is often assumed that men can and will do both. Even today, the reason given for women not entering some careers and professions is that this kind of work is not suitable for women because of their physical capabilities or assumed child-care responsibilities.

Different gender socialization may occur as people reach their forties and enter "middle age." A double standard of aging exists that affects women more than men. Often, men are considered to be at the height of their success as their hair turns gray and their face gains a few wrinkles. By contrast, not only do other people in society make middle-age

# GENDER IN THE MEDIA

## Celebrity Weight Battles Versus "Thinspiration" in Media Representations

When you have famous people turning their weight tribulations into mass-media extravaganzas, they're contributing to a culture where passing comments on strangers' bodies is considered O.K.

>—Lesley Kinzel, who writes the blog *Two Whole Cakes* (formerly *Fatshionista.com*), has argued that shows such as *The Biggest Loser* encourage people to believe that it is acceptable to see other individuals' attempts at weight loss as entertainment (qtd. in Hoffman, 2009).

Stories about today's female Hollywood stars and other celebrities remain the same: How do they look? Celebrities who maintain an average body weight receive little attention about this issue, but those who are judged as overweight or underweight receive unprecedented coverage about their physical appearance. Negative media framing of stories about women's weight is further encouraged by celebrities who willingly make derogatory remarks about their own appearance when they are overweight: "Hideous!" (Kirstie Alley) or "Embarrassing" (Oprah Winfrey). By contrast, journalists who refer to female celebrities who are thin or underweight as "dangerously skinny or anorexic" put them on the defensive.

If the media are setting the standards for beauty and appropriate weight, does this harm anyone? Yes, it may have a negative influence on younger and older persons alike. Research has shown that 80 percent of ten-year-olds fear that they are fat, and many of these perceptions are based on media representations. More than half of teenage girls and nearly one-third of teenage boys engage in unhealthy weight-control behaviors, including skipping meals, fasting, smoking cigarettes, vomiting, and taking laxatives, some of which they also learned from media coverage (ABCNews.com, 2006).

Moreover, "fat bashing" and "thinspiration" in the media may affect people of all ages. In the words of one journalist, media accounts of celebrity weight battles and their "mortification-of-the flesh" narratives may be "toxic, undermining [heavy women's] hard-won self-esteem and exacerbating the

Singer Sheryl Crow displays her thin and highly toned body on the cover of a magazine. Can such an emphasis on thin and sculpted bodies be harmful to women who are not celebrities?

derision they face" (Hoffman, 2009). Similarly, media framing of stories about "thinspiration" and "rapidly shrinking" stars may influence people to believe that you can never be too thin (or too rich).

To counter this perception, celebrities such as Kate Bosworth and Kate Winslet have gone public with their concern that the wrong idea is being conveyed to young girls about eating and weight. These stars emphasize that they do not have an eating disorder but instead spend large amounts of time exercising and eating nutritious food. Some even go so far as to appear on TV talk shows describing how they eat calorie-laden foods such as macaroni and cheese, and have lavish desserts while on vacation. Even this "I'm just like everyone" approach may discourage overweight women, who wonder, "Why can't I eat like that without gaining ten pounds?"

### reflect & analyze

Are media stories about celebrity weight gain or loss just entertainment for audiences? Should we be concerned that media representations about weight and appearance may influence how people, particularly girls and women, think about eating and weight? What do you think?

women feel as if they are "over the hill," but multimillion-dollar advertising campaigns continually call attention to women's every weakness, every pound gained, and every bit of flabby flesh, wrinkle, or gray hair. Increasingly, both women and men have turned to "miracle" products, and sometimes to cosmetic surgery, to reduce the visible signs of aging. However, the vast majority (92 percent) of all cosmetic surgery is performed on female patients (American Society for Aesthetic Plastic Surgery, 2010).

Knowledge of how we develop a gender-related self-concept and learn to feel, think, and act in feminine or masculine ways is important for an understanding of ourselves. Examining gender socialization makes us aware of the effect of our parents, siblings, teachers, friends, and the media on our perspectives about gender. However, the gender-socialization perspective has been criticized on several accounts. Childhood gender-role socialization may not affect people as much as some analysts have suggested. For example, the types of jobs that people take as adults may have less to do with how they were socialized in childhood than with how they are treated in the workplace. From this perspective, women and men will act in ways that bring them the most rewards and produce the fewest punishments. Also, gender-socialization theories can be used to blame women for their own subordination by not taking into account structural barriers that perpetuate gender inequality. We will now examine a few of those structural forces.

# Contemporary Gender Inequality

**LO6** According to feminist scholars, women experience gender inequality as a result of past and present economic, political, and educational discrimination. Women's position in the U.S. workforce reflects the years of subordination that women have experienced in society.

## Gendered Division of Paid Work in the United States

Where people are located in the occupational structure of the labor market has a major impact on their earnings. The workplace may be a gendered institution if jobs are segregated by gender and by race/ethnicity. In a comprehensive study, sociologists Kevin Stainback and Donald Tomaskovic-Devey (2012) describe how data from five million private-sector workplaces that they examined confirm that white men still dominate the management ranks and that workplace segregation, based on both gender and race, is increasing in many employment sectors. Consider, for example, that white men are 68 percent more likely to be in management positions than to be regular staffers, white women are 28 percent less likely to be in management, African American (black) men are 53 percent less likely to be in leadership positions, and African American (black) women are 73 percent less likely to be in management positions as compared to being

regular staffers in the private sector (Stainback and Tomaskovic-Devey, 2012).

*Gender-segregated work* refers to the concentration of women and men in different occupations, jobs, and places of work. The 2010 Census reported that 96 percent of all secretaries in the United States were women while 93.3 percent of all mechanical engineers were men (U.S. Census Bureau, 2011f). To eliminate gender-segregated jobs in the United States, more than half of all men or all women workers would have to change occupations. Moreover, women are severely underrepresented at the top of major U.S. corporations. Slightly less than 17 percent of corporate board seats in the largest companies, such as Walmart and Sysco, were held by women in 2013. Women of color held only 3.2 percent of all board seats, and more than 66 percent of companies had no women of color serving as directors. Out of the top Fortune 500 companies, 135 had no female executive officers, and only 21 had CEOs who were women (Catalyst, 2013). According to a report by Catalyst (2013), when there are few, or no, women in top leadership roles in business, young women lack role models to encourage them to enter the business world.

Although the degree of gender segregation in the professional labor market (including physicians, dentists, lawyers, accountants, and managers) has declined since the 1970s, racial–ethnic segregation has remained deeply embedded in the social structure. Although some change has occurred in recent years, women of color are more likely than their white counterparts to be concentrated in public-sector employment (as public schoolteachers, welfare workers, librarians, public defenders, and faculty members at public colleges, for example) rather than in the private sector (for example, in large corporations, major law firms, and private educational institutions). And it appears that resegregation is occurring in the private sector. According to the study of fifty-eight industries by Stainback and Tomaskovic-Devey (2012), seven had a rise in gender segregation between 2001 and 2005. These include airlines, railroads, and the mining sectors. Similarly, the research found an increase in racial segregation in eighteen industries, including transportation and the lumber and leather manufacturing industries. Across all categories of occupations, white women and all people of color are not evenly represented, as shown in Table 10.3.

*Labor market segmentation*—the division of jobs into categories with distinct working conditions—results in women having separate and unequal jobs. Why does gender-segregated work matter? Although we look more closely at the issue of the pay gap in the following section, it is important to

## table 10.3

### Percentage of the Workforce Represented by Women, African Americans, Hispanics, and Asian Americans in Selected Occupations

The U.S. Census Bureau accumulates data that show what percentage of the total workforce is made up of women, African Americans, Asian Americans, and Hispanics. As used in this table, *women* refers to females in all racial–ethnic categories, whereas *African Americans, Hispanics,* and *Asian Americans* refer to both women and men.

| | Women | African Americans | Hispanics | Asian Americans |
|---|---|---|---|---|
| All occupations | 47.2 | 10.8 | 14.3 | 4.8 |
| Managerial, professional, and related occupations | 51.5 | 8.4 | 7.3 | 6.1 |
| Management occupations | 38.2 | 6.4 | 7.8 | 4.4 |
| Professional and related occupations | 57.4 | 9.2 | 7.1 | 7.0 |
| Architecture and engineering | 12.9 | 5.2 | 6.8 | 9.0 |
| Lawyers | 31.5 | 4.3 | 3.4 | 3.4 |
| Physicians and surgeons | 32.3 | 5.8 | 6.8 | 15.7 |
| Service occupations (all) | 56.8 | 15.3 | 21.3 | 4.9 |
| Food preparation and serving | 55.1 | 11.3 | 22.2 | 5.6 |
| Cleaning and building-service workers | 40.6 | 13.6 | 35.2 | 3.1 |
| Health care service support occupations | 88.9 | 25.5 | 15.2 | 4.1 |
| Grounds maintenance workers | 5.8 | 6.3 | 43.8 | 1.3 |

Source: U.S. Bureau of Labor Statistics, 2013f.

**FIGURE 10.9** What stereotypes are associated with men in female-oriented positions? With women in male-oriented occupations? Do you think such stereotypes will change in the near future?

note here that the pay gap between men and women is the best-documented consequence of gender-segregated work. Most women work in lower-paying, less-prestigious jobs, with less opportunity for advancement than their male counterparts.

Gender-segregated work affects both men and women (Figure 10.9). Men are often kept out of certain types of jobs. Those who enter female-dominated occupations frequently have to justify themselves and prove that they are "real men." Even if these concerns do not push men out of female-dominated occupations, they affect how the men manage their gender identity at work. For example, men in occupations such as nursing tend to emphasize their masculinity, attempt to distance themselves from female colleagues, and try to move quickly into management and supervisory positions.

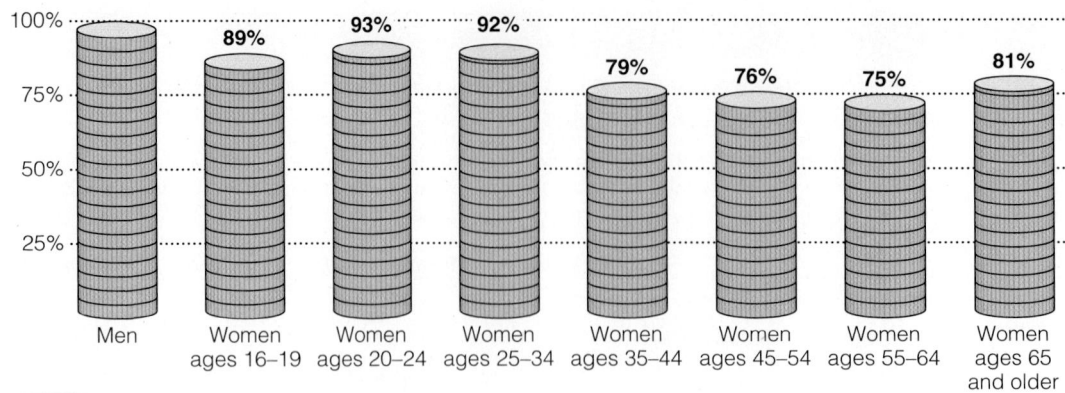

**FIGURE 10.10** THE WAGE GAP, 2012

Source: AAUW, 2013.

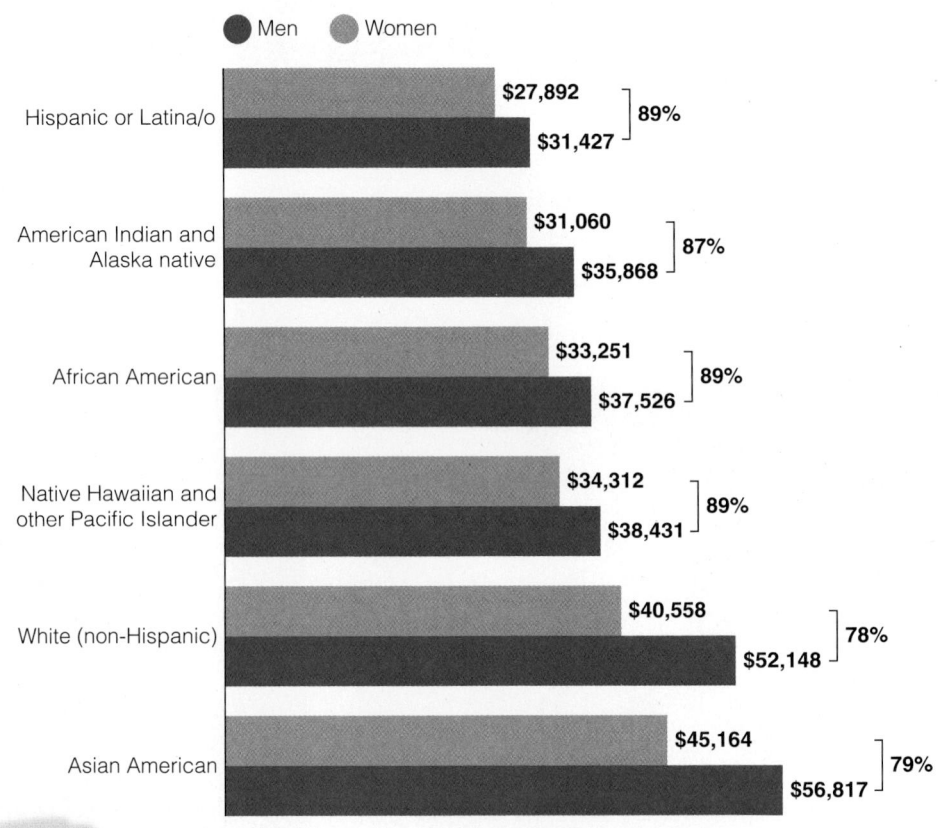

**FIGURE 10.11** WOMEN'S ANNUAL WAGES AS A PERCENTAGE OF MEN'S IN EACH RACIAL–ETHNIC CATEGORY, 2012

Source: AAUW, 2013.

Occupational gender segregation contributes to stratification in society. Job segregation is structural; it does not occur simply because individual workers have different abilities, motivations, and material needs. As a result of gender and racial segregation, employers are able to pay many men of color and all women less money, promote them less often, and provide fewer benefits.

## Pay Equity (Comparable Worth)

Occupational segregation contributes to a **pay gap**—the disparity between women's and men's **earnings.** The pay gap is calculated by dividing women's earnings by men's earnings to yield a percentage, also known as the earnings ratio. In 2012, for example, women who were classified as "full-time wage and salary workers" earned about 77 percent (or 77 cents for every dollar) of the amount earned by men in the same category. However, as Figure 10.10 shows, women in all age categories also receive less pay than men, with the disparity growing wider in the older age brackets.

Earnings differences between women and men in various racial–ethnic categories are the widest for white Americans and Asian Americans. White

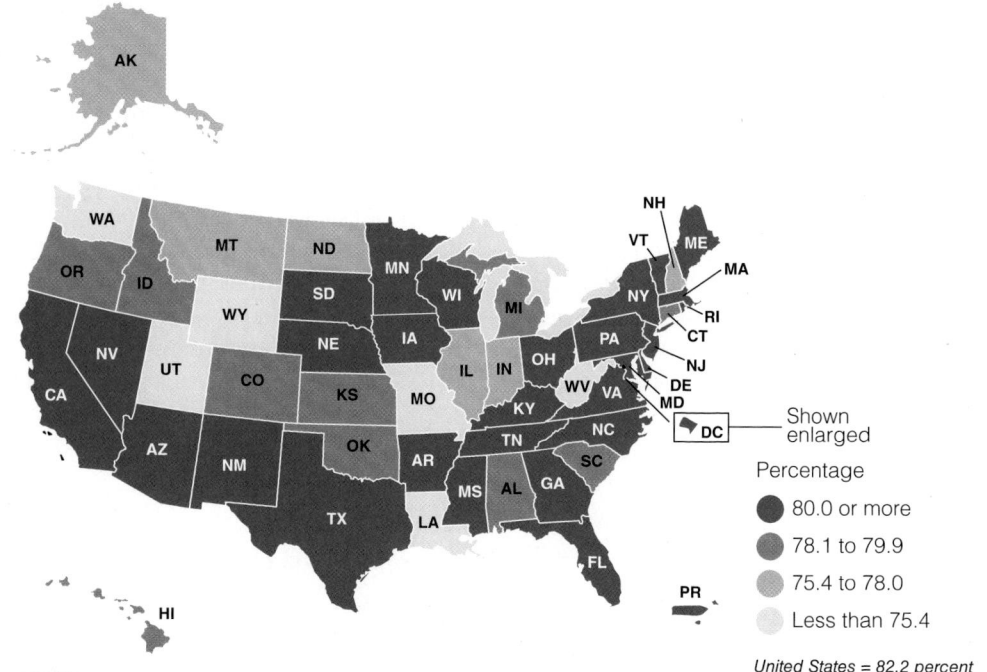

**FIGURE 10.12** WOMEN'S EARNINGS AS A PERCENTAGE OF MEN'S BY STATE AND PUERTO RICO, 2012

Source: U.S. Bureau of Labor Statistics, 2012e.

Percentage

- 80.0 or more
- 78.1 to 79.9
- 75.4 to 78.0
- Less than 75.4

*United States = 82.2 percent*

(non-Hispanic) women's earnings were about 78 percent of their white male counterparts in 2012, while Asian American women earned slightly more: about 79 percent as much as their male counterparts (see Figure 10.11). By comparison, Hispanic women earned about 89 percent as much as their Hispanic male counterparts, American Indian and Alaska Native women earned 87 percent as much as their male counterparts, and African American women earned about 89 percent as much as their male counterparts (AAUW, 2013). It should also be noted that when these figures are compared to white (non-Hispanic) men's earnings, Hispanic women earn only 53 percent as much while American Indian and Alaska Native women, African American women, and Native Hawaiian and other Pacific Islander women are in the 60–69 percent range. For white (non-Hispanic) women, the ratio of women's earnings compared to white men's increases to 78 percent, and for Asian American women, as compared to white men's earnings, the ratio reaches the highest of any category, at 87 percent. Figure 10.12 shows women's median earnings as compared to men's median earnings in each of the fifty states, the District of Columbia, and Puerto Rico.

*Pay equity* or **comparable worth** is the belief that wages ought to reflect the worth of a job, not the gender or race of the worker. How can the comparable worth of different kinds of jobs be determined? One way is to compare the actual work of women's and men's jobs and see if there is a disparity in the salaries paid for each. To do this, analysts break a job into components—such as the education, training, and skills required, the extent of responsibility for others' work, and the working conditions—and then allocate points for each (Lorber, 2005). For pay equity to exist, men and women in occupations that receive the same number of points should be paid the same. However, pay equity exists for very few jobs.

## Paid Work and Family Work

As previously discussed, the first big change in the relationship between family and work occurred with the Industrial Revolution and the rise of capitalism. The cult of domesticity kept many middle- and upper-class women out of the workforce during this period. Primarily, working-class and poor women were the ones who had to deal with the work/family conflict. Today, however, the issue spans the entire economic spectrum. The typical married woman in the United States combines paid work in the labor force and family work as a homemaker. Although

**pay gap** the disparity between women's and men's earnings.

**comparable worth** (or *pay equity*) the belief that wages ought to reflect the worth of a job, not the gender or race of the worker.

this change has occurred at the societal level, individual women bear the brunt of the problem.

Even with dramatic changes in women's workforce participation, the sexual division of labor in the family remained essentially unchanged for many years. Most married women share responsibility for the breadwinner role, yet some men do not accept their full share of domestic responsibilities. Consequently, women may have a "double day" or "second shift" because of their dual responsibilities for paid and unpaid work (Hochschild, 1989, 2003). However, recent studies by the Pew Research Center have shown that the way that mothers and fathers spend their time has changed significantly, with fathers now performing more housework and childcare activities and women being more involved in paid employment than in the past. For both men and women, juggling work and family life constitutes a major concern that may produce stress. Among working mothers, 56 percent reported that they found it difficult to balance work and family life; 50 percent of fathers reported a similar problem (K. Parker, 2013). According to the Pew Research Center report, "Fathers have by no means caught up with mothers in terms of time spent caring for children and doing household chores, but there has been some gender convergence in the way they divide their time between work and home" (K. Parker, 2013). Among dual-income couples, fathers spend about 42 hours each week on paid work, as compared to 31 hours of paid work for mothers. Housework took up an average of 16 hours per week of mothers' time, as compared to 9 hours of fathers' time. Child care accounted for an average of 12 hours per week of mothers' time, as compared to 7 hours per week of fathers' time (K. Parker, 2013).

Problems from the past remain in many households: Working women have less time to spend on housework; if husbands do not participate in routine domestic chores, some chores simply do not get done or get done less often. Although the income that many women earn is essential to the economic survival of their families, they still must spend part of their earnings on family maintenance, such as day-care centers, fast-food restaurants, and laundries, in an attempt to keep up with their obligations.

Especially in families with young children, domestic responsibilities consume a great deal of time and energy. Although some kinds of housework can be put off, the needs of children often cannot be ignored or delayed. When children are ill or school events cannot be scheduled around work, parents (especially mothers) may experience stressful role conflicts ("Shall I be a good employee or a good mother?"). Many working women care not only for themselves, their husbands, and their children but

also for elderly parents or in-laws. Some analysts refer to these women as the "sandwich generation"—caught between the needs of their young children and their elderly relatives. Many women try to solve their time crunch by forgoing leisure time and sleep. When Arlie Hochschild interviewed working mothers, she found that they talked about sleep "the way a hungry person talks about food" (1989: 9). Perhaps this is one reason that in later research, Hochschild (1997) learned that some married women with children found more fulfillment at work and that they worked longer hours because they liked work better than facing the pressures of home.

# Perspectives on Gender Stratification

Sociological perspectives on gender stratification vary in their approach to examining gender roles and power relationships in society. Some focus on the roles of women and men in the domestic sphere; others note the inequalities arising from a gendered division of labor in the workplace. Still others attempt to integrate both the public and private spheres into their analyses.

## Functionalist and Neoclassical Economic Perspectives

**LO7** As seen earlier, functionalist theory views men and women as having distinct roles that are important for the survival of the family and society. The most basic division of labor is biological: Men are physically stronger, and women are the only ones able to bear and nurse children. Gendered belief systems foster assumptions about appropriate behavior for men and women and may have an impact on the types of work that women and men perform.

**The Importance of Traditional Gender Roles** According to functional analysts such as Talcott Parsons (1955), women's roles as nurturers and caregivers are even more pronounced in contemporary industrialized societies. While the husband performs the *instrumental* tasks of providing economic support and making decisions, the wife assumes the *expressive* tasks of providing affection and emotional support for the family. This division of family labor ensures that important societal tasks will be fulfilled; it also provides stability for family members.

This view has been adopted by a number of politically conservative analysts who assert that relationships between men and women are damaged when changes in gender roles occur, and family life suffers as a consequence. From this perspective, the traditional division of labor between men and women is the natural order of the universe.

**The Human Capital Model** Functionalist explanations of occupational gender segregation are similar to neoclassical economic perspectives, such as the human capital model. According to this model, individuals vary widely in the amount of human capital they bring to the labor market. *Human capital* is acquired by education and job training; it is the source of a person's productivity and can be measured in terms of the return on the investment (wages) and the cost (schooling or training).

From this perspective, what individuals earn is the result of their own choices (the kinds of training, education, and experience they accumulate, for example) and of the labor market need (demand) for and availability (supply) of certain kinds of workers at specific points in time. For example, human capital analysts might argue that women diminish their human capital when they leave the labor force to engage in childbearing and child-care activities (Figure 10.13). While women are out of the labor force, their human capital deteriorates from nonuse. When they return to work, women earn lower wages than men because they have fewer years of work experience and have "atrophied human capital" because their education and training may have become obsolete. One study found that over a fifteen-year period, married women with children worked fewer years in paid employment than men living in similar family arrangements. As a result, the women were more likely to work fewer hours in the labor market and be low earners (Rose and Hartman, 2008).

**Evaluation of Functionalist and Neoclassical Economic Perspectives** Although Parsons and other functionalists did not specifically endorse the gendered division of labor, their analysis suggests that it is natural and perhaps inevitable. However, critics argue that problems inherent in traditional gender roles, including the personal role strains of men and women and the social costs to society, are minimized by this approach. For example, men are assumed to be "money machines" for their families when they might prefer to spend more time in child-rearing activities. Also, the woman's place is assumed to be in the home, an assumption that ignores the fact that many women hold jobs because of economic necessity.

Peter Dazeley/The Image Bank/Getty Images

**FIGURE 10.13** According to the human capital model, women may earn less in the labor market because of their child-rearing responsibilities. What other sociological explanations are offered for the lower wages that women receive?

In addition, the functionalist approach does not take a critical look at the structure of society (especially the economic inequalities) that makes educational and occupational opportunities more available to some than to others. Furthermore, it fails to examine the underlying power relations between men and women or to consider the fact that the tasks assigned to women and to men are unequally valued by society. Similarly, the human capital model is rooted in the premise that individuals are evaluated based on their human capital in an open, competitive market where education, training, and other job-enhancing characteristics are taken into account. From this perspective, those who make less money (often men of color and all women) have no one to blame but themselves.

Critics note that instead of blaming people for their choices, we must acknowledge other realities. Wage discrimination occurs in two ways: (1) the wages are higher in male-dominated jobs, occupations, and segments of the labor market, regardless of whether women take time for family duties, and (2) in any job, women and people of color will be paid less.

## Conflict Perspectives

According to many conflict analysts, the gendered division of labor within families and in the workplace results from male control of and dominance over women and resources. Differentials between men and women may exist in terms of economic, political,

**FIGURE 10.14** Although the demographic makeup of the U.S. Senate has been gradually changing in recent decades, men still dominate it, a fact that the conflict perspective attributes to a very old pattern in human societies.

physical, and/or interpersonal power (Figure 10.14). The importance of a male monopoly in any of these arenas depends on the significance of that type of power in a society. In hunting and gathering and horticultural societies, male dominance over women is limited because all members of the society must work in order to survive. In agrarian societies, however, male sexual dominance is at its peak. Male heads of household gain a monopoly not only on physical power but also on economic power, and women become sexual property.

Although men's ability to use physical power to control women diminishes in industrial societies, men still remain the head of household and control the property. In addition, men gain more power through their predominance in the most highly paid and prestigious occupations and the highest elected offices. By contrast, women have the ability in the marriage market to trade their sexual resources, companionship, and emotional support for men's financial support and social status. As a result, women as a group remain subordinate to men.

All men are not equally privileged; some analysts argue that women and men in the upper classes are more privileged, because of their economic power, than men in lower-class positions and all people of color. In industrialized societies, persons who occupy elite positions in corporations, universities, the mass media, and government or who have great wealth have the most power. Most of these are men, however.

Conflict theorists in the Marxist tradition assert that gender stratification results from private ownership of the means of production; some men not only gain control over property and the distribution of goods but also gain power over women. According to Friedrich Engels and Karl Marx, marriage serves to enforce male dominance. Men of the capitalist class instituted monogamous marriage (a gendered institution) so that they could be certain of the paternity of their offspring, especially sons, whom they wanted to inherit their wealth. Feminist analysts have examined this theory, among others, as they have sought to explain male domination and gender stratification.

## Feminist Perspectives

**LO8** *Feminism*—the belief that women and men are equal and should be valued equally and have equal rights—is embraced by many men as well as women. It holds in common with men's studies the view that gender is a socially constructed concept that has important consequences for the lives of all people. According to sociologists, both women and men can be feminists and propose feminist theories because they have much in common as they seek to gain a better understanding of the causes and consequences of gender inequality.

Feminist theory seeks to identify ways in which norms, roles, institutions, and internalized expectations limit women's behavior. It also seeks to demonstrate how women's personal control operates even within the constraints of relative lack of power. In the twenty-first century, feminist theory focuses more on global issues, as shown in "Sociology in Global Perspective."

**Liberal Feminism** In liberal feminism, gender equality is equated with equality of opportunity. The roots of women's oppression lie in their lack of equal civil rights and educational opportunities. Only when these constraints on women's participation are removed will women have the same chance for success as men. This approach notes the importance of gender-role socialization and suggests that changes need to be made in what children learn from their families, teachers, and the media about appropriate masculine and feminine attitudes and behavior. Liberal feminists fight for better child-care options, a woman's right to choose an abortion, and the elimination of sex discrimination in the workplace.

**Radical Feminism** According to radical feminists, male domination causes all forms of human oppression, including racism and classism. Radical feminists often trace the roots of patriarchy to women's childbearing and child-rearing responsibilities, which make them dependent on men. In the radical feminist view, men's oppression of women is deliberate, and ideological justification for this subordination is provided by other institutions such as the media and religion. For women's lives to improve, radical feminists claim, patriarchy must be abolished. If institutions are currently gendered, alternative institutions—such as women's organizations seeking better health care, day care, and shelters for victims of domestic violence and rape—should be developed to meet women's needs.

**Socialist Feminism** Socialist feminists argue that the oppression of women results from their dual roles as paid *and* unpaid workers in a capitalist economy. In the workplace, women are exploited by capitalism; at home, they are exploited by patriarchy. Women are easily exploited in both sectors; they are paid low wages and have few economic resources. According to some feminist scholars, gender-segregated work is a central way in which men remain dominant over women in capitalist economies, primarily because most women have lower wages and fewer opportunities than men. As a result, women must do domestic labor either to gain a better-paid man's economic support or to stretch their own wages. According to socialist feminists, the only way to achieve gender equality is to eliminate capitalism and develop a socialist economy that would bring equal pay and rights to women.

**Multicultural Feminism** Recently, academics and activists have been rethinking the experiences of women of color from a feminist perspective. The experiences of African American women and Latinas/Chicanas have been of particular interest to some social analysts. Building on the civil rights and feminist movements of the late 1960s and early 1970s, some contemporary black feminists have focused on the cultural experiences of African American women. A central assumption of this analysis is that race, class, and gender are forces that simultaneously oppress African American women. The effects of these three statuses cannot be adequately explained as "double" or "triple" jeopardy (race + class + gender = a poor African American woman) because these ascribed characteristics are not simply added to one another. Instead, they are multiplicative in nature (race × class × gender); different characteristics may be more significant in one situation than another. For example, a well-to-do white woman (class) may be in a position of privilege when compared to people of color (race) and men from lower socioeconomic positions (class), yet be in a subordinate position when compared to a white man (gender) from the capitalist class (Andersen and Collins, 2010). In order to analyze the complex relationship among these characteristics, the lived experiences of African American women and other previously "silenced people" must be heard and examined within the context of particular historical and social conditions.

A classic example of multicultural feminist studies is the work of the psychologist Aida Hurtado (1996), who explored the cultural identification of Latina/Chicana women. According to Hurtado, distinct differences exist between the worldviews of the white (non-Latina) women who participate in the women's movement and many Chicanas, who have a strong sense of identity with their own communities. From this perspective, women of color do not possess the "relational privilege" that white women have because of their proximity to white patriarchy through husbands, fathers, sons, and others. To change this situation, there must be a "politics of inclusion," which might create social structures that lead to positive behavior and bring more people

---

**feminism** the belief that men and women are equal and should be valued equally and have equal rights.

# SOCIOLOGY *in Global Perspective*

## The Rise of Islamic Feminism in the Middle East?

I would like for all of the young Muslim girls to be able to relate to Iman, whether they wear the hijab [head scarf] or not. Boys will also enjoy Iman's adventures because she is one tough, smart girl! Iman gets her super powers from having very strong faith in Allah, or God. She solves many of the problems by explaining certain parts of the Koran that relate to the story.

—Rima Khoreibi, an author from Dubai (United Arab Emirates) (qtd. in Kristof, 2006)

**A**ccording to some gender-studies specialists, a newer form of feminist thinking is emerging among Muslim women. Often referred to as "feminist Islam" or "Islamic feminism," this approach is based on the belief that greater gender equality may be possible in the Muslim world if the teachings of Islam, as set forth in the Koran (or Qur'an)—the Islamic holy book—are followed more closely. Islamic feminism is based on the principle that Muslim women should retain their allegiance to Islam as an essential part of their self-determination and identity but that they should also work to change patriarchal control over the basic Islamic worldview (Wadud, 2002). According to the journalist Nicholas D. Kristof (2006), both Islam and evangelical Christianity have been on the rise in recent years because both religions provide "a firm moral code, spiritual reassurance and orderliness to people vexed by chaos and immorality around them, and . . . dignity to the poor."

Islamic feminists believe that the rise of Islam might contribute to greater, rather than less, equality for women. From this perspective, stories about characters such as Iman may help girls and young women realize that they can maintain their deep religious convictions and their head scarf (hijab) while, at the same time, working for greater equality for women and more opportunities for themselves. In *The Adventures of Iman,* the female hero always wears a pink scarf around her neck, and she uses the scarf to cover her hair when she is praying to Allah. Iman quotes the Qur'an when she is explaining to others that Muslims are expected

into a dialogue about how to improve social life and reduce inequalities.

**Evaluation of Conflict and Feminist Perspectives** Conflict and feminist perspectives provide insights into the structural aspects of gender inequality in society. These approaches emphasize factors external to individuals that contribute to the oppression of white women and people of color; however, they have been criticized for emphasizing the differences between men and women without taking into account the commonalities they share. Feminist approaches have also been criticized for their emphasis on male dominance without a corresponding analysis of the ways in which some men may also be oppressed by patriarchy and capitalism. The Concept Quick Review outlines the key aspects of each sociological perspective on gender socialization.

## [ concept quick review ]

### Sociological Perspectives on Gender Stratification

| Perspective | Focus | Theory/Hypothesis |
|---|---|---|
| Functionalist | Macrolevel analysis of women's and men's roles | Traditional gender roles ensure that expressive and instrumental tasks will be performed.<br><br>Human capital model |
| Conflict | Power and economic differentials between men and women | Unequal political and economic power heightens gender-based social inequalities. |
| Feminist Approaches | Feminism should be embraced to reduce sexism and gender inequality. | 1. Liberal feminism<br><br>2. Radical feminism<br><br>3. Socialist feminism<br><br>4. Multicultural feminism |

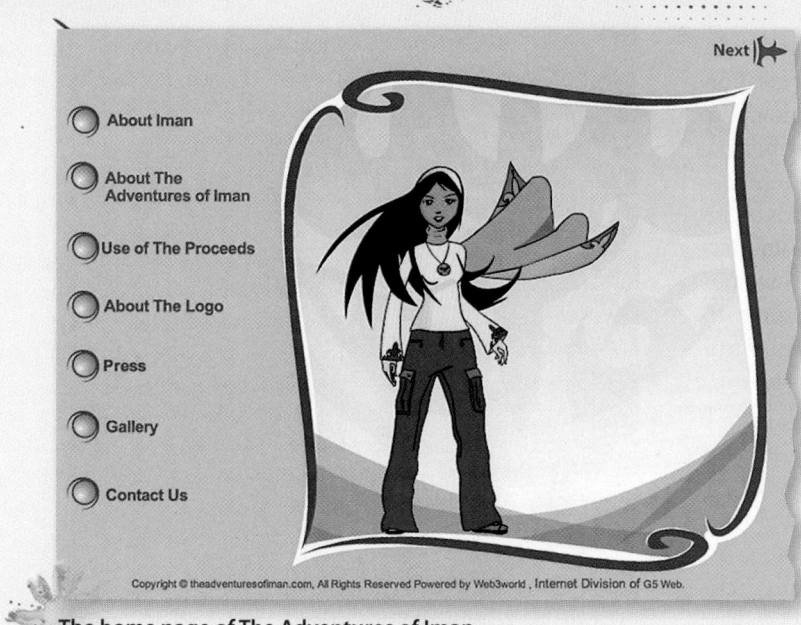

Courtesy of Rima Khorebi

**The home page of The Adventures of Iman.**

to be tolerant, kind, and righteous. For Iman, religion is a form of empowerment, not an extension of patriarchy.

Islamic feminism is quite different from what most people think of as Western feminism (particularly in regard to issues such as the wearing of the *hijab* or the fact that in Saudi Arabia, a woman may own a motor vehicle but may not legally drive it). However, change is clearly under way in many regions of the Middle East and other areas of the world as rapid economic development and urbanization quickly change the lives of many people.

## reflect & analyze

Why is women's inequality a complex issue to study across nations? What part does culture play in defining the roles of women and men in various societies? How do religious beliefs influence what we perceive as "appropriate" or "inappropriate" behaviors for men, women, and children? What do you think?

## Gender Issues in the Future

Over the past century, women made significant progress in the labor force. Laws were passed to prohibit sexual discrimination in the workplace and school. Affirmative action programs helped make women more visible in education, government, and the professional world. More women entered the political arena as candidates and elected officials instead of as volunteers in the campaign offices of male candidates.

Many men joined movements to raise their consciousness, realizing that what is harmful to women may also be harmful to men. For example, women's lower wages in the labor force suppress men's wages as well; in a two-paycheck family, women who are paid less contribute less to the family's finances, thus placing a greater burden on men to earn more money. In the midst of these changes, however, many gender issues remain unresolved in the second decade of the twenty-first century. For example, recent national surveys have shown that the movement toward attitudes of greater gender equality in the United States has slowed and that more people are embracing a new cultural framework of "egalitarian essentialism," which is a blend of feminism equality and

traditional motherhood roles (Hermsen, Cotter, and Vanneman, 2011).

In the labor force, gender segregation and the wage gap are still problems. As the United States attempts to climb out of the worst economic recession in decades, job loss has affected both women and men. However, data from the Bureau of Labor Statistics show that the wages of the typical woman who has a job have risen slightly faster than those of the typical man. Rather than this being considered a gain for women, some analysts suggest that it is a situation where everyone is losing but that men are simply losing more because of job insecurity or loss, declining real wages, and the loss of benefits such as health care and pension funds.

In the United States and other nations of the world, gender equity, political opportunities, education, and health care remain pressing problems for women. Gender issues and imbalances can contribute not only to individual problems but also to societal problems, such as the destabilization of nations in the global economy. Gender inequality is also an international problem because it is related to violence against women, trafficking, and other crimes against girls and women. To bring about social change for women, it is important for them to be equal players in the economy and the political process (International Foundation for Electoral

Systems, 2011) (Figure 10.15). According to Hillary Rodham Clinton (2011), former U.S. secretary of state,

> Governments and business leaders worldwide should view investing in women as a strategy for job creation and economic growth. And while many are doing so, the pool of talented women remains underutilized, underpaid and underrepresented overall in business and society. Worldwide, women do two-thirds of the work, yet they earn just one-third of the income and own less than 2 percent of the land. . . . If we invest in women's education and give them the opportunity to access credit or start a small business, we add fuel to a powerful engine for progress for women, their families, their communities and their countries.

As Clinton suggests, an investment in girls and women, whether in the United States or in other nations of the world, will strengthen other efforts to deal with social problems such as violence against women, inequality, and poverty. However, we must ask this: How will economic problems around the world affect gender inequality in the twenty-first century? Sadly, the fallout from the economic crisis that began in 2007 still lingers and is most strongly felt by people who are already in the greatest social and financial peril. What do you think might be done to provide more equal opportunities for girls and women in changing economic times?

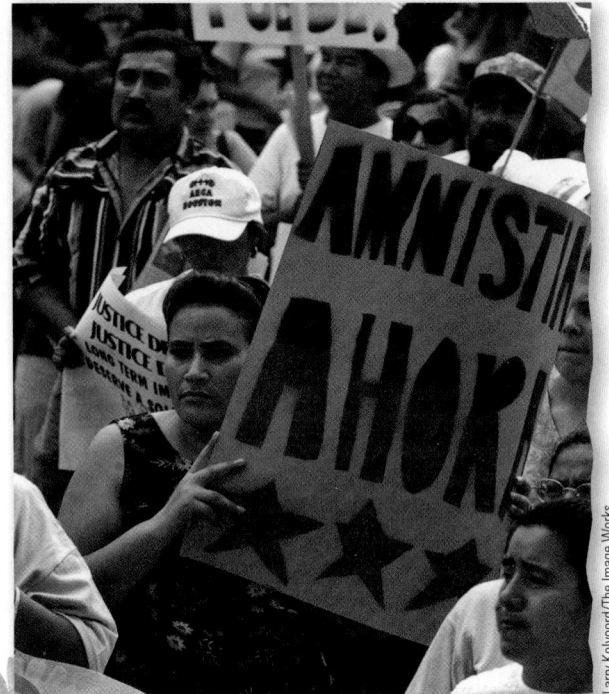

**FIGURE 10.15** Latinas have become increasingly involved in social activism for causes that they believe are important. This woman is showing her support for amnesty for undocumented workers in the United States.

*Larry Kolvoord/The Image Works*

# chapter REVIEW

**LO1 How do sex and gender differ?**

Sex refers to the biological categories and manifestations of femaleness and maleness; gender refers to the socially constructed differences between females and males. In short, sex is what we (generally) are born with; gender is what we acquire through socialization.

**LO2 What kind of discrimination occurs on the basis of sexual orientation?**

Homophobia refers to extreme prejudice and sometimes discriminatory actions directed at gays, lesbians, bisexuals, and others who are perceived as not being heterosexual. Discrimination occurs in many forms, including marital and parenting rights, housing, health care, bank lending policies, and other rights and privileges taken for granted by heterosexual persons.

**LO3 How are *gender roles*, *gender identity*, *body consciousness*, and *sexism* defined?**

*Gender roles* encompass the attitudes, behaviors, and activities that are socially assigned to each sex and that are learned through socialization. *Gender identity* is an individual's perception of self as either female or male. *Body consciousness* is how a person perceives and feels about his or her body. *Sexism* is the subordination of one sex, usually female, based on the assumed superiority of the other sex.

**LO4 How does the division of labor between women and men differ in various kinds of societies?**

In most hunting and gathering societies, fairly equitable relationships exist between women and men because neither sex has the ability to provide all of the food necessary for survival. In horticultural societies, a fair degree of gender equality exists because neither sex controls the food supply. In agrarian societies, male dominance is overt; agrarian tasks require more labor and physical strength, and females are often excluded from these tasks because they are viewed as too weak or too tied to child-rearing activities. In industrialized societies, a gap exists between nonpaid work performed by women at home and paid work performed by men and women. A wage gap also exists between women and men in the marketplace.

## LO5 What are the primary agents of gender socialization, and what is their role in socializing people throughout life?

Parents, peers, teachers and schools, sports, and the media are agents of socialization that tend to reinforce stereotypes of appropriate gender behavior. From birth, parents act differently toward children on the basis of the child's sex. Peers help children learn gender-role stereotypes, as well as gender-appropriate and gender-inappropriate behavior. Schools operate as a gendered institution, and teachers provide messages about gender through the formal content of assignments and informal interactions. In terms of sports, boys are socialized to participate in highly competitive, rule-oriented games, whereas girls have traditionally been socialized to participate in activities that involve less competitiveness. Recently, however, more girls have started to participate in sports formerly regarded as "male" activities.

## LO6 In what ways does the contemporary workplace reflect gender stratification?

Many women work in lower-paying, less-prestigious jobs than men. This occupational segregation leads to a disparity, or pay gap, between women's and men's earnings. Even when women are employed in the same job as men, on average they do not receive the same, or comparable, pay.

## LO7 How do functionalists and conflict theorists differ in their perspectives on gender inequality?

According to functionalist analysts, women's roles as caregivers in contemporary industrialized societies are crucial in ensuring that key societal tasks are fulfilled. While the husband performs the instrumental tasks of economic support and decision making, the wife assumes the expressive tasks of providing affection and emotional support for the family. According to conflict analysts, the gendered division of labor within families and the workplace—particularly in agrarian and industrial societies—results from male control and dominance over women and resources.

## LO8 What are the feminist perspectives on gender inequality?

Feminist perspectives provide insights into the structural aspects of gender inequality in society. In liberal feminism, gender equality is equated with equality of opportunity. Radical feminists often trace the roots of patriarchy to women's childbearing and child-rearing responsibilities, which make them dependent on men. Socialist feminists argue that the oppression of women results from their dual roles as paid and unpaid workers in a capitalist economy. Academics and activists have been rethinking the experiences of women of color from a feminist perspective. The experiences of African American women and Latinas/Chicanas have been of particular interest to some social analysts.

## key TERMS

body consciousness 304

comparable worth 319

cross-dresser 300

feminism 322

gender 303

gender bias 312

gender identity 304

gender role 304

homophobia 303

intersexed person 298

matriarchy 307

patriarchy 306

pay gap 318

primary sex characteristics 298

secondary sex characteristics 298

sex 298

sexism 306

sexual orientation 300

sexualization 297

transgendered person 299

## questions for CRITICAL THINKING

1. Do the media reflect societal attitudes on gender, or do the media determine and teach gender behavior? (As a related activity, watch television for several hours, and list the roles for women and men depicted in programs and those represented in advertising.)

2. Examine the various academic departments at your college. What is the gender breakdown of the faculty in selected departments? What is the gender breakdown of undergraduates and graduate students in those departments? Are there major differences among various academic areas of teaching and study? What hypothesis can you come up with to explain your observations?

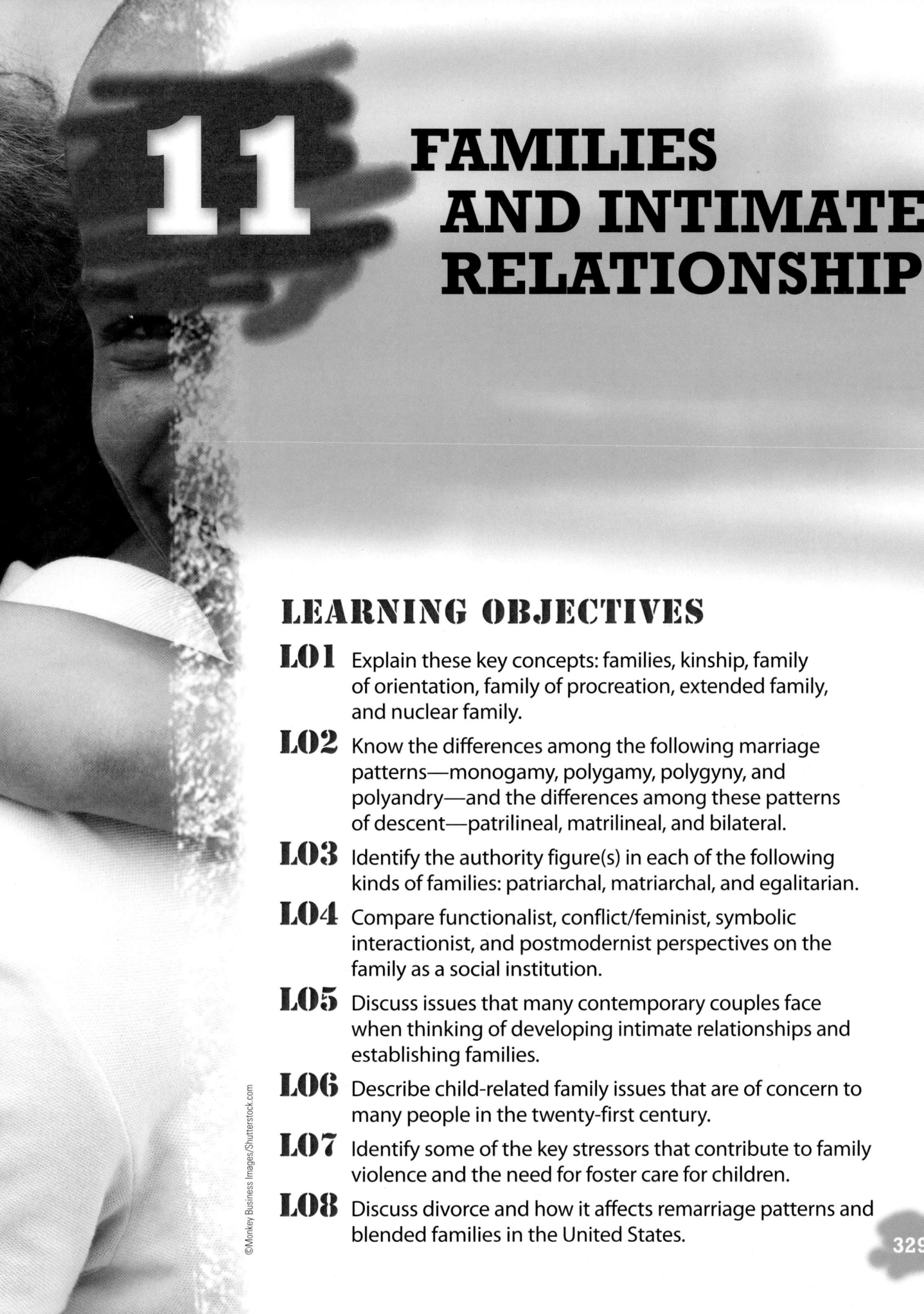

# 11 FAMILIES AND INTIMATE RELATIONSHIPS

## LEARNING OBJECTIVES

**LO1** Explain these key concepts: families, kinship, family of orientation, family of procreation, extended family, and nuclear family.

**LO2** Know the differences among the following marriage patterns—monogamy, polygamy, polygyny, and polyandry—and the differences among these patterns of descent—patrilineal, matrilineal, and bilateral.

**LO3** Identify the authority figure(s) in each of the following kinds of families: patriarchal, matriarchal, and egalitarian.

**LO4** Compare functionalist, conflict/feminist, symbolic interactionist, and postmodernist perspectives on the family as a social institution.

**LO5** Discuss issues that many contemporary couples face when thinking of developing intimate relationships and establishing families.

**LO6** Describe child-related family issues that are of concern to many people in the twenty-first century.

**LO7** Identify some of the key stressors that contribute to family violence and the need for foster care for children.

**LO8** Discuss divorce and how it affects remarriage patterns and blended families in the United States.

©Monkey Business Images/Shutterstock.com

# Diverse Family Landscapes in the Twenty-First Century

**It definitely felt a little weird to be back [living at home].** I enjoy spending time with my mom, but I also felt like by now I should be established and not depending on her to help me.

— Amanda Seals, age 30, describing how it feels to be part of the "boomerang generation"—adult children living at home with their parents while looking for a job or otherwise getting back on their feet (qtd. in Trounson, 2012)

Contemporary families are more diverse than in the past, including an increasing number of households made up of young people who, at least temporarily, have returned to live in their parents' homes. What larger societal factors contribute to this living arrangement?

© Iakov Filimonov/Shutterstock.com

**Kelly is 33, and if dreams were winds, you might say that hers have shifted.** She believes that every household needs one primary caretaker, that women are, broadly speaking, better at that job than men, and that no amount of professional success could possibly console her if she felt her two children were not being looked after the right way. The maternal instinct is a real thing, Kelly argues.

—Kelly Makino, featured in an article titled "The Retro Wife" in *New York* magazine, telling journalist Lisa Miller (2013: 22) why she prefers to be a stay-at-home mom even if less money comes into her two-parent household

Although the personal narratives about the families of Amanda, Kelly, and Bruce may initially seem unrelated, a common thread weaves through them: Contemporary families are diverse, ranging from "boomerang" families in which adult children return home until they can find jobs, to families with a stay-at-home parent who takes care of the children, to extended families in which individuals may not always see things in the same way. The United States is a highly diverse nation with many family patterns. Parents socialize and interact with their children in various ways, often linked to their cultural background, socioeconomic status, race/ethnicity, religion, national origin, and/or other personal characteristics, attributes, and preferences.

In this chapter we examine the increasing complexity of family life in the United States and other nations. Pressing family and societal issues such as interpersonal communications, financial hardships, teenage childbearing, divorce, and child-care issues will be used as examples of how families and intimate relationships continue to change over time and place. Before reading on, test your knowledge about trends in family life by taking the "Sociology and Everyday Life" quiz.

## Families in Global Perspective

**LO1** As the nature of family life has changed in high-, middle-, and low-income nations, the issue of what constitutes a "family" continues to be widely debated. In the "Universal Declaration of Human Rights," Article 16, adopted by the United Nations (1948), the family is defined as follows:

■ Men and women of full age, without any limitation due to race, nationality, or religion, have the right to marry and to found a family. They are entitled to equal rights as to marriage, during marriage and at its dissolution.

I hit the breaking point as a parent a few years ago. It was the week of my extended family's annual gathering in August, and we were struggling with assorted crises. . . . Sure enough, one night all the tensions boiled over. At dinner, I noticed my nephew texting under the table. I knew I shouldn't say anything, but I couldn't help myself and asked him to stop. Ka-boom! My sister snapped at me to not discipline her child. My dad pointed out that my girls were the ones balancing spoons on their noses. My mom said none of the grandchildren had manners. Within minutes, everyone had fled to separate corners.

—Journalist Bruce Feiler (2013) describes how easily tensions can mount at extended-family gatherings. He further argues that families need a shared narrative about their history across generations so each person will develop an "intergenerational self" to provide personal identity and help in difficult times.

## How Much Do You Know About Contemporary Trends in U.S. Family Life?

| True | False | |
|---|---|---|
| T | F | 1. Most U.S. adults view having a baby outside of marriage as "morally wrong." |
| T | F | 2. More than four in ten U.S. adults have at least one step-relative in their family. |
| T | F | 3. The percentage of children living with two parents, regardless of their marital status, differs by race and Hispanic origin. |
| T | F | 4. The birth rate among U.S. teenagers is much lower than the birth rate among teens in other Western industrialized countries. |
| T | F | 5. Couples who cohabit before they get married are more likely to stay married. |

| True | False | |
|---|---|---|
| T | F | 6. Stay-at-home parents have become less common since the 1990s, when more two-paycheck families were established to help pay the bills. |
| T | F | 7. The percentage of U.S. households that contain only one person has continued to increase since the 1970s. |
| T | F | 8. Among U.S. married couples, it is very rare for the wife to earn more money than the husband. |

Answers on page 332.

■ Marriage shall be entered into only with the free and full consent of the intending spouses.

■ The family is the natural and fundamental group unit of society and is entitled to protection by society and the States.

According to this declaration, the social institution of family must be protected in all societies because family is the "natural" and "fundamental" group unit of society. Although families differ widely around the world, they also share certain common concerns in their everyday lives. Food, clothing, shelter, and child care are necessities important to all people.

In the United States the Census Bureau defines a family as consisting of two or more people who are related by birth, marriage, or adoption, and residing in the same housing unit. (The Census Bureau specifies that one person in the household unit will be identified as the "householder.") For many years the standard sociological definition of *family* has been a group of people who are related to one another by bonds of blood, marriage, or adoption and who live together, form an economic unit, and bear and raise children. Some people believe that this definition should not be expanded—that social approval should not be extended to other relationships simply because the persons in those relationships wish to consider themselves to be a family. However, other people challenge this definition because it simply does not match the reality of family life in contemporary society, particularly at a time when only about half of adults ages eighteen and older are married in the legal sense of the term.

Today's families include many types of living arrangements and relationships, including single-parent households; unmarried heterosexual couples; lesbian, gay, bisexual, and transgendered couples with or without children; and multiple generations (such as grandparent, parent, and child) living in the same household. To accurately reflect these changes in family life, some sociologists believe that we need a more encompassing definition of what constitutes

# ANSWERS to the Sociology Quiz on Contemporary Trends in U.S. Family Life

**1. False.** A Gallup "Values and Beliefs" poll found that more than half (54 percent) of U.S. adults say that having a baby outside of marriage is "morally acceptable." Slightly less than half (41 percent) view out-of-wedlock childbearing as "morally wrong" (Saad, 2011).

**2. True.** A Pew Research Center report revealed that more than four in ten U.S. adults have at least one step-relative in their family, such as a stepparent, a step- or half-sibling, or a stepchild.

**3. True.** Among persons reporting a single race on the U.S. Census, 85 percent of Asian American children lived with two parents compared with 77 percent of white (non-Hispanic) children, 66 percent of Hispanic (Latino/a) children, and 38 percent of black (African American) children.

**4. False.** Recent research indicates that couples who cohabit before they get married are less likely to stay married; however, their chances of remaining married improve if they were already engaged when they began living together.

**5. False.** The U.S. teen birth rate is up to nine times *higher* than it is in other Western industrialized countries.

**6. False.** There has been an *increase* in the number of stay-at-home parents, particularly mothers. In 2012, among married-couple families with children younger than 15, the percentage of stay-at-home mothers rose to 24 percent, up from 20 percent in the 1990s.

**7. True.** The percentage of U.S. households that consist of one person living alone has risen from 17 percent in 1970 to 27 percent in 2012.

**8. False.** According to a study by the Pew Research Center, more than one in five (22 percent) of U.S. husbands are in marriages in which the wife earns more than the husband—up from only 4 percent in 1970.

Sources: Based on Newport, 2011; Parker, 2011; Roberts, 2010; Saad, 2011; U.S. Centers for Disease Control and Prevention, 2011d; and Vespa, Lewis, and Kreider, 2013. © Cengage Learning.

a family. Accordingly, we will define *families* **as relationships in which people live together with commitment, form an economic unit and care for any young, and consider their identity to be significantly attached to the group.** Sexual expression and parent–child relationships are a part of most, but not all, family relationships.

How do sociologists approach the study of families? In our study of families we will use our sociological imagination to see how our personal experiences are related to the larger happenings in society. At the microlevel, each of us has a "biography," based on our experience within our family; at the macrolevel, our families are embedded in a specific cultural and social context that has a major effect on them. We will examine the institution of the family at both of these levels, starting with family structure and characteristics.

## Family Structure and Characteristics

In preindustrial societies the primary form of social organization is through kinship ties. **Kinship refers to a social network of people based on common ancestry, marriage, or adoption.** Through kinship networks, people cooperate so that they can acquire the basic necessities of life, including food and shelter. Kinship systems can also serve as a means by which property is transferred, goods are produced and distributed, and power is allocated.

In industrialized societies other social institutions fulfill some of the functions previously taken care of by the kinship network. For example, political systems provide structures of social control and authority, and economic systems are responsible for the production and distribution of goods and services. Consequently, families in industrialized societies serve fewer and more-specialized purposes than do families in preindustrial societies. Contemporary families are primarily responsible for regulating sexual activity, socializing children, and providing affection and companionship for family members.

## Families of Orientation and Procreation

During our lifetime, many of us will be members of two different types of families—a family of orientation and a family of procreation. The *family of orientation* **is the family into which a person is born** and in which early socialization usually takes place. Although most people are related to members of their family of orientation by blood ties, those who are adopted have a legal tie that is patterned after a blood relationship. The *family of procreation* is the family that a person forms by having or adopting children. Both legal and blood ties are found in most families of procreation. The relationship between a husband and wife is based on legal ties; however, the relationship between a parent and child may be based on either blood ties or legal ties, depending on whether the child has been adopted (Figure 11.1).

Some sociologists have emphasized that "family of orientation" and "family of procreation" do not encompass all types of contemporary families. Instead, many gay men, lesbians, bisexuals, and transgendered persons have *families we choose*—social arrangements that include intimate relationships between couples and close familial relationships among other couples and other adults and children. According to the sociologist Judy Root Aulette (1994), "families we choose" include blood ties and legal ties, but they also include *fictive kin*—persons who are not actually related by blood but who are accepted as family members.

**Extended and Nuclear Families** Sociologists distinguish between extended families and nuclear families based on the number of generations that live within a household. An *extended family* is a family unit composed of relatives in addition to parents and children who live in the same household. These families often include grandparents, uncles, aunts, or other relatives who live close to the parents and children, making it possible for family members to share resources. In horticultural and agricultural societies, extended families are extremely important; having a large number of family members participate in food production may be essential for survival. Today, extended-family patterns are found in Latin America, Africa, Asia, and some parts of eastern and southern Europe. With the

advent of industrialization and urbanization, maintaining the extended-family pattern becomes more difficult. Increasingly, young people move from rural to urban areas in search of employment in the industrializing sector of the economy. At that time, some extended families remain, but the nuclear family typically becomes the predominant family form in the society.

A *nuclear family* is a family composed of one or two parents and their dependent children, all of whom live apart from other relatives. A traditional definition specifies that a nuclear family is made up of a "couple" and their dependent children; however, this definition became outdated when a significant shift occurred in the family structure. A comparison of Census Bureau data from 1970 to 2012 shows that there has been a significant decline (from slightly over 40 percent to less than 20 percent) in the percentage of U.S. households comprising a married couple with their own children under eighteen years of age (see "Census Profiles: Household

**FIGURE 11.1** Whereas the relationship between spouses is based on legal ties, relationships between parents and children may be established by either blood tie or legal tie.

Myrleen Pearson/Alamy

**families** relationships in which people live together with commitment, form an economic unit and care for any young, and consider their identity to be significantly attached to the group.

**kinship** a social network of people based on common ancestry, marriage, or adoption.

**family of orientation** the family into which a person is born and in which early socialization usually takes place.

**family of procreation** the family that a person forms by having or adopting children.

**extended family** a family unit composed of relatives in addition to parents and children who live in the same household.

**nuclear family** a family composed of one or two parents and their dependent children, all of whom live apart from other relatives.

## Household Composition, 1970 and 2012

The Census Bureau asks a representative sample of the U.S. population about their marital status and also asks those individuals questions about other individuals residing in their household. Based on the most recent data, the Census Bureau reports that the percentage distribution of nonfamily and family households has changed substantially during the past four decades. The most noticeable trend is the decline in the number of married-couple households with their own children living with them, which decreased from about 40 percent of all households in 1970 to slightly less than 20 percent in 2012. The distribution shown below reflects this and other changes in household structure.

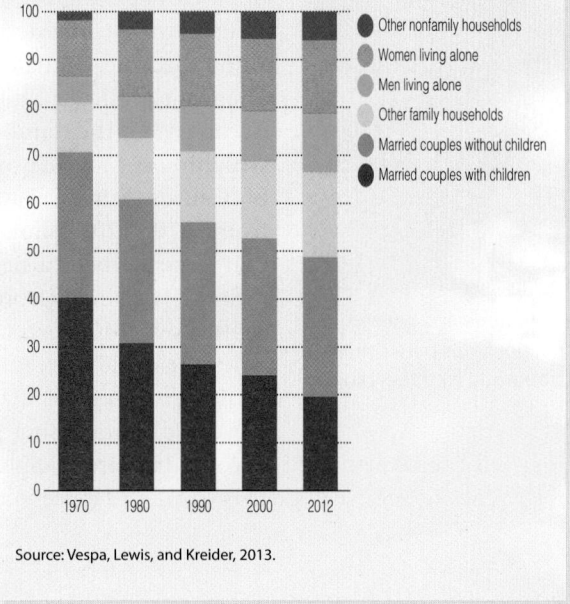

● Other nonfamily households
● Women living alone
● Men living alone
○ Other family households
● Married couples without children
● Married couples with children

Source: Vespa, Lewis, and Kreider, 2013.

Composition, 1970 and 2012"). Conversely, there has been an increase in the percentage of households in which either a woman or a man lives alone.

## Marriage Patterns

**LO2** Across cultures, different forms of marriage characterize families. **Marriage is a legally recognized and/or socially approved arrangement between two or more individuals that carries certain rights and obligations and usually involves sexual activity.** In most societies, marriage involves a mutual commitment by each partner, and linkages between two individuals and families are publicly demonstrated.

In the United States the only legally sanctioned form of marriage is **monogamy—a marriage between two partners, usually a woman and a man.** For some people, marriage is a lifelong commitment that ends only with the death of a partner. For others, marriage is a commitment of indefinite duration. Through a pattern of marriage, divorce, and remarriage, some people practice *serial monogamy*—a succession of marriages in which a person has several spouses over a lifetime but is legally married to only one person at a time.

**Polygamy** is the concurrent marriage of a person of one sex with two or more members of the opposite sex.** The most prevalent form of polygamy is *polygyny*—the concurrent marriage of one man with two or more women** (Figure 11.2). Polygyny has been practiced in a number of societies, including parts of Europe until the Middle Ages. More recently, some marriages in Islamic societies in Africa and Asia have been polygynous; however, the cost of providing for multiple wives and numerous children makes the practice impossible for all but the wealthiest men. In addition, because roughly equal numbers of women and men live in these areas, this nearly balanced sex ratio tends to limit polygyny.

The second type of polygamy is *polyandry*—the concurrent marriage of one woman with two or more men.** Polyandry is very rare; when it does occur, it is typically found in societies where men greatly outnumber women because of high rates of female infanticide.

## Patterns of Descent and Inheritance

Even though a variety of marital patterns exist across cultures, virtually all forms of marriage establish a system of descent so that kinship can be determined and inheritance rights established. In preindustrial societies, kinship is usually traced through one parent (unilineally). The most common pattern of unilineal descent is *patrilineal descent*—a system of tracing descent through the father's side of the family.** Patrilineal systems are set up in such a manner that a legitimate son inherits his father's property and sometimes his position upon the father's death. In nations such as India, where boys are seen as permanent patrilineal family members but girls are seen as only temporary family members, girls tend to be considered more expendable than boys.

Even with the less common pattern of *matrilineal descent*—a system of tracing descent through the mother's side of the family**—women may not control property. However, inheritance of property and position is usually traced from the maternal uncle (mother's brother) to his nephew (mother's son). In some cases, mothers may pass on their property to daughters.

By contrast, kinship in industrial societies is usually traced through both parents (bilineally). The most common form is **bilateral descent—a system of tracing descent through both the mother's and father's sides of the family.** This pattern is used in the United States for the purpose of determining kinship and inheritance rights; however, children typically take the father's last name.

## Power and Authority in Families

**LO3** Descent and inheritance rights are intricately linked with patterns of power and authority in families. The most prevalent forms of familial power and authority are patriarchy, matriarchy, and egalitarianism. A **patriarchal family** is a **family structure in which authority is held by the eldest male (usually the father).** The male authority figure acts as head of the household and holds power and authority over the women and children, as well as over other males. A **matriarchal family is a family structure in which authority is held by the eldest female (usually the mother).** In this case the female authority figure acts as head of the household. Although there has been a great deal of discussion about matriarchal families, scholars have found no historical evidence to indicate that true matriarchies ever existed.

The most prevalent pattern of power and authority in families is patriarchy. Across cultures, men are the primary (and often sole) decision makers regarding domestic, economic, and social concerns facing the family. The existence of patriarchy may give men a sense of power over their own lives, but it can also create an atmosphere in which some men feel greater freedom to abuse women and children.

An **egalitarian family** is a **family structure in which both partners share power and authority equally.** Recently, a trend toward more-egalitarian relationships has been evident in a number of countries as women have sought changes in their legal status and increased educational and employment opportunities. Some degree of economic independence makes it possible for women to delay marriage or to terminate a problematic marriage. Recent cross-national studies have found that larger increases in the proportion of women who have higher levels of education, who hold jobs with higher wages, who have more commitment

AP Images/Rick Bowmer

**FIGURE 11.2** Polygamy is the concurrent marriage of a person of one sex with two or more persons of the opposite sex. Although most people do not practice this pattern of marriage, some men are married to more than one wife.

to careers outside the family, and who have greater interest in gender equality all contribute to the support of egalitarian gender values in the larger society as these ideas eventually spread to others (Pampel, 2011).

**marriage** a legally recognized and/or socially approved arrangement between two or more individuals that carries certain rights and obligations and usually involves sexual activity.

**monogamy** a marriage between two partners, usually a woman and a man.

**polygamy** the concurrent marriage of a person of one sex with two or more members of the opposite sex.

**polygyny** the concurrent marriage of one man with two or more women.

**polyandry** the concurrent marriage of one woman with two or more men.

**patrilineal descent** a system of tracing descent through the father's side of the family.

**matrilineal descent** a system of tracing descent through the mother's side of the family.

**bilateral descent** a system of tracing descent through both the mother's and father's sides of the family.

**patriarchal family** a family structure in which authority is held by the eldest male (usually the father).

**matriarchal family** a family structure in which authority is held by the eldest female (usually the mother).

**egalitarian family** a family structure in which both partners share power and authority equally.

## Residential Patterns

Residential patterns are interrelated with the authority structure and the method of tracing descent in families. *Patrilocal residence* **refers to the custom of a married couple living in the same household (or community) as the husband's parents.** Across cultures, patrilocal residency is the most common pattern. Patrilocal residency can be found in countries where it is to the distinct advantage of young men to remain close to their parents' household.

Few societies have residential patterns known as *matrilocal*—**the custom of a married couple living in the same household (or community) as the wife's parents.** In industrialized nations such as the United States, most couples hope to live in a *neolocal residence*—**the custom of a married couple living in their own residence apart from both the husband's and the wife's parents.**

Up to this point, we have examined a variety of marriage and family patterns found around the world. Even with the diversity of these patterns, most people's behavior is shaped by cultural rules pertaining to endogamy and exogamy. *Endogamy* **is the practice of marrying within one's own group.** In the United States, for example, most people practice endogamy: They marry people who come from the same social class, racial–ethnic group, religious affiliation, and other categories considered important within their own social group. *Exogamy* **is the practice of marrying outside one's own group.** Depending on the circumstances, exogamy may not be noticed at all, or it may result in a person being ridiculed or ostracized by other members of the "in" group. The three most important sources of positive or negative sanctions for intermarriage are the family, the church, and the state. Participants in these social institutions may look unfavorably on the marriage of an in-group member to an "outsider" because of the belief that it diminishes social cohesion in the group. However, educational attainment is also a strong indicator of marital choice. Higher education emphasizes individual achievement, and college-educated people may be less likely than others to identify themselves with their social or cultural roots and thus more willing to marry outside their own social group or category if their potential partner shares a similar level of educational attainment.

## Theoretical Perspectives on Family

**LO4** The *sociology of family* **is the subdiscipline of sociology that attempts to describe and explain patterns of family life and variations in family structure.** Functionalist perspectives emphasize the functions that families perform at the macrolevel of society, whereas conflict and feminist perspectives focus on families as a primary source of social inequality. Symbolic interactionists examine microlevel interactions that are integral to the roles of different family members. Postmodern analysts view families as being permeable, capable of being diffused or invaded so that their original purpose is modified.

## Functionalist Perspectives

Functionalists emphasize the importance of the family in maintaining the stability of society and the well-being of individuals. According to Emile Durkheim, marriage is a microcosmic replica of the larger society; both marriage and society involve a mental and moral fusion of physically distinct individuals. Durkheim also believed that a division of labor contributes to greater efficiency in all areas of life—including marriages and families—even though he acknowledged that this division imposes significant limitations on some people.

In the United States, Talcott Parsons was a key figure in developing a functionalist model of the family. According to Parsons (1955), the husband/father fulfills the *instrumental role* (meeting the family's economic needs, making important decisions, and providing leadership), whereas the wife/mother fulfills the *expressive role* (running the household, caring for children, and meeting the emotional needs of family members).

Contemporary functionalist perspectives on families derive their foundation from Durkheim. Division of labor makes it possible for families to fulfill a number of functions that no other institution can perform as effectively. In advanced industrial societies, families serve four key functions:

1. *Sexual regulation.* Families are expected to regulate the sexual activity of their members and thus control reproduction so that it occurs within specific boundaries. At the macrolevel, incest taboos prohibit sexual contact or marriage between certain relatives. For example, virtually all societies prohibit sexual relations between parents and their children and between brothers and sisters.
2. *Socialization.* Parents and other relatives are responsible for teaching children the necessary knowledge and skills to survive. The smallness and intimacy of families make them best suited for providing children with the initial learning experiences they need.
3. *Economic and psychological support.* Families are responsible for providing economic and psychological support for members. In preindustrial

societies, families are economic production units; in industrial societies, the economic security of families is tied to the workplace and to macrolevel economic systems. In recent years, psychological support and emotional security have been increasingly important functions of the family.

4. *Provision of social status.* Families confer social status and reputation on their members. These statuses include the ascribed statuses with which individuals are born, such as race/ethnicity, nationality, social class, and sometimes religious affiliation. One of the most significant and compelling forms of social placement is the family's class position and the opportunities (or lack thereof) resulting from that position. Examples of class-related opportunities are access to quality health care, higher education, and a safe place to live.

## Conflict and Feminist Perspectives

Conflict and feminist analysts view functionalist perspectives on the role of the family in society as idealized and inadequate. Rather than operating harmoniously and for the benefit of all members, families are sources of social inequality and conflict over values, goals, and access to resources and power (Figure 11.3).

According to some classical conflict theorists, families in capitalist economies are similar to the work environment of a factory. Men in the home dominate women in the same manner that capitalists and managers in factories dominate their workers (Engels, 1970/1884). Although childbearing and care for family members in the home contribute to capitalism, these activities also reinforce the subordination of women through unpaid (and often devalued) labor. Other conflict analysts are concerned with the effect that class conflict has on the family. The exploitation of the lower classes by the upper classes contributes to family problems such as high rates of divorce and overall family instability.

Some feminist perspectives on inequality in families focus on patriarchy rather than class. From this viewpoint, men's domination over women existed long before capitalism and private ownership of property. Women's subordination is rooted in patriarchy and men's control over women's labor power. According to one scholar, "Male power in our society is expressed in economic terms even if it does not originate in property relations; women's activities in the home have been undervalued at the same time as their labor has been controlled

**FIGURE 11.3** Functionalist theorists believe that families serve a variety of functions that no other social institution can adequately fulfill. In contrast, conflict and feminist theorists believe that families may be a source of conflict over values, goals, and access to resources and power. Children in upper-class families have many advantages and opportunities that are not available to other children.

by men" (Mann, 1994: 42). In addition, men have benefited from the privileges they derive from their status as family breadwinners.

## Symbolic Interactionist Perspectives

Early symbolic interactionists such as Charles Horton Cooley and George Herbert Mead provided key insights on the roles that we play as family members and how we modify or adapt our roles

**patrilocal residence** the custom of a married couple living in the same household (or community) as the husband's parents.

**matrilocal residence** the custom of a married couple living in the same household (or community) as the wife's parents.

**neolocal residence** the custom of a married couple living in their own residence apart from both the husband's and the wife's parents.

**endogamy** the practice of marrying within one's own group.

**exogamy** the practice of marrying outside one's own group.

**sociology of family** the subdiscipline of sociology that attempts to describe and explain patterns of family life and variations in family structure.

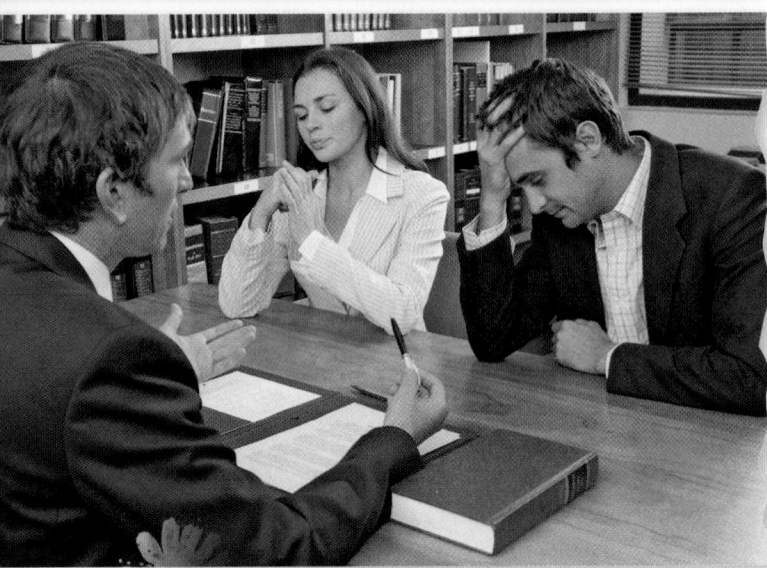

**FIGURE 11.4** Marriage is a complicated process involving rituals and shared moments of happiness. When marriage is followed by divorce, couples must abandon a shared reality and then reestablish individual ones.

to the expectations of others—especially significant others such as parents, grandparents, siblings, and other relatives. How does the family influence the individual's self-concept and identity? In order to answer questions such as this one, contemporary symbolic interactionists examine the roles of husbands, wives, and children as they act out their own parts and react to the actions of others. From such a perspective, what people think, as well as what they say and do, is very important in understanding family dynamics.

Some symbolic interactionist theorists focus on how interaction between marital partners contributes to a shared reality (Berger and Kellner, 1964). Although newlyweds bring separate identities to a

marriage, over time they construct a shared reality as a couple. In the process the partners redefine their past identities to be consistent with new realities. Development of a shared reality is a continuous process, taking place not only in the family but in any group in which the couple participates together. Divorce is the reverse of this process; couples may start with a shared reality and, in the process of uncoupling, gradually develop separate realities (Figure 11.4).

Symbolic interactionists explain family relationships in terms of the subjective meanings and everyday interpretations that people give to their lives. As the sociologist Jessie Bernard (1982/1973) pointed out, women and men experience marriage differently. Although the husband may see *his* marriage very positively, the wife may feel less positive about *her* marriage, and vice versa. Researchers have found that husbands and wives may give very different accounts of the same event and that their "two realities" frequently do not coincide.

## Postmodernist Perspectives

According to postmodern theories, we have experienced a significant decline in the influence of the family and other social institutions. As people have pursued individual freedom, they have been less inclined to accept the structural constraints imposed on them by institutions. Given this assumption, how might a postmodern perspective view contemporary family life? For example, how might this approach answer the question "How is family life different in the information age"?

The postmodern family has been described as *permeable*—capable of being diffused or invaded in such a manner that the family's original nature is modified or changed (Elkind, 1995). From this approach, if the nuclear family is a reflection of the age of modernity, the permeable family reflects the postmodern assumptions of difference and irregularity. This is evident in the fact that the nuclear family is now only one of many family forms. Similarly, under modernity the idea of romantic love has given way to the idea of consensual love: Individuals agree to have sexual relations with others whom they have no intention of marrying or, if they marry, do not necessarily see the marriage as having permanence. Maternal love has also been transformed into shared parenting, which includes not only mothers and fathers but also caregivers who may either be relatives or nonrelatives.

Urbanity is another characteristic of the postmodern family. The boundaries between the public sphere (the workplace) and the private sphere (the home) are becoming much more open and flexible.

## Theoretical Perspectives on Families

| | Focus | Key Points | Perspective on Family Problems |
|---|---|---|---|
| Functionalist | Role of families in maintaining stability of society and individuals' well-being | In modern societies, families serve the functions of sexual regulation, socialization, economic and psychological support, and provision of social status. | Family problems are related to changes in social institutions such as the economy, religion, education, and law/government. |
| Conflict/ Feminist | Families as sources of conflict and social inequality | Families both mirror and help perpetuate social inequalities based on class and gender. | Family problems reflect social patterns of dominance and subordination. |
| Symbolic Interactionist | Family dynamics, including communication patterns and the subjective meanings that people assign to events | Interactions within families create a shared reality. | How family problems are perceived and defined depends on patterns of communication, the meanings that people give to roles and events, and individuals' interpretations of family interactions. |
| Postmodernist | Permeability of families | In postmodern societies, families are diverse and fragmented. Boundaries between workplace and home are blurred. | Family problems are related to cyberspace, consumerism, and the hyperreal in an age increasingly characterized by high-tech "haves" and "have-nots." |

In fact, family life may be negatively affected by the decreasing distinction between what is work time and what is family time. As more people are becoming connected "24/7" (twenty-four hours per day, seven days per week), the boss who in the past would not call at 11:30 P.M. may send an e-mail asking for an immediate response to some question that has arisen while the person is away on vacation with family members.

The Concept Quick Review summarizes sociological perspectives on the family. Taken together, these perspectives reflect various ways in which familial relationships may be viewed in contemporary societies. Now we shift our focus to love, marriage, intimate relationships, and family issues in the United States.

# Developing Intimate Relationships and Establishing Families

**LO5** The United States has been described as a "nation of lovers"; it has been said that we are "in love with love." Why is this so? Perhaps the answer lies in the fact that our ideal culture emphasizes *romantic love,* which refers to a deep emotion, the satisfaction of significant needs, a caring for and acceptance of the person we love, and involvement in an intimate relationship (Lamanna and Riedmann, 2012) (Figure 11.5).

## Love and Intimacy

In the late nineteenth century, during the Industrial Revolution, people came to view work and home as separate spheres in which different feelings and emotions were appropriate (Coontz, 2000). The public sphere of work—men's sphere—emphasized self-reliance and independence. By contrast, the private sphere of the home—women's sphere—emphasized the giving of services, the exchange of gifts, and love. Accordingly, love and emotions became the domain of women, and work and rationality became the domain of men (Lamanna and Riedmann, 2012). Although the roles of women and men have changed dramatically over the past hundred years, women and men may still not share the same perceptions about romantic love today, and

**FIGURE 11.5** In the United States the notion of romantic love is deeply intertwined with our beliefs about how and why people develop intimate relationships and establish families. Not all societies share this concern with romantic love. However, in this country the number of opportunities for romance is sometimes increased by online matching services, such as this one, which caters to farmers and people who love nature.

women may express their feelings verbally, whereas men may express their love through nonverbal actions; however, in other cases, the situation may be just the opposite.

Love and intimacy are closely intertwined. Intimacy may be psychic ("the sharing of minds"), sexual, or both. Although sexuality is an integral part of many intimate relationships, perceptions about sexual activities vary from one culture to the next and from one time period to another. For example, kissing has traditionally been found primarily in Western cultures; many African and Asian cultures have viewed kissing negatively, although some change has occurred among younger people in recent years.

For decades, the work of the biologist Alfred C. Kinsey was considered to be the definitive research on human sexuality, even though some of his methodology had serious limitations. However, in the 1990s the work of Kinsey and his associates

was superseded by the National Health and Social Life Survey, which was conducted by the National Opinion Research Center at the University of Chicago (see Laumann et al., 1994; Michael et al., 1994). Based on interviews with more than 3,400 men and women ages 18 to 59, this random survey tended to reaffirm the significance of the dominant sexual ideologies. Most respondents reported that they engaged in heterosexual relationships, although 9 percent of the men said they had had at least one homosexual encounter resulting in orgasm. Although 6.2 percent of men and 4.4 percent of women said that they were at least somewhat attracted to others of the same gender, only 2.8 percent of men and 1.4 percent of women identified themselves as gay or lesbian. According to the study, persons who engaged in extramarital sex found their activities to be more thrilling than those with a marital partner, but they also felt more guilt. Persons in sustained relationships such as marriage or cohabitation found sexual activity to be the most satisfying emotionally and physically.

## Cohabitation and Domestic Partnerships

Attitudes and actions regarding cohabitation have changed dramatically over the past five decades. *Cohabitation* **refers to two people who live together, and think of themselves as a couple, without being legally married.** The U.S. Census Bureau uses the terms *unmarried partner, cohabiting partner,* and *cohabiter* interchangeably when referring to individuals who cohabit (Vespa, Lewis, and Kreider, 2013). The number of unmarried opposite-sex partners living together has increased from about 2.9 million in 1996 to nearly 8.1 million in 2013 (see Figure 11.6). Forty percent of all cohabiting partners had one or more children younger than age 18 residing in their household in 2013, an increase from approximately 1.2 million cohabitating partners with children under 18 in the late 1990s to about 3.1 million in 2013 (U.S. Census Bureau, 2013a).

As compared with children who reside with married parents, more children in cohabiting households experience economic disadvantages that are related to lower levels of parental education and lack of stability often afforded by home ownership. Among cohabiters, women are better educated than men, with more than half (55 percent) of all female cohabiters having some college or a bachelor's degree, compared with 46 percent of male cohabiters. Those who are statistically most likely to cohabit are younger individuals (typically under

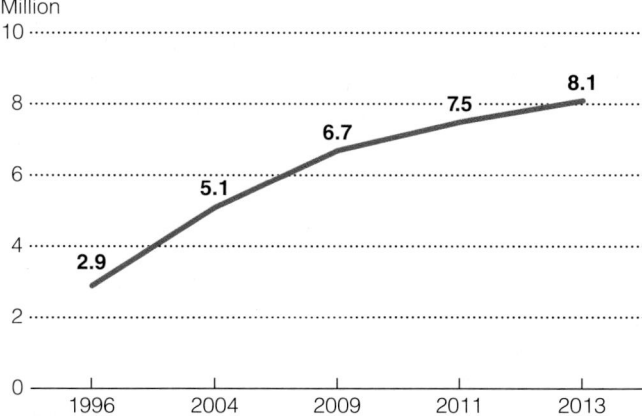

**FIGURE 11.6** ESTIMATED NUMBER OF OPPOSITE-SEX COUPLES COHABITING IN THE UNITED STATES IN SELECTED YEARS, 1996–2013

Source: U. S. Census Bureau, 2011d.

age 45), people who have been married before, or older persons who do not want to lose retirement benefits that are contingent upon not remarrying. Overall, cohabiting couples have fared less well in the second decade of the twenty-first century than they did in the first decade. Some problems were linked to national and international economic conditions, such as lack of new job creation and high rates of long-term unemployment. About 18 percent of male cohabiters and 27 percent of female cohabiters had no earnings in 2013, as compared with 2003, when 11 percent of males and 20 percent of females reported no earned income for that year (Vespa, Lewis, and Kreider, 2013).

How common is cohabitation among women? A study by the National Center for Health Statistics found that nearly one-half (48 percent) of all U.S. women between the ages of 15 and 44 had cohabited before marriage. Between 1995 and 2010, the percentage of women who cohabited as a first union increased to 57 percent of Hispanic (Latina) women, 43 percent of white (non-Hispanic) women, and 39 percent of African American (black) women (Copen, Daniels, and Mosher, 2013). According to researchers in this study, 40 percent of first premarital cohabitations among women transitioned to marriage within 3 years, about 33 percent of these cohabiting relationships continued, and 27 percent dissolved (Copen, Daniels, and Mosher, 2013). This study looked only at *first premarital cohabitation,* so it did not include those women who had cohabited with other partners.

To what extent do cohabiting couples have children together? Over the past decade, childbearing has become more frequent among cohabiting couples. Among all U.S. women ages 15–44, about 23 percent of births occurred within cohabiting relationships, up from 14 percent in 2002. According to research participants, about 50 percent of these births were unintended (Copen, Daniels, and Mosher, 2013).

Does cohabitation serve as a step toward marriage, or is it an alternative to marriage? Research shows that cohabitation is more likely to serve as a transition toward marriage among women with higher levels of education and income than for cohabiting women with lower levels (Copen, Daniels, and Mosher, 2013).

Among heterosexual couples, many reasons exist for cohabitation; for LGBT couples, however, no alternatives to cohabitation have existed for many years. For that reason, many lesbians, gays, bisexuals, and transgendered persons have sought recognition of their civil unions or **domestic partnerships—household partnerships in which an unmarried couple lives together in a committed, sexually intimate relationship and is granted some of the same rights and benefits as those accorded to married heterosexual couples.** Civil unions—which are available in some states to both same-sex and opposite-sex couples—provide legal recognition of the couple's relationship and afford legal rights to the partners similar to those accorded to spouses in marriage. Some states, such as Delaware and Rhode Island, replaced civil union provisions with same-sex marriage. It should be noted that domestic partnerships vary from state to state and that they are in a period of fluctuation triggered by changes in legal provisions pertaining to same-sex marriage. Rights typically granted to those domestic partners who meet state requirements include the following: hospital visitation rights, child custody and visitation rights, health insurance coverage if it is provided for heterosexual spouses by the partner's employer, and similar family-related rights that are granted to opposite-sex families. State and federal laws are in a period of transition regarding civil unions and domestic partnerships.

**cohabitation** a situation in which two people live together, and think of themselves as a couple, without being legally married.

**domestic partnerships** household partnerships in which an unmarried couple lives together in a committed, sexually intimate relationship and is granted some of the same rights and benefits as those accorded to married heterosexual couples.

## Marriage

Why do people get married? Couples get married for a variety of reasons. Some do so because they are "in love," desire companionship and sex, want to have children, feel social pressure, are attempting to escape from a bad situation in their parents' home, or believe that they will have more money or other resources if they get married. These factors notwithstanding, the selection of a marital partner is actually fairly predictable. As previously discussed, most people in the United States tend to choose marriage partners who are similar to themselves. *Homogamy* **refers to the pattern of individuals marrying those who have similar characteristics, such as race/ethnicity, religious background, age, education, or social class.** However, homogamy provides only the general framework within which people select their partners; people are also influenced by other factors. For example, some researchers claim that people want partners whose personalities match their own in significant ways. Thus, people who are outgoing and friendly may be attracted to other people with those same traits. However, other researchers claim that people look for partners whose personality traits differ from but complement their own.

The number of married households in the United States has been in a downward trend from the 1970s, when married couples made up 71 percent of all households, to 2012, when under half (49 percent) of all U.S. households were composed of married couples. The median age at first marriage in 2013 was 29.8 for men and 26.9 for women, as compared to 23.2 for men and 20.8 for women in 1970 (Roy, 2013).

## Same-Sex Marriages

As discussed in previous chapters, controversy continues over the legal status of gay and lesbian couples, particularly those who seek to make their relationship a legally binding commitment through marriage. Rapid changes have occurred in recent years regarding the legality of same-sex marriages in more states. One case regarding same-sex marriage, *United States v. Windsor,* challenged Section 3 of the 1996 Defense of Marriage Act (DOMA), which explicitly defined marriage for all purposes under federal law as the legal union of one man and one woman as husband and wife. The issue before the court was whether DOMA deprives same-sex couples, who are lawfully married in states that permit it, of the equal protection of the laws guaranteed by the Fifth Amendment to the Constitution. In 2011 President Obama declared this act to be

unconstitutional and ordered the U.S. Justice Department to stop defending the law in court. His highly controversial decision was applauded by gay and lesbian rights advocates but sharply denounced by conservative political leaders. The case moved through various courts until it reached the highest court in the nation.

The U.S. Supreme Court ruled that DOMA is unconstitutional because it amounts to a deprivation of the equal liberty of persons that is protected by the Fifth Amendment. This ruling struck down the central provisions of DOMA that denied federal benefits to same-sex couples who were married in jurisdictions that permit same-sex unions. However, this decision did not address the larger issue of whether there is a nationwide right of all same-sex couples to marry regardless of where they live.

In the second Supreme Court case, *Hollingsworth v. Perry,* the Court examined the issue of whether the Equal Protection Clause of the Fourteenth Amendment prohibits the state of California from enforcing Proposition 8, a voter-approved measure defining marriage as the union of a man and woman and banning same-sex marriage. Prior to passage of this proposition, in a five-month period in 2008 same-sex marriages were legally performed in California. Ultimately, the Supreme Court ruled that the proponents who intervened to defend Proposition 8 did not have legal standing based on the U.S. Constitution and that their appeal should be dismissed. The Court's decision cleared the way for same-sex marriages to resume in California. It also meant that same-sex couples married in jurisdictions that permit such unions could not be denied federal benefits.

Although legal, political, and social support for same-sex marriage has increased in recent years, many questions remain unanswered, including (1) how state-by-state laws will be applied in various situations, (2) what happens when a same-sex couple legally married in one state moves across state lines to another state where same-sex marriage is not recognized, and (3) how the patchwork of state laws affects divorce; as well, many other issues that the Supreme Court did not address when it reached decisions in these two cases still remain (Figure 11.7).

Overall, research indicates that there is more support for same-sex marriage among the general U.S. population than in the past. In March 2013, 48 percent of persons responding to a survey by the Pew Research Center (2013a) indicated that they favored allowing gays and lesbians to marry legally, while 43 percent opposed the notion. By contrast, in a 2001 survey 57 percent of Americans opposed gay

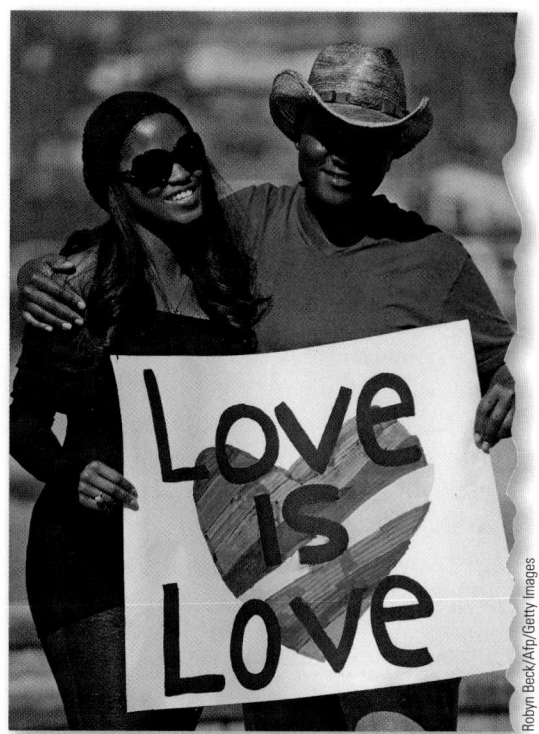

**FIGURE 11.7** The issue of same-sex marriage was frequently in the headlines in 2013 because of two major legal challenges that came before the U.S. Supreme Court.

marriage (Pew Research Center, 2013a). Researchers attribute this change in attitude to the millennial generation (people born after 1980), which is made up of young adults who are more open to gay rights than previous generations. The most frequently cited reason for changing attitudes on same-sex marriage is having friends, family members, or acquaintances who are lesbian or gay (Pew Research Center, 2013a).

## Housework and Child-Care Responsibilities

Taking care of housework, child-care responsibilities, and paid employment is a continuing challenge for families. In 2012, among the 34.6 million families with children in the United States, nearly 88 percent had at least one employed parent. In families with children maintained by women with no spouse present, the mother was employed in 67 percent of the households. By contrast, in families with children maintained by men with no spouse present, 82 percent of the fathers were employed. And in married-couple families with children, 96.3 percent of the households had at least one employed parent. The share of married-couple families with children where both parents work is 59 percent (U.S. Bureau of Labor Statistics, 2014a). What this means is that many marriages in the United States

are *dual-earner marriages*—**marriages in which both spouses are in the labor force.** Even when their children are under age six, nearly two-thirds of mothers are employed in the labor force (U.S. Bureau of Labor Statistics, 2014a). Moreover, many married women leave their paid employment at the end of the day and then go home to perform hours of housework and child care. Sociologist Arlie Hochschild (2012) refers to this as the *second shift*—**the domestic work that employed women perform at home after they complete their workday on the job.** Thus, many married women contribute to the economic well-being of their families and also meet many of the domestic needs of family members by cooking, cleaning, shopping, taking

**homogamy** the pattern of individuals marrying those who have similar characteristics, such as race/ethnicity, religious background, age, education, or social class.

**dual-earner marriages** marriages in which both spouses are in the labor force.

**second shift** Arlie Hochschild's term for the domestic work that employed women perform at home after they complete their workday on the job.

care of children, and managing household routines (Figure 11.8). According to Hochschild, the unpaid housework that women do on the second shift amounts to an extra month of work each year. For fathers with children and no spouse present, a similar second shift may also exist.

In recent years, married and cohabiting couples with more-egalitarian ideas about women's and men's roles have tended to share more equally in food preparation, housework, and child-care responsibilities (Lachance-Grzela and Bouchard, 2010). Although some studies show that when husbands share household responsibilities, they spend less time in these activities than do their wives, other studies shown that the higher a woman's educational resources and earnings potential, the more help from her partner that she actually gets with child care and housework (Sullivan, 2011).

In the United States, millions of parents rely on outside child care so that they can work. Relatives (father, grandparent, sibling, or other relative) take care of almost half of all children between birth and four years of age who have full-time employed mothers. Another 25 percent of children with full-time employed mothers spend most of their time away from home in a center-based arrangement, such as day care, nursery school, preschool, or Head Start. An additional 13 percent are primarily cared for by a nonrelative, such as a babysitter, nanny, au pair, or family day-care provider, in a home-based environment (Federal Interagency Forum on Child and Family Statistics, 2013). The cost of child-care programs often makes it difficult for families to find a high-quality environment for their children. Church-sponsored programs, such as "Mothers' Day Out," have become increasingly popular with young working mothers who use these facilities for one or more days per week as a form of relatively inexpensive day care for their children while they are at work.

Although organized after-school programs have become more numerous, the percentage of children staying home alone has remained steady in recent years. About 25 percent of school-age children stay alone after the school day ends until a parent returns home from work. Child-care specialists are concerned about this because children need productive and safe activities to engage in while their parents are working, but many home-alone children spend time eating junk food, watching television, talking on their cell phone, or playing computer and video games. Many of these children are under the supervision of an older brother or sister who may not be particularly interested in taking care of them (McClure, 2010). Older children are more likely than younger ones to care for themselves: 33 percent of children ages 12–14 are regularly in self-care situations, as compared to 10 percent of children ages 9–11 and 2 percent of children ages 5–8 (Federal Interagency Forum on Child and Family Statistics, 2013).

Ariel Skelley/Taxi/Getty Images

**FIGURE 11.8** Juggling housework, child care, and a job in the paid workforce is all part of the average day for many women. Why does sociologist Arlie Hochschild believe that many women work a "second shift"?

## Child-Related Family Issues and Parenting

**LO6** Not all couples become parents. Those who decide not to have children often consider themselves to be "child-free," whereas those who do not produce children through no choice of their own may consider themselves "childless."

### Deciding to Have Children

Cultural attitudes about having children and about the ideal family size began to change in the United States in the late 1950s. Women, on average, are now having (statistically speaking) 1.88 children, less than the average of 2.0 children in the past, and also less than what is referred to as the "replacement level" of 2.1 children (see "Sociology Works!"). For example, the birth rate for women ages 20–24 declined from 85.3 births in 2011 to 83.1 births per 1,000 women in 2012. However, the birth rates for women in their thirties and early forties rose in that

# Sociology WORKS!

## The Economics of Fertility Rates

**M**any people who are not looking to purchase a house anytime soon enjoy watching the popular HGTV series *House Hunters*. A scenario found in numerous episodes is a young couple seeking a home where they can start a family. Finding a "nest," in this case a residence that appeals to both partners and is within their budget, is viewed as a central factor in not only the couple's personal happiness but also in their ability to procreate and socialize a child or children in the right setting. In other words, *social factors* are important in fertility rates and procreation.

Demographers and other sociologists who specialize in population trends have sought to identify how biological and social factors affect fertility rates in various nations. Although biological factors such as general health and levels of nutrition in a region clearly affect the number of children that a couple may produce, social factors are also important in determining the estimated number of children that a woman will have in her lifetime (see Chapter 15 for additional information).

A key social factor is the housing market where a couple lives: The ability to buy a house and have a relatively large home may influence a couple's decision about how many children to have. Early in the 2000s, the housing market in some areas of the United States provided an opportune time for more young couples to purchase their own home. Mortgages were readily available, and it typically did not take much cash (in the way of a down payment) up front to sign the contract and move into one's dream home. The availability of such loans to people who really couldn't afford the

home they were buying has since come back to haunt many underfunded homebuyers. By 2007, more babies were being born, and the U.S. fertility rate reached a historic high of 69.3 per 1,000 women between the ages of 15 and 44. However, by 2013, the fertility rate for women in that age category declined to 62.7 births per 1,000 women, at least partly because of continuing economic concerns across the nation (Hamilton and Sutton, 2013).

Sociological insights on the social aspects of fertility, which at first might appear to be primarily a biological phenomenon, have provided us with new information on why people decide to have children and how many children they might have. However, much remains unknown about the relationship among the housing market, the maternity ward, and other social factors, including how income and feelings of optimism or pessimism about the local and national economy might affect a couple's decisions regarding parenting.

### reflect & analyze

**How might sociological findings about factors that influence a couple's decision to have children be useful in your community? For example, why is information about the availability of housing and local fertility trends important to school board members and administrators when they are making enrollment projections or deciding where to build a new school in the future?**

---

same year, suggesting that more women may be deferring childbearing rather than not producing children at all. In 2012 more than half of all U.S. women between the ages of 15 and 44 were mothers, and 81 percent of women had become mothers by age 45. This figure is somewhat lower than the 90 percent of women who had given birth by that age in the 1970s.

Today, more than 40 percent of all births in the United States are to unmarried women, and this number rose by 1 percent from 2011 to 2012, the latest year for which data are available (National Vital Statistics Reports, 2013a).

Advances in birth control techniques over the past four decades—including the birth control pill and contraceptive patches and shots—now make it possible for people to decide whether they want to have children, how many they wish to have, and to determine (at least somewhat) the spacing of the children's births. However, sociologists suggest that

fertility is linked not only to reproductive technologies but also to women's beliefs that they do or do not have other opportunities in society that are viable alternatives to childbearing (Lamanna and Riedmann, 2012).

Today, the concept of reproductive freedom includes both the desire *to have* or *not to have* one or more children. According to some sociologists, many U.S. women spend up to one-half of their life attempting to control reproduction while other women and men choose to be child-free. Among U.S. women ages 15 to 44, 6 percent are voluntarily childless. The percentage of women who are voluntarily childless (and have never given birth) is higher among women who hold professional degrees (a master's or equivalent and higher). One in four women with a master's degree and nearly that many women with doctorates have no biological children by the time the women reach the ages of 40 to 44. Of course, some of these women may

# SOCIOLOGY *in Global Perspective*

## Wombs-for-Rent: Commercial Surrogacy in India

I wanted to be a surrogate mother because I wanted to deposit money into an account for my children for their future. I also wanted to help parents who cannot have children. I am proud to have given birth to a beautiful baby. . . . I feel like part of the family.

> —Thapa, a 31-year-old Indian woman who became a surrogate for an Australian couple, explains why, in addition to her own children, she is willing to help other people become parents (qtd. in AFP, 2013). Thapa works with a New Delhi, India, surrogacy center, where she and other surrogate mothers earn about $6,000 for carrying a child for a foreign couple.

**W**hy do some infertile couples in the United States, Britain, and elsewhere want to "hire" a woman in India to have their child? Most couples who engage in this practice have made numerous attempts to have a child through in vitro fertilization and other assisted reproductive technologies. If they have been unsuccessful in their efforts, the couple may first attempt to find a surrogate in the United States, but they quickly learn that a U.S. gestational surrogate costs more than $50,000—far more than they would pay for a surrogate in India (Kohl, 2007; United Press International, 2007). According to infertility specialist Dr. Nayna Patel, earning money through surrogacy helps uplift Indian women: It provides money for their household and makes them more independent. For example, the typical woman might earn more for one surrogate pregnancy than she would earn in 15 years from other kinds of employment (*CBS News,* 2007).

Are there any problems with global "rent-a-womb"? If there is an agreement between a surrogate mother and a couple who badly wants a child, some analysts believe that "offer and acceptance" is nothing more than capitalism at work—where there is a demand (for infants by infertile couples), there will be a supply (from low-income surrogate

---

have children through adoption, by marriage to a partner who already had one or more children, or other factors (Wade, 2012). When some people decide not to have children, their wishes come into conflict with our society's *pronatalist bias,* which assumes that having children is the norm and can be taken for granted, whereas those who choose not to have children believe they must justify their decision to others (Lamanna and Riedmann, 2012).

Some couples experience the condition of *involuntary infertility,* whereby they want to have a child but find that they are physically unable to do so. *Infertility* is defined as an inability to conceive after a year of unprotected sexual relations. Women who are able to get pregnant but who are not able to stay pregnant may also be defined as infertile. Research suggests that fertility problems originate in females in approximately one-third of the cases, with males in about one-third of the cases, and with both the male and female in about one-third of the cases (Mayo Clinic, 2011). Leading causes of male infertility are abnormal sperm production or function, sexual problems, general health and lifestyle issues, overexposure to certain environmental factors (such as pesticides and other chemicals or heat), and age. The most common causes of female infertility include fallopian tube damage or blockage, endometriosis, ovulation disorders, early menopause, and other health-related disorders (Mayo Clinic, 2011).

It is estimated that about half of infertile couples who seek treatments such as medication, behavioral approaches, fertility drugs, artificial insemination, and surgery can be helped; however, some are unable to overcome infertility despite expensive treatments such as *assisted reproductive technology (ART),* which includes medical procedures such as *in vitro fertilization,* whereby medical professionals help infertile couples achieve pregnancy (Mayo Clinic, 2011). Some people who are involuntarily childless may choose surrogacy or adoption as alternative ways of becoming a parent (see "Sociology in Global Perspective").

### Adoption

*Adoption* is a legal process through which the rights and duties of parenting are transferred from a child's biological and/or legal parents to a new legal parent or parents. This procedure gives the adopted child all the rights of a biological child. In most adoptions a new birth certificate is issued, and the child has no future contact with the biological parents; however, some states have "right-to-know" laws under which adoptive parents must grant the biological parents visitation rights. In 2011–2012, approximately 67 percent of all adoptive households were made up of married couples, followed by single women at 27 percent, single men at 2.9 percent, and unmarried it, not because she is being paid to give birth to someone else's baby. Other social critics are concerned about the potential mistreatment of low-income women who may be exploited or may suffer long-term emotional damage from functioning as a surrogate (Dunbar, 2007). For the time being, in clinics such as the one in India, hopeful parents just provide the egg, the sperm, and the money, and all the rest is done for them by the clinic and the surrogates, who live in a spacious house where they are taken care of by maids, cooks, and doctors.

A surrogate mother (left) has delivered a baby for Karen Kim (center), with the help of infertility specialist Dr. Nayna Patel (right). This practice, sometimes called "rent-a-womb," remains controversial.

mothers). However, some ethicists raise troubling questions about the practice of commercial surrogacy: A mother should give birth to her child because it is hers and she loves

## reflect & analyze

What are your thoughts on surrogacy? Is there any difference between surrogacy when it occurs in high-income nations such as the United States and Britain as compared to situations in which the parents live in a high-income nation and the surrogate mother lives in a lower-income nation? How might we relate the specific issue of outsourced surrogacy to some larger concerns about families and intimate relationships that we have discussed in this chapter?

couples at 2.3 percent (U.S. Department of Health and Human Services, 2013).

Matching children who are available for adoption with prospective adoptive parents can be difficult. The available children have specific needs, and the prospective parents often set specifications on the type of child they want to adopt. Some adoptions are by relatives of the child; others are by infertile couples (although many fertile couples also adopt). Increasing numbers of gays, lesbians, and people who are single are adopting children. Although thousands of children are available for adoption each year in the United States, many prospective parents seek out children in nations such as China, Russia, and India. The primary reason is that the available children in the United States are thought to be "unsuitable." They may have disabilities, or they may be sick, nonwhite (most of the prospective parents are white), or too old. In addition, fewer infants are available for adoption today than in the past because better means of contraception exist, abortion is more readily available, and more unmarried teenage parents decide to keep their babies.

## Teenage Childbearing

Teenage childbearing is a popular topic in the media and political discourse, and the United States has the highest rate of teen pregnancy in the Western

industrialized world (National Vital Statistics Report, 2013a). However, as shown in Figure 11.9, the U.S. teen birth rate (between the ages of 15 and 19) has generally been in a decline from 1960 to 2012, briefly increased in periods such as 2005–2007, and then resumed its long-term downward trend. In 2012 the birth rate for teenagers between the ages of 15 and 19 dropped 6 percent (to 29.4 births per 1,000 teenagers ages 15–19) as compared with 2011. This was the lowest rate ever reported in the United States. Rates were also down for ages 15 through 19 and for nearly all races and Hispanic-origin groups (National Vital Statistics Report, 2013).

Although teen pregnancy rates have continued to decline, concern remains about the number of teenagers who are producing children and the fact that unmarried teenagers accounted for 17 percent of all nonmarital births in 2012 (National Vital Statistics Report, 2013). What are the primary reasons for teenage pregnancy? At the microlevel, several issues are most important: (1) many sexually active teenagers do not use contraceptives; (2) teenagers—especially those from low-income families and/or subordinate racial and ethnic groups—may receive little accurate information about the use of, and problems associated with, contraception; (3) some teenage males (because of a double standard based on the myth that sexual promiscuity is acceptable among males but not females) believe that females

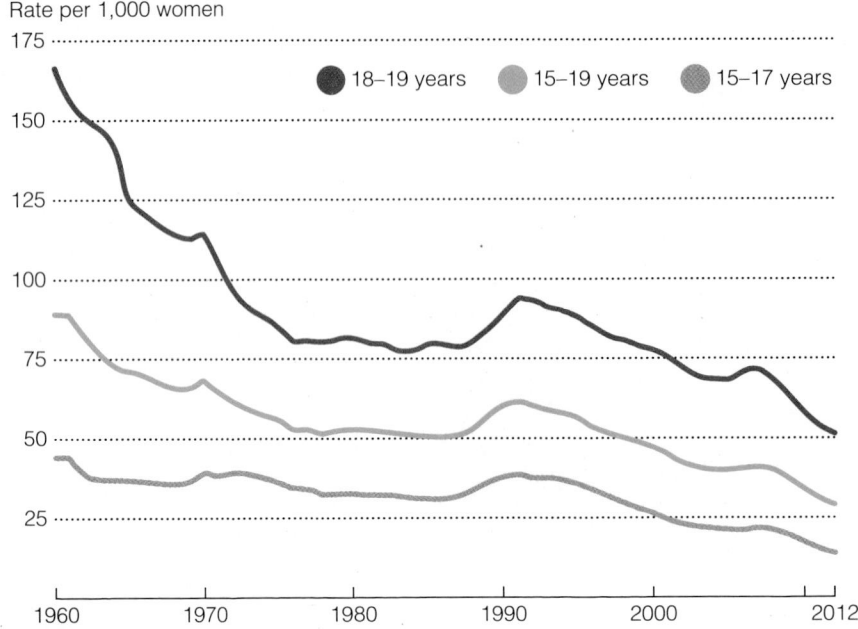

Rate per 1,000 women

● 18–19 years　● 15–19 years　● 15–17 years

**FIGURE 11.9** BIRTH RATES FOR TEENAGERS AGES 15–19 YEARS, BY AGE, UNITED STATES, 1960–2012

Source: National Vital Statistics Reports, 2013a.

should be responsible for contraception; and (4) some teenagers view pregnancy as a sign of male prowess or as a way to gain adult status.

At the macrolevel, structural factors also contribute to teenage pregnancy rates. Lack of education and employment opportunities may discourage young people's thoughts of upward mobility. Likewise, religious and political opposition has resulted in issues relating to reproductive responsibility not being dealt with as openly in the United States as in some other nations. Finally, advertising, films, television programming, magazines, music, and other forms of media often flaunt the idea of being sexually active without showing the possible consequences of such behavior. However, recent television series have been both praised and blamed for their portrayals of teen pregnancy, even as producers and sponsors have claimed to send a message about fighting teenage pregnancy (see "Teen Pregnancy in the Media").

Teen pregnancies are of concern to analysts who argue that teenage mothers and their children experience strong socioeconomic disadvantages (Mollborn and Dennis, 2011). Studies show that teen mothers may be less skilled at parenting, are less likely to complete high school than their counterparts without children, and possess few economic and social supports other than their relatives. In addition, these births may have negative long-term consequences for these mothers and their children, who may also have limited educational and employment opportunities, and a high likelihood of living in poverty.

Teenage fathers have largely been left out of the picture in most studies of teen pregnancy and parenting. How does having a teen father affect a child's health and development? One study found that the father–child relationship does not differ significantly between having a teen or adult father when factors such as marital status and economic disadvantage are held constant. For example, in families where the father, regardless of his age, is nonresidential (lives elsewhere) or cohabits occasionally, children are placed at a social disadvantage when compared to children whose father is married to the teen mother and resides in a more permanent family arrangement (Mollborn and Lovegrove, 2010). What about the effects of teen parenting on the father? According to a study on the effects of teenage fatherhood, teen pregnancy led to a decrease in the number of years of schooling among teen fathers who sought early full-time employment, enrolled in the military, or acquired high school equivalency diplomas once they became fathers (Sciencedaily. com, 2011).

## Single-Parent Households

Since 1970, there has been a significant increase in single-parent or one-parent households with children under age 18 because of divorce and births outside of marriage (Figure 11.10). In 2013 more

PART 4 | SOCIAL INSTITUTIONS

than one in four (27 percent) U.S. children under age 18 lived in a single-parent household. As shown in Figure 11.11, the percentage of mother-only family groups continues to rise while the percentage of father-only family groups with children under age 18 remains consistently much lower (Vespa, Lewis, and Kreider, 2013).

Does living in a single-parent family put children at risk? According to research by sociologist Paul R. Amato (2005), changes in family structure such as an increase in the proportion of single-parent households tends to place children in situations where they experience a lower standard of living, receive less-effective parenting, experience less-cooperative co-parenting, are less emotionally close to both parents, and are subjected to more stressful events and circumstances than children who grow up in stable, two-parent families. Why does this occur? Because of factors such as economic hardships that force single-parent families to do without books, without computers, and without homes in better neighborhoods and school districts. When this problem is coupled with the lack of time for parenting while the single parent struggles to make ends meet, and the fact that many children lose contact with their fathers after separation or divorce, the quality of parenting is often less than that found in supportive, co-parenting relationships. Even for a person with a stable income and a network of friends and family to help with child care, raising a child alone can be an emotional and financial burden.

Because of the nature of marriage laws in some states, lesbian mothers and gay fathers are counted in some studies as single parents even when they share parenting responsibilities with a same-sex partner. More research has been conducted in recent years on parenting by lesbian, gay, bisexual, and transgendered (LGBT) couples, and the results are favorable; living in these families appears in some respects to be better for children than living in heterosexual families. One of the areas that appears to be better is the division of parenting and household labor, which has a distinct pattern of equality and sharing among LGBT

**FIGURE 11.10** Mothers and fathers in single-parent households are confronted with the necessity of meeting most of their children's daily needs without help from others. However, even in two-parent households, children are not guaranteed a happy childhood simply because both parents live in the same household.

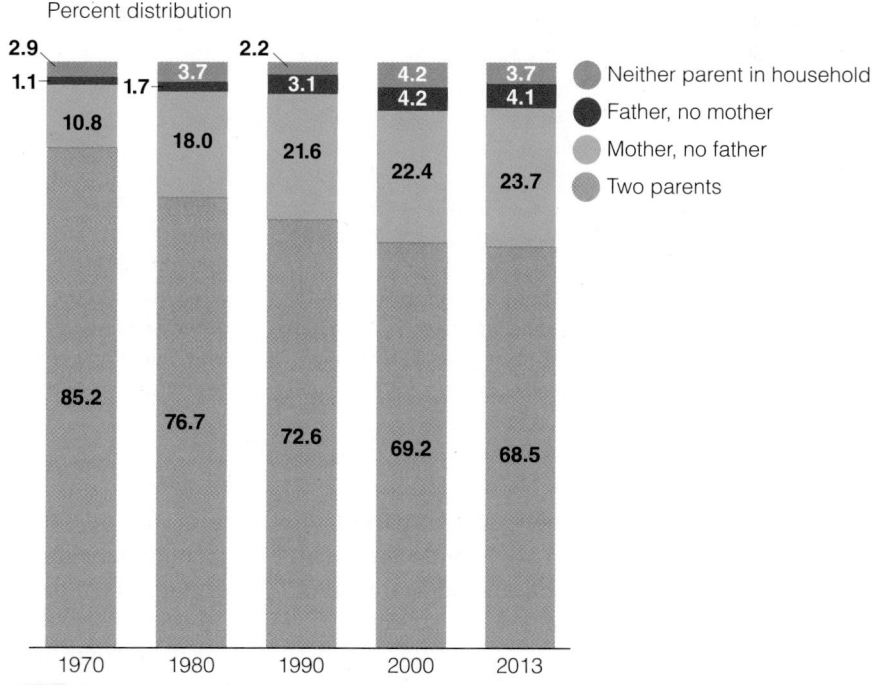

Percent distribution

Neither parent in household
Father, no mother
Mother, no father
Two parents

**FIGURE 11.11** LIVING ARRANGEMENTS OF CHILDREN UNDER 18 YEARS OLD FOR SELECTED YEARS, 1970–2012

Source: Vespa, Lewis, and Kreider, 2013.

# TEEN PREGNANCY IN THE MEDIA

## Sex Ed on TV? Mixed Messages on Teen Pregnancy

Chelsea, Jenelle, Kailyn, and Leah were like many other happy-go-lucky teenagers. Chelsea was a popular softball star until she became pregnant and dropped out of high school during her senior year to have her daughter. Jenelle was a carefree beach bunny until she gave birth to her son. Kailyn was a highly motivated high school senior until she became the mother of her son. Leah was a former cheerleader until she became the teen mother of twins.

Although this sounds like an episode of *Glee* or a soap opera, this is actually the cast of one season of MTV's *Teen Mom* franchise, where young girls are shown going through the process of finding out that they are pregnant, making adjustments to their daily lives, and entering the world of motherhood and complex relationships with their parents and boyfriends as their lives swiftly change around them. For some girls this means giving up their child for adoption and dealing with the consequences of that decision.

*16 and Pregnant* and the *Teen Mom* franchise have been very popular with many adolescent girls. Like *Teen Mom*, *16 and Pregnant* also follows high school girls as they cope with being pregnant and becoming parents. These programs are so popular among some teens that rumors have spread that girls were becoming pregnant so that they could "star" on the shows. Although this rumor was never confirmed, it does bring up an interesting question: What kind of media message is being sent to young women about becoming pregnant while in high school?

A content analysis of these shows reveals several standard framing devices: (1) "What would it be like?" framing, (2) "Glad it's not me" framing, and (3) "cautionary tale" framing. In the series *16 and Pregnant*, viewers follow a different girl each week as she has her baby and moves into her new life. Viewers have an opportunity to see "what it would be like" because each girl describes her feelings and how she made the decisions she did along the way. Viewers see her friends, family, and sometimes the child's father as they respond to the situation. "Glad it's not me" framing comes into play in *16 and Pregnant* and *Teen Mom* when viewers see the many obstacles that a young mother faces as she tries to have a life of her own, parent an infant, and deal with interpersonal relationships. "Glad it's not me" framing provides a voyeuristic view of the often-troubled relationship between the new mother and the baby's father, and this is sometimes the focus of the show, more than other issues—such as the baby's future—that might also be discussed.

Finally, "cautionary tale" framing of these shows is what many parents and educators like most about *16 and Pregnant* and *Teen Mom:* They believe that these programs open up the topic of teen pregnancy and parenting for discussion with adolescents who are rarely interested in talking about such matters with adults. It is the hope of some parents and educators that girls will be drawn into the lives of the young women featured in these reality shows and that they will come away with the message that they do not want to be in a similar situation. For example, some teachers ask their students to think about what it would be like to have schoolwork, babies who are ill, lack of sleep, and a boyfriend who does not help them out, all at the same time.

However, one of the major criticisms of this type of show as a cautionary tale is the fact that these shows do not provide useful sex education for young people. In some cases the shows may glamorize teen pregnancy and parenting for young women who have no other way of getting attention from their parents, peers, and boyfriends. Overall, though, programs such as these have been applauded for framing a pressing social concern in such a manner that it gains the attention of young women, the very audience that sex educators and parents have been trying to reach for many years (Hoffman, 2011).

## reflect & analyze

Clearly, getting pregnant and having a child is a major milestone, but should teenage mothers share their thoughts and feeling about this event with millions of viewers? Do programs such as *Teen Mom* have the potential to generate constructive conversations about the choices that teens make and how they deal with the consequences of their actions? What do you think?

---

couples as compared with heterosexual parents. It also appears that lesbian and gay parents tend to be more responsive to their children and more child-oriented in their outlook (Short et al., 2007).

## Two-Parent Households

Between 1970 and 2012, the share of all U.S. households comprising married couples with children under age 18 halved from 40 percent to slightly less than 20 percent. During that time period, the percentage of children living in two-parent households dropped from 85.2 percent to 68.5 percent while the percentage living with a single parent increased. In computing these statistics for "parents," the U.S. Census Bureau includes not only biological parents but also stepparents who adopt their children. However, foster parents are considered nonrelatives.

For families in which a couple truly shares parenting, children have two primary caregivers. Some parents share parenting responsibilities by choice; others share out of necessity because both hold full-time jobs. Some studies have found that men's taking an active part in raising the children is beneficial not only for mothers (who then have a little more time for other activities) but also for the men and the children. The men benefit through increased access to children and greater opportunity to be nurturing parents (Coltrane, 2010).

## Remaining Single

Some never-married people remain single by choice. Reasons include opportunities for a career (especially for women), the availability of sexual partners without marriage, the belief that the single lifestyle is full of excitement, and the desire for self-sufficiency and freedom to change and experiment. Some scholars have concluded that individuals who prefer to remain single hold more-individualistic values and are less family oriented than those who choose to marry. Friends and personal growth tend to be valued more highly than marriage and children.

Other never-married individuals remain single out of necessity. Being single is an economic necessity for those who cannot afford to marry and set up their own household. Structural changes in the economy have limited the options of many young working-class people. Even some college graduates have found that they cannot earn enough money to set up a household separate from that of their parents.

The proportion of singles varies significantly by racial and ethnic group, as shown in Figure 11.12. Among persons ages 18 and over in 2012, slightly over 43 percent of African Americans had never married, compared with 35 percent of Latinos/as, nearly 27 percent of Asian and Pacific Islander Americans, and slightly less than 23 percent of non-Hispanic whites (ProQuest, 2014).

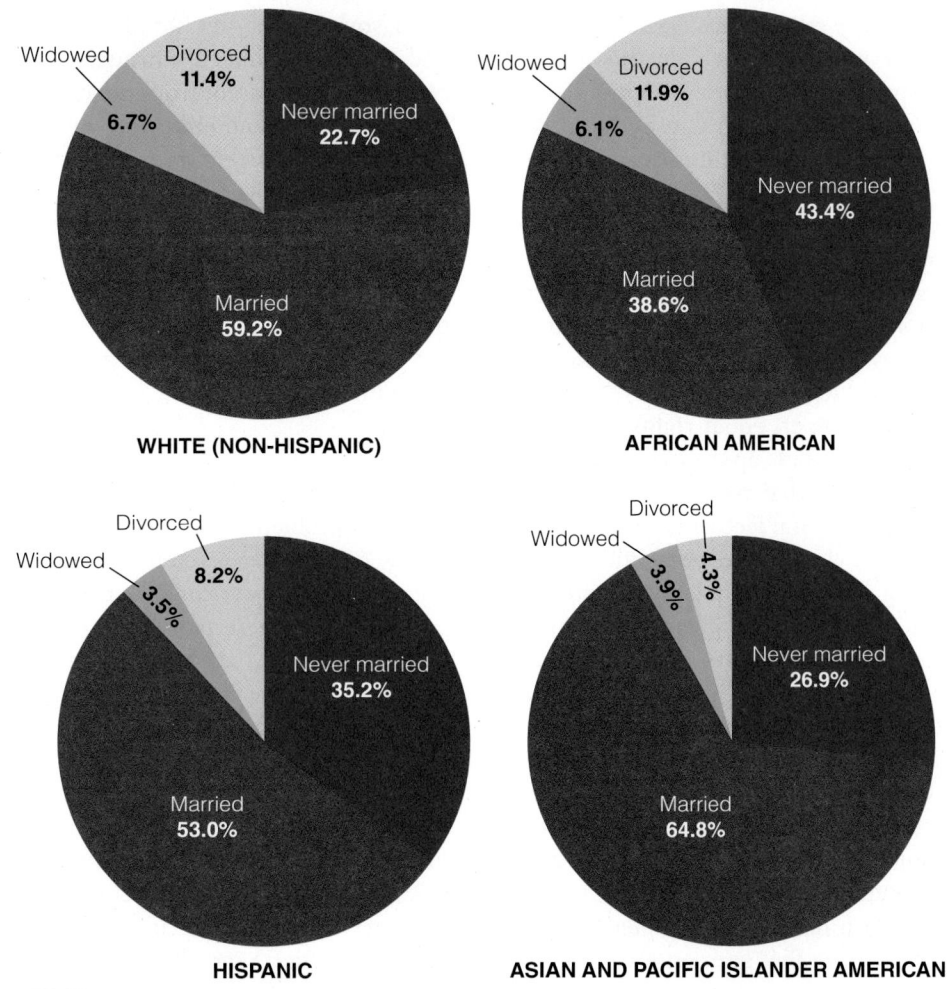

**FIGURE 11.12** MARITAL STATUS OF U.S. POPULATION AGES 18 AND OVER BY RACE/ETHNICITY, 2012

Source: ProQuest, 2014.

# Transitions and Problems in Families

**LO7** Families go through many transitions and experience a wide variety of problems, ranging from high rates of divorce and teen pregnancy to domestic abuse and family violence. These all-too-common experiences highlight two important facts about families: (1) for good or ill, families are central to our existence, and (2) the reality of family life is far more complicated than the idealized image of families found in the media and in many political discussions. Moreover, as people grow older, transitions inevitably occur in family life.

## Family Violence

*Family violence* refers to various forms of abuse that take place among family members, including child abuse, spousal abuse, and elder abuse. We will primarily focus on domestic violence—also referred to as spousal abuse or intimate-partner violence—and elder abuse. *Domestic violence* refers to any intentional act or series of acts—whether physical, emotional, or sexual—by one or both partners in an intimate relationship that causes injury to either person. An intimate relationship might include marriage or cohabitation, as well as people who are separated or living apart from a former partner or spouse.

There are numerous causes of domestic violence, and many factors are interrelated. Factors contributing to unequal power relations in families include economic inequality, legal and political sanctions that deny girls and women equal rights, and cultural sanctions that dictate appropriate sex roles and reinforce the belief that males are inherently superior to females. Cultural factors that perpetuate domestic violence include gender-specific socialization that establishes dominant–subordinate sex roles. Economic factors include poverty or limited financial resources within families that contribute to tension and sometimes to violence. Economic factors are intertwined with women's access to education, employment, and sufficient income so that they can take care of themselves and their children. Regardless of the factors that contribute to domestic violence, control is central to all forms of abuse: Gaining and maintaining control over the victim is the key factor. As a result, family violence often involves a cycle of abuse that goes on for extended periods of time.

How much do we really know about family violence? Women, as compared with men, are more likely to be the victims of violence perpetrated by intimate partners. Recent statistics indicate that more than 25 million women in the United States experience domestic violence at some time during their lives. However, we cannot know the true extent of family violence because much of it is not reported to police. For example, it is estimated that only about one-half of the intimate-partner violence against women is reported. The reasons that many victims of family violence do not report the violence to police include (1) belief that such violence is a private or personal matter, (2) fear of retaliation, (3) view of the violence as a "minor" crime, (4) desire to protect the offender, and (5) belief that the police will not help or will be ineffective.

Although everyone in a household where family violence occurs is harmed psychologically, whether or not they are the victims of violence, children are especially affected. It is estimated that between 3 million and 10 million children witness some form of domestic violence in their homes each year, and there is evidence to suggest that domestic violence and child maltreatment often take place in the same household (Children's Defense Fund, 2011).

In some situations, family violence can be reduced or eliminated through counseling, the removal of one parent from the household, or other steps that are taken either by the family or by social service or law enforcement officials. However, children who witness violence in the home may display certain emotional and behavioral problems that adversely affect their school life and communication with other people. In some families the problems of family violence are great enough that the children are removed from the household and placed in foster care.

## Children in Foster Care

A special problem in families is when children must be placed in foster care either voluntarily or involuntarily. Many foster children have been in dysfunctional homes where parents or other relatives lack the ability to meet the children's daily needs. *Foster care* refers to institutional settings or residences where adults other than a child's own parents or biological relatives serve as caregivers. States provide financial aid to foster parents, and the intent of such programs is that the children will either return to their own families or be adopted by other families. However, this is often not the case for "difficult to place" children, particularly those who are over ten years of age, have illnesses or disabilities, or are perceived as suffering from "behavioral problems." More than 400,000 children are in foster care at any given time, though the number has declined each year since 2005. More than one-half of all children in foster care are in just nine states: California, Florida, Illinois, Indiana, Michigan, New York, Ohio, Pennsylvania, and Texas. About one-third of children in foster care are age five or younger, one-third are between the ages of six and thirteen, and one-third are age fourteen or older. Those who enter foster care at a younger age tend to remain in the system for a longer period of time. Forty-five percent or more of all children in foster care in one-fourth of the states and the District of Columbia are African American, and in twenty-one states the percentage of African American children in foster care is more than twice their proportion in the general child population (Children's Bureau, 2013).

Problems in the family contribute to the large numbers of children who are in foster care. Such factors include parents' illness, unemployment, or death; violence or abuse in the family; and high rates of divorce.

## Divorce

**LO8** *Divorce* is the legal process of dissolving a marriage that allows former spouses to remarry if they so choose. Most divorces today are granted on the grounds of *irreconcilable differences*, meaning that there has been a breakdown of the marital relationship for which neither partner is specifically blamed. Prior to the passage of more-lenient divorce laws, many states required that the partner seeking the divorce prove misconduct on the part of the other spouse. Under *no-fault divorce laws*, however, proof of "blameworthiness" is generally no longer necessary.

Over the past 100 years, the U.S. divorce rate (number of divorces per 1,000 population) has varied from a low of 0.7 in 1900 to an all-time high of 5.3 in 1981; by 2012, it had decreased to slightly less than 4.2 percent (Vespa, Lewis, and Kreider, 2013). Divorce statistics may vary based on the source because some organizations include annulments in their count, and a number of states no longer provide divorce statistics to national reporting agencies. Figure 11.13 shows the latest available U.S. divorce rates for each state so that you can see how your state compares with others.

Although many people believe that marriage should last for a lifetime, others believe that marriage is a commitment that may change over time. Between 40 and 50 percent of new U.S. marriages will end in divorce (Cherlin, 2010). According to various studies on divorce in the United States, there are significant differences in the rates of divorce for first, second, and third marriages: The divorce rate for first marriages is between 40 and 50 percent, the rate for second marriages is from 60 to 67 percent,

and the rate for third marriages is from 73 to 75 percent. Research has also found that couples with children have a slightly lower rate of divorce as compared to couples without children.

Financial stressors are a contributing factor in some divorces (National Marriage Project, 2010). During times of a national recession such as the one that United States experienced beginning in 2007–2008, some people decide to remain married only until conditions change so that they can sell a house or have better financial stability when they part. However, others gain a deeper commitment to their marriage as they struggle through adversity (National Marriage Project, 2010).

As previously stated, we often hear that 50 percent of all marriages end in divorce. But this figure is misleading in some ways, even though it is true for the U.S. population as a whole. Although there is a 40–50 percent chance that a first marriage will end in either divorce or separation before one partner dies, divorce rates vary widely based on certain factors. There is a decrease in risk of divorce for people in the following categories:

- Making over $50,000 annually (as compared with under $25,000)
- Having graduated from college (as opposed to not completing high school)
- Having a baby 7 months or more after marriage (as opposed to before marriage)
- Marrying over 25 years of age (as opposed to under 18)
- Coming from an intact family of origin (as opposed to having divorced parents)
- Having a religious affiliation (as opposed to having none) (National Marriage Project, 2010)

**Consequences of Divorce** In some families, divorce may have a dramatic economic and emotional impact on family members. In others, the effect may be more marginal. In most families where one or more children are present, the children remain with their mother and live in a single-parent household for a period of time. In recent years, there has been a debate over whether children who live with a same-sex parent after divorce are better off than their peers who live with an opposite-sex parent. However, virtually no evidence has been found to support the belief that children are better off living with a same-sex parent.

Although divorce decrees frequently provide for parental joint custody of children, this arrangement may create unique problems. Furthermore, some children experience more than one divorce during their childhood because one or both of their parents may remarry and subsequently divorce again.

| | Rates per 1,000 population[a, b] | | | | | | Rates per 1,000 population[a, b] | | | |
|---|---|---|---|---|---|---|---|---|---|---|
| | 1990 | 2000 | 2005 | 2011 | | | 1990 | 2000 | 2005 | 2011 |
| United States[c] | 4.7 | 4.0 | 3.6 | 3.6 | | Missouri | 5.1 | 4.5 | 3.6 | 3.9 |
| Alabama | 6.1 | 5.5 | 4.9 | 4.3 | | Montana | 5.1 | 4.2 | 4.5 | 4.0 |
| Alaska | 5.5 | 3.9 | 4.3 | 4.8 | | Nebraska | 4.0 | 3.7 | 3.3 | 3.5 |
| Arizona | 6.9 | 4.6 | 4.2 | 3.9 | | Nevada | 11.4 | 9.9 | 7.4 | 5.6 |
| Arkansas | 6.9 | 6.4 | 6.0 | 5.3 | | New Hampshire | 4.7 | 4.8 | 3.9 | 3.8 |
| California | 4.3 | --- | --- | --- | | New Jersey | 3.0 | 3.0 | 2.9 | 2.9 |
| Colorado | 5.5 | 4.7 | 4.4 | 4.4 | | New Mexico | 4.9 | 5.1 | 4.6 | 3.3 |
| Connecticut | 3.2 | 3.3 | 3.0 | 3.1 | | New York | 3.2 | 3.0 | 2.9 | 2.9 |
| Delaware | 4.4 | 3.9 | 3.8 | 3.6 | | North Carolina | 5.1 | 4.5 | 4.1 | 3.7 |
| District of Columbia | 4.5 | 3.2 | 2.0 | 2.9 | | North Dakota | 3.6 | 3.4 | 2.9 | 2.7 |
| Florida | 6.3 | 5.1 | 4.6 | 4.5 | | Ohio | 4.7 | 4.2 | 3.5 | 3.4 |
| Georgia | 5.5 | 3.3 | --- | --- | | Oklahoma | 7.7 | --- | 5.6 | 5.2 |
| Hawaii | 4.6 | 3.9 | --- | --- | | Oregon | 5.5 | 4.8 | 4.2 | 3.8 |
| Idaho | 6.5 | 5.5 | 5.0 | 4.9 | | Pennsylvania | 3.3 | 3.1 | 2.3 | 2.8 |
| Illinois | 3.8 | 3.2 | 2.6 | 2.6 | | Rhode Island | 3.7 | 2.9 | 3.0 | 3.2 |
| Indiana | --- | --- | --- | --- | | South Carolina | 4.5 | 3.8 | 2.9 | 3.2 |
| Iowa | 3.9 | 3.3 | 2.7 | 2.4 | | South Dakota | 3.7 | 3.5 | 2.8 | 3.3 |
| Kansas | 5.0 | 3.6 | 3.1 | 3.9 | | Tennessee | 6.5 | 5.9 | 4.6 | 4.3 |
| Kentucky | 5.8 | 5.1 | 4.6 | 4.4 | | Texas | 5.5 | 4.0 | 3.3 | 3.2 |
| Louisiana | --- | --- | --- | --- | | Utah | 5.1 | 4.3 | 4.1 | 3.7 |
| Maine | 4.3 | 5.0 | 4.1 | 4.2 | | Vermont | 4.5 | 4.1 | 3.6 | 3.6 |
| Maryland | 3.4 | 3.3 | 3.1 | 2.9 | | Virginia | 4.4 | 4.3 | 4.0 | 3.8 |
| Massachusetts | 2.8 | 2.5 | 2.2 | 2.7 | | Washington | 5.9 | 4.6 | 4.3 | 4.1 |
| Michigan | 4.3 | 3.9 | 3.4 | 3.4 | | West Virginia | 5.3 | 5.1 | 5.1 | 5.2 |
| Minnesota | 3.5 | 3.2 | --- | --- | | Wisconsin | 3.6 | 3.2 | 2.9 | 2.9 |
| Mississippi | 5.5 | 5.0 | 4.4 | 4.0 | | Wyoming | 6.6 | 5.8 | 5.2 | 4.8 |

**FIGURE 11.13** U.S. DIVORCE RATES BY STATE, 1990–2011

[a]Based on total population residing in area; population enumerated as of April 1 for 1990; estimated as of July 1 for all other years.
[b]Includes annulments.
[c]U.S. totals for the numbers of divorces are an estimate that includes states not reporting, such as California, Colorado, Indiana, and Louisiana.

Source: ProQuest, 2014.

But divorce does not have to be always negative. For some people, divorce may be an opportunity to terminate destructive relationships. For others, it may represent a means to achieve personal growth by managing their lives and social relationships and establishing their own social identity. Still others choose to remarry one or more times.

## Remarriage

Most people who divorce get remarried. Over the past decade, more than 40 percent of all marriages have been between previously married brides and/ or grooms. Most divorced people remarry others who have been divorced. However, remarriage rates vary by gender and age. At all ages, a greater proportion of men than women remarry, often relatively soon after the divorce. In the 2010s, among divorced men ages 25 and above, 52 percent of them remarried. Among women, the older a woman is at the time of divorce, the lower her likelihood of remarrying. Women who have not graduated from high school and who have young children tend to remarry relatively quickly; by contrast, women with at least a college degree and without children are less likely to remarry.

Unfortunately, the National Center for Health Statistics at the Centers for Disease Control and Prevention discontinued some of its research and statistical analysis on remarriage early in the 2000s, so extensive data are not available on the nature and extent of remarriage in the United States. It is assumed that there is a relatively high failure rate for remarriages and that reasons for the failure include a history of psychological stress or general anxiety disorder among one or more of the partners, financial problems associated with the lingering recession and slow recovery, and lower family incomes (edivorcestatistics.com, 2013).

As a result of divorce and remarriage, complex family relationships are often created. Some people become part of stepfamilies or **blended families, which consist of a husband and wife, children from previous marriages, and children (if any) from the new marriage** (Figure 11.14). At least initially, levels of family stress may be fairly high because of rivalry among the children and hostilities directed toward stepparents or babies born into the family. In spite of these problems, however, many blended families succeed. The family that results from divorce and remarriage is typically a complex, binuclear family in which children may have a biological parent and a stepparent, biological siblings and stepsiblings, and an array of other relatives, including aunts, uncles, and cousins.

The norms governing divorce and remarriage are ambiguous. Because there are no clear-cut guidelines, people must make decisions about family life (such as whom to invite for a birthday celebration or wedding) based on their beliefs and feelings about the people involved. This adds to individuals' insecurity and confusion about how to interact with others on social occasions and has contributed to the use of informal social networking technology to communicate rather than having face-to-face discussions about plans that must be made involving children, finances, or other shared concerns.

Kayte Deioma/PhotoEdit

**FIGURE 11.14** Remarriage and blended families create new opportunities and challenges for parents and children alike.

## Family Issues in the Future

As we have seen, families and intimate relationships changed dramatically as we moved into the twenty-first century. Some people believe that the family as we know it is doomed. Others believe that a return to traditional family values will save this important social institution and create greater stability in society.

Current economic conditions constitute a threat for many families in the United States and other nations. Throughout good and bad economic times, families are important to people because the family is a vital social institution in society and often serves as the major source of support for individuals (Kreider and Elliott, 2009). However, in periods like the 1930s depression or the recent recession, families are particularly affected by problems in the economy. Worldwide, the economies in various nations have been undergoing transformation, and in the United States alone, since the onset of the 2007 recession, there has been a net loss of 7.5 million jobs (Zuckerman, 2011). From 2011 to 2013, the rate of unemployment averaged about 9 percent, after having peaked at over 10 percent in parts of 2010, and this counts only individuals who are actively seeking

**blended family** a family consisting of a husband and wife, children from previous marriages, and children (if any) from the new marriage.

employment, not those who have become discouraged and have given up looking for work.

Many families experience long-term repercussions from a national or global economic crisis and the financial and emotional havoc that such crises wreak on workers and their families. For example, studies have found that individuals are feeling more tension and having more arguments with other family members because of personal fears that people have regarding unemployment, loss of housing values, mortgage foreclosures, and other personal and financial problems (Luo and Thee-Brenan, 2009). It is not surprising to sociologists that families and personal relationships are affected by financial insecurity: Individuals are forced to deal with problems of low self-esteem associated with job loss and seeking new employment, and they must also focus on how to take care of their family in tough economic times.

People in the lower tiers of the U.S. class structure are confronted with problems such as these all of the time; however, this level of anxiety is somewhat new to many individuals in the middle and upper-middle classes. And family problems related to economic crises are not limited to the United States: People in many countries are affected by changes in the global economy. The present economy casts a shadow over family issues in the future. How national and global political and business leaders deal with economic issues will no doubt have an important effect on the future of families worldwide.

In addition, how we view the concept of "family" is continuing to evolve in the twenty-first century. Some of these changes are already becoming evident. For example, as discussed earlier, many men are taking an active role in raising their children and helping with household chores. More individuals rely on others who are not necessarily immediate family for friendship, emotional support, and help in time of emergencies. As well, more people are cohabiting or living in domestic partnerships or other relationships that have not been legally formalized.

Regardless of problems facing families today, many people still demonstrate their faith in the future by getting married and having children. It will be interesting to see what people in the future decide to do about marriage and parenthood. How many children will people have? What will family life be like in 2030 or 2040? What will your own family be like?

# chapter REVIEW

## LO1 How are the following key concepts defined: families, kinship, family of orientation, family of procreation, extended family, and nuclear family?

Today, families may be defined as relationships in which people live together with commitment, form an economic unit and care for any young, and consider their identity to be significantly attached to the group. Through kinship networks, people cooperate so that they can acquire the basic necessities of life, including food and shelter. Kinship systems can also serve as a means by which property is transferred, goods are produced and distributed, and power is allocated. Although most people are related to members of their family of orientation by blood ties, those who are adopted have a legal tie that is patterned after a blood relationship. The family of orientation is the family into which a person is born; the family of procreation is the family that a person forms by having or adopting children. An extended family is a family unit composed of relatives in addition to parents and children who live in the same household. A traditional definition specifies that a nuclear family is made up of a "couple" and their dependent children; however, this definition became outdated when a significant shift occurred in the family structure.

## LO2 What are the differences among the following marriage patterns—monogamy, polygamy, polygyny, and polyandry—and the differences among these patterns of descent—patrilineal, matrilineal, and bilateral?

Marriage is a legally recognized and/or socially approved arrangement between two or more individuals that carries certain rights and obligations and usually involves sexual activity. In the United States the only legally sanctioned form of marriage is monogamy—a marriage between two partners, usually a woman and a man. Polygamy is the concurrent marriage of a person of one sex with two or more members of the opposite sex. The most prevalent form of polygamy is polygyny—the concurrent marriage of one man with two or more women. The second type of polygamy is polyandry—the concurrent marriage of one woman with two or more men. Virtually all forms of marriage establish a system of descent so that kinship can be determined and inheritance rights established. In preindustrial societies, kinship is usually traced through one parent (unilineally). The most common pattern of unilineal descent is patrilineal descent. Matrilineal descent is the less common pattern. Kinship in industrial societies is usually traced through both parents (bi-lineally). The most common form of descent is bilateral.

## LO3 What are the kinds of authority figures in patriarchal, matriarchal, and egalitarian families?

The most prevalent forms of familial power and authority are patriarchy, matriarchy, and egalitarianism. A patriarchal family is a family structure in which authority is held by the eldest male (usually the father). A matriarchal family is a family structure in which authority is held by the eldest female (usually the mother). An egalitarian family is a family structure in which both partners share power and authority equally.

## LO4 What are the primary sociological perspectives on the family as a social institution?

Functionalists emphasize the importance of the family in maintaining the stability of society and the well-being of individuals. Conflict and feminist perspectives view the family as a source of social inequality and an arena for conflict. Symbolic interactionists explain family relationships in terms of the subjective meanings and everyday interpretations that people give to their lives. Postmodern analysts view families as being permeable, capable of being diffused or invaded so that their original purpose is modified.

## LO5 What issues do many contemporary couples face when thinking of developing intimate relationships and establishing families?

Families are changing dramatically in the United States. Cohabitation has increased significantly in the past three decades. Among heterosexual couples, many reasons exist for cohabitation; for gay and lesbian couples, however, no alternatives to cohabitation exist in most U.S. states. For that reason, many lesbians and gays seek recognition of their domestic partnerships—household partnerships in which an unmarried couple lives together in a committed, sexually intimate relationship and is granted some of the same rights and benefits as those accorded to married heterosexual couples. With the increase in dual-earner marriages, women have become larger contributors to the financial well-being of their families, but some have become increasingly burdened by the second shift—the domestic work that employed women perform at home after they complete their workday on the job.

## LO6 What child-related family issues are of concern to many people in the twenty-first century?

Cultural attitudes about having children and about the ideal family size have changed dramatically in the United States. Today, the concept of reproductive freedom includes both the desire *to have* or *not to have* one or more children. Issues of concern include teenage childbearing. Many single-parent families also exist today.

## LO7 What are some of the key stressors that contribute to family violence and the need for foster care for children?

Factors contributing to unequal power relations in families include economic issues, legal and political sanctions that deny girls and women equal rights, and cultural factors that perpetuate domestic violence. Regardless of the factors that contribute to domestic violence, control is central to all forms of abuse. Women and children are most strongly affected by family violence, although domestic violence is also perpetrated against men. However, everyone in a household where family violence occurs is harmed psychologically, and children are especially harmed. Foster care is often used as a safe place for children who have been in dysfunctional families, some of which are the sites of family violence, others of which are not.

## LO8 What is divorce, and how does it affect remarriage patterns and blended families in the United States?

Divorce is the legal process of dissolving a marriage. At the macrolevel, changes in social institutions may contribute to an increase in divorce rates; at the microlevel, factors contributing to divorce include age at marriage, length of acquaintanceship, economic resources, education level, and parental marital happiness.

# key TERMS

# questions for CRITICAL THINKING

1. In your opinion, what constitutes an ideal family? How might functionalist, conflict, feminist, and symbolic interactionist perspectives describe this family?

2. Suppose that you wanted to find out about people's perceptions about love and marriage. What specific issues might you examine? What would be the best way to conduct your research?

3. You have been appointed to a presidential commission on child-care problems in the United States. How to provide high-quality child care at affordable prices is a key issue for the first meeting. What kinds of suggestions would you take to the meeting? How do you think your suggestions should be funded? How does the future look for children in high-, middle-, and low-income families in the United States?

# 12

# EDUCATION AND RELIGION

Michael Ventura/PhotoEdit, Inc.

## LEARNING OBJECTIVES

**LO1** Explain why education and religion are important areas of sociological inquiry.

**LO2** Identify the key assumptions of functionalist, conflict, and symbolic interactionist perspectives on education.

**LO3** Describe the major problems faced by U.S. elementary and secondary schools today.

**LO4** Discuss ways in which problems in U.S. higher education are linked to problems in the United States and the world.

**LO5** Define *religion* and identify its key components.

**LO6** Compare and contrast the major theoretical perspectives on religion.

**LO7** Identify and briefly describe the major types of religious organizations.

**LO8** Discuss major trends in U.S. education and religion in the twenty-first century.

## Sociology and Everyday Life

# *An Ongoing Debate*

**As someone who has spent her career in science,** I have no delusion that the debate over

For many years, people have argued about what should (or should not) be taught in U.S. public schools, including creationism or intelligent design as contrasted with evolution. Reverend. Jim Grove (shown here) introduces a video on creationism to a group in Dover, Pennsylvania.

teaching evolution in public schools will go away . . . [, but] I hope we can put at least one issue to rest. Intelligent design is not science. . . . Scientific theories must change in light of new data. . . . The lack of this characteristic in intelligent design is well-illustrated by a conversation I had with the tour guide at a creationist museum. . . . I posited to the fellow: "Look, you and I are both trying to learn the truth. But in science, theories must be modified to account for new data, and you are never going to change your theory. So creationism—by any name—is not science!" He calmly replied, "But I already know the truth and the truth doesn't change." . . . That statement explains precisely why creationism, intelligent design, or whatever other label comes along 10 years from now, is not science. The theory refuses to change.

—Elizabeth Cooper (2011), a college lecturer, stating why she believes that intelligent design, or creationism, should not be taught in public school science courses

**W**hat is all the controversy about? Why do lawsuits about teaching intelligent design in schools arise so often? The argument over intelligent design—an assertion that the universe is so complex that an intelligent, supernatural power must have created it—as an alternative to the theory of evolution is the latest debate in a lengthy battle over the teaching of creationism versus evolutionism in public schools, and it is only one of many arguments that will continue to take place regarding the appropriate relationship between public education and religion in the United States. About ninety years ago, evolutionism versus creationism was hotly debated in the famous "Scopes monkey trial," so named because of Charles Darwin's assertion that human beings had evolved from lower primates. In this case, John Thomas Scopes, a substitute high school biology teacher in Tennessee, was found guilty of teaching evolution, which denied the "divine creation of man as taught in the Bible." Although an appeals court later overturned Scopes's conviction and $100 fine (on

the grounds that the fine was excessive), teaching evolution in Tennessee's public schools remained illegal until 1967. By contrast, more-recent U.S. Supreme Court rulings have looked unfavorably on the teaching of creationism in public schools, based on a provision in the Constitution that requires a "wall of separation" between church (religion) and state (government). Initially, this wall of separation was erected to protect religion from the state, not the state from religion.

As these examples show, religion can be a highly controversial topic. One group's deeply held beliefs or cherished religious practices may be a source of irritation to another. Today, religion is a source of both stability and conflict not only in the United States but also throughout the world. In this chapter we examine how religion influences life in the United States and in other areas of the world. Before reading on, test your knowledge about how religion affects public education in this country by taking the "Sociology and Everyday Life" quiz.

**We are teaching our children a theory** [evolution] that most of us don't believe in. I don't think God creates everything on a day-to-day basis, like the color of the sky. But I do believe he created Adam and Eve—instantly.

—Steve Farrell, a Dover, Pennsylvania, resident explaining why he approved of the Dover school board's decision to require biology teachers to teach intelligent design, before a U.S. District Court ruled (in *Kitzmiller v. Dover*) that such teaching violates the Establishment Clause of the First Amendment to the U.S. Constitution (qtd. in Powell, 2004: A1)

## How Much Do You Know About the Effects of Religion on U.S. Education?

| True | False | |
|------|-------|--|
| T | F | 1. The U.S. Constitution originally specified that religion should be taught in the public schools. |
| T | F | 2. Virtually all sociologists have advocated the separation of moral teaching from academic subject matter. |
| T | F | 3. The federal government has limited control over how funds are spent by school districts because most of the money comes from the state and local levels. |
| T | F | 4. Parochial schools have decreased in enrollment as interest in religion has waned in the United States. |

| True | False | |
|------|-------|--|
| T | F | 5. Studies have found that the religious affiliation of schoolchildren may be unrelated to their religious affiliation as adults. |
| T | F | 6. Debates over textbook content focus only on elementary education because of the vulnerability of young children. |
| T | F | 7. More parents are instructing their own children through home schooling because of their concerns about what public schools are (or are not) teaching their children. |
| T | F | 8. The U.S. Congress has the ultimate authority over whether religious education can be included in public school curricula. |

*Answers on page 364.*

## An Overview of Education and Religion

**LO1** Education and religion are powerful and influential forces in contemporary societies. Both institutions impart values, beliefs, and knowledge considered essential to the social reproduction of individual personalities and entire cultures (Bourdieu and Passeron, 1990). Education and religion both grapple with issues of societal stability and social change, reflecting society even as they attempt to shape it. Education and religion also share certain commonalities as objects of sociological study; for example, both are socializing institutions. Whereas early socialization is primarily informal and takes place within our families and friendship networks, as we grow older, socialization passes to the more formalized organizations created for the specific purposes of education and religion.

Areas of sociological inquiry that specifically focus on those institutions are (1) the *sociology of education,* which primarily examines formal education or schooling in industrial societies, and (2) the *sociology of religion,* which focuses on religious groups and organizations, on the behavior of individuals within those groups, and on ways in which religion is intertwined with other social institutions. Let's start our examination by looking at sociological perspectives on education.

## Sociological Perspectives on Education

**LO2** *Education* is the social institution responsible for the systematic transmission of knowledge, skills, and cultural values within a formally organized structure. In all societies, people

**education** the social institution responsible for the systematic transmission of knowledge, skills, and cultural values within a formally organized structure.

# ANSWERS to the Sociology Quiz on Religion and Education

**1. False.** Because of the diversity of religious backgrounds of the early settlers, no mention of religion was made in the original Constitution. Even the sole provision that currently exists (the establishment clause of the First Amendment) does not speak directly of the issue of religious learning in public education.

**2. False.** Obviously, contemporary sociologists hold strong beliefs and opinions on many subjects; however, most of them do not think that it is their role to advocate specific stances on a topic. Early sociologists were less inclined to believe that they had to be "value-free." For example, Durkheim strongly advocated that education should have a moral component and that schools had a responsibility to perpetuate society by teaching a commitment to the common morality.

**3. True.** Most public school revenue comes from local funding through property taxes and state funding from a variety of sources, including sales taxes, personal income taxes, and, in some states, oil revenues or proceeds from lotteries.

**4. False.** Just the opposite has happened. As parents have felt that their children were not receiving the type of education they desired in public schools, parochial schools have flourished. Christian schools have grown to over five thousand in number; Jewish parochial schools have also grown rapidly over the past decade.

**5. True.** More than one-quarter of U.S. adults have left the faith in which they were raised (and might have participated while in school) in favor of another religion or no religion at all (referred to as "unaffiliated" in surveys).

**6. False.** Attempts to remove textbooks occur at all levels of schooling. For example, cases have involved removing Chaucer's "The Miller's Tale" and Aristophanes' *Lysistrata* from high school curricula.

**7. True.** Some parents choose home schooling for religious reasons; others embrace it for secular reasons, including fear for their children's safety and concerns about the quality of public schools.

**8. False.** Ultimately, issues relating to the separation of church and state, including religious instruction in public schools, are constitutional issues that are decided by the U.S. Supreme Court in the absence of a constitutional amendment.

Sources: Based on Ballantine and Hammack, 2012; and Pew Forum on Religion and Public Life, 2007a, 2007b. © Cengage Learning

---

must acquire certain knowledge and skills in order to survive. In less-developed societies, these skills might include hunting, gathering, fishing, farming, and self-preservation. In contemporary, developed societies, knowledge and skills are often related to the requirements of a highly competitive job market.

Sociologists have divergent perspectives on the purpose of education in contemporary society. Functionalists suggest that education contributes to the maintenance of society and provides people with an opportunity for self-enhancement and upward social mobility. Conflict theorists argue that education perpetuates social inequality and benefits the dominant class at the expense of all others. Symbolic interactionists focus on classroom dynamics and the effect of self-concept on grades and aspirations.

## Functionalist Perspectives on Education

Functionalists view education as one of the most important components of society. According to Emile Durkheim, education is the "influence exercised by adult generations on those that are not yet ready for social life" (1956: 28). Durkheim asserted that moral values are the foundation of a cohesive social order and that schools have the responsibility to teach a commitment to the common morality.

From this perspective, students must be taught to put the group's needs ahead of their individual desires and aspirations. Contemporary functionalists suggest that education is responsible for teaching U.S. values. However, not all analysts agree on what those values should be or what functions that education should serve in contemporary societies. In analyzing the values and functions of education, sociologists using a functionalist framework distinguish between manifest and latent functions. Manifest functions and latent functions of education are compared in Figure 12.1.

**Manifest Functions of Education** Some functions of education are *manifest functions*, previously defined as the open, stated, and intended goals or

**Manifest functions**—open, stated, and intended goals or consequences of activities within an organization or institution. In education, these are:

- socialization
- transmission of culture
- social control
- social placement
- change and innovation

**Latent functions**—hidden, unstated, and sometimes unintended consequences of activities within an organization. In education, these include:

- matchmaking and production of social networks
- restricting some activities
- creating a generation gap

**FIGURE 12.1   MANIFEST AND LATENT FUNCTIONS OF EDUCATION**
© Cengage Learning

consequences of activities within an organization or institution. Examples of manifest functions in education include teaching specific subjects, such as science, mathematics, reading, history, and English. Education serves five major manifest functions in society:

1. *Socialization.* From kindergarten through college, schools teach students the student role, specific academic subjects, and political socialization. In primary and secondary schools, students are taught specific subject matter appropriate to their age, skill level, and previous educational experience. At the college level, students focus on more-detailed knowledge of subjects that they have previously studied while also being exposed to new areas of study and research.

2. *Transmission of culture.* Schools transmit cultural norms and values to each new generation and play an active part in the process of assimilation, whereby recent immigrants learn dominant cultural values, attitudes, and behavior so that they can be productive members of society.

3. *Social control.* Schools are responsible for teaching values such as discipline, respect, obedience, punctuality, and perseverance. Schools teach conformity by encouraging young people to be good students, conscientious future workers, and law-abiding citizens (see "Sociology Works!").

4. *Social placement.* Schools are responsible for identifying the most qualified people to fill available positions in society. As a result, students are channeled into programs based on individual ability and academic achievement. Graduates receive the appropriate credentials for entry into the paid labor force.

5. *Change and innovation.* Schools are a source of change and innovation. As student populations change over time, new programs are introduced to meet societal needs; for example, sex education, drug education, and multicultural studies have been implemented in some schools to help students learn about pressing social issues. Innovation in the form of new knowledge is required of colleges and universities. Faculty members are encouraged, and sometimes required, to engage in research and to share the results with students, colleagues, and others.

**Latent Functions of Education**   In addition to manifest functions, all social institutions, including education, have some *latent functions,* which, as you will recall, are the hidden, unstated, and sometimes unintended consequences of activities within an organization or institution. Education serves at least three latent functions:

1. *Restricting some activities.* Early in the twentieth century, all states passed *mandatory education laws* that require children to attend school until they reach a specified age or until they complete a minimum level of formal education. Out of these laws grew one latent function of education, which is to

# Sociology WORKS!

## Stopping Bullying: Character Building and Social Norms

**News Item:** Fifteen-year-old Phoebe Prince spent her last day enduring relentless bullying before she walked home from school and hanged herself. The South Hadley, Massachusetts, ninth-grader suffered a "nearly three-month campaign of verbally assaultive behavior and threats of physical harm," Northwestern District Attorney Elizabeth Scheibel said when she announced the results of her investigation into Phoebe's death. (Green, 2010)

The nation was shocked when a teenager whose family had recently emigrated from Ireland took her own life after enduring months of verbal assaults and threats after she briefly dated a popular boy. Phoebe was harassed in school, on Facebook, and in text messages. Nine students at her school were charged in the bullying of Phoebe, and some analysts believe that school authorities should be held accountable because they did not do enough to stop the bullying. Unfortunately, Phoebe Prince's case is not an isolated incident of bullying: Studies show that numerous students are bullied, and some studies even find that nearly one-fifth of students are willing to admit that they bully others with some frequency.

What is bullying? By definition, *bullying* is aggressive behavior that is intentional and that involves an imbalance of power or strength. Bullying is often repeated, and the child who is routinely the object of the bullying usually has a difficult time defending himself or herself (U.S. Department of Health and Human Services, 2008b).

What have sociologists learned that might be useful for eliminating negative behavior and maintaining social control in schools? One method of reducing bullying is rooted in character education, which focuses on the reinforcement of acceptable character traits, such as respect and responsibility. From character-education courses and online studies, students learn problem-solving techniques, conflict-resolution approaches, and communication skills to help them interact with one another in a more positive manner.

However, an alternative approach has been suggested by such sociologists as H. Wesley Perkins, who emphasizes that the best practice in bullying prevention and intervention is to focus on the *social environment* of the school. According to Perkins, to reduce bullying it is important to change the climate of the school and social norms regarding bullying. As you will recall from Chapter 2, norms are established rules of behavior or standards of conduct; therefore, a school's social norms must establish standards of conduct that make it "uncool" for one person to bully another but "cool" for individuals to help out those who are bullied. Perkins developed a social norms approach to preventing bullying which assumes that a person's behavior may be strongly influenced by the incorrect perceptions the individual holds about how other members

Sociologist H. Wesley Perkins sees the problem of bullying as an issue involving norms: If bullying becomes uncool, young people will be less likely to take part in it and more likely to help someone who is a victim of this very harmful activity.

of his or her social group think and act. From this approach, the first step to reducing the problem of bullying is to correct students' misperceptions about how prevalent this type of behavior really is: If students think that bullying is very common in their school and that "everybody is doing it," they may believe that the social norms support such behavior. If this misperception can be corrected, bullying will decrease as students come to see that the "nonbully status" is the social norm (Teicher, 2006).

Emphasizing the social environment of education and its norms is an important sociological contribution to our understanding of both positive and negative actions that take place in schools. It also makes us aware that social institutions fulfill a variety of important functions in society and that one of those functions for education is to maintain social control and to encourage civility among participants, whether or not they subscribe to moral values that teach them to "Love thy neighbor. . . ." Although it is also important to focus on character education for students and to teach them to respect others and have a sense of responsibility for their own actions, all of us have a responsibility for maintaining a social environment in which everyone is treated with kindness and respect. Bullying isn't just a matter of "kids being kids": It is a serious sociological problem and must be dealt with as such.

### reflect & analyze

**What do you think about the problem of bullying? How can sociology help us to reduce problems such as this?**

keep students off the streets and out of the full-time job market for a number of years, thus helping keep unemployment within reasonable bounds.

2. *Matchmaking and production of social networks.* Because schools bring together people of similar ages, social class, and race/ethnicity, young people often meet future marriage partners and develop social networks that may last for many years.

3. *Creating a generation gap.* Students may learn information in school that contradicts beliefs held by their parents or their religion. When education conflicts with parental attitudes and beliefs, a generation gap is created if students embrace the newly acquired perspective.

As we have seen, education fulfills both manifest and latent functions in society; however, some aspects of this important social institution may be impaired and generate problems not only for students but also for others in the larger society (such as lagging educational standards that result in employees coming into the workforce with fewer marketable skills).

**Dysfunctions of Education** Functionalists acknowledge that education has certain dysfunctions. Some analysts argue that U.S. education is not promoting the high-level skills in reading, writing, science, and mathematics that are needed in the workplace and the global economy. For example, mathematics and science education in the United States does not compare favorably with that found in many other industrialized countries. Are U.S. schools dysfunctional as a result of lower test scores? Analysts do not agree on what these score differences mean. For most functionalist thinkers, lagging test scores are a sign that dysfunctions exist in the nation's educational system. According to this approach, improvements will occur only when more-stringent academic requirements are implemented for students and when teachers receive sufficient training. Overall, functionalists typically advocate the importance of establishing a more rigorous academic environment in which students are required to learn the basics that will make them competitive in school and job markets.

## Conflict Perspectives on Education

In contrast with the functionalist perspective, conflict theorists argue that schools often perpetuate class, racial–ethnic, and gender inequalities as some groups seek to maintain their privileged position at the expense of others (Ballantine and Hammack, 2012).

**Cultural Capital and Class Reproduction** Although many factors—including intelligence, motivation, and previous achievement—are important

Cleve Bryant/PhotoEdit

**FIGURE 12.2** Children who are able to visit museums, libraries, and musical events may gain cultural capital that other children do not possess. What is cultural capital? Why is it important in the process of class reproduction?

in determining how much education a person will attain, conflict theorists argue that access to quality education is closely related to social class. From this approach, education is a vehicle for reproducing existing class relationships. According to the French sociologist Pierre Bourdieu, the school legitimates and reinforces the social elites by engaging in specific practices that uphold the patterns of behavior and the attitudes of the dominant class. Bourdieu asserts that students from diverse class backgrounds come to school with differing amounts of *cultural capital*—social assets that include values, beliefs, attitudes, and competencies in language and culture (Bourdieu and Passeron, 1990). Cultural capital involves "proper" attitudes toward education, socially approved dress and manners, and knowledge about books, art, music, and other forms of high and popular culture (Figure 12.2).

**cultural capital** Pierre Bourdieu's term for people's social assets, including values, beliefs, attitudes, and competencies in language and culture.

Middle- and upper-income parents endow their children with more cultural capital than do working-class and poverty-level parents. Because cultural capital is essential for acquiring an education, children with less cultural capital have fewer opportunities to succeed in school. For example, standardized tests that are used to group students by ability and to assign them to classes often measure students' cultural capital rather than their "natural" intelligence or aptitude. Thus, a circular effect occurs: Students with dominant cultural values are more highly rewarded by the educational system. In turn, the educational system teaches and reinforces those values that sustain the elite's position in society.

**Tracking and Detracking** Closely linked to the issue of cultural capital is how tracking in schools is related to social inequality. *Tracking refers to the practice of assigning students to specific curriculum groups and courses on the basis of their test scores, previous grades, or other criteria.* Conflict theorists believe that tracking seriously affects many students' educational performance and their overall academic accomplishments. In elementary schools, tracking is often referred to as *ability grouping* and is based on the assumption that it is easier to teach a group of students who have similar abilities. However, class-based factors also affect which children are most likely to be placed in "high," "middle," or "low" groups, often referred to by such innocuous terms as "Blue Birds," "Red Birds," and "Yellow Birds." This practice is described by Ruben Navarrette Jr. (1997: 274–275), who tells us about his own experience with tracking:

One fateful day, in the second grade, my teacher decided to teach her class more efficiently by dividing it into six groups of five students each. Each group was assigned a geometric symbol to differentiate it from the others. There were the Circles. There were the Squares. There were the Triangles and Rectangles.

I remember being a Hexagon. . . . I remember something else, an odd coincidence. The Hexagons were the smartest kids in the class. These distinctions are not lost on a child of seven. . . . Even in the second grade, my classmates and I knew who was smarter than whom. And on the day on which we were assigned our respective shapes, we knew that our teacher knew, too. As Hexagons, we would wait for her to call on us, then answer by hurrying to her with books and pencils in hand. We sat around a table in our "reading group," chattering excitedly to one another and basking in the intoxication of positive

learning. We did not notice, did not care to notice, over our shoulders, the frustrated looks on the faces of Circles and Squares and Triangles who sat quietly at their desks, doodling on scratch paper or mumbling to one another. We knew also that, along with our geometric shapes, our books were different and that each group had different amounts of work to do. . . . The Circles had the easiest books and were assigned to read only a few pages at a time. . . . Not surprisingly, the Hexagons had the most difficult books of all, those with the biggest words and the fewest pictures, and we were expected to read the most pages.

The result of all of this education by separation was exactly what the teacher had imagined that it would be: Students could, and did, learn at their own pace without being encumbered by one another. Some learned faster than others. Some, I realized only [later], did not learn at all.

As Navarrette suggests, tracking does make it possible for students to work together based on their perceived abilities and at their own pace; however, it also extracts a serious toll from students who are labeled as "underachievers" or "slow learners." However, as he also points out, race, class, language, gender, and many other social categories may determine the placement of children in elementary tracking systems as much or more than their actual academic abilities and interests. Today, Ruben Navarrette Jr. is a nationally recognized syndicated newspaper columnist and blogger who writes on many important issues facing our nation and the world.

The practice of tracking often continues in middle school/junior high and high school. Although schools in some communities bring together students from diverse economic and racial–ethnic backgrounds, the students do not necessarily take the same courses, move on the same academic career paths, or have the same opportunities even when they attend the same school (Gilbert, 2010).

The *detracking movement*—which emphasizes that students should be deliberately placed in classes of mixed ability to improve their academic performance and test scores—has influenced a growing number of educators. Detracking is based on the assumption that intensifying the secondary school curricula may help close the achievement gap among students, particularly those dimensions that are based on class or race/ethnicity. For example, a growing proportion of U.S. students are being enrolled in higher-level academic math courses than in the past so that they will be exposed

**FIGURE 12.3** As Ruben Navarrette Jr. so powerfully describes, school is extremely tedious for underachieving students, who may find themselves "tracked" in such a way as to deny them upward mobility in the future.

to more-complex course content and be influenced by students who perform at a higher level in such courses.

Detracking is a major concern for parents of high-achieving students: They often believe their children are losing out because lower-achieving students are in their courses. According to their perspective, high-achieving students should have classes that maximize their potential rather than hold them back with less-able or less-talented students. According to the sociologist Maureen Hallinan (2005), who has extensively studied detracking, tracking is not the answer: Schools should provide more-engaging lessons for all students, alter teachers' assumptions about students, and raise students' performance requirements (Figure 12.3).

### The Hidden Curriculum and Inequality

According to conflict theorists, the *hidden curriculum* **is the transmission of cultural values and attitudes, such as conformity and obedience to authority, through implied demands found in the rules, routines, and regulations of schools.** In other words, through the experience of being in school, students pick up on subtle messages about attitudes, beliefs, values, and behavior that are either "appropriate" or "inappropriate" from teachers and other school personnel. These messages are not part of the official curriculum or the school's mission to educate students for the future.

Although all students are subjected to the hidden curriculum, students who are from low-income families and/or are African American or Hispanic (Latino/a) may be affected the most adversely by educational settings that have been established on the basis of upper- and middle-class white (non-Hispanic) values, attitudes, and behavior (AAUW, 2008). When teachers from middle- and upper-middle-class backgrounds instruct students from lower-income families, the teachers often establish a more structured classroom and a more controlling environment for students (Figure 12.4). These teachers may also have lower expectations for students' academic achievements. Schools with many students from low-income families often emphasize procedures and rote memorization without focusing on decision making and choice, or on providing explanations of why something is done a particular way. Schools for middle-class students stress the

**tracking** the assignment of students to specific curriculum groups and courses on the basis of their test scores, previous grades, or other criteria.

**hidden curriculum** the transmission of cultural values and attitudes, such as conformity and obedience to authority, through implied demands found in rules, routines, and regulations of schools.

© iStockphoto.com/Svetlana Braun

**CLASSROOM RULES**

1. Be Prepared
   ★ Come ready to learn
   ★ Have materials

2. Be Respectful
   ★ Show respect to others and their property
   ★ Respect YOURSELF!

3. Be Safe
   ★ DO NOT verbally or physically abuse or hurt anyone!
   ★ Think before you do!!!

4. Raise Your Hand
   ★ DO NOT Speak when others are speaking... Classmates, teacher

5. Do Your Best!!!
   ★ Always try
   ★ Give 100%
   ★ Have pride in all you do.

**FIGURE 12.4** Signs such as this found in elementary classrooms list the rules, and sometimes the rewards and consequences, of different types of student behavior. According to conflict theorists, schools impose rules on working-class and poverty-level students so that they will learn to follow orders and to be good employees in the workplace.

**370**

processes (such as figuring and decision making) involved in getting the right answer. Schools for affluent students focus on creative activities in which students express their own ideas and apply them to the subject under consideration, as well as building students' analytical and critical-thinking skills.

Over time, students may become frustrated with the educational system and drop out or become very marginal students, making it even more difficult for them to attend college and gain the appropriate credentials for gaining better-paying jobs. Educational credentials are extremely important in a nation such as ours that emphasizes *credentialism*—**a process of social selection in which class advantage and social status are linked to the possession of academic qualifications** (Collins, 1979; Marshall, 1998). Credentialism is closely related to *meritocracy*—a social system in which status is assumed to be acquired through individual ability and effort. Persons who acquire the appropriate credentials for a job are assumed to have gained the

position through what they know, not who they are or whom they know. According to conflict theorists, the hidden curriculum determines in advance that the most valued credentials will primarily stay in the hands of the middle and upper classes, so the United States is not actually as meritocratic as some might claim.

The hidden curriculum is also related to gender bias. For many years, the focus in education was on how gender bias harmed girls and women: Reading materials, classroom activities, and treatment by teachers and peers contributed to a feeling among many girls and young women that they were less important than male students. The accepted wisdom was that, over time, differential treatment undermines females' self-esteem and discourages them from taking certain courses, such as math and science, that have been dominated by male teachers and students. In the early 1990s the American Association of University Women issued *The AAUW Report: How Schools Shortchange Girls,* which highlighted inequalities in women's education and started a national debate on gender equity (AAUW, 1995). Over the past twenty years, improvements have occurred in girls' educational achievement, as females have attended and graduated from high school and college at a higher rate than their male peers. More females have enrolled in advanced placement or honors courses and in academic areas, such as math and science, where they had previously lagged (AAUW, 2008). However, some traditional gender differences persist at some grade levels, with boys generally outscoring girls on math tests by a small margin and girls outscoring boys on reading tests by a small margin.

Ironically, after many years of discussion about how the hidden curriculum and other problems in schools served to disadvantage female students, the emphasis has now shifted to the question of whether girls' increasing accomplishments from elementary school to college and beyond have come at the expense of boys and young men. But this is not true, according to research by the AAUW (2008: 2): "Educational achievement is not a zero-sum game, in which a gain for one group results in a corresponding loss for the other. If girls' success comes at the expense of boys, one would expect to see boys' scores decline as girls' scores rise, but this has not been the case."

Regardless of gender, large differences remain in scores on academic tests among students by race/ethnicity. Studies have shown that white children are more likely to graduate from high school and college than are their African American and Hispanic peers. Likewise, children from higher-income families are more likely to graduate from high school

**Question 2:** Consider the following two statements: all farmers who are also ranchers cannot come near town; and most of the ranchers who are also farmers cannot surf. Which of the following statements MUST be true?

○ Most of the farmers who cannot come near town can surf

○ Only some farmers who ranch can surf near town

○ A surfer who ranches and farms cannot surf near town

○ Some ranchers who farm can come to town to learn to surf

○ Any farmer who cannot surf also ranches

**FIGURE 12.5** IQ TEST SAMPLE QUESTION

IQ tests containing questions such as this are often used to place students in ability groups. Placement can set the course of a person's entire education.

© Cengage Learning

than are children from lower-income families, who are also less likely to attend college (AAUW, 2008).

The conflict theorists' focus on the hidden curriculum calls our attention to the fact that students learn far more—both positively and negatively—than just the subject matter that is being taught in classrooms. Students are exposed to a wide range of beliefs, values, attitudes, and behavioral expectations that are not directly related to specific subject matter.

## Symbolic Interactionist Perspectives on Education

Unlike functionalist analysts, who focus on the functions and dysfunctions of education, and conflict theorists, who focus on the relationship between education and inequality, symbolic interactionists focus on classroom communication patterns and educational practices, such as labeling, that affect students' self-concept and aspirations.

### Labeling and the Self-Fulfilling Prophecy

According to symbolic interactionists, the process of labeling is directly related to the power and status of those persons who do the labeling and those who are being labeled. Chapter 6 explains that *labeling* is the process whereby a person is identified by others as possessing a specific characteristic or exhibiting a certain pattern of behavior (such as being deviant). In schools, teachers and administrators are empowered to label children in various ways, including grades, written comments on deportment (classroom behavior), and placement in classes. For some students, labeling amounts to a *self-fulfilling prophecy*—previously defined as an unsubstantiated belief or prediction resulting in behavior that makes the originally false belief come true (Merton, 1968).

A classic form of labeling and the self-fulfilling prophecy may occur through the use of IQ (intelligence quotient) tests, which claim to measure a person's inherent intelligence, apart from any family or school influences on the individual. In many school systems, IQ tests are used as one criterion in determining student placement in classes and ability groups (see Figure 12.5). The way in which IQ test scores may become a self-fulfilling prophecy was revealed in the 1960s when two social scientists conducted an experiment in an elementary school during which they intentionally misinformed teachers that some of the students had extremely high IQ test scores whereas others had average to below-average scores. As the researchers observed, the teachers began to teach "exceptional" students in a different manner from other students. In turn, the "exceptional" students began to outperform their "average" peers and to excel in their classwork. This study called attention to the labeling effect of IQ scores.

Today, so-called IQ fundamentalists continue to label students and others on the basis of IQ tests, claiming that these tests measure some identifiable trait that predicts the quality of people's thinking and their ability to perform. Critics of IQ tests continue to argue that these exams measure a number of factors—including motivation, home environment, type of socialization at home, and quality of schooling—not intelligence alone (Yong, 2011).

## Problems in Elementary and Secondary Schools

**LO3** Education in kindergarten through high school is a microcosm of many of the issues and problems facing the United States. The problems we address in this section include unequal funding of public schools, high dropout rates, racial segregation and resegregation, and competition for public schools in the form of school choice, charter schools, and home schooling.

### Unequal Funding of Public Schools

Why does unequal funding exist in public schools? Most educational funds come from state legislative

**credentialism** a process of social selection in which class advantage and social status are linked to the possession of academic qualifications.

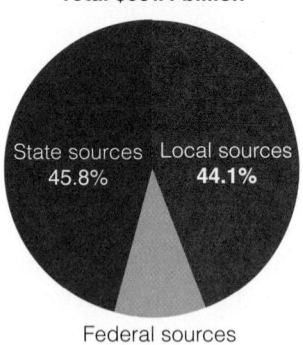

**Total $601.4 billion**

State sources
45.8%

Local sources
**44.1%**

Federal sources
10.1%

**FIGURE 12.6** PERCENTAGE DISTRIBUTION OF TOTAL PUBLIC ELEMENTARY–SECONDARY SCHOOL SYSTEM REVENUE, 2012–2013

Source: National Education Association, 2012. © Cengage Learning

appropriations and local property taxes. As shown in Figure 12.6, in the 2012–2013 school year, state sources contributed 45.8 percent of public elementary–secondary school system revenue, 44.1 percent came from local sources, and 10.1 percent came from federal sources (National Education Association, 2012). Much of the money from federal sources is earmarked for special programs for students who are disadvantaged (e.g., the Head Start program) or who have a disability. Expenditures per pupil for public and secondary public schools vary from state to state, with the U.S. average being $10,669 (National Education Association, 2012).

Spending on public and secondary education varies widely from state to state. In part, this is because the local property-tax base has been eroding in central cities as major industries have relocated or gone out of business. Many middle- and upper-income families have moved to suburban areas with their own property-tax base so their children can attend relatively new schools equipped with the latest textbooks and state-of-the-art computers—advantages that schools in central cities and poverty-ridden rural areas lack (Ballantine and Hammack, 2012).

## School Dropouts

High dropout rates are a major problem facing contemporary schools. Dropout rates are computed in various ways, but one of the most telling is the status dropout rate—the percentage of people in a specific age range who are not currently enrolled in high school and who do not have a high school degree or its equivalent. In recent years slightly more than three million 16- to 24-year-olds were not enrolled in high school and had not earned a high school diploma or its equivalent.

Status dropout rates vary by gender, race/ethnicity, and region of the country (see Figure 12.7). Males (7.0 percent) have a higher status dropout rate than females (6.0 percent). Status dropout rates also vary by race/ethnicity: Hispanics/ Latinos/as (13.0 percent), American Indians/Alaska Natives (13.0 percent), and blacks/African Americans (8.0 percent) have higher status dropout rates than persons reporting two or more races (6.5 percent), whites (4.0 percent), and Asians/Pacific Islanders (3.0 percent) (National Center for Educational Statistics, 2014). Finally, region is also an issue in status dropout rates: The Northeastern United States has the lowest status dropout rates, while the South and West have the highest.

Using a second approach to determine dropout rates, the *event dropout rate*—which estimates the percentage of both public and private high school students who left high school between the beginning of one school year and the beginning of the next without earning a high school diploma or an alternative credential such as a GED—we find that every school day, at least 7,000 U.S. students (on average) leave high school and never return. What this means is that, on average, 3.5 percent of students who were enrolled in public or private high schools in October of one year left school before October of the next year without completing a high school program. Perhaps the most telling statistic when using the event dropout rate is that students living in low-income families are about four-and-one-half times more likely to drop out in any given year than are students living in high-income families (National Center for Educational Statistics, 2012a).

Dropping out of school produces serious economic and social consequences for individuals and nations. There is substantial evidence that education levels are related to earnings. Young people without a high school education not only have a difficult time finding work that will pay a living wage, but when they are employed, they make far less than individuals with at least a high school diploma. The median weekly earnings of full-time workers ages 25 and over who had not completed high school (in 2012) was $478, as compared to earnings of $647 for persons with a high school diploma, GED, or equivalent (no college) and $1,168 for those holding at least a bachelor's degree (U.S. Bureau of Labor Statistics, 2013e).

A higher percentage of dropouts (ages 25 and over) are also unemployed or holding temporary or part-time jobs while seeking full-time work. Cities and states suffer because tax revenues are lower

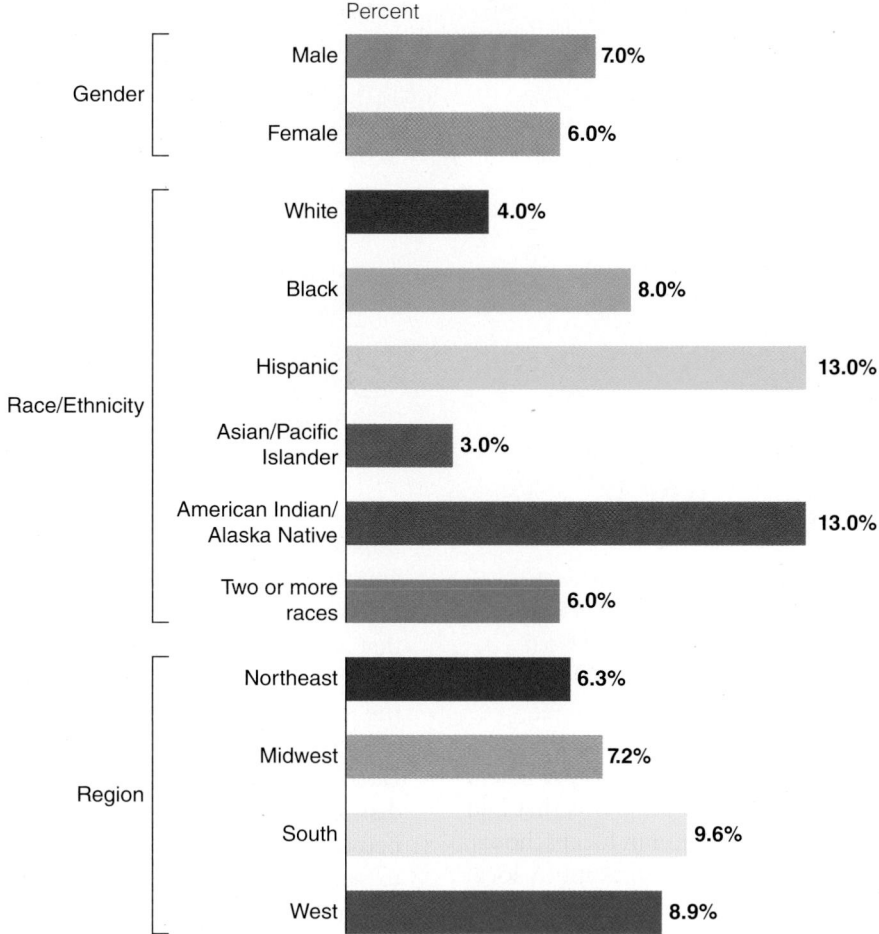

Percent

**Gender**
- Male — **7.0%**
- Female — **6.0%**

**Race/Ethnicity**
- White — **4.0%**
- Black — **8.0%**
- Hispanic — **13.0%**
- Asian/Pacific Islander — **3.0%**
- American Indian/ Alaska Native — **13.0%**
- Two or more races — **6.0%**

**Region**
- Northeast — **6.3%**
- Midwest — **7.2%**
- South — **9.6%**
- West — **8.9%**

**FIGURE 12.7** STATUS DROPOUT RATES FOR 16- TO 24-YEAR-OLDS, BY RACE/ETHNICITY, GENDER, AND REGION

Source: National Center for Educational Statistics, 2014.

when many people are unemployed, and the societal costs for public assistance, crime control, and health care are higher.

Why do students drop out of school? Some students believe that their classes are boring; others are skeptical about the value of schooling and think that completing high school will not increase their job opportunities. Upon leaving school, many dropouts have high hopes of making money and enjoying newfound freedom; however, many find that few jobs are available and that they do not have the minimum education required for any "good" jobs that exist.

## Racial Segregation and Resegregation

Although some people believe that the issue of racial segregation has long been solved in America's schools, in many areas of the United States schools remain racially segregated or have become

resegregated (Figure 12.8). In 1954 the U.S. Supreme Court ruled (in *Brown v. Board of Education of Topeka, Kansas*) that "separate but equal" segregated schools are unconstitutional because they are inherently unequal. Today, racial segregation remains a fact of life in education: Efforts to bring about *desegregation*—the abolition of legally sanctioned racial–ethnic segregation—or *integration*—the implementation of specific action to change the racial–ethnic and/or class composition of the student body—have failed in many districts. Some school systems have bused students across town to achieve racial integration. Others have changed school attendance boundaries or introduced magnet schools with specialized programs such as science or the fine arts to change the racial–ethnic composition of schools. But school segregation does not exist in isolation: Racially segregated housing patterns contribute to high rates of school segregation.

**FIGURE 12.8** Although many people believe that the United States is a racially integrated nation, a look at schools throughout the country reveals that many of them remain segregated or have become largely resegregated in recent decades.

Resegregation is also an issue because some school districts have abandoned programs that had produced greater racial integration in local schools. Raleigh, North Carolina, is a case in point: A local school board decided to end consideration of race and socioeconomic status in determining school assignments and stopped the district's busing-for-diversity program. Those who opposed this return to the "neighborhood school" concept argued that resegregation would quickly occur throughout the district (Mooney, 2011).

How segregated are U.S. schools? According to one study by the Civil Rights Project at the University of California, approximately 40 percent of African American and Hispanic (Latino/a) students are enrolled in U.S. public schools where less than 10 percent of their classmates are white (non-Hispanic) Americans. Despite declining residential segregation for African Americans and extensive movement to the suburbs, school segregation remains very high for many African American students. The problem involves double segregation by both race and poverty: The typical African American or Latina/o student attends a school with almost twice as many low-income students as a school attended by the typical white American or Asian American student (Orfield, Kucsera, and Siegel-Hawley, 2012).

## Competition for Public Schools

Public schools do not have a monopoly on K–12 education. Today, parents have more choices of where to send their children than in the past.

**School Choice and School Vouchers** School choice is a persistent issue in education. Much of the discussion about school choice focuses on school voucher programs in which public funds (tax dollars) are provided to parents so that they can pay their child's tuition at a private school of their choice. Many parents praise the voucher system because it provides them with options for schooling their children. Some political leaders applaud vouchers and other school-choice policies for improving public school performance. However, voucher programs are controversial: Some critics believe that giving taxpayer money to parents so that they can spend it at private (often religious) schools violates constitutional requirements for the separation of church and state. However, the U.S. Supreme Court ruled in *Zelman v. Simmons-Harris,* a case involving a Cleveland, Ohio, school district, that voucher policies are constitutional because parents have a choice and are not required to send their children to church-affiliated schools. Other critics claim that voucher programs are less effective in educating children than public schools. According to studies in the District of Columbia, Milwaukee, and Cleveland, public school students outperformed voucher students in both reading and math on state proficiency tests; however, neither group reached state proficiency requirements (Ott, 2011). In sum, advocates like the choice factor in voucher programs while critics believe that vouchers undermine public education, lack accountability, and may contribute to the collapse of the public school system.

**Charter Schools** Charter schools (or "schools of choice") are primary or secondary schools that receive public money but are free from some of the day-to-day bureaucracy of a larger school district that may limit classroom performance. These schools operate under a charter contract negotiated by the school's organizers (often parents or teachers) and a sponsor (usually a local school board, a state board of education, or a university) that oversees the provisions of the contract. Some school districts "contract out" by hiring for-profit companies on a contract basis to manage charter schools, but the schools themselves are nonprofit. Among the largest educational management organizations are Imagine Schools, National Heritage Academies, the Leona Group, Edison Learning, White Hat

Management, and Mosaica Education (Molnar, Miron, and Urschel, 2010).

In the 2012–2013 academic year, more than 500 new public charter schools opened; however, in the same time period, about 150 public charter schools closed their doors because of low enrollment, financial concerns, and low academic performance. Advocates for charter schools suggest that these figures show that there is both demand for charter schools and evidence that schools that do not meet the needs of their students will be closed (National Alliance for Public Charter Schools, 2013).

What are the unique claims for charter schools? Charter schools are supposed to provide more autonomy for individual students and teachers, and to provide a large number of minority students with a higher-quality education than they would receive in the public schools in their area. Public charter schools enroll a greater percentage of low-income students (46 percent as compared to 41 percent in traditional public schools), African American students (27 percent versus 15 percent), Hispanic/Latino/a students (26 percent versus 22 percent), and students with lower standardized assessments before their transfer to charter schools (National Alliance for Public Charter Schools, 2013). Charter schools attempt to maintain an organizational culture that motivates students and encourages achievement rather than having a negative school environment where minority students are ridiculed for "acting white" or making good grades. Some charter schools offer college-preparatory curriculums and help students of color achieve their goal of enrolling in the college or university of their choice.

However, charter schools have numerous challenges. Some schools have high turnover rates, perhaps partly because of family instability, students' socioeconomic status, or other factors not under the direct control of the schools. A number of charter-school officials have been accused of misappropriating school funds or other financial irregularities, but many analysts believe that the positives seem to outweigh the negatives when it comes to charter schools addressing the academic gap among minority students.

**Home Schooling** Another alternative, home schooling, has been chosen by some parents who hope to avoid the problems of public schools while providing a quality education for their children. It is estimated that about 1.5 million children are home-schooled in grades K through 12 (Kerkman, 2011). This is a significant increase from the estimated 1.1 million students who were home-schooled in 2003. The primary reasons that parents indicate for preferring to home-school their children are (1) concern about the school environment, (2) the desire to provide religious or moral instruction, and (3) dissatisfaction with the academic instruction available at traditional schools. Typically, the parents of homeschoolers are better educated, on average, than other parents; however, their income is about the same. Researchers have found that boys and girls are equally likely to be home-schooled.

Parents who educate their children at home believe that their children are receiving a better education at home because instruction can be individualized to the needs and interests of their children. Some parents also indicate religious reasons for their decision to home-school their children. An association of homeschoolers now provides communication links for parents and children, and technological advances in computers and the Internet have made it possible for homeschoolers to gain information and communicate with one another. In some states, parents organize athletic leagues, proms, and other social events so that their children will have an active social life without being part of a highly structured school setting. According to advocates, home-schooled students typically have high academic achievements and high rates of employment.

Critics of home schooling question how much parents know about school curricula and how competent they are to educate their own children at home, particularly in rapidly changing subjects such as science and computer technology. Some states have passed accountability laws that must be met by parents who teach their children at home. For example, Florida requires that parents register their children with their school districts or with umbrella schools that maintain and submit records to the state.

## School Safety and Violence at All Levels

Today, officials in schools from the elementary years to two-year colleges and four-year universities are focusing on how to reduce or eliminate violence. In many schools, teachers and counselors are instructed in anger management and peer mediation, and they are encouraged to develop classroom instruction that teaches values such as respect and responsibility (National Center for Educational Statistics, 2012b). Some schools create partnerships with local law enforcement agencies and social service organizations to link issues of school safety to larger concerns about safety in the community and the nation.

**FIGURE 12.9** In the aftermath of the 2012 mass school shooting in Newtown, Connecticut, many people sought more-stringent gun laws to prevent future occurrences of similar horrendous events. Do you believe that there is sufficient public concern, from this and other horrendous mass shootings, to bring about stronger gun control legislation in the United States?

In fact, violent incidents are reported each year by nearly 75 percent of all public schools, grades K through 12, and 40 percent of these incidents are reported to the local police. Estimates of the most common incidents reported include physical attack or fight without a weapon (about 70 percent), threat of physical attack without a weapon (about 46 percent), vandalism (about 45 percent), theft/larceny (about 44 percent), possession of a knife or other sharp object (about 40 percent), and distribution, possession, or use of illegal drugs (about 25 percent) (National Center for Education Statistics, 2012b). Yet despite these numbers, statistics related to school safety continue to show that U.S. schools are among the safest places for young people. According to "Indicators of School Crime and Safety," jointly released by the National Center for Education Statistics and the U.S. Department of Justice's Bureau of Justice Statistics, young people are more likely to be victims of violent crime at or near their home, on the streets, at commercial establishments, or at parks than they are at school (National Center for Educational Statistics, 2012b). However, these statistics do not keep many people from believing that schools are becoming more dangerous with each passing year and that all schools should have high-tech surveillance equipment and police officers in every school to help maintain a safe environment.

Even with safety measures in place, violence and fear of violence continue to be pressing problems in schools throughout the United States.

This concern extends from kindergarten through grade 12 because violent acts have resulted in deaths in a number of communities throughout the United States (Figure 12.9). The Newtown, Connecticut, school massacre occurred on December 14, 2012, when Adam Lanza fatally shot twenty children and six adult staff members at an elementary school before committing suicide. Considered the second-deadliest mass murder at an elementary school in U.S. history (following the 1927 Bath School bombings in Michigan), this violent attack prompted extensive debate about gun control and whether school officials should be armed to prevent future occurrences of this kind or whether police officers should be stationed in every school.

So far, the evidence is unclear on the effect of placing police officers in every school. Do they deter crime or provide safety if an armed person seeks to harm students and school personnel? Advocates say that the answer is an unequivocal "yes." However, critics note that the recent increase in police officers and armed guards in schools has also brought about an increase in the number of students who are arrested for minor behavior problems, thus pushing children whose detrimental behavior might best be handled by the school's discipline system into the criminal courts for relatively minor offenses (Eckholm, 2013).

Like public elementary and secondary schools, college and university campuses are not immune to violence, as deranged individuals have engaged in acts of personal terrorism at the expense of students, professors, and other victims. In the aftermath of tragedies such as the one that occurred at Virginia Tech in 2007, in which thirty-two people were killed by a student, there was a massive outpouring of public sympathy and calls for greater campus security. As usual, gun-control advocates called for greater control over the licensing and ownership of firearms and for heightened police security on college campuses, whereas pro-gun advocates argued that people should be allowed to carry firearms on campus for their own protection. Lawmakers in a number of states introduced measures seeking to relax concealed-weapons restrictions on college and university campuses. As of this writing, legislatures in a number of states and the U.S. Congress were debating issues pertaining to gun control, particularly in light of the Newtown violence. Many people have serious concerns about arming schoolteachers and allowing concealed weapons on schools grounds and college campuses because having weapons readily available might make a bad situation even worse.

# Opportunities and Challenges in Colleges and Universities

**LO4** Who attends college? What sort of college or university do they attend? We will explore these and other questions in this section.

## Community Colleges

One of the fastest-growing areas of U.S. higher education today is the community college; however, the history of two-year colleges goes back more than a century, with the establishment of Joliet Junior College in Illinois (Figure 12.10). Following World War II, the GI Bill of Rights provided the opportunity for more people to attend college, and in 1948 a presidential commission report called for the establishment of a network of public community colleges that would charge little or no tuition, serve as cultural centers, be comprehensive in their program offerings, and serve the area in which they were located.

Hundreds of community colleges were opened across the nation during the 1960s, and the number of such institutions has steadily increased since that time as community colleges have responded to the needs of their students and local communities. Community colleges offer a variety of courses, some of which are referred to as "transfer courses," in which students earn credits that are fully transferable to a four-year college or university. Other courses are in technical/occupational programs, which provide formal instruction in fields such as nursing, emergency medical technology, plumbing, carpentry, and computer information technology. Many community colleges are now also offering four-year degrees, sometimes in conjunction with four-year colleges located in the same state.

Community colleges educate about half of the nation's undergraduates. According to the American Association of Community Colleges (2013), the 1,132 community colleges (including public, private, and tribal colleges) in the United States institutions enroll about 13 million students in credit and non-credit courses. Women make up a slight majority

Joliet Junior College

Phil Roche/Miami Dade College

**FIGURE 12.10** Joliet Junior College (Illinois) is the oldest two-year college in the United States, having opened its doors in 1901. The bottom photo shows a scene from graduation day at the nation's largest two-year school, Miami Dade College (Florida), which recently began offering four-year degrees, part of a national trend among two-year colleges. Today, Joliet, Miami Dade, and other schools like them fulfill many needs in the competitive world of higher education.

(57 percent) of community college students, and for working women and mothers of young children, these schools provide a unique opportunity to attend classes on a part-time basis as their schedule permits. Men also benefit from flexible scheduling because they can work part time or full time while

enrolled in school. About 59 percent of all community college students are enrolled part time, while 41 percent are full-time students (taking 12 or more credit hours each semester). Community colleges are also important for underrepresented minority student enrollment: 55 percent of all Native American college students attend a community college, as do 56 percent of all Hispanic students, 49 percent of African American students, and 44 percent of Asian American/Pacific Islander students (American Association of Community Colleges, 2013).

One of the greatest challenges facing community colleges today is money. Across the nation, state and local governments struggling to balance their budgets have slashed funding for community colleges. In a number of regions, these cuts have been so severe that schools have been seriously limited in their ability to meet the needs of their students. In some cases, colleges have terminated programs, slashed course offerings, reduced the number of faculty, and eliminated essential student services. Many people were hopeful that President Obama's "American Graduation Initiative" of 2009 would strengthen community colleges, offer greater financial support for students, and help increase the college graduation rate for the nation by 2020. However, only a small portion of the $12 billion in grants to community colleges and universities for innovative training programs has been distributed. Grants such as these may help more students gain an education that will help them develop their life and workplace skills and thus offer them a better future.

## Four-Year Colleges and Universities

More than 18 million undergraduate students and nearly 3 million graduate students attend public or private degree-granting colleges or universities in the United States (*Chronicle of Higher Education*, 2012). Four-year schools typically offer a general education curriculum that gives students exposure to multiple disciplines and ways of knowing, along with more in-depth study (known as a "major") in at least one area of concentration. However, many challenges are faced by four-year institutions, ranging from the cost of higher education to racial and ethnic differences in enrollment and lack of faculty diversity.

## The High Cost of a College Education

What does a college education cost? According to the *Chronicle of Higher Education* (2013), in the 2012–2013 academic year the average total cost of attendance for full-time students living on campus and paying in-state tuition was $22,261 at public four-year schools and $43,289 at private nonprofit institutions. The lowest total costs were for students living with family and paying in-state tuition and fees at public two-year institutions ($3,131). More than half of students at public colleges pay less than $9,000 in tuition and fees. Although community colleges and public state colleges and universities typically have lower tuition and overall costs—because they are funded primarily by tax dollars—than private colleges have, even the cost of attending public institutions has increased significantly over the past decade.

According to some social analysts, a college education is a bargain and a means of upward mobility. However, other analysts believe that the high cost of a college education reproduces the existing class system: Students who lack money may be denied access to higher education, and those who are able to attend college receive different types of education based on their ability to pay. For example, a community college student who receives an associate's degree or completes a certificate program may be prepared for a position in the middle of the occupational status range, such as a dental assistant, computer programmer, or auto mechanic. In contrast, university graduates with four-year degrees are more likely to find initial employment with firms where they stand a chance of being promoted to high-level management and executive positions.

Given the necessity of getting a college education in the twenty-first century, is any financial assistance available? The Obama administration passed a 2010 student-loan bill to aid colleges and students. The legislation was designed to "cut out the middle person" by ending the bank-based system of distributing federally subsidized student loans and instead have the Department of Education give loan money directly to colleges and their students. With the savings from this approach, more money was to be put into the Pell Grant program. Unlike a loan, a federal Pell Grant does not have to be repaid. The maximum award for the 2014–2015 academic year was $5,730, although many students do not qualify for this amount (for example, the minimum award was $587 in the same time period). Many questions remain about student loans and the possible long-term effects of high student debt on individuals after they complete their college education (Figure 12.11).

## Racial and Ethnic Differences in Enrollment

How does college enrollment differ by race and ethnicity? People of color (who are more likely than the average white student to be from lower-income

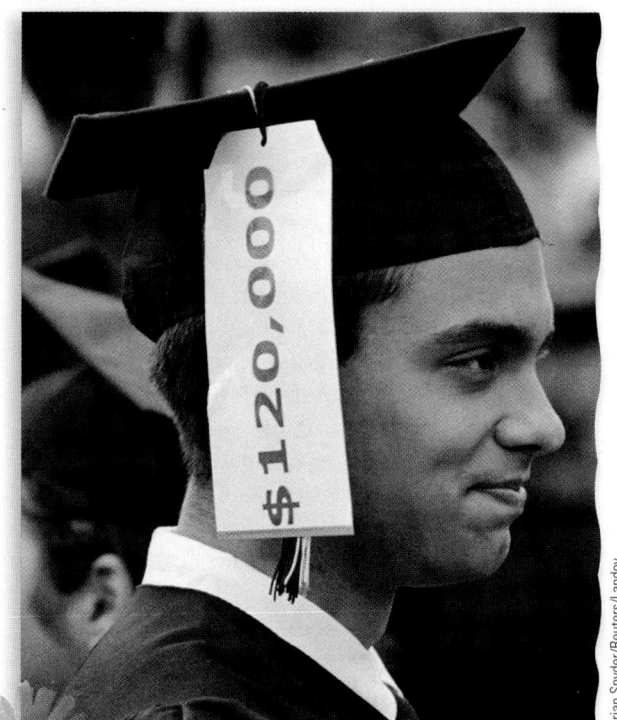

Brian Snyder/Reuters/Landov

**FIGURE 12.11** Soaring costs of both public and private institutions of higher education are a pressing problem for today's college students and their parents. What factors have contributed to the higher overall costs of obtaining a college degree?

families) are underrepresented in higher education. White Americans make up about 61 percent of all college students, as compared to African American enrollment at 14 percent, Hispanic/Latina/o enrollment at 13 percent, Asian American enrollment at 6 percent, and American Indian/Alaska Native enrollment at 1 percent. States with the largest share of underrepresented minorities enrolled are New Mexico, Texas, Mississippi, Florida, Georgia, California, Maryland, Louisiana, District of Columbia, and Arizona (*Chronicle of Higher Education,* 2012). Although Native American/Alaska Native enrollment rates have remained stagnant at about 1.0 percent, tribal colleges on reservations have experienced growth in student enrollment. Founded to overcome racism experienced by Native American students in traditional four-year colleges and to shrink the high dropout rate among Native American college students, thirty-one colleges are now chartered and run by the Native American nations (American Association of Community Colleges, 2013). Unlike other community colleges, the tribal colleges receive no funding from state and local governments and, as a result, are often short of funds to fulfill their academic mission.

The proportionately low number of people of color enrolled in colleges and universities is reflected in the educational achievement of people ages 25 and over, as shown in "Census Profiles." If we focus on persons who receive doctorate degrees, the underrepresentation of persons of color is even more striking. According to the *Chronicle of Higher Education* (2013), of the 31,789 doctoral degrees conferred in 2011 on recipients who were U.S. citizens or permanent residents, African Americans earned slightly more than 6 percent, Hispanics earned 6.3 percent, and Native Americans/American Indians earned 0.4 percent. By contrast, whites (non-Hispanic) earned 74.1 percent of these degrees, and Asian Americans earned 9 percent. Persons with two or more races and those indicating "other or unknown" races accounted for an additional 4 percent of the total doctoral degrees awarded.

Underrepresentation is not the only problem faced by students of color: Problems of prejudice and discrimination continue on some college campuses. Some problems are overt and highly visible; others are more covert and hidden from public view. Examples of overt racism include mocking Black History Month or a Latino celebration on campuses, referring to individuals by derogatory names, tying nooses on doorknobs of dorm rooms or faculty offices, and having "parties" where guests dress in outfits that ridicule people from different cultures or nations. A study by the sociologists Leslie Houts Picca and Joe R. Feagin (2007) found that many blatantly racist events, ranging from private jokes and conversations to violent incidents, occurred in the presence of 600 white students at 28 colleges and universities across the country who were asked to keep diaries and record any racist events that they observed. In addition to overt patterns of discrimination, other signs of racism included numerous conversations that took place "backstage" (in whites-only spaces where no person of color was present) and involved derogatory comments, skits, or jokes about persons of color. According to Picca and Feagin, most of the racial events were directed at African Americans, but Latinos/as and Asian Americans were also objects of some negative comments.

E-mail and social networking sites offer additional avenues for college students and others to engage in backstage racism. However, racist statements made by white persons in "private" digital communications are sometimes made public and shared in the front stage by individuals who do not share their views. As a result, cyber racism brings embarrassment not only to the perpetrators but also to persons of color who experience emotional distress as a result of the behavior of others (Daniels, 2009, 2010).

## Educational Achievement of Persons Ages 25 and Over

The Census Bureau asks people to indicate the highest degree or level of schooling they have completed. Categories range from "no schooling completed" to "doctorate degree"; however, we are looking only at the categories of high school graduate and above. As shown below, census data reflect that Asian Americans and non-Hispanic whites hold the highest levels of educational attainment. For these statistics to change significantly, greater educational opportunities and more-affordable higher education would need to be readily available to African Americans and Latinos/as, who historically have experienced racial discrimination and inadequately funded public schools with high dropout rates and low graduation rates.

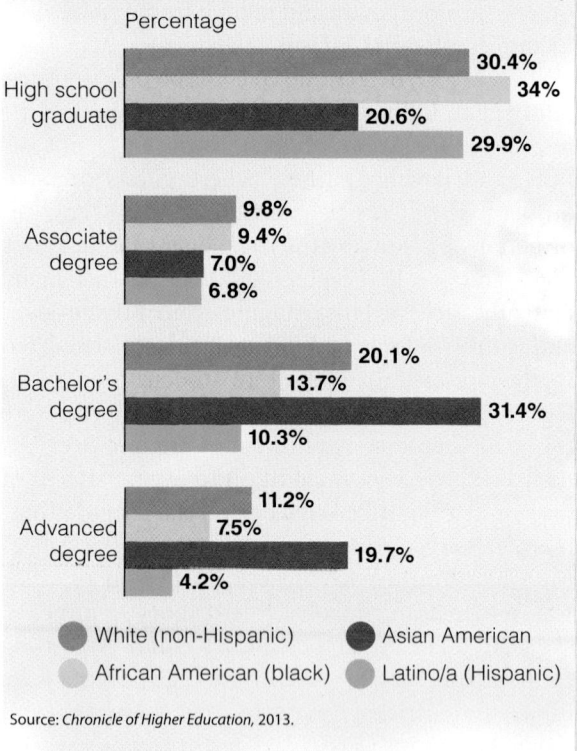

Percentage

High school graduate
30.4%
34%
20.6%
29.9%

Associate degree
9.8%
9.4%
7.0%
6.8%

Bachelor's degree
20.1%
13.7%
31.4%
10.3%

Advanced degree
11.2%
7.5%
19.7%
4.2%

● White (non-Hispanic)   ● Asian American
● African American (black)   ● Latino/a (Hispanic)

Source: *Chronicle of Higher Education,* 2013.

## Religion in Historical Perspective

**LO5** *Religion* is a social institution composed of a unified system of beliefs, symbols, and rituals—based on some sacred or supernatural realm—that guides human behavior, gives meaning to life, and unites believers into a community. For many people, religious beliefs

provide the answers for seemingly unanswerable questions about the meaning of life and death.

## Religion and the Meaning of Life

Religion seeks to answer important questions such as why we exist, why people suffer and die, and what happens when we die. Whereas science and medicine typically rely on existing scientific evidence to respond to these questions, religion seeks to explain suffering, death, and injustice in the realm of the sacred. According to Emile Durkheim (1995/1912), *sacred refers to those aspects of life that are extraordinary or supernatural*—in other words, those things that are set apart as "holy." People feel a sense of awe, reverence, deep respect, or fear for that which is considered sacred. Across cultures and in different eras, many things have been considered sacred, including invisible gods, spirits, specific animals or trees, altars, crosses, holy books, and special words or songs that only the initiated could speak or sing. Those things that people do not set apart as sacred are referred to as *profane—the everyday, secular, or "worldly" aspects of life.* Thus, whereas sacred beliefs are rooted in the holy or supernatural, secular beliefs have their foundation in scientific knowledge or everyday explanations. In the debate between creationists and evolutionists, for example, advocates of creationism view it as a belief founded in sacred (Biblical) teachings, whereas advocates of evolutionism assert that their beliefs are based on provable scientific facts.

In addition to beliefs, religion also comprises symbols and rituals (Figure 12.12). According to the anthropologist Clifford Geertz (1966), religion is a set of cultural symbols that establishes powerful and pervasive moods and motivations to help people interpret the meaning of life and establish a direction for their behavior. People often act out their religious beliefs in the form of *rituals*—regularly repeated and carefully prescribed forms of behaviors that symbolize a cherished value or belief (Kurtz, 1995). Rituals range from songs and prayers to offerings and sacrifices that worship or praise a supernatural being, an ideal, or a set of supernatural principles. For example, Muslims bow toward Mecca, the holy city of Islam, five times a day at fixed times to pray to God, whereas Christians participate in the celebration of communion (or the "Lord's Supper") to commemorate the life, death, and resurrection of Jesus. Rituals differ from everyday actions in that they involve very strictly determined behavior. The rituals involved in praying or in observing communion are carefully orchestrated and must be followed with precision. According to the sociologist Randall Collins (1982: 34), "In rituals, it is the forms that count. Saying prayers, singing

**FIGURE 12.12** One of the rituals in most religions is the identification of a holy place. For Jews (clockwise from top left), one such place is the Western Wall in Jerusalem. Buddhists might travel to the Mahaparinirvan Stupa, a shrine in India. For Muslims, an example of a holy place is the Blue Mosque in Turkey. Historical Native Americans believed in the ritual of creating burial mounds. And for Christians, Bethlehem is a historic pilgrimage.

a hymn, performing a primitive sacrifice or a dance, marching in a procession, kneeling before an idol or making the sign of the cross—in these, the action must be done the right way."

**Categories of Religion** Although it is difficult to establish exactly when religious rituals first began, anthropologists have concluded that all known societies over the past 100,000 years have had some form of religion (Haviland et al., 2013). The original locations of the world's major religions are shown in Figure 12.13.

**religion** a social institution composed of a unified system of beliefs, symbols, and rituals—based on some sacred or supernatural realm—that guides human behavior, gives meaning to life, and unites believers into a community.

**sacred** those aspects of life that are extraordinary or supernatural.

**profane** the everyday, secular, or "worldly" aspects of life.

**FIGURE 12.13** ORIGINAL LOCATIONS OF THE WORLD'S MAJOR RELIGIONS
© Cengage Learning

Religions have been classified into four main categories based on their dominant belief: simple supernaturalism, animism, theism, and transcendent idealism. In very simple preindustrial societies, religion often takes the form of *simple supernaturalism*—the belief that supernatural forces affect people's lives either positively or negatively. This type of religion does not acknowledge specific gods or supernatural spirits but focuses instead on impersonal forces that may exist in people or natural objects. By contrast, **animism is the belief that plants, animals, or other elements of the natural world are endowed with spirits or life forces having an effect on events in society.** Animism is identified with early hunting and gathering societies and with many Native American societies, in which everyday life was not separated from the elements of the natural world (Albanese, 2007).

The third category of religion is *theism*—a belief in a god or gods. Horticultural societies were among the first to practice *monotheism*—a belief in a single, supreme being or god who is responsible for significant events such as the creation of the world. Three of the major world religions—Christianity, Judaism, and Islam—are monotheistic. By contrast, Shinto and a number of indigenous religions of Africa are forms of *polytheism*—a belief in more than one god

(see Table 12.1). The fourth category of religion, transcendent idealism, is *nontheistic* because it does not focus on worship of a god or gods. *Transcendent idealism* is a belief in sacred principles of thought and conduct. Principles such as truth, justice, affirmation of life, and tolerance for others are central tenets of transcendent idealists, who seek an elevated state of consciousness in which they can fulfill their true potential.

## Religion and Scientific Explanations

During the Industrial Revolution, scientific explanations began to compete with religious views of life. Rapid growth in scientific and technological knowledge gave rise to the idea that science would ultimately answer questions that had previously been in the realm of religion. Science is the accumulated knowledge of the physical world that is systematically obtained through observation, experimentation, and other scientific methods. Many scholars believed that increases in scientific knowledge would result in *secularization*—**the process by which religious beliefs, practices, and institutions lose their significance in sectors of society**

PART 4 | SOCIAL INSTITUTIONS

## table 12.1

### Major World Religions

| | | Current Followers | Founder/Date | Beliefs |
|---|---|---|---|---|
| ✝ | **Christianity** | 2.2 billion | Jesus—1st century C.E. | Jesus is the Son of God. Through good moral and religious behavior (and/or God's grace), people achieve eternal life with God. |
| ☪ | **Islam** | 1.6 billion | Muhammad—ca. 600 C.E. | Muhammad received the Qur'an (scriptures) from God. On Judgment Day, believers who have submitted to God's will, as revealed in the Qur'an, will go to an eternal Garden of Eden. |
| ॐ | **Hinduism** | 1 billion | No specific founder—ca. 1500 B.C.E. | Brahma (creator), Vishnu (preserver), and Shiva (destroyer) are divine. Union with ultimate reality and escape from eternal reincarnation are achieved through yoga, adherence to scripture, and devotion. |
| ☸ | **Buddhism** | 488 million | Siddhartha Gautama—500 to 600 B.C.E. | Through meditation and adherence to the Eight-Fold Path (correct thought and behavior), people can free themselves from desire and suffering, escape the cycle of eternal rebirth, and achieve nirvana (enlightenment). |
| ✡ | **Judaism** | 14 million | Abraham, Isaac, and Jacob—ca. 2000 B.C.E. | God's nature and will are revealed in the Torah (Hebrew scripture) and in His intervention in history. God has established a covenant with the people of Israel, who are called to a life of holiness, justice, mercy, and fidelity to God's law. |
| ☯ | **Confucianism** | 6.3 million | K'ung Fu-Tzu (Confucius)—ca. 500 B.C.E. | The sayings of Confucius (collected in the *Analects*) stress the role of virtue and order in the relationships among individuals, their families, and society. |

**and culture** (Berger, 1967). Secularization involves a decline of religion in everyday life and a corresponding increase in organizations that are highly bureaucratized, fragmented, and impersonal.

## Sociological Perspectives on Religion

**LO6** Religion as a social institution is a powerful, deeply felt, and influential force in human society. Sociologists study the social institution of religion because of the importance that religion holds for many people; they also want to know more about the influence of religion on society, and vice versa. For example, some people believe that the introduction of prayer or religious instruction in public schools would have a positive effect on the teaching of values such as honesty, compassion, courage, and tolerance because these values could be given a moral foundation. However, society has strongly influenced the practice of religion in the United States as a result of court rulings and laws that have limited religious activities in public settings, including schools (see "Sociology and Social Policy").

The major sociological perspectives have different outlooks on the relationship between religion and society. Functionalists typically emphasize the ways in which religious beliefs and rituals can bind people together. Conflict explanations suggest that religion can be a source of false consciousness in society. Symbolic interactionists focus on the meanings that people give to religion in their everyday life.

**animism** the belief that plants, animals, or other elements of the natural world are endowed with spirits or life forces having an effect on events in society.

**secularization** the process by which religious beliefs, practices, and institutions lose their significance in sectors of society and culture.

# SOCIOLOGY and Social Policy

## Prayer in Public Schools? The Issue of Separation of Church and State

ALL THINGS ARE POSSIBLE!

"Only Believe"

The faculty and staff are working hard to prepare our students for the MSA.

Let's come together, as one, in prayer (teachers, students and parents) and ask God to bless our school to pass the MSA (Jeremiah 33:3). He will do it again! Everyone is invited.

Time: Saturday, 10:00–10:30 a.m.

Where: Tench Tilghman

    —This poster announcing a Baltimore, Maryland, principal's invitation to students, teachers, and parents to join together in prayer at a public school for success on the Maryland School Assessments test for third through eighth graders created a stir and brought an attorney for the American Civil Liberties Union forward to argue that such activity was a violation of the U.S. Constitution (Green, 2011).

**W**hy is the issue of prayer or other religious observances in public schools an ongoing concern for many people? As we think about this question, it is necessary to understand how social policy has been historically used to establish a division between church and state. When the U.S. Constitution was ratified, in 1789, the colonists who made up the majority of the population of the original states were of many different faiths. Because of this diversity, there was no mention of religion in the original Constitution. However, in 1791 the First Amendment added the following provision (to this day, the only reference to religion in that document): "Congress shall make no law respecting an establishment of religion, or prohibiting the free exercise thereof." The ban was binding only on the federal government; however, in 1947 the U.S. Supreme Court held that the Fourteenth Amendment ("No State shall make or enforce any law which shall abridge the privileges or immunities of citizens") had the effect of making the First Amendment's separation of church and state applicable to state governments as well.

Historically, the Supreme Court has been called on to define the boundary between permissible and impermissible governmental action with regard to religion. In 1947 the Court held that the ban on establishing a religion made it unconstitutional for a state to use tax revenues to support an institution that taught religion. In 1962 and 1963 the Court expanded this ruling to include many types of religious activities at schools or in connection with school activities, including group prayer, invocations at sporting events, and distribution of religious materials at school. However, in 1990 the Court ruled that religious groups could meet on school property if certain conditions were met, including that attendance must be voluntary, the meetings must be organized and run by students, and the activities must occur outside of regular class hours.

Rob Crandall/Stock Connection Blue/Alamy

**Should prayer be permitted in the classroom? On the school grounds? At school athletic events?** Given the diversity of beliefs that U.S. people hold, arguments and court cases over activities such as prayer around the school flagpole will no doubt continue in the future.

Compromises regarding prayer in schools have been attempted in some states as legislators passed laws that either permit or mandate a daily moment of silence in public schools; however, the effect of such compromises has often been that neither side in the debate is appeased. Advocates of allowing more religious activities in public education believe that the constitutional dictate prohibiting "establishment of religion" was not intended to keep religion out of the public schools and that students need greater access to religious and moral training. Opponents believe that any entry of religious training and religious observances into public education and taxpayer-supported school facilities or events violates the Constitution and might be used by some people to promote their religion over others, or over a person's right to have no religion at all. No doubt, this debate will continue in the future as some political leaders in state legislatures look for creative ways to authorize school prayer or other religious functions in public schools that will pass constitutional muster while, at the same time, staunch opponents of such measures will continue to argue that any form of religious expression should not be allowed in this nation's public schools.

Sources: Based on Albanese, 2007; and Green, 2011.

## reflect & analyze

**Where should the line be drawn on religious observances and public education? What do you think?**

## Functionalist Perspectives on Religion

The functionalist perspective on religion finds its roots in the works of Emile Durkheim, who emphasized that religion is essential to the maintenance of society. He suggested that religion is a cultural universal found in all societies because it meets basic human needs and serves important societal functions.

For Durkheim, the central feature of all religions is the presence of sacred beliefs and rituals that bind people together in a collectivity. In his studies of the religion of the Australian aborigines, for example, Durkheim found that each clan had established its own sacred totem, which included kangaroos, trees, rivers, rock formations, and other animals or natural creations. To clan members, their totem was sacred; it symbolized some unique quality of their clan. People developed a feeling of unity by performing ritual dances around their totem, which caused them to abandon individual self-interest. Durkheim suggested that the correct performance of the ritual gives rise to religious conviction. Religious beliefs and rituals are *collective representations*—group-held meanings that express something important about the group itself. Because of the intertwining of group consciousness and society, functionalists suggest that religion has three important functions in any society:

1. *Meaning and purpose.* Religion offers meaning for the human experience. Some events create a profound sense of loss on both an individual basis (such as injustice, suffering, and the death of a loved one) and a group basis (such as famine, earthquake, economic depression, or subjugation by an enemy). Inequality may cause people to wonder why their own situation is no better than it is. Most religions offer explanations for these concerns. Explanations may differ from one religion to another, yet each tells the individual or group that life is part of a larger system of order in the universe. Some (but not all) religions even offer hope of an afterlife for persons who follow the religion's tenets of morality in this life. Such beliefs help make injustices in this life easier to endure.

2. *Social cohesion and a sense of belonging.* Religious teachings and practices, by emphasizing shared symbolism, help promote social cohesion. An example is the Christian ritual of communion, which not only commemorates a historical event but also allows followers to participate in the unity ("communion") of themselves with other believers. All religions have some form of shared experiences that rekindle the group's consciousness of its own unity.

3. *Social control and support for the government.* All societies attempt to maintain social control through systems of rewards and punishments. Sacred symbols and beliefs establish powerful, pervasive, long-lasting motivations based on the concept of a general order of existence. In other words, if individuals consider themselves to be part of a larger order that holds the ultimate meaning in life, they will feel bound to one another (and past and future generations) in a way that otherwise might not be possible.

Religion also helps maintain social control in society by conferring supernatural legitimacy on the norms and laws of society. In some societies, social control occurs as a result of direct collusion between the dominant classes and the dominant religious organizations.

In the United States the separation of church and state reduces religious legitimation of political power. Nevertheless, political leaders often use religion to justify their decisions, stating that they have prayed for guidance in deciding what to do. This informal relationship between religion and the state has been referred to as *civil religion*—**the set of beliefs, rituals, and symbols that makes sacred the values of the society and places the nation in the context of the ultimate system of meaning.** Civil religion is not tied to any one denomination or religious group; it has an identity all its own. For example, many civil ceremonies in the United States have a marked religious quality. National values are celebrated on "high holy days" such as Memorial Day and the Fourth of July. Both political inaugurations and courtroom trials require people to place their hand on a Bible while swearing to do their duty or tell the truth, as the case may be. The U.S. flag is the primary sacred object of our civil religion, and the pledge of allegiance has included the phrase "one nation under God" for many years now. U.S. currency bears the inscription "In God We Trust."

## Conflict Perspectives on Religion

Many functionalists view religion, including civil religion, as serving positive functions in society, but some conflict theorists view religion negatively.

**Karl Marx on Religion** For Marx, *ideologies*—systematic views of the way the world ought

---

**civil religion** the set of beliefs, rituals, and symbols that makes sacred the values of the society and places the nation in the context of the ultimate system of meaning.

to be—are embodied in religious doctrines and political values (Turner, Beeghley, and Powers, 2002). These ideologies also serve to justify the status quo and retard social change. The capitalist class uses religious ideology as a tool of domination to mislead the workers about their true interests. For this reason, Marx wrote his now-famous statement that religion is the "opiate of the masses." People become complacent because they have been taught to believe in an afterlife in which they will be rewarded for their suffering and misery in this life. Although these religious teachings soothe the masses' distress, any relief is illusory. Religion unites people under a "false consciousness" that they share common interests with members of the dominant class.

**Max Weber on Religion** Whereas Marx believed that religion retarded social change, Weber argued just the opposite. For Weber, religion could be a catalyst to produce social change. In *The Protestant Ethic and the Spirit of Capitalism* (1976/1904–1905), Weber asserted that the religious teachings of John Calvin were directly related to the rise of capitalism. Calvin emphasized the doctrine of *predestination*— the belief that even before they are born, all people are divided into two groups, the saved and the damned, and only God knows who will go to heaven (the elect) and who will go to hell. Because people cannot know whether they will be saved, they tend to look for earthly signs that they are among the elect. According to the Protestant ethic, those who have faith, perform good works, and achieve economic success are more likely to be among the chosen of God. As a result, people work hard, save their money, and do not spend it on worldly frivolity; instead, they reinvest it in their land, equipment, and labor.

The spirit of capitalism grew in the fertile soil of the Protestant ethic. Even as people worked ever harder to prove their religious piety, structural conditions in Europe led to the Industrial Revolution, free markets, and the commercialization of the economy—developments that worked hand in hand with Calvinist religious teachings. From this viewpoint, wealth was an unintended consequence of religious piety and hard work. With the contemporary secularizing influence of wealth, people often think of wealth and material possessions as the major (or only) reason to work. Although it is no longer referred to as the "Protestant ethic," many people still refer to the "work ethic" in somewhat the same manner that Weber did. For example, political and business leaders in the United States often claim that "the work ethic is dead."

Like Marx, Weber was acutely aware that religion could reinforce existing social arrangements, especially the stratification system (Figure 12.14). The wealthy can use religion to justify their power and privilege: It is a sign of God's approval of their hard work and morality. As for the poor, if they work hard

Devendra M Singh/AFP/Getty Images

**FIGURE 12.14** According to Marx and Weber, religion serves to reinforce social stratification in a society. For example, according to Hindu belief, a person's social position in his or her current life is a result of behavior in a former life.

and live a moral life, they will be richly rewarded in another life.

From a conflict perspective, religion tends to promote conflict between groups and societies. According to conflict theorists, conflict may be *between* religious groups (for example, anti-Semitism), *within* a religious group (for example, when a splinter group leaves an existing denomination), or between a religious group and the *larger society* (for example, the conflict over religion in the classroom). Conflict theorists assert that in attempting to provide meaning and purpose in life while at the same time promoting the status quo, religion is used by the dominant classes to impose their own control over society and its resources. Many feminists object to the patriarchal nature of most religions; some advocate a break from traditional religions, whereas others seek to reform religious language, symbols, and rituals to eliminate the elements of patriarchy.

## Symbolic Interactionist Perspectives on Religion

Thus far, we have been looking at religion primarily from a macrolevel perspective. Symbolic interactionists focus their attention on a microlevel analysis which examines the meanings that people give to religion in their everyday life.

**Religion as a Reference Group** For many people, religion serves as a reference group to help them define themselves. For example, religious symbols have meaning for large bodies of people. The Star of David holds special significance for Jews, just as the crescent moon and star do for Muslims and the cross does for Christians. For individuals as well, a symbol may have a certain meaning beyond that shared by the group. For instance, a symbol given to a child may have special meaning when he or she grows up and faces war or other crises. It may not only remind the adult of a religious belief but also create a feeling of closeness with a relative who is now deceased. It has been said that the symbolism of religion is so very powerful because it "expresses the essential facts of our human existence" (Collins, 1982: 37).

**Her Religion and His Religion** Early feminist Charlotte Perkins Gilman (1976/1923) believed that "men's religion" taught people to submit and obey rather than to think about and realistically confront situations. Consequently, she asserted, the monopolization of religious thoughts and doctrines by men contributed to intolerance and the subordination of women. Along with other feminist thinkers, Gilman concluded that "God the Mother" images could be useful as a means of encouraging people to be more cooperative and compassionate, rather than competitive and violent.

Not all people interpret religion in the same way. In virtually all religions, women have much less influence in establishing social definitions of appropriate gender roles both within the religious community and in the larger community. Therefore, women and men may belong to the same religious group, but their individual religion will not necessarily be a carbon copy of the group's entire system of beliefs. In fact, women's versions of any given religion probably differ markedly from men's versions.

Religious symbolism and language typically create a social definition of the roles of men and women. For example, religious symbolism may depict the higher deities as male and the lower deities as female. Sometimes, females are depicted as spiritual forces that are negative or evil. For instance, the Hindu goddess Kali represents men's eternal battle against the evils of materialism (Daly, 1973). Historically, language has defined women as being nonexistent in the world's major religions. Phrases such as "for all men" in Catholic and Episcopal services gradually have been changed to "for all"; however, some churches retain the traditional liturgy. Although there has been resistance, especially by women, to some of the terms, inclusive language is less common than older male terms for God.

## Rational Choice Perspectives on Religion

In terms of religion, rational choice theory is based on the assumption that religion is essentially a rational response to human needs; however, the theory does not claim that any particular religious belief is necessarily true or more rational than another. The rational choice perspective views religion as a competitive marketplace in which religious organizations (suppliers) offer a variety of religions and religious products to potential followers (consumers), who shop around for the religious theologies, practices, and communities that best suit them.

According to this approach, people need to know that life has a beneficial supernatural element, such as that there is meaning in life or that there is life after death, and they seek to find these rewards in various religious organizations. The rewards include explanations of the meaning of life and reassurances about overcoming death. However, because religious organizations cannot offer religious certainties, they instead offer compensators—a body of language and practices that compensates for some physical lack or frustrated goal. According to the sociologists Rodney Stark and William Bainbridge (1985),

## Sociological Perspectives on Education and Religion

| | Education | Religion |
|---|---|---|
| **Functionalist Perspective** | One of the most important components of society: Schools teach students not only content but also to put group needs ahead of the individual's. | Sacred beliefs and rituals bind people together and help maintain social control. |
| **Conflict Perspective** | Schools perpetuate class, racial–ethnic, and gender inequalities through what they teach to whom. | Religion may be used to justify the status quo (Marx) or to promote social change (Weber). |
| **Symbolic Interactionist Perspective** | Labeling and the self-fulfilling prophecy are an example of how students and teachers affect one another as they interpret their interactions. | Religion may serve as a reference group for many people, but because of race, class, and gender, people may experience it differently. |
| **Rational Choice Perspective** | | Religious persons and organizations, interacting within a competitive market framework, offer a variety of religions and religious products to consumers, who shop around for religious theologies, practices, and communities that best suit them. |

all religions offer compensators, such as a belief in heaven, personal fulfillment, and control over evil influences in the world, to offset the fact that they cannot offer certainty of an afterlife or other valued resources that potential followers and adherents might desire. Rational choice theory focuses on the process by which actors—individuals, groups, and communities—settle on one optimal outcome out of a range of possible choices (a cost–benefit analysis). These compensators provide a range of possible choices for people in the face of a limited (or non-existent) supply of the choice (certainty, for example) that they truly desire.

Some sociologists of religion have applied rational choice theory to an examination of the very competitive U.S. religious marketplace and have found that people are actively shopping around for beliefs, practices, and religious communities that best suit them. For example, in recent years some religious followers have been attracted to churches that preach the so-called prosperity gospel, which is based on the assumption that if you give your money to God, He will bless you with more money and other material possessions (such as a larger house and a luxury vehicle) that you desire. Several megachurches, including Joel Osteen's Lakewood in Houston, T. D. Jakes's Potter's House in South Dallas, and Creflo Dollar's World Changers near Atlanta, are partly based on teaching that suggests God wants people to be prosperous if they are "right" with Him.

Based on the diverse teaching and practices of various religious bodies, adherents and prospective followers move among various religious organizations, with every major religious group simultaneously gaining and losing adherents. Although some find the religious home they seek, others decide to consider themselves unaffiliated with any specific faith (Pew Forum on Religion and Public Life, 2008). Based on the movement of possible adherents, religious groups challenge one another for followers, emphasize specific moral values, and create a civil society that offers followers a religious faith that does not unduly burden them (Stark and Tobin, 2008).

The Concept Quick Review summarizes the major sociological perspectives on education and religion.

# Types of Religious Organization

**LO7** Religious groups vary widely in their organizational structure. Although some groups are large and somewhat bureaucratically organized, others are small and have a relatively informal authority structure. Some require total commitment from their members; others expect members to have only a partial commitment (see Table 12.2). Sociologists have developed typologies or ideal types of religious organization to enable

## table 12.2

### Characteristics of Churches and Sects

| Characteristic | Church | Sect |
|---|---|---|
| **Organization** | Large, bureaucratic organization, led by a professional clergy | Small, faithful group, with high degree of lay participation |
| **Membership** | Open to all; members usually from upper and middle classes | Closely guarded membership, usually from lower classes |
| **Type of Worship** | Formal, orderly | Informal, spontaneous |
| **Salvation** | Granted by God, as administered by the church | Achieved by moral purity |
| **Attitude Toward Other Institutions and Religions** | Tolerant | Intolerant |

them to study a wide variety of religious groups. The most common categorization includes four types: ecclesia, church, sect, and cult.

## Ecclesia

An *ecclesia* **is a religious organization that is so integrated into the dominant culture that it claims as its membership all members of a society.** Membership in the ecclesia occurs as a result of being born into the society rather than by any conscious decision on the part of individual members. The linkages between the social institutions of religion and government are often very strong in such societies. Although no true ecclesia exists in the contemporary world, the Anglican church (the official church of England), the Lutheran church in Sweden and Denmark, the Roman Catholic church in Italy and Spain, and Islamic mosques in Iran and Pakistan come fairly close.

## Churches, Denominations, and Sects

Unlike an ecclesia, a church is not considered to be a state religion; however, it may still have a powerful influence on political and economic arrangements in society. A *church* **is a large, bureaucratically organized religious organization that tends to seek accommodation with the larger society in order to maintain some degree of control over it.** Church membership is largely based on birth; children of church members are typically baptized as infants and become lifelong members of the church. Older children and adults may choose to join the church, but they are required to go through an extensive training program that culminates in a ceremony similar to the one that infants go through. Churches have a bureaucratic structure, and leadership is

hierarchically arranged. Usually, the clergy have many years of formal education. Churches have very restrained services that appeal to the intellect rather than the emotions. Religious services are highly ritualized; they are led by clergy who wear robes, enter and exit in a formal processional, administer sacraments, and read services from a prayer book or other standardized liturgical format. The Lutheran church and the Episcopal church are two examples.

Midway between the church and the sect is a *denomination*—**a large, organized religion characterized by accommodation to society but frequently lacking in the ability or intention to dominate society** (Niebuhr, 1929). Denominations have a trained ministry, and although involvement by lay members is encouraged more than in the church, their participation is usually limited to particular activities, such as readings or prayers. Denominations tend to be more tolerant and less likely than churches to expel or excommunicate members. This form of organization is most likely to thrive in societies characterized by *religious pluralism*—a situation in which many religious groups exist because they have a special appeal to specific segments of the population. Perhaps because of its diversity,

**ecclesia** a religious organization that is so integrated into the dominant culture that it claims as its membership all members of a society.

**church** a large, bureaucratically organized religious organization that tends to seek accommodation with the larger society in order to maintain some degree of control over it.

**denomination** a large, organized religion characterized by accommodation to society but frequently lacking in the ability or intention to dominate society.

the United States has more denominations than any other country. Table 12.3 shows this diversity.

A *sect* **is a relatively small religious group that has broken away from another religious organization to renew what it views as the original version of the faith.** Unlike churches, sects offer members a more personal religion and an intimate relationship with a supreme being, depicted as taking an active interest in the individual's everyday life. Whereas churches use formalized prayers, often from a prayer book, sects have informal prayers composed at the time they are given. Typically, religious sects appeal to those who might be characterized as lower class, whereas denominations primarily appeal to the middle and upper-middle classes, and churches focus on the upper classes.

According to the church–sect typology, as members of a sect become more successful economically and socially, their religious organization is also likely to focus more on this world and less on the next. If some members of the sect do not achieve financial success, they may feel left behind as other members and the ministers shift their priorities. Eventually, this process will weaken some organizations, and people will split off to create new, less worldly versions of the group that will be more committed to "keeping the faith." Those who defect to form a new religious organization may start another sect or form a cult (Stark and Bainbridge, 1981).

## Cults (New Religious Movements)

Previously, sociologists defined a *cult* **as a loosely organized religious group with practices and teachings outside the dominant cultural and religious traditions of a society.** Because the term *cult* has assumed a negative and sometimes offensive meaning because of the beliefs and actions of a few highly publicized cults, some researchers now use the term *new religious movement* (NRM) and point out that a number of major world religions (including Judaism, Islam, and Christianity) and some denominations (such as the Mormons) started as cults. Also, most cults or NRMs do not exhibit the bizarre behavior or have the unfortunate ending that a few notorious groups have had in the past.

NRMs usually have a leader who exhibits charismatic characteristics (personal magnetism or mystical leadership) and possesses a unique ability to communicate and form attachments with others. An example was the late Reverend Sun Myung Moon, a former Korean electrical engineer who founded the Unification church, or "Moonies," claiming God revealed to him that the Judgment Day was rapidly approaching. This former cult identified itself as "comprised of families striving to embody the

### table 12.3

Top 25 U.S. Denominations That Self-Identify as Christian in 2012

| Religious Body | Members |
| --- | --- |
| Catholic Church | 68,202,492 |
| Southern Baptist Convention | 16,136,044 |
| United Methodist Church | 7,679,850 |
| Church of Jesus Christ of Latter Day Saints | 6,157,238 |
| Church of God in Christ | 5,499,875 |
| National Baptist Convention U.S.A. | 5,197,512 |
| Evangelical Lutheran Church in America | 4,274,855 |
| National Baptist Convention of America | 3,500,000 |
| Assemblies of God | 3,030,944 |
| Presbyterian Church (U.S.A.) | 2,675,873 |
| African Methodist Episcopal Church | 2,500,000 |
| National Missionary Baptist Convention of America | 2,500,000 |
| Lutheran Church—Missouri Synod | 2,278,586 |
| Episcopal Church | 1,951,906 |
| Pentecostal Assemblies | 1,800,000 |
| Churches of Christ | 1,639,495 |
| Greek Orthodox Church | 1,500,000 |
| African Methodist Episcopal Zion Church | 1,400,000 |
| American Baptist Churches in U.S.A. | 1,308,054 |
| Jehovah's Witnesses | 1,184,249 |
| Church of God (Cleveland, Tenn.) | 1,074,047 |
| Christian Churches/Churches of Christ | 1,071,616 |
| Seventh-Day Adventist Church | 1,060,386 |
| United Church of Christ | 1,058,423 |
| Progressive National Baptist Convention | 1,010,000 |

Source: National Council of Churches, 2012.

ideal of true love and to establish a world of peace and unity among all peoples, races, and religions as envisioned by Reverend Sun Myung Moon. Members of the Unification Church accept and follow Reverend Moon's particular religious teaching, the Divine Principle" (*Unification Church News*, 2011).

Reverend Moon died at the age of 92 in 2012. Initially, his movement flourished because it recruited new members through their personal attachments to present members. In the twenty-first century, it became more institutionalized. Other cult leaders did not fare so well for so long, including Jim Jones, whose ill-fated cult ended up committing mass suicide in Guyana in 1978, and Marshall Herff Applewhite ("Do"), who led his thirty-eight Heaven's Gate followers to commit mass suicide at their Rancho Santa Fe, California, mansion after convincing them that the comet Hale-Bopp, which swung by Earth in late March 1997, would be their celestial chariot taking them to a higher level.

What eventually happens to cults or NRMs? Over time, some disappear; however, others gradually transform into other types of religious organizations, such as sects or denominations. An example is Mary Baker Eddy's Christian Science church, which started as a cult but became an established denomination with mainstream methods of outreach, such as Christian Science Reading Rooms placed in office buildings or shopping malls, where individuals can learn of the group's beliefs while going about their routine activities. Other cults or new religious movements lose their "newness" as they are embraced, as Scientology has been, by celebrities such as Tom Cruise, John Travolta, Kirstie Alley, and Brandy.

## Trends in Religion in the United States

**LO8** Religion in the United States is very diverse. Pluralism and religious freedom are among the cultural values most widely espoused, and no state church or single denomination predominates. As shown in Figure 12.15, Protestants

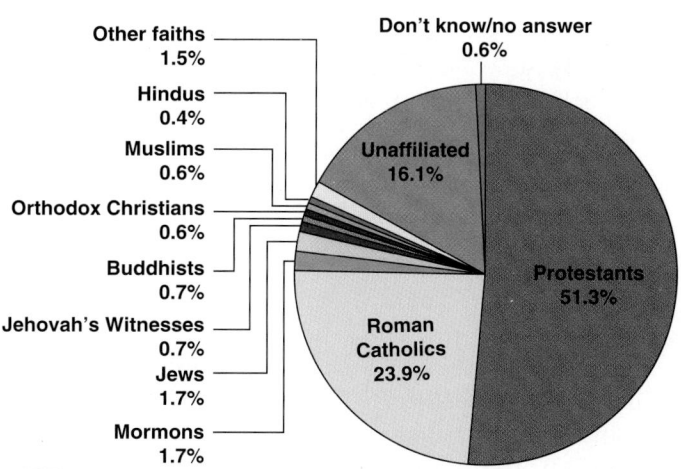

**FIGURE 12.15** U.S. RELIGIOUS TRADITIONS' MEMBERSHIP

Source: Pew Research Religion and Public Life Project, 2014. © Cengage Learning

constitute the largest religious body in the United States, followed by Roman Catholics, Jews, Mormons, and others.

## The Secularization Debate

As previously defined, *secularization* is the process by which religious beliefs, practices, and institutions lose their significance in society and nonreligious values, principles, and institutions take their place. Secularization has two components: (1) a decline in religious values and institutions in everyday life and (2) a corresponding increase in nonreligious values or principles and greater significance given to secular institutions. Although secularization has been widely studied and hotly debated, many scholars argue that levels of religiosity are not declining in the United States and other high-income nations. In fact, sociologist Peter Berger, who proposed the secularization theory, reconsidered his theory when he observed that modernity, rather than secularizing society, had contributed to a counter-secularization movement. Berger also found that higher levels of religiosity accompanied modernity—the way that people are influenced by religious beliefs and shape their social reality accordingly. *Religiosity* may also be referred to as religious commitment or "religiousness." Social scientists use various measures to determine religiosity, including the extent to which a person does one or more of the following: (1) believes in and "feels" or experiences certain aspects of religion, (2) becomes involved in religious activities such as attending church or reading sacred texts, (3) believes in the teachings of the church, and (4) lives in accordance with those teachings and beliefs.

So is secularization occurring in the twenty-first century? Although the U.S. religious landscape has fluid and diverse patterns, some trends have been identified that might support the secularization thesis. First, the number of people in the United States who do not identify with any religion has grown at a rapid rate. In 2012 one-fifth of the U.S. public, and one-third of adults under age thirty, indicated that they were religiously unaffiliated (Pew Forum on Religion and Public Life, 2012b). When the

**sect** a relatively small religious group that has broken away from another religious organization to renew what it views as the original version of the faith.

**cult** a loosely organized religious group with practices and teachings outside the dominant cultural and religious traditions of a society.

Pew Research Center looked at trends from 2007 to 2012, it found that over the five-year-period, the number of people who considered themselves to be unaffiliated increased from slightly over 15 percent to just under 20 percent of all U.S. adults. These percentages included nearly 33 million individuals who stated that they had no particular religious affiliation and 13 million who self-described as atheists and agnostics (Pew Forum for Religion and Public Life, 2012b).

However, these figures do not necessarily mean that Americans are losing their religion. According to recent surveys, many of the 46 million unaffiliated adults say that they are religious or spiritual in some way. For example, 68 percent state that they believe in God, and more than 50 percent indicate that they feel a deep connection with nature and the Earth. In addition, one in five of the so-called "unaffiliated" say that they pray every day (Pew Forum on Religion and Public Life, 2012b). One of the reasons that many unaffiliated persons indicate that they are not looking for a church or other religious institution to join is that they believe these organizations are too concerned with money and power, as well as overly emphasizing rules and becoming too involved in politics (Pew Forum on Religion and Public Life, 2012b). Issues of secularization and the growing number of unaffiliated people in major national religion surveys will continue to be of concern to persons who identify themselves as "fundamentalists."

### The Rise of Religious Fundamentalism

*Fundamentalism* **is a traditional religious doctrine that is conservative, is typically opposed to modernity, and rejects "worldly pleasures" in favor of otherworldly spirituality.** In the past, traditional fundamentalism primarily appealed to people from lower-income, rural, southern backgrounds; however, newer fundamentalist movements have had a much wider appeal to people from all socioeconomic levels, geographical areas, and occupations in the United States. One reason for the rise of fundamentalism has been a reaction against modernization and secularization. Around the world, those who adhere to fundamentalism—whether they are Muslims, Christians, or followers of one of the other world religions—believe that sacred traditions must be revitalized. For example, public education in the United States has been the focus of some who follow the tenets of Christian fundamentalism. Various religious and political leaders vow to bring the Christian religion "back" into the public life of this country. They have been especially critical of educators who teach what they perceive to be *secular humanism*—the belief that human beings can become better through their own efforts rather than through belief in God and a religious conversion. According to some Christian fundamentalists, elementary schoolchildren do not receive a fair and balanced picture of the Christian religion, but instead are taught that their parents' religion is inferior and perhaps irrational (Carter, 1994). But how might students and teachers who come from the diverse religious heritages feel about Christian religious instruction or organized prayer in public schools? Many social analysts believe that such practices would cause conflict and perhaps discrimination on the basis of religion.

## Education and Religion in the Future

Education and religion are powerful and influential forces because these social institutions instill in individuals the values, beliefs, and knowledge that many people believe are essential for the survival of individuals and entire cultures. Because education and religion are both socializing institutions, children and young people are particularly affected by the ideas that teachers and religious leaders impart to them. Through formal education or schooling, people learn about some aspects of the society in which they live and the dominant culture of which they are a part. In religious groups, individuals of all ages learn about the institutionalized beliefs and rituals of specific organizations that have been established to serve people's religious needs and other purposes.

What will be the future of education in the United States? The answer to this question depends on how successful elected officials are in getting their agendas through state legislatures and the U.S. Congress. In regard to public elementary and secondary schools, provisions of the Elementary and Secondary Education Act—or No Child Left Behind (NCLB)—implemented in 2001 by the Bush administration remain in effect as of this writing, but most of the states have received waivers that extend the amount of time for reaching certain prescribed deadlines or exempt schools from specific provisions of the law. The Obama administration has consistently opposed No Child Left Behind and has established a mechanism by which states can get relief from some provisions by implementing state-led efforts to close academic achievement gaps. Overall, however, the effects of No Child Left Behind are still being felt in the nation's schools. A primary purpose of NCLB was to close the achievement gap between rich and poor students by holding schools accountable for students' learning. The law required states to test every student's progress toward meeting specific standards that were established for each level of schooling. School districts were required to report students' results to demonstrate that they were making progress toward meeting these standards. Schools that closed the education gap received additional federal dollars, but schools and districts that did not show adequate progress lost funding and pupils: In some school districts, parents were able to move their children from lower- to higher-performing schools.

During the NCLB era, improvements occurred in fourth- and eighth-grade reading and math scores nationwide, particularly among African American and Hispanic students. However, critics believed that NCLB did not accurately define the main problem facing U.S. education: What the schools really needed was more money and more incentives for teaching and learning, not more testing of students or teachers. As testing became the focal point in schools and pressure increased on teachers and schools to improve students' achievement test scores, investigations in a number of states uncovered scandals in which educators had tampered with children's standardized tests in an effort to improve scores and enhance their own performance reviews. Although scandals such as these were rare overall, they pointed to the kind of pressure that the federal No Child Left Behind law put on students, teachers, and school administrators as performance requirements moved higher annually and students and schools were penalized if they were unable to reach the level of educational attainment that was expected.

Plans still in the process of implementation for the future include a "Race to the Top" in education, which was designed by the Obama administration to reshape the U.S. educational system to better prepare students for success. Race to the Top would provide additional competitive grants to states to expand innovations in education, support effective charter schools, promote public-school choice, and provide assistance to magnet schools. Forty-six states and the District of Columbia submitted blueprints for reform under this competition, and nineteen states received Race to the Top funding, which was supposed to benefit 22 million students (White House, 2013). Advocates for this program believe that it improves the future of U.S. schools because it includes an "Early Learning Challenge" in which states are asked to raise their standards and outcomes in early education with the overall goal of preparing all students to enter kindergarten ready to succeed. Race to the Top is also supposed to make it possible for more states to gain waivers from having to adhere to some guidelines in the NCLB law. Other issues addressed by this measure include

attracting outstanding teachers, creating conditions in schools that support effective teaching and learning, and modernizing outdated schools. However, some critics point out that such initiatives reward a few schools but leave the many without the necessary resources to improve. Other critics argue that many plans have been drawn up for education but that without the necessary financial backing at local, state, and federal levels, it is unlikely that many of the idealistic goals of plans such as Race to the Top will become a reality.

President Obama issued "A Blueprint for Reform" and set the four following goals to be reached by 2020:

1. Improving teacher and principal effectiveness.
2. Providing information to educators and families to help them evaluate and improve schools and students' learning.
3. Implementing college- and career-ready standards and developing improved assessments aligned with those standards.
4. Improving student learning and achievement in the lowest-performing schools by providing support and interventions.

Overall, the plan's purpose is to produce greater equity and opportunity for students.

Just as early schooling is faced with many challenges today, colleges and universities are confronted with many pressing issues, particularly given the major financial constraints that many institutions must confront. Among the Obama administration's initiatives for higher education are linking community colleges to employers to provide pathways to jobs, making major federal investments in higher education through increasing Pell Grants, helping students manage student loan debt, and expanding education tax credits to assist families with the costs of college (White House, 2013).

At the university level, institutions continue to expand their focus while undergoing strenuous budget cuts coupled with increasing demands to meet the needs of diverse student populations. Tightening of financial resources available to colleges and universities will lead to even more schools seeking alternative ways to fund their operations. Some will further raise tuition; others will seek different sources of funding. Some will move beyond the United States to find ways

**fundamentalism** a traditional religious doctrine that is conservative, is typically opposed to modernity, and rejects "worldly pleasures" in favor of otherworldly spirituality.

base of operation. For example, some U.S. universities are expanding their educational operations to emerging nations where demand is high for certain kinds of curricula, such as advanced business and petroleum engineering courses in Qatar and other Middle Eastern countries. Experts suggest that "university globalization" is here to stay, with both the export of students from countries such as India and China to other countries to study and the development of top-tier research universities in countries, including China, Singapore, and Saudi Arabia, where students may study without living abroad (Figure 12.16). Increasing numbers of U.S. students may go to school in these countries if their institutions offer opportunities, possibly at a lower cost, than do schools in the United States.

As discussed throughout this chapter, one of the major issues—present and future—in colleges and universities in the United States and other nations is the part that higher education will play in reducing or maintaining and perpetuating social inequality.

As we shift our focus to religion, we ask this: What significance will religion have in the future? In the twenty-first century, organized religion has seen a significant rise in power and importance in many regions of the world. It is interesting to note that religion has been empowered by three trends that secularization theory believed would be the cause of its death: (1) modernization, (2) democratization, and (3) globalization. Rather than modernization spelling the end of religion, more than 80 percent of people around the world claim that religion is important to them. Even for advanced industrialized nations, the figure is more than 60 percent (Toft, Philpott, and Shah, 2011). With the growth of democracies in numerous countries, including those Arab nations currently seeking change, religious actors have played a leading part in the political arena that was prohibited by former authoritarian regimes. Finally, globalization has produced a surge in the number of people and ideas that now travel around the world at much faster speeds than in the past, and this has made it possible for transnational actors such as religious groups to make themselves known to others and to use newer technologies such as the Internet, text messaging, and Twitter to bring people together quickly for collective action. How will these three trends play out in the future? According to Toft, Philpott, and Shah (2011),

Ted Pink/Alamy

**FIGURE 12.16** The term *university globalization* refers to the export of students from countries such as India and China to nations where they study and immerse themselves in another culture. How does this process affect higher education in the United States and other countries?

Religion is far from being the only or even the most decisive factor in global politics. But it has played—and will continue to play—a key role. The 21st century has brought us a world radically different from the one secularization theory promised. We have no choice but to build new theories and devise fresh policy strategies for the religious age we live in, God's Century, not the secular age that never came.

As discussed in this chapter, religion can be a unifying force in societies and around the world, or it can be a source of conflict and violence. It remains to be seen what the relationship between religion and peace and/or conflict will be in this century. Throughout history, people with religious beliefs have contributed to both peace and terrorism, democracy and authoritarianism, and reconciliation and civil war (Toft, Philpott, and Shah, 2011).

Debates over religion around the world, particularly issues such as secularization, fundamentalism, and violence perpetrated by religious extremists, will no doubt continue. Some studies show that the more-conservative or fundamental expressions of religious traditions are the most likely to present major challenges to peaceful accords among individuals and nations. Why? It is because conservative or fundamentalist factions primarily focus on those issues that bring about discord, conflict, and even the use of violence to resolve perceived differences. It is important to consider ways in which the leaders and teachings of the world's religions can best help cope with the opportunities and challenges that we face in the twenty-first century. However, the challenge is greatest when we acknowledge that some peoples of the world may believe that their religion will be furthered only if they are able to remain aloof from, and perhaps even banish, Western civilization as we know it.

Regarding the future of religion in America, studies have shown that we have become more polarized religiously over the past fifty years, but this may not remain true in the future. According to the political scientists Robert Putnam and David Campbell (2010), the United States is not only religiously devout and diverse, but it is also more religiously tolerant than we initially might think. To support this point, they point to intermarriage across religious lines, as well as to the extent that people today are far more likely to change religions or list themselves as "unaffiliated" or "none" when asked about religious preference than they were fifty years ago. In the future, more people may turn away from organized religions, creating a drop in U.S. religiosity, but that does not necessarily mean that these individuals have no religion at all: Many young people who indicate that they have no religious affiliation do state that they believe in God or provide other indicators of their religious beliefs. The extent to which religion is linked to conservative politics and/or discussions of sexual morality and rights regarding abortion, gay marriage, and other "hot topics" will play a central part in whether younger people become more (or less) involved in organized religion in the future.

In the twenty-first century we are likely to see not only the creation of new religious forms but also a dramatic revitalization of traditional forms of religious life. Religious fundamentalism has found a new audience among young people between the ages of eighteen and twenty-four. Some are Christian fundamentalists who will continue to emphasize Biblical literalism, which means that every word of the Bible is literally true (inerrancy). Other Christians are creating their own modified version of what it means to be a Christian, a fundamentalist, or an evangelical (a person actively involved in sharing the "Good News" of his or her religion with others). Ultimately, one thing we probably can expect is considerable religious tension between Americans with differing beliefs and people around the globe who do not share our worldviews. All of these factors are going to make it imperative that we become more tolerant of others and hope that they will do likewise toward us.

As we look to the future, we can expect even more fluidity in religion and the expansion of religious pluralism. In the words of Putnam and Campbell (2010: 550),

> How has America solved the puzzle of religious pluralism—the coexistence of religious diversity and devotion? And how has it done so in the wake of growing religious polarization? By creating a web of interlocking personal relationships among people of many different faiths. This is America's grace.

This is an interesting and optimistic view of the future of religion. As we have seen in this chapter, the debate continues over what education and religion should do and what the relationship between these two important social institutions should be. It will be up to your generation to work for greater understanding among the diverse people who make up our nation and the world.

# chapter REVIEW

## LO1 Why are education and religion important areas of sociological inquiry?

Education and religion are powerful and influential forces in society. Both institutions impart values, beliefs, and knowledge considered essential to the social reproduction of individual personalities and entire cultures. For this reason, analysis of education and religion is central to a comprehensive understanding of sociology.

## LO2 What are the key assumptions of functionalist, conflict, and symbolic interactionist perspectives on education?

According to functionalists, education has both manifest functions (socialization, transmission of culture, multicultural education, social control, social placement, and change and innovation) and latent functions (keeping young people off the streets and out of the job market, matchmaking and producing social networks, and creating a generation gap). From a conflict perspective, education is used to perpetuate class, racial–ethnic, and gender inequalities through tracking, ability grouping, and a hidden curriculum that teaches subordinate groups conformity and obedience. Symbolic interactionists examine classroom dynamics and study ways in which practices such as labeling may become a self-fulfilling prophecy for some students.

## LO3 What are some major problems in U.S. elementary and secondary schools?

Most educational funds come from state legislative appropriations and local property taxes. In difficult economic times, this means that schools must do without the necessary funds to provide students with teachers, supplies, and the best educational environment for learning. High dropout rates, racial segregation and resegregation, and equalizing educational opportunities for students are also among the many pressing issues facing U.S. public education today.

## LO4 How are problems in U.S. higher education related to larger issues in the United States and the world?

The funding of higher education is a central problem linked to crises in national and global economies. The problem of increasing costs of higher education for students is compounded by state budget shortfalls, which have caused funding for public higher education to be slashed. Declining state and federal support has become a major concern for colleges and universities because as enrollments drop, along with financial support, these institutions will have to find new sources of revenue, sharply reduce expenses, and rework administrative costs. The global economy also has a significant influence on the nature of higher education in contemporary societies. Other issues, such as the underrepresentation of minorities as students, a lack of faculty diversity, and problems of prejudice and discrimination are also linked to larger inequalities in society.

## LO5 What is religion, and what are its key components?

Religion is a social institution composed of a unified system of beliefs, symbols, and rituals, based on some sacred or supernatural realm, that guides human behavior, gives meaning to life, and unites believers into a community.

## LO6 What are the major theoretical perspectives on religion?

According to functionalists, religion has three important functions in any society: (1) providing meaning and purpose to life, (2) promoting social cohesion and a sense of belonging, and (3) providing social control and support for the government. From a conflict perspective, religion can have negative consequences in that the capitalist class uses religion as a tool of domination to mislead workers about their true interests. However, Weber believed that religion could be a catalyst for social change. Symbolic interactionists focus on a micro-level analysis of religion, examining the meanings that people give to religion and that they attach to religious symbols in their everyday life. According to the rational choice perspective, religious persons and organizations offer a variety of religions and religious products to consumers who shop around for religious theologies, practices, and communities that best suit them.

## LO7 What are the major types of religious organization?

Religious organizations can be categorized as ecclesia, churches, denominations, sects, and cults (now frequently referred to as new religious movements or NRMs).

## LO8 What are some major trends in U.S. education and religion in the twenty-first century?

Major trends in education include continuing debates about how schools should be funded, what should be taught, and how student learning and other outcomes should be assessed. The future of education may hinge on which elected officials are able to get their agendas for pre-kindergarten through college passed by state legislatures and the U.S. Congress. For-profit institutions will also play an increasing role in the future of this important social institution. One of the most important trends in religion has been the ongoing debate over secularization—the process by which religious beliefs, practices, and institutions lose their significance in society and nonreligious values, principles, and institutions take their place. Because religion meets many needs that no other institution can provide, it will continue as a major social institution.

# key TERMS

animism 382

church 389

civil religion 385

credentialism 370

cult 390

cultural capital 367

denomination 389

ecclesia 389

education 363

fundamentalism 392

hidden curriculum 369

profane 380

religion 380

sacred 380

sect 390

secularization 382

tracking 368

# questions for CRITICAL THINKING

1. Why does so much controversy exist over what should be taught in U.S. public schools?

2. How are the values and attitudes you learned from your family reflected in your beliefs about education and religion?

3. How would you design a research project to study the effects of civil religion on everyday life? What kind of data would be most accessible?

4. If Durkheim, Marx, and Weber were engaged in a discussion about education and religion, on what topics might they agree? On what topics would they disagree?

# 13 POLITICS AND THE ECONOMY IN GLOBAL PERSPECTIVE

## LEARNING OBJECTIVES

**LO1** Define the terms *power* and *authority*, and distinguish among traditional, charismatic, and rational–legal authority.

**LO2** Compare the major political systems found around the world.

**LO3** Describe the pluralist model of the U.S. power structure and note the part that special interest groups and political action committees play in this approach.

**LO4** Explain how political parties shape the government and identify recent trends in political participation and voter apathy.

**LO5** Describe the primary function of the economy and distinguish among preindustrial, industrial, and postindustrial economies.

**LO6** Discuss the key characteristics of capitalism, socialism, and mixed economies.

**LO7** Describe professions and explain how they differ from other occupations.

**LO8** Explain problems that are associated with contingent work and the underground (informal) economy.

© iStockphoto.com/best-photo

# Experiencing Politics and History at a Presidential Inauguration

**Tens of thousands of college students were present in Washington, D.C.,** when Barack Obama was publicly sworn in for his second term as president of the United States on January 21, 2013.

Thousands of students attended the 2013 presidential inauguration. What do you think college students like you might learn about the political process by observing such a historic event?

Millions of others watched the inauguration live on TV, computers, or smartphones and tablets that were using the latest apps. Among those was George Washington University senior Jamie Blynn, who blogged about how much more exciting it was to attend the event in person, as compared to watching the previous (2009) inauguration on an outdated TV in her twelfth-grade English classroom:

Luckily, we pushed our way through the hordes of people . . . [to] the entrances to the National Mall, and somehow stumbled upon the perfect location just minutes before Sen. Chuck Schumer kicked off the ceremony. . . . In the crowd, we stood among hundreds of thousands of people from [all] around the country, all of us waving the same American flags proudly in the air. [The patriotism] was like nothing I have ever experienced before; we all came from different races, ethnicities, and all-around different backgrounds, but we could all find common ground behind the flag. . . . This was a great end to . . . my college career.

—Jamie Blynn (2013)

College students Jamie Blynn and Kevin Kucharshi both believed that it was important to be at a presidential inauguration—to be a part of history and to see the results of the democratic process of electing public officials. Such student involvement in politics and government stands in sharp contrast to what we often hear about voter apathy and lack of engagement among young people. But, media reports aside, everyone is not apathetic: Many college students care about their nation and what their own future holds in the United States and the larger global community.

The political landscape has been vastly modified in a very short period of time as a result of instantaneous social networking and websites. Only a few years ago, voters learned about political candidates from stump speeches, newspapers, and the evening news on television. In the second decade of the twenty-first century, millions of people of all ages sign on to Facebook, Twitter, and other social media sites to follow their favorite candidates and learn more about their messages. Do people actually vote for specific candidates or engage in political activism because they have been influenced by the media? This is an even more significant question in light of the nature of both traditional and social media. And the line between traditional and more-contemporary media has blurred rapidly. Most traditional media outlets, including television, radio, newspaper, and other print media, have adopted social media into their reporting so that the two flow together and are often intertwined. Television anchors give play-by-play accounts of a political event, while viewers comment on Twitter and/or Facebook. Like these rapid changes in the political-media landscape, the U.S. economy has also seen dramatic changes that have affected the lives of people and changed the economy as a social institution.

Student media representatives were also present in Washington: Multimedia reporter Kevin Kucharshi, Yale class of 2015, provided students with eyewitness accounts of the inauguration and students' perceptions about the event and the future of the United States in his YouTube documentary *A Part of History: College Students and the 2013 Inauguration.*

## How Much Do You Know About Politics and the Media?

| True | False | |
|---|---|---|
| T | F | 1. More than 80 percent of American adults use the Internet, and many of them also participate in social networking sites where political information is available. |
| T | F | 2. Studies have found that most users of social networking sites find these sites to be "very important" or "somewhat important" in keeping up with political news or discussing political issues with others. |
| T | F | 3. Persons who describe their political beliefs as conservative are more likely to use social networking sites than those who describe themselves as "moderate" or "liberal." |
| T | F | 4. Persons using social networking sites are often surprised about what their friends have posted about politics because they assumed the friends' political beliefs were different from what their statements indicate. |
| T | F | 5. Young adults and African Americans are more likely than white Americans |

| True | False | |
|---|---|---|
| | | (non-Hispanic) and Hispanics to see social networking sites as important for keeping up with political news, recruiting people to get involved, finding others with similar views, and debating or discussing political issues with others. |
| T | F | 6. Many social networking site users indicate that they have changed their political views based on reading posts about specific issues. |
| T | F | 7. Because more media users have turned to social media for information about politics, traditional print and television journalists have taken a more active role in conducting investigations into the backgrounds and political platforms of candidates. |
| T | F | 8. The number of viewers of local TV news programming has decreased each year since 2007, even in presidential election years. |

Answers on page 402.

In twenty-first-century America, the issue of politics and the global economy is a hot topic for concerned people because we live in an age of discord regarding some decisions that have been made by our country's political and business leaders. Sociologists are concerned about how the social institutions of politics and the economy operate and how the decisions made in these social arenas affect people's everyday lives. In this chapter we discuss the intertwining nature of contemporary politics, the economy, and the media. Before reading on, test your knowledge of politics and the media by taking the "Sociology and Everyday Life" quiz.

## Politics, Power, and Authority

*Politics* **is the social institution through which power is acquired and exercised by some people and groups.** In contemporary societies the government is the primary political system. *Government* **is the formal organization that has the legal and political authority to regulate the relationships among members of a society and between the society and those outside its borders.** Some social analysts refer to the government as the *state*—**the political entity that possesses a**

**politics** the social institution through which power is acquired and exercised by some people and groups.

**government** the formal organization that has the legal and political authority to regulate the relationships among members of a society and between the society and those outside its borders.

**state** the political entity that possesses a legitimate monopoly over the use of force within its territory to achieve its goals.

401

# ANSWERS to the Sociology Quiz on Politics and the Media

**1. True.** Despite the "digital divide" (which means unequal access to the Internet based on the interactivity of race and income), 80 percent of adults in the United States indicate in studies that they use the Internet, and many are also social networking site users.

**2. False.** Although social networking sites contain political news, only about one-third of users indicated such sites were "very" or "somewhat" important in helping them keep up with information about the 2012 presidential election; only about one-quarter of users surveyed indicated that they used these sites to debate or discuss political issues with others.

**3. False.** Persons who describe their political beliefs as conservative were less likely to use social media than those who describe themselves as "moderate" or "liberal." Only 49 percent of conservatives use social networking sites, as compared to 60 percent of liberals and 61 percent of moderates. Although the figures change slightly, the differences hold true for the overall U.S. population generally and Internet users specifically.

**4. True.** About 38 percent of social networking site users indicate that they were surprised about their friends' political beliefs, and 18 percent have blocked, unfriended, or hidden someone on a site because of what the other person posted about politics.

**5. True.** Young adults between the ages of 18 and 29 and African Americans are more likely to indicate that social media are important for politics than white (non-Hispanic) Americans and Hispanics.

**6. False.** The vast majority of social networking users do not indicate that they have changed their political views or activity based on information from the sites. However, some who listed themselves as Democrats or liberals stated that they became more actively interested in a political issue as a result of involvement in social media.

**7. False.** Partly as a result of substantial cuts in newsroom personnel, media campaign reporters are now less involved in investigative journalism and instead primarily broadcast statements and assertions set forth by political candidates and their managers without interpreting or contextualizing these messages.

**8. True.** Audiences for local TV news programs have continued to decline over the past decade, even in years (such as 2012) in which a presidential election is held.

Sources: Based on Pew Research Center, 2012b, 2013c.

---

**legitimate monopoly over the use of force within its territory to achieve its goals.**

Whereas political science focuses primarily on power and its distribution in different types of political systems, *political sociology* is the area of sociology that examines politics and the government. Political sociology primarily focuses on the *social circumstances* of politics and explores how the political arena and its actors are intertwined with social institutions such as the economy, religion, education, and the media.

What is the relationship between politics and media? Recent research suggests that this relationship is dramatically changing in an era where various forms of media are omnipresent twenty-four hours a day, seven days a week, or "24/7." As early as 1998, sociologist Michael Parenti argued that the media distort—either intentionally or unintentionally—the information they provide to citizens. According to Parenti, the media use their power to influence public opinion in favor of management over labor, corporations over those who criticize them, affluent whites over racial and ethnic minorities, political officials over protestors, and free-market capitalists over those who are in favor of public-sector development. Nearly two decades later, if we look at the overall relationship among politics, government, and both the "traditional" media and the rapidly advancing social media, we can see that some of Parenti's assertions remain relevant regarding the use, distribution, and possible abuse of power. According to a Pew Research Center Project study (2013c), the role of social media such as Twitter and Facebook grew exponentially in influencing voters in the 2012 U.S. presidential election while the part journalists played was much smaller than in previous elections.

## Power and Authority

**LO1** *Power* **is the ability of persons or groups to achieve their goals despite opposition from others** (Weber, 1968/1922). Through the use of persuasion, authority, or force, some people are able to get others to acquiesce to their demands. Consequently, power is a *social relationship* that involves both leaders and followers. Power is also a dimension in the structure of social stratification. Persons in positions of power control valuable resources of society—including wealth, status, comfort, and safety—and are able to direct the actions of others while protecting and enhancing the privileged social position of their class (Domhoff, 2014). And, according to the sociologist G. William Domhoff (2002), the media tend to reflect "the biases of those with access to them—corporate leaders, government officials, and policy experts." The Pew Research Center (2013c) study concluded that in the 2012 presidential election, campaign reporters acted primarily as "megaphones," rather than as "investigators," in reporting the assertions that were put forward by candidates and other political participants. As a result, there was less effort on the part of journalists to interpret, contextualize, and sometimes verify statements that were made by the opposing parties. Reports such as this raise an important question: What effects might power shifts in the media have on our perceptions of what is taking place in U.S. politics and government?

Shifting to the bigger picture: What about power on a global basis? Although the most basic form of power is physical violence or force, most political leaders do not want to base their power on force alone. Instead, they seek to legitimize their power by turning it into *authority*—**power that people accept as legitimate rather than coercive.**

## Ideal Types of Authority

Who is most likely to accept authority as legitimate and adhere to it? People have a greater tendency to accept authority as legitimate if they are economically or politically dependent on those who hold power. They may also accept authority more readily if it reflects their own beliefs and values. Weber's outline of three *ideal types* of authority—traditional, charismatic, and rational–legal—shows how different bases of legitimacy are tied to a society's economy.

**Traditional Authority** According to Weber, *traditional authority* **is power that is legitimized on the basis of long-standing custom.** In preindustrial societies the authority of traditional leaders, such as kings, queens, pharaohs, emperors, and religious dignitaries, is usually grounded in religious beliefs and custom. For example, British kings and queens have historically traced their authority from God. Members of subordinate classes obey a traditional leader's edicts out of economic and political dependency and sometimes personal loyalty. However, as societies industrialize, traditional authority is challenged by a more complex division of labor and by the wider diversity of people who now inhabit the area as a result of high immigration rates.

Gender, race, and class relations are closely intertwined with traditional authority. Political scientist Zillah R. Eisenstein (1994) suggests that *racialized patriarchy*—the continual interplay of race and gender—reinforces traditional structures of power in contemporary societies. According to Eisenstein (1994: 2), "Patriarchy differentiates women from men while privileging men. Racism simultaneously differentiates people of color from whites and privileges whiteness. These processes are distinct but intertwined."

**Charismatic Authority** *Charismatic authority* **is power legitimized on the basis of a leader's exceptional personal qualities or the demonstration of extraordinary insight and accomplishment that inspire loyalty and obedience from followers.** Charismatic leaders may be politicians, soldiers, or entertainers, among others.

Charismatic authority tends to be temporary and relatively unstable; it derives primarily from individual leaders (who may change their minds, leave, or die) and from an administrative structure usually limited to a small number of faithful followers. For this reason, charismatic authority often becomes routinized. The *routinization of charisma* **occurs when charismatic authority is succeeded**

---

**power** according to Max Weber, the ability of persons or groups to achieve their goals despite opposition from others.

**authority** power that people accept as legitimate rather than coercive.

**traditional authority** power that is legitimized on the basis of long-standing custom.

**charismatic authority** power legitimized on the basis of a leader's exceptional personal qualities or the demonstration of extraordinary insight and accomplishment that inspire loyalty and obedience from followers.

**routinization of charisma** the process by which charismatic authority is succeeded by a bureaucracy controlled by a rationally established authority or by a combination of traditional and bureaucratic authority.

by a bureaucracy controlled by a rationally established authority or by a combination of traditional and bureaucratic authority (Turner, Beeghley, and Powers, 2007). According to Weber (1968/1922: 1148), "It is the fate of charisma to recede . . . after it has entered the permanent structures of social action."

**Rational–Legal Authority** According to Weber, *rational–legal authority* is power legitimized by law or written rules and regulations. Rational–legal authority—also known as *bureaucratic authority*—is based on an organizational structure that includes a clearly defined division of labor, hierarchy of authority, formal rules, and impersonality. Power is legitimized by procedures; if leaders obtain their positions in a procedurally correct manner (such as by election or appointment), they have the right to act (Figure 13.1).

Rational–legal authority is held by elected or appointed government officials and by officers in a formal organization. However, authority is invested in the *office,* not in the *person* who holds the office. For example, although the U.S. Constitution grants rational–legal authority to the office of the presidency, a president who fails to uphold the public trust may be removed from office. In contemporary society the media may play an important role in bringing to light allegations about presidents or other elected officials, including President Richard M. Nixon in the 1970s Watergate investigation, the late-1990s sex scandal involving President Bill Clinton, and the more-recent ethics charges of money laundering involving members of Congress during the Obama administration.

In a rational–legal system, the governmental bureaucracy is the apparatus responsible for creating and enforcing rules in the public interest. Weber believed that rational–legal authority was the only means to attain efficient, flexible, and competent regulation under a rule of law.

**FIGURE 13.1** Max Weber's three types of global authority are shown here in global perspective. Pope Francis is an example of traditional authority sanctioned by custom. Mother Teresa exemplifies charismatic authority, for her leadership was based on personal qualities. The U.S. Supreme Court represents rational–legal authority, which depends upon established rules and procedures.

**404**

## Weber's Three Types of Authority

|  | Description | Examples |
|---|---|---|
| **Traditional** | Legitimized by long-standing custom | Patrimony (authority resides in traditional leader supported by larger social structures, as in old British monarchy) |
|  | Subject to erosion as traditions weaken | Patriarchy (rule by men occupying traditional positions of authority, as in the family) |
| **Charismatic** | Based on leader's personal qualities | Napoleon |
|  |  | Adolf Hitler |
|  | Temporary and unstable | Martin Luther King Jr. |
|  |  | César Chávez |
|  |  | Mother Teresa |
| **Rational–Legal** | Legitimized by rationally established rules and procedures | Modern British Parliament |
|  | Authority residing in the office, not the person | U.S. presidency, Congress, federal bureaucracy |

Weber's three types of authority are summarized in the Concept Quick Review.

# Political Systems in Global Perspective

**LO2** Political systems as we know them today have evolved slowly. In the earliest societies, politics was not an entity separate from other aspects of life. Political institutions first emerged in agrarian societies as they acquired surpluses and developed greater social inequality. Elites took control of politics and used custom or traditional authority to justify their position. When cities developed circa 3500–3000 B.C.E., the *city-state*—a city whose power extended to adjacent areas—became the center of political power.

Nation-states as we know them began to develop in Europe between the twelfth and fifteenth centuries (see Tilly, 1975). A *nation-state* is a unit of political organization that has recognizable national boundaries and whose citizens possess specific legal rights and obligations. Nation-states emerge as countries develop specific geographic territories and acquire greater ability to defend their borders. Improvements in communication and transportation make it possible for people in a larger geographic area to share a common language and culture. As charismatic and traditional authority are superseded by rational–legal authority, legal standards come to prevail in all areas of life, and the nation-state claims a monopoly over the legitimate use of force.

The U.S. State Department recognizes 195 independent nation-states throughout the world (U.S. Department of State, 2013). Today, everyone is born, lives, and dies under the auspices of a nation-state. Four main types of political systems are found in nation-states: monarchy, authoritarianism, totalitarianism, and democracy.

## Monarchy

*Monarchy* **is a political system in which power resides in one person or family and is passed from generation to generation through lines of inheritance** (Figure 13.2). Monarchies are most common in agrarian societies and are associated with traditional authority patterns. However, the relative power of

**rational–legal authority** power legitimized by law or written rules and regulations.

**monarchy** a political system in which power resides in one person or family and is passed from generation to generation through lines of inheritance.

**FIGURE 13.2** Through its many ups and downs, the British royal family has remained a symbol of Great Britain's monarchy, today headed by Queen Elizabeth. Monarchies typically pass power from generation to generation, and Queen Elizabeth's grandsons, Prince Harry and Prince William (shown here, at his wedding), represent the future of the royal family's rule.

monarchs has varied across nations, depending on religious, political, and economic conditions.

*Absolute monarchs* claim a hereditary right to rule (based on membership in a noble family) or a divine right to rule (a God-given right to rule that legitimizes the exercise of power). In limited monarchies, rulers depend on powerful members of the nobility to retain their thrones. Unlike absolute monarchs, *limited monarchs* are not considered to be above the law. In *constitutional monarchies,* the royalty serve as symbolic rulers or heads of state while actual authority is held by elected officials in national parliaments. In present-day monarchies such as the United Kingdom, Sweden, Spain, and the Netherlands, members of royal families primarily perform ceremonial functions. In the United Kingdom, for example, the media often focus large amounts of time and attention on the royal family but concentrate on the personal lives of its members.

## Authoritarianism

***Authoritarianism* is a political system controlled by rulers who deny popular participation in government.** A few authoritarian regimes have been absolute monarchies whose rulers claimed a hereditary right to their position and where the final decision on issues rests with the monarch. Today,

Saudi Arabia, Qatar, and Swaziland are examples of absolute monarchies. In *dictatorships,* power is gained and held by a single individual. Pure dictatorships are rare; all rulers need the support of the military and the backing of business elites to maintain their position. *Military juntas* result when military officers seize power from the government, as has happened in recent decades in Argentina, Chile, and Haiti. Currently, authoritarian regimes exist in Cuba and in the People's Republic of China. Authoritarian regimes seek to control the media and to suppress coverage of any topics or information that does not reflect upon the regime in a favorable light.

## Totalitarianism

***Totalitarianism* is a political system in which the state seeks to regulate all aspects of people's public and private lives.** Totalitarianism relies on modern technology to monitor and control people; mass propaganda and electronic surveillance are widely used to influence people's thinking and control their actions. The World War II Nazi Party in Germany is an example of a totalitarian regime: Military leaders sought to control all aspects of national life, not just government operations. Other examples include the former Soviet Union, with vestiges of this approach remaining in contemporary Russian leadership.

To keep people from rebelling, totalitarian governments enforce conformity. People are denied the right to assemble for political purposes, access to information is strictly controlled, and secret police enforce compliance, creating an environment of constant fear and suspicion.

Many nations do not recognize totalitarian regimes as being the legitimate government of a particular country. Afghanistan in the year 2001 was an example. As the war on terrorism began in the aftermath of the September 11 terrorist attacks on the United States, many people developed a heightened awareness of the Taliban regime, which ruled most of Afghanistan and was engaged in fierce fighting to capture the rest of the country. The Taliban regime maintained absolute control over the Afghan people in most of that nation, including requiring that all Muslims take part in prayer five times each day and that women wear the hijab (veil). Since military action commenced in Afghanistan, most of what U.S. residents have known about the Taliban, the war, and its aftermath has been based on media reports and "expert opinions" expressed on television and the Internet.

## Democracy

**Democracy is a political system in which the people hold the ruling power either directly or through elected representatives.** The literal meaning of *democracy* is "rule by the people" (from the Greek words *demos,* meaning "the people," and *kratein,* meaning "to rule"). In an ideal-type democracy, people would actively and directly rule themselves. *Direct participatory democracy* requires that citizens be able to meet together regularly to debate and decide the issues of the day. Because there are approximately 319 million people in the United States today, it would be impossible for everyone to come together in one place for a meeting.

In countries such as the United States, Canada, Australia, and the United Kingdom, people have a voice in the government through *representative democracy,* **whereby citizens elect representatives to serve as bridges between themselves and the government.** The U.S. Constitution requires that each state have two senators and a minimum of one member in the House of Representatives. The number of voting representatives in the House (435 seats) has not changed since the apportionment following the 1910 census; however, those 435 seats are reapportioned based on an increase or decrease in a state's population as shown in census data gathered every ten years.

In a representative democracy, elected representatives are supposed to convey the concerns and interests of those they represent, and the government is expected to be responsive to the wishes of the people. Elected officials are held accountable to the people through elections. However, representative democracy is not always equally accessible to all people in a nation. Throughout U.S. history, members of subordinate racial–ethnic groups have been denied full participation in the democratic process. Gender and social class have also limited some people's democratic participation.

Even representative democracies are not all alike. As compared to the winner-takes-all elections in the United States, which are usually decided by the candidate who wins the most votes, the majority of European elections are based on a system of proportional representation, meaning that each party is represented in the national legislature according to the proportion of votes that party received. For example, a party that won 40 percent of the vote would receive 40 seats in a 100-seat legislative body, and a party receiving 20 percent of the votes would receive 20 seats.

# Perspectives on Power and Political Systems

Is political power in the United States concentrated in the hands of the few or distributed among the many? Sociologists and political scientists have suggested many different answers to this question; however, two prevalent models of power have emerged: pluralist and elite.

## Functionalist Perspectives: The Pluralist Model

**LO3** The pluralist model is rooted in a functionalist perspective which assumes that people share a consensus on central concerns, such as freedom and protection from harm, and

**authoritarianism** a political system controlled by rulers who deny popular participation in government.

**totalitarianism** a political system in which the state seeks to regulate all aspects of people's public and private lives.

**democracy** a political system in which the people hold the ruling power either directly or through elected representatives.

**representative democracy** a form of democracy whereby citizens elect representatives to serve as bridges between themselves and the government.

(a)

(b)

**FIGURE 13.3** GOVERNMENT FROM A FUNCTIONALIST PERSPECTIVE

From the functionalist perspective, government serves important functions that no other institution can fulfill. Contemporary functionalists identify four main functions: (a) maintaining law and order, (b) planning and directing society, (c) meeting social needs, and (d) handling international relations, including warfare.

© Cengage Learning

(c)

(d)

that the government serves important functions no other institution can fulfill. According to Emile Durkheim (1933/1893), the purpose of government is to socialize people to be good citizens, to regulate the economy so that it operates effectively, and to provide necessary services for citizens. Contemporary functionalists state the four main functions as follows: (1) maintaining law and order, (2) planning and directing society, (3) meeting social needs, and (4) handling international relations, including warfare (see Figure 13.3).

But what happens when people do not agree on specific issues or concerns? Functionalists suggest that divergent viewpoints lead to a system of political pluralism, in which the government functions as an arbiter between competing interests and viewpoints. According to the ***pluralist model,*** **power in political systems is widely dispersed throughout many competing interest groups** (Dahl, 1961).

In the pluralist model the diverse needs of women and men, people of all religions and racial–ethnic backgrounds, and the wealthy, middle class, and poor are met by political leaders who engage in a process of bargaining, accommodation, and compromise. Competition among leadership groups in government, business, labor, education, law, medicine, and consumer organizations, among others, helps prevent abuse of power by any one group. Everyday people can influence public policy by

voting in elections, participating in existing special interest groups, or forming new ones to gain access to the political system. In sum, power is widely dispersed, and leadership groups that wield influence on some decisions are not the same groups that may be influential in other decisions (Dye, Zeigler, and Schubert, 2012).

**Special Interest Groups** *Special interest groups* are political coalitions made up of individuals or groups that share a specific interest they wish to protect or advance with the help of the political system. Examples of special interest groups include the AFL-CIO (representing the majority of labor unions) and public interest or citizens' groups such as the American Conservative Union and Population Connection (formerly Zero Population Growth).

What purpose do special interest groups serve in the political process? According to some analysts, special interest groups help people advocate their own interests and further their causes. Broad categories of special interest groups include banking, business, education, energy, the environment, health, labor, persons with a disability, religious groups, retired persons, women, and those espousing a specific ideological viewpoint; obviously, many groups overlap in interests and membership. Special interest groups are also referred to as *pressure groups* (because they put pressure on political leaders) or *lobbies.* Lobbies are often referred to in terms of the organization they represent or the single issue on which they focus—for example, the "gun lobby" and the "dairy lobby." The people who are paid to influence legislation on behalf of specific clients are referred to as *lobbyists.*

Over the past fifty years, special interest groups have become more involved in "single-issue politics," in which political candidates are often supported or rejected solely on the basis of their views on a specific issue—such as abortion, gun control, LGBT rights, or the environment. Single-issue groups derive their strength from the intensity of their beliefs; leaders have little room to compromise on issues.

**Political Action Committees** For many years, the funding of lobbying efforts has been a hotly debated issue. Numerous attempts have been made to limit campaign contributions and expenditures to ensure that wealthy and influential individuals and organizations are not able to silence the voices of people who do not have equal resources. Reforms in campaign finance laws in the 1970s set limits on direct contributions to political candidates and led to the creation of *political action committees* **(PACs)—organizations of special interest groups that solicit contributions from donors and fund campaigns to help elect (or defeat) candidates based on their stances on specific issues.** As the cost of running for political office has skyrocketed, candidates have relied more on PACs for financial assistance. Advertising, staff, direct-mail operations, telephone banks, computers, consultants, travel expenses, office rentals, and other expenses incurred in political campaigns make PAC money vital to candidates.

Some PACs represent the "public interest" and ideological interest groups such as LGBT rights or the National Rifle Association. Other PACs represent the capitalistic interests of large corporations. Realistically, PACs do not represent members of the least-privileged sectors of society, such as food stamp PACs.

Various efforts have been made to curb excessive spending in political elections. As an outgrowth of record-setting campaign spending in the 1996 national election, Congress passed the 2002 McCain–Feingold campaign finance law prohibiting "soft money" contributions (which are made outside the limits imposed by federal election law). This law pertains to federal elections only and does not include state or local elections.

But a shift occurred in 2010, when the U.S. Supreme Court ruled in *Citizens United v. Federal Election Commission* that corporations and other organizations could bypass existing spending limits by giving unlimited amounts to "independent groups" that support candidates (but not to the candidates themselves). In this decision the Court struck down a provision of the McCain–Feingold Act that prohibited both for-profit and not-for-profit corporations and unions from broadcasting "electioneering communications," defined as a broadcast, cable, or satellite communication that mentioned a candidate within sixty days of a general election or thirty days of a primary. The Court's decision was made in terms of First Amendment rights, based on the assumption that a decision to spend money in support of a political cause or candidate was similar to giving a speech or carrying a campaign sign and thus protected by the First Amendment (Toobin, 2011). This controversial decision was criticized for granting lobbyists for special interests even more power than they previously held in Washington, while average Americans were further downgraded in their efforts to support a political candidate. Clearly, the Court's decision contributed to the rise of *super PACs,* which were major contributors in the 2012 elections.

**Rise of the Super PACs** Super PACs came into existence following two important legal cases. In *SpeechNow.org v. Federal Election Commission,* a federal court ruled that it was a violation of the First Amendment to establish limitations on individual contributions to independent organizations that seek to influence elections. In the second case, as previously discussed, the U.S. Supreme Court decided that it was a violation of the U.S. Constitution to establish spending limits for corporations and unions. This decision opened the door for corporations, unions, and other organizations to bypass existing spending limits and give unlimited amounts of money to various "independent groups" that support candidates, but not to candidates or specific political parties.

What is the difference between a PAC and a super PAC? Simply stated, PACs have restrictions on who may contribute and how much they can give, whereas super PACs are unrestricted in who can give and how much they may donate. As a result, large sums of money flow into candidates' campaign

**pluralist model** an analysis of political systems that views power as being widely dispersed throughout many competing interest groups.

**political action committees** organizations of special interest groups that solicit contributions from donors and fund campaigns to help elect (or defeat) candidates based on their stances on specific issues.

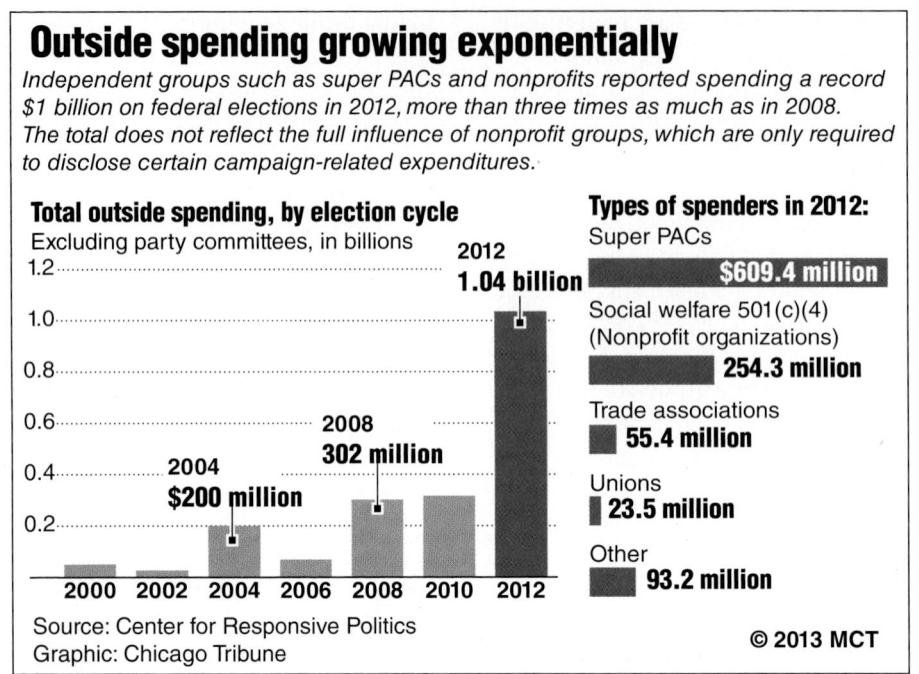

**FIGURE 13.4** INFLUENCE OF SUPER PACS IN THE 2012 PRESIDENTIAL ELECTION

Source: Center for Responsive Politics, 2013.

coffers from entities that have specific agendas. Super PACs can contribute money to candidates who will uphold the best interests of specific interest groups, corporations, or unions, or they can provide funds to defeat candidates whose political platforms will not benefit the major contributors of the super PAC. The presidential campaign in 2012 was the first election in which super PACs were involved in supporting campaigns, as shown in Figure 13.4. It remains to be seen what the longer-term effects of these massive contributions from corporations, unions, and other major donors will be on U.S. elections and the political process.

Based on a functionalist approach, some special interest groups and political action committees continue to represent the broader needs and interests of many individuals, groups, and organizations in society. However, what happens when they become narrowly focused on the issues that benefit only a very small percentage of elites at the top of the pyramid, as has recently happened with some super PACs? At this juncture we must turn to conflict perspectives and elite models for a better explanation of how politics and government operate when power is more concentrated.

## Conflict Perspectives: Elite Models

Although conflict theorists acknowledge that the government serves a number of important purposes in society, they assert that government exists for the benefit of wealthy or politically powerful elites who use the government to impose their will on the masses. According to the **elite model, power in political systems is concentrated in the hands of a small group of elites, and the masses are relatively powerless.** The pluralist model and the elite model are compared in Figure 13.5.

Contemporary elite models are based on the assumption that decisions are made by the elites, who agree on the basic values and goals of society. However, the needs and concerns of the masses are not often given full consideration by those in the elite. According to this approach, power is highly concentrated at the top of a pyramid-shaped social hierarchy, and public policy reflects the values and preferences of the elite, not the preferences of the people (Dye, Zeigler, and Schubert, 2012).

**C. Wright Mills and the Power Elite** Who makes up the U.S. power elite? According to the sociologist C. Wright Mills (1959a), the *power elite* comprises leaders at the top of business, the executive branch of the federal government, and the military. Of these three, Mills speculated that the "corporate rich" (the highest-paid officers of the biggest corporations) are the most powerful because of their unique ability to parlay the vast economic resources at their disposal into political power (see "Sociology Works!"). At the middle level of the pyramid, Mills placed the legislative branch of government, special interest groups, and local opinion leaders. The

- Decisions are made on behalf of the people by leaders who engage in bargaining, accommodation, and compromise.

- Competition among leadership groups makes abuse of power by any one group difficult.

- Power is widely dispersed, and people can influence public policy by voting.

- Public policy reflects a balance among competing interest groups.

- Decisions are made by a small group of elite people.

- Consensus exists among the elite on the basic values and goals of society.

- Power is highly concentrated at the top of a pyramid-shaped social hierachy.

- Public policy reflects the values and preferences of the elite.

**FIGURE 13.5** PLURALIST AND ELITE MODELS
© Cengage Learning

bottom (and widest) layer of the pyramid is occupied by the unorganized masses, who are relatively powerless and are vulnerable to economic and political exploitation.

**G. William Domhoff and the Ruling Class** Sociologist G. William Domhoff (2002) asserts that, in fact, this nation has a *ruling class*—the corporate rich, who make up less than 1 percent of the U.S. population. Domhoff uses the term *ruling class* to signify a relatively fixed group of privileged people who wield sufficient power to constrain political processes and serve underlying capitalist interests. Although the power elite controls the everyday operation of the political system, who *governs* is less important than who *rules.*

According to Domhoff (2005), the upper class and the corporate rich influence politics in the following ways:

- "The rich" coalesce into a social upper class that has developed institutions by which the children of its members are socialized into an upper-class worldview and newly wealthy people are assimilated.
- Members of this upper class control corporations, which have been the primary mechanisms for

generating and holding wealth in the United States for upwards of 150 years now.
- In a network of nonprofit organizations, members of the upper class and hired corporate leaders not yet in the upper class shape policy debates.
- Members of the upper class, with the help of their high-level employees in for-profit and nonprofit institutions, are able to dominate the federal government.
- The rich, and corporate leaders, nonetheless claim to be relatively powerless.
- Working people have less power than in many other democratic countries, so they have little chance to influence the political process.

However, Domhoff emphasizes that people in the United States do have options because of their rights to free speech and to vote. Although the U.S. government may not always be responsive to the will of the majority of voters, citizens have been able to

**elite model** an analysis of political systems that views power in political systems as being concentrated in the hands of a small group of elites, and the masses are relatively powerless.

# Sociology WORKS!

## C. Wright Mills's Ideas About the Media: Ahead of His Time?

In a mass society the dominant type of communication is the formal media, and the [various audiences] become mere *media markets:* all those exposed to the contents of a given mass media.

>—In *The Power Elite*, the influential sociologist C. Wright Mills (1959a: 299) provided this early critique of the U.S. mass media and its effect on the general public.

During the mid-twentieth century, when Mills wrote these words, the press in the United States was often described as a "watchdog for the public." It was widely believed that journalists would reliably report what the public needed to know to stay informed, including any government and/or corporate dishonesty or ineptitude. However, Mills rejected this notion because he believed the media industry lulled individuals into complacency and persuaded them to accept the status quo rather than encouraging the audience member to become a "public" person who formulates his or her own opinion on a given topic.

Do the contemporary media inform people or lull them into complacency? Mills hoped that individuals would develop the capacity for critical judgment based on information they received from the media industry; however, he was doubtful that this would occur for two important sociological reasons: (1) media communication involved a limited number of people who communicated to a great number of others ("the masses"), and (2) audiences had no effective way of answering back, making mass communication a one-way process.

Mills made an important point when he stated that critical thinking is more likely to occur if public communications are organized in such a way that people have a chance to immediately and effectively respond to opinions expressed in public. When individuals are able to respond immediately and provide their own opinions on a topic, media communications become a two-way street. Clearly, Mills was ahead of his time. Social network sites, smartphones, text messaging, e-mail, and other rapid forms of communication now provide us with opportunities for media interactivity through

Alex Segre/Alamy

Karl Marx's beliefs about capitalism, socialism, and communism are widely known today, although it has been more than 100 years since his famous proclamation—as emblazoned on his tombstone outside London—"Workers of all lands unite."

which we can respond to media discourse and make our own thoughts known.

### reflect & analyze

Do you use social media to respond to ideas presented by political candidates and elected officials? How might we relate Mills's ideas to the contemporary role of the media industries in shaping political and economic "realities" around the world?

---

place restraints on some actions of wealthy elites to keep them from having an even greater influence on policy than they already possess.

Power elite models call our attention to a central concern in contemporary U.S. society: the ability of democracy and its ideals to survive in the context of the increasingly concentrated power held by capitalist oligarchies such as the media giants that we discuss in this chapter.

## The U.S. Political System

The U.S. political system is made up of formal elements, such as the legislative process and the duties of the president, and informal elements, such as the role of political parties in the election process. We now turn to an examination of these informal elements, including political parties, political socialization, and voter participation.

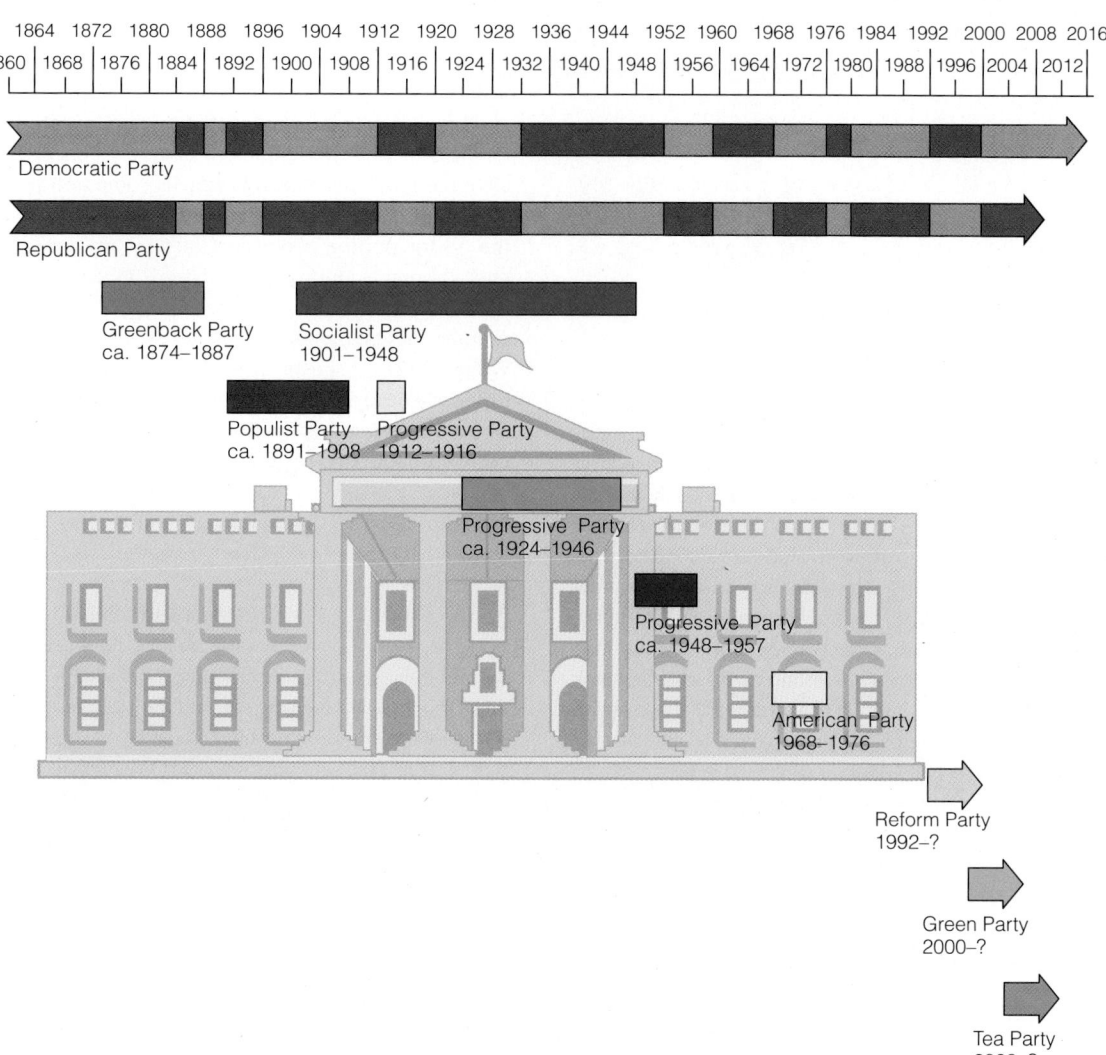

Party in control of White House

1864 1872 1880 1888 1896 1904 1912 1920 1928 1936 1944 1952 1960 1968 1976 1984 1992 2000 2008 2016
1860 1868 1876 1884 1892 1900 1908 1916 1924 1932 1940 1948 1956 1964 1972 1980 1988 1996 2004 2012

Democratic Party

Republican Party

Greenback Party
ca. 1874–1887

Socialist Party
1901–1948

Populist Party
ca. 1891–1908

Progressive Party
1912–1916

Progressive Party
ca. 1924–1946

Progressive Party
ca. 1948–1957

American Party
1968–1976

Reform Party
1992–?

Green Party
2000–?

Tea Party
2009–?

**FIGURE 13.6** MAJOR U.S. POLITICAL PARTIES

Despite recurring attempts by other groups to organize third parties, the Democratic and Republican parties have dominated national politics in the United States. Control of the presidency has alternated between these two parties since the Civil War.

Note: Three different "third parties" have gone by the name Progressive Party.

© Cengage Learning 2015

## Political Parties and Elections

**LO4** A *political party* is an organization whose purpose is to gain and hold legitimate control of government; it is usually composed of people with similar attitudes, interests, and socioeconomic status. A political party (1) develops and articulates policy positions, (2) educates voters about issues and simplifies the choices for them, and (3) recruits candidates who agree with those policies, helps those candidates win office, and holds the candidates responsible for implementing the party's policy positions. In carrying out these functions, a party may try to modify the demands of special interests, build a consensus that could win majority support, and provide simple and identifiable choices for the voters on election day. Political parties create a *platform*, a formal statement of the party's political positions on various social and economic issues.

Since the Civil War, the Democratic and Republican parties have dominated the U.S. political system (see Figure 13.6). Although one party may control the presidency for several terms, at some point the voters elect the other party's nominee, and control shifts.

**political party** an organization whose purpose is to gain and hold legitimate control of government.

413

## "The Life of Julia": The Internet, Twitter, and Continuing Political Battles

**D**id you meet "Julia" during the 2012 presidential campaign? If not, here's the rundown: The Obama reelection campaign created a new interactive Web slideshow that followed "Julia," a fictional character, to show how President Obama's vision for this country would benefit millions of people, particularly women. The slideshow starts when Julia is three years old and going to a government Head Start program (because Obama supported the program while Mitt Romney, his opponent, had threatened it with drastic cuts). Julia's high school participates in the president's "Race to the Top" program, which helps her enter college, get cheaper student loans, get a Pell Grant, and receive a tax credit. At age 22, Julia can count on being covered by her parents' health insurance because of President Obama's Affordable Care Act, and when she reaches age 67, Julia is happily retired and receiving her Social Security and Medicare benefits (which the president has protected for Julia and millions of other Americans like her) (based on Gavora, 2012, and Scherer, 2012b).

This slideshow ad was created to demonstrate sharp differences in what people might expect if President Obama were reelected to a second term, as compared to the election of his opponent, Mr. Romney. Some people applauded the ad for putting a personal face on the president's accomplishments during his first term in office and for reminding potential voters that they might lose important programs and benefits if a major sea change occurred in the White House. However, the ad also produced a backlash, especially from Republicans, who tweeted to #Julia that she was a bad example of people who are dependent on government handouts. For example, Michelle Malkin wrote, "Hi @Barack Obama, I will read Life of #Julia to my kids to show them how NOT to live their lives—tethered to Nanny State" (qtd. in Scherer, 2012b). Others tweeted that "Julia needs a job" (Caldwell, 2012).

The Internet and social media such as Twitter and Facebook have drastically changed politics, how candidates run for office, and how people respond to candidates and the political process. During much of the twentieth century, political parties were more extensively involved in announcing potential candidates and ensuring that their voices were heard. Supporters walked from door to door in neighborhoods encouraging people to vote; other grassroots activities included

How well do the parties measure up to the ideal-type characteristics of a political party? Although both parties have been successful in getting their candidates elected at various times, they generally do not meet the ideal characteristics for a political party because they do not offer voters clear policy alternatives. Moreover, the two parties are oligarchies, dominated by active elites who hold views that are further from the center of the political spectrum than are those of a majority of members of their party. As a result, voters in primary elections (in which the nominees of political parties for most offices other than president and vice president are chosen) may select nominees whose views are closer to the center of the political spectrum and further away from the party's own platform. Likewise, party loyalties appear to be declining among voters, who may vote in one party's primary but then cast their ballot in general elections without total loyalty to that party, or cast a "split-ticket" ballot (voting for one party's candidate in one race and another party's candidate in the next one).

Finally, the media have replaced the party as a means of political communication. Often, the candidate who wins does so as a result of media presentation, not the political party's platform. Candidates no longer need political parties to carry their message to the people (see "Politics in the Media").

### Discontent with the Current Political System and Parties

Although many individuals identify themselves as either Republicans (GOP) or Democrats, a growing number of people have expressed discontent with politicians and the existing U.S. political parties that have dominated the political system and may view themselves as independent from the traditional parties. In one national survey, for example, persons who self-identified as "independents" accounted for 39 percent of all those surveyed (Holyk, 2012). In 2012 the Democratic Party was near a record low in national poll ratings, as compared to prior presidential election years, and the

How has Facebook changed the face of politics?

town hall meetings and campaign buses traveling from town to town.

The popularity of television dramatically changed these campaign strategies, bringing about more-extensive efforts to mass "sell" political candidates to voters while soundly dismissing their opponents as unworthy to hold office. Then social media arrived, and election campaigns started to be played out in tweets—character messages that often become a war of words—or on Facebook pages where people could digitally connect to a campaign. In the 2012 election the Obama administration offered a free Facebook application that allowed "targeted sharing"—access to the "friends" of more than one million Obama backers. Supporters were asked to click a button to send their Facebook friends online content that asked them to register to vote, give money to the campaign, watch a video, or go to the polls and vote (Scherer, 2012a). Rather than having people travel from door to door making an appeal, which is slow and sometimes unproductive, with Facebook it is now possible to connect a political campaign through one person to more than 500 friends!

## reflect & analyze

**How will new media communications options affect political elections in the United States and other nations in the future? What problems can you foresee as social media such as Twitter and Facebook become even more prevalent in political elections?**

GOP was even further behind in ratings. Although the Tea Party movement initially came into the political arena with gusto, when its leaders tried to turn it into a full-fledged third political party, the Tea Party continued to slip in favorability in national polls (Holyk, 2012). Let's look briefly at the background of the Tea Party and the Green Party as case studies in discontent with current political parties.

**Tea Party: Past, Present, Future** In 2009 the Tea Party movement emerged to support more constitutionally limited government and to oppose various stimulus and bailout programs that use federal monies. The protesters refer to themselves as the "Tea Party" based on the Boston Tea Party, a 1773 protest by American colonists against "taxation without representation" by the British government because the colonists were not represented in the British Parliament but were required to pay taxes to that government. A demographic analysis reveals that the typical Tea Party supporter is white, male, married, and over

forty-five years of age. Most are registered Republicans, but they disagree with party leadership about various issues. Clearly, the Tea Party movement is pessimistic about the direction that its members think the United States has taken under the Obama administration, and the group wants to do something about it (Zernike and Thee-Brenan, 2010). However, the Tea Party suffered a setback in the 2012 presidential election when polls showed that only 32 percent rated the Tea Party favorably, as compared to a 49 percent favorable rating for the Democratic Party. By contrast, the Republican Party received only a 39 percent favorable rating (Holyk, 2012). As the Tea Party drifted downward, Barack Obama was reelected to a second term, and Republican candidates suffered losses in the U.S. Senate.

What does the future hold for the Tea Party? Some political analysts believe that the movement will grow in the future, largely populated by people who have strong beliefs about tax cuts, balanced budgets, and gun control. The Tea Party has created strong roots through online activism and strong fund-raising activities. Overall, it appears that the Tea Party is most popular among very conservative voters, and this group may continue to influence the GOP because a few of its key players were elected to the U.S. Congress.

**Green Party: Past, Present, Future** Although the roots of the Green Party started as early as the 1970s with the Values Party in New Zealand, the Green Party as we know it in the United States today was founded in 1996, when a number of state green parties decided to join together to express their discontent with the U.S. political process and provide a viable alternative to the existing parties. The Green Party's first presidential candidate was the consumer advocate Ralph Nader, who ran unsuccessfully in 1996 and 2000. In 2001 the Green Party of the United States was formed out of the older association of state organizations to help establish parties in all fifty states.

Today, the Green Party focuses on these ten key values: (1) grassroots democracy; (2) social justice and equal opportunity; (3) ecological wisdom; (4) nonviolence; (5) decentralization of wealth and power in social, political, and economic institutions; (6) community-based economics and economic justice; (7) feminism and gender equity; (8) respect for diversity; (9) personal and global responsibility; and (10) future focus and sustainability to protect valuable natural resources and develop sustainable economies that are not dependent on continual expansion for survival (GreenParty.org, 2013) (Figure 13.7). The stated purposes of the organization are to renew democracy without the support of corporate donors, to take a stand against super PACs and the big-money system of government, and to find viable solutions for real-world problems such as health care, corporate globalization, alternative energy, election reform, and wages for workers (GreenParty.org, 2013).

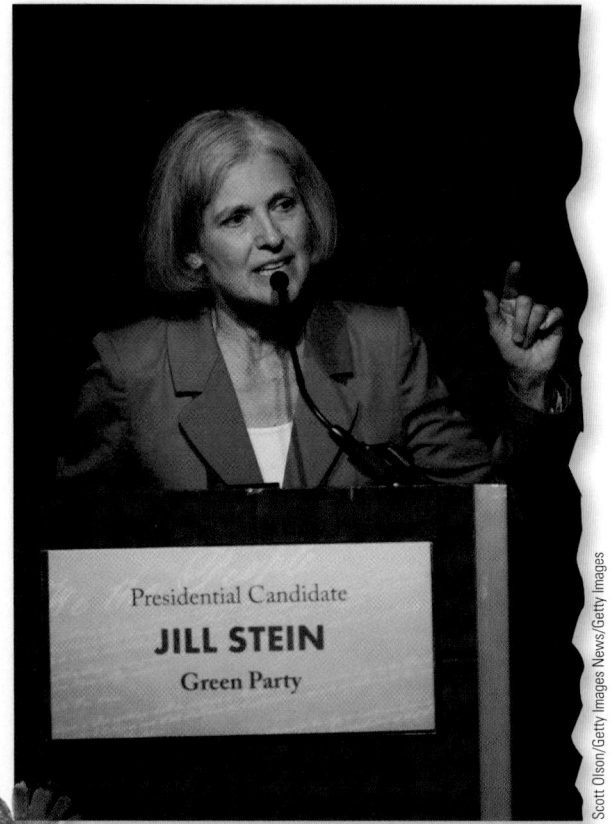

**FIGURE 13.7** Jill Stein was the Green Party candidate for president in the most recent U.S. presidential election. How does the Green Party differ from other political parties?

Scott Olson/Getty Images News/Getty Images

What does the future hold for the Green Party? In national elections the party has not fared well as of 2013, receiving very few votes for its candidate, Jill Stein, in the 2012 presidential election and currently holding no seats in the U.S. Congress. At the state level, the Green Party holds one seat in a state house of representatives but no state senate seats or governorships. In 2012 Greens held more than 130 positions in local offices, ranging from mayor or member of the city council to school board members. Today, most Green Party supporters live on the Pacific Coast, around the upper Great Lakes, and in the northeastern United States. In the future the Green Party, like other third parties, will probably continue to struggle in overcoming ballot access laws, such as the requirement of many states that candidates get petitions signed by a certain percentage of registered voters before they can run for office.

## Political Participation and Voter Apathy

Why do some people vote and others not? How do people come to think of themselves as being conservative, moderate, or liberal? Key factors include individuals' political socialization and attitudes.

*Political socialization* **is the process by which people learn political attitudes, values, and behavior.** For young children the family is the primary agent of political socialization, and children tend to learn and hold many of the same opinions held by their parents. By the time children reach school age, they typically identify with the political party (if any) of their parents. As we grow older, other agents of socialization begin to affect our political beliefs, including our peers, teachers, and the media. If we grow up around family members and friends who vote and discuss politics, we are more likely to be interested in the political process and to vote.

In addition to the socialization process, people's socioeconomic status affects their political attitudes, values, and beliefs. For example, individuals who are very poor or are unable to find employment may believe that society has failed them and therefore tend to be indifferent toward the political system. Believing that casting a ballot would make no difference to their circumstances, they do not vote.

Democracy in the United States has been defined as a government "of the people, by the people, and for the people." Accordingly, it would stand to reason that "the people" would actively participate in their government at any or all of four levels: (1) voting, (2) attending and taking part in political meetings, (3) actively participating in political campaigns, and (4) running for and/or holding political office. At most, about 10 percent of the voting-age population in this country participates at a level higher than simply voting, and over the past fifty years less than half of the voting-age population has voted in nonpresidential elections. Even in presidential election years, voter turnout is often relatively low.

**Voter Turnout and Political Preferences**
In the 2012 presidential election, 58.7 percent of the 219 million eligible voters cast ballots. This percentage was down from the 2008 presidential election, when 62.3 percent of the 208.3 million eligible voters actually voted. The number of ballots cast in 2008 was the highest in history at 131 million votes, as compared to 126 million in 2012. However, more than half (57.4 percent) of the drop-off in voter turnout was in three states: New York and New Jersey had experienced the devastation of Hurricane Sandy just prior to the election, and California continued a downward decline in voter turnout, from ranking thirty-third in turnout in the 2008 election to forty-first in turnout in 2012. Overall, Minnesota had the highest voter turnout (76.1 percent) while Hawaii had the lowest (44.5 percent).

Among young voters between the ages of 18 and 29, 23 million people cast ballots. The presidential election in 2012 was the third election in a row in which about half of the eligible younger voting population actually voted, and their vote was a determining factor in the swing states of Florida, Ohio, Pennsylvania, and Virginia. These voters were the most racially and ethnically diverse: Latinos/as, African Americans, and other so-called nonwhite voters made up 42 percent of the voting-eligible population, as compared to only 24 percent

among voters ages 30 and older (Pillsbury and Johannesen, 2012).

For the first time, in 2012 Latinos/as (across age categories) constituted one in ten voters (10 percent of the electorate). This increase is attributed to changing demographics; it is estimated that the percentage of Latino/a voters will double in size by 2030, at which time they will account for 40 percent of the growth in the eligible electorate in the United States. By contrast, increases in the numbers of African American voters in the 2012 election are attributed by political analysts to higher voter turnout among African Americans rather than demographic shifts in the population. In 2012 African Americans voted at a higher rate than in the past and accounted for an estimated 13 percent of all votes cast (Taylor, 2012). Barack Obama's candidacy in 2008 and 2012 is believed to be a key reason why African Americans constituted a higher percentage of all voters than in the past, but other factors may also be significant, such as increasing diversity of the electorate and a declining white (non-Hispanic) turnout rate (Taylor, 2012). Figure 13.8 shows the percentage of voter participation in the 2012 presidential election by race and ethnicity.

State-by-state differences in voting preferences are also highly visible in what political analysts refer to as the "red states" and the "blue states." These terms refer to those states whose residents predominantly vote for Republican Party (red) or Democratic Party (blue) presidential candidates. Today, the terms are often used to indicate that

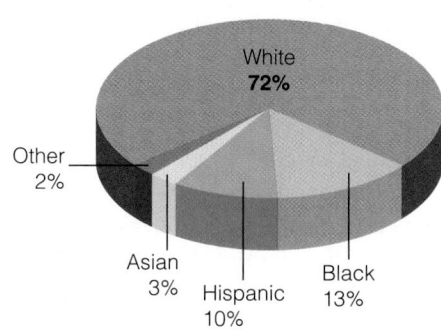

**FIGURE 13.8** VOTER PARTICIPATION IN THE 2012 PRESIDENTIAL ELECTION BY RACE AND ETHNICITY

Source: Based on Taylor, 2012.

**political socialization** the process by which people learn political attitudes, values, and behavior.

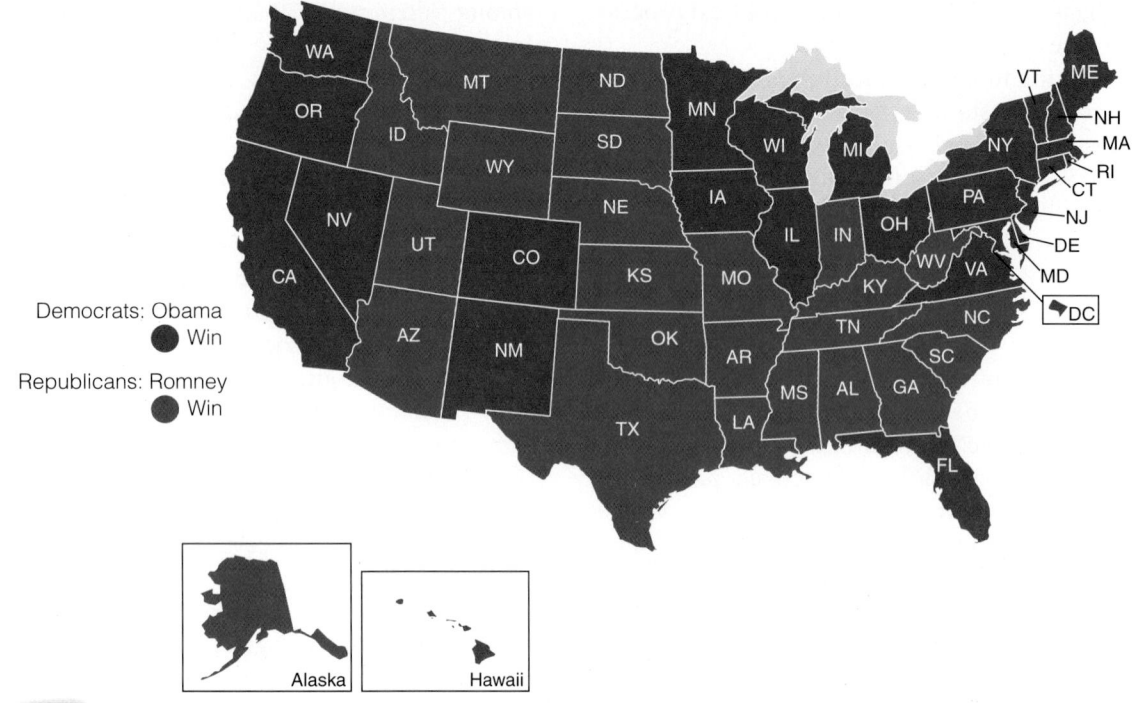

Democrats: Obama
● Win

Republicans: Romney
● Win

Alaska    Hawaii

**FIGURE 13.9**  2012 PRESIDENTIAL ELECTION: STATE BY STATE

Staff/MCT/Newscom

voters residing there are either conservative or liberal but not necessarily to indicate their political party affiliation (see Figure 13.9 for a "red" and "blue" state breakdown of the 2012 presidential election).

## Governmental Bureaucracy

When most people think of political power, they overlook one of its major sources—the governmental bureaucracy. Negative feelings about bureaucracy are perhaps strongest when people are describing the "faceless bureaucrats" and "red tape" with which they must deal in government. But who are these "faceless bureaucrats," and what do they do?

Bureaucratic power tends to take on a life of its own. During the nineteenth century, the government had a relatively limited role in everyday life. In the 1930s, however, the scope of government was extended greatly during the Great Depression to deal with labor–management relations, public welfare, and the regulation of the securities markets. With dramatic increases in technology and increasing demands from the public that the government "do something" about the many problems facing our nation, the federal government now employs about 2.1 million people in civilian positions, with 1.8 million of those positions offering full-time permanent employment (U.S. Office of Personnel Management, 2013).

Because much of the actual functioning of the government is carried on by its bureaucracy, even the president, the White House staff, and cabinet officials have difficulty establishing control over the bureaucracy (Dye, Zeigler, and Schubert, 2012). Many employees in the federal bureaucracy have seen several presidents "come and go," but the *permanent government* in Washington, made up of top-tier civil service bureaucrats who have a major power base, has remained more constant. Figure 13.10 shows characteristics of the "typical" federal civilian employee.

The governmental bureaucracy has been able to perpetuate itself and expand because many of its employees have highly specialized knowledge and skills and cannot be easily replaced by "outsiders." In addition, as the United States has grown in size and complexity, public policy is increasingly made by bureaucrats rather than by elected officials. For example, offices and agencies have been established to create rules, policies, and procedures for dealing with complex issues such as nuclear power, environmental protection, and drug safety; bureaucracies announce an estimated twenty rules or regulations for every one law passed by Congress (Dye, Zeigler, and Schubert, 2012).

The federal budget is the central ingredient in the bureaucracy. Preparing the annual federal budget is a major undertaking for the president and the Office of Management and Budget, one of the most

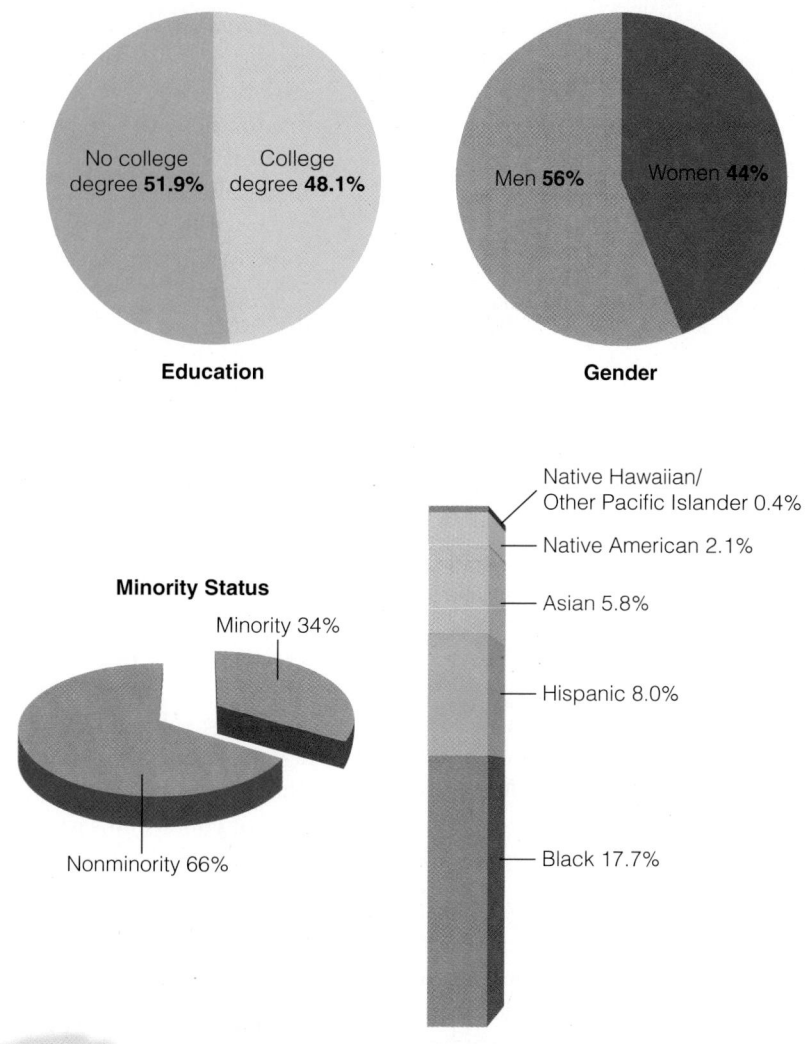

**FIGURE 13.10** THE "TYPICAL" FEDERAL CIVILIAN EMPLOYEE

Source: U.S. Office of Personnel Management, 2013.

important agencies in Washington. Getting the budget approved by Congress is an even more monumental task; however, even with the highly publicized wrangling over the budget by the president and Congress, the final congressional appropriations are usually within 2–3 percent of the budget originally proposed by the president.

In part, this is caused by the *iron triangle of power*—a three-way arrangement in which a private interest group (usually a corporation), a congressional committee or subcommittee, and a bureaucratic agency make the final decision on a political issue that is to be decided by that agency. Figure 13.11 illustrates the alliance among the Defense Department (Pentagon), private military (or defense) contractors, and members of Congress. According to the sociologist Joe Feagin,

The Iron Triangle has a revolving door of money, influence, and jobs among these three sets of actors, involving trillions of dollars.

Military contractors who receive contracts from the Defense Department serve on the advisory committees that recommend what weapons they believe are needed. Many people move around the triangle from job to job, serving in the military, then in the Defense Department, then in military industries. (Feagin and Feagin, 1994: 405)

The Iron Triangle is also referred to as the ***military-industrial complex*—the mutual interdependence of the military establishment and private military contractors.** According to the sociologist C. Wright Mills (1976), this alliance of economic, military, and political power amounts to a "permanent war economy" or "military economy." However, the economist

**military–industrial complex** the mutual interdependence of the military establishment and private military contractors.

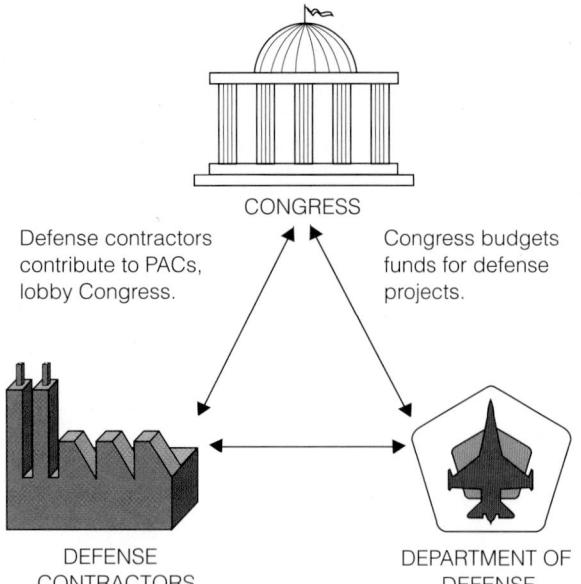

CONGRESS

Defense contractors contribute to PACs, lobby Congress.

Congress budgets funds for defense projects.

DEFENSE CONTRACTORS

DEPARTMENT OF DEFENSE

Defense Department awards contracts to contractors; personnel move back and forth between employment by Defense Department and by defense contractors.

**FIGURE 13.11** EXAMPLE OF THE IRON TRIANGLE OF POWER

© Cengage Learning

John Kenneth Galbraith (1985) has argued that the threat of war is good for the economy because government money spent on military preparedness stimulates the private sector of the economy, creates jobs, and encourages consumer spending. Issues regarding the military–industrial complex are intricately linked to problems in the economy, which we will examine now.

# Economic Systems in Global Perspective

The *economy* **is the social institution that ensures the maintenance of society through the production, distribution, and consumption of goods and services.** *Goods* are tangible objects that are necessary (such as food, clothing, and shelter) or desired (such as DVDs and electric toothbrushes). *Services* are intangible activities for which people are willing to pay (such as dry cleaning, a movie, or medical care).

## Preindustrial, Industrial, and Postindustrial Economies

**L05** In all societies the specific method of producing goods is related to the technoeconomic base of the society, as discussed in Chapter 10. In each society people develop an economic system, ranging from simple to very complex, for the sake of survival.

*Preindustrial economies* include hunting and gathering, horticultural and pastoral, and agrarian societies. Most workers engage in *primary sector production*—**the extraction of raw materials and natural resources from the environment.** These materials and resources are typically consumed or used without much processing. The production units in hunting and gathering societies are small; most goods are produced by family members. The division of labor is by age and by gender. The potential for producing surplus goods increases as people learn to domesticate animals and grow their own food. In horticultural and pastoral societies, the economy becomes distinct from family life. The distribution process becomes more complex, with the accumulation of a surplus such that some people can engage in activities other than food production. In agrarian societies, production is related primarily to producing food. However, workers have a greater variety of specialized tasks, such as warlord or priest; for example, warriors are necessary to protect the surplus goods from plunder by outsiders.

*Industrial economies* result from sweeping changes to the system of production and distribution of goods and services during industrialization. Drawing on new forms of energy (such as steam, gasoline, and electricity) and machine technology, factories proliferate as the primary means of producing goods. Most workers engage in *secondary sector production*—**the processing of raw materials (from the primary sector) into finished goods.** For example, steel workers process metal ore; auto workers then convert the ore into automobiles, trucks, and buses. In industrial economies, work becomes specialized and repetitive, activities become bureaucratically organized, and workers primarily work with machines instead of with one another. With the emergence of mass production, larger surpluses are generated, typically benefiting some people and organizations but not others.

A *postindustrial economy* is based on *tertiary sector production*—**the provision of services rather than goods**—as a primary source of livelihood for workers and profit for owners and corporate shareholders. Tertiary sector production includes a wide range of activities, such as fast-food service, transportation, communication, education, real estate, advertising, sports, and entertainment. Yahoo! and Google are examples of postindustrial corporations that primarily provide services rather than selling products (Figure 13.12). As shown in "Census Profiles," a majority of U.S. jobs are in tertiary sector employment, as contrasted with primary or secondary sector employment, and some of the employment sectors are segregated by gender and/or race/ethnicity.

Five characteristics are central to the postindustrial economy:

1. Service industries dominate over manufacturing.
2. Information and technological innovation displace property as the central preoccupations in the economy.
3. Professional and technical classes grow more predominant, and workplace culture shifts from factories to diversified work settings.
4. Traditional boundaries between work and home (public and private life) no longer exist because digital technologies such as cell phones and computers make global communication possible.
5. High levels of urbanization and a decline in population occur in many rural areas.

Capitalism and socialism are the principal economic models in industrial and postindustrial countries. As we examine these two models, keep in mind that no society has a purely capitalist or socialist economy.

iStockphoto.com/4x6

**FIGURE 13.12** Yahoo! CEO Marissa Mayer surprised many when she announced that Yahoo employees would no longer be allowed to work from home. What might employees gain from working in the office rather than at home or other locations such as the local Starbucks?

## Capitalism

**LO6** *Capitalism* **is an economic system characterized by private ownership of the means of production, from which personal profits can be derived through market competition and without government intervention.** Most of us think of ourselves as "owners" of private property because we own a car, a television, or other possessions. However, most of us are not capitalists; we *spend money* on the things we own rather than *make money* from them. Only a relatively few people own income-producing property from which a profit can be realized by producing and distributing goods and services; everyone else is a consumer. "Ideal" capitalism has four distinctive features: (1) private ownership of the means of production, (2) pursuit of personal profit, (3) competition, and (4) lack of government intervention.

**Private Ownership of the Means of Production** Capitalist economies are based on the right of individuals to own income-producing property, such as land, water, mines, and factories, and the right to "buy" people's labor. However, under early monopoly capitalism (1890–1940), most ownership shifted from individuals to huge *corporations*—**organizations that have legal powers, such as the ability to enter into contracts and buy and sell property, separate from their individual owners.** In advanced monopoly capitalism

(1940–present), ownership and control of major industrial and business sectors have become increasingly concentrated, and many corporations have become more global in scope. *Transnational corporations* **are large corporations that**

**economy** the social institution that ensures the maintenance of society through the production, distribution, and consumption of goods and services.

**primary sector production** the sector of the economy that extracts raw materials and natural resources from the environment.

**secondary sector production** the sector of the economy that processes raw materials (from the primary sector) into finished goods.

**tertiary sector production** the sector of the economy that is involved in the provision of services rather than goods.

**capitalism** an economic system characterized by private ownership of the means of production, from which personal profits can be derived through market competition and without government intervention.

**corporations** organizations that have legal powers, such as the ability to enter into contracts and buy and sell property, separate from their individual owners.

**transnational corporations** large corporations that are headquartered in one country but sell and produce goods and services in many countries.

# census PROFILES

## Civilian Occupations

*The Statistical Abstract of the United States*, compiled until 2013 by the U.S. Census Bureau, reports data from the Bureau of Labor Statistics about the occupations of people who are age 16 or over. (This information is now reported in *ProQuest Statistical Abstract of the United States,* available in print and online.) Here is a summary of recent findings: The majority of people who are not full-time military employees hold jobs in the tertiary sector, including management, professional, service, and sales occupations, as contrasted with primary sector and secondary sector employment. As shown in the figures below, job segregation by gender and/or race/ethnicity exists in several employment sectors.

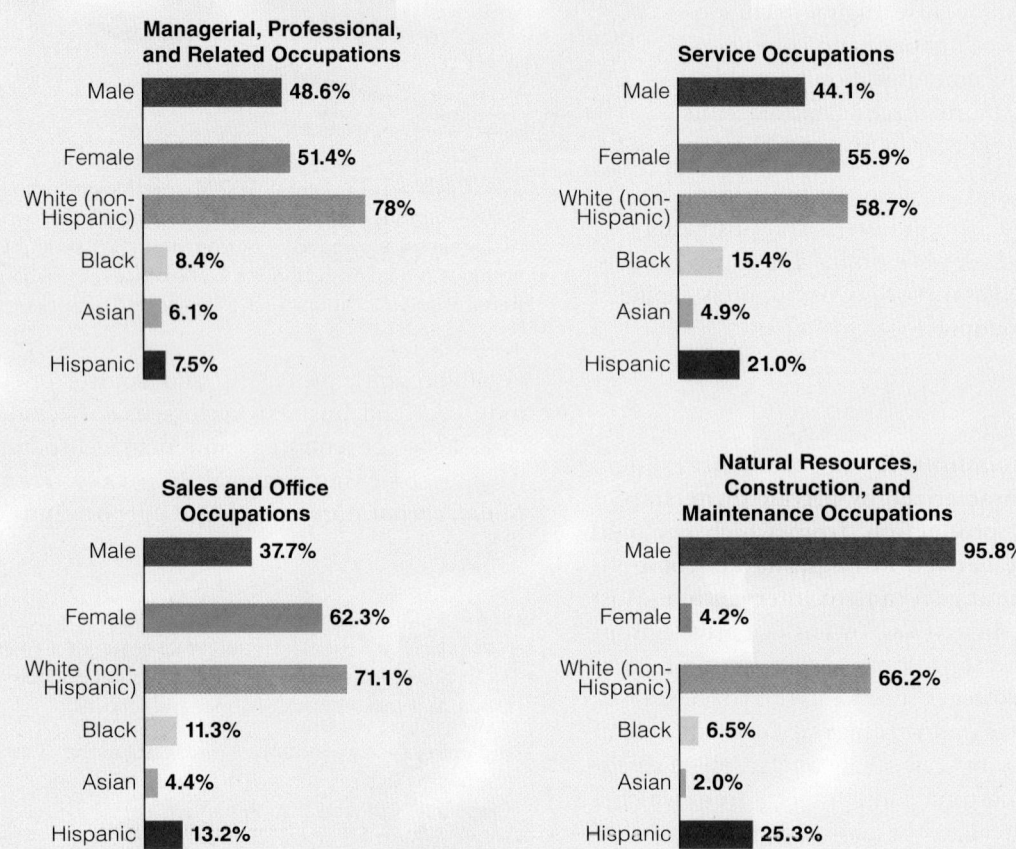

**Managerial, Professional, and Related Occupations**

- Male: 48.6%
- Female: 51.4%
- White (non-Hispanic): 78%
- Black: 8.4%
- Asian: 6.1%
- Hispanic: 7.5%

**Service Occupations**

- Male: 44.1%
- Female: 55.9%
- White (non-Hispanic): 58.7%
- Black: 15.4%
- Asian: 4.9%
- Hispanic: 21.0%

**Sales and Office Occupations**

- Male: 37.7%
- Female: 62.3%
- White (non-Hispanic): 71.1%
- Black: 11.3%
- Asian: 4.4%
- Hispanic: 13.2%

**Natural Resources, Construction, and Maintenance Occupations**

- Male: 95.8%
- Female: 4.2%
- White (non-Hispanic): 66.2%
- Black: 6.5%
- Asian: 2.0%
- Hispanic: 25.3%

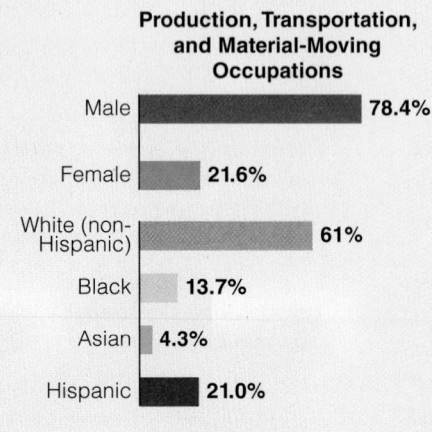

**Production, Transportation, and Material-Moving Occupations**

- Male: 78.4%
- Female: 21.6%
- White (non-Hispanic): 61%
- Black: 13.7%
- Asian: 4.3%
- Hispanic: 21.0%

Source: ProQuest, 2014.

## table 13.1

### Revenues of the World's 20 Largest Public and Private Corporations, 2012

| Company and Location | Revenues/GDP (in billions) |
|---|---|
| Exxon Mobil—U.S. (Oil and gas) | $486.4 |
| Royal Dutch Shell—UK (Oil and gas) | $470.1 |
| Walmart—U.S. (Retail) | $446.9 |
| Sinopec Group—China (Oil and gas) | $397.7 |
| China National Petroleum—China (Oil and gas) | $378.0 |
| BP—UK (Oil and gas) | $375.5 |
| Saudi Aramco—Saudi Arabia (Oil and gas) | $356 |
| Vitol—Netherlands (Commodities) | $297 |
| State Grid Corporation of China—China (Electric utility) | $265.9 |
| Chevron—U.S. (Oil and gas) | $253.7 |
| ConocoPhillips—U.S. (Oil and gas) | $251.2 |
| Toyota—Japan (Automotive) | $226.1 |
| Total—France (Oil and gas) | $215.5 |
| Volkswagen Group—Germany (Automotive) | $206.2 |
| Japan Post Holdings—Japan (Conglomerate) | $201.1 |
| Glencore—Switzerland (Commodities) | $186.1 |
| Gazprom—Russia (Oil and gas) | $158.1 |
| Apple—U.S. (Electronics) | $156.5 |
| General Motors—U.S. (Automotive) | $150.2 |
| General Electric—U.S. (Conglomerate) | $147.3 |

Sources: Compiled by the author and by DeCarlo, 2013; Murphy, 2012.

are headquartered in one country but sell and produce goods and services in many countries. These corporations play a major role in the economies and governments of many nations. Revenues for the top twenty public and private transnational or global corporations worldwide are shown in Table 13.1, where their concentration in only a few industries, such as oil and gas, is evident.

**Pursuit of Personal Profit** A tenet of capitalism is the belief that people are free to maximize their individual gain through personal profit; in the process, the entire society will benefit from their activities (Smith, 1976/1776). Economic development is assumed to benefit both capitalists and workers, and the general public also benefits from public expenditures (such as for roads, schools, and parks) made possible through an increase in business tax revenues.

During the period of industrial capitalism, however, specific individuals and families (not the general public) were the primary recipients of profits. For many generations, descendants of some of the early industrial capitalists have benefited from the economic deeds (and misdeeds) of their ancestors. In early monopoly capitalism, some stockholders derived massive profits from companies that held near-monopolies on specific goods and services.

In advanced (late) monopoly capitalism, profits have become even more concentrated. Although some people own *some* stock, they do not own *control*; in other words, they are unable to participate in establishing the policies that determine the size of the profit or the rate of return on investments (which affects the profits they derive).

**Competition** In theory, competition acts as a balance to excessive profits. When producers vie with one another for customers, they must be able to offer innovative goods and services at competitive prices. However, from the time of early industrial capitalism, the trend has been toward less, rather than more, competition among companies. In early monopoly capitalism, competition was diminished by increasing concentration *within* a particular industry, a classic case being the virtual monopoly on oil held by John D. Rockefeller's Standard Oil Company (Tarbell, 1925/1904; Lundberg, 1988). Today, Microsoft Corporation has been the subject of federal investigations and lawsuits because it so dominates certain areas of the computer software industry that it has virtually no competitors in those areas. In other situations, just a few companies may dominate certain industries. An *oligopoly,* which in Greek means "few sellers," **refers to an industry dominated by just a few companies.** An example is the music industry, in which a few

**oligopoly** an industry dominated by just a few companies.

## table 13.2

The Music Industry's Big Three

| Company | Country | Leading Artists |
|---|---|---|
| Universal Music Group (EMI Recorded Music) | France | Taylor Swift<br>Lady Gaga<br>Justin Bieber<br>Lil Wayne<br>Eminem |
| Sony Music Entertainment (EMI Music Publishing) | United States | Avril Lavigne<br>Foo Fighters<br>Pink<br>Kings of Leon |
| Warner Music Group (Atlantic, Elektra) | United States | Lupe Fiasco<br>Death Cab for Cutie<br>Rick Ross<br>Linkin Park |

© Cengage Learning

giant companies are behind many of the labels and artists (see Table 13.2). In fact, what was formerly known as the Big Four—Universal, Sony, Warner, and EMI—further consolidated into the Big Three in 2011, leaving only Universal, Sony, and Warner. This would be considered a **shared monopoly— when four or fewer companies supply 50 percent or more of a particular market.** Other industries that are dominated by just a few companies include those that manufacture automobiles, breakfast cereals, cigarettes, oil, and personal computers and tablets. However, as we have seen in recent years, even some corporations, such as automobile manufacturers, that have been considered "too big to fail" have indeed had problems serious enough that government bailouts have been required to keep them in business and some of their workers employed.

In advanced monopoly capitalism, mergers also occur *across* industries: Corporations gain near-monopoly control over all aspects of the production and distribution of a product by acquiring both the companies that supply the raw materials and the companies that are the outlets for the product. For example, an oil company may hold leases on the land where the oil is pumped out of the ground, own the plants that convert the oil into gasoline, and own the individual gasoline stations that sell the product to the public.

Corporations with control both within and across industries are often formed by a series of mergers and acquisitions. These corporations are referred to as **conglomerates—combinations of businesses in different commercial areas, all of which are owned by one holding company.** Media ownership is a case in point; companies such as Time Warner have extensive holdings in radio and television stations, cable television companies, book publishing firms, and film production and distribution companies, to name only a few.

Competition is reduced over the long run by **interlocking corporate directorates—members of the board of directors of one corporation who also sit on the board(s) of other corporations.** Although the Clayton Antitrust Act of 1914 made it illegal for a person to sit simultaneously on the boards of directors of two corporations that are in *direct* competition with each other, a person may serve simultaneously on the board of a financial institution (a bank, for example) and the board of a commercial corporation (a computer manufacturing company or a furniture store chain, for example) that borrows money from the bank. Directors of competing corporations may also serve together on the board of a third corporation that is not in direct competition with the other two. An example of interlocking directorates is depicted in Figure 13.13. Compensation for members of the boards of top corporations can be more than $1 million per year when stock, stock options, and pensions are taken into account.

Interlocking directorates diminish competition by producing interdependence. Individuals who serve on multiple boards are often able to forge cooperative arrangements that benefit their corporations but not necessarily the general public. When several corporations are controlled by the same financial interests, they are more likely to cooperate with one another than to compete.

**Lack of Government Intervention** Ideally, capitalism works best without government intervention in the marketplace. The policy of *laissez-faire* (les-ay-FARE, which means "leave alone") was advocated by the economist Adam Smith in his

**FIGURE 13.13** THE GENERAL MOTORS BOARD OF DIRECTORS

The 2013 General Motors Board of Directors shows the nature of interlocking directorates. On the chair representing each of the directors is the name of another entity the director is connected with, and his or her position with that entity.

Source: General Motors, 2013. © Cengage Learning

1776 treatise *An Inquiry into the Nature and Causes of the Wealth of Nations.* Smith argued that when people pursue their own selfish interests, they are guided "as if by an invisible hand" to promote the best interests of society (see Smith, 1976/1776). Today, terms such as *market economy* and *free enterprise* are often used, but the underlying assumption is the same: Free market competition, not the government, should regulate prices and wages. However, the "ideal" of unregulated markets benefiting all citizens has seldom been realized. Individuals and companies in pursuit of higher profits have run roughshod over weaker competitors, and small businesses have grown into large, monopolistic corporations. Accordingly, government regulations were implemented in an effort to curb the excesses of the marketplace brought about by laissez-faire policies.

However, much of what is referred to as government intervention has been in the form of aid to business. Between 1850 and 1900, corporations received government assistance in the form of public subsidies and protection from competition by tariffs, patents, and trademarks. Government intervention in the twenty-first century has included the Emergency Economic Stabilization Act, which created a $700 billion Troubled Assets Relief Program (TARP). This program allowed the government to purchase failed bank assets that resulted primarily from the subprime mortgage crisis (in which people were encouraged to purchase homes that many of them could not afford). General Motors, Bank of America, and American International Group (AIG) were among the companies that received taxpayers' money from the TARP bailout. Overall, most corporations have gained much more than they have lost as a result of government involvement in the economy.

## Socialism

***Socialism* is an economic system characterized by public ownership of the means of production, the pursuit of collective goals, and centralized decision making.** Like "pure" capitalism, "pure" socialism does not exist. Karl Marx described socialism as a temporary stage en route to an ideal communist society. Although the terms *socialism* and *communism* are associated with Marx and are often used interchangeably, they are not identical. Marx defined *communism* as an economic system characterized by common ownership of all economic resources. In the *Communist Manifesto* and *Das Kapital,* he predicted that the working class would become increasingly impoverished and alienated under capitalism. As a result, the workers would become aware of their own class interests, revolt against the capitalists, and overthrow the entire system (see Turner, Beeghley, and Powers, 2007). After the revolution, private property would be abolished, and capital would be controlled by collectives of workers who

**shared monopoly** a situation that occurs when four or fewer companies supply 50 percent or more of a particular market.

**conglomerate** a combination of businesses in different commercial areas, all of which are owned by one holding company.

**interlocking corporate directorates** members of the board of directors of one corporation who also sit on the board(s) of other corporations.

**socialism** an economic system characterized by public ownership of the means of production, the pursuit of collective goals, and centralized decision making.

would own the means of production. The government (previously used to further the interests of the capitalists) would no longer be necessary. People would contribute according to their abilities and receive according to their needs (Marx and Engels, 1967/1848; Marx, 1967/1867). "Ideal" socialism has three distinctive features: (1) public ownership of the means of production, (2) pursuit of collective goals, and (3) centralized decision making.

**Public Ownership of the Means of Production** In a truly socialist economy, the means of production are owned and controlled by a collectivity or the state, not by private individuals or corporations. For example, prior to the early 1990s the state owned all the natural resources and almost all the capital in what was then the Soviet Union. At least in theory, goods were produced to meet the needs of the people. Access to housing and medical care was considered to be a right. The leaders of the former Soviet Union and some Eastern European nations decided to abandon government ownership and control of the means of production because the system was unresponsive to the needs of the marketplace and offered no incentive for increased efficiency (Boyes and Melvin, 2011). Over the past two decades, Russia and other states in the former Soviet Union have attempted to privatize most industries, with the exceptions of the energy and defense-related sectors. In *privatization*, resources are converted from state ownership to private ownership; the government takes an active role in developing, recognizing, and protecting private property rights (Boyes and Melvin, 2011). Today, the state-owned Russian oil company Rosneft makes billions of dollars annually (Kramer, 2011).

China—previously the world's other major communist economy—has privatized many state industries. In the second decade of the twenty-first century, China has a hybrid political economy made up of both capitalism and an autocratic form of Communist Party governance. Double-digit economic growth has brought about an increase in annual urban income, life expectancy has increased by more than six years, and the rate of illiteracy has dropped significantly. With these improvements for middle-class Chinese, it is likely that the combination of communism and a modified form of capitalism will remain for the foreseeable future (Jacobs, 2011). Both Russia and China have undergone significant changes in recent years, moving from globally isolated, centrally controlled economies to more market-based and globally integrated economies.

**Pursuit of Collective Goals** Socialism is based on the pursuit of collective goals, rather than on personal profits. Equality in decision making replaces hierarchical relationships (such as between owners and workers or between political leaders and citizens). Everyone shares in the goods and services of society—especially necessities such as food, clothing, shelter, and medical care—based on need, not on ability to pay. In reality, in nations such as China, members of the Communist Party are able to obtain low-interest loans from state-owned and state-operated banks as long as they play by party rules (Jacobs, 2011). In sum, even though pursuit of collective goals is one of the ideals of socialism, few societies can or do pursue purely collective goals.

**Centralized Decision Making** Another tenet of socialism is centralized decision making. In theory, economic decisions are based on the needs of society; the government is responsible for aiding the production and distribution of goods and services. Central planners set wages and prices to ensure that the production process works. When problems such as shortages and unemployment arise, they can be dealt with quickly and effectively by the central government (Boyes and Melvin, 2011).

## Mixed Economies

As we have seen, no economy is truly capitalist or socialist; most economies are mixtures of both. A *mixed economy* **combines elements of a market economy (capitalism) with elements of a command economy (socialism).** Sweden, Great Britain, and France have mixed economies, sometimes referred to as *democratic socialism*—**an economic and political system that combines private ownership of some of the means of production, governmental distribution of some essential goods and services, and free elections.** For example, government ownership in Sweden is limited primarily to railroads, mineral resources, a public bank, and liquor and tobacco operations (Figure 13.14). Compared with capitalist economies, however, the government in a mixed economy plays a larger role in setting rules, policies, and objectives.

The government is also heavily involved in providing services such as medical care, child care, and transportation. In Sweden, for example, all residents have health insurance, housing subsidies, child allowances, paid parental leave, and day-care subsidies. Recently, some analysts have suggested that the United States has assumed many of the characteristics of a *welfare state*, **a state in which there is extensive government action to provide support and services to the citizens,** as it has attempted to meet the basic needs of older persons, young children, unemployed people, and persons with a disability.

**FIGURE 13.14** In Sweden, liquor stores are owned by the government. How are employees affected by employment in a nation with a mixed economy such as Sweden, where elements of capitalism and socialism are both present?

# Work in the Contemporary United States

The economy in the United States and other contemporary societies is partially based on the work (purposeful activity, labor, or toil) that people perform. However, work in high-income nations is highly differentiated and often fragmented because people have many kinds of occupations. Some occupations are referred to as professions.

## Professions

**L07** Although sociologists do not always agree on exactly which occupations are professions, most of them agree that the term *professionals* includes most doctors, natural scientists, engineers, computer scientists, certified public accountants, economists, social scientists, psychotherapists, lawyers, policy experts of various sorts, professors, at least some journalists and editors, some clergy, and some artists and writers.

**Characteristics of Professions** *Professions* **are high-status, knowledge-based occupations that have five major characteristics** (Freidson, 1970, 1986; Larson, 1977):

1. *Abstract, specialized knowledge.* Professionals have abstract, specialized knowledge of their field based on formal education and interaction with colleagues.

2. *Autonomy.* Professionals are autonomous in that they can rely on their own judgment in selecting the relevant knowledge or the appropriate technique for dealing with a problem.

3. *Self-regulation.* In exchange for autonomy, professionals are theoretically self-regulating. All professions have licensing, accreditation, and regulatory associations that set professional standards and that require members to adhere to a code of ethics as a form of public accountability.

4. *Authority.* Because of their authority, professionals expect compliance with their directions and advice. Their authority is based on mastery of the body of specialized knowledge and on their profession's autonomy.

5. *Altruism.* Ideally, professionals have concern for others, not just their own self-interest. The word *altruism* implies some degree of self-sacrifice whereby professionals go beyond their self-interest or personal comfort so that they can help a patient or client.

**Social Reproduction of Professionals** Although higher education is one of the primary qualifications for a profession, the emphasis on education gives children whose parents are professionals a disproportionate advantage early in life. For example, there is a direct linkage between parental education/family income and children's scores on college admissions tests such as the SAT. In turn, test scores are directly related to students' ability to gain admission to colleges and universities, which serve as springboards to most professions. For this reason, the practice of heavy reliance on SAT, ACT, and other admissions tests for determining who is admitted to certain colleges is being widely debated.

**Deprofessionalization** Certain professions are undergoing a process of *deprofessionalization,* in which some of the characteristics of a profession

**mixed economy** an economic system that combines elements of a market economy (capitalism) with elements of a command economy (socialism).

**democratic socialism** an economic and political system that combines private ownership of some of the means of production, governmental distribution of some essential goods and services, and free elections.

**welfare state** a state in which there is extensive government action to provide support and services to the citizens.

**professions** high-status, knowledge-based occupations.

are eliminated. Some physicians believe that insurance companies, political leaders, and lobbyists have deprofessionalized their occupation by referring to them as "health care providers" rather than as "medical doctors" or "physicians."

## Other Occupations

*Occupations* **are categories of jobs that involve similar activities at different work sites.** More than 600 different occupational categories and 35,000 occupation titles, ranging from motion picture cartoonist to drop-hammer operator, are currently listed by the U.S. Census Bureau. Historically, occupations have been classified as blue collar and white collar. Blue-collar workers were primarily factory and craft workers who did manual labor; white-collar workers were office workers and professionals. However, contemporary workers in the service sector do not easily fit into either of these categories; neither do the so-called pink-collar workers, primarily women, who are employed in occupations such as preschool teacher, dental assistant, secretary, and clerk. (The term refers to an era when some female restaurant employees were required to wear uniforms with a pink collar.)

Sociologists establish broad occupational categories by distinguishing between employment in the primary labor market and in the secondary labor market. The *primary labor market* **consists of high-paying jobs with good benefits that have some degree of security and the possibility of future advancement.** By contrast, the *secondary labor market* **consists of low-paying jobs with few benefits and very little job security or possibility for future advancement** (Bonacich, 1972).

**Upper-Tier Jobs: Managers and Supervisors** Managers are essential in contemporary bureaucracies, where work is highly specialized and authority structures are hierarchical. Workers at each level of the hierarchy take orders from their immediate superiors and perhaps give orders to a few subordinates. Upper-level managers are typically responsible for coordination of activities and control of workers.

**Lower-Tier and Marginal Jobs** Positions in the lower tier of the service sector are part of the secondary labor market, characterized by low wages, little job security, few chances for advancement, higher unemployment rates, and very limited (if any) unemployment benefits. Typical lower-tier positions include janitor, waitperson, messenger, sales clerk, typist, file clerk, migrant laborer, and textile worker. Large numbers of young people, people of color, recent immigrants, and white women are employed in this sector (Figure 13.15).

**FIGURE 13.15** Occupational segregation by race and gender is clearly visible in personal service industries, such as restaurants and fast-food chains. Women and people of color are disproportionately represented in marginal jobs such as waitperson or fast-food server—jobs that typically do not meet societal norms for benefits and security.

*Marginal jobs* **differ from the employment norms of the society in which they are located.** Examples of marginal jobs in the U.S. labor market include private household workers—such as launderers, cooks, maids, housekeepers, gardeners, babysitters, and nannies—and personal service employees in eating and drinking places, hotels, laundries, and beauty and nail salons, for example. This kind of work is typically not covered by government work regulations—such as minimum standards of pay, working conditions, and safety standards—and does not offer sufficient hours of work each week to provide a living. The jobs are excluded from most labor legislation, employers often pay cash in order to avoid payroll taxes and Social Security, and the jobs typically provide no insurance or retirement benefits.

## Contingent Work

**LO8** *Contingent work* **is part-time work, temporary work, or subcontracted work that offers advantages to employers but that can be detrimental to the welfare of workers.** Contingent work is found in every segment of the workforce. The federal government is part of this trend, as is private enterprise. In the health care field, physicians, nurses, and other workers are increasingly employed through temporary agencies. Employers benefit by hiring workers on a part-time or temporary basis; they are able to cut costs, maximize profits, and have workers available only when they need them. Temporary workers are the fastest-growing segment of the contingent workforce, and agencies that "place" them have increased dramatically in number in the last decade.

Subcontracted work is another form of contingent work that often cuts employers' costs at the

Peter Hvizdak/The Image Works

expense of workers. Instead of employing a large workforce, many companies have significantly reduced the size of their payrolls and benefit plans by *subcontracting*—**an agreement in which a corporation contracts with other (usually smaller) firms to provide specialized components, products, or services to the larger corporation.** Hiring and paying workers become the responsibility of the subcontractor, not of the larger corporation.

## The Underground (Informal) Economy

Some social analysts make a distinction between the legitimate and the underground (informal) economies in the United States. For the most part, the occupations previously described in this chapter operate within the *legitimate economy:* Taxes on income are paid by employers and employees, and individuals who hold jobs requiring a specialized license (such as craftspeople or taxi drivers) possess the appropriate credentials for their work. By contrast, the *underground economy* is made up of a wide variety of activities through which people make money that they do not report to the government, and in some cases, their endeavors may involve criminal behavior (Venkatesh, 2006). Sometimes referred to as the "shadow economy," one segment of the underground economy is made up of workers who are paid "off the books," which means that they are paid in cash, their earnings are not reported, and no taxes are paid. Lawful jobs, such as nannies, construction workers, and landscape/yard workers, are often part of the shadow economy because workers and bosses make under-the-table deals so that both can gain through the transaction: Employers pay less for workers' services, and workers have more money to take home than if they paid taxes on their earnings.

The underground economy also involves trade in lawful goods that are sold "off the books" because no taxes are paid on the sales. In his study of the underground economy of the urban poor, the sociologist Sudhir Venkatesh (2006: 9) explains how these transactions take place:

> Most underground exchanges are short-term efforts to make a buck, but they can nevertheless follow strict patterns. Individuals know where to meet one another to trade off the books; there are usually particular places where this trading occurs and particular people who are known to be involved. People will have a rough idea of prices or rates of barter and trade before initiating the exchange. And it is not difficult to predict when conflict may arise; nor are people entirely unaware of the means for addressing disputes

over quality, pricing, and service. In other words, while there are endless reasons to participate in the underground (or to stop doing so), there are always rules to be obeyed, codes to be followed, and likely consequences of actions.

The selling of goods in the underground economy may have increased in recent years because of the popularity of online commerce. Although individual sellers are responsible for paying taxes on items sold on websites such as eBay, it is unknown how many of the sellers actually report income from their sales.

According to one way of thinking, operating a business in the underground economy reveals capitalism at its best because it shows how the "free market" might work without government intervention. However, from another perspective, selling goods or services in the underground economy borders on—or moves into—criminal behavior. For some individuals the underground economy offers the only alternative to unemployment, particularly in poor communities where people often feel alienated from the wider world and believe that they must use shady means to survive (see Venkatesh, 2006).

## Unemployment

There are three major types of unemployment—cyclical, seasonal, and structural. *Cyclical unemployment* occurs as a result of lower rates of production during recessions in the business cycle; a recession is a decline in an economy's total production that lasts at least six months. Although massive layoffs initially

**occupations** categories of jobs that involve similar activities at different work sites.

**primary labor market** the sector of the labor market that consists of high-paying jobs with good benefits that have some degree of security and the possibility of future advancement.

**secondary labor market** the sector of the labor market that consists of low-paying jobs with few benefits and very little job security or possibility for future advancement.

**marginal jobs** jobs that differ from the employment norms of the society in which they are located.

**contingent work** part-time work, temporary work, or subcontracted work that offers advantages to employers but that can be detrimental to the welfare of workers.

**subcontracting** an agreement in which a corporation contracts with other (usually smaller) firms to provide specialized components, products, or services to the larger corporation.

occur, some of the workers will eventually be rehired, largely depending on the length and severity of the recession. *Seasonal unemployment* results from shifts in the demand for workers based on conditions such as the weather (in agriculture, the construction industry, and tourism) or the season (holidays and summer vacations). Both of these types of unemployment tend to be relatively temporary.

By contrast, structural unemployment may be permanent. *Structural unemployment* arises because the skills demanded by employers do not match the skills of the unemployed or because the unemployed do not live where the jobs are located. This type of unemployment often occurs when a number of plants in the same industry are closed or when new technology makes certain jobs obsolete. Structural unemployment often results from capital flight—the investment of capital in foreign facilities. This type of unemployment leads to long-term unemployment for some former employees and higher rates of unemployment for the nation.

The ***unemployment rate*** **is the percentage of unemployed persons in the labor force actively seeking jobs.** The second decade of the twenty-first century has seen a significant increase in unemployment. The U.S. unemployment rate in 2000 was 4 percent. By 2011, the overall rate hovered around 9 percent before falling to 6.7 percent in February 2014. However, this drop may mean that some workers have stopped actively seeking employment. In 2014 the overall U.S. unemployment rate for adult men was 6.4 percent, as compared to 5.9 percent for women. However, the breakdown across racial–ethnic and age categories tells a more complete story: The unemployment rate for African Americans was 12 percent, and for Hispanics (Latinos/as) it was 8.1 percent. By contrast, white Americans (non-Hispanics) had an unemployment rate of 5.8 percent, and Asian Americans had a rate of 6.0 percent. Among teenagers, the story is even more disparate: Teenagers of all racial–ethnic categories (and both sexes) had a 21.4 percent unemployment rate (U.S. Bureau of Labor Statistics, 2014b).

Like other types of "official" statistics, unemployment rates may be misleading. Individuals who become discouraged in their attempt to find work and no longer actively seek employment are not counted as unemployed. Some analysts believe that unemployment rates may drop for a period of time because people either do not seek work or they accept part-time or temporary jobs when they cannot find full-time employment.

## Labor Unions and Worker Activism

In their individual and collective struggles to improve their work environment and gain some degree of control over their work-related activities, workers have used a number of methods to resist workplace alienation. Many have joined labor unions to gain strength through collective action. Others have engaged in various forms of worker activism.

**Labor Unions** U.S. labor unions came into being in the mid-nineteenth century. Unions have been credited with gaining an eight-hour workday, a five-day workweek, health and retirement benefits, sick leave and unemployment insurance, and workplace health and safety standards for many employees. Most of these gains have occurred through *collective bargaining*—negotiations between employers and labor union leaders on behalf of workers. In some cases, union leaders have called strikes to force employers to accept the union's position on wages and benefits. While on strike, workers may picket in front of the workplace to gain media attention, to fend off "scabs" (nonunion workers) who might take over their jobs, and in some cases to discourage customers from purchasing products made or sold by their employer. In recent years, strike activity has diminished significantly because many workers have feared losing their jobs. In 2013 only 11.3 percent of wage and salary workers were union members, compared with 20.1 percent in 1983, the first year for which the federal government compiled such data (U.S. Bureau of Labor Statistics, 2014e). About 14.5 million workers belonged to unions in 2013, with the rate of membership for public-sector workers being more than five times higher (35.3 percent) than that for private-sector workers (6.7 percent). The highest rate of unionization (35.3 percent for each occupation group) is among workers in education, training, library occupations, and protective service occupations (U.S. Bureau of Labor Statistics, 2014e).

A 2012 study by the Pew Research Center found that nearly two-thirds (64 percent) of Americans agree that labor unions are necessary to protect the working person, but 33 percent of the respondents disagreed. Fifty-four percent of Republicans in this study stated they did not believe labor unions were necessary to protect workers while 80 percent of Democrats indicated the opposite (Pew Research Center, 2012b).

## Employment Opportunities for Persons with a Disability

For many years, people with disabilities have been steered away from occupations by teachers, parents, prospective employers, and others who tended to focus more on what they could not do rather than what they are capable of doing. In 1990 the United States became the first nation to formally address the issue of equality for persons with a disability when Congress passed the Americans with Disabilities Act (ADA) to prohibit discrimination on the basis of disability. Combined with previous disability rights laws (such as those that provide for the elimination of architectural barriers from new, federally funded buildings and for the maximum integration of schoolchildren with disabilities), the ADA is a legal mandate for the full equality of people with disabilities. The federal law defines a person with a disability as an individual with at least one of the following conditions: He or she is deaf or has serious difficulty hearing; is blind or has serious difficulty seeing even when wearing glasses; has serious difficulty concentrating, remembering, or making decisions because of a physical, mental, or emotional condition; has serious difficulty walking or climbing stairs; has difficulty dressing or bathing; or has difficulty doing errands alone such as visiting a doctor's office or shopping because of a physical, mental, or emotional condition.

Despite the ADA and other laws regarding disability rights, many persons with a disability remain unemployed or have been tracked into disability-related service roles (such as helping other persons with a disability). The economic recession and the slow recovery thereafter also hit workers with disabilities hard, particularly those individuals who have mobility impairments or difficulty performing routine daily activities. As of February 2014, only 19.1 percent of persons ages 16 and over with one or more disabilities were employed (as compared with 68.5 percent of people without disabilities). The 2014 unemployment rate for people with disabilities was 14.3 percent, as compared with 6.8 percent for people without disabilities (U.S. Bureau of Labor Statistics, 2014d). Many other persons with disabilities were not in the labor force at all, nor were they actively seeking employment. Many persons with a disability believe that they could work if they were offered the opportunity. However, even when persons with a disability are able to find jobs, they earn less than persons without a disability. On average, workers with a disability receive lower pay, have less job security, and have less access to health insurance, pension plans, and training than their nondisabled counterparts (Society for Human Resource Management, 2010).

## Politics and the Economy in the Future

Thinking about U.S. politics and government in the future is very much like the old story about optimists and pessimists. According to the story, an eight-ounce cup containing exactly four ounces of water is placed on a table. The optimist comes in, sees the cup, and says, "The cup is half full." The pessimist comes in, sees the cup, and says, "The cup is half empty." Clearly, both the pessimist and the optimist are looking at the same cup containing the same amount of water, but their perspective on what they see is quite different. For some analysts, looking at the future of the U.S. government and the economy is very much like this.

First, views of the future of politics and government relate to specific concerns about the United States:

■ What will be the future of political parties? What did the last presidential election tell us about the nature of political parties and politics? Does social networking reduce the need for political parties?
■ Are global corporate interests and the concerns of the wealthy and powerful overshadowing the needs and interests of everyday people?
■ How will elected politicians and appointed government officials handle the challenges facing the United States in a global economy and rapidly shrinking world?

Second, how will the U.S. economy look in the future? What about the global economy? Although sociologists do not have a crystal ball with which to predict the future, some general trends can be suggested.

Many of the trends we examined in this chapter will produce dramatic changes in the organization of the economy and work in the future. U.S. workers may find themselves fighting for a larger piece of an ever-shrinking economic pie that includes a trade deficit and a huge national debt. Even as the gender and racial–ethnic composition of the labor force continues to diversify, many workers will remain in occupations and industries segregated by gender and/or race/ethnicity. At the same time, workers may increasingly be fragmented into two major labor market divisions: (1) those who work in the innovative, primary sector and (2) those whose jobs are located in the growing secondary, marginal sector. In the innovative sector, increased productivity will be the watchword as corporations respond to heightened international competition. In the marginal sector, alienation will grow as temporary workers, sometimes including professionals such as lawyers and health care specialists, look for avenues of upward mobility or at least a chance to make their work life more tolerable.

**unemployment rate** the percentage of unemployed persons in the labor force actively seeking jobs.

One of the greatest concerns for the future is the employment of U.S. workers. There is also cause for concern because many jobs that do exist do not provide sufficient income for people to pay for fundamentals such as housing, utilities, food, health care, and transportation. However, despite these problems, many analysts predict that the United States will remain a major player in the world economy. Changes in the global economy, instantaneous worldwide communications and social networking, and rapidly shifting political boundaries require people in all nations—including the United States—to make changes in the way that things are done on the individual, regional, national, and international levels. Will we be sufficiently flexible and well informed to make those changes in a timely manner? We must address this issue now and in the decades that follow.

# chapter REVIEW

## LO1 How are the terms *power* and *authority* defined, and what are the differences among traditional, charismatic, and rational–legal authority?

Power is the ability of persons or groups to carry out their will even when opposed by others. Authority is power that people accept as legitimate rather than coercive. Max Weber identified these types of authority: traditional, charismatic, and rational–legal. Traditional authority is based on long-standing custom. Charismatic authority is power based on a leader's personal qualities. Rational–legal authority is based on law or written rules and regulations, as found in contemporary bureaucracies.

## LO2 What are the main types of political systems found around the world?

The main types of political systems are monarchies, authoritarian systems, totalitarian systems, and democratic systems. In a monarchy, one person is the hereditary ruler of the nation. In authoritarian systems, rulers tolerate little or no public opposition and generally cannot be removed from office by legal means. In totalitarian systems, the state seeks to regulate all aspects of society and to monopolize all societal resources in order to exert complete control over both public and private life. In democratic systems, the powers of government are derived from the consent of all the people.

## LO3 How does the pluralist perspective view power in the United States, and what part do special interest groups and political action committees play in this approach?

According to the pluralist (functionalist) model, power is widely dispersed throughout many competing interest groups. People influence policy by voting, joining special interest groups and political action campaigns, and forming new groups. Special interest groups are political coalitions made up of individuals or groups that share a specific interest they wish to protect or advance with the help of the political system. Political action committees (PACs) are organizations of special interest groups that solicit contributions from donors and fund campaigns to help elect (or defeat) candidates based on their stances on specific issues.

## LO4 How do political parties shape the government, and what are some recent trends in political participation and voter apathy?

A political party (1) develops and articulates policy positions, (2) educates voters about issues and simplifies the choices for them, and (3) recruits candidates who agree with those policies, helps those candidates win office, and holds the candidates responsible for implementing the party's policy positions. Some people respond when they are displeased with current political parties and form their own organization; however, others simply do not participate and contribute to high rates of voter apathy in many elections. In the 2012 presidential election, 58.7 percent of the 219 million eligible voters cast ballots. This percentage was down from the 2008 presidential election, when 62.3 percent of the 208.3 million eligible voters actually voted.

## LO5 What is the primary function of the economy, and how do preindustrial, industrial, and postindustrial economies compare?

The economy is the social institution that ensures the maintenance of society through the production, distribution, and consumption of goods and services. Preindustrial economies include hunting and gathering, horticultural and pastoral, and agrarian societies. Most workers engage in primary sector production—the extraction of raw materials and natural resources from the environment. Industrial societies engage in secondary sector production, which is based on the processing of raw materials (from the primary sector) into finished goods. Postindustrial societies engage in tertiary sector production by providing services rather than goods.

## LO6 What are the key characteristics of capitalism, socialism, and mixed economies?

Capitalism is characterized by private ownership of the means of production, the pursuit of personal profit, competition, and limited government intervention. By contrast, socialism is characterized by public ownership of the means of production, the pursuit of collective goals, and centralized decision making. In mixed economies, elements of a capitalist, market economy are combined with elements of a command, socialist economy. These mixed economies are often referred to as democratic socialism.

## LO7 What are the characteristics of professions, and how do professions differ from other occupations?

Professions are high-status, knowledge-based occupations characterized by abstract, specialized knowledge; autonomy; self-regulation; authority over clients and subordinate occupational groups; and a degree of altruism. Occupations are categories of jobs that involve similar activities at different work sites.

## LO8 What problems are associated with contingent work and the underground (informal) economy?

Contingent work is part-time work, temporary work, or subcontracted work that offers advantages to employers but may be detrimental to workers. Through the use of contingent workers, employers are able to cut costs and maximize profits, but workers have little or no job security. The underground economy is made up of a wide variety of activities through which people make money that they do not report to the government or through endeavors that may involve criminal behavior.

## key TERMS

authoritarianism 406

authority 403

capitalism 421

charismatic authority 403

conglomerate 424

contingent work 428

corporations 421

democracy 407

democratic socialism 426

economy 420

elite model 410

government 401

interlocking corporate
    directorates 424

marginal jobs 428

military–industrial complex 419

mixed economy 426

monarchy 405

occupations 428

oligopoly 423

pluralist model 408

political action committees 409

political party 413

political socialization 416

politics 401

power 403

primary labor market 428

primary sector production 420

professions 427

rational–legal authority 404

representative democracy 407

routinization of charisma 403

secondary labor market 428

secondary sector production 420

shared monopoly 424

socialism 425

state 401

subcontracting 429

tertiary sector production 420

totalitarianism 406

traditional authority 403

transnational corporations 421

unemployment rate 430

welfare state 426

## questions for CRITICAL THINKING

1. Who is ultimately responsible for decisions and policies that are made in a democracy such as the United States: the people or their elected representatives? Why is it important to examine how economic influences affect the political process in the United States and other nations?

2. How would you design a research project that studies the effects of social media on people's views about political candidates and issues? What would be your hypothesis? What kinds of data would you need to gather? How would you gather accurate data?

# 14 HEALTH, HEALTH CARE, AND DISABILITY

## LEARNING OBJECTIVES

**LO1** Discuss the relationship between the social environment and health and illness.

**LO2** Define *social epidemiology* and identify key demographic factors that are studied by social epidemiologists.

**LO3** Explain how the profession of medicine emerged in the United States.

**LO4** Discuss how health care is paid for in the United States and in other nations.

**LO5** Describe how advanced medical technology has changed the practice of medicine and the cost of health services.

**LO6** Explain how functionalist, conflict, symbolic interactionist, and postmodern approaches differ in their analysis of health and health care.

**LO7** Discuss what is meant by the term *mental illness*, and explain why it is a difficult topic for sociological research.

**LO8** Discuss the concept of disability and identify key sociological perspectives on disability.

Michael Keller/AGE Fotostock

435

# Medicine as a Social Institution

**Medicine is, I have found, a strange and in many ways disturbing business.** The stakes are high, the liberties taken tremendous. We drug people, put needles and tubes into them, manipulate their chemistry, biology, and physics, lay them unconscious and open their bodies up to the world.

We do so out of an abiding confidence in our know-how as a profession. What you find when you get in close, however—close enough to see the furrowed brows, the doubts and missteps, the failures as well as the successes—is how messy, uncertain, and also surprising medicine turns out to be.

The thing that still startles me is how fundamentally human an endeavor it is. Usually, when we think about medicine and its remarkable abilities, what comes to mind is the science and all it has given us to fight sickness and misery: the tests, the machines, the drugs, the procedures. And without question, these are at the center of virtually everything medicine achieves. But we rarely see how it all actually works. You have a cough that won't go away—and then? It's not science you call upon but a doctor. A doctor with good days and bad days. A doctor with a weird laugh and a bad

Erik Jacobs/The New York Times/Redux

Dr. Atul Gawande (center) has written movingly about the differences between people's expectations of physicians and the medical establishment and the realities that they find in health care today. Sociologists study these contradictions to better understand a very complex and important part of U.S. social life.

---

The everyday life of a doctor like Atul Gawande is filled with its high points and low points: Some patients benefit from medical treatments they receive from physicians, whereas others have sustained injuries or developed illnesses that are too severe or are beyond the scope of current knowledge and practice in the health care system to be successfully resolved. Physicians are human beings just like the patients they treat; however, much more is expected of them because of the availability of health care in the United States and other high-income nations and because the dominant role of doctors in modern high-tech medicine has led many individuals to believe that virtually anything should be possible when it comes to one's health and longevity. However, this assumption is often not an accurate reflection of how health, illness, and health care actually work.

What does the concept of health mean to you? At one time, health was considered to be simply the absence of disease. However, the World Health Organization defines *health* as **a state of complete physical, mental, and social well-being.** According to this definition, health involves not only the absence of disease but also a positive sense of wellness. In other words, health is a multidimensional phenomenon: It includes physical, social, and psychological factors.

What is illness? *Illness* refers to an interference with health; like health, illness is socially defined and may change over time and between cultures. For example, in the United States and Canada, obesity is viewed as unhealthy, whereas in other times and places, obesity indicated that a person was prosperous and healthy.

What happens when a person is perceived to have an illness or disease? Healing involves both personal and institutional responses to perceived illness and disease. One aspect of institutional healing is health care and the health care delivery system in a society. **Health care is any activity intended to improve health.** When people experience illness, they often seek medical attention in hopes of having their health restored. A vital part of haircut. A doctor with three other patients to see and, inevitably, gaps in what he knows and skills he's still trying to learn. . . . We look for medicine to be an orderly field of knowledge and procedure. But it is not. It is an imperfect science, an enterprise of constantly changing knowledge, uncertain information, fallible individuals, and at the same time lives on the line. There is science in what we do, yes, but also habit, intuition, and sometimes plain old guessing. The gap between what we know and what we aim for persists. And this gap complicates everything we do.

—Atul Gawande, M.D. (2002: 4, 5, 7), a surgeon at Brigham and Women's Hospital in Boston, was a surgical resident when he wrote these words describing how he feels about the power and the limits of medicine.

## How Much Do You Know About Health, Illness, and Health Care?

**True  False**

T    F    1. The idea that everyone should have guaranteed health insurance coverage is accepted by nearly all Americans.

T    F    2. Worldwide, most HIV infections are spread by homosexual contact.

T    F    3. The primary reason that African Americans have shorter life expectancies than whites is the high rate of violence in central cities and the rural South.

T    F    4. The most common cause of death among Americans ages 15 to 24 is unintentional injuries.

**True  False**

T    F    5. Health care in most high-income, developed nations is organized on a fee-for-service basis as it is in the United States.

T    F    6. The medical–industrial complex has operated in the United States with virtually no regulation, and allegations of health care fraud have largely been overlooked by federal and state governments.

T    F    7. Media coverage of chronic depression and other mental conditions focuses primarily on these problems as "women's illnesses."

T    F    8. About two-thirds of U.S. adults are either overweight or obese.

Answers on page 439.

health care is *medicine*—**an institutionalized system for the scientific diagnosis, treatment, and prevention of illness.**

In this chapter we will explore the dynamics of health, health care, and disability from a sociological perspective, as well as look at issues through the eyes of those who have experienced medical problems. Before reading on, test your knowledge about health, illness, and health care by taking the "Sociology and Everyday Life" quiz.

## Health in Global Perspective

**LO1** Studying health and health care issues around the world offers insights on illness and how political and economic forces shape health care in nations. Disparities in health are glaringly apparent between high-income and low-income nations when we examine factors such as the prevalence of life-threatening diseases, rates of life expectancy and infant mortality, and access to health services. In regard to global health, for example, approximately 2.3 million people worldwide became newly infected with HIV in 2012 and about 1.6 million died from AIDS-related causes (UNAIDS, 2013). Twenty-six countries saw a 50 percent or greater drop in new HIV infections between 2001 and 2012. Likewise, AIDS-related deaths have fallen by 30 percent since the peak year of 2005. In 2012, 1.5 million people died from AIDS-related causes worldwide, compared to 2.3 million in 2005. Overall, more than 36 million have died from AIDS-related illnesses since the start of the epidemic (UNAIDS, 2013).

**health** a state of complete physical, mental, and social well-being.

**health care** any activity intended to improve health.

**medicine** an institutionalized system for the scientific diagnosis, treatment, and prevention of illness.

**FIGURE 14.1** Japan has one of the lowest infant mortality rates in the world. Why? Finding the answer to this question would benefit people in every nation.

*Life expectancy* **refers to an estimate of the average lifetime of people born in a specific year.** However, AIDS is not the only disease reducing life expectancy in some nations. Most deaths in low- and middle-income nations are linked to infectious and parasitic diseases that are now rare in high-income, industrialized nations. Among these diseases are tuberculosis, polio, measles, diphtheria, meningitis, hepatitis, malaria, and leprosy. Life expectancy ranges from a low of 48.7 years in Chad to 89.7 years in Monaco (*CIA Factbook,* 2013).

The *infant mortality rate* **is the number of deaths of infants under 1 year of age per 1,000 live births in a given year.** In 2012 the infant mortality rate ranged from a low of 1.8 in high-income nations such as Monaco and Japan to a high of 117 in low-income nations such as Afghanistan (Figure 14.1). Consider the tragic difference in the deaths of 117 infants out of every 1,000 infants before their first birthday in a given year, as compared with the deaths of slightly less than 2 infants (statistically speaking) out of every 1,000. A large proportion of these deaths occur during the infants' *first month* of life. There are many reasons for these differences in life expectancy and infant mortality. Many people in low-income countries have insufficient or contaminated food; lack access to pure, safe water; and do not have adequate sewage and refuse disposal. Added to these hazards is a lack of information about how to maintain good health. Many of these nations also lack qualified physicians and health care facilities with up-to-date equipment and medical procedures.

Nevertheless, tremendous progress has been made in saving the lives of children and adults over the past 20 years. The average life expectancy at birth is more than 70 years in the following regions: the Arab States, East Asia and the Pacific, Europe and Central Asia, and Latin America and the Caribbean. By contrast, life expectancy at birth remains at 54.9 years in sub-Saharan Africa (United Nations Development Programme, 2013). One of the key indicators of the level of human development in a nation or region is the rate at which children survive. Referred to as "child survival," this is often measured by the child's fifth birthday. However, even for children who survive past this birthday, poor health is still a major concern because it can permanently damage their cognitive development and ability to fully function as an adult (United Nations Development Programme, 2013).

Despite the progress that has been made in life expectancy and medical care, many challenges remain. Consider these current global health concerns:

- Although raised blood pressure is a known risk factor for deaths from stroke and coronary heart disease, many people worldwide have undiagnosed or untreated hypertension. In Africa it is estimated that more than one-third of the population has high blood pressure, and this problem is growing worse.
- Being overweight or obese is a problem that increases the risk of coronary heart disease, ischemic stroke, type 2 diabetes, and some cancers, and the worldwide prevalence of these conditions has doubled over the past two decades.
- Maternal deaths are a continuing problem in lower-income nations, where women face fifteen times the risk of dying during pregnancy or from childbirth-related complications than women in higher-income nations.
- Worldwide, the amount of money spent on health care varies widely, and the extent to which persons derive improved medical attention or quality of life through higher expenditures (such as in the United States) is not easily measured. However, lack of adequate funds to pay for appropriate health care is also a pressing concern. (World Health Organization, 2012)

With all of these global health care issues, pressing problems remain regarding the costs and availability of advanced diagnostic and surgical technologies and lifesaving drugs around the world. Advanced technologies such as the da Vinci surgery system (physician-directed robotic surgery) and MRI machines cost in the range of $1.5 million to $3 million, not including expenses involved in creating specially designed suites to house the equipment.

Similarly, drugs to reduce pain and suffering or to save lives are very costly for individuals and medical institutions. The major pharmaceutical

# ANSWERS to the Sociology Quiz on Health, Illness, and Health Care

**1. False.** The battle over passage of the historic health care reform law in 2010 was a reflection of how divided the United States is over universal health coverage. Debates continue over trying to end "Obamacare," as this law is now referred to by some.

**2. False.** In the United States, more than half of HIV infections occur among gay and bisexual men and other men who have sex with men. However, globally, most cases of HIV infection are transmitted through heterosexual contact.

**3. False.** The lower life expectancy of African Americans as a category is caused by a higher prevalence of life-threatening illnesses, such as cancer, heart disease, hypertension, and AIDS. However, it should be noted that African American males do have the highest death rates from homicide of any racial–ethnic category in the United States.

**4. True.** Unintentional injuries account for nearly half of all deaths among Americans age 15–24, making unintentional injuries the most common cause of death in this age group. The majority of unintentional injury deaths in this age group result from motor vehicle accidents.

**5. False.** Most high-income, developed nations have some form of universal health coverage, which is either provided by the government or purchased by the government.

**6. False.** A number of government investigations have focused on rising health care payments and allegations of fraud in the health care delivery system. Billing fraud has been found in Medicare and Medicaid payments to physicians, hospitals, nursing homes, home health agencies, medical labs, and medical equipment manufacturers.

**7. False.** Until recently, chronic depression and other mental conditions were most often depicted as "female" problems. However, male depression has become more widely publicized through documentaries on individuals' lives and through advertising for antidepressants.

**8. True.** Overweight and obesity rates are alarmingly high in the United States, with two-thirds of U.S. adults being either overweight or obese.

Sources: Based on Kaiser Family Foundation, 2012; National Center for Health Statistics, 2012; and Weiss and Lonnquist, 2012. © Cengage Learning

companies that hold patents on the most widely used drugs believe that their name-brand products need to be protected by law, whereas people in international human relief agencies believe that the most important concern is providing needed medication to the one-third of the world's population that does not have access to essential medicines. Transnational pharmaceutical companies fear that if they provide their name-brand drugs (if these are the only ones available to treat a specific condition) at lower prices in low-income countries, this will undercut their major sales base in high-income countries. For this reason the companies are reluctant to provide inexpensive or free drugs to people in low-income countries while, at the same time, they have been aggressively marketing name-brand drugs to patients in high-income nations through the media (see "Health Issues in the Media"). If we are to see a significant improvement in life expectancy and health among people of all nations, improvements are needed in accessibility to advanced medical technologies and effective drugs.

How about improvements in health and health care within one nation? Is there a positive relationship between the amount of money that a society spends on health care and the overall physical, mental, and social well-being of its people? Not necessarily. If there were such a relationship, people in the United States would be among the healthiest and most physically fit people in the world. Some estimates suggest that we spend as much as $2.8 trillion annually on health care (Brill, 2013). In 2012 health care spending accounted for 17.7 percent of the gross domestic product (GDP) in the United States, while it accounted for only 11.2 percent in Canada, 9.1 in Italy, and 7.4 in Korea (OECD, 2013).

**life expectancy** an estimate of the average lifetime of people born in a specific year.

**infant mortality rate** the number of deaths of infants under 1 year of age per 1,000 live births in a given year.

# HEALTH ISSUES IN THE MEDIA

## It's Right for You! The Framing of Drug Ads

JB Reed/Landov

You know when you feel the weight of sadness. You may feel exhausted, hopeless, anxious. Whatever you do you feel lonely and do not enjoy the things you once loved. Things just don't feel like they used to. These are some symptoms of depression, a serious medical condition affecting over 20 million Americans. . . . You just shouldn't have to feel this way anymore. . . . When you know more about what's wrong, you can help make it right.

—ad for Zoloft, an antidepression drug

This ad for a prescription antidepression drug is only one of thousands of pharmaceutical advertisements that constantly bombard us. Drug advertising aimed at consumers has accounted for at least $5 billion in annual advertising dollars for a number of years (Japsen, 2012). Most drug ads use *sympathetic and intuitive framing* to help viewers and readers believe that they are not alone if they have feelings of depression, patterns of sexual dysfunction, or whatever else may be bothering them. In many ads a "real person" gives his or her testimonial about how a particular drug has helped with the problem. The undertone of the person's statement is to offer a *sympathetic* message to viewers ("I know what you're going through because I've been there myself!"). Then a narrator or voiceover (a person who is heard but not seen) makes several understanding (*intuitive*) statements that suggest to individuals in the media audience, "We know you don't want to discuss this with anyone, but we can help." Ads such as this offer a simple solution (in the form of a pill) to help people solve their problems if they will only follow the narrator's advice: "Ask your doctor if [this product] is right for you."

What are the strengths and limitations of framing prescription drug ads in this manner and addressing consumers directly rather than pitching the products only to the physicians who prescribe the drugs? An obvious strength of direct advertising is that patients become aware of products and learn about illnesses or conditions of which they were unaware. However, major limitations exist when it comes to direct advertising of such products. Many drug ads do not tell the whole story: Some play down the fact that other drugs,

**Increasingly, advertisements for prescription drugs are aimed at consumers, although they cannot purchase such products until a doctor prescribes them. Do you agree with this approach to informing patients? Why or why not?**

or a nondrug option, may be more effective in treating the condition than the pill being marketed. Other ads encourage people who are not good candidates for a particular drug to insist that their physician prescribe it for them anyway.

Although more prescription drug ads today inform media audiences about possible side effects than those in the past, the narrators usually read the list of negative effects very quickly and then—in a highly reassuring voice—state that most side effects are rare. One example is an ad for Mirapex, a restless-leg-syndrome drug, which states, "Tell your doctor . . . if you experience increased gambling, sexual, or other intense urges." Comments such as this may produce laughter from viewers rather than giving them a reason to seriously consider the possible negative consequences of certain prescription medications.

Today, the U.S. Congress and various governmental agencies continue to ponder the future of prescription drug ads that directly target consumers, but little is likely to come of these inquiries because spending on consumer drug advertising is so high and pressure from the pharmaceutical industry to continue these lucrative ads is so great. However, Mo Rocca (2007), of CBS News' *Sunday Morning,* offered the following humorous, short-term solution to dealing with the side effects of TV drug ads:

> If there's one side effect to all these ads, it's heightened anxiety. . . . Maybe the best solution is to tune out the ads. Or at least ask your doctor if watching them is right for you. And in no instance should you watch them while you drink or operate heavy machinery. Otherwise, you're likely to experience nausea, dizziness . . . and an uncontrollable urge to throw your TV out the window.

## reflect & analyze

**As a sociology student, what effect (if any) do you think that prescription drug ads have on people you know?**

# Health in the United States

Even if we limit our discussion (for the moment) to people in the United States, why are some of us healthier than others? Is it biology—our genes—that accounts for this difference? Does the environment within which we live have an effect? How about our own individual lifestyle?

## Social Epidemiology

**LO2** The field of social epidemiology attempts to answer questions such as these. *Social epidemiology* is the study of the causes and distribution of health, disease, and impairment throughout a population (Weiss and Lonnquist, 2012). Typically, the target of the investigation is disease agents, the environment, and the human host. *Disease agents* include biological agents such as insects, bacteria, and viruses that carry or cause disease; nutrient agents such as fats and carbohydrates; chemical agents such as gases and pollutants in the air; and physical agents such as temperature, humidity, and radiation. The *environment* includes the physical (geography and climate), biological (presence or absence of known disease agents), and social (socioeconomic status, occupation, and location of home) environments (Figure 14.2). The human *host* takes into account demographic factors (age, sex, and race/ethnicity), physical condition, habits and customs, and lifestyle (Weiss and Lonnquist, 2012). Let's look briefly at some of these factors.

**Age** Rates of illness and death are highest among the old and the young. Mortality rates drop shortly after birth and begin to rise significantly during middle age. After age 65, rates of chronic diseases and mortality increase more rapidly. *Chronic diseases are illnesses that are long term or lifelong and that develop gradually or are present from birth;* in contrast, *acute diseases are illnesses that strike suddenly and cause dramatic incapacitation and sometimes death* (Weitz, 2013). Two of the most common sources of chronic disease and premature death are tobacco use, which increases mortality among both smokers and people who breathe the tobacco smoke of others, and alcohol abuse, both of which are discussed later in this chapter.

The fact that rates of chronic diseases increase rapidly after age 65 has obvious implications not only for people reaching that age (and their families) but also for society. Between 2010 and 2015, the percentage of the U.S. population ages 65 and

**FIGURE 14.2** Access to quality health care is much greater for some people than for others. The factors that are involved vary not only for people within one nation but also across the nations of the world.

**social epidemiology** the study of the causes and distribution of health, disease, and impairment throughout a population.

**chronic diseases** illnesses that are long term or lifelong and that develop gradually or are present from birth.

**acute diseases** illnesses that strike suddenly and cause dramatic incapacitation and sometimes death.

over will increase from 13 percent to 21 percent, and the percentage of those ages 80 and over is expected to double from 4 percent to 8 percent. The cost of caring for the older population—especially those who must live in institutionalized settings—will increase in at least direct proportion to their numbers.

**Sex** Prior to the twentieth century, women had lower life expectancies than men because of high mortality rates during pregnancy and childbirth. Preventive measures have greatly reduced this cause of female mortality, and women now live longer than men. For babies born in the United States in 2013, for example, the estimated life expectancy at birth was 78.7 years, with 76.3 years for males and 81.1 years for females (Minino, 2013). Females have a slight biological advantage over males in this regard from the beginning of life, as can be seen in the fact that they have lower mortality rates both in the prenatal stage and in the first month of life.

However, gender roles and gender socialization also contribute to the difference in life expectancy: Men are more likely to work in dangerous occupations such as commercial fishing, mining, construction, and public safety/firefighting. As a result of gender roles, males may be more likely than females to engage in risky behavior such as drinking alcohol, smoking cigarettes (there is more social pressure on women not to smoke), using drugs, driving dangerously, and engaging in fights. Finally, women are more likely to use the health care system, with the result that health problems are identified and treated earlier (while there is a better chance of a successful outcome), whereas many men are more reluctant to consult doctors.

Because women on average live longer than men, it is easy to jump to the conclusion that they are healthier than men. However, although men at all ages have higher rates of fatal diseases, women have higher rates of chronic illness.

**Race/Ethnicity and Social Class** Racial–ethnic differences are also visible in statistics pertaining to life expectancy. Projections for infants born in 2013 reveal this sobering fact: Even with improvements in life expectancy among persons of color in recent years, life expectancy for African American males is estimated at 71.6 years as compared to 78.9 for Latino (Hispanic) males and 76.4 for white (non-Hispanic) males. For African American females, life expectancy is 77.8 years compared to 83.7 years for Latina (Hispanic) females and 81.1 years for white females (Minino, 2013). For both sexes, the widest life expectancy gap is 6.6 years between Hispanics at 81.4 years and African Americans (non-Hispanic) at 74.8 years.

Although race/ethnicity and social class are related to issues of health and mortality, research continues to show that income and the neighborhood in which a person lives may be equally or more significant than race or ethnicity with respect to these issues. How is it possible that the neighborhood you live in may significantly affect your risk of dying during the next year? Numerous studies have found that people have a higher survival rate if they live in better-educated or wealthier neighborhoods than if the neighborhood is low income and has low levels of education. Among the reasons that researchers believe that neighborhoods make a difference are the availability (or lack thereof) of safe areas to exercise, grocery stores with nutritious foods, and access to transportation, education, and good jobs. Many low-income neighborhoods are characterized by fast-food restaurants, liquor stores, and other facilities that do not afford residents healthy options (Figure 14.3).

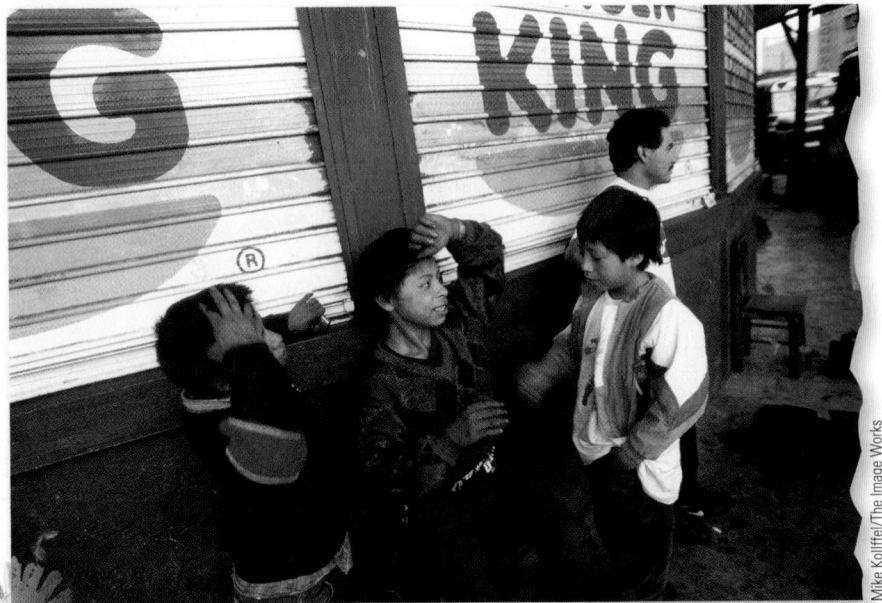

**FIGURE 14.3** Can your neighborhood be bad for your health? According to recent research, it can indeed, especially if it predominantly contains fast-food restaurants, liquor stores, and similarly unhealthy lifestyle options.

As discussed in prior chapters, people of color are more likely to have incomes below the poverty line, and the poorest people typically receive less preventive care and less-optimal management of chronic diseases than do other people. People living in central cities, where there are high levels of poverty and crime, or in remote rural areas generally have greater difficulty in getting health care because most doctors prefer to locate their practice in a "safe" area, particularly one with a patient base that will produce a high income.

© iStockphoto.com/ConorCrowePhoto

**FIGURE 14.4** Occupation and life expectancy may be related. Men are overrepresented in high-risk jobs, such as die-casting molten metal, that may affect their life expectancies.

Although rural Americans make up 20 percent of the U.S. population, only about 10 percent of the nation's physicians practice in rural areas, and fewer specialists such as cardiologists are available there.

Another factor is occupation (see Figure 14.4). People with lower incomes are more likely to be employed in jobs that expose them to danger and illness—working in the construction industry or around heavy equipment in a factory, for example, or holding a job as a convenience store clerk or other position that exposes a person to the risk of armed robbery. Finally, people of color and poor people are more likely to live in areas that contain environmental hazards.

However, although Latinas/os are more likely than non-Latino/a whites to live below the poverty line, they have lower death rates from heart disease, cancer, accidents, and suicide, and they have a higher life expectancy. One explanation may be dietary factors and the strong family life and support networks found in many Latina/o families (Weiss and Lonnquist, 2012). Obviously, more research is needed on this point, for the answer might be beneficial to all people.

## Lifestyle Factors

As noted previously, social epidemiologists also examine lifestyle choices as a factor in health, disease, and impairment. We will examine three lifestyle factors as they relate to health: drugs, sexually transmitted diseases, and diet and exercise.

**Drug Use and Abuse** What is a drug? There are many different definitions, but for our purposes a *drug* is any substance—other than food and water—that, when taken into the body, alters its functioning in some way. Drugs are used for either therapeutic or recreational purposes. *Therapeutic* use occurs when a person takes a drug for a specific purpose such as reducing a fever or controlling a cough. In contrast, *recreational* drug use occurs when a person takes a drug for no purpose other than achieving a pleasurable feeling or psychological state. Alcohol and tobacco are examples of drugs that are primarily used for recreational purposes; their use by people over a fixed age (which varies from time to time and place to place) is lawful. Other drugs—such as some antianxiety drugs or tranquilizers—may be used legally only if prescribed by a physician for therapeutic use but are frequently used illegally for recreational purposes.

**Alcohol** The use of alcohol is considered an accepted part of the culture in the United States. According to a Gallup survey, the majority of adults in the United States consume alcoholic beverages, averaging four drinks per week (Saad, 2012). When survey respondents were asked how many alcoholic drinks they had consumed in the past week, 52 percent stated that they had from one to seven drinks. However, 12 percent had consumed eight or more drinks during the same week.

**drug** any substance—other than food and water—that, when taken into the body, alters its functioning in some way.

Overall, 22 percent of the respondents indicated that they sometimes drink too much. Men tended to drink more than women, reporting that they consumed an average of 6.2 drinks as compared with 2.2 drinks for women. Men were more likely to drink beer, followed by liquor and then wine, while women were more likely to drink wine, followed by liquor and then beer (Saad, 2012).

Age and gender are both factors in regard to drinking behavior: Younger men between the ages of 18 to 49 are the heaviest drinkers of any age group or gender. The Gallup study found that 36 percent of respondents in this category indicated that they sometimes drink too much, as compared with 18 percent of men ages 50 and older, 20 percent of women ages 18 to 49, and 8 percent of women ages 50 and older (Saad, 2012).

Race is also a factor in drinking behavior: The Gallup survey found that white Americans are more likely to drink than African Americans and Latinos/as (Hispanic Americans). This study also found that when white Americans drink, they consume more alcohol than do nonwhite Americans: White Americans consumed on average 4.5 drinks in the past week before the study was conducted, as compared with 3.3 drinks for African Americans and Latinas/os (Hispanic Americans) (Saad, 2012).

Is alcohol consumption a problem? Even short-term alcohol use may have negative effects, including motor vehicle crashes, violence against others, sexual aggression, spread of HIV and sexually transmitted diseases (STDs), unplanned pregnancy, and binge drinking and the risky behavior associated with such activity. *Binge drinking* is defined as men drinking five or more alcoholic drinks within a short period of time or women drinking four or more drinks within a short period of time (U.S. CDC, 2012b). More than

38 million U.S. adults engage in binge drinking, but the extent of binge drinking varies from state to state (see Figure 14.5). The average largest number of drinks within a short period of time among binge drinkers ranges from six drinks in the District of Columbia to nine drinks in Wisconsin (see Figure 14.6).

Many long-term problems are associated with alcohol consumption as well. Chronic heavy drinking or alcoholism can cause permanent damage to the brain or other parts of the body. For alcoholics, the long-term negative health effects include *nutritional deficiencies* resulting from poor eating habits (chronic heavy drinking contributes to high caloric consumption but low nutritional intake); *cardiovascular problems* such as inflammation and enlargement of the heart muscle, high blood pressure, and stroke; and eventually to *alcoholic cirrhosis*—a progressive development of scar tissue that chokes off blood vessels in the liver and destroys liver cells by interfering with their use of oxygen. The social consequences of heavy drinking are not always limited to the person doing the drinking. For example, abuse of alcohol and/or other drugs by a pregnant woman can damage her unborn fetus.

Can alcoholism be overcome? Many people get and stay sober. Some rely on organizations such as Alcoholics Anonymous or church support groups to help them overcome their drinking problem. Others undergo medical treatment and therapy sessions to learn more about the root causes of their problem.

**Nicotine (Tobacco)** The nicotine in tobacco is a toxic, dependency-producing psychoactive drug that is more addictive than heroin. It is classified as a stimulant because it stimulates central nervous system receptors and activates them to release adrenaline, which raises blood pressure, speeds

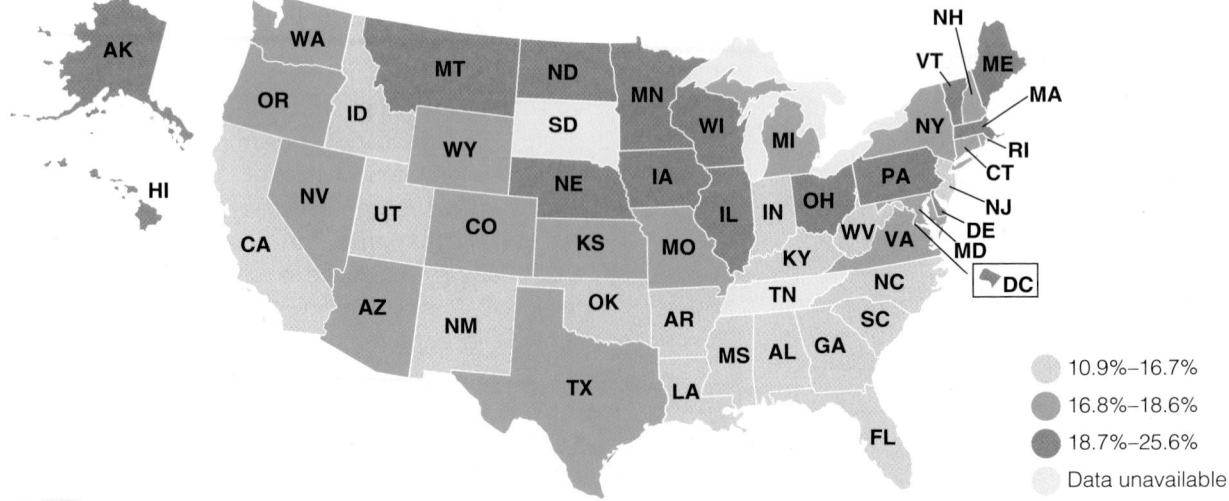

10.9%–16.7%
16.8%–18.6%
18.7%–25.6%
Data unavailable

**FIGURE 14.5** PERCENTAGE OF ADULTS WHO BINGE DRINK
Source: U.S. CDC, 2012b.

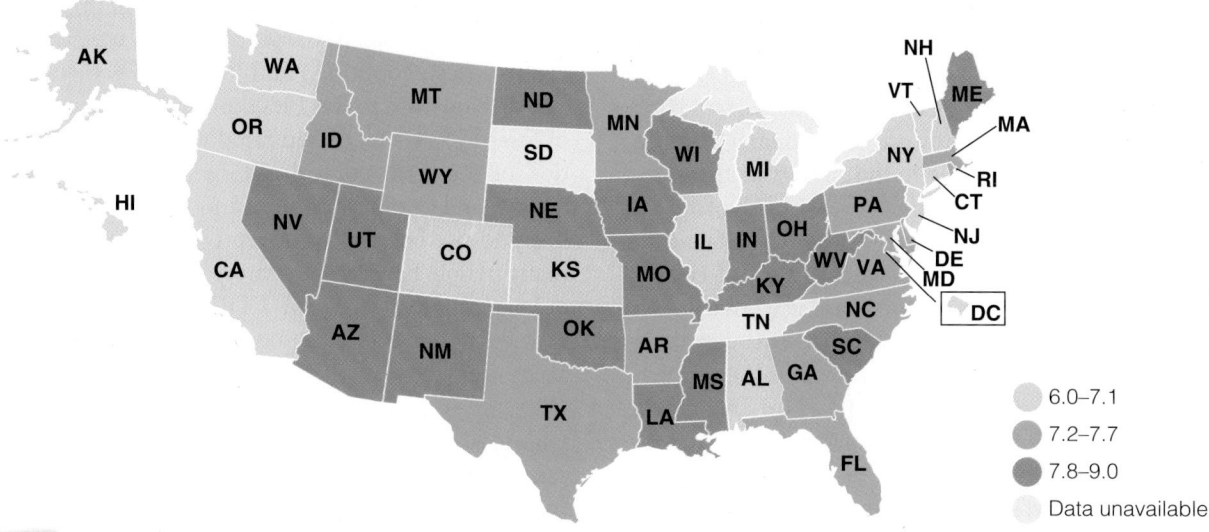

| | |
|---|---|
| ● | 6.0–7.1 |
| ● | 7.2–7.7 |
| ● | 7.8–9.0 |
| | Data unavailable |

**FIGURE 14.6** THE AVERAGE LARGEST NUMBER OF DRINKS CONSUMED BY BINGE DRINKERS ON A SINGLE OCCASION

Source: U.S. CDC, 2012b.

up the heartbeat, and gives the user a temporary sense of alertness. Although the overall proportion of smokers in the general population has declined somewhat since the 1964 Surgeon General warning that smoking is linked to cancer and other serious diseases, tobacco is still the single most preventable cause of disease, disability, and death in the United States. Tobacco is responsible for about one in every five deaths in this country, and it is estimated that more than 450,000 people die prematurely from smoking or exposure to secondhand smoke each year (Figure 14.7). Even people who never light up a cigarette can be harmed by *environmental tobacco smoke*—the smoke in the air inhaled by nonsmokers as a result of other people's tobacco smoking and the residue of smoke on garments and furniture, for example. Researchers have found that environmental smoke is especially hazardous for nonsmokers who carpool or work with heavy smokers. Another 8.5 to 9 million people have a serious illness that is related to smoking (U.S. CDC, 2013c).

**Illegal Drugs** In the United States, marijuana is the most extensively used drug that is illegal in most states. Colorado and Washington were the first states to allow the sale of recreational marijuana (as compared to marijuana for medicinal purposes), in 2014. In Colorado, for example, individuals over age 21 are allowed to purchase a certain quantity of cannabis, but they are not permitted to smoke it in public. Currently, federal law prohibits recreational use of marijuana, but the U.S. Justice Department has indicated that the agency will not challenge states that pass laws legalizing the drug (Martinez, 2014).

The use of this drug has grown exponentially among young people in the nation's middle and

**FIGURE 14.7** Since 2012, the U.S. Food and Drug Administration has required that all cigarette packages carry a vivid color image about tobacco-related cancer disease and a warning about the negative consequences of smoking. Do you think that warnings such as these will help reduce the rate of smoking and tobacco-related deaths in the United States?

high schools. In 2012 researchers found that a high rate of marijuana use by eighth-, tenth-, and twelfth-grade students, combined with a drop in perceptions about the potential harm of such drug use, had contributed to regular or daily use of marijuana by more young people. According to the survey, as teens get older, their perception about the risk involved in smoking marijuana diminishes despite the fact that studies funded by the National Institutes of Health (NIH) have found that persons who used marijuana heavily in their teens and continued through their adulthood showed a significant drop in IQ between the ages of 13 and 38 (NIH, 2012). Moreover, even young people who used marijuana heavily before age 18 (when the brain is still developing) and quit taking it showed impaired mental abilities even after they no longer took the drug.

High doses of marijuana smoked during pregnancy can disrupt the development of a fetus and result in congenital abnormalities and neurological disturbances. Furthermore, some studies have found an increased risk of cancer and other lung problems associated with marijuana because its smokers are believed to inhale more deeply than tobacco users.

Another widely used illegal drug is cocaine. Cocaine may be either inhaled, injected intravenously, or smoked ("crack cocaine"). According to the National Institute on Drug Abuse, nearly five million Americans ages twelve and older have used cocaine, and one million have used crack at least once. People who use cocaine over extended periods of time have higher rates of infection, heart problems, internal bleeding, hypertension, stroke, and other neurological and cardiovascular disorders than do nonusers. Intravenous cocaine users who share contaminated needles are also at risk for contracting HIV.

Each of the drugs discussed in these few paragraphs represents a lifestyle choice that affects health. Whereas age, race/ethnicity, sex, and—at least to some degree—social class are ascribed characteristics, taking drugs is a voluntary action on a person's part.

**Sexually Transmitted Diseases** The circumstances under which a person engages in sexual activity constitute another lifestyle choice with health implications. Although most people find consensual sexual activity enjoyable, it can result in transmission of certain *sexually transmitted diseases (STDs)*, including HIV/AIDS, chlamydia, gonorrhea, and syphilis. The latest estimates from the U.S. Centers for Disease Control and Prevention (CDC) show that there are about 20 million new STD infections in the United States each year. These new infections alone cost the health care system nearly $16 billion in direct medical costs.

Prior to 1960, the incidence of STDs in this country had been reduced sharply by barrier-type contraceptives (e.g., condoms) and the use of penicillin as a cure. However, in the 1960s and 1970s the number of cases of STDs increased rapidly with the introduction of the birth control pill, which led to women having more sexual partners and couples being less likely to use barrier contraceptives. Since that time, many people have become more aware of STDs, and organizations such as the CDC have mounted aggressive public service campaigns to make individuals aware of choices that contribute to sexual health.

**Chlamydia** Chlamydia is the most commonly reported STD in terms of the number of cases (more than 1.4 million) reported annually to the U.S. Centers for Disease Control and Prevention (2014a). However, the CDC estimates that more than half of new cases remain undiagnosed and unreported each year. This is an issue of great concern because chlamydial infections in women, which are usually asymptomatic, can result in pelvic inflammatory disease (PID)—a major cause of infertility, ectopic pregnancy, and chronic pelvic pain. Like other STDs, chlamydia can also contribute to the transmission of HIV. Women are at higher risk of chlamydia, as shown in Figure 14.8. However, in recent years more men have also been tested for chlamydial infection, resulting in an increase in the number of male cases reported. Rates of this disease vary widely among racial and ethnic populations: The rate for African Americans in 2012 was seven times the rate for white Americans (U.S. CDC, 2014a).

**Gonorrhea** Gonorrhea, the second most commonly reported STD in the United States, is caused by a bacterium that can grow and multiply easily in the warm, moist areas of the reproductive tract. It can also grow in the mouth, throat, eyes, and anus. Untreated gonorrhea may lead to serious outcomes such as tubal infertility, ectopic pregnancy, and chronic pelvic pain. Gonorrhea can help the transmission of HIV infection, and in advanced cases it can spread to the blood or joints and can be life threatening.

Specific populations are at higher risk for gonorrhea. Among African Americans, rates are nearly fifteen times higher than rates among white Americans. The South has the highest gonorrhea rate among the four regions of the country (U.S. CDC, 2014b). Although antibiotics can cure gonorrhea, some strains of the infection have developed resistance to each of the antibiotics used for treatment.

**Syphilis** Syphilis is a genital ulcerative disease that can be acquired not only by sexual intercourse but also by kissing or coming into intimate bodily

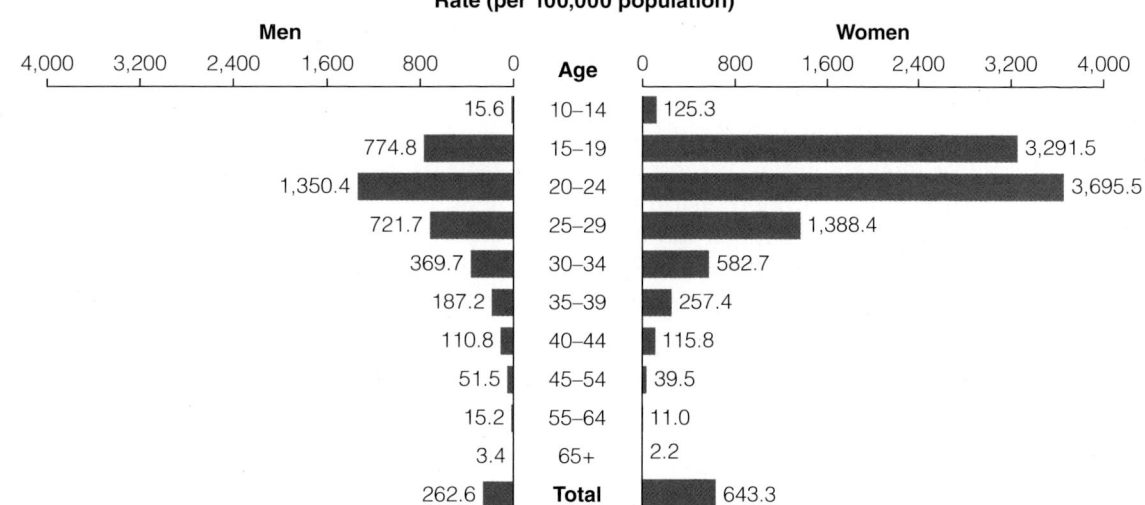

**Rate (per 100,000 population)**

| | Men | | | | | | Age | | | | Women | | |
|---|---|---|---|---|---|---|---|---|---|---|---|---|---|
| 4,000 | 3,200 | 2,400 | 1,600 | 800 | 0 | | | 0 | 800 | 1,600 | 2,400 | 3,200 | 4,000 |

Men / Age / Women values:

- 10–14: Men 15.6 | Women 125.3
- 15–19: Men 774.8 | Women 3,291.5
- 20–24: Men 1,350.4 | Women 3,695.5
- 25–29: Men 721.7 | Women 1,388.4
- 30–34: Men 369.7 | Women 582.7
- 35–39: Men 187.2 | Women 257.4
- 40–44: Men 110.8 | Women 115.8
- 45–54: Men 51.5 | Women 39.5
- 55–64: Men 15.2 | Women 11.0
- 65+: Men 3.4 | Women 2.2
- Total: Men 262.6 | Women 643.3

**FIGURE 14.8** CHLAMYDIA: RATES BY AGE AND SEX, UNITED STATES, 2012

Source: U.S. CDC, 2014a.

contact with an infected person. Untreated early syphilis in pregnant women results in perinatal death in up to 40 percent of cases and may lead to infection of the fetus in up to 80 percent of cases if syphilis was acquired during the four years prior to the pregnancy (U.S. CDC, 2014c). Untreated syphilis can, over time, cause cardiovascular problems, brain damage, and even death. Penicillin can cure most cases of syphilis as long as the disease has not spread.

Like gonorrhea, syphilis is a major health problem in the South and in urban areas in other regions. High-risk populations include young African American men, all men who have sex with other men, and men who have sex with both men and women (U.S. CDC, 2012c).

**AIDS** AIDS (acquired immunodeficiency syndrome), which is caused by HIV (human immunodeficiency virus), is among the most significant health problems that this nation—and the world—faces today. Although AIDS almost inevitably ends in death, no one actually dies *of* AIDS. Rather, AIDS reduces the body's ability to fight diseases, making a person vulnerable to many diseases—such as pneumonia—that result in death.

AIDS was first identified in 1981, and the total number of AIDS-related deaths in the United States through 1985 was only 12,493; however, the numbers rose rapidly and precipitously after that. In recent years the CDC has developed new estimates of HIV prevalence, meaning the total number of people living with HIV. Current estimates suggest that 1.14 million people in the United States ages 13 years and older are in this category. Nearly 16 percent (180,900 people) are not aware that they are infected with HIV. The majority of new HIV infections are attributed to male-to-male sexual behavior, followed by heterosexual transmission and intravenous drug use. Gay, bisexual, and other men who have sex with men (MSM) of all races and ethnicities are the population most profoundly affected by HIV (U.S. CDC, 2013b). Although African Americans make up about 12 percent of the total U.S. population, they account for nearly half (44 percent) of new HIV infections each year. Hispanics/Latinos/as represent 16 percent of the U.S. population but account for 21 percent of new HIV infections. Heterosexuals make up about 25 percent of estimated new HIV infections each year. Most new HIV infections among women are attributed to heterosexual contact or injection drug use (U.S. CDC, 2013b).

According to the U.S. Centers for Disease Control and Prevention (2012a), racial disparities in HIV/AIDS are driven by factors such as the higher HIV prevalence (proportion of people living with HIV) in many African American and Latino communities, which means that individuals who reside in these communities have a higher risk of infection with every sexual encounter. Other factors include the stigma associated with HIV/AIDS and homophobia, which may prevent people from seeking prevention or medical attention.

Worldwide, the number of people with HIV/AIDS continues to increase, but progress has been made in some countries in addressing this epidemic. An estimated 35.3 million people worldwide are living with HIV/AIDS, including 2.1 million between the ages of 10 and 19 (AIDS.gov, 2013). HIV is the world's leading infectious killer.

According to the World Health Organization (WHO), about 2.3 million new cases of HIV were reported worldwide in 2012. Almost all (95 percent) of new infections are reported in low- and middle-income countries, particularly sub-Saharan Africa, where nearly 1 in every 20 adults lives with HIV. WHO also reports that 3.34 million children worldwide are infected. Most of the children were infected by their HIV-positive mothers during pregnancy, childbirth, or breastfeeding. Even more alarming is the fact that more than 700 children per day are newly infected with HIV (AIDS.gov, 2013).

HIV is transmitted through unprotected (or inadequately protected) sexual intercourse with an infected partner (either male or female), by sharing a contaminated hypodermic needle with someone who is infected, by exposure to blood or blood products (usually from a transfusion), or, as previously stated, by an HIV-inflected woman who passes the virus on to her child. It is not transmitted by casual contact such as shaking hands.

Over the past decade, global efforts have been increased to reduce the HIV/AIDS epidemic. More people are receiving standard antiretroviral therapy (ART), which is a combination of at least three antiretroviral (ARV) drugs to maximally suppress the HIV virus and stop the progression of the disease. Massive reductions in rates of death and suffering have occurred when the potent ART regimen is used, particularly in early stages of the disease. ART is also used for the prevention of HIV infection, particularly among pregnant women, children, and other high-risk populations.

**Staying Healthy: Diet and Exercise** Up to this point we have primarily been looking at problematic aspects of lifestyle that have a negative effect on health; however, lifestyle choices also include positive actions such as a healthy diet and good exercise. Over the past several decades, a dramatic improvement in our understanding of food and diet has taken place, and many people in the United States have begun to improve their dietary habits. A significant portion of the population now eats larger amounts of vegetables, fruits, and cereals, substituting unsaturated fats and oils for saturated fats. However, many people in the United States still eat diets that are deficient in fruits and vegetables, and their meals contain more fats and added sugars than are recommended by current dietary guidelines. Eating behavior such as this contributes to obesity, which may shorten the average U.S. life expectancy by 2 to 5 years, reversing the steady increase in life expectancy that has occurred over the past 200 years. Figure 14.9 shows adult obesity rates across the United States.

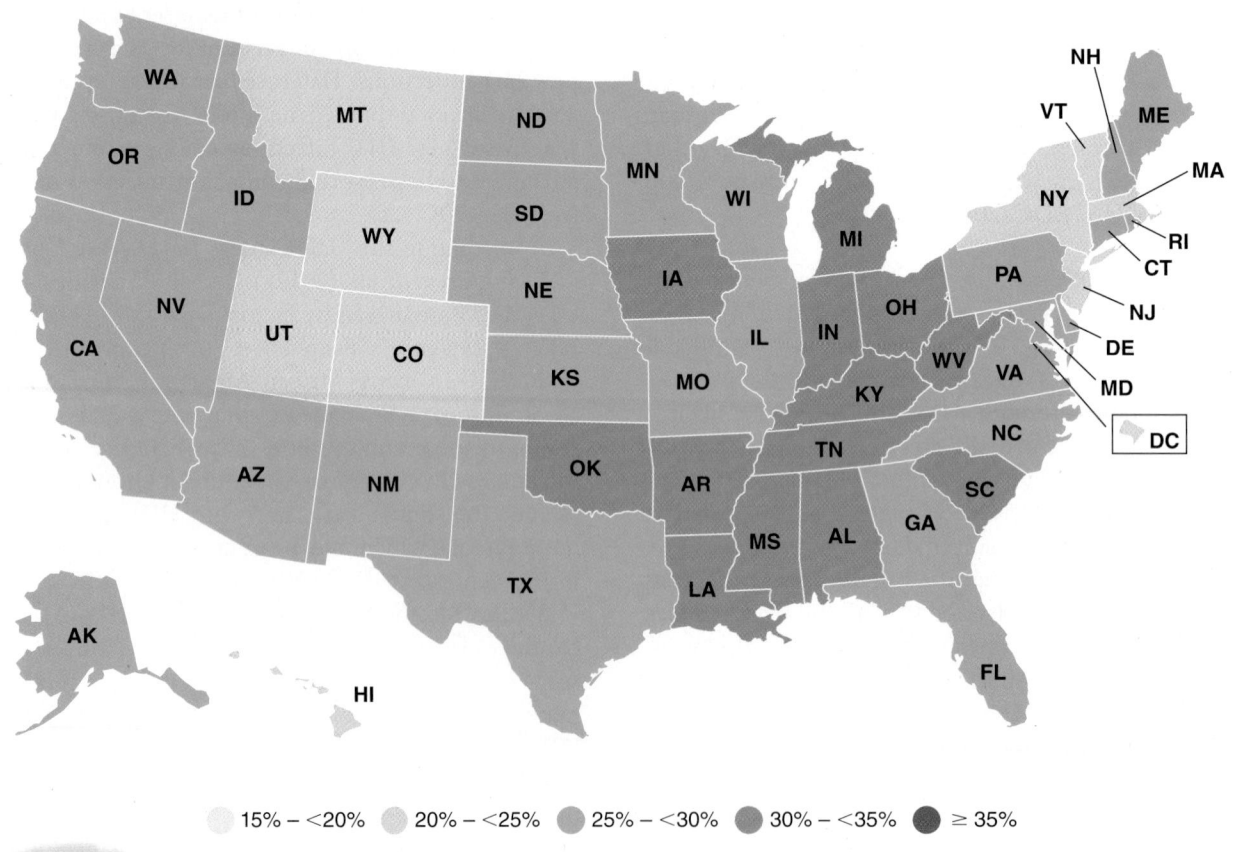

15% – <20%    20% – <25%    25% – <30%    30% – <35%    ≥ 35%

**FIGURE 14.9** ADULT OBESITY IN THE UNITED STATES, 2012
Source: U.S. CDC, 2013a.

PART 4 | SOCIAL INSTITUTIONS

Obesity increases the risk of health problems such as hypertension, type 2 diabetes, and certain types of cancer. It is estimated that more than one-third of all U.S. adults are obese and that many others are overweight. One of the important ways to reduce obesity is through exercise. Regular exercise (at least three times a week) keeps the heart, lungs, muscles, and bones in good health and slows the aging process.

# Health Care in the United States

Understanding health care as it exists in the United States today requires a brief examination of its history. During the nineteenth century, people became doctors in this country either through apprenticeships, purchasing a mail-order diploma, completing high school and attending a series of lectures, or obtaining bachelor's and M.D. degrees and studying abroad for a number of years. At that time, medical schools were largely proprietary institutions, and their officials were often more interested in acquiring students than in enforcing standards. The state licensing boards established to improve medical training and stop the proliferation of "irregular" practitioners failed to slow the growth of medical schools, and their number increased from 90 in 1880 to 160 in 1906. Medical school graduates were largely poor and frustrated because of the overabundance of doctors and quasi-medical practitioners, so doctors became highly competitive and anxious to limit the number of new practitioners. The obvious way to accomplish this was to reduce the number of medical schools and set up licensing laws to eliminate unqualified or irregular practitioners (Kendall, 1980).

## The Rise of Scientific Medicine and Professionalism

**LO3** Although medicine had been previously viewed more as an art than as a science, several significant discoveries during the nineteenth century in areas such as bacteriology and anesthesiology began to give medicine increasing credibility as a science. At the same time that these discoveries were occurring, the ideology of science was being advocated in all areas of life, and people came to believe that almost any task could be done better if the appropriate scientific methods were used. To make medicine in the United States more scientific (and more profitable), the Carnegie Foundation (at the request of the American Medical Association and the forerunner of the Association of American Medical Colleges) commissioned an official study of medical education. The "Flexner report" that resulted from this study has been described as the catalyst of modern medical education but has also been criticized for its lack of objectivity.

**The Flexner Report** To conduct his study, Abraham Flexner met with the leading faculty at the Johns Hopkins University School of Medicine to develop a model of how medical education should take place; he next visited each of the medical schools then in existence, comparing them with the model. Included in the model was the belief that a medical school should be a full-time, research-oriented, laboratory facility that devoted all of its energies to teaching and research, not to the practice of medicine (Kendall, 1980). It should employ "laboratory men" to train students in the "science" of medicine, and the students should then apply the principles they had learned in the sciences to the illnesses of patients (Brown, 1979). Only a few of the schools Flexner visited were deemed to be equipped to teach scientific medicine; nonetheless, his model became the standard for the profession (Figure 14.10).

As a result of the Flexner report (1910), all but two of the African American medical schools then in existence were closed, and only one of the medical schools for women survived. As a result, white women and people of color were largely excluded from medical education for the first half of the twentieth century. Until the civil rights movement and the women's movement of the 1960s and 1970s, virtually all physicians were white, male, and upper or upper-middle class.

**The Professionalization of Medicine** Despite its adverse effect on people of color and women who might desire a career in medicine, the Flexner report did help *professionalize* medicine. When we compare post-Flexner medicine with the characteristics of professions, we find that it meets those characteristics:

1. *Abstract, specialized knowledge.* Physicians undergo a rigorous education that results in a theoretical understanding of health, illness, and medicine. This education provides them with the credentials, skills, and training associated with being a professional.
2. *Autonomy.* Physicians are autonomous and (except as discussed subsequently in this chapter) rely on their own judgment in selecting the appropriate technique for dealing with a problem. They expect patients to respect that autonomy.

**449**

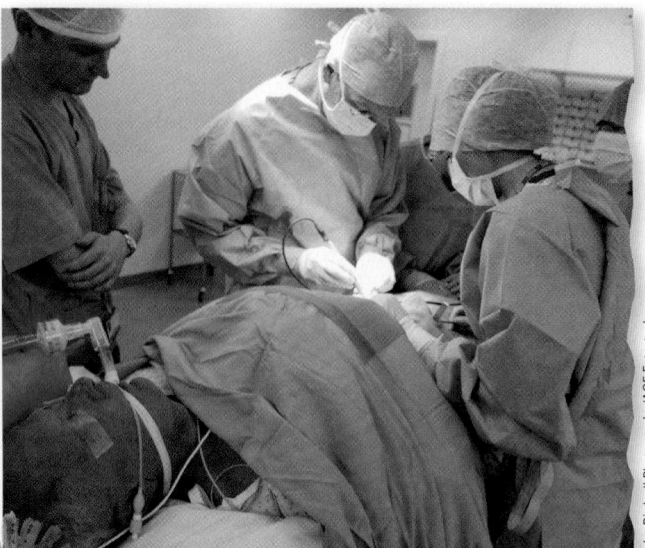

**FIGURE 14.10** A sight similar to the one on the left might have greeted Abraham Flexner as he prepared his report on medical education in the United States: students observing while their professor demonstrates surgical techniques. And although today's health care facilities look very different, many of the same teaching techniques are still employed.

3. *Self-regulation.* Theoretically, physicians are self-regulating. They have licensing, accreditation, and regulatory boards and associations that set professional standards and require members to adhere to a code of ethics as a form of public accountability.

4. *Authority.* Because of their authority, physicians expect compliance with their directions and advice. They do not expect clients to argue about the advice rendered (or the price to be charged).

5. *Altruism.* Physicians perform a valuable service for society rather than acting solely in their own self-interest. Many physicians go beyond their self-interest or personal comfort so that they can help a patient.

However, with professionalization, licensed medical doctors gained control over the entire medical establishment, a situation that has continued until the present and—despite current efforts at cost control by insurance companies and others—may continue into the future.

## Medicine Today

Throughout its history in the United States, medical care has been on a *fee-for-service* basis: Patients have been billed individually for each service they receive, including treatment by doctors, laboratory work, hospital visits, prescriptions, and other health-related expenses. Fee for service is an expensive way to deliver health care because there are few restrictions on the fees charged by doctors, hospitals, and other medical providers. For this reason the Affordable Care Act, passed in 2010, proposed to change how health care is paid for and the ways in which some services are provided.

There are both good and bad sides to the fee-for-service approach. The good side is that in the "true spirit" of capitalism, coupled with the hard work and scholarship of many people, this approach has resulted in remarkable advances in medicine. The bad side of fee-for-service medicine is its inequality of distribution. In effect, the United States has a two-tier system of medical care. For the most part, those who can afford it are able to get good medical treatment and comfortable surroundings (Bernstein, 2012).

However, top-tier medical care is not within the budget of many people. The annual cost of health care per person in the United States rose from $141 in 1960 to $8,233 in 2012 (pbs.org, 2012). Figure 14.11 reflects how expenditures for health care have increased from 1993 to 2012. Keeping in mind the issues that have been raised throughout this text regarding income disparity in the United States, the questions to be considered at this point are "Who pays for medical care, and how?" and "What about the people who simply cannot afford adequate medical care?"

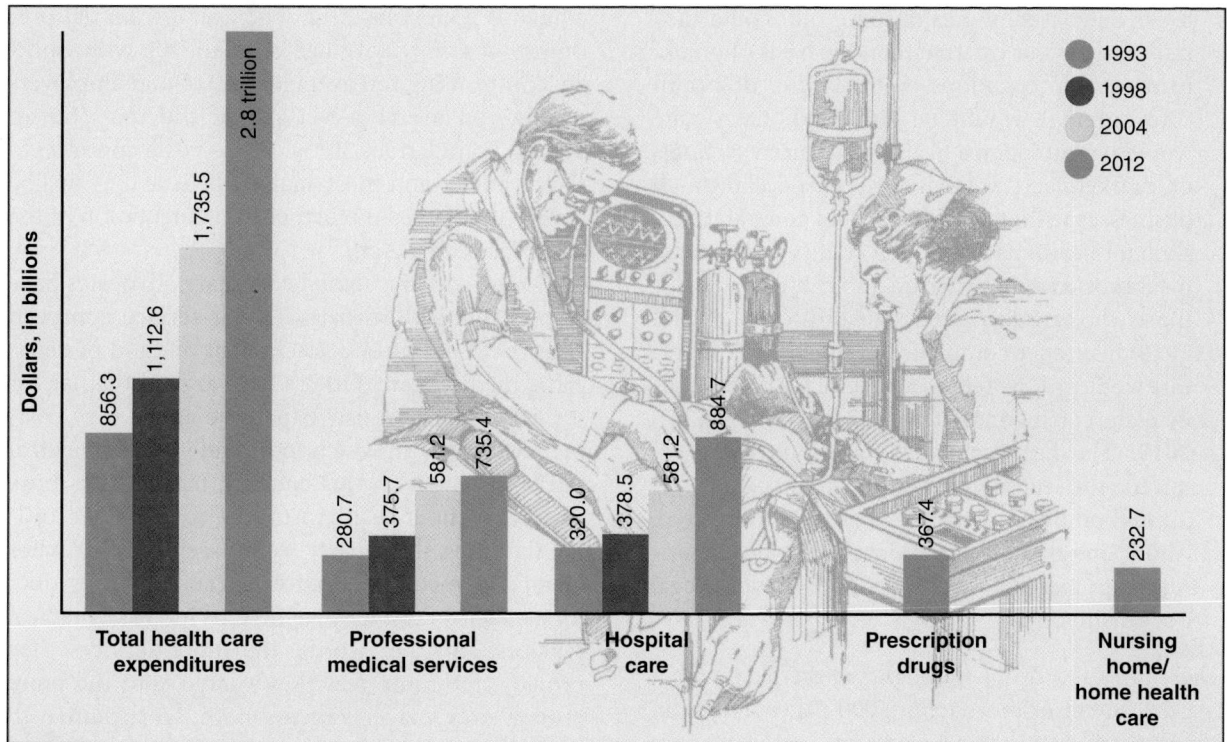

Dollars, in billions

- 1993
- 1998
- 2004
- 2012

**Total health care expenditures:** 856.3 | 1,112.6 | 1,735.5 | 2.8 trillion

**Professional medical services:** 280.7 | 375.7 | 581.2 | 735.4

**Hospital care:** 320.0 | 378.5 | 581.2 | 884.7

**Prescription drugs:** 367.4

**Nursing home/home health care:** 232.7

**FIGURE 14.11** INCREASE IN COST OF HEALTH CARE, 1993–2012

Source: Plunkett Research, 2013. © Cengage Learning

## Paying for Medical Care in the United States

**LO4** Until recently, the United States was the only high-income nation without some form of universal health coverage for all citizens. As previously mentioned, the U.S. Congress passed a sweeping health care reform bill in 2010 that would gradually bring about some changes in how health care is funded. Let's look first at the health reform legislation and then compare its provisions with previous and current methods of funding of health care in the United States.

**The Affordable Care Act of 2010** After a lengthy struggle in the U.S. Congress, a major health care reform bill was signed into law in 2010. One of the central tenets in the law was the creation of a new insurance marketplace that made it possible for individuals and families without coverage and small business owners to pool their resources to increase their buying power in order to make health insurance more affordable. Private insurance companies would compete for their business based on cost and quality. Advocates of the law believed that it might be a first step in curbing abuses in the insurance industry.

Health care reform was originally scheduled to occur in the following stages:

- **2010:** Adults who had been unable to get coverage because of a preexisting condition could join

a high-risk insurance pool (as a stopgap measure until the competitive health insurance marketplace was scheduled to begin in 2014). Insurance companies would have to cover children with preexisting conditions. Policies could not be revoked when people got sick. Preventive services would be fully covered without co-pays or deductibles. Dependent children could remain on their parents' insurance plans until they reached the age of twenty-six.

- **2011:** Medicare recipients would have access to free annual wellness visits with no cost for preventive care, and those recipients who had to pay out of pocket for prescription drugs would receive substantial discounts.

- **2012:** The federal government would provide additional money for primary-care services, and new incentives would be offered to encourage doctors to join together in accountability-care organizations. Hospitals with high readmission rates would face stiff penalties.

- **2013:** Households with incomes above $250,000 would be subject to higher taxes to help pay for health care reform. Medicare would launch "payment bundling" so that hospitals, doctors, and other health care providers could be paid on the basis of patient outcome, not services provided.

- **2014:** Most people would be required to buy health insurance or pay a penalty for not having it. Insurance companies could not deny a policy

**451**

to anyone based on health status, nor could they refuse to pay for treatment on the basis of preexisting health conditions. Annual limits on health care coverage would be abolished. Each state would have to open a health insurance exchange, or marketplace, so that individuals and small businesses without coverage could comparatively shop for health packages. Tax credits would make insurance and health care more affordable for those who earn too much to qualify for Medicaid.

- **2018:** Insurance companies and plan administrators would pay a 40 percent excise tax on all family plans costing more than $27,500 per year.
- **2019:** The health reform law should have reduced the number of uninsured people by 32 million, leaving about 23 million uninsured. About one-third of the uninsured would be immigrants residing in the country without legal documentation.

These are a few of the highlights of the plan, which is a document of more than 2,000 pages in length. However, in 2012, after a series of legal challenges and an appeal of a decision of the U.S. Court of Appeals of the Eleventh Circuit, the U.S. Supreme Court heard a challenge to the 2010 health care law on the issue of whether Congress had overstepped its constitutional authority by requiring most people in the United States to purchase health insurance or be assessed a penalty. According to the Obama administration, Congress is justified in this action by the constitutional power granted to this legislative body "to regulate commerce" and "to lay and collect taxes." After extensive deliberations, the U.S. Supreme Court ruled that Congress had not exceeded its constitutional authority, and the Affordable Care Act was further implemented following the Court's decision. If this is the future of health care funding in the United States, how has it looked in the past and present? Let's look first at how private health insurance works.

**Private Health Insurance** Private health insurance is largely paid for by businesses and households. Beginning in the 1960s, medical insurance programs began to expand, and third-party providers (public and private insurers) began picking up large portions of doctor and hospital bills for insured patients. With third-party fee-for-service payment, patients pay premiums into a fund that in turn pays doctors and hospitals for each treatment the patient receives.

In the 2010s, private health insurance premiums have continued to increase for both individuals and families. The average annual premium for an individual in an employer-sponsored health insurance plan was $5,615 in 2012. The average annual premium for family coverage was $15,745, with workers contributing (on average) $4,316 and employers paying (on average) $11,429 of the cost (Kaiser Family Foundation, 2013). These costs continue to rise annually, and most analysts believe they will be largely unaffected by further implementation of the Affordable Care Act.

Over the years, some health care advocates have argued that a third-party fee-for-service approach is the best and most cost-efficient method of delivering medical care. Others have argued that fee for service is outrageously expensive and a very cost-ineffective way in which to provide for the medical needs of people in this country, particularly those who are without health insurance coverage. According to critics, third-party fee for service contributes greatly to medical inflation because it gives doctors and hospitals an incentive to increase medical services. In other words, the more services they provide, the more fees they charge, and the more money they make. Patients have no incentive to limit their visits to doctors or hospitals because they have already paid their premiums and feel entitled to medical care, regardless of the cost. This is one of the spiraling costs that advocates of health care reform hope will be reduced.

**Public Health Insurance** Since the 1960s, the United States has had two nationwide public health insurance programs, Medicare and Medicaid. In 1965 Congress enacted Medicare, a federal program for people ages 65 and over (who are covered by Social Security or railroad retirement insurance or who have been permanently and totally disabled for two years or more). This program was primarily funded through Social Security taxes paid by current workers. Medicare has several components, referred to as Part A, Part B, Part C, and Part D. Medicare Part A (hospital insurance) helps cover inpatient care in hospitals, skilled-nursing facilities, hospice, and home health care. Part B (medical insurance) helps cover doctors' services and outpatient care. It covers some of the services of physical and occupational therapists, and some home health care. Most people pay a monthly premium for Part B. Medicare Part C is an insurance plan run by private companies approved by Medicare. Medicare Part D provides prescription drug coverage.

Medicaid is the federal government's health care program for low-income and disabled persons and certain groups of seniors in nursing homes. Medicaid is jointly funded by federal–state–local monies, and various factors are taken into account when determining whether a person is eligible for

Medicaid. Among these are age, disability, blindness, and pregnancy. Income and resources are also taken into consideration, as well as whether the person is a U.S. citizen or a lawfully admitted immigrant. Each state has its own rules regarding who may be covered under Medicaid, and some provide time-limited coverage for specific categories of individuals, such as uninsured women with breast or cervical cancer or those individuals diagnosed with TB (tuberculosis). As compared to Medicare, Medicaid has had a more tarnished image throughout its history. Unlike Medicare recipients, who are often seen as "worthy" of their health care benefits, Medicaid recipients have been stigmatized by politicians and media outlets for their participation in a "welfare program." Today, many physicians refuse to take Medicaid patients because the administrative paperwork is burdensome and reimbursements are low—typically less than one-half of what private insurance companies pay for the same services.

When the health care reform law passed in 2010, both the Medicaid program and the Medicare program were in financial difficulty. In 2011 these two programs cost $769 billion, or 21 percent of the federal budget, according to the Center on Budget and Policy Priorities (Leonard, 2012). An influx of retiring baby boomers and rising health care costs have contributed to increases in Medicare spending to the extent that the Centers for Medicare and Medicaid Services predicts that the cost of these two plans will rise to $1.8 trillion by 2020. This is especially likely to happen because under the Affordable Care Act, eligibility for "free" Medicaid coverage will significantly expand the number of persons covered under Medicaid, perhaps by as many as 30 million additional people, so cost containment will be difficult in this area. However, preventive health care services could improve the health of recipients while reducing costs.

**Health Maintenance Organizations (HMOs)** Created in an effort to provide workers with health coverage by keeping costs down, *health maintenance organizations (HMOs)* **provide, for a set monthly fee, total care with an emphasis on prevention to avoid costly treatment later.** The doctors do not work on a fee-for-service basis, and patients are encouraged to get regular checkups and to practice good health practices (e.g., exercise and eat right). However, research shows that preventive care is good for the individual's health but does not necessarily lower total costs. As long as patients use only the doctors and hospitals that are affiliated with their HMO, they pay no fees, or only small co-payments, beyond their insurance premiums.

Recent concerns about physicians being used as gatekeepers who might prevent some patients from obtaining referrals to specialists or from getting needed treatment have resulted in changes in the policies of some HMOs, which now allow patients to visit health care providers outside an HMO's network or to receive other previously unauthorized services by paying a higher co-payment. However, critics charge that those HMOs whose primary-care physicians are paid on a capitation basis—meaning that they receive only a fixed amount per patient whom they see, regardless of how long they spend with that patient—in effect encourage doctors to undertreat patients.

**Managed Care** Another approach to controlling health care costs in the United States is known as *managed care*—**any system of cost containment that closely monitors and controls health care providers' decisions about medical procedures, diagnostic tests, and other services that should be provided to patients.** One type of managed care in the United States is a *preferred provider organization* (PPO), which is an organization of medical doctors, hospitals, and other health care providers that enters into a contract with an insurer or a third-party administrator to provide health care at a reduced rate to patients who are covered under specific insurance plans. In most managed-care programs, patients choose a primary-care physician from a list of participating doctors. Unlike many of the HMOs, when a patient covered under a PPO plan needs medical services, he or she may contact any one of a number of primary-care physicians or specialists who are "in-network" providers. Like HMOs, most PPO plans do contain a pre-certification requirement in which scheduled (nonemergency) hospital admissions and certain kinds of procedures must be approved in advance. Through measures such as this, these insurance plans have sought (unsuccessfully) to curb the rapidly increasing costs of medical care and to reduce the extensive paperwork and bureaucracy involved in the typical medical visit.

**health maintenance organizations (HMOs)** companies that provide, for a set monthly fee, total care with an emphasis on prevention to avoid costly treatment later.

**managed care** any system of cost containment that closely monitors and controls health care providers' decisions about medical procedures, diagnostic tests, and other services that should be provided to patients.

For the foreseeable future, HMOs and PPOs are supposed to remain somewhat the same. After the passage of the health care reform measure, the Obama administration widely publicized a statement that people in plans such as these would not be affected by the new law. Rather, they would have assurance that they could get health care coverage even if they lost their job, changed jobs, moved out of state, got divorced, or were diagnosed with a serious illness.

**The Uninsured** Despite existing public and private insurance programs, about 48 million people in the United States did not have health insurance in 2012. Although full-time workers are more likely to have health insurance than those who work less than full time and persons who are unemployed, having a full-time job is still no guarantee of health insurance coverage: Slightly more than 15 percent of all full-time workers lacked health insurance in 2012 (DeNavas-Walt, Proctor, and Smith, 2013). In a worst-case scenario, some who are uninsured (despite being employed full time) make too little to afford health insurance but too much to qualify for Medicaid. Even if their employers offer group health insurance, some employees cannot afford their share of the premiums.

U.S. children without health insurance in 2012 totaled 6.6 million (DeNavas-Walt, Proctor, and Smith, 2013). Figure 14.12 shows that the uninsured rate for children in poverty (12.9 percent) was greater than the rate for all children (8.9 percent). Statistically speaking, as this figure shows, children who are more likely to be uninsured are ones who live in poverty, are Hispanic (any race), and are not citizens of the United States. The rate of uninsured children would be even higher if there were not state-sponsored programs such as CHIP, a health insurance plan for children that is offered on a sliding scale with premiums (down to zero) based on family income. Nearly 28 percent of young adults between the ages of 19 and 34 were uninsured in 2012—the highest uninsured rate of all age categories.

Race and ethnicity are important factors in health insurance coverage. Although every racial and ethnic group is affected, Latinos/as and African Americans are more likely to be uninsured than are white (non-Hispanics) Americans. For Hispanic Americans the uninsured rate was 29 percent, as compared to African Americans at 19 percent, Asian Americans at 15 percent, and white (non-Hispanic) Americans at 11 percent. Across all racial–ethnic categories, 17.7 percent of people under the age of 65 were uninsured for some portion, or all, of 2012.

Under the 2010 Affordable Care Act, the health insurance marketplace was implemented in 2014. This system was designed to make buying health coverage easier and more affordable by allowing individuals and small businesses to compare health plans, get answers to questions, find out if they are eligible for tax credits for private insurance or

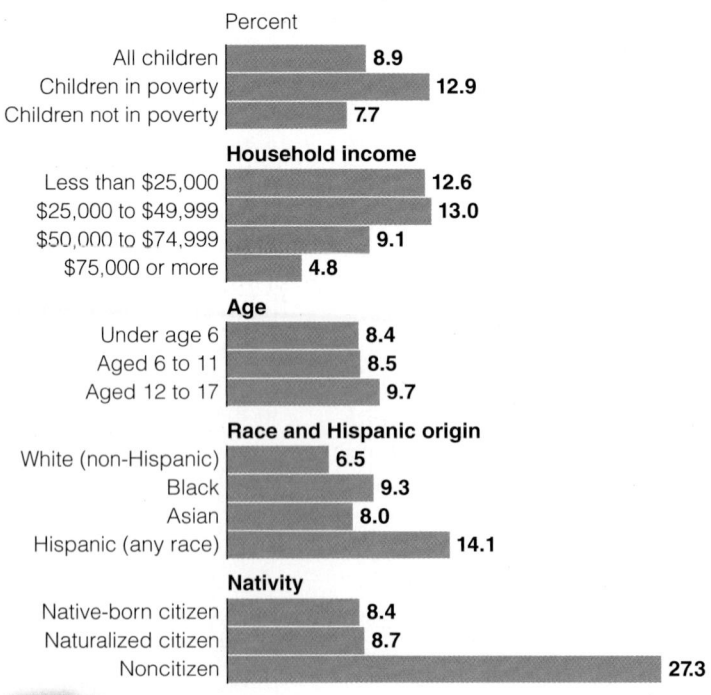

**FIGURE 14.12** UNINSURED CHILDREN BY POVERTY STATUS, HOUSEHOLD INCOME, AGE, RACE, HISPANIC ORIGIN, AND NATIVITY, 2012

Source: DeNavas-Walt, Proctor, and Smith, 2013.

programs such as the Children's Health Insurance Program (CHIP), and enroll in the health plan that best meets their needs. Every health insurance plan in the new marketplace will offer comprehensive coverage, from doctors to medications to hospital visits. Persons wishing to purchase health insurance will be able to compare prices, benefits, quality, and other features that may be of importance to them so that they can make an informed choice about which plan they want to purchase. At the time of this writing (September 2014) more than eight million people have signed up for "Obamacare," but there is widespread disagreement about how many have actually enrolled and paid for the plan. It remains to be seen how effective implementation of the Affordable Care Act will be in reducing the number of people without health insurance coverage and increasing the quality of health care in this country.

**Preventive Health Care Services** Under the Affordable Care Act, some preventive services are provided for individuals who are enrolled in job-related health plans or individual health insurance policies created after March 23, 2010. Among the preventive services covered at no cost are blood pressure, diabetes, and cholesterol tests; many cancer screenings, including mammograms and colonoscopies; counseling on quitting smoking, losing weight, eating healthfully, treating depression, and reducing alcohol use; and certain types of routine vaccinations and flu and pneumonia shots, as well as well-baby and well-child visits to a physician. It is hoped that early detection and preventive measures will cut down the overall cost of health care.

## Paying for Medical Care in Other Nations

Other nations have various ways in which they provide health care for their citizens. Let's examine how other nations pay for health care.

**Canada** Health care in Canada is delivered through a publicly funded health care system. Services are provided by private entities and are mostly free to patients. Information remains private between physician and patient, and the government is not involved in patient care. Each citizen receives a health card, and all patients receive the same level of care. As long as a person's premiums are paid up, health coverage is not affected by losing or changing jobs.

How did the contemporary Canadian health care system get started? In 1962 the government of the province of Saskatchewan implemented a health insurance plan despite opposition from doctors, who went on strike to protest the program. The strike was not successful, as the vast majority of citizens supported the government, which maintained health services by importing doctors from Great Britain. The Saskatchewan program proved itself viable in the years following the strike, and by 1972 all Canadian provinces and territories had coverage for medical and hospital services (Murray, Linden, and Kendall, 2012). As a result, Canada has **universal health care**—a health care system in which all citizens receive medical services paid for by tax revenues. In Canada these revenues are supplemented by insurance premiums paid by all taxpaying citizens.

One major benefit of the Canadian system is a significant reduction in administrative costs. Whereas more than 20 percent of the U.S. health care dollar represents administrative costs, in Canada the corresponding figure is 10 percent (Weiss and Lonnquist, 2012). However, the system is not without its critics, who claim that it is costly and often wasteful.

The Canadian health care system does not constitute what is referred to as **socialized medicine**—a health care system in which the government owns the medical care facilities and employs the physicians. Rather, Canada has maintained the private nature of the medical profession. Although the government pays most health care costs, the physicians are not government employees and have much greater autonomy than do physicians in the health care system in Great Britain.

**Great Britain** Great Britain has a centralized, single-payer health care system that is funded by general revenues. The National Health Service Act of 1946 provides for all health care services to be available at no charge to the entire population. Although physicians work out of offices or clinics—as in the United States or Canada—the government sets health care policies, raises funds and controls the medical care budget, owns health care facilities, and directly employs physicians and other health care personnel (Weiss and Lonnquist, 2012).

Unlike the Canadian model, the health care system in Great Britain *does* constitute socialized medicine. Physicians receive capitation payments from the government: a fixed annual fee for each patient in their practice regardless of how many

**universal health care** a health care system in which all citizens receive medical services paid for by tax revenues.

**socialized medicine** a health care system in which the government owns the medical care facilities and employs the physicians.

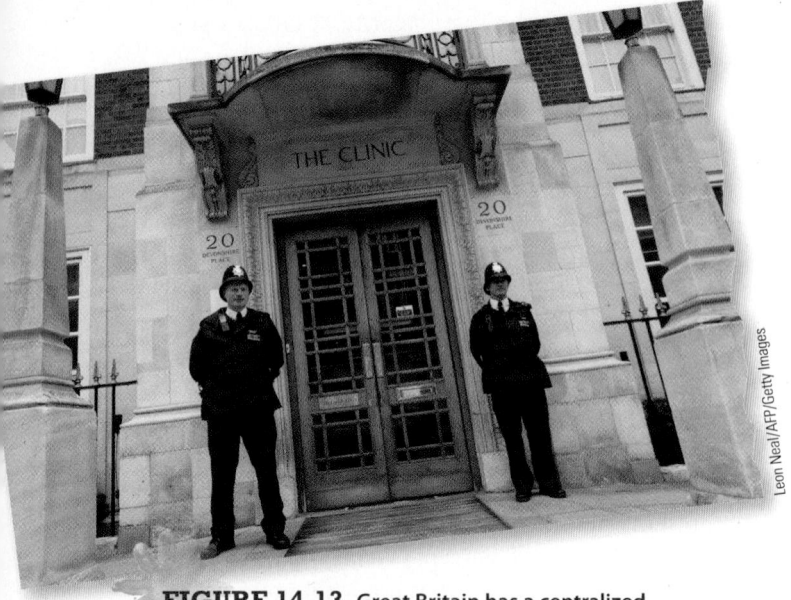

**FIGURE 14.13** Great Britain has a centralized, single-payer health care system that is funded by general revenues, but private patients can pay more to avoid long waits. Here, police officers stand outside the private London Clinic, where Prince Philip, husband of Queen Elizabeth, recovered from surgery in 2013.

times they see the patient or how many procedures they perform. They also receive supplemental payments for each low-income or elderly patient in their practice, to compensate for the extra time such patients may require; bonus payments if they meet targets for providing preventive services, such as immunizations against disease; and financial incentives if they practice in medically underserved areas. Physicians may accept private patients, but such patients rarely constitute more than a small fraction of a physician's practice; hospitals reserve a small number of beds for private patients (Weiss and Lonnquist, 2012) (Figure 14.13). Why would anyone want to be a private patient who pays for her or his own care or hospital bed? The answer is primarily found in the desire to avoid the long waits ("queues") that the general population encounters and the fact that private patients can enter the hospital for surgery at times convenient to the consumer rather than wait upon the convenience of the system.

**People's Republic of China** In recent years the People's Republic of China has seen many changes in the delivery and financing of medical care. When most U.S. people speak of "Chinese medicine," they are referring to treatments such as acupuncture, herbal remedies, or massage. However, health care in China is much more complex than the use of alternative remedies.

During the past two decades the health care system in China has become a complex mix of market-driven capitalism, communism, and massive government spending. The profit-driven system is based on fee-for-service practice by physicians and the sale of highly expensive pharmaceutical products that are marketed by doctors and hospitals. Much of the profit is through drug sales, particularly intravenous (IV) treatments, which are used much more widely than in the United States. Until reforms occurred in about 2008–2009, pharmaceutical products were very costly; however, they have become somewhat more affordable since that time.

Ironically, as the Chinese government stepped up funding for health care, social disease epidemics, such as obesity and/or illnesses related to excessive use of tobacco, alcohol, and salt, took their toll on many in the population: Type 2 diabetes, hypertension, and environmentally induced heart and lung diseases have remained on the upswing for a number of years (Mills, 2010).

What is the history of health care delivery and funding in China? After a lengthy civil war, in 1949 the Communist Party won control of mainland China but found itself in charge of a vast nation with a population of one billion people, most of whom lived in poverty and misery. Malnutrition was prevalent, life expectancies were short, and infant and maternal mortality rates were high. In the cities, only the elite could afford medical care; in the rural areas, where most of the population resided, Western-style health care barely existed (Weitz, 2013). With a lack of financial resources and not enough trained health care personnel, China adopted a policy to create a large number of physician extenders who could educate the public about health and the treatment of illness and disease. Referred to as *street doctors* in urban areas and *village doctors* (formerly "barefoot doctors") in the countryside, these individuals had little formal training and worked under the supervision of trained physicians (Weitz, 2013).

Over the past six decades, medical training has become more rigorous. All doctors receive training in both Western and traditional Chinese medicine. Most doctors who work in hospitals receive a salary; all other doctors now work on a fee-for-service basis. In urban areas, about 94 percent of the working population has health insurance coverage paid for by employers, but individuals often pay large out-of-pocket fees because of gaps in health care coverage. Today, 60 percent of the population of mainland China lives in rural areas, where large numbers of migrant farmworkers must provide for their own health care. New government initiatives established in the 2010s call for the establishment of a clinic in each rural community; however, many more physicians will be needed to provide adequate care for the millions of people living in these areas.

In urban areas the latest reports show that it is difficult to get into a hospital even with insurance. Increasing demands for services have placed the already overburdened system under greater stress, and many people line up outside the more-prestigious hospitals early in the morning on the day before they want treatment to get their name on lengthy waiting lists. Although local doctors might be able to take care of their needs, many Chinese patients want to be treated in the more-prestigious hospitals, where they believe they will receive higher-quality care. To help meet the growing demands on the health care delivery system, the Chinese government is attempting to improve thousands of medical centers and to focus on preventive care, especially for infants, children, pregnant women, and those in need of mental health care. Despite these efforts, both services and funding remain quite different when comparing urban and rural areas: The health care disparities remain great, and more money and medical personnel are needed throughout the system, but particularly in rural areas, to meet the needs of a rapidly growing population.

Regardless of which approach to health care delivery and financing that a nation uses, health care providers, hospitals, governmental agencies, political leaders, and the general public all face many difficult issues about the best way to meet patients' needs and not bankrupt the medical system and the nation. One area that may increase costs and generate complex concerns is advanced medical technology.

## Social Implications of Advanced Medical Technology

**LO5** Advances in medical technology are occurring at a speed that is almost unbelievable; however, sociologists and other social scientists have identified specific social implications of some of the new technologies (see Weiss and Lonnquist, 2012):

1. *Advanced technologies create options for people and for society, but these options alter human relationships.* An example is the ability of medical personnel to sustain a life that in earlier times would have ended as the result of disease or an accident. Although this can be beneficial, technologically advanced equipment (that can sustain life after consciousness is lost and there is no likelihood that the person will recover) can create a difficult decision for the family of that person if he or she has not left a *living will*—a document

**FIGURE 14.14** High-tech medical imaging devices such as the one shown here are very expensive. Some analysts question whether an expenditure of this size could be better used to help more people in different ways.

stating the person's wishes regarding the medical circumstances under which his or her life should be terminated. Federal law requires all hospitals and other medical facilities to honor the terms of a living will.

2. *Advanced technologies increase the cost of medical care.* For example, the computerized axial tomography (CT or CAT) scanner—which combines a computer with X-rays that are passed through the body at different angles—produces clear images of the interior of the body that are invaluable in investigating disease. The cost of such a scanner is around $1 million. Magnetic resonance imaging (MRI) equipment that allows pictures to be taken of internal organs ranges in cost from $1 million to $2.5 million. Can the United States afford such equipment in every hospital for every patient (Figure 14.14)? The money available for health care is not unlimited, and when it is spent on high-tech equipment and treatment, it is being reallocated from other health care programs that might be of greater assistance to more people.

3. *Advanced technologies such as cloning and stem cell research raise provocative questions about the very nature of life.* In 1997 Dr. Ian Williams and his associates in Scotland cloned a lamb (that they named Dolly) from the DNA of an adult sheep. Subsequently, scientists have cloned other animals in the same manner, raising a number of profound questions: If scientists can duplicate mammals from adult DNA, is it possible to clone a human being? If it is possible, would it be ethical? Like cloning, stem cell research has been an important and controversial issue in medicine. Stem cell research is important because stem cells—which are accessible in the skin and through

**457**

**FIGURE 14.15** The use of herbal therapies is a form of alternative medicine that is increasing in popularity in the United States. How does this approach to health care differ from a more traditional medical approach?

*(Xie Guang Hui/Redlink/Corbis)*

extraction from umbilical-cord blood and human embryos—can be used to generate virtually any type of specialized cell in the human body and to replace diseased or damaged human tissue. However, opponents of embryonic stem cell research believe that a human life is taken when a human embryo is destroyed in this research. Proponents of stem cell research respond that these studies do not always require the use of embryos.

At the same time that high-tech medicine is becoming a major part of overall health care, many people are turning to holistic medicine and alternative healing practices.

## Holistic Medicine and Alternative Medicine

When examining the subject of medicine, it is easy to think only in terms of conventional (or mainstream) medical treatment. By contrast, **holistic medicine is an approach to health care that focuses on prevention of illness and disease and is aimed at treating the whole person—body and mind—rather than just the part or parts in which symptoms occur.** Under this approach it is important that people not look solely to medicine and doctors for their health but also that people engage in health-promoting behavior. Likewise,

medical professionals must not only treat illness and disease but also work with the patient to promote a healthy lifestyle and self-image.

Many practitioners of *alternative medicine*—healing practices inconsistent with dominant medical practice—take a holistic approach, and today many people are turning to alternative medicine either in addition to or in lieu of traditional medicine (Figure 14.15). However, some medical doctors are opposed to alternative medicine. In understanding the medical establishment's reaction to alternative medicine, it is important to keep in mind the philosophy of scientific medicine—that medicine is a science, not an art. Thus, to the extent to which alternative medicine is "nonscientific," it must be quackery and therefore something that is undoubtedly worthless and possibly harmful. Undoubtedly, self-interest is also involved in mainstream medicine's reaction to alternative medicine: If the public can be persuaded that scientific medicine is the only legitimate healing practice, fewer health care dollars will be spent on a form of medical treatment that is (at least to some extent) in competition with the medical establishment. But if all forms of alternative medicine (including chiropractic, massage, and spiritual) are taken into account, people spend more money on unconventional therapies than they do for all hospitalizations.

# Sociological Perspectives on Health and Medicine

**LO6** Functionalist, conflict, symbolic interactionist, and postmodernist perspectives focus on different aspects of health and medicine; each provides us with significant insights on the problems associated with these pressing social concerns.

## A Functionalist Perspective: The Sick Role

According to the functionalist approach, if society is to function as a stable system, it is important for people to be healthy and to contribute to their society. Consequently, sickness is viewed as a form of deviant behavior that must be controlled by society. This view was initially set forth by the sociologist Talcott Parsons (1951) in his concept of the *sick role*—the set of patterned expectations that defines the norms and values appropriate for individuals who are sick and for those who interact with them. According to Parsons, the sick role has four primary characteristics:

1. People who are sick are not responsible for their condition. It is assumed that being sick is not a deliberate and knowing choice of the sick person.
2. People who assume the sick role are temporarily exempt from their normal roles and obligations. For example, people with illnesses are typically not expected to go to school or work.
3. People who are sick must want to get well. The sick role is considered to be a temporary one that people must relinquish as soon as their condition improves sufficiently. Those who do not return to their regular activities in a timely fashion may be labeled as hypochondriacs or malingerers.
4. People who are sick must seek competent help from a medical professional to hasten their recovery.

As these characteristics show, Parsons believed that illness is dysfunctional for both individuals and the larger society. Those who assume the sick role are unable to fulfill their necessary social roles, such as being parents or employees. Similarly, people who are ill lose days from their productive roles in society, thus weakening the ability of groups and organizations to fulfill their functions.

According to Parsons, it is important for the society to maintain social control over people who enter the sick role. Physicians are empowered to determine who may enter this role and when patients are ready to exit it. Because physicians spend many years in training and have specialized knowledge about illness and its treatment, they are certified by the society to be "gatekeepers" of the sick role. When patients seek the advice of a physician, they enter into the doctor–patient relationship, which does not contain equal power for both parties. The patient is expected to follow the "doctor's orders" by adhering to a treatment regime, recovering from the malady, and returning to a normal routine as soon as possible.

What are the major strengths and weaknesses of Parsons's model and, more generally, of the functionalist view of health and illness? Parsons's analysis of the sick role was pathbreaking when it was introduced. Some social analysts believe that Parsons made a major contribution to our knowledge of how society explains illness-related behavior and how physicians have attained their gatekeeper status. In contrast, other analysts believe that the sick-role model does not take into account racial–ethnic, class, and gender variations in the ways that people view illness and interpret this role. For example, this model does not take into account the fact that many individuals in the working class may choose not to accept the sick role unless they are seriously ill—because they cannot afford to miss time from work and lose a portion of their earnings. Moreover, people without health insurance may not have the option of assuming the sick role.

## A Conflict Perspective: Inequalities in Health and Health Care

Unlike the functionalist approach, conflict theory emphasizes the political, economic, and social forces that affect health and the health care delivery system. Among the issues of concern to conflict theorists are the ability of all people to obtain health care; how race, class, and gender inequalities affect health and health care; power relationships between doctors and other health care workers; the dominance of the medical model of health care; and the role of profit in the health care system.

**holistic medicine** an approach to health care that focuses on prevention of illness and disease and is aimed at treating the whole person—body and mind—rather than just the part or parts in which symptoms occur.

**sick role** the set of patterned expectations that defines the norms and values appropriate for individuals who are sick and for those who interact with them.

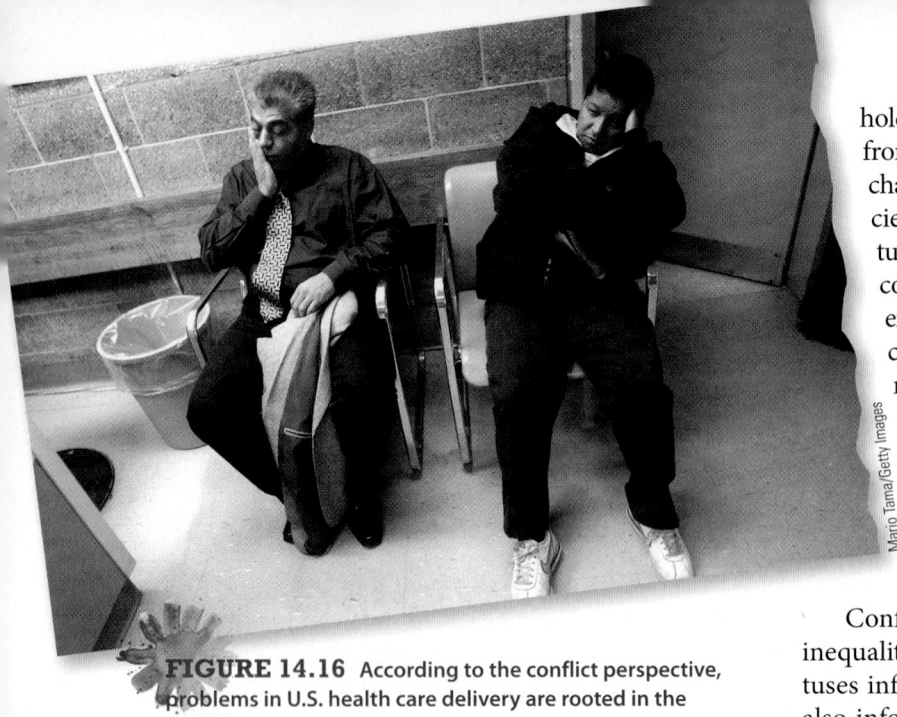

**FIGURE 14.16** According to the conflict perspective, problems in U.S. health care delivery are rooted in the capitalist economy, which views medicine as a commodity that is produced and sold by the medical–industrial complex.

Who is responsible for problems in the U.S. health care system? According to many conflict theorists, problems in U.S. health care delivery are rooted in the capitalist economy, which views medicine as a commodity that is produced and sold by the medical–industrial complex (Figure 14.16). The *medical–industrial complex* **encompasses local physicians and hospitals as well as global health-related industries such as insurance companies and pharmaceutical and medical supply companies** (Relman, 1992).

The United States is one of the few industrialized nations that has relied almost exclusively on the medical–industrial complex for health care delivery and has not had any form of universal health coverage to provide some level of access to medical treatment for all people. Consequently, access to high-quality medical care has been linked to people's ability to pay and to their position within the class structure. Those who are affluent or have good medical insurance may receive high-quality, state-of-the-art care in the medical–industrial complex because of its elaborate technologies and treatments. However, people below the poverty level and those just above it have greater difficulty gaining access to medical care. Referred to as the *medically indigent,* these individuals do not earn enough to afford private medical care but earn just enough money to keep them from qualifying for Medicaid. In the profit-oriented capitalist economy, these individuals are said to "fall between the cracks" in the health care system.

Who benefits from the existing structure of medicine? According to conflict theorists, physicians—who hold a legal monopoly over medicine—benefit from the existing structure because they can charge inflated fees. Similarly, clinics, pharmacies, laboratories, hospitals, supply manufacturers, insurance companies, and many other corporations derive excessive profits from the existing system of payment in medicine. In recent years, large drug companies and profit-making hospital corporations have come to occupy a larger and larger part of health care delivery. As a result, medical costs have risen rapidly, and the federal government and many insurance companies have placed pressure for cost containment on other players in the medical–industrial complex.

Conflict theorists increase our awareness of inequalities of race, class, and gender as these statuses influence people's access to health care. They also inform us about the problems associated with health care becoming "big business." However, some analysts believe that the conflict approach is unduly pessimistic about the gains that have been made in health status and longevity—gains that are at least partially the result of large investments in research and treatment by the medical–industrial complex.

## A Symbolic Interactionist Perspective: The Social Construction of Illness

Symbolic interactionists attempt to understand the specific meanings and causes that we attribute to particular events. In studying health, symbolic interactionists focus on the meanings that social actors give their illness or disease and how these affect people's self-concept and relationships with others. According to symbolic interactionists, we socially construct "health" and "illness" and how both should be treated. For example, some people explain disease by blaming it on those who are ill. If we attribute cancer to the acts of a person, we can assume that we will be immune to that disease if we do not engage in the same behavior. Nonsmokers who learn that a lung cancer victim had a two-pack-a-day habit feel comforted that they are unlikely to suffer the same fate. Similarly, victims of AIDS are often blamed for promiscuous sexual conduct or intravenous drug use, regardless of how they contracted HIV. In this case the social definition of the illness leads to the stigmatization of individuals who suffer from the disease.

Although biological characteristics provide objective criteria for determining medical conditions such as heart disease, tuberculosis, or cancer, there is also a subjective component to how illness is defined. This subjective component is very important when we look at conditions such as childhood hyperactivity, mental illness, alcoholism, drug abuse, cigarette smoking, and overeating, all of which have been medicalized. The term **medicalization refers to the process whereby nonmedical problems become defined and treated as illnesses or disorders.** Medicalization may occur on three levels: (1) the conceptual level (e.g., the use of medical terminology to define the problem), (2) the institutional level (e.g., physicians are supervisors of treatment and gatekeepers to applying for benefits), and (3) the interactional level (e.g., when physicians treat patients' conditions as medical problems). For example, the sociologists Deborah Findlay and Leslie Miller (1994: 277) explain how gambling has been medicalized (Figure 14.17):

> Habitual gambling . . . has been regarded by a minority as a sin, and by most as a leisure pursuit—perhaps wasteful but a pastime nevertheless. Lately, however, we have seen gambling described as a psychological illness—"compulsive gambling." It is in the process of being medicalized. The consequences of this shift in discourse (that is, in the way of thinking and talking) about gambling are considerable for doctors, who now have in gamblers a new market for their services or "treatment"; perhaps for gambling halls, which may find themselves subject to new regulations, insofar as they are deemed to contribute to the "disease"; and not least, for gamblers themselves, who are no longer treated as sinners or wastrels, but as patients, with claims on our sympathy, and to our medical insurance plans as well.

Sociologists often refer to this form of medicalization as the *medicalization of deviance* because it gives physicians and other medical professionals greater authority to determine what should be considered "normal" and "acceptable" behavior and to establish the appropriate mechanisms for controlling "deviant behaviors."

According to symbolic interactionists, medicalization is a two-way process: Just as conditions can be medicalized, so can they be demedicalized. **Demedicalization refers to the process whereby a problem ceases to be defined as an illness or a disorder.** Examples include the removal of certain behaviors (such as homosexuality) from the list of mental disorders compiled by the American Psychiatric Association and the deinstitutionalization of mental health patients. The process of demedicalization also continues in women's health as advocates seek to redefine childbirth and menopause as natural processes rather than as illnesses.

In addition to how health and illness are defined, symbolic interactionists examine how doctors and

Stockbyte/Jupiter Images

**FIGURE 14.17** Is gambling a moral issue or a medical one? According to sociologists, the recent trend toward viewing compulsive gambling as a health care issue is an example of the medicalization of deviance.

patients interact in health care settings (see "Sociology Works!"). Some physicians may hesitate to communicate certain kinds of medical information to patients, such as why they are prescribing certain medications or what side effects or drug interactions may occur.

Symbolic interactionist perspectives on health and health care provide us with new insights on the social construction of illness and how health and illness cannot be strictly determined by medical criteria. Symbolic interactionists also make us aware of the importance of communication between physicians and patients, including factors that may reduce effective medical treatment for some individuals. However, these approaches have been criticized for suggesting that few objective medical criteria exist for many illnesses and for overemphasizing microlevel issues without giving adequate recognition to macrolevel issues such as the effects on health care of managed care, health maintenance organizations, and for-profit hospital chains.

**medical–industrial complex** local physicians, local hospitals, and global health-related industries such as insurance companies and pharmaceutical and medical supply companies.

**medicalization** the process whereby nonmedical problems become defined and treated as illnesses or disorders.

**demedicalization** the process whereby a problem ceases to be defined as an illness or a disorder.

**461**

# Sociology WORKS!

## Sociology Sheds Light on the Physician–Patient Relationship

DOCTOR: What's the problem?

(Chair noise).

PATIENT: . . . had since last Monday evening so it's a week of sore throat.

DOCTOR: . . . hm . . . hm . . .

PATIENT: . . . which turned into a cold . . . and then a cough.

DOCTOR: A cold you mean what? Stuffy nose?

PATIENT: uh stuffy nose yeah not a chest . . . cold . . .

DOCTOR: . . . hm . . . hm . . . And a cough.

PATIENT: And a cough . . . which is the most irritating aspect. . . .

DOCTOR: Okay. Uh, any fever?

PATIENT: Not that I know of . . . I took it a couple of times in the beginning but haven't felt like—

DOCTOR: How 'bout your ears? . . . (Mishler, 2005: 322)

In this brief excerpt from the transcript of a discussion between a doctor and patient, the patient responds to the doctor's request for information by telling him what her symptoms are and how they began to change when her illness "turned into a cold . . . and then a cough." According to sociologists who study the social organization of health care, this transcript indicates that the physician wants the patient to continue speaking when he makes sounds such as "hm . . . hm," but he also wants to remain in control of the conversation. When the patient mentions that she has a "cold," for example, the doctor asks for further clarification of her specific symptoms so that he can determine, efficiently and in a short period of time, what kind of cold she has and what the treatment plan should be (Mishler, 2005). This is one brief passage from many pages of medical transcripts that researchers have used to study what they refer to as the struggle between the *voice of medicine* and the *voice of the lifeworld* (Mishler, 1984, 2005).

What are the voices of medicine and of the lifeworld? In this context the voice of medicine refers to the technical, scientific attitude adopted by many doctors in their communication with patients. This type of discourse is generally abstract, neutral, and somewhat distant. By contrast, the voice of the lifeworld refers to the natural, everyday attitudes that are expressed by patients when they talk to their physician in the hope of gaining additional insights about their medical condition. Some sociologists believe that a constant struggle exists between these two "voices" in the doctor–patient relationship and that this struggle affects the outcome of each medical encounter. The voice of medicine makes it difficult for patients to believe that their concerns are being heard when the physician is visibly in a hurry, does not listen, interrupts frequently, and/or talks down to the patient.

To minimize the voice of medicine, some sociologists advocate *therapeutic communication* between doctors and patients. In therapeutic communication, (1) the physician engages in full and open communication with the patient and feels free to ask questions about psychosocial as well as physical conditions, (2) the patient provides full and open information to the physician and feels free to ask questions and seek clarifications, and (3) a genuine rapport develops between physician and patient (Weiss and Lonnquist, 2012). According to some social analysts, it takes a doctor no additional time to use a positive communication style that conveys friendliness, empathy, genuineness, candor, an openness to conversation, and a nonjudgmental attitude toward the patient. Such a communication style certainly helps physicians establish positive relationships with their patients. Of course, patients should also try to communicate in a positive manner with physicians, people whom the patients hope will be able to help them remain healthy or help them resolve an existing medical problem.

### reflect & analyze

**Sociologists who study the social organization of medicine will continue to look for new insights on the physician–patient relationship. What other sociological perspectives do you believe might be useful in explaining the dynamics of doctor–patient communications or other social interactions (such as between physicians and other health professionals) that routinely take place within the health care system?**

## A Postmodernist Perspective: The Clinical Gaze

In *The Birth of the Clinic* (1994/1963), postmodern theorist Michel Foucault questioned existing assumptions about medical knowledge and the power that doctors have gained over other medical personnel and everyday people. Foucault asserted that truth in medicine—as in all other areas of life—is a social construction, in this instance one that doctors have created. Foucault believed that doctors gain power through the *clinical* (or "observing") *gaze*, which they use to gather information. Doctors develop the clinical gaze through their observation of patients; as the doctors begin to diagnose and treat medical conditions, they also start to speak "wisely" about everything. As a result, other people start to believe that doctors can "penetrate illusion and see . . . the hidden truth" (Shawver, 1998).

## Sociological Perspectives on Health and Medicine

| | |
|---|---|
| A Functionalist Perspective: The Sick Role | People who are sick are temporarily exempt from normal obligations but must want to get well and seek competent help. |
| A Conflict Perspective: Inequalities in Health and Health Care | Problems in health care are rooted in the capitalist system, exemplified by the medical–industrial complex. |
| A Symbolic Interactionist Perspective: The Social Construction of Illness | People socially construct both "health" and "illness," and how both should be treated. |
| A Postmodernist Perspective: The Clinical Gaze | Doctors gain power through observing patients to gather information, thus appearing to speak "wisely." |

According to Foucault, the prestige of the medical establishment was further enhanced when it became possible to categorize all illnesses within a definitive network of disease classification under which physicians can claim that they know why patients are sick. Moreover, the invention of new tests made it necessary for physicians to gaze upon the naked body, to listen to the human heart with an instrument, and to run tests on the patient's body fluids. Patients who objected were criticized by doctors for their "false modesty" and "excessive restraint" (Foucault, 1994/1963: 163). As the new rules allowed for the patient to be touched and prodded, the myth of the doctor's diagnostic wisdom was further enhanced, and "medical gestures, words, gazes took on a philosophical density that had formerly belonged only to mathematical thought" (Foucault, 1994/1963: 199). For Foucault, the formation of clinical medicine was merely one of the more-visible ways in which the fundamental structures of human experience have changed throughout history.

Foucault's work provides new insights on medical dominance, but it has been criticized for its lack of attention to alternative viewpoints. Among these is the possibility that medical breakthroughs and new technologies actually help physicians become wiser and more scientific in their endeavors. Another criticism is that Foucault's approach is based on the false assumption that people are passive individuals who simply comply with doctors' orders—he does not take into account that people (either consciously or unconsciously) may resist the myth of the "wise doctor" and not follow "doctors' orders" (Lupton, 1997). Foucault's analysis was not limited to doctors who treat bodily illness; he also critiqued psychiatrists and the treatment of insanity (1988/1961).

The Concept Quick Review summarizes the major sociological perspectives on health and medicine.

## Mental Illness

**LO7** Mental illness affects many people; however, it is a difficult topic for sociological research. An earlier social analyst, Thomas Szasz (1984), argued that mental illness is a myth. According to Szasz, "mental illnesses" are actually individual traits or behaviors that society deems unacceptable, immoral, or deviant. As a result, labeling individuals as "mentally ill" harms them because they often come to accept the label and are then treated accordingly by others.

Is mental illness a myth? After decades of debate on this issue, social analysts are no closer to reaching a consensus than they were when Szasz originally introduced his ideas. However, many scholars believe that mental illness is a reality that has biological, psychological, environmental, social, and other factors involved. Many medical professionals distinguish between a *mental disorder*—a condition that makes it difficult or impossible for a person to cope with everyday life—and *mental illness*—a condition in which a person has a severe mental disorder requiring extensive treatment with medication, psychotherapy, and sometimes hospitalization.

How many people are affected by mental illness? Data from the Substance Abuse and Mental Health Services Administration suggest that 45.9 million adults age 18 or older, or 20 percent of this age group, experienced mental illness in 2011–2012 (SAMHSA, 2012). The rate of mental illness is more than twice as high among those ages 18 to 25 (nearly 30 percent) than among persons ages 50 and older (14.3 percent). Adult women are more likely than men to have had mental illness in the past year prior to the survey. By definition, mental illness means that a person has a diagnosable mental, behavioral, or emotional disorder. In any given year, about 5 percent of the adult population (or approximately 11.5 million individuals) has serious mental illness, which means that a person has serious functional impairment that substantially interferes with or limits life activities (SAMHSA, 2012).

The most widely accepted classification of mental disorders is the American Psychiatric Association's (2013) *Diagnostic and Statistical Manual of Mental Disorders (DSM–5),* which is illustrated in Table 14.1. Mental disorders are very costly to the nation. Direct costs associated with mental disorders include the price of medication, clinic visits, and hospital stays. However, many indirect costs are

incurred as well. These include the loss of earnings by individuals, the costs associated with homelessness and incarceration, and other indirect costs that exist but are difficult to document.

## The Treatment of Mental Illness

In *Madness and Civilization,* Michel Foucault (1988/1961) examined the "archaeology of madness" from 1500 to 1800 to determine how ideas of mental illness have changed over time and to describe the

### table 14.1

Criteria for Psychiatric Disorders Identified by the American Psychiatric Association

| | |
|---|---|
| **Neurodevelopmental disorders** | Intellectual disability (intellectual developmental disorder), communication disorders, and autism spectrum disorder |
| **Schizophrenia spectrum and other psychotic disorders** | Disorders with symptoms such as delusions or hallucinations |
| **Bipolar and related disorders** | Mood swings that range from depression to mania |
| **Depressive disorders** | Includes major depressive disorder (clinical depression), disruptive mood dysregulation disorder in children up to age 18 years, and premenstrual dysphoric disorder |
| **Anxiety disorders** | Disorders characterized by anxiety that is manifested in phobias or panic attacks |
| **Obsessive-compulsive and related disorders** | Behavior that is obsessive and compulsive in nature, including excoriation (skin-picking) disorder, hoarding disorder, substance-/medication-induced obsessive-compulsive and related disorder, and obsessive-compulsive and related disorder caused by another medical condition |
| **Trauma- and stressor-related disorders** | Includes posttraumatic stress disorder (PTSD), reactive attachment disorder, and disinhibited social engagement disorder |
| **Dissociative disorders** | Depersonalization/derealization disorder and other problems involving disassociation with normal consciousness such as dissociative amnesia |
| **Somatic symptom and related disorders** | Including psychological problems that present themselves as symptoms of physical disease |
| **Feeding and eating disorders** | Including avoidant/restrictive food intake disorder, anorexia nervosa, bulimia nervosa, and binge eating disorder |
| **Sleep–wake disorders** | Including insomnia disorder, narcolepsy, some breathing-related sleep disorders, and other problems associated with sleep |
| **Sexual dysfunctions** | Gender-specific sexual dysfunctions of a duration of 6 months or more |
| **Gender dysphoria** | Intense anxiety about one's birth gender as compared to one's perceived gender |
| **Disruptive, impulse-control, and conduct disorders** | Including the inability to control disruptive behavior, lack of impulse control, and conduct disorders such as such as kleptomania and pyromania |
| **Substance-related and addictive disorders** | Disorders resulting from abuse of alcohol and/or drugs; gambling disorder and tobacco use disorder |
| **Neurocognitive disorders (NCDs)** | Including dementia and amnestic disorder and substantive/medication-induced NCDs |
| **Paraphilic disorders** | Sexual masochism disorder rooted in atypical sexual interests that cause persons to feel distressed about their interests and/or to cause another person to have psychological distress, injury, or death as a result of this disorder |

Source: Adapted from American Psychiatric Association, 2013.

"birth of the asylum." According to Foucault, early in this period insanity was considered part of every-day life, and people with mental illnesses were free to walk the streets; however, beginning with the Renaissance and continuing into the seventeenth and eighteenth centuries, the mentally ill were viewed as a threat to others. During that time, asylums were built, and a clear distinction was drawn between the "insane" and the rest of humanity. According to Foucault (1988/1961: 252), people came to see "madness" as a minority status that does not have the right to autonomy:

> Madness is childhood. Everything at the [asylum] is organized so that the insane are transformed into minors. They are regarded as children who have an overabundance of strength and make dangerous use of it. They must be given immediate punishments and rewards; whatever is remote has no effect on them.

In the second decade of the twenty-first century, many people with mental disorders do not receive professional treatment. U.S. adults who receive care for mental health problems are more likely to receive such care from primary-care and general-medical-care providers who are not specialists in mental health care. The lack of available mental health providers, as well as inadequate insurance coverage, means that many people with mental health problems fall through the cracks. In 2011–2012 about four in ten people experiencing any mental illness received mental health services. For those with "serious mental illness," about 60 percent received some form of treatment, leaving an additional 40 percent who had received no mental health care in the past year (SAMHSA, 2012).

Many people seeking psychiatric assistance are treated with medications or psychotherapy—which is believed to help patients understand the underlying reasons for their problem—and sometimes treated in psychiatric wards of local hospitals or in private psychiatric hospitals. However, the introduction of new psychoactive drugs to treat mental disorders and the deinstitutionalization movement in the 1960s have created dramatic changes in how people with mental disorders are treated. *Deinstitutionalization* **refers to the practice of rapidly discharging patients from mental hospitals into the community.** Originally devised as a solution for the problem of "warehousing" mentally ill patients in large, prison-like mental hospitals in the first half of the twentieth century, deinstitutionalization is now viewed as a problem by many social scientists. The theory behind this process was that patients' rights were being violated because many patients experienced involuntary commitment

(i.e., without their consent) to the hospitals, where they remained for extended periods of time. Instead, some professionals believed that the patients' mental disorders could be controlled with proper medications and treatment from community-based mental health services. Advocates of deinstitutionalization also believed that this practice would relieve the stigma attached to mental illness and hospitalization. However, critics of deinstitutionalization argue that this process exacerbated long-term problems associated with inadequate care for people with mental illness.

Admitting people to psychiatric hospitals on an involuntary basis ("involuntary commitment") has always been controversial; however, it remains the primary method by which police officers, judges, social workers, and other officials deal with people—particularly the homeless—whom they have reason to believe are mentally ill and imminently dangerous to others or themselves if not detained. State psychiatric hospitals continue to provide most of the chronic inpatient care for poor people with mental illnesses; these institutions tend to serve as a revolving door to poverty-level board-and-care homes, nursing homes, or homelessness, as contrasted with the situation of patients who pay their bills at private psychiatric facilities through private insurance coverage or Medicare.

According to the sociologist Erving Goffman's (1961a) classical work on this topic, mental hospitals are an excellent example of a *total institution*, previously defined as a place where people are isolated from the rest of society for a period of time and come under the complete control of the officials who run the institution.

## Disability

**LO8** What is a disability? *Disability* **refers to a physical or mental impairment that substantially limits one or more major activities that a person would normally do at a given stage of life and that may result in stigmatization or discrimination against the person with the disability.**

**deinstitutionalization** the practice of rapidly discharging patients from mental hospitals into the community.

**disability** a physical or mental impairment that substantially limits one or more major activities that a person would normally do at a given stage of life and that may result in stigmatization or discrimination against the person with the disability.

Some disabilities involve physical conditions, while others involve mental abilities. However, according to disability rights advocates, social attitudes and the social and physical environments in which people live are contributing factors to the extent to which a person may be considered "disabled." An example of a disabling environment is a school or office building in which elevator buttons and faucets on public restroom sinks are located beyond the reach of a person using a wheelchair. In such a setting the person's disability derives from the fact that necessary objects in everyday life have been made inaccessible. According to advocates, disability must be thought of in terms of how society causes or contributes to the problem—not in terms of what is "wrong" with the person with a disability (Albrecht, Seelman, and Bury, 2001).

A second crucial factor in understanding disability is how a person with a disability is viewed and treated by other people. Many examples could be given of persons with a disability who have been made uncomfortable by "able-bodied" individuals who think they are "helping" the person with a disability when they make some comment about the disability or the perceived inadequacies it causes. However, Dr. Adrienne Asch (2001), director of the Center for Ethics at Yeshiva University, speaks volumes about what her life has been like as an outstanding academic and as a person with a visual impairment—blindness:

> [A] growing number of professionals with disabilities, including myself, can point to professional recognition and the joys of doing work we love as well as its relative financial security and social status. Yet like . . . others, we have all-too-frequent reminders that we are unanticipated participants in workshops or conferences or unexpected guests at social gatherings. Sitting beside a stranger waiting for a lecture to begin at an academic conference, the stranger whispers loudly not "Hello, my name is Carol," but "Let me know how I can help you." What help do I need while waiting for the speaker to begin? Why not introduce herself, rather than assume that the only sociability I could possibly want is her help? When I respond by saying that she can let me know if I can help her, she does not get the point and I am all too well aware that the point is subtle; instead she needs to be thanked for her offer and reassured that I will accept it—and then many pleasantries later perhaps we can discuss why we are at the lecture and whether we like it and what workshop we will attend that afternoon.

Asch emphasizes that the incident she described is not an isolated one in her life: Many people who have known her for years do not feel comfortable treating her the same way they would a person without a disability. Even more important, countless other persons with a disability have shared similar experiences in which they been stigmatized or marginalized because of how another individual responded to them based on their disability.

Estimates suggest that about 56.7 million Americans (about 18.7 percent of the civilian noninstitutionalized U.S. population) have some level of disability, and nearly 13 percent have a severe disability (Brault, 2012). Although anyone can become disabled, some people are more likely to be or to become disabled than others. People who work in hazardous settings have higher rates of disability than workers in seemingly safer occupations. African Americans have higher rates of disability than whites, especially more-serious disabilities; persons with lower incomes also have higher rates of disability (Weitz, 2013). As age increases, the chances of disability also rise. But no one is immune to disability. Many people believe that most disabilities are caused by catastrophic events such as serious accidents; however, illness causes more than 90 percent of all disabilities.

Environment, lifestyle, and working conditions may all contribute to either temporary or chronic disability. For example, air pollution in automobile-clogged cities leads to a higher incidence of chronic respiratory disease and lung damage, which may result in severe disability for some people. Eating certain types of food and smoking cigarettes increase the risk for coronary and cardiovascular diseases. In contemporary industrial societies, workers in the second tier of the labor market (primarily recent immigrants, white women, and people of color) are at the greatest risk for certain health hazards and disabilities. Employees in data processing and service-oriented jobs may also be affected by work-related disabilities. The extensive use of computers has been shown to harm some workers' vision; to produce joint problems such as arthritis, low-back pain, and carpal tunnel syndrome; and to place employees under high levels of stress that may result in neuroses and other mental health problems.

As shown in Table 14.2, more than one out of five people in the United States (21.3 percent) has a chronic health condition that, given the physical, attitudinal, and financial barriers built into the social system, makes it difficult to perform one or more activities generally considered appropriate for persons of their age. This percentage refers only to the noninstitutionalized population. If people living in nursing homes and other institutional settings are considered, the percentage is significantly higher.

## table 14.2

### Percentage of Noninstitutionalized U.S. Population with Disabilities

| Characteristic | Percentage[a] |
|---|---|
| With a disability | 21.3 |
|     Severe | 14.8 |
|     Not severe | 6.5 |
| Has difficulty or is unable to: | |
|     See | 4.8 |
|     Hear | 3.1 |
|     Have speech understood | 1.2 |
|     Lifting | 7.1 |
|     Using stairs | 9.2 |
|     Walking | 9.9 |
| Has difficulty or needs assistance with: | |
|     Getting around | 1.9 |
|     Getting into bed | 2.5 |
|     Taking a bath or shower | 2.3 |
|     Dressing | 1.8 |
|     Eating | 0.8 |
|     Getting to or using the toilet | 1.2 |
| Has difficulty or needs assistance with: | |
|     Going outside the home alone | 4.2 |
|     Managing money | 2.4 |
|     Preparing meals | 2.4 |
|     Doing housework | 3.2 |
|     Using the phone | 1.2 |

[a]Percentage of persons ages 15 and older.

Source: Brault, 2012. © Cengage Learning

For infants born with a disability, living with a disability is a long-term process for parents and child alike. Parents report that more than 289,000 children under age three have a disability, either a developmental delay or difficulty moving their arms or legs. In addition, about 465,000 children between the ages of three and five have a disability in which they experience a developmental delay or have difficulty walking, running, or playing (Brault, 2012).

Among persons who acquire disabilities later in life, through disease or accidents, the social significance of their disability can be seen in how they initially respond to their symptoms and diagnosis, how they view the immediate situation and their future, and how the illness and disability affect their lives. When confronted with a disability, most people adopt one of two strategies—avoidance or vigilance.

Those who use the avoidance strategy deny their condition in order to maintain hopeful images of the future and elude depression; for example, some individuals refuse to participate in rehabilitation following a traumatic injury because they want to pretend that the disability does not exist. By contrast, those using the vigilance strategy actively seek knowledge and treatment so that they can respond appropriately to the changes in their bodies (Weitz, 2013).

## Sociological Perspectives on Disability

How do sociologists view disability? Those using the functionalist framework often apply Parsons's sick-role model, which is referred to as the medical model of disability. According to the medical model, people with disabilities become, in effect, chronic patients under the supervision of doctors and other medical personnel, subject to a doctor's orders or a program's rules and not to their own judgment. From this perspective, disability is deviance.

The deviance framework is also apparent in some symbolic interactionist perspectives. According to symbolic interactionists, people with a disability experience role ambiguity because many people equate disability with deviance. By labeling individuals with a disability as "deviant," other people can avoid them or treat them as outsiders. Society marginalizes people with a disability because they have lost old roles and statuses and are labeled as "disabled" persons (Figure 14.18). According to the sociologist Eliot Freidson (1965), how the people are labeled results from three factors: (1) their degree of responsibility for their impairment, (2) the apparent seriousness of their condition, and (3) the perceived legitimacy of the condition. Freidson concluded that the definitions of and expectations for people with a disability are socially constructed factors.

Finally, from a conflict perspective, persons with a disability are members of a subordinate group in conflict with persons in positions of power in the government, in the health care industry, and in the rehabilitation business, all of whom are trying to control their destinies. Those in positions of power have created policies and artificial barriers that keep people with disabilities in a subservient position (Asch, 1986; Hahn, 1987). Moreover, in a capitalist economy, disabilities are big business. When people with disabilities are defined as a social problem and public funds are spent to purchase goods and services for them, rehabilitation becomes a commodity that can be bought and sold by the medical–industrial complex. From this perspective, persons with a disability are objectified. They have an

**FIGURE 14.18** Disabilities are often the result of violent activity. Many veterans of the wars in Iraq and Afghanistan now suffer physical and emotional disabilities that may be with them for the rest of their lives and involve ongoing therapies.

economic value as consumers of goods and services that will allegedly make them "better" people. Many persons with a disability endure the same struggle for resources faced by people of color, women, and older persons. Individuals who hold more than one of these ascribed statuses, combined with experiencing disability, are doubly or triply oppressed by capitalism.

Today, many working-age persons with a disability in the United States are unemployed (see "Census Profiles: Disability and Employment Status"). Most of them believe that they could and would work if offered the opportunity. However, even when persons with a severe disability are able to find jobs, they typically earn less than persons without a disability.

Employment, poverty, and disability are related. On the one hand, people may become economically disadvantaged as a result of chronic illness or disability. On the other hand, poor people are less likely to be educated and more likely to be malnourished and have inadequate access to health care—all of which contribute to risk of chronic illness, physical and mental disability, and the inability to participate in the labor force. As previously mentioned, the type of employment available to people with limited resources increases their chances of becoming disabled. They may work in hazardous places such as mines, factory assembly lines, and chemical plants, or in the construction industry, where the chance of becoming seriously disabled is much higher.

## Health Care in the Future

Central questions regarding the future of health care are how to provide coverage for the largest number of people and how to do this without bankrupting the entire nation. The implementation of the Affordable Care Act will affect health care in this nation more than any other factor. Overall, our best hope for good medical care is a payment system that will result in the best outcomes for patients at reasonable costs and thus an overall transformation of the current U.S. health care system.

A key issue in the United States is how to prevent, reduce, and best treat epidemics that affect all Americans. If we do not have a specific illness or health condition, many of us do not see that disease as being "our problem." In the future, however, we must come to see problems such as HIV/AIDS as everyone's concern. The Obama administration has developed a national HIV/AIDS strategy to help reduce the number of people who become infected with HIV, to increase access to care and improve health outcomes for people living with HIV, and to reduce HIV-related health disparities with more and better community-level approaches that integrate HIV prevention and care with other social service

## Disability and Employment Status

One of the questions that the U.S. Bureau of the Census asks respondents is about long-lasting conditions such as physical, mental, or emotional conditions that substantially limit important basic activities. The survey also asks if they are working at a job or business. The answers allow the Census Bureau to determine how many people have a disability and how many people without a disability were employed at the time the census was conducted. As you can see from the figure below, for the population age 18–64 years (the period during which people are most likely to be employed), about 30 percent of persons with a disability were employed, compared with 75 percent of persons without a disability who were employed. Is employment among persons with a disability related primarily to their disability status, or do other factors—such as prejudice or lack of willingness to make the necessary accommodations that would allow such persons to hold a job—play a significant part in the high rate of unemployment of persons with a disability?

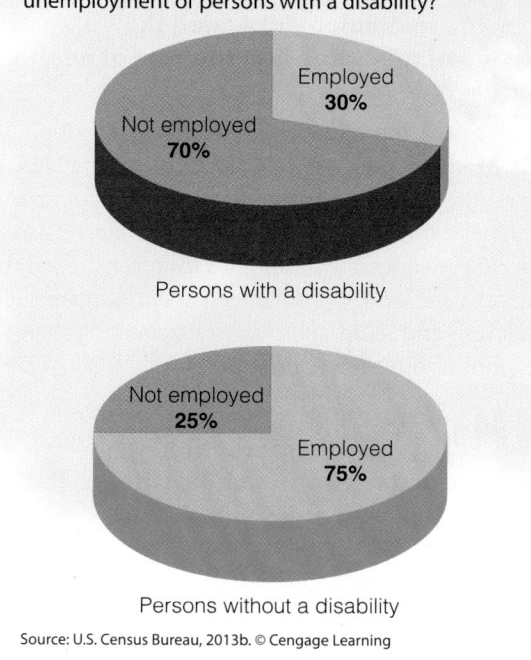

Employed
**30%**

Not employed
**70%**

Persons with a disability

Not employed
**25%**

Employed
**75%**

Persons without a disability

Source: U.S. Census Bureau, 2013b. © Cengage Learning

rising costs of medical care and their usefulness as major tools for diagnosis and treatment. Technology is a major stimulus for social change, and the health care systems in high-income nations such as the United States reflect the rapid rate of technological innovation that has occurred in the last few decades. In the future, advanced health care technologies will no doubt provide even more accurate and quicker diagnoses, effective treatment techniques, and increased life expectancy.

However, technology alone cannot solve many of the problems confronting us in health and health care delivery. In fact, some aspects of technological innovation may be dysfunctional for individuals and society. As we have seen, some technological "advances" raise new ethical concerns, such as the moral and legal issues surrounding the cloning of human life. Some "advances" also may fail: A new prescription drug may be found to cause side effects that are more serious than the illness that it was supposed to remedy. Whether advanced technology succeeds or fails in some areas, it will probably continue to increase the cost of health care in the future. As a result, the gap between the rich and the poor in the United States will contribute to inequalities of access to vital medical services. On a global basis, new technologies may lower the death rate in some low-income countries, but it will primarily be the wealthy in those nations who will have access to the level of health care that many people in higher-income countries take for granted.

In the developing nations of the world, preventive health care and more-effective and efficient health care delivery are crucial as the world's population continues to increase and as some regions of the world are plagued by high rates of disease and poverty, a deadly combination in any setting. The concerns of the World Health Organization and other organizations must be heeded to prevent global pandemics: epidemics of infectious disease that spread through human populations across a wide region, country, continent, or the whole world. In the future it will be necessary to direct more money and attention toward preventing major health crises rather than trying to find some way to deal with them after they develop.

It goes without saying that health, illness, and health care will continue to change in the future. To a degree, health care in the future will be up to each of us. What measures will we take to safeguard ourselves against illness and disorders? How can we help others who are the victims of acute and chronic diseases or disabilities? Will we seek to have a voice in how our community, state, and nation deal with health care issues now and in the future?

needs. Along with meeting these goals, it will be important to reduce stigma and the discrimination against people living with HIV. This HIV/AIDS national initiative is only one of many possible examples which show that in the future, we must have a coordinated national response to preventive health care and to the equitable distribution of health care services throughout the country.

Another key issue in contemporary health care is the role that advanced technologies play in the

# chapter REVIEW

## LO1 What is the relationship between the social environment and health and illness?

According to the World Health Organization, health is a state of complete physical, mental, and social well-being. Illness refers to an interference with health; like health, illness is socially defined and may change over time and between cultures. In other words, health and illness are not only biological issues but also social issues. Studying health and health care issues around the world offers insights on illness and how political and economic forces shape health care in nations.

## LO2 What is social epidemiology, and what key demographic factors do social epidemiologists study?

Social epidemiology is the study of the causes and distribution of health, disease, and impairment throughout a population. Typically, the target of the investigation is disease agents, the environment, and the human host (age, sex, race/ethnicity, physical condition, habits and customs, and lifestyle).

## LO3 How did the profession of medicine emerge in the United States?

During the nineteenth century, medical schools were largely proprietary institutions, and their officials were often more interested in acquiring students than in enforcing standards. Gradually, the number of medical schools was reduced, and licensing laws were established to eliminate unqualified or irregular practitioners. Although medicine had been previously viewed more as an art than as a science, several significant discoveries during the nineteenth century in areas such as bacteriology and anesthesiology began to give medicine increasing credibility as a science.

## LO4 How is health care paid for in the United States and other nations?

Throughout most of the past hundred years, medical care in the United States has been paid for on a fee-for-service basis. This approach to paying for medical services is expensive because few restrictions are placed on the fees that doctors, hospitals, and other medical providers can charge patients. Recently, there have been efforts at cost containment, and HMOs and managed care have produced both positive and negative results in the contemporary practice of medicine. Health maintenance organizations (HMOs) provide, for a set monthly fee, total care with an emphasis on prevention to avoid costly treatment later. Managed care is any system of cost containment that closely monitors and controls health care providers' decisions about medical procedures, diagnostic tests, and other services that should be provided to patients. Other nations have various ways in which they provide health care for their citizens. Some nations (such as Canada) have a universal health care system in which all citizens receive medical services paid for by tax revenues. Nations such as Great Britain have a socialized health care system in which the government owns the medical care facilities and employs the physicians.

## LO5 How has advanced medical technology changed the practice of medicine and the cost of health services?

Sociologists have identified specific social implications of some new medical technologies: (1) Advanced technologies create options for people and for society, but these options alter human relationships; (2) advanced technologies increase the cost of medical care; and (3) advanced technologies such as cloning and stem cell research raise provocative questions about the very nature of life.

## LO6 How do functionalist, conflict, symbolic interactionist, and postmodern approaches differ in their analysis of health and health care?

According to the functionalist approach, if society is to function as a stable system, it is important for people to be healthy and to contribute to their society. Consequently, sickness is viewed as a form of deviant behavior that must be controlled by society. Conflict theory tends to emphasize the political, economic, and social forces that affect health and the health care delivery system. Among these issues are the ability of all people to obtain health care; how race, class, and gender inequalities affect health and health care; power relations between doctors and other health care workers; the dominance of the medical model of health care; and the role of profit in the health care system. In studying health, symbolic interactionists focus on the fact that the meaning that social actors give their illness or disease will affect their self-concept and their relationships with others. Symbolic interactionists also examine medicalization—the process whereby non-medical problems become defined and treated as illnesses or disorders. Postmodern theorists argue that doctors and the medical establishment have gained control over illness and patients at least partly because of the physicians' clinical gaze, which replaces all other systems of knowledge.

## LO7 What is meant by the term *mental illness,* and why is it a difficult topic for sociological research?

Mental illness affects many people; however, it is a difficult topic for sociological research. An earlier social analyst, Thomas Szasz, argued that mental illness is a myth. After decades of debate on this issue, social analysts are no closer to reaching a consensus than they were when Szasz originally introduced his ideas.

## LO8 What is a disability, and what are some key sociological perspectives on disability?

Disability is a physical or health condition that stigmatizes or causes discrimination. In viewing disability, sociologists using the functionalist framework often apply the medical model of disability. According to the medical model, people with disabilities become chronic patients under the supervision of medical personnel, subject to a doctor's orders or a program's rules and not to their own judgment. According to symbolic interactionists, people with a disability experience role ambiguity because many people equate disability with deviance. From a conflict perspective, persons with a disability are members of a subordinate group in conflict with persons in positions of power in the government, the health care industry, and the rehabilitation business, all of whom are trying to control their destinies.

## key TERMS

acute diseases  441

chronic diseases  441

deinstitutionalization  465

demedicalization  461

disability  465

drug  443

health  436

health care  436

health maintenance organizations (HMOs)  453

holistic medicine  458

infant mortality rate  438

life expectancy  438

managed care  453

medical–industrial complex  460

medicalization  461

medicine  437

sick role  459

social epidemiology  441

socialized medicine  455

universal health care  455

## questions for CRITICAL THINKING

1. Why is it important to explain the social, as well as the biological, aspects of health and illness in societies?
2. In what ways are race, class, and gender intertwined with physical and mental disorders?
3. How would functionalists, conflict theorists, and symbolic interactionists suggest that health care delivery might be improved in the United States?
4. Based on this chapter, how do you think illness and health care will be handled in the United States in the future? Are there things that we can learn from other nations regarding the delivery of health care? Why or why not?

# 15 POPULATION AND URBANIZATION

© iStockphoto.com/Holger Mette

## LEARNING OBJECTIVES

**LO1** Define *demography* and explain the processes that produce population changes.

**LO2** Explain what is meant by population composition and describe how it is measured.

**LO3** Discuss Malthusian, Marxist, and neo-Malthusian perspectives on population as well as demographic transition theory.

**LO4** Identify the key points in international migration theories.

**LO5** Explain how cities develop through preindustrial, industrial, and postindustrial stages.

**LO6** Compare ecological/functionalist models with political economy/conflict models in their explanations of urban growth.

**LO7** Describe what is meant by the experience of urban life and indicate how sociologists seek to explain this experience.

**LO8** Identify the best-case and worst-case scenarios regarding population and urban growth in the twenty-first century and discuss how some of the worst-case scenarios might be averted.

# The Immigration Debate

**After being born and reared in the Philippines,** my mother wanted to give me a better life. So she sent me to live with my grandparents in Silicon Valley. It was 1993, and I was 12 years old. I loved America the moment I got here, and embraced the language, the culture, and the people. . . .

At 16, I rode my bike to the DMV [Department of Motor Vehicles] to get my driver's permit. I brought my green card with me. The woman at the DMV looked at it, leaned over and whispered, "This is fake. Don't come back here again."

I went home and confronted my grandfather. He confirmed it. That was the first time I realized I am an undocumented immigrant—what some people call an "illegal." I decided then that people could never doubt that I am an American. Speak English well. Write English well. Contribute to this society. If I worked hard enough, if I achieved enough, I felt I could earn what it means to be an American.

Jose Antonio Vargas

Justin Sullivan/Getty Images

Immigration is one of the most politically and emotionally charged issues in the twenty-first century. Terms such as *illegal alien* and *undocumented immigrant* have been tossed around by the media and political candidates in the United States for decades, but concerns about the effects of immigration on cities have also escalated as major shifts in the global population have occurred.

Around the world, people move from one location to another for many reasons. Some individuals and families move for economic opportunities, while others move because they fear for their life. In the United States and many other high-income countries, there is a lack of consensus about the causes and consequences of immigration: Some people are adamantly opposed to immigration; others believe that it is acceptable under certain circumstances. The result has been a divisive battle over immigration reform. A number of states have passed stringent immigration laws, causing extensive political strife and litigation.

Illegal immigration has declined in the aftermath of the U.S. economic crisis of 2007–2008 and the slowed recovery continuing into the second decade of the 2000s. However, many people remain concerned that rapid population growth and the changing demographic characteristics of new arrivals will further harm the quality of life in the United States, and they demand more-restrictive immigration laws (Preston, 2012).

Clearly, immigration is an important factor in understanding the dynamics of population and urbanization; however, immigration is only one factor associated with population growth and urban change. Birth rates and death rates are also important but have received relatively little attention. In this chapter we explore the dynamics of population and urbanization, with a focus on how birth rates, death rates, and migration affect growth and change in societies such as ours. Before reading on, test your knowledge about current migration and U.S. immigration issues by taking the "Sociology and Everyday Life" quiz.

—Jose Antonio Vargas (2012), an award-winning journalist who informed the world of his undocumented immigrant status in the *New York Times* (see Vargas, 2011), describes how he learned that he was in the United States illegally. Vargas founded Define American (see its website) to encourage others to use social media to support immigration reform.

## How Much Do You Know About Migration and U.S. Immigration?

| True | False | |
|------|-------|---|
| T | F | 1. All "unauthorized immigrants" in the United States entered the country illegally. |
| T | F | 2. Most people migrate to another country because of "push" factors that make them unhappy in their country of origin, such as poor economic conditions, political unrest, or war. |
| T | F | 3. U.S. immigrants are just as likely to have college degrees as are native-born Americans. |
| T | F | 4. Immigrants increase unemployment and lower wages among native workers. |
| T | F | 5. Most children living in undocumented immigrant families were born in the United States and are, therefore, U.S. citizens. |

| True | False | |
|------|-------|---|
| T | F | 6. When a Gallup poll asked Americans if illegal immigrants mostly take jobs that Americans want, or if they mostly take low-paying jobs that Americans don't want, the majority of Americans indicated that they believe that most undocumented immigrants take low-paying jobs that Americans don't want. |
| T | F | 7. To become a naturalized U.S. citizen, immigrants must pass a basic test on English and U.S. civics (history and government). |
| T | F | 8. Foreign-born non-U.S. citizens who marry a U.S. citizen are automatically granted U.S. citizenship. |

Answers on page 477.

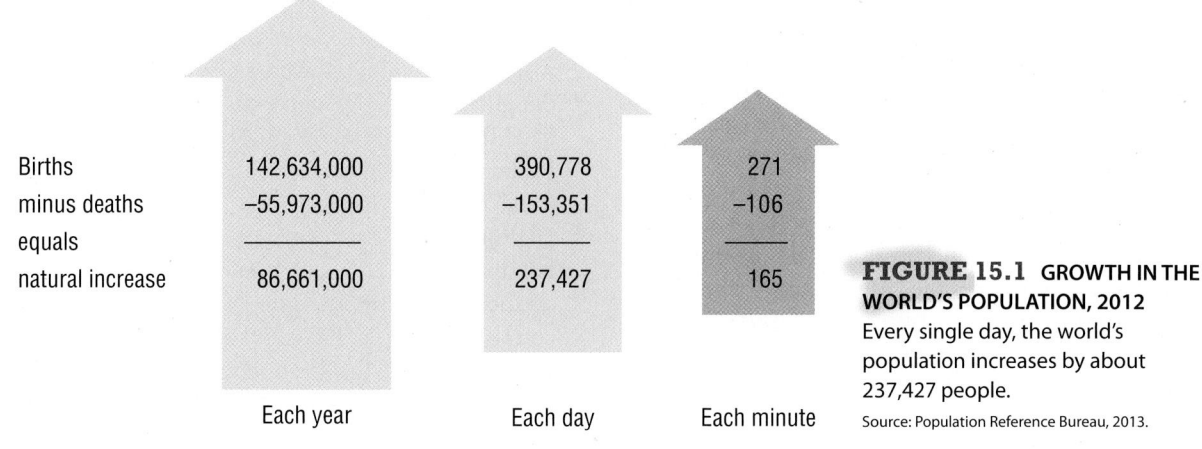

| | Each year | Each day | Each minute |
|---|---|---|---|
| Births | 142,634,000 | 390,778 | 271 |
| minus deaths | −55,973,000 | −153,351 | −106 |
| equals | | | |
| natural increase | 86,661,000 | 237,427 | 165 |

**FIGURE 15.1** GROWTH IN THE WORLD'S POPULATION, 2012
Every single day, the world's population increases by about 237,427 people.
Source: Population Reference Bureau, 2013.

## Demography: The Study of Population

**LO1** How large is the world's population? The world population has reached more than 7.2 billion people and continues to grow rapidly. Every year, the population increases by about 86,661,000 people when we calculate the *natural increase* in population: births minus deaths (see Figure 15.1). If we break down the natural increase even further, 237,427 people are added to the world's population each day, and 165 persons are added every minute (Population Reference Bureau, 2013).

What causes the population to grow so rapidly? This question is of interest to scholars who specialize in the study of *demography*—**a subfield of sociology that examines population size, composition, and distribution.** Many sociological studies use demographic analysis as a component of the research design because all aspects of social life are affected by demography. For example, an important relationship exists between population size and the availability of food, water, energy, and housing. Population size, composition, and distribution are also connected to issues such as poverty, racial and ethnic diversity, shifts in the age structure of society, and concerns about environmental degradation (Weeks, 2012).

In the twenty-first century, nearly all (97 percent) of the world population growth is occurring in developing nations, such as the 51 countries in sub-Saharan Africa, because of high birth rates combined with younger populations. By contrast, birth rates in developed nations are barely exceeding death rates because of a combination of low birth rates and much older populations (Population Reference Bureau, 2013).

Increases or decreases in population can have a powerful impact on the social, economic, and political structures of societies. As used by demographers, a *population* is a group of people who live in a specified geographic area. Changes in populations occur as a result of three processes: fertility (births), mortality (deaths), and migration.

## Fertility

*Fertility* **is the actual level of childbearing for an individual or a population.** The level of fertility in a society is based on biological and social factors, the primary biological factor being the number of women of childbearing age (usually between ages 15 and 45). Other biological factors affecting fertility include the general health and level of nutrition of women of childbearing age. Social factors influencing the level of fertility include the roles available to women in a society and prevalent viewpoints regarding what constitutes the "ideal" family size.

Based on biological capability alone, most women could produce twenty or more children during their childbearing years. *Fecundity* is the potential number of children who could be born if every woman reproduced at her maximum biological capacity. Fertility rates are not as high as fecundity rates because people's biological capabilities are limited by social factors such as practicing voluntary abstinence and refraining from sexual intercourse until an older age, as well as by contraception, voluntary sterilization, abortion, and infanticide. Additional social factors affecting fertility include significant changes in the number of available partners for sex and/or marriage (as a result of war, for example), increases in the number of women of childbearing age in the workforce, and high rates of unemployment. In some countries, governmental policies also affect the fertility rate. For example, until recently the family-planning restrictions established by the People's Republic of China allowed only one child per family in order to limit population growth in light of a population explosion that occurred between 1949 and 1976, when the population almost doubled. In 2013 these limits were relaxed for couples in which one or both partners are only-children themselves.

The most basic measure of fertility is the *crude birth rate*—**the number of live births per 1,000 people in a population in a given year.** The crude birth rate in the United States was 12 per 1,000 in 2013, as compared with an all-time high rate of 27 per 1,000 in 1947 (following World War II). This measure is referred to as a "crude" birth rate because it is based on the entire population and is not "refined" to incorporate significant variables affecting fertility, such as age, marital status, religion, and race/ethnicity.

The United Nations divides countries into three categories based on fertility levels:

1. *Low-fertility countries,* where women are not having enough children to ensure that, on average, each woman is replaced by a daughter who survives to the age when she will be old enough to have children. Forty-two percent of the world's population lives in low-fertility countries, which include all countries in Europe except Iceland and Ireland; 19 out of 51 countries in Asia; 14 out of 39 in the Americas; Mauritius and Tunisia in Africa; and Australia in Oceania.
2. *Intermediate-fertility countries,* where each woman is having, on average, between 1 and 1.5 daughters. Forty percent of the world's population lives in intermediate-fertility countries, such as India, the United States, Indonesia, Bangladesh, Mexico, and Egypt.
3. *High-fertility countries,* where the average woman has more than 1.5 daughters. Eighteen percent of the world's population lives in high-fertility countries, such as Pakistan, Nigeria, the Philippines, Ethiopia, the Democratic Republic of the Congo, the United Republic of Tanzania, Sudan, Kenya, Uganda, Iraq, Afghanistan, Ghana, Yemen, Mozambique, and Madagascar.

Obviously, high-fertility countries will potentially add the most to future population growth. However, it is difficult to estimate how much the world's population will *actually* increase over the next century because even small differences in fertility levels that are sustained over long periods of time can have a major effect on population size.

# ANSWERS to the Sociology Quiz on Migration and U.S. Immigration

**1. False.** Although the term *unauthorized immigrant* refers to a U.S. resident who is not a citizen of this country, who has not been admitted for permanent residence, or who does not have an authorized temporary status that permits longer-term residence and work, some "unauthorized immigrants" originally entered the country with valid visas but overstayed their visas' expiration or otherwise violated the terms of their admission.

**2. False.** Although "push" factors are important in migrating to another country, "pull" factors are even more important. Pull factors include better job opportunities, family or friends who already reside in the country of destination, and other favorable conditions that draw them to their new destination.

**3. True.** The percentage of U.S. immigrants (ages 25 and older) with at least a bachelor's degree (27 percent) is nearly identical to that of native-born U.S. adults (28 percent).

**4. False.** Research typically has not shown that immigrants increase unemployment in a region or lower wages among native workers. In fact, some studies have found that immigration produces modest gains in wages for native workers.

**5. True.** The majority (82 percent) of children living in undocumented immigrant families (with at least one undocumented immigrant parent) in the United States were born in this country and have U.S. citizenship status.

**6. True.** In a 2011 Gallup Organization survey, the majority (79 percent) of Americans said they believe that unauthorized immigrant workers mostly take low-paying jobs that Americans don't want.

**7. True.** To become a U.S. citizen, immigrants must go through an application process that includes taking a test on English and civics. To pass this test, applicants must demonstrate an ability to read, write, and speak simple words and phrases in English, and must demonstrate a knowledge and understanding of U.S. history and government.

**8. False.** A non-U.S. citizen who marries a U.S. citizen is *not* automatically granted U.S. citizenship. To become a U.S. citizen, a foreign-born spouse of a U.S. citizen must meet certain requirements (e.g., must have been married to and living with his or her U.S. citizen spouse for the past three years) and must go through the citizenship application process. (Requirements are different for noncitizen spouses of military personnel who have died in service.)

Sources: Based on Congressional Budget Office, 2011; Gallup Organization, 2011a; Gibson and Jung, 2006; Gryn and Larsen, 2010; Kandel, 2011; Passel and Cohn, 2011; Shierholz, 2010; and U.S. Citizenship and Immigration Services, 2011. © Cengage Learning

---

In most areas of the world, women are having fewer children. Women who have six or seven children tend to live in agricultural regions of the world, where children's labor is essential to the family's economic survival and child mortality rates are very high. For example, Niger has a crude birth rate of 50 per 1,000, as compared with 12 per 1,000 in the United States (Population Reference Bureau, 2013). However, in Niger and some other African nations, families need to have many children in order to ensure that one or two will live to adulthood because of high rates of poverty, malnutrition, and disease.

## Mortality

The primary cause of world population growth in recent years has been a decline in **mortality—the incidence of death in a population.** The simplest measure of mortality is the **crude death rate—the number of deaths per 1,000 people in a population in a given year.** The world mortality rate is about 8 per 1,000. The Population Reference Bureau (2013) has estimated that the U.S. mortality rate is also 8 per 1,000. In high-income, developed nations, mortality rates have declined dramatically as diseases such as malaria, polio, cholera, tetanus, typhoid, and measles have been virtually eliminated

**demography** a subfield of sociology that examines population size, composition, and distribution.

**fertility** the actual level of childbearing for an individual or a population.

**crude birth rate** the number of live births per 1,000 people in a population in a given year.

**mortality** the incidence of death in a population.

**crude death rate** the number of deaths per 1,000 people in a population in a given year.

by vaccinations and improved sanitation and personal hygiene (Weeks, 2012). The leading causes of deaths in most regions of the world are noncommunicable conditions such as cardiovascular diseases, cancer, diabetes, and chronic lung diseases. (The ten leading causes of death in the United States in 1900 and 2010–2013 are shown in Table 15.1.) In regions such as sub-Saharan Africa, infectious diseases remain the leading cause of death because mortality rates remain high as a result of HIV/AIDS.

Many children do not survive long enough to contract communicable diseases. On a global basis, large numbers of newborn infants do not live to see their first birthday. The measure of these deaths is referred to as the *infant mortality rate,* which is the number of deaths of infants under 1 year of age per 1,000 live births in a given year. The world infant mortality rate is about 40 per 1,000 live births each year (Population Reference Bureau, 2013). The infant mortality rate is an important reflection of a society's level of preventive (prenatal) medical care, maternal nutrition, childbirth procedures, and neonatal care for infants. The infant mortality rate varies widely: Sierra Leone has 128 deaths in the first year per 1,000 live births, compared to 1.2 deaths in the first year per 1,000 live births in China, Hong Kong SAR (Special Administrative Region).

In the United States the infant mortality rate has not changed significantly in recent years. In 2012 there were approximately 6 infant deaths per 1,000 live births. However, the mortality rate for African American (black) infants has remained 2.2 times as high as the rate for white American (non-Hispanic) infants for several years: 12.4 deaths per 1,000 live births for black (African American) infants, as compared to 8.47 deaths for American Indian/Alaska Native infants, 5.33 deaths for white (non-Hispanic) infants, 5.29 for Hispanic infants, and 4.40 for Asian and Pacific Islander infants per 1,000 live births in each category. Among other factors, differential levels of access to prenatal counseling and medical services are often reflected in the divergent infant mortality rates.

*Life expectancy* is an estimate of the average lifetime in years of people born in a specific year. For persons born in the United States in 2014, for example, the estimated life expectancy at birth was 79.6 years, with 77 years for males and 81.9 years for females (CIA *Factbook,* 2014). However, disparities exist among racial groups and divergent educational categories. Studies have found that adult men and women in the United States with fewer than twelve years of education had life expectancies that were about the same as adults in the 1950s and 1960s. When combined with race, the disparity grows even wider: White (non-Hispanic) women and men with 16 years or more of schooling had life expectancies far greater

### table 15.1

The Ten Leading Causes of Death in the United States, 1900 and 2010–2013

| Cause of Death—1900 | Rank | Cause of Death—2010–2013 |
|---|---|---|
| Influenza/pneumonia | 1 | Heart disease |
| Tuberculosis | 2 | Cancer (malignant neoplasms) |
| Stomach/intestinal disease | 3 | Chronic lower respiratory diseases |
| Heart disease | 4 | Stroke (cerebrovascular diseases) |
| Cerebral hemorrhage | 5 | Accidents (unintentional injuries) |
| Kidney disease | 6 | Alzheimer's disease |
| Accidents | 7 | Diabetes mellitus |
| Cancer | 8 | Influenza and pneumonia |
| Diseases in early infancy | 9 | Nephritis, nephritic syndrome, and nephrosis |
| Diphtheria | 10 | Suicide (intentional self-harm) |

Source: National Vital Statistics Reports, 2013b. © Cengage Learning

than African Americans with fewer than 12 years of education. Between white men and African American men, the difference was 14.2 additional years of life expectancy for white men. Between white women and African American women, the difference was 10.3 years more for white women than for African American women (Health Affairs, 2012).

## Migration

*Migration* **is the movement of people from one geographic area to another for the purpose of changing residency.** Migration affects the size and distribution of the population in a given area. *Distribution* refers to the physical location of people throughout a geographic area. In the United States, people are not evenly distributed throughout the country; many of us live in densely populated areas. *Density* is the number of people living in a specific geographic area. In urbanized areas, density may be measured by the number of people who live per room, per block, or per square mile.

Migration may be either international (movement between two nations) or internal (movement within national boundaries). Internal migration has occurred throughout U.S. history and has significantly changed the distribution of the population over time.

In essence, migration involves two types of movement: immigration and emigration. *Immigration* is the movement of people into a geographic area to take up residency. Although immigration to the United States has continued steadily since the 1970s, as noted earlier, economic conditions in this century have reduced the flow of both legal and unauthorized immigration to the United States. Each year, more than one million people obtain legal permanent residence (LPR) in the United States. In 2012, for example, a total of 1,031,631 persons became legal permanent residents of this country, and the majority (53 percent) had already lived in the United States. Nearly 66 percent were granted permanent resident status based on a family relationship with a U.S. citizen or with a legal permanent resident of the United States. The leading countries of origin of new LPRs are Mexico (14 percent), China (nearly 8 percent), and India (6.4 percent) (Monger and Yankay, 2013).

Immigration rates are not an accurate reflection of the actual number of immigrants who enter a country. The U.S. Immigration and Naturalization Service records only legal immigration based on entry visas and change-of-immigration-status forms. Similarly, few records are maintained regarding *emigration*—the movement of people out of a geographic area to take up residency elsewhere. To determine the net migration in a geographic area, the number of people leaving that area to take up permanent or semipermanent residence elsewhere (emigrants) is subtracted from the number of people entering that area to take up residence there (immigrants), unless more people are moving out of the area than into it, in which case the mathematical process is reversed. It is estimated that the net immigration rate in the United States is about 3.64 migrants per 1,000 population (cia.gov, 2013). In 2012 the net migration from Mexico fell to zero and perhaps even less because immigration to the United States from that country came to a halt. Factors such as the weakened U.S. economy, increased law enforcement along the U.S.–Mexico border, growing fears about drug wars and crossing the border illegally, and slightly improved economic conditions in Mexico contributed to at least a temporary reduction in new immigration to the United States (Passel, Cohn, and Gonzalez-Barrera, 2012). Improved economic conditions in the United States may bring new waves of immigration, but even if this does not happen, nearly one-third of all current U.S. immigrants are persons born in Mexico (Passel, Cohn, and Gonzalez-Barrera, 2012).

People migrate either voluntarily or involuntarily. *Pull* factors at the international level, such as a democratic government, religious freedom, employment opportunities, or a more temperate climate, may draw voluntary immigrants into a nation. Within nations, people from large cities may be pulled to rural areas by lower crime rates, more space, and

**FIGURE 15.2** Political unrest, violence, and war are "push" factors that encourage people to leave their country of origin. By contrast, job opportunities, such as construction work in the United States, are a major "pull" factor for people from low-income countries.

a lower cost of living. Some people are drawn by pull factors such as greater economic opportunities at their destination and are pushed by factors such as low wages and few employment opportunities in their previous place of residence (Figure 15.2). *Push* factors at the international level, such as political unrest, violence, war, famine, plagues, and natural

**migration** the movement of people from one geographic area to another for the purpose of changing residency.

# IMMIGRATION IN THE MEDIA

## Language Matters: "Illegal Immigrants" No Longer

The [Associated Press] Stylebook no longer sanctions the term "illegal immigrant" or the use of "illegal" to describe a person. Instead, it tells users that "illegal" should describe only an action, such as living in or immigrating to a country illegally.

—Paul Colford (2013), Associated Press (AP) senior vice president, along with Kathleen Carroll, executive editor, issued this statement, along with an explanation of why the *AP Stylebook*, which is used by journalists as a guide for correct capitalization, punctuation, and news writing rules, was being changed.

Several top voices at the Associated Press have made a significant change in how journalists should refer to persons who have entered, or are residing in, a country in violation of civil or criminal law. As indicated in the passage above, the word *illegal* is to be used only to refer to an action ("illegal immigration"), not to a person ("illegal immigrant"). According to the *AP Stylebook*, it is also acceptable to state that a person is "living in" or "entering a country illegally" or "without legal permission" (Colford, 2013). This change is part of a much larger issue of how the media frame discussions of immigration.

How should immigration be represented in the media? In the past, workers and their families who were living in the country illegally, particularly young children who worked in the fields or saw some of their family members deported, were shown in a sympathetic manner. *Sympathetic framing* refers to news writing that focuses on the human interest side of a story and shows that the individuals involved are caring people who are representative of a larger population. In this situation the focus would be on the issues and concerns of undocumented workers and why they feel a compelling need to come into the United States for work and what they perceive to be a "better quality of life." Sympathetic coverage has also been given to young immigrants who were

eager to finish their education and become part of the American Dream through attaining a college education and remaining in the United States (Opportunity Agenda, 2013).

In stark contrast to sympathetic framing are news reports that employ negative framing to describe immigrants from countries such as Mexico and others in Central and South America. Negative framing sometimes describes "illegal" immigrants as a drain on education, health care, and the U.S. welfare system. Other negative framing in stories focuses on these workers as a source of cheap labor that primarily benefits employers who do not have to pay a living wage. Negative framing of immigration in the media is not new in the United States. As far back as 1904, the *San Francisco Chronicle* carried lengthy articles describing how Japanese laborers were taking jobs away from U.S. workers, reflecting a pattern of reporting that continues in the twenty-first century (Puette, 1992).

How the media frame stories about immigration may influence social policy. Will we further restrict immigration for national security reasons? Will we develop new laws to allow limited entry of immigrant workers who meet the needs of certain employers? Not only are these important legal and social policy questions, but they are also issues that journalists and other news analysts face when they identify people, frame stories on immigration, and explain how U.S. policy makers choose to include or exclude immigrants from this country.

How do media representations affect your views on legal and illegal immigration in the United States?

## reflect & analyze

As this chapter has emphasized, population changes occur because of births, deaths, and migration. How important has migration been to social change in the United States? What part have the media played in our perceptions of immigration in this country?

disasters, may encourage people to leave one area and relocate elsewhere. Push factors in regional U.S. migration include unemployment, harsh weather conditions, a high cost of living, inadequate school systems, and high crime rates.

Involuntary, or forced, migration usually occurs as a result of political oppression, such as when Jews fled Nazi Germany in the 1930s or when Afghans left their country to escape oppression there in the early 2000s. Slavery is the most striking example of involuntary migration; for example, the 10–20 million Africans forcibly transported to the Western Hemisphere prior to 1800 did not come by choice.

Where do we stand in the United States in the twenty-first century? In the 2010s a number of states, including Alabama, Arizona, Georgia, Indiana, South Carolina, and Utah, passed stringent laws regarding illegal immigration. These laws caused strife and legal contests throughout the nation. As an example, the original Arizona law required that immigrants carry necessary identity documents to show that they were legally in the United States, police were empowered to detain persons suspected of being in the country illegally, and employers would be legally sanctioned for hiring undocumented workers. However, in *United States of America v. Arizona* a district court judge blocked some controversial provisions of the law, and in 2011 Arizona Governor Jan Brewer appealed the case to the U.S. Supreme Court. In a 5–3 ruling, the Court ruled that the federal government has significant power to regulate immigration and struck down most of the key provisions of the Arizona law. However, the Court let stand the provision of the Arizona law involving police checks on people's immigration status while the police are enforcing other laws if "reasonable suspicion" exists that the person is in the United States illegally.

In 2012 and 2013, members of Congress continued to propose immigration legislation to break the gridlock on this pressing issue. Leaders have engaged in sparring matches in the media that have been framed in divergent ways regarding immigration and border security. Some research has found that how members of the media frame their coverage on immigration may influence public opinion on the topic (see "Immigration in the Media"). Priorities of immigration bills often include one or more of these provisions:

1. *Securing the border:* The Department of Homeland Security would be required to monitor 100 percent of the southwest border with Mexico

**FIGURE 15.3** This fence between Mexico and the United States has been the subject of extensive media attention and public controversy. Although not a continuous fence, the barriers—combined with a "virtual fence" of sensors and cameras monitored by the U.S. Border Patrol—are designed to prevent illegal migration between the two countries.

*Spencer Platt/Getty Images*

and intercept 90 percent of people trying to cross it illegally (Figure 15.3).

2. *Path to citizenship:* Unauthorized immigrants who were in the United States before December 31, 2011, would be allowed to apply for temporary legal status. After ten years, they could apply for a green card (a visa for legal permanent residents), and three years later they could apply to become U.S. citizens.

3. *Interior enforcement:* U.S. business owners would be required, within five years, to verify the legal status of new employees to ensure that they are not hiring undocumented immigrants. Homeland Security would also have to track all immigrants each time they enter and exit the United States.

4. *Immigration overhaul:* The legal immigration system would be overhauled, reducing the percentage of visas that go to immigrants based on family ties from 75 percent to 50 percent, with the other 50 percent based on occupational specializations in science, technology, math and engineering, or low-skilled jobs.

These immigration initiatives would also provide an opportunity for young undocumented immigrants who arrived in the United States as children, often referred to as "Dreamers," to become citizens in five years after they apply to do so. Many college students with undocumented immigrant status had been negatively affected by the failure of a previous "Dream Act" bill proposed by the

Obama administration, which was supposed to provide citizenship for them. However, legal analysts at the time of this writing suggest that this, and other immigration bills that follow it, will no doubt produce more debates about whether millions of undocumented immigrants should receive citizenship and about whether there should be further so-called militarization of the U.S.–Mexico border (CNN.com, 2012a; A. Parker, 2013; Preston, 2013).

Although our discussion has focused on how immigration affects the United States, many other nations face similar issues, including the United Kingdom and Spain, both of which have announced that they are passing more-stringent laws to regulate immigration and make it easier to remove undocumented persons from within their nations' borders. Various factors, such as lack of jobs and poor economic conditions, fear of violence and terrorism, and concerns about how to provide an adequate social safety net for citizens, contribute to the belief that immigration is harmful to a nation's population and damages the existing social order.

## Population Composition

**LO2** Changes in fertility, mortality, and migration affect the *population composition*—the biological and social characteristics of a population, including age, sex, race, marital status, education, occupation, income, and size of household.

One measure of population composition is the *sex ratio*—the number of males for every hundred females in a given population. A sex ratio of 100 indicates an equal number of males and females in the population. If the number is greater than 100, there are more males than females; if it is less than 100, there are more females than males. In the United States the sex ratio in 2013 was 97, which means there were 97 males per 100 females. Although approximately 124 males are conceived for every 100 females, male fetuses miscarry at a higher rate. At birth, there are 105 males for every 100 females, but because males at all ages generally have higher mortality rates than females, the sex ratio declines with age. In the 25–54 age category, the sex ratio in 2013 was 100, meaning that for every 100 men in that age category, there were 100 women. However, by age 68 and older, the sex ratio was 77, meaning that for every 100 women ages 68 and older, there were only 77 men in this age category.

For demographers, sex and age are significant population characteristics; they are key indicators of fertility and mortality rates. The age distribution of a population has a direct bearing on the demand for schooling, health, employment, housing, and

pensions. The current distribution of a population can be depicted in a ***population pyramid***—a graphic **representation of the distribution of a population by sex and age.** Population pyramids are a series of bar graphs divided into five-year age cohorts; the left side of the pyramid shows the number or percentage of males in each age bracket; the right side provides the same information for females. The age/sex distribution in the United States and other high-income nations does not have the appearance of a classic pyramid, but rather is more rectangular or barrel-shaped. By contrast, low-income nations, such as Mexico and Iran, which have high fertility and mortality rates, do fit the classic population pyramid. Figure 15.4 compares the demographic composition of Mexico, Iran, the United States, and France.

## Population Growth in Global Context

What are the consequences of global population growth? Scholars do not agree on the answer to this question. Some biologists have warned that Earth is a finite ecosystem that cannot support the global population of 9.3 billion people predicted by 2050 or the population of more than 10 billion people predicted by the end of this century (UNFPA, 2011). However, some economists have emphasized that free-market capitalism is capable of developing innovative ways to solve such problems. The debate is not a new one; for several centuries, strong opinions have been voiced about the effects of population growth on human welfare.

### The Malthusian Perspective

**LO3** English clergyman and economist Thomas Robert Malthus (1766–1834) was one of the first scholars to systematically study the effects of population. Displeased with societal changes brought about by the Industrial Revolution in England, Malthus (1965/1798: 7) anonymously published *An Essay on the Principle of Population, as It Affects the Future Improvement of Society,* in which he argued that "the power of population is infinitely greater than the power of the earth to produce subsistence [food] for man."

According to Malthus, the population, if left unchecked, would exceed the available food supply. He argued that the population would increase in a geometric (exponential) progression (2, 4, 8, 16 . . .) while the food supply would increase only by an arithmetic progression (1, 2, 3, 4 . . .). In other words, a *doubling effect* occurs: Two parents can have four children, sixteen grandchildren, and so on, but food production increases by only one acre at a time.

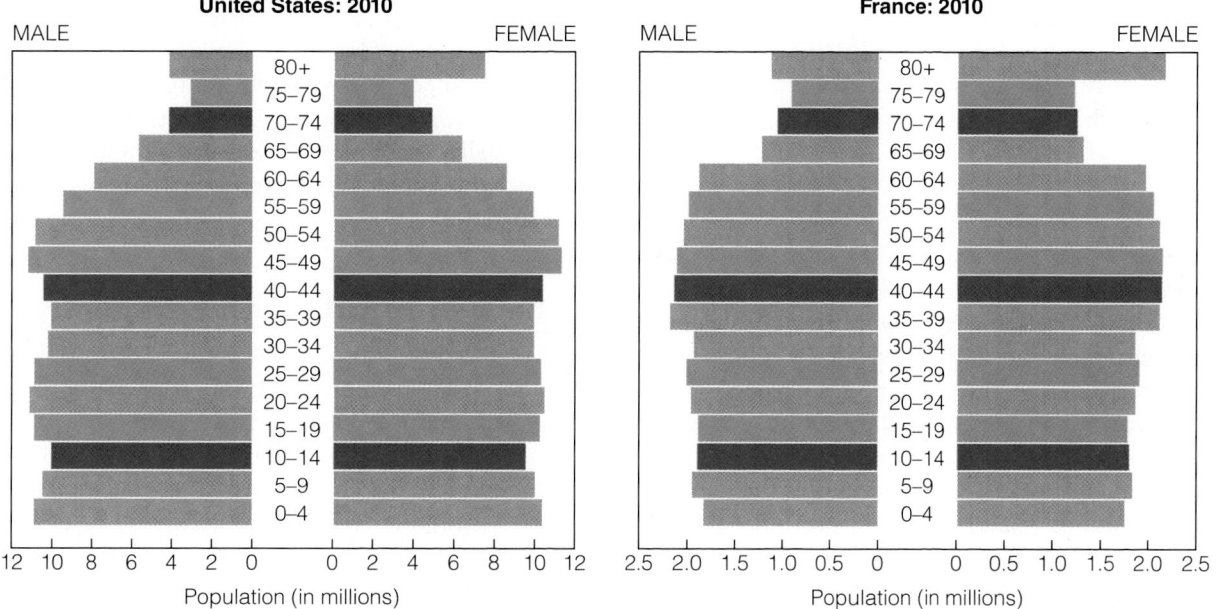

**FIGURE 15.4** POPULATION PYRAMIDS FOR MEXICO, IRAN, THE UNITED STATES, AND FRANCE

Source: U.S. Census Bureau International Data Base, 2011. © Cengage Learning

Thus, population growth inevitably surpasses the food supply, and the lack of food ultimately ends population growth and perhaps eliminates the existing population (Weeks, 2012). Even in a best-case scenario, overpopulation results in poverty.

However, Malthus suggested that this disaster might be averted by either positive or preventive checks on population. *Positive checks* are mortality risks, such as famine, disease, and war; *preventive checks* are limits to fertility. For Malthus, the only acceptable preventive check was *moral restraint*; people should practice sexual abstinence before marriage and postpone marriage as long as possible in order to have only a few children.

Malthus has had a lasting impact on the field of population studies. Most demographers refer

**population composition** the biological and social characteristics of a population, including age, sex, race, marital status, education, occupation, income, and size of household.

**sex ratio** the number of males for every hundred females in a given population.

**population pyramid** a graphic representation of the distribution of a population by sex and age.

to his dire predictions when they examine the relationship between fertility and subsistence needs. Overpopulation is still a daunting problem that capitalism and technological advances thus far have not solved, especially in middle- and low-income nations with rapidly growing populations and very limited resources.

In 2014 some economists and journalists revisited Malthus because they believed that his prediction was based on a realistic premise that the Earth's carrying capacity definitely had a limit and that the world might be closer than many people thought to reaching that limit, not only in terms of food and water but also climate-related risks. A United Nations panel issued a warning that global warming was challenging the world's food supply. For example, it is believed that the ozone accumulations and high temperatures will negatively affect the production of all the major grains. As Craig Hanson, Director of Food, Forests, and Water Programs at the World Resources Institute, stated, "This is a wake-up call for the agriculture sector. Climate change is a food security issue. It's not just an environmental issue" (qtd. in Porter, 2014).

## The Marxist Perspective

Among those who attacked the ideas of Malthus were Karl Marx and Frederick Engels. According to Marx and Engels, the food supply is not threatened by overpopulation; technologically, it is possible to produce the food and other goods needed to meet the demands of a growing population. Marx and Engels viewed poverty as a consequence of the exploitation of workers by the owners of the means of production. For example, they argued that England had poverty because the capitalists skimmed off some of the workers' wages as profits. The labor of the working classes was used by capitalists to earn profits, which, in turn, were used to purchase machinery that could replace the workers rather than supply food for all. From this perspective, overpopulation occurs because capitalists desire to have a surplus of workers (an industrial reserve army) in order to suppress wages and force workers concerned about losing their livelihoods to be more productive.

According to some contemporary economists, the greatest crisis today facing low-income nations is capital shortage, not food shortage. Through technological advances, agricultural production has reached the level at which it can meet the food needs of the world if food is distributed efficiently. *Capital shortage* refers to the lack of adequate money or property to maintain a business; it is a problem because the physical capital of the past no longer meets the needs of modern economic development. In the past, self-contained rural economies survived on local labor, using local materials to produce the capital needed for other laborers. For example, in a typical village a carpenter made the loom needed by the weaver to make cloth. Today, in the global economy the one-to-one exchange between the carpenter and the weaver is lost. With an antiquated, locally made loom, the weaver cannot compete against electronically controlled, mass-produced looms. Therefore, the village must purchase capital from the outside, using its own meager financial resources. In the process the complementary relationship between labor and capital is lost; modern technology brings with it steep costs and results in village noncompetitiveness and underemployment.

Marx and Engels made a significant contribution to the study of demography by suggesting that poverty, not overpopulation, is the most important issue with regard to food supply in a capitalist economy. Although Marx and Engels offer an interesting counterpoint to Malthus, some scholars argue that the Marxist perspective is self-limiting because it attributes the population problem solely to capitalism. In actuality, nations with socialist economies also have demographic trends similar to those in capitalist societies.

## The Neo-Malthusian Perspective

More recently, *neo-Malthusians* (or "new Malthusians") have reemphasized the dangers of overpopulation. To neo-Malthusians, Earth is "a dying planet" with too many people and too little food, compounded by environmental degradation and overconsumption. The 1968 publication of Paul R. Ehrlich's *The Population Bomb* launched a worldwide discussion about the effects of overpopulation and rapid population growth. "The Population Bomb Revisited," published forty years later, reasserted the growing importance of the demographic element in the human predicament: "[t]he Earth's finite capacity to sustain human civilization" (Ehrlich and Ehrlich, 2009: 63). In other words, overpopulation and rapid population growth result in global environmental problems, ranging from global warming and rain-forest destruction to famine and vulnerability to epidemics (Ehrlich and Ehrlich, 2009). Unless significant changes are made, including improving the status of women, reducing racism and religious prejudice, reforming the agriculture system, and shrinking the growing gap between rich and poor, the consequences will be dire (Ehrlich and Ehrlich, 2009).

Early neo-Malthusians published birth control handbooks, and widespread acceptance of birth control eventually reduced the connection between people's sexual conduct and fertility (Weeks, 2012). Later neo-Malthusians have encouraged people to be part of the solution to the problem of overpopulation by having only one or two children in order to bring about *zero population growth*—**the point at which no population increase occurs from year to year** because the number of births plus immigrants is equal to the number of deaths plus emigrants (Weeks, 2012).

Today, the Ehrlich proposal remains the same: "Adopt policies that gradually reduce birthrates and eventually start a global decline toward a human population size that is sustainable in the long run" (Ehrlich and Ehrlich, 2009).

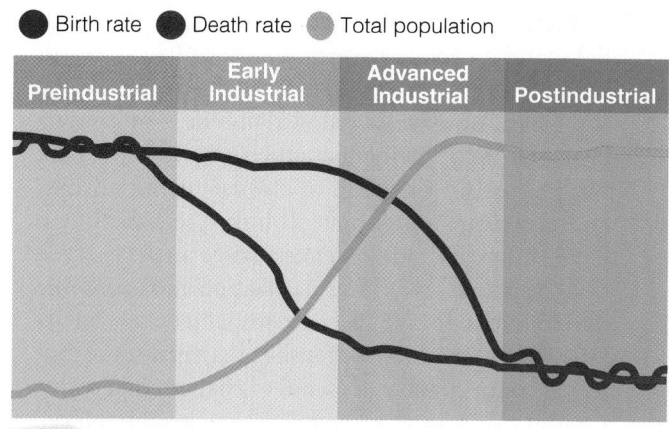

● Birth rate ● Death rate ○ Total population

**FIGURE 15.5** **THE DEMOGRAPHIC TRANSITION**
© Cengage Learning

## Demographic Transition Theory

Some scholars who disagree with the neo-Malthusian viewpoint suggest that the theory of demographic transition offers a more accurate picture of future population growth. **Demographic transition is the process by which some societies have moved from high birth and death rates to relatively low birth and death rates as a result of technological development.** Demographic transition is linked to four stages of economic development (see Figure 15.5):

- *Stage 1: Preindustrial societies.* Little population growth occurs because high birth rates are offset by high death rates. Food shortages, poor sanitation, and lack of adequate medical care contribute to high rates of infant and child mortality.
- *Stage 2: Early industrialization.* Significant population growth occurs because birth rates are relatively high whereas death rates decline. Improvements in health, sanitation, and nutrition produce a substantial decline in infant mortality rates. Overpopulation is likely to occur because more people are alive than the society has the ability to support.
- *Stage 3: Advanced industrialization and urbanization.* Very little population growth occurs because both birth rates and death rates are low. The birth rate declines as couples control their fertility through contraceptives and become less likely to adhere to religious directives against their use. Children are not viewed as an economic asset; they consume income rather than produce it. Societies in this stage attain zero population growth, but the actual number of births per year may still rise because of an increased number of women of childbearing age.
- *Stage 4: Postindustrialization.* Birth rates continue to decline as more women gain full-time

employment and the cost of raising children continues to increase. The population grows very slowly, if at all, because the decrease in birth rates is coupled with a stable death rate.

Debate continues about whether this evolutionary model accurately explains the stages of population growth in all societies. Advocates note that demographic transition theory highlights the relationship between technological development and population growth, thus making Malthus's predictions obsolete. Scholars also point out that demographic transitions occur at a faster rate in now-low-income nations than they previously did in the nations that are already developed. For example, nations in the process of development have higher birth rates and death rates than the now-developed societies did when they were going through the transition. The death rates declined in the now-developed nations as a result of internal economic development—not, as is the case today, through improved methods of disease control (Weeks, 2012). Critics suggest that this theory best explains development in Western societies.

## Other Perspectives on Population Change

In recent decades, other scholars have continued to develop theories about how and why changes in population growth patterns occur. Some have

**zero population growth** the point at which no population increase occurs from year to year.

**demographic transition** the process by which some societies have moved from high birth and death rates to relatively low birth and death rates as a result of technological development.

studied the relationship between economic development and a decline in fertility; others have focused on the process of secularization—the decline in the significance of the sacred in daily life—and how a change from believing that otherworldly powers are responsible for one's life to a sense of responsibility for one's own well-being is linked to a decline in fertility. Based on this premise, some analysts argue that the processes of industrialization and economic development are typically accompanied by secularization but that the relationship between these factors is complex when it comes to changes in fertility.

Shifting from the macrolevel to the microlevel, educational and social psychological factors also play into the decisions that individuals make about how many children to have. Family planning information is more readily available to people with more years of formal education and may cause them to engage in decision making in accord with *rational choice theory,* which is based on the assumption that people make decisions based on a calculated cost–benefit analysis ("What do I gain and lose from a specific action?"). In low-income countries or other settings in which children are identified as an economic resource for their parents throughout life, fertility rates are higher than in higher-income countries. However, as modernization and urbanization occur in such societies, the positive economic effects of having more children may be offset by the cost of caring for those children and the lowered economic advantage gained from having children in an industrialized nation.

As demographers have reformulated the demographic transition theory, they have highlighted additional factors that are likely to be causes of fertility decline, and they have suggested that demographic transition is not just one process but rather a set of intertwined transitions. One is the epidemiological transition—the shift from deaths at younger ages caused by acute, communicable diseases. Another is the fertility transition—the shift from natural fertility to controlled fertility, resulting in a decrease in the fertility rate. Other transitions include the migration transition, the urban transition, the age transition, and the family and household transition, which occur as a result of lower fertility, longer life, an older age structure, and predominantly urban residence.

# A Brief Glimpse at International Migration Theories

**LO4** Why do people relocate from one nation to another? Several major theories have been developed in an attempt to explain international migration. The *neoclassical economic approach* assumes that migration patterns occur based on geographic differences in the supply of and demand for labor. The United States and other high-income countries that have had growing economies and a limited supply of workers for certain types of jobs have paid higher wages than are available in areas with a less developed economy and a large labor force. As a result, people move to gain higher wages and sometimes better living conditions. They also may take jobs in other countries so that they can send money to their families in their country of origin.

Unlike the neoclassical explanation of migration, which focuses on individual decision making, the *new households economics of migration approach* emphasizes the part that entire families or households play in the migration process. From this approach, Mexican workers' temporary migration to the United States would be examined not only from the perspective of the individual worker but also in terms of what the entire family gains from the process of having one or more migrant family members work in another country. By having a diversity of family income (originating from more than one source), the family is cushioned from the economic woes of the nation that most of the family members think of as "home."

Two conflict perspectives on migration add to our knowledge of why people migrate. Split-labor-market theory suggests that immigrants from low-income countries are often recruited for secondary labor market positions: dead-end jobs with low wages, unstable employment, and sometimes hazardous working conditions. By contrast, migrants from higher-income countries may migrate for primary-sector employment—jobs in which well-educated workers are paid high wages and receive benefits such as health insurance and a retirement plan. The global migration of some high-tech workers is an example of this process, whereas the migration of farm workers and construction helpers is an example of secondary labor market migration.

Finally, world systems theory (discussed later in this chapter) views migration as linked to the problems caused by capitalist development around the world (Massey et al., 1993). As the natural resources, land, and workforce in low-income countries with little or no industrialization have come under the influence of international markets, there has been a corresponding flow of migrants from those nations to the highly industrialized, high-income countries, especially those with which the poorer nations have had the most economic, political, or military contact.

After flows of migration commence, the pattern may continue because potential migrants have personal ties with relatives and friends who now live in the country of destination and can serve as a source of stability when the potential migrants relocate to the new country. Known as *network theory*, this approach suggests that once migration has begun, it takes on a life of its own and that the migration pattern which ensues may be different from the original push or pull factors that produced the earlier migration. Another approach, *institutional theory*, suggests that migration may be fostered by groups—such as humanitarian aid organizations relocating refugees or smugglers bringing people into a country illegally—and that the actions of these groups may produce a larger stream of migrants than would otherwise be the case.

As you can see from these diverse approaches to explaining contemporary patterns of migration, the reasons that people migrate are numerous and complex, involving processes occurring at the individual, family, and societal levels.

# Urbanization in Global Perspective

**LO5** *Urban sociology* is a subfield of sociology that examines social relationships and political and economic structures in the city. According to urban sociologists, a *city* is a relatively dense and permanent settlement of people who secure their livelihood primarily through nonagricultural activities. Although cities have existed for thousands of years, only about 3 percent of the world's population lived in cities 200 years ago, as compared with more than 50 percent today. Current estimates suggest that 69 percent of the global population will live in urban areas by 2050 (United Nations DESAPD, 2010).

## Emergence and Evolution of the City

Cities are a relatively recent innovation when compared with the length of human existence. The earliest humans are believed to have emerged anywhere from 40,000 to 1,000,000 years ago, and permanent human settlements are believed to have begun first about 8000 B.C.E. However, some scholars date the development of the first city between 3500 and 3100 B.C.E., depending largely on whether a formal writing system is considered as a requisite for city life (Sjoberg, 1965; Weeks, 2012).

According to the sociologist Gideon Sjoberg (1965), three preconditions must be present in order for a city to develop:

1. *A favorable physical environment,* including climate and soil conducive to the development of plant and animal life and an adequate water supply to sustain both.

2. *An advanced technology* (for that era) that could produce a social surplus in both agricultural and nonagricultural goods.
3. *A well-developed social organization,* including a power structure, in order to provide social stability to the economic system.

Based on these prerequisites, Sjoberg places the first cities in the Middle Eastern region of Mesopotamia or in areas immediately adjacent to it at about 3500 B.C.E. However, not all scholars concur; some place the earliest city in Jericho (located in present-day Jordan) at about 8000 B.C.E., with a population of about 600 people (see Kenyon, 1957).

The earliest cities were not large by today's standards. The population of the larger Mesopotamian centers was between 5,000 and 10,000 (Sjoberg, 1965). The population of ancient Babylon (probably founded around 2200 B.C.E.) may have grown as large as 50,000 people; Athens may have held 80,000 people (Weeks, 2012). Four to five thousand years ago, cities with at least 50,000 people existed in the Middle East (in today's Iraq and Egypt) and Asia (in today's Pakistan and China), as well as in Europe. About 3,500 years ago, cities began to reach this size in Central and South America.

## Preindustrial Cities

The largest preindustrial city was Rome; by 100 C.E., it may have had a population of 650,000. With the fall of the Roman Empire in 476 C.E., the nature of European cities changed. Seeking protection and survival, those persons who lived in urban settings typically did so in walled cities containing no more than 25,000 people. For the next 600 years, the urban population continued to live in walled enclaves as competing warlords battled for power and territory during the "dark ages." Slowly, as trade increased, cities began to tear down their walls.

Preindustrial cities were limited in size by a number of factors. For one thing, crowded housing conditions and a lack of adequate sewage facilities increased the hazards from plagues and fires, and death rates were high. For another, food supplies were limited. In order to generate food for each city resident, at least fifty farmers had to work in the fields, and animal power was the only means of bringing food to the city. Once foodstuffs arrived in the city, there was no effective way to preserve them. Finally, migration to the city was difficult. Many people were in serf, slave, and caste systems whereby they were bound to the land. Those able to escape such restrictions still faced several weeks of travel to reach the city, thus making it physically and financially impossible for many people to become city dwellers.

**FIGURE 15.6** Since the 1950s, postindustrial cities have emerged in which families do not live close to a central business district and more homes are located in suburban or outlying residential enclaves such as the neighborhood shown here.

technology, including new forms of transportation and agricultural production, made it easier for people to leave the countryside and move to the city. Between 1700 and 1900, the population of many European cities mushroomed. For example, the population of London increased from 550,000 to almost 6.5 million. Although the Industrial Revolution did not start in the United States until the mid-nineteenth century, the effect was similar. Between 1870 and 1910, for example, the population of New York City grew by 500 percent. In fact, New York City became the first U.S. *metropolis*—one or more central cities and their surrounding suburbs that dominate the economic and cultural life of a region. Nations such as Japan and Russia, which became industrialized after England and the United States, experienced a delayed pattern of urbanization, but this process moved quickly once it began in those countries.

In spite of these problems, many preindustrial cities had a sense of *community*—a set of social relationships operating within given spatial boundaries or locations that provided people with a sense of identity and a feeling of belonging. The cities were full of people from all walks of life, both rich and poor, and they felt a high degree of social integration. You will recall that Ferdinand Tönnies (1940/1887) described such a community as *Gemeinschaft*—a society in which social relationships are based on personal bonds of friendship and kinship and on intergenerational stability, such that people have a commitment to the entire group and feel a sense of togetherness. By contrast, industrial cities were characterized by Tönnies as *Gesellschaft*—societies exhibiting impersonal and specialized relationships, with little long-term commitment to the group or consensus on values. In *Gesellschaft* societies, even neighbors are "strangers" who perceive that they have little in common with one another.

### Industrial Cities

The Industrial Revolution changed the nature of the city. Factories sprang up rapidly as production shifted from the primary, agricultural sector to the secondary, manufacturing sector. With the advent of factories came many new employment opportunities not available to people in rural areas. Emergent

### Postindustrial Cities

Since the 1950s, postindustrial cities have emerged in nations such as the United States as their economies have gradually shifted from secondary (manufacturing) production to tertiary (service and information-processing) production. Postindustrial cities increasingly rely on an economic structure that is based on scientific knowledge rather than industrial production, and, as a result, a class of professionals and technicians grows in size and influence. Postindustrial cities are dominated by "light" industry, such as software manufacturing; information-processing services, such as airline and hotel reservation services; educational complexes; medical centers; convention and entertainment centers; and retail trade centers and shopping malls. Most families do not live close to a central business district (Figure 15.6). Technological advances in communication and transportation make it possible for middle- and upper-income individuals and families to have more work options and to live greater distances from the workplace; however, these options are not often available to people of color and those at the lower end of the class structure.

On a global basis, cities such as New York, London, and Tokyo appear to fit the model of the postindustrial city. These cities have experienced a rapid growth in knowledge-based industries such as financial services. London, Tokyo, and New York have—at least until recently—experienced an increase in the number of highly paid professional jobs, and more workers have been in high-income categories. Many people have benefited for a number of years from these high incomes and have created a lifestyle that is based on materialism and the gentrification of urban spaces. Meanwhile, those persons outside the growing professional categories have seen their own quality of life further deteriorate and their job opportunities become increasingly restricted to secondary labor markets in their respective "global" cities.

What is next for the postindustrial city? Various social analysts have proposed that cities are far from dead because of the many benefits they offer people. According to urban economist Edward Glaeser (2011), the future of nations and the world relies on cities that bring people together in a setting that is "healthier, greener, and richer" than urban myths would have us believe. From this perspective, incomes are higher in metropolitan areas; cities are more energy efficient than suburban areas, where people commute great distances, often in heavy traffic congestion with stop-and-go traffic, between work and home. Cities are also centers of consumption; however, this benefit primarily accrues to the wealthy. However, the poor fare better than many think because they have inexpensive mass transit, the ability to "cram into small apartments in the outer boroughs," and "plenty of entry-level service-sector jobs with wages that beat those in Ghana or Guatemala" (Glaeser, 2011). People in the middle-income category often have a harder time because of the costs of housing in good neighborhoods and a quality education for their children (Glaeser, 2011).

A second perspective on the postmodern city has been offered by John D. Kasarda, who coined the term *aerotropolis* to describe a new urban pattern in which cities are built around airports rather than airports being built around cities. An aerotropolis is a combination of giant airport, planned city, shipping facility, and business hub. According to this approach, the pattern of the twentieth century was city in the center, airport on the periphery. However, this pattern has shifted, with the airport now in the center and the city on the periphery, because of extensive growth in jet travel, 24/7 workdays, overnight shipping, and global business networks (Kasarda and Lindsay, 2011). Aerotropoli are now found in Seoul, Amsterdam, Dallas, Memphis, Washington, D.C., and other cities where globalization has forever changed the nature of urban life.

# Perspectives on Urbanization and the Growth of Cities

Urban sociology follows in the tradition of early European sociological perspectives that compared social life with biological organisms or ecological processes. For example, Auguste Comte pointed out that cities are the "real organs" that make a society function. Emile Durkheim applied natural ecology to his analysis of *mechanical solidarity,* characterized by a simple division of labor and shared religious beliefs such as are found in small, agrarian societies, and *organic solidarity,* characterized by interdependence based on the elaborate division of labor found in large, urban societies. These early analyses became the foundation for ecological models/functionalist perspectives in urban sociology.

## Functionalist Perspectives: Ecological Models

**LO6** Functionalists examine the interrelations among the parts that make up the whole; therefore, in studying the growth of cities, they emphasize the life cycle of urban growth. Like the social philosophers and sociologists before him, the University of Chicago sociologist Robert Park (1915) based his analysis of the city on *human ecology*—the study of the relationship between people and their physical environment. According to Park (1936), economic competition produces certain regularities in land-use patterns and population distributions. Applying Park's idea to the study of urban land-use patterns, the sociologist Ernest W. Burgess (1925) developed the concentric zone model, an ideal construct that attempts to explain why some cities expand radially from a central business core.

**The Concentric Zone Model** Burgess's *concentric zone model* is a description of the process of urban growth that views the city as a series of circular areas or zones, each characterized by a different type of land use, which developed from a central core (see Figure 15.7a). *Zone 1* is the central business district and cultural center. In *Zone 2,* houses formerly occupied by wealthy families are divided into rooms and rented to recent immigrants and poor persons; this zone also contains light manufacturing and marginal businesses (such as second-hand stores, pawnshops, and taverns). *Zone 3* contains working-class residences and shops and ethnic enclaves.

**a.** Concentric Zone Model   **b.** Sector Model   **c.** Multiple Nuclei Model

- **1.** Central business district
- **2.** Wholesale light manufacturing
- **3.** Working-class residential
- **4.** Middle-class residential
- **5.** High-class residential
- **6.** Heavy manufacturing
- **7.** Outlying business district
- **8.** Suburb

**FIGURE 15.7** THREE MODELS OF THE CITY
Source: Adapted from Harris and Ullman, 1945. © Cengage Learning

*Zone 4* comprises homes for affluent families, single-family residences of white-collar workers, and shopping centers. *Zone 5* is a ring of small cities and towns populated by persons who commute to the central city to work and by wealthy people living on estates.

Two important ecological processes are involved in the concentric zone theory: invasion and succession. ***Invasion*** **is the process by which a new category of people or type of land use arrives in an area previously occupied by another group or type of land use** (McKenzie, 1925). For example, Burgess noted that recent immigrants and low-income individuals "invaded" Zone 2, formerly occupied by wealthy families. ***Succession*** **is the process by which a new category of people or type of land use gradually predominates in an area formerly dominated by another group or type of land use** (McKenzie, 1925). In Zone 2, for example, when some of the single-family residences were sold and subsequently divided into multiple housing units, the remaining single-family owners moved out because the "old" neighborhood had changed. As a result of their move, the process of invasion was complete, and succession had occurred.

Invasion and succession theoretically operate in an outward movement: Those who are unable to "move out" of the inner rings are those without upward social mobility, so the central zone ends up being primarily occupied by the poorest residents— except when gentrification occurs. ***Gentrification*** **is the process by which members of the middle and upper-middle classes, especially whites, move into a central-city area and renovate existing properties.** Centrally located, naturally attractive areas

are the most likely candidates for gentrification. To urban ecologists, gentrification is the solution to revitalizing the central city. To conflict theorists, however, gentrification creates additional hardships for the poor by depleting the amount of affordable housing available and by "pushing" them out of the area.

The concentric zone model demonstrates how economic and political forces play an important part in the location of groups and activities, and it shows how a large urban area can have internal differentiation. However, the model is most applicable to older cities that experienced high levels of immigration early in the twentieth century and to a few midwestern cities such as St. Louis. No city, including Chicago (on which the model is based), entirely conforms to this model.

**The Sector Model** In an attempt to examine a wider range of settings, urban ecologist Homer Hoyt (1939) studied the configuration of 142 cities. Hoyt's *sector model* emphasizes the significance of terrain and the importance of transportation routes in the layout of cities. According to Hoyt, residences of a particular type and value tend to grow outward from the center of the city in wedge-shaped sectors, with the more-expensive residential neighborhoods located along the higher ground near lakes and rivers or along certain streets that stretch in one direction or another from the downtown area (see Figure 15.7b). By contrast, industrial areas tend to be located along river valleys and railroad lines. Middle-class residential zones exist on either side of the wealthier neighborhoods. Finally, lower-class residential areas occupy the remaining space, bordering the central business area and the industrial areas. Hoyt (1939) concluded that the sector model applied to cities such as Seattle, Minneapolis, San Francisco, Charleston (South Carolina), and Richmond (Virginia).

**The Multiple Nuclei Model** According to the *multiple nuclei model* developed by urban ecologists Chauncey Harris and Edward Ullman (1945), cities do not have one center from which all growth radiates but rather have numerous centers of development based on specific urban needs or activities (see Figure 15.7c). As cities began to grow rapidly, they annexed formerly outlying and independent townships that had been communities in their own right. In addition to the central business district, other nuclei developed around entities such as an educational institution, a medical complex, or a government center. Residential neighborhoods may exist close to or far away from these nuclei. A wealthy residential enclave may be located near a high-priced shopping center, for instance, whereas less-expensive housing must locate closer to industrial and transitional areas of town. This model may be applicable to cities such as Boston (Figure 15.8). However, critics suggest that it does not provide insights about the uniformity of land-use patterns among cities and relies on an after-the-fact explanation of why certain entities are located where they are.

**Contemporary Urban Ecology** Urban ecologist Amos Hawley (1950) revitalized the ecological tradition by linking it more closely with functionalism. According to Hawley, urban areas are complex and expanding social systems in which growth patterns are based on advances in transportation and communication. For example, commuter railways and automobiles led to the decentralization of city life and the movement of industry from the central city to the suburbs (Hawley, 1981).

Other urban ecologists have continued to refine the methodology used to study the urban environment. *Social area analysis* examines urban populations in terms of economic status, family status, and ethnic classification (Shevky and Bell, 1966). For example, middle- and upper-middle-class parents with school-age children tend to cluster together in "social areas" with a "good" school district; young single professionals may prefer to cluster in the central city for entertainment and nightlife.

The influence of human ecology on the field of urban sociology is still very strong today (see Frisbie and Kasarda, 1988). Contemporary research on European and North American urban patterns is often based on the assumption that spatial arrangements in cities conform to a common, most efficient design. However, some critics have noted that ecological

Education Images/Universal Images Group Limited/Alamy

**FIGURE 15.8** Boston's South End has recently experienced gentrification. Why do some social theorists disagree on the positive and negative effects of gentrification on communities?

models do not take into account the influence of powerful political and economic elites on the development process in urban areas (Feagin and Parker, 2002).

## Conflict Perspectives: Political Economy Models

Conflict theorists argue that cities do not grow or decline by chance. Rather, they are the product of specific decisions made by members of the capitalist class and political elites. These far-reaching decisions regarding land use and urban development benefit the members of some groups at the expense of others (see Castells, 1977/1972). Karl Marx suggested that cities are the arenas in which the intertwined processes of class conflict and capital accumulation take place; class consciousness and worker revolt are more likely to develop when workers are concentrated in urban areas.

According to the sociologists Joe R. Feagin and Robert Parker (2002), three major themes prevail

**invasion** the process by which a new category of people or type of land use arrives in an area previously occupied by another group or type of land use.

**succession** the process by which a new category of people or type of land use gradually predominates in an area formerly dominated by another group or type of land use.

**gentrification** the process by which members of the middle and upper-middle classes, especially whites, move into a central-city area and renovate existing properties.

in political economy models of urban growth. First, both economic *and* political factors affect patterns of urban growth and decline. Economic factors include capitalistic investments in production, workers, workplaces, land, and buildings. Political factors include governmental protection of the right to own and dispose of privately held property as owners see fit and the role of government officials in promoting the interests of business elites and large corporations.

Second, urban space has both an exchange value and a use value. *Exchange value* refers to the profits that industrialists, developers, bankers, and others make from buying, selling, and developing land and buildings. By contrast, *use value* is the utility of space, land, and buildings for everyday life, family life, and neighborhood life. In other words, land has purposes other than simply for generating profits—for example, for homes, open spaces, and recreational areas (see Figure 15.9). Today, class conflict exists over the use of urban space, as is evident in battles over the rental costs, safety, and development of large-scale projects.

Third, both structure and agency are important in understanding how urban development takes place. *Structure* refers to institutions such as state bureaucracies and capital investment circuits that are involved in the urban development process. *Agency* refers to human actors, including developers, business elites, and activists protesting development, who are involved in decisions about land use.

**Capitalism and Urban Growth in the United States** According to political economy models, urban growth is influenced by capital investment decisions, power and resource inequality, class and class conflict, and government subsidy programs. Members of the capitalist class choose corporate locations, decide on sites for shopping centers and factories, and spread the population that can afford to purchase homes into sprawling suburbs located exactly where the capitalists think they should be located (Feagin and Parker, 2002).

Today, a few hundred financial institutions and developers finance and construct most major and many smaller urban development projects around the country, including skyscrapers, shopping malls, and suburban housing projects. These decision makers set limits on the individual

choices of the ordinary citizen with regard to real estate, just as they do with regard to other choices (Feagin and Parker, 2002). They can make housing more affordable or totally unaffordable for many people. Ultimately, their motivation rests not in benefiting the community but rather in making a profit; the cities that they produce reflect this mindset.

One of the major results of these urban development practices is *uneven development*—the tendency of some neighborhoods, cities, or regions to grow and prosper whereas others stagnate and decline. Conflict theorists argue that uneven development reflects inequalities of wealth and power in society. The problem not only affects areas in a state of decline but also produces external costs, even in "boom" areas, that are paid by the entire community. Among these costs are increased pollution, increased traffic congestion, and rising rates of crime and violence. According to the sociologist Mark Gottdiener (1985: 214), these costs are "intrinsic to the very core of capitalism, and those who profit the most from development are not called upon to remedy its side effects."

**The Gated Community in the Capitalist Economy** The growth of *gated communities*—subdivisions or neighborhoods surrounded by

**Exchange Value**
Profits from buying, selling, and developing urban land

**Use Value**
Utility of urban land, space, and buildings for everyday personal and community life

**Examples of who profits**
- Industrialists
- Developers
- Bankers
- Tax collectors

**Examples of uses**
- Affordable housing
- Open spaces
- Recreational areas
- Public services

**FIGURE 15.9** THE VALUE OF URBAN SPACE
© Cengage Learning

barriers such as walls, fences, gates, or earth banks covered with bushes and shrubs, along with a secured entrance—is an example to many people of how developers, builders, and municipalities have encouraged an increasing division between public and private property in capitalist societies (Figure 15.10). Many gated communities are created by developers who hope to increase their profits by offering potential residents a semblance of safety, privacy, and luxury that they might not have in nongated residential areas. Other gated communities have been developed after the fact in established neighborhoods by adding walls, gates, and sometimes security guard stations. In the past, for example, residents of elite residential enclaves, such as the River Oaks area of Houston or the "Old Enfield" area of Austin, Texas, were able to gain approval from the city to close certain streets and create cul-de-sacs, or to erect other barriers to discourage or prevent outsiders from driving through the neighborhood (Kendall, 2002). Gated communities for upper-middle-class and upper-class residents convey the idea of exclusivity and privilege, whereas such communities for middle- and lower-income residents typically focus on such features as safety for children and the ability to share amenities such as a "community" swimming pool or recreational center with other residents.

Regardless of the social and economic reasons given for the development of gated communities, many analysts agree that these communities reflect a growing divide between public and private space in urban areas. According to a qualitative study by the anthropologist Setha Low (2003), gated communities do more than simply restrict access to the residents' homes: They also limit the use of public spaces, making it impossible for others to use the roads, parks, and open space contained within the enclosed community. Low (2003) refers to this phenomenon as the "fortressing of America."

**Gender Regimes in Cities** Some feminist perspectives focus on urbanization as a reflection not only of the workings of the political economy but also of patriarchy. According to the sociologist Lynn M. Appleton (1995), different kinds of cities have different *gender regimes*—prevailing ideologies of how women and men should think, feel, and act; how access to social positions and control of resources should be managed; and how relationships between men and women should be conducted. The higher density and greater diversity found in central cities such as New York City serve as a challenge to the private patriarchy found in the home and workplace in lower-density, homogeneous areas such as suburbs and rural areas.

**FIGURE 15.10** How do gated communities increase the division between public and private property in the United States?

*Private patriarchy* is based on a strongly gendered division of labor in the home, gender-segregated paid employment, and women's dependence on men's income.

At the same time, cities may foster *public patriarchy* in the form of women's increasing dependence on paid work and the government for income and their decreasing emotional interdependence with men. At this point, gender often intersects with class and race as a form of oppression because lower-income women of color often reside in central cities. Public patriarchy may be perpetuated by cities through policies that limit women's access to paid work and public transportation. However, such cities may also be a forum for challenging patriarchy; all residents who differ in marital status, paternity, sexual orientation, class, and/or race/ethnicity tend to live close to one another and may hold a common belief that both public and private patriarchy should be eliminated (Appleton, 1995).

## Symbolic Interactionist Perspectives: The Experience of City Life

**LO7** Symbolic interactionists examine the *experience* of urban life. How does city life affect the people who live in a city? Some analysts answer this question positively; others are cynical about the effects of urban living on the individual.

**Simmel's View of City Life** According to the German sociologist Georg Simmel (1950/1902–1917), urban life is highly stimulating, and it shapes people's thoughts and actions. Urban residents are

**493**

influenced by the quick pace of the city and the pervasiveness of economic relations in everyday life. Because of the intensity of urban life, people have no choice but to become somewhat insensitive to events and individuals around them. Many urban residents avoid emotional involvement with one another and try to ignore events taking place around them. Urbanites feel wary toward other people because most interactions in the city are economic rather than social. Simmel suggests that attributes such as punctuality and exactness are rewarded but that friendliness and warmth in interpersonal relations are viewed as personal weaknesses. Some people act reserved to cloak their deeper feelings of distrust or dislike toward others. However, Simmel did not view city life as completely negative; he also pointed out that urban living could have a liberating effect on people because they have opportunities for individualism and autonomy.

**Urbanism as a Way of Life** Based on Simmel's observations on social relations in the city, the early Chicago School sociologist Louis Wirth (1938) suggested that urbanism is a "way of life." *Urbanism* refers to the distinctive social and psychological patterns of life typically found in the city. According to Wirth, the size, density, and heterogeneity of urban populations typically result in an elaborate division of labor and in spatial segregation of people by race/ethnicity, social class, religion, and/or lifestyle. In the city, primary-group ties are largely replaced by secondary relationships; social interaction is fragmented, impersonal, and often superficial. Even though people gain some degree of freedom and privacy by living in the city, they pay a price for their autonomy, losing the support and reassurance that come from primary-group ties.

From Wirth's perspective, people who live in urban areas are alienated, powerless, and lonely. A sense of community is obliterated and replaced by the "mass society"—a large-scale, highly institutionalized society in which individuality is supplanted by mass messages, faceless bureaucrats, and corporate interests.

**Gans's Urban Villagers** In contrast to Wirth's gloomy assessment of urban life, the sociologist Herbert Gans (1982/1962) suggested that not everyone experiences the city in the same way. Based on research in the west end of Boston in the late 1950s, Gans concluded that many residents develop strong loyalties and a sense of community in central-city areas that outsiders may view negatively. According to Gans, there are five major categories of adaptation among urban dwellers. *Cosmopolites* are students, artists, writers, musicians, entertainers, and professionals who choose to live in the city because they want to be close to its cultural facilities. *Unmarried people* and *childless couples* live in the city because they want to be close to work and entertainment. *Ethnic villagers* live in ethnically segregated neighborhoods; some are recent immigrants who feel most comfortable within their own group. The *deprived* are poor individuals with dim future prospects; they have very limited education and few, if any, other resources. The *trapped* are urban dwellers who can find no escape from the city; this group includes persons left behind by the process of invasion and succession, downwardly mobile individuals who have lost their former position in society, older persons who have nowhere else to go, and individuals addicted to alcohol or other drugs. Gans concluded that the city is a pleasure and a challenge for some urban dwellers and a nightmare for others (see "Sociology Works!").

**Cities and Persons with a Disability** Disability rights advocates believe that structural barriers create a "disabling" environment for many people, particularly in large urban settings (Figure 15.11). Many cities have made their streets and sidewalks more user friendly for persons in wheelchairs and for individuals with visual disability by constructing concrete ramps with slide-proof surfaces at intersections or installing traffic lights with sounds designating when to "Walk." However, both urban and rural areas have a long way to go before many persons with disabilities will have the access to the things they need to become productive members of the community: educational and employment opportunities. Some persons with disabilities cannot navigate

**FIGURE 15.11** Most U.S. cities are laid out for motorized vehicles, not people. Traveling urban areas via wheelchair can be a daunting proposition.

Bildagentur Waldhaeus/AGE Fotostock

# Sociology WORKS!

## Herbert Gans and Twenty-First-Century Urban Villagers

I moved to Austin because it's a high-tech city with a small-town feel. Kinda my own "urban village" where I can cycle around town when I want but still own a nice car to go out in. I chose my neighborhood because it's centrally located to downtown eating and live entertainment. Austin calls itself the "Live Music Capital of the World," and I have plenty of opportunities to hear the music I like. Overall, I'd say that I'm compatible with Austin, and Austin's compatible with me.

> —"Brad," a twenty-four-year-old college graduate, explaining why he chose to become an "urban villager" in Austin, Texas (author's files, 2007)

In the more than five decades since the urban sociologist Herbert J. Gans examined life in Boston's west end and identified five major categories of adaptation among urban dwellers, we continue to find that many residents think of themselves as living in an "urban village" despite the differences in high-rise buildings and the smaller settings in which the Boston west enders lived. Today, many younger urban residents think of themselves as having strong loyalties to a specific segment of their community with which they share interests and common experiences. Although many contemporary studies of urban life have emphasized the problems of poverty, crime, racial and ethnic discrimination, inadequate health care, and poor schools in our nation's major cities, it is useful for us to examine positive aspects of urban life as well. We can also gain important insights from examining the experiences of middle- and upper-income residents of our cities.

Herbert Gans believed that the people he referred to as *cosmopolites* chose to live in the city so that they could be close to cultural facilities, while he thought that *unmarried people* and *childless couples* chose to live there because they wanted to be close to work and entertainment. Among some affluent residents in high-tech cities such as Austin, Texas, married people and families with children have increasingly joined the ranks of individuals who live in or near the city's downtown area. According to contemporary urban villagers such as "Brad," they can find other people who are like themselves, who participate in activities they enjoy, and who support one another much like the members of an extended family when they need friendship or assistance. From this perspective, Gans's ideas about the urban village work because they show us that an important way to understand city life is through the experiences of people who live there. In the final analysis, of course, all people—including lower-income individuals, who have been further disadvantaged or even displaced by gentrification, and the poor and

The people shown here enjoying a walk down East 6th Street, also known as Old Pecan Street, in Austin, Texas, may consider themselves to be "urban villagers" because they can enjoy small-town amenities such as little shops, neighborhood-like restaurants, and live music venues while residing in a larger city. Do you consider yourself to be an urban villager?

homeless—must be included in any thorough sociological examination of city life in the twenty-first century. However, "urban villagers," as coined by Gans, has staying power as a concept because it encourages us to look at urban life as a kaleidoscope of diversity that includes the wealthy and the merely affluent, as well as those who are "just getting by" or who are poor, because they live near one another as contemporary urban dwellers.

## reflect & analyze

Can you identify categories of urban villagers in a city with which you are familiar? To what extent do people live in certain areas of the city based on personal choice? What factors appear to be beyond their control?

the streets and sidewalks of their communities, and some face obstacles getting into buildings that marginally, at best, meet the accessibility standards of the Americans with Disabilities Act; thus, many persons with a disability are unemployed.

Political scientist Harlan Hahn (1997: 177–78) traces the problem of lack of access to the beginnings of industrialism:

> The rise of industrialism produced extensive changes in the lives of disabled as well as nondisabled people. As factories replaced private dwellings as the primary sites of production, routines and architectural configurations were standardized to suit nondisabled workers. Both the design of worksites and of the products that were manufactured gave virtually no attention to the needs of people with disabilities. As a result, patterns of aversion and avoidance toward disabled persons were embedded in the construction of commodities, landscapes, and buildings that would remain for centuries. . . .
>
> The social and economic changes fostered by industrialization may have been exacerbated by the accompanying process of urbanization. As workers increasingly moved from farms and rural villages to live near the institutions of mass production, the character of community life appeared to shift perceptibly. Deviant or atypical personal characteristics that may have gradually become familiar in a small community seemed bizarre or disturbing in an urban milieu.

As Hahn's statement suggests, historical patterns in the dynamics of industrial capitalism contributed to discrimination against persons with disabilities, and this legacy remains evident in contemporary cities. Structural barriers are further intensified when other people do not respond favorably toward persons with disabilities. A classic life experience was told by scholar and disability rights advocate Sally French (1999: 25–26), who described her experience as a person with a visual disability:

> I have lived in the same house for 16 years and yet I cannot recognize my neighbors. I know nothing about them at all; which children belong to whom, who has come and gone, who is old or young, ill or well, black or white. . . . On moving to my present house I informed several neighbors that, because of my inability to recognize them, I would doubtless pass them by in the street without greeting them. One neighbor, who had previously seen me striding confidently down the road, refused to believe me, but the others said they understood and would talk to me if our paths crossed. For the first couple of weeks it worked and I was surprised how often we met,

but after that their greetings rapidly decreased and then ceased altogether. Why this happened I am not sure, but I suspect that my lack of recognition strained the interaction and limited the social reward they received from the encounter.

The Concept Quick Review examines the multiple perspectives on urban growth and urban living.

## Problems in Global Cities

Although people have lived in cities for thousands of years, the time is rapidly approaching when more people worldwide will live in or near a city than live in a rural area. In the middle-income and low-income regions of the world, Latin America is becoming the most urbanized: Four megacities—Mexico City, Buenos Aires, Lima, and Santiago—already contain more than half of the region's population and continue to grow rapidly. Soon, Rio de Janeiro and São Paulo are expected to have a combined population of about 35 million people living in a 350-mile-long megalopolis.

Rapid population growth will have a major effect on cities throughout the world. Essential services such as health, education, transportation, and sanitation are already strained in many cities, and the problem will only grow worse as the world's population moves upward toward a projected 10 billion or more by the end of the twenty-first century. In China alone it is estimated that 220 cities will have more than 1 million people by 2025 and that more than 350 million rural residents will move to the cities from rural areas. Requirements for new and expanded infrastructure in these cities will be tremendous, particularly in the building of high-rise buildings, mass-transit systems, and other amenities that will be needed to support this colossal population shift.

Today, some social analysts look beyond the city proper, which is defined as a locality with legally fixed boundaries and an administratively recognized urban status that is usually characterized by some form of local government, to see the larger picture of what takes place in urban agglomerations. An *urban agglomeration* is defined as comprising the city or town proper and also the suburban fringe or thickly settled territory lying outside of, but adjacent to, the city boundaries, as a more accurate reflection of population composition and density in a given region. Figure 15.12 shows the populations of the world's fourteen largest urban agglomerations.

Natural increases in population (higher birth rates than death rates) account for two-thirds of new urban growth. In recent years, fewer deaths and more births have occurred than demographers had anticipated. High fertility brings about booming

## Perspectives on Urbanism and the Growth of Cities

| | | |
|---|---|---|
| Functionalist Perspectives: Ecological Models | Concentric zone model | Because of invasion, succession, and gentrification, cities are a series of circular zones, each characterized by a particular land use. |
| | Sector model | Cities consist of wedge-shaped sectors, based on terrain and transportation routes, with the most-expensive areas occupying the best terrain. |
| | Multiple nuclei model | Cities have more than one center of development, based on specific needs and activities. |
| Conflict Perspectives: Political Economy Models | Capitalism and urban growth | Members of the capitalist class choose locations for skyscrapers and housing projects, limiting individual choices by others. |
| | Gender regimes in cities | Different cities have different prevailing ideologies regarding access to social positions and resources for men and women. |
| | Gated communities in the capitalist economy | A belief in private ownership of property under capitalism promotes real estate developments such as gated communities that contribute to a growing divide between public and private space in urban areas. |
| Symbolic Interactionist Perspectives: The Experience of City Life | Simmel's view of city life | Because of the intensity of city life, people become somewhat insensitive to individuals and events around them. |
| | Urbanism as a way of life | The size, density, and heterogeneity of urban population result in an elaborate division of labor and space. |
| | Gans's urban villagers | Five categories of adaptation occur among urban dwellers, ranging from cosmopolites to trapped city dwellers. |
| | Disability and city life | Cities provide unique challenges for persons with disabilities because urban areas often have barriers that make equal access difficult, despite rights that were gained under the Americans with Disabilities Act. |

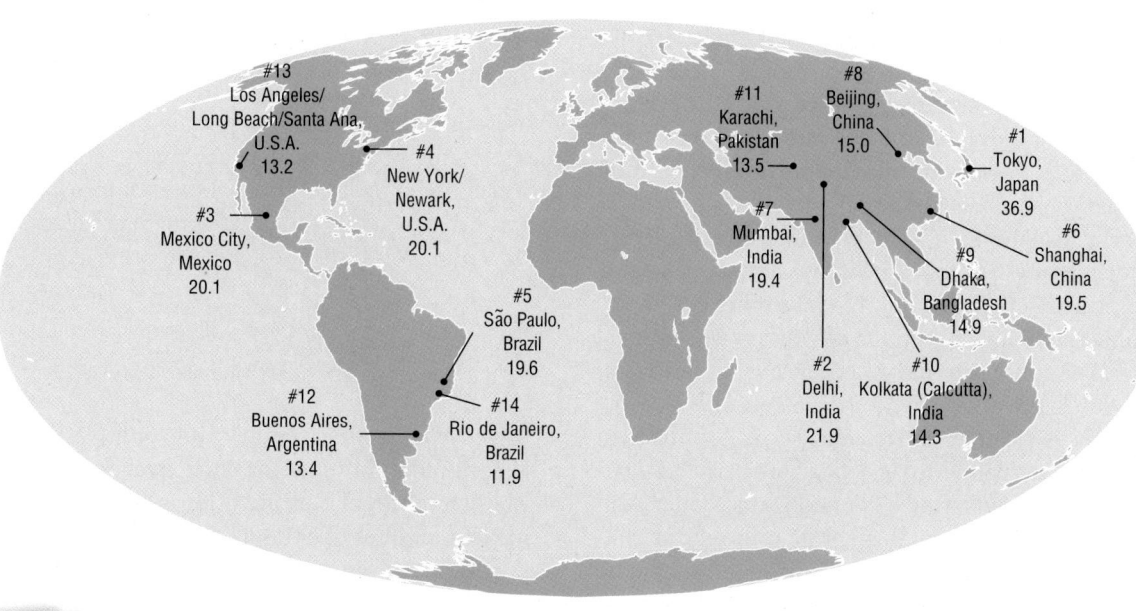

**FIGURE 15.12** THE WORLD'S FOURTEEN LARGEST AGGLOMERATIONS

Note: Estimation of the 2011 population in millions (rounded).

Source: United Nations World Urbanization Prospects On-Line Data, 2012. © Cengage Learning populations and a corresponding strain on food and other resources; however, low fertility contributes to an aging population and stress on social services.

The other component of new urban growth is rural-to-urban migration. Some people move from rural areas to urban areas because they have been displaced from their land. Others move because they are looking for a better life. No matter what the reason, migration has caused rapid growth in cities in China, sub-Saharan Africa, India, Algeria, and Egypt. At the same time that the population is growing rapidly, the amount of farmland available for growing crops to feed people is decreasing. In Egypt, for example, land that was previously used for growing crops is now used for petroleum refineries, food-processing plants, and other factories. Issues such as this contributed to a political uprising in 2011 and to continued unrest in the following years.

Rapid global population growth in Latin America and other regions is producing a variety of urban problems, including overcrowding, environmental pollution, and the disappearance of farmland. In fact, many cities in middle- and low-income nations are quickly reaching the point at which food, housing, and basic public services are available to only a limited segment of the population.

As global urbanization has continued to increase, differences in urban areas based on economic development at the national level have become apparent. Some cities in what Immanuel Wallerstein's (1984) world systems theory describes as core nations are referred to as *global cities*—interconnected urban areas that are centers of political, economic, and cultural activity. New York City, London, Paris, Tokyo, Hong Kong, Los Angeles, Seoul, and Singapore are among the largest global cities today. These cities are the sites of new and innovative product development and marketing, and they are often the "command posts" for the world economy. But economic prosperity is not shared equally by all of the people in the core-nation global cities. Sometimes the living conditions of workers in low-wage service-sector jobs or in assembly production jobs more closely resemble the living conditions of workers in semiperipheral nations than they resemble the conditions of middle-class workers in their own country.

Many African countries and some countries in South America and the Caribbean are *peripheral* nations, previously defined as nations that depend on core nations for capital, have little or no industrialization (other than what may be brought in by core nations), and have uneven patterns of urbanization. According to Wallerstein (1984), the wealthy in peripheral nations support the exploitation of poor workers by core-nation capitalists in return for maintaining their own wealth and position. Poverty is thus perpetuated, and the problems worsen because of the unprecedented population growth in these countries. Like peripheral nations, semiperipheral nations—such as India, Iran, and Mexico—are confronted with unprecedented population growth. In addition, a steady flow of rural migrants to large cities is creating enormous urban problems.

More recently, Wallerstein suggested that a new world system is emerging that will retain some basic features of the existing one, but it will not be based on capitalism as we know it. According to Wallerstein, the new system will be either hierarchical and exploitative or relatively democratic and relatively egalitarian. In either scenario, national and local governments will be faced with difficult choices, including how to deal with major fiscal crises and immigration. In the United States and throughout Europe, for example, a growing cry demands that governments "do something" about evicting "foreigners"; however, the effects of such an action may create great turmoil and economic instability in cities. Wallerstein's original theory that work migrates from core nations to workers in semiperipheral and peripheral nations has shifted: Workers are now migrating to cities where they hope to find employment to support themselves and their families, and this process is not likely to stop in the near future.

# Urban Problems in the United States

Even the most optimistic observers tend to agree that cities in the United States have problems brought on by years of neglect and deterioration. As we have seen in previous chapters, poverty, crime, racism, sexism, homelessness, inadequate public school systems, alcoholism and other drug abuse, gangs and guns, and other social problems are most visible and acute in urban settings. Issues of transportation and urban growth and development are intertwined with many of these problems (Figure 15.13).

## Divided Interests: Cities and Suburbs

In the twenty-first century the composition of the U.S. population in major cities is changing rapidly. The white population in cities is aging and declining in a number of cities, while the minority population is continuing to grow and disperse. According to the 2010 census, Hispanics (Latinos/as) represent the largest minority group in major U.S. cities. The Hispanic share of urban population rose in all of the largest 100 metropolitan areas to the extent

**FIGURE 15.13** Despite an increase in telecommuting and more-diverse employment opportunities in the high-tech economy, our highways have grown increasingly congested. Can we implement measures to reduce the problems of urban congestion and environmental pollution, or will these problems grow worse with each passing year?

that Hispanics now outnumber African Americans (blacks) as the largest minority group in major U.S. cities. The data show that Hispanics make up 26 percent of primary city populations, while African Americans (blacks) make up 22 percent. Across all cities in 2010, white Americans account for 41 percent of residents in primary cities—continuing a downward trend from 53 percent in 1990 to 45 percent in 2000 and then reaching the current 41 percent (Frey, 2011).

The story of divided interests between cities and suburbs has continued for more than a century. Following World War II, a dramatic population shift occurred as thousands of families moved from cities to suburbs. Postwar suburban growth was fueled by aggressive land developers, inexpensive real estate and construction methods, better transportation, abundant energy, government subsidies such as liberal lending policies by federal agencies including the Veterans Administration and the Federal Housing Authority, and racial stress in the cities. This pattern of postwar suburbanization and decades of white flight from cities produced a pattern in which the majority of today's white Americans who live in large urban areas are in the suburbs. By 2010, 78 percent of all white Americans who resided in large metro areas lived in suburbs, up from 74 percent in 1990 (Frey, 2011).

Suburbanization created a territorial division of interests between cities and suburban areas. City services and school districts have continued to languish for lack of funds. Affluent families living in "gentrified" properties in the city typically send their children to elite private schools, whereas the children of poor families living in racially segregated public housing projects attend underfunded (and often substandard) public schools. For decades the wealthy and the poor have lived in different spheres, even when they reside in close proximity in urban areas.

**Race, Class, and Suburbs** Sharp racial and ethnic divisions between cities and suburbs have become more blurred in the 2010s; however, class lines remain more distinct, particularly in some wealthy suburban areas. In the past, most suburbs were predominantly white and middle class. Today, some upper-middle-class and upper-class suburbs still have a white majority, but the old image of "chocolate city and vanilla suburbs" is now obsolete (Frey, 2011: 2). According to the 2010 census, minorities represent 35 percent of suburban residents in the 100 largest metropolitan areas in the United States, similar to their share of the overall population. Hispanics make up 17 percent of suburbanites (as compared to 16 percent of the overall population); blacks make up 10 percent of suburbanites (as compared to 12 percent of the U.S. population) (Frey, 2011).

What most data do not reveal is how class and other demographic factors are intertwined with

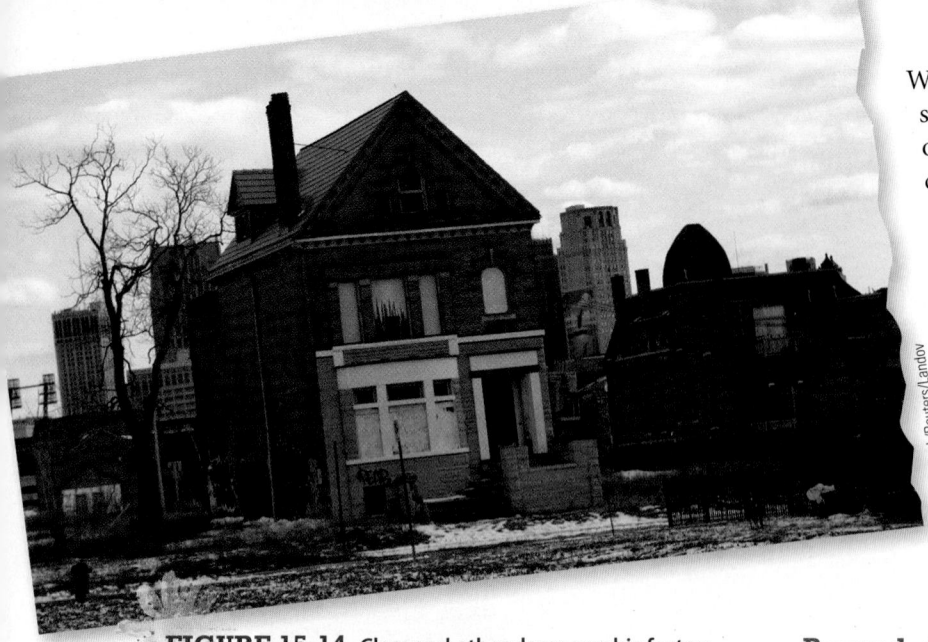

Washington, where a sharp rise in black suburbanization has occurred as a result of economic progress among younger college-educated blacks (Frey, 2011).

However, Hispanics account for almost half (49 percent) of the overall growth in population in suburban areas in the 2000s. In the 100 largest metropolitan areas, Hispanics contributed more to the growth than any other racial and ethnic group in 49 of those areas. Hispanics had the largest gains in areas surrounding cities such as New York City, Houston, Miami, Los Angeles, and Riverside, California.

**FIGURE 15.14** Class and other demographic factors are intertwined with suburban integration. Some suburbs, such as this one near Detroit, have become racially resegregated and provide few job opportunities and public services for residents who live there.

suburban integration. In some suburbs, people of color (especially African Americans) become resegregated. An example is the Detroit metropolitan area, where Census Bureau data show a 25 percent drop in the city's population over the last decade and a corresponding increase in minority population in the suburbs. According to a study by historian and sociologist Thomas J. Sugrue (2011), many African Americans are moving into so-called secondhand suburbs in that city and others. Sugrue defines *secondhand suburbs* as "established communities with deteriorating housing stock that are falling out of favor with younger white homebuyers." He argues that if history holds true, these suburban areas will soon look like Detroit itself, with "resegregated schools, dwindling tax bases and decaying public services." (See Figure 15.14.)

Although some analysts claim that the location of one's residence is a matter of personal choice, African Americans and other persons of color have looked for the same things in suburban properties that white Americans have desired, namely safe streets and low crime rates, the best housing they can afford, quality schools for their children, and the amenities of life outside the hubbub of the central city. However, what African Americans have often found in areas such as Detroit are fewer job opportunities, poorer services, older houses, and run-down shopping districts (Sugrue, 2011). Fortunately, the overall picture is more positive in some suburbs located near Atlanta, Houston, Dallas, and

**Beyond the Suburbs: Edge Cities and Exurban Areas** In the past, urban fringes (referred to as *edge cities*) developed beyond central cities and suburbs (Garreau, 1991). The Massachusetts Turnpike corridor west of Boston and the Perimeter area north of Atlanta are examples. Edge cities initially develop as residential areas; then retail establishments and office parks move into the area, creating the unincorporated edge city. Commuters from the edge city are able to travel around (rather than in and out of) the metropolitan region's center and can avoid its rush-hour traffic quagmires. Edge cities may not have a governing body or correspond to municipal boundaries; however, they drain taxes from central cities and older suburbs. Many businesses and industries have moved physical plants and tax dollars to these areas: Land is cheaper, and utility rates and property taxes are lower.

Eventually, edge cities in some states have become commuter towns or exurban areas with more structure. Fast-growing exurban areas in Kentucky, Georgia, Virginia, Missouri, Tennessee, Texas, North Carolina, Alabama, Oklahoma, and Pennsylvania have remained mostly white and have depended overwhelmingly on whites for growth in the twenty-first century (Frey, 2011). Of the 22 exurban areas, 16 are more than 75 percent white, with white Americans accounting for at least 80 percent of the population in 15 of the 20 areas. Examples include Spencer, Kentucky, in the Louisville/Jefferson County Kentucky–Indiana metro area; Dawsonville, Georgia, in the Atlanta metro area; New Kent, Virginia, in the Richmond metro area; and Lincoln County, Missouri, in the St. Louis metro area (Frey, 2011). According to a recent study, these exurban areas are a current reflection of what suburbia was in the past: "new housing, growth, and demographic detachment

Rebecca Cook/Reuters/Landov

from the more urban portions of their metropolitan areas" (Frey, 2011: 11). And, to some extent, this situation includes and reflects the continuing financial woes of the cities.

## The Continuing Fiscal Crises of the Cities

The largest cities in the United States have faced periodic fiscal crises for many years; however, in the 2010s, cities of all sizes and their adjoining suburbs are experiencing even greater financial problems partially linked to a major downturn in national and international economic trends. Economic recoveries in cities take at least two years longer than a national recovery, which means that financial problems brought about by the Great Recession may last for years to come. An example is the decline in the U.S. housing market, which affects city tax revenues: Housing prices fell by 4 percent during 2011, which reduced household wealth by $500 billion nationwide and further reduced local governments' property tax base (U.S. Conference of Mayors, 2012).

Why have national and international economic downturns hurt U.S. cities so drastically? What are cities doing about it? Cities have experienced extensive shortfalls in revenue because states have reduced the amount of money that they provide for cities, and the cities have had decreased revenue from sales taxes, corporate taxes, and personal income taxes. Funds from the federal government to states and cities have also been limited and are often earmarked for specific projects rather than for use in the general operating budget. These budget crises have forced states to cut funding to already cash-strapped cities. Vital services, including police, firefighting, and public works, have been cut drastically, and the slashing of city budgets and programs may continue in some areas for years to come. As cities lose revenue, officials must decide to lay off or furlough employees, charge higher fees for services, and cancel major projects such as street repairs or infrastructure improvements (building a new water treatment facility, for example).

A study by the Pew Charitable Trusts (2013) found that thirty cities in the most populous metropolitan areas of the United States faced more than $192 billion in unpaid commitments for pensions and other retiree benefits, such as health care, and that was as far back as 2009. The majority of the shortfall was in New York City, followed by Philadelphia, Chicago, and Portland, Oregon. It is a major concern not only for persons expecting to receive pensions or to be covered by health care

benefits that these cities do not have the money needed to fully fund their pension plans, but it also is a major problem for everyone, particularly taxpayers. Although the pension systems and state insurance plans may be able to return to higher levels if the economy improves, the fiscal crisis of the cities will be deepened by the problem that this situation presents because it uniquely falls on cities to provide for many of their public-sector employees.

City officials continue to urge leaders at the state and federal levels to create new programs that will help cities meet their residents' needs. Demands are specifically being made for more federal aid through job-creation programs and other economic stimulus packages. Some analysts believe that inaction at the state and federal levels may create even greater financial chaos by forcing some cities into bankruptcy. Local officials emphasize that the state of America's cities continues to threaten the long-term national economic recovery (National League of Cities, 2010). It remains to be seen what the eventual effects of these continuing fiscal crises will be on various cities throughout the nation.

# Rural Community Issues in the United States

Although most people think of the United States as highly urbanized, about 20 percent of the U.S. population resides in rural areas, identified as communities of 2,500 people or less by the U.S. Census Bureau. Sociologists typically identify *rural communities* as small, sparsely settled areas that have a relatively homogeneous population of people who primarily engage in agriculture. However, rural communities today are more diverse than this definition suggests.

Unlike the standard migration patterns from rural to urban places in the past, recently more people have moved from large urban areas and suburbs into rural areas. Many of those leaving urban areas today want to escape the high cost of living, crime, traffic congestion, and environmental pollution that make daily life difficult. Technological advances make it easier for people to move to outlying rural areas and still be connected to urban centers if they need to be. The proliferation of computers, cell phones, commuter airlines, and highway systems has made previously remote areas seem much more accessible to many people. However, many recent immigrants to rural areas do not face some traditional problems experienced by long-term rural residents, particularly farmers, small-business owners, teachers, doctors, and other medical personnel.

For many people in rural areas who have made their livelihood through farming and other agricultural endeavors, recent decades have been very difficult, both financially and emotionally. Rural crises such as droughts, crop failures, and the loss of small businesses in the community have had a negative effect on many adults and their children. Like their urban counterparts, rural families have experienced problems of divorce, alcoholism, abuse, and other crises, but these issues have sometimes been exacerbated by such events as the loss of the family farm or business. Because home is also the center of work in farming families, the loss of the farm may also mean the loss of family and social life, and the loss of things dear to children such as their 4-H projects—often an animal that a child raises to show and sell. Some rural children and adolescents are also subject to injuries associated with farm work, such as from being kicked or crushed by livestock, from falling out of a tractor or a pickup, or from operating machinery designed for adults, that are not typically experienced by their urban counterparts.

Economic opportunities are limited in many rural areas, and average salaries are typically lower than in urban areas, based on the assumption that a family can live on less money in rural communities than in cities. An example is rural teachers, who earn substantially less than their urban and suburban counterparts. Some rural areas have lost many teachers and administrators to higher-paying districts.

Although many of the problems we have examined in this book are intensified in rural areas, one of the most pressing is the availability of health services and doctors. Recently, some medical schools have established clinics and practices in outlying rural regions of the states in which they are located in an effort to increase the number of physicians available to rural residents. Typically, physicians who have just started to practice medicine have chosen to work in large urban centers with accessible high-tech medical facilities. Because of the pressing time constraints of tending to patients with life-threatening problems, the availability of community clinics and hospitals in rural areas may be a life-or-death matter for some residents. The loss of these facilities can have a devastating effect on people's health and life chances.

In addition to the movement of some urban dwellers to rural areas, two other factors have changed the face of rural America in some regions. One is the proliferation of superstores, such as Walmart, PetSmart, Lowe's, and Home Depot. In some cases these superstores have effectively doomed small businesses such as local hardware stores and pet shops because local merchants cannot meet the prices established by these large-volume discount chains. The development of superstores and outlet malls along the rural highways of this country has raised new concerns about environmental issues such as air and water pollution, and has brought about new questions regarding whether these stores benefit the rural communities where they are located.

A second factor that has changed the face of some rural areas (and is sometimes related to the growth of superstores and outlet malls) is an increase in tourism in rural America. The vast majority of adults in the United States have taken a trip to a rural destination, usually for leisure purposes, over the past few years. Tourism produces jobs; however, many of the positions are for food servers, retail clerks, and hospitality workers, which are often low-paying, seasonal jobs that have few benefits. Tourism may improve a community's tax base, but this does not occur when the outlet malls, hotels, and fast-food restaurants are located outside of the rural community's taxing authority, as frequently occurs when developers decide where to locate malls and other tourist amenities.

## Population and Urbanization in the Future

**LO8** In the future, rapid population growth is inevitable. World population is projected to grow from more than 7 billion in 2013 to 9.3 billion in 2050, and to 10 billion in 2085 (see Figure 15.15). Nearly all of this growth in

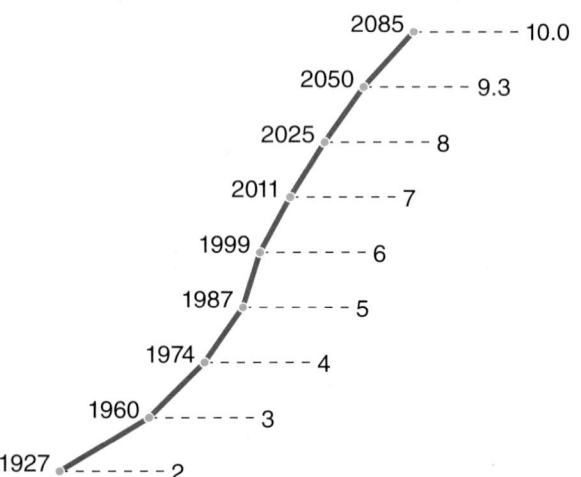

**FIGURE 15.15** INCREASE IN THE WORLD'S POPULATION IN BILLIONS OF PEOPLE

Source: United Nations Population Division, 2011. © Cengage Learning

PART 5 | SOCIAL DYNAMICS AND SOCIAL CHANGE

population is projected to occur in low-income nations, mostly in Africa and Asia. Higher population growth in low-income countries stems from the higher total fertility rates—the average lifetime number of births per woman in a population. In sub-Saharan Africa, for example, women have an average of nearly 5 children in their lifetime. Worldwide, total fertility rates have declined significantly from nearly 5 in the 1950s to 2.5 in 2012 (*CIA Factbook,* 2013).

According to *World Urbanization Prospects* (United Nations, 2012), the urban population worldwide will grow by 2.6 billion people between 2011 and 2050. In 2011 approximately 3.6 billion people resided in urban areas, and it is estimated that 6.3 billion will be urban dwellers by 2050. North America, Latin America and the Caribbean, and Europe, which are already highly urbanized, will continue the urbanization process and, by 2050, will be more than 84 percent urban. In contrast, Africa and Asia remain mostly rural, with slightly more than 40 percent of their populations living in urban areas in 2011. But by 2050, about two-thirds of the population in Africa and Asia are expected to live in urban areas.

In a worst-case scenario, by the 2020s or the 2030s central cities and nearby suburbs in the United States will have experienced bankruptcy exacerbated by sporadic race- and class-oriented violence. Sporadic urban violence brought about by mass killings in public locations such as schools and movie theaters, combined with random and terrorist attacks such as the Boston Marathon bombing and the Fort Hood (Killeen, Texas) military base shootings, may also become more frequent occurrences, based on the nature of recent events. The infrastructure of cities will be beyond the possibility of repair. Families and businesses with the ability to do so will have long since moved to "new cities," where they will inevitably diminish the quality of life that they originally sought there. Areas that we currently think of as being relatively free from such problems will be characterized by depletion of natural resources and by greater air and water pollution.

By contrast, in a best-case scenario the problems brought about by rapid population growth in low-income nations will be remedied by new technologies that make goods readily available to people. International trade agreements are removing trade barriers and making it possible for all nations to fully engage in global trade. People in low-income nations will benefit by gaining jobs and opportunities to purchase goods at lower prices. Of course, the opposite may also occur: People may be exploited as inexpensive labor, and their country's natural resources may be depleted as transnational corporations buy up raw materials without contributing to the long-term economic stability of the nation.

In the United States a best-case scenario for the future might include improvements in how taxes are collected and spent. Some analysts suggest that regional governments should be developed that would be responsible for water, wastewater (sewage), transportation, schools, and other public services over a wider area. Others believe that we should put more effort into smart growth and/or green movements to improve our cities. "Smart Growth" is the name given to development decisions that cities are making across the United States. Examples of Smart Growth planning include working to conserve resources, preserving natural lands and critical environmental areas, protecting water and air quality, and reusing already-developed land by reinvesting in existing infrastructure and reclaiming historic buildings. The following are Smart Growth principles (EPA.gov, 2013):

- Encourage mixed land use
- Take advantage of compact building design
- Create a range of housing opportunities and choices
- Create walkable neighborhoods
- Foster distinctive, attractive communities with a strong sense of place
- Preserve open spaces, farmland, natural beauty, and critical environmental areas
- Strengthen and direct development toward existing communities
- Provide a variety of transportation choices
- Make development decisions predictable, fair, and cost effective
- Encourage community and stakeholder collaboration in development decisions

Although some critics argue that cities that have tried Smart Growth have not brought about profound changes through their efforts, goals such as these seem desirable for bringing about a better future for our communities. At the macrolevel we may be able to do little about population and urbanization; however, at the microlevel we may be able to exercise some degree of control over our communities and our lives. Reclaiming public space for daily life would be an important start, along with making neighborhoods more sustainable and helping inhabitants feel that they have a vested interest in their own community. Hopefully, we will also have found a way to curb random violence and acts of terrorism that produce great harm and keep people living in a state of fear and consternation rather than moving forward for the betterment of themselves and future generations.

# chapter REVIEW

## LO1 What is demography, and what are the processes that produce population changes?

Demography is a subfield of sociology that examines population size, composition, and distribution. Populations change as the result of fertility (births), mortality (deaths), and migration.

## LO2 What is meant by population composition, and how is it measured?

Changes in fertility, mortality, and migration affect the population composition—the biological and social characteristics of a population, including age, sex, race, marital status, education, occupation, income, and size of household. One measure of population composition is the sex ratio—the number of males for every hundred females in a given population.

## LO3 What are the Malthusian, Marxist, and neo-Malthusian perspectives, and what are the components of demographic transition theory?

More than 200 years ago, Thomas Malthus warned that overpopulation would result in poverty, starvation, and other major problems that would limit the size of the population. According to Malthus, the population would increase geometrically while the food supply would increase only arithmetically, resulting in a critical food shortage and poverty. According to Karl Marx, poverty is the result of capitalist greed, not overpopulation. More recently, neo-Malthusians have reemphasized the dangers of overpopulation and encouraged zero population growth. Demographic transition theory links population growth to four stages of economic development: (1) the preindustrial stage, with high birth rates and death rates; (2) early industrialization, with relatively high birth rates and a decline in death rates; (3) advanced industrialization and urbanization, with low birth rates and death rates; and (4) postindustrialization, with additional decreases in the birth rate coupled with a stable death rate.

## LO4 What are some of the key points in international migration theories?

The neoclassical economic approach assumes that migration patterns occur based on geographic differences in the supply of and demand for labor. The new households economics of migration approach emphasizes the part that families or households play in migration. Split-labor-market theory suggests that immigrants from low-income countries are often recruited for secondary labor market positions, whereas migrants from higher-income countries may migrate for primary-sector employment. World systems theory views migration as linked to the problems caused by capitalist development around the world.

## LO5 How do cities develop through preindustrial, industrial, and postindustrial stages?

According to Gideon Sjoberg, three preconditions must be present in order for a city to develop: a favorable physical environment, an advanced technology, and a well-developed social organization. The largest preindustrial city was Rome; by 100 C.E., it may have had a population of 650,000. The Industrial Revolution changed the nature of the city. Emergent technology, including new forms of transportation and agricultural production, made it easier for people to leave the countryside and move to the city. Postindustrial cities increasingly rely on an economic structure that is based on scientific knowledge rather than industrial production.

## LO6 How do ecological/functionalist models compare with political economy/conflict models of urban growth?

Functionalists view urban growth in terms of ecological models. The concentric zone model sees the city as a series of circular areas, each characterized by a different type of land use. The sector model describes urban growth in terms of terrain and transportation routes. The multiple nuclei model views cities as having numerous centers of development from which growth radiates. According to political economy models/conflict perspectives, urban growth is influenced by capital investment decisions, class and class conflict, and government subsidy programs. At the global level, capitalism also influences the development of cities in core, peripheral, and semiperipheral nations.

## LO7 What is meant by the experience of urban life, and how do sociologists seek to explain this experience?

Symbolic interactionist perspectives focus on how people experience urban life. Some analysts view the urban experience positively; others believe that urban dwellers become insensitive to events and to people around them.

## L08 What are some best-case and worst-case scenarios regarding population and urban growth in the twenty-first century?

In a worst-case scenario, central cities and nearby suburbs in the United States will have experienced bankruptcy, and the infrastructure of cities will be beyond repair. Areas that we currently think of as being relatively free from such problems will be characterized by depletion of natural resources and by greater air and water pollution. In a best-case scenario, the problems brought about by rapid population growth in low-income nations will be remedied by new technologies that make goods readily available to people.

## key TERMS

crude birth rate  476

crude death rate  477

demographic transition  485

demography  476

fertility  476

gentrification  490

invasion  490

migration  478

mortality  477

population composition  482

population pyramid  482

sex ratio  482

succession  490

zero population growth  485

## questions for CRITICAL THINKING

1. What impact does a relatively high rate of immigration have on a nation's culture? Does it affect personal identity? Why or why not?

2. If you were designing a study of growth patterns for the city where you live (or one you know well), which theoretical model(s) would provide the most useful framework for your analysis?

3. What do you think that everyday life in U.S. cities, suburbs, and rural areas will be like twenty years from the date you are reading this? Where would you prefer to live? What does your answer reflect about the future of U.S. cities?

# 16 COLLECTIVE BEHAVIOR, SOCIAL MOVEMENTS, AND SOCIAL CHANGE

## LEARNING OBJECTIVES

**LO1** Define collective behavior and list factors that contribute to it.

**LO2** Describe the most common types of crowd behavior.

**LO3** Discuss these explanations of crowd behavior: contagion theory, social unrest and circular reaction, convergence theory, and emergent norm theory.

**LO4** Explain how mass behavior differs from other forms of collective behavior.

**LO5** Compare fads and fashions, and evaluate the "trickle-down" approach as an explanation for the origin of contemporary fashions.

**LO6** Identify and explain the major types of social movements.

**LO7** Compare the following theories about conditions that are likely to produce social movements: relative deprivation theory, value-added theory, and resource mobilization theory.

**LO8** Describe how research based on social constructionist theory—frame analysis, political opportunity theory, and new social movement theory—variously explains how people are drawn into social movements.

Glowimages/Age Fotostock

# Many Voices Needed for Social Change

Imagine for a moment that you live in a community that is poor.... You want to send your child to a school that can challenge her to help reach her potential. But, you can't because of your limited income.... This is how one mother described her situation to me recently in Detroit, Michigan ... the area around her home and school had lead levels, left over from former lead smelter activities, which were three times the legal standard.... When people hear about the struggles in environmental justice communities, they often only think about the immediate pollution and health impacts in a low wealth community. But to understand it one level deeper, you need to understand that families living in these communities are really trapped. ... Those who have the power to change this cycle of poison and poverty choose not to....

I'm not sure how to change this situation. It is a larger societal crisis that will take the majority of people to demand change. Today it is only the voices of the desperate parents, frustrated teachers that sound the alarm and cry for justice. This must change.

—Lois Gibbs (2013), founder and executive director of the Center for Health, Environment and Justice, describes an environmental "circle of poison and poverty" that has persisted for more than 35 years while Gibbs has been involved in organizing social movements to fight pollution in locations such as Love Canal, New York, that were damaged by corporate environmental pollution.

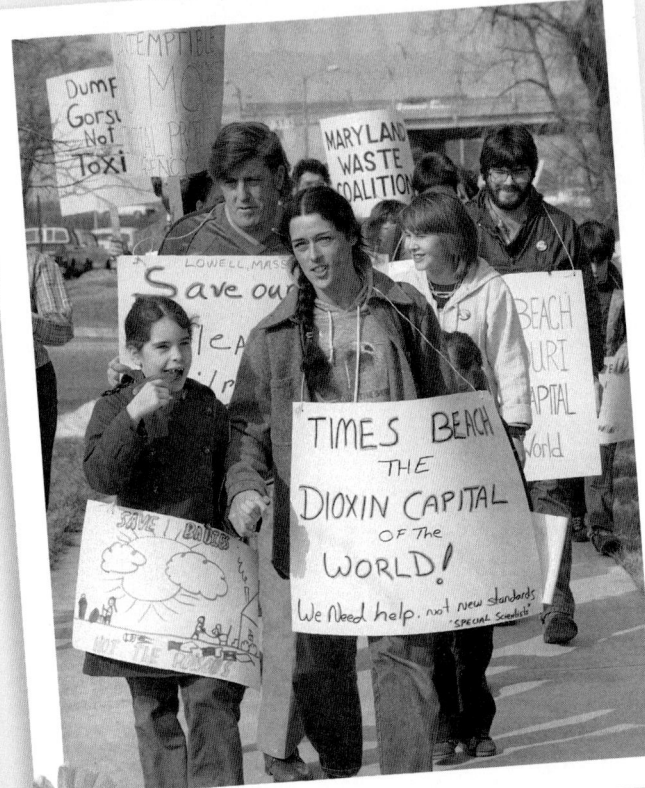

AP Images/Scott Stewart

Here, Lois Gibbs (center) leads members of toxic-waste protest groups as they demonstrate in 1983 over the Environmental Protection Agency's handling of the Superfund to clean up toxic waste sites.

Demonstrations and protests are a daily part of life in the twenty-first century. Almost daily, social media, the Internet, TV, and newspapers inform us of new or unresolved social or environmental problems. As the chapter's opening narrative indicates, the message of social movements is that people should act collectively and immediately to help reduce these problems. In other words, social change is essential. Sociologists define **social change as the alteration, modification, or transformation of public policy, culture, or social institutions** over time. Social change is usually brought about by collective behavior and social movements. But how much social change can the individual bring about?

In this chapter we will examine collective behavior, social movements, and social change from a sociological perspective. We will use the Occupy movement and environmental activism (also known as the Green movement) as examples of how people try to use social movements as a form of mass mobilization and social transformation. For example, the Occupy movement, which began in 2011 as Occupy Wall Street in

## How Much Do You Know About Collective Behavior and Environmental Issues?

| True | False | |
|------|-------|---|
| T | F | 1. Scientists are forecasting a global warming of between 2 and 11 degrees Fahrenheit over the next century. |
| T | F | 2. The environmental movement in the United States started in the 1960s. |
| T | F | 3. People who hold strong attitudes regarding the environment are very likely to be involved in social movements to protect the environment. |
| T | F | 4. Environmental groups may engage in civil disobedience or use symbolic gestures to call attention to their issue. |

| True | False | |
|------|-------|---|
| T | F | 5. People are most likely to believe rumors when no other information is readily available on a topic. |
| T | F | 6. Influencing public opinion is a very important activity for many social movements. |
| T | F | 7. Most social movements in the United States seek to improve society by changing some specific aspect of the social structure. |
| T | F | 8. Sociologists have found that people in a community respond very similarly to natural disasters and to disasters caused by technological failures. |

Answers on page 511.

Manhattan, referred to itself as a leaderless resistance movement comprising diverse people from all backgrounds and political affiliations. The group claimed on its website to represent 99 percent of the U.S. population while standing up against the wealthiest 1 percent in the nation, which has been labeled as greedy and corrupt and a sure sign that capitalism has run amok. Similarly, the "Go-Green" movement has become popular in recent years, but it should be noted that the environmental movement can be traced back to the 1960s and 1970s, when many people were actively involved in collective behavior, making demands that additional efforts be put forth to protect the environment and to save our planet. In fact, some ideas expressed by the "Go-Green" movement can be traced back to the 1930s or earlier.

Thus, collective behavior and social movements may come and go over lengthy periods of time, but the themes of concern may remain similar. Before reading on, test your knowledge about collective behavior and environmental issues by taking the "Sociology and Everyday Life" quiz.

## Collective Behavior

**LO1** *Collective behavior* is voluntary, often spontaneous activity that is engaged in by a large number of people and typically violates dominant-group norms and values. Unlike the *organizational behavior* found in corporations and voluntary associations (such as labor unions and environmental organizations), collective behavior lacks an official division of labor, hierarchy of authority, and established rules and procedures.

Unlike *institutional behavior* (in education, religion, or politics, for example), it lacks institutionalized norms to govern behavior. Collective behavior can take various forms, including crowds, mobs, riots, panics, fads, fashions, and public opinion.

Early sociologists studied collective behavior because they lived in a world that was responding to the processes of modernization, including urbanization, industrialization, and the proletarianization of workers. Contemporary forms of collective behavior, such as the Occupy Wall Street movement and similar social protests, are variations on the theme that originated during the transition from feudalism to capitalism and the rise of modernity in Europe. Some forms of collective behavior and social movements are directed toward public issues such as air pollution, water pollution, and the exploitation of workers in global sweatshops by transnational corporations.

### Conditions for Collective Behavior

Collective behavior occurs as a result of some common influence or stimulus that produces a response from a collectivity. A *collectivity* is a number of

**social change** the alteration, modification, or transformation of public policy, culture, or social institutions over time.

**collective behavior** voluntary, often spontaneous activity that is engaged in by a large number of people and typically violates dominant-group norms and values.

**509**

people who act together and may mutually transcend, bypass, or subvert established institutional patterns and structures. Three major factors contribute to the likelihood that collective behavior will occur: (1) structural factors that increase the chances of people responding in a particular way, (2) timing, and (3) a breakdown in social control mechanisms and a corresponding feeling of normlessness.

A common stimulus is an important factor in collective behavior. For example, the publication of *Silent Spring* (1962) by former Fish and Wildlife Service biologist Rachel Carson is credited with triggering collective behavior directed at demanding a clean environment and questioning how much power large corporations should have. Carson described the dangers of pesticides such as DDT, which was then being promoted by the chemical industry as the miracle that could give the United States the unchallenged position as food supplier to the world (Griswold, 2012). Carson's activism has been described in this way:

> Carson was not a wild-eyed reformer intent on bringing the industrial age to a grinding halt. She wasn't even opposed to pesticides per se. She was a careful scientist and brilliant writer whose painstaking research on pesticides proved that the "miraculous" bursts of agricultural productivity had long-term costs undisclosed in the chemical industry's exaggerated puffery. Americans were losing things—their health, many birds and fishes, and the purity of their waterways—that they should value more than modest savings at the grocery store. (Cronin and Kennedy, 1999: 151)

Timing is another significant factor in bringing about collective behavior. For example, in the 1960s smog had started staining the skies in this country; in Europe, birds and fish were dying from environmental pollution; and oil spills from tankers were provoking public outrage worldwide. People in this country were ready to acknowledge that problems existed. By writing *Silent Spring*, Carson made people aware of the hazards of chemicals in their foods and the destruction of wildlife. However, that is not all she accomplished: As a consequence of her careful research and writing, she also produced anger in people at a time

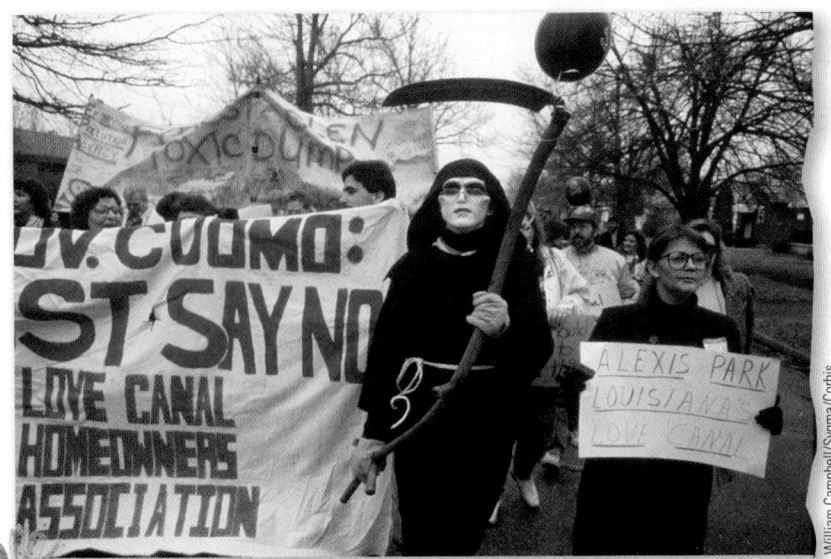

**FIGURE 16.1** The Love Canal area of Niagara Falls, New York, has been the site of protests and other forms of collective behavior because of hazardous environmental pollution. Original protests in the 1970s, demanding a cleanup of the site, were followed in the 1990s by new protests, this time over the proposed resettlement of the area.

when they were beginning to wonder if they were being deceived by the very industries that they had entrusted with their lives and their resources. Once aroused to action, many people began demanding an honest, comprehensive accounting of where pollution was occurring and how it might be endangering public health and environmental resources (Figure 16.1). Public outcries also led to investigations in courts and legislatures throughout the United States as people began to demand legal recognition of the right to a clean environment. Now, more than fifty years and two million copies later, no other single work has had the impact of Carson's *Silent Spring* on environmental issues and activism (Griswold, 2012).

A breakdown in social control mechanisms has been a powerful force in triggering collective behavior regarding environmental protection and degradation. During the 1970s, people in the "Love Canal" area of Niagara Falls, New York, became aware that their neighborhood and their children's school had been built over a canal where tons of poisonous waste had been dumped by a chemical company between 1930 and 1950. After the company closed the sixteen-acre site, covered it with soil, and sold it (for $1) to the city of Niagara Falls, homes and a school were built on it. Over the next two decades an oily black substance began oozing into the homes in the area and killing the trees and grass on the lots; schoolchildren reported mysterious illnesses and feelings of malaise. Tests indicated that the dump site contained more than 200 different chemicals, many of which

# ANSWERS to the Sociology Quiz on Collective Behavior and Environmental Issues

**1. True.** Global surface temperatures have increased about 0.4 degrees Fahrenheit during the past 25 years, and scientists believe that this trend will grow more pronounced during the next 100 years.

**2. False.** The environmental movement in the United States is the result of more than 135 years of collective action. The first environmental organization, the American Forestry Association (now American Forests), originated in 1875.

**3. False.** Public opinion polls continue to show that the majority of people in the United States have favorable attitudes regarding protection of the environment and banning nuclear weapons; however, far fewer individuals are actually involved in collective action to further these causes.

**4. True.** Environmental groups have held sit-ins, marches, boycotts, and strikes, which sometimes take the form of civil disobedience. Others have hanged political leaders in effigy or held officials hostage. Still others have dressed as grizzly bears to block traffic in Yellowstone National Park or created a symbolic "crack" (made of plastic) on the Glen Canyon Dam on the Colorado River to denounce development in the area.

**5. True.** Rumors are most likely to emerge and circulate when people have very little information on a topic that is important to them. For example, rumors abound in times of technological disasters, when people are fearful and often willing to believe a worst-case scenario.

**6. True.** Many social movements, including grassroots environmental activism, attempt to influence public opinion so that local decision makers will feel obliged to correct a specific problem through changes in public policy.

**7. True.** Most social movements are reform movements that focus on improving society by changing some specific aspect of the social structure. Examples include environmental movements and the disability rights movement.

**8. False.** Most sociological studies have found that people respond differently to natural disasters, which usually occur very suddenly, than to technological disasters, which may occur gradually. One of the major differences is the communal bonding that tends to occur following natural disasters, as compared with the extreme social conflict that may follow technological disasters.

could cause cancer or other serious health problems. Upon learning this information, Lois Gibbs, a mother of one of the schoolchildren (and the woman who we featured in this chapter's opening vignette), began a grassroots campaign to force government officials to relocate community members injured by seepages from the chemical dump. The collective behavior of neighborhood volunteers was not only successful in eventually bringing about social change but also inspired others to engage in collective behavior regarding environmental problems in their communities.

## Dynamics of Collective Behavior

To better understand the dynamics of collective behavior, let's briefly examine several questions. First, how do people come to transcend, bypass, or subvert established institutional patterns and structures? Some environmental activists have found that they cannot get their point across unless they

go outside established institutional patterns and organizations. For example, Lois Gibbs and other Love Canal residents initially tried to work within established means through the school administration and state health officials to clean up the problem. However, they quickly learned that their problems were not being solved through "official" channels. As the problem appeared to grow worse, "official" responses became more defensive and obscure. Accordingly, some residents began acting outside of established norms by holding protests and strikes (Gibbs, 1982). Some situations are more conducive to collective behavior than others. When people can communicate quickly and easily with one another, spontaneous behavior is more likely. When people are gathered together in one general location (whether lining the streets or assembled in a massive stadium), they are more likely to respond to a common stimulus.

Second, how do people's actions compare with their attitudes? People's attitudes (as expressed in

**511**

public opinion surveys, for instance) are not always reflected in their political and social behavior. Issues pertaining to the environment are no exception. For example, people may indicate in survey research that they believe the quality of the environment is very important, but the same people may not turn out on election day to support propositions that protect the environment or candidates who promise to focus on environmental issues. Likewise, individuals who indicate on a questionnaire that they are concerned about increases in ground-level ozone—the primary component of urban smog—often drive single-occupant, oversized vehicles that government studies have shown to be "gas guzzlers" that contribute to lowered air quality in urban areas. As a result, smog levels increase, contributing to human respiratory problems and dramatically reduced agricultural crop yields.

Third, why do people act collectively rather than singly? Many people believe that there is strength in numbers, whether they are attending a rock concert or protesting an injustice in society. Individuals may act as a collectivity when they believe it is the only way to fight those with greater power and resources. Collective behavior is not just the sum total of a large number of persons acting at the same time; rather, it reflects people's joint response to some common influence or stimulus.

## Distinctions Regarding Collective Behavior

People engaging in collective behavior may be divided into crowds and masses. A **crowd is a relatively large number of people who are in one another's immediate vicinity.** Examples of crowds include the audience in a movie theater or people at a pep rally for a sporting event. By contrast, a **mass is a number of people who share an interest in a specific idea or issue but who are not in one another's immediate vicinity.** An example is the popularity of Facebook and Twitter. Through these forms of instantaneous communication, people express their views on everyday life and on larger social issues such as economic crises and environmental problems. Individuals who read what someone has posted and make comments in response usually share a common interest even if these individuals have not met in a face-to-face encounter and may not agree on certain issues.

Collective behavior may also be distinguished by the dominant emotion expressed. The *dominant emotion* refers to the "publicly expressed feeling perceived by participants and observers as the most prominent in an episode of collective behavior"

(Lofland, 1993: 72). Fear, hostility, and joy are three fundamental emotions found in collective behavior; however, grief, disgust, surprise, or shame may also predominate in some forms of collective behavior (Lofland, 1993). Can you think of other emotions that people might express when they engage in collective behavior?

## Types of Crowd Behavior

**LO2** When we think of a crowd, many of us think of *aggregates,* previously defined as a collection of people who happen to be in the same place at the same time but who share little else in common. For example, think of thousands of people stranded in an airport when harsh weather or other conditions make it impossible for them to board flights and head for their destinations. Although stranded businesspeople and tourists are together in the airport, they do not necessarily share anything in common. Moreover, the presence of a relatively large number of people in the same location does not necessarily produce collective behavior. To help explain this phenomenon, early symbolic interactionist sociologist Herbert Blumer (1946) developed a typology that divides crowds into four categories: casual, conventional, expressive, and acting. Other scholars have added a fifth category, protest crowds.

**Casual and Conventional Crowds** *Casual crowds* are relatively large gatherings of people who happen to be in the same place at the same time; if they interact at all, it is only briefly. People in a shopping mall or a subway car are examples of casual crowds (Figure 16.2). Other than sharing a momentary interest, such as a clown's performance or a small child's fall, a casual crowd has nothing in common. The casual crowd plays no active part in the event—such as the child's fall—which likely would have occurred whether or not the crowd was present; the crowd simply observes.

*Conventional crowds* are made up of people who come together for a scheduled event and thus share a common focus. Examples include religious services, graduation ceremonies, concerts, and college lectures. Each of these events has preestablished schedules and norms. Because these events occur regularly, interaction among participants is much more likely; in turn, the events would not occur without the crowd, which is essential to the event.

**Expressive and Acting Crowds** *Expressive crowds* provide opportunities for the expression of some strong emotion (such as joy, excitement, or grief). People release their pent-up emotions in conjunction with other persons experiencing similar emotions. Examples include worshippers at religious revival services; mourners lining the streets when a celebrity, a public official, or religious leader has died; and revelers assembled at Mardi Gras or on New Year's Eve at Times Square in New York.

*Acting crowds* are collectivities so intensely focused on a specific purpose or object that they may erupt into violent or destructive behavior. Mobs, riots, and panics are examples of acting crowds, but casual and conventional crowds may become acting crowds under some circumstances. A *mob* **is a highly emotional crowd whose members engage in, or are ready to engage in, violence against a specific target—a person, a category of people, or physical property.** Mob behavior in this country has included lynchings, fire bombings, effigy hangings, and hate crimes. Mob violence tends to dissipate relatively quickly once a target has been injured, killed, or destroyed. Sometimes, actions such as an effigy hanging are used symbolically by groups that are not otherwise violent. For example, Lois Gibbs and other Love Canal residents called attention to their problems with the chemical dump site by staging a protest in which they "burned in effigy" the governor and the health commissioner to emphasize their displeasure with the lack of response from these public officials.

Compared with mob actions, riots may be of somewhat longer duration. A *riot* **is violent crowd behavior that is fueled by deep-seated emotions but not directed at one specific target.** Riots are often triggered by fear, anger, and hostility; however, not all riots are caused by deep-seated hostility and hatred—people may be expressing joy and exuberance when rioting occurs. Examples include celebrations after sports victories that turn into riots such as when some of the fans rioted in the streets of Baltimore following the Ravens' victory over the San Francisco 49ers in the 2013 Super Bowl.

A *panic* **is a form of crowd behavior that occurs when a large number of people react to a real or perceived threat with strong emotions and self-destructive behavior.** The most common type of panic occurs when people seek to escape from a perceived danger, fearing that few (if any) of them will be able to get away from that danger. Examples include passengers on a sinking cruise ship or persons in a burning nightclub. Panics can also arise in response to larger social, financial, or political conditions that people believe are beyond their control—such as a major disruption in the economy. A "bank run" in

**FIGURE 16.2** Crowds of people come together for a variety of reasons. The people pictured here are gathering in Times Square in New York City for a New Year's Eve celebration. How does a crowd such as this differ from other types of crowds?

which hundreds or thousands of customers seek to take out all of their money at the same time, fearing that the financial institution is becoming insolvent, is an example. Although panics are relatively rare, they receive massive media coverage because they provoke strong feelings of fear in readers and viewers, and the number of casualties may be large. An example was the 2013 tragic garment factory blaze in Bangladesh in which more than 100 workers died as fire gutted the large manufacturing warehouse and panic-stricken workers jammed a stairwell trying to escape their workplace.

**crowd** a relatively large number of people who are in one another's immediate vicinity.

**mass** a number of people who share an interest in a specific idea or issue but who are not in one another's immediate vicinity.

**mob** a highly emotional crowd whose members engage in, or are ready to engage in, violence against a specific target—a person, a category of people, or physical property.

**riot** violent crowd behavior that is fueled by deep-seated emotions but not directed at one specific target.

**panic** a form of crowd behavior that occurs when a large number of people react to a real or perceived threat with strong emotions and self-destructive behavior.

**FIGURE 16.3** Protest crowds are often confronted by law enforcement officials who are responsible for maintaining order and safety in the community. Standoffs such the one shown here are not an unusual sight on a global basis as people seek to make their concerns known to others even as officials attempt to fulfill their duties.

**Protest Crowds** Sociologists Clark McPhail and Ronald T. Wohlstein (1983) added protest crowds to the four types of crowds identified by Blumer. *Protest crowds* engage in activities intended to achieve specific political goals. Examples include sit-ins, marches, boycotts, blockades, and strikes. Some protests take the form of *civil disobedience*—**nonviolent action that seeks to change a policy or law by refusing to comply with it.** Acts of civil disobedience may become violent, as in a confrontation between protesters and police officers (Figure 16.3); in this case a protest crowd becomes an *acting crowd*. In the 1960s, African American students and sympathetic whites used sit-ins to call attention to racial injustice and demand social change. Some of these protests could escalate into violent confrontations even when violence was not the intent of the organizers. Some analysts have questioned whether the Occupy movement is a new form of civil disobedience, but others have argued that this is not accurate. From their perspective, civil disobedience views political institutions as legitimate but does not see some laws (such as those that once discriminated against persons of color) as being legitimate. By contrast, the Occupy movement might be understood as political disobedience because participants resist the current system of governing in the United States, particularly partisan politics and ideologies (Harcourt, 2011). At the bottom line, protest crowds typically seek to change some aspect of the status quo even if they do not state specific policy demands, as was the case with a large segment of the initial Occupy movement.

However, some protest crowds do state specific policy demands, such as protesters who sought to call attention to the environmental impact of offshore drilling, particularly in the aftermath of the BP explosion at a Gulf of Mexico oil rig. These protesters held rallies in front of the White House and other public areas where they hoped that political leaders would observe them and take their demands for change into account. In 2013 protesters with signs and American flags demanded a change in how the Internal Revenue Service operates in response to the acknowledgment that the IRS had targeted U.S. citizens based on their political or religious beliefs. Similarly, immigrant rights activists have rallied to protest how certain political leaders had voted on immigration reform legislation and to demand different ways of dealing with border security and citizenship for persons without legal documentation who are already residing in the United States. People with many different issues take their concerns to the streets, as well as to mainstream and social media, to have their voices heard about problems that they believe need to be addressed.

## Explanations of Crowd Behavior

**LO3** What causes people to act collectively? How do they determine what types of action to take? One of the earliest theorists to provide an answer to these questions was Gustave Le Bon, a French scholar who focused on crowd psychology in his contagion theory.

**Contagion Theory** *Contagion theory* focuses on the social–psychological aspects of collective behavior; it attempts to explain how moods, attitudes, and behavior are communicated rapidly and why they are accepted by others. Le Bon (1841–1931) argued that people are more likely to engage in antisocial behavior in a crowd because they are anonymous and feel invulnerable. Le Bon (1960/1895) suggested that a crowd takes on a life of its own that is larger than the beliefs or actions of any one person. Because of its anonymity, the crowd transforms individuals from rational beings into a single organism with a collective mind. In essence, Le Bon asserted that emotions such as fear and hate are contagious in crowds because people experience a decline in personal responsibility; they will do things as a collectivity that they would never do when acting alone.

Le Bon's theory is still used by many people to explain crowd behavior. However, critics argue that the "collective mind" has not been documented by systematic studies.

LOUISA GOULIAMAKI/AFP/Getty Images/Newscom

**Social Unrest and Circular Reaction** Sociologist Robert E. Park was the first U.S. sociologist to investigate crowd behavior. Park believed that Le Bon's analysis of collective behavior lacked several important elements. Intrigued that people could break away from the powerful hold of culture and their established routines to develop a new social order, Park added the concepts of social unrest and circular reaction to contagion theory. According to Park, social unrest is transmitted by a process of *circular reaction*—the interactive communication between persons such that the discontent of one person is communicated to another, who, in turn, reflects the discontent back to the first person (Park and Burgess, 1921).

**Convergence Theory** *Convergence theory* focuses on the shared emotions, goals, and beliefs that many people may bring to crowd behavior. Because of their individual characteristics, many people have a predisposition to participate in certain types of activities. From this perspective, people with similar attributes find a collectivity of like-minded persons with whom they can express their underlying personal tendencies. Although people may reveal their "true selves" in crowds, their behavior is not irrational; it is highly predictable to those who share similar emotions or beliefs.

Convergence theory has been applied to a wide array of conduct, from lynch mobs to environmental movements. In a now-classic study of lynching, social psychologist Hadley Cantril (1941) found that the participants shared certain common attributes: They were poor and working-class whites who felt that their status was threatened by the presence of more-successful African Americans. Consequently, the characteristics of these individuals made them susceptible to joining a lynch mob even if they did not know the target of the lynching.

Convergence theory adds to our understanding of certain types of collective behavior by pointing out how individuals may have certain attributes—such as racial hatred or fear of environmental problems that directly threaten them—that initially bring them together. However, this theory does not explain how the attitudes and characteristics of individuals who take some collective action differ from those who do not.

**Emergent Norm Theory** Unlike contagion and convergence theories, *emergent norm theory* emphasizes the importance of social norms in shaping crowd behavior. Drawing on the symbolic interactionist perspective, the sociologists Ralph Turner and Lewis Killian (1993: 12) asserted that crowds develop their own definition of a situation and establish norms for behavior that fit the occasion:

Some shared redefinition of right and wrong in a situation supplies the justification and coordinates the action in collective behavior. People do what they would not otherwise have done when they panic collectively, when they riot, when they engage in civil disobedience, or when they launch terrorist campaigns, because they find social support for the view that what they are doing is the right thing to do in the situation.

According to Turner and Killian (1993: 13), emergent norms occur when people define a new situation as highly unusual or see a long-standing situation in a new light.

Sociologists using the emergent norm approach seek to determine how individuals in a given collectivity develop an understanding of what is going on, how they construe these activities, and what type of norms are involved. For example, in a study of audience participation, the sociologist Steven E. Clayman (1993) found that members of an audience listening to a speech applaud promptly and independently but wait to coordinate their booing with other people; they do not wish to "boo" alone.

Some emergent norms are permissive—that is, they give people a shared conviction that they may disregard ordinary rules, such as waiting in line, taking turns, or treating a speaker courteously. Collective activity such as mass looting may be defined (by participants) as taking what rightfully belongs to them. In the aftermath of the 2010 Haiti earthquake, when relief aid was slow in coming, looting was commonplace, but so too was "mob justice" for those who were caught stealing other people's possessions.

Emergent norm theory points out that crowds are not irrational. Rather, new norms are developed in a rational way to fit the immediate situation. However, critics note that proponents of this perspective fail to specify exactly what constitutes a norm, how new ones emerge, and how they are so quickly disseminated and accepted by a wide variety of participants. One variation of this theory suggests that no single dominant norm is accepted by everyone in a crowd; instead, norms are specific to the various categories of actors rather than to the collectivity as a whole (Snow, Zurcher, and Peters, 1981). For example, in a study of football victory celebrations sociologists found that each week, behavioral patterns were changed in the postgame revelry, with some being modified, some added, and some deleted (Snow, Zurcher, and Peters, 1981).

**civil disobedience** nonviolent action that seeks to change a policy or law by refusing to comply with it.

## Mass Behavior

**LO4** Not all collective behavior takes place in face-to-face collectivities. ***Mass behavior*** **is collective behavior that takes place when people (who are often geographically separated from one another) respond to the same event in much the same way.** For people to respond in the same way, they typically have common sources of information that provoke their collective behavior. The most frequent types of mass behavior are rumors, gossip, mass hysteria, public opinion, fashions, and fads. Under some circumstances, social movements constitute a form of mass behavior. However, we will examine social movements separately because they differ in some important ways from other types of dispersed collectivities.

**Rumors and Gossip** *Rumors* **are unsubstantiated reports on an issue or subject.** Whereas a rumor may spread through an assembled collectivity, rumors may also be transmitted among people who are dispersed geographically, including people spreading rumors on Twitter, posting messages on Facebook, or talking on the phone. Although rumors may initially contain a kernel of truth, they may be modified as they spread to serve the interests of those repeating them. Rumors thrive when tensions are high and when little authentic information is available on an issue of great concern. For example, a week after the terrorist explosion at the 2013 Boston Marathon, a fake Twitter feed purporting to be from the Associated Press stating that President Barack Obama had been injured in an explosion at the White House caused the stock market to take a sharp decline, wiping out $130 billion in stock value in a matter of seconds. Although the market recovered the value, the power of rumor in an age of rapid communication and social media was proven in a new and frightening manner (Matthews, 2013).

Why do people believe rumors? People are often willing to give rumors credence when no opposing information is available. Once a rumor begins to circulate, it seldom stops unless compelling information comes to the forefront that either proves the rumor false or makes it obsolete. In contemporary societies with sophisticated technology, rumors come from a wide variety of sources and may be difficult to trace. Print media (newspapers and magazines) and electronic media (radio and television), fax machines, cellular networks, satellite systems, the Internet, and social media aid the rapid movement of rumors around the globe. In addition, modern communications technology makes anonymity much easier. In a split second, messages (both factual and fictitious) can be disseminated to millions of people worldwide.

Whereas rumors deal with an issue or a subject, *gossip* **refers to rumors about the personal lives of individuals.** Charles Horton Cooley (1963/1909) viewed gossip as something that spread among a small group of individuals who personally knew the person who was the object of the rumor. Today, this is frequently not the case; many people enjoy gossiping about people whom they have never met. Tabloid newspapers and magazines, such as the *National Enquirer* and *People,* along with television "news" programs, websites, Facebook, and Twitter, provide "inside" information on the lives of celebrities. We are constantly bombarded with information and gossip, much of which has not been checked for authenticity.

**Mass Hysteria and Panic** *Mass hysteria* is a form of dispersed collective behavior that occurs when a large number of people react with strong emotions and self-destructive behavior to a real or perceived threat. Does mass hysteria actually occur? Although the term has been widely used, many sociologists believe that this behavior is best described as a panic with a dispersed audience.

A classic example of mass hysteria or a panic with a widely dispersed audience was actor Orson Welles's 1938 Halloween eve radio dramatization of H. G. Wells's science fiction classic *The War of the Worlds.* A CBS radio dance-music program was interrupted suddenly by a news bulletin informing the audience that Martians had landed in New Jersey and were in the process of conquering Earth. Some listeners became extremely frightened even though an announcer had indicated before, during, and after the performance that the broadcast was a fictitious dramatization (Figure 16.4). According to some reports, as many as 1 million of the estimated 10 million listeners believed that this astonishing event had occurred. Thousands were reported to have hidden in their storm cellars or to have gotten in their cars so that they could flee from the Martians (see Brown, 1954). In actuality, the program probably did not generate mass hysteria, but rather a panic among gullible listeners. Others switched stations to determine if the same "news" was being broadcast elsewhere. When they discovered that it was not, they merely laughed at the joke being played on listeners by CBS. In 1988, on the fiftieth anniversary of the broadcast, a Portuguese radio station rebroadcast the program; once again, a panic ensued.

Today, panics such as this tend to occur among segments of the population but not the entire population. For example, in 2013 some gun enthusiasts

**FIGURE 16.4** Although a spokesperson for CBS Radio stated to listeners that they were hearing a dramatization of a novel, the 1938 presentation of H. G. Wells's *The War of the Worlds*, as presented by Orson Welles and his Mercury Theatre, terrified untold numbers of people. Here Welles talks to interviewers the day after the event caused a nationwide panic.

became concerned that there was going to be a shortage of ammunition and started purchasing all the bullets they could find. The panic started following the reelection of President Barack Obama; the Newtown, Connecticut, school shootings and demands for more-stringent federal gun control laws; and persistent rumors that the Department of Homeland Security and the federal government were stockpiling ammunition. Some people could not be persuaded that ammunition would still be available on store shelves when they needed it for their weapons.

# LO5 Fads and Fashions

As you will recall from Chapter 3, a *fad* is a temporary but widely copied activity enthusiastically followed by large numbers of people. Fads can be embraced by widely dispersed collectivities: TV, the Internet, and social media bring the latest fads—such as top memes, YouTube videos of cats or dogs doing funny things, zombie films and TV series like *The Walking Dead,* and Ben & Jerry's latest ice cream flavor—to the attention of audiences around the world.

Unlike fads, fashions tend to be longer lasting. In Chapter 3, *fashion* is defined as a currently valued style of behavior, thinking, or appearance. Fashion also applies to art, music, drama, literature, architecture, interior design, and automobiles, among other things. However, most sociological research on fashion has focused on clothing, especially women's apparel and how it may relate to culture and social customs in a specific era.

In preindustrial societies, clothing styles remained relatively unchanged. With the advent of industrialization, items of apparel became readily available at low prices because of mass production. Fashion became more important as people embraced the "modern" way of life and as advertising encouraged "conspicuous consumption."

Georg Simmel, Thorstein Veblen, and Pierre Bourdieu have all viewed fashion as a means of status differentiation among members of different social classes. Simmel (1957/1904) suggested a classic "trickle-down" theory (although he did not use those exact words) to describe the process by which members of the lower classes emulate the fashions of the upper class (Figure 16.5). As the fashions descend through the status hierarchy, they are watered down and "vulgarized" so that they are no longer recognizable to members of the upper class, who then regard them as unfashionable and in bad taste (Davis, 1992). Veblen (1967/1899) asserted that fashion serves mainly to institutionalize conspicuous consumption among the wealthy. More than eighty years later, Bourdieu (1984) similarly (but more subtly) suggested that "matters of taste," including fashion sensibility, constitute a large share of the "cultural capital" possessed by members of the dominant class.

Herbert Blumer (1969) disagreed with the trickle-down approach, arguing that "collective selection" best explains fashion. Blumer suggested that people in the middle and lower classes follow fashion because it is fashion, not because they desire to emulate members of the elite class. Blumer thus shifted the focus on fashion to collective mood, tastes, and choices: "Tastes are themselves a product of experience. . . . They are formed in the context of social interaction, responding to the definitions and affirmation given by others. People thrown into areas of common interaction and having similar runs of experience develop common tastes" (qtd. in Davis, 1992: 116). Perhaps one of the best refutations of the trickle-down approach is the way in which fashion today often originates among people in the lower social classes and is mimicked by the elites.

**mass behavior** collective behavior that takes place when people (who are often geographically separated from one another) respond to the same event in much the same way.

**rumor** an unsubstantiated report on an issue or subject.

**gossip** rumors about the personal lives of individuals.

**FIGURE 16.5** Georg Simmel suggested a "trickle-down" theory such as this one in Paris to describe the process by which the lower classes emulate the fashions of the upper class. Do runway shows provide evidence for such a theory?

© ostill/Shutterstock.com

**Public Opinion** *Public opinion* **consists of the attitudes and beliefs communicated by ordinary citizens to decision makers.** It is measured through polls and surveys, which use research methods such as interviews and questionnaires, as described in Chapter 1. Many people are not interested in all aspects of public policy but are concerned about issues that they believe are relevant to them. Even on a single topic, public opinion will vary widely based on race/ethnicity, religion, region, social class, education level, gender, age, and so on.

Scholars who examine public opinion are interested in the extent to which the public's attitudes are communicated to decision makers and the effect (if any) that public opinion has on policy making (Turner and Killian, 1993). Some political scientists argue that public opinion has a substantial effect on decisions at all levels of government; others strongly disagree. Table 16.1 shows a public opinion survey

of the top policy priorities of the U.S. public in 2014. As you will note, the environment is not listed among the ten items that the Pew Research Center (2014) survey identified as the public's top priorities. Slightly over half (52 percent) of Americans indicated that they believed the environment should be a top priority for the president and the Congress. Of course, other pressing issues such as the economy, jobs, and the budget deficit were also on their minds when they were taking this survey.

Today, people attempt to influence elites, and vice versa. Consequently, a two-way process occurs with the dissemination of *propaganda*—**information provided by individuals or groups that have a vested interest in furthering their own cause or damaging an opposing one.** Although many of us think of propaganda in negative terms, the information provided can be correct and can have a positive effect on decision making.

Grassroots environmental activists (including the Love Canal residents) have attempted to influence public opinion for many years. However, it is less clear that public opinion translates into action by either decision makers in government and industry or by individuals (such as a willingness to adopt a more ecologically sound lifestyle).

Initially, most grassroots environmental activists attempt to influence public opinion so that local decision makers will feel the necessity of correcting a specific problem through changes in public policy. Although activists usually do not start out seeking broader social change, they often move in that direction when they become aware of how widespread the problem is in the larger society or on a global basis. One of two types of social movements often develops at this point—one focuses on NIMBY ("not in my backyard"), whereas the other focuses on NIABY ("not in anyone's backyard").

## table 16.1

### Top 10 Policy Priorities: 2014

| Issue | Percentage |
| --- | --- |
| Strengthening the nation's economy | 80 |
| Improving the jobs situation | 74 |
| Defending against terrorism | 73 |
| Improving education | 69 |
| Securing Social Security | 66 |
| Reducing the budget deficit | 63 |
| Securing Medicare | 61 |
| Reducing health care costs | 59 |
| Reducing crime | 55 |
| Helping the poor and needy | 49 |

Sources: Pew Research Center, 2014. © Cengage Learning

# Social Movements

Although collective behavior is short-lived and relatively unorganized, social movements are longer lasting, are more organized, and have specific goals. A *social movement* is an organized group that acts consciously to promote or resist change through collective action. Because social movements initially are not institutionalized and are outside the political mainstream, they offer "outsiders" an opportunity to have their voices heard.

Social movements are more likely to develop in industrialized societies than in preindustrial societies, where acceptance of traditional beliefs and practices makes such movements unlikely. Diversity and a lack of consensus (hallmarks of industrialized nations) contribute to demands for social change, and people who participate in social movements typically lack the power and other resources to bring about change without engaging in collective action. Social movements are most likely to spring up when people come to see their personal troubles as public issues that cannot be solved without a collective response. Although the government is most frequently the target of social movement activity, other organizations—such as schools, corporations, or financial institutions—are also the targets of social activism.

Social movements make democracy more available to excluded groups. Historically, people in the United States have worked at the grassroots level to bring about changes even when elites sought to discourage activism. For example, the civil rights movement brought into its ranks African Americans in the South who had never before been allowed to participate in politics. The women's suffrage movement gave voice to women who had been denied the right to vote. Similarly, a grassroots environmental movement gave the working-class residents of Love Canal a way to "fight city hall" and Hooker Chemicals, as Lois Gibbs (1982: 38–40) explains:

> People were pretty upset. They were talking and stirring each other up. I was afraid there would be violence. We had a meeting at my house to try to put everything together [and] decided to form a homeowners' association. We got out the word as best we could and told everyone to come to the Frontier Fire Hall on 102d Street. . . . The firehouse was packed with people, and more were outside. . . .
>
> I was elected president. . . . I took over the meeting but I was scared to death. It was only the second time in my life I had been in front of a microphone or a crowd. . . . We set four goals right at the beginning—(1) get all the residents within the Love Canal area who wanted to be evacuated, evacuated and relocated, especially during the construction and repair of the canal; (2) do something about

propping up property values; (3) get the canal fixed properly; and (4) have air sampling and soil and water testing done throughout the whole area, so we could tell how far the contamination had spread. . . .

Most social movements rely on volunteers like Lois Gibbs to carry out the work. Traditionally, women have been strongly represented in both the membership and the leadership of many grassroots movements.

The Love Canal activists set the stage for other movements that have grappled with the kind of issues that the sociologist Kai Erikson (1994) refers to as a "new species of trouble." Erikson was a pioneer in the field of studying this new species of trouble, which is no longer new but by which he meant environmental problems that contaminate, pollute, befoul, taint, and scare human beings in new ways that produce uncanny fear (Erikson, 1991). The chaos that Erikson describes is the result of technological disasters caused by system failures, human error, faulty designs, and other problems that wreak havoc on people and things. An example is the 2011 disaster in Japan, where, in the aftermath of a major earthquake and massive flood, thousands of people living near Fukushima Daiichi and Daini nuclear power stations were ordered to evacuate because it was likely that small amounts of radioactive material would leak from the plants (Wald, 2011). The health effects of the radiation leak on residents and workers who sought to save the plant may not be known for years as problems such as this continue to be a "new species of trouble."

Social movements provide people who otherwise would not have the resources to enter the game of politics a chance to do so. We are most familiar with those movements that develop around public policy issues considered newsworthy by the media, ranging from abortion and women's rights to gun control and environmental justice. However, a number of other types of social movements exist as well.

## Types of Social Movements

**LO6** Social movements are difficult to classify; however, sociologists distinguish among movements on the basis of their *goals* and

**public opinion** the attitudes and beliefs communicated by ordinary citizens to decision makers.

**propaganda** information provided by individuals or groups that have a vested interest in furthering their own cause or damaging an opposing one.

**social movement** an organized group that acts consciously to promote or resist change through collective action.

**FIGURE 16.6** Martin Luther King Jr., a leader of the civil rights movement in the 1950s and 1960s, advocated nonviolent protests that sometimes took the form of civil disobedience. Here he marches alongside his wife, Coretta Scott King, who for many years took over Dr. King's activities after he was assassinated.

the *amount of change* they seek to produce. Some movements seek to change people, whereas others seek to change society.

**Reform Movements** Grassroots environmental movements are an example of *reform movements*, which seek to improve society by changing some specific aspect of the social structure. Members of reform movements usually work within the existing system to attempt to change public policy so that it more adequately reflects their own value system. Examples of reform movements (in addition to the environmental movement) include labor movements, animal rights movements, antinuclear movements, Mothers Against Drunk Driving, and the disability rights movement.

Some social movements arise specifically to alter negative stereotypes and reduce stigma associated with specific categories of people. Such social movements may not only bring about changes in societal attitudes and practices but also produce changes in participants' social emotions. For example, the civil rights and gay rights movements helped replace shame with pride (Figure 16.6).

**Revolutionary Movements** Movements seeking to bring about a total change in society are referred to as *revolutionary movements*. These movements usually do not attempt to work within the existing system; rather, they aim to remake the system by replacing existing institutions with new ones. This

was apparently the goal of the Occupy Wall Street movement, although it was unclear because it was leaderless and functioned with spokespersons who declined to express specific objectives for their activism. Revolutionary movements range from utopian groups seeking to establish an ideal society to radical terrorists who use fear tactics to intimidate those with whom they disagree ideologically. In the twenty-first century, people in Muslim countries around the world, including Tunisia, Egypt, and Iran, have participated in revolutionary movements and risen up against what they perceived to be tyrannical regimes (Figure 16.7). For example, people in Tunisia gathered in 2013 to call for unity and support for the resistance movement against tyranny and injustices perpetrated by the former regime of Zine El Abidine Ben Ali, the ousted president of that nation who had banned all public meetings and engaged in other forms of oppression.

Movements based on terrorism often use tactics such as bombings, kidnappings, hostage taking, hijackings, and assassinations. A number of movements in the United States have engaged in terrorist activities or supported a policy of violence. However, the terrorist attacks in New York City and Washington, D.C., on September 11, 2001, and the events that followed those attacks proved to all of us that terrorism within this country can originate from the activities of revolutionary terrorists from outside the country as well. Some terrorist events do not appear to be as fully organized or to include as many people as revolutionary movements are typically thought to involve.

**Religious Movements** Social movements that seek to produce radical change in individuals are typically based on spiritual or supernatural belief

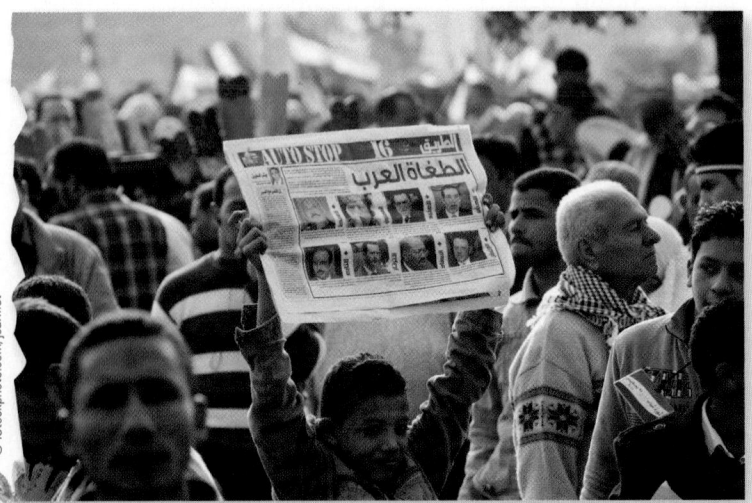

**FIGURE 16.7** Revolutionary movements have taken place in Egypt and other Arab nations in recent years because of a strong belief that leaders are oppressive and governments are not benefiting the people.

systems. Also referred to as *expressive movements*, *religious movements* are concerned with renovating or renewing people through "inner change." Fundamentalist religious groups seeking to convert nonbelievers to their belief system are an example of this type of movement. Some religious movements are *millenarian*—that is, they forecast that "the end is near" and assert that an immediate change in behavior is imperative. Examples include Hare Krishnas, the Unification Church, Scientology, and the Divine Light Mission, all of which tend to appeal to the psychological and social needs of young people seeking meaning in life that mainstream religions have not provided for them.

**Alternative Movements** Movements that seek limited change in some aspect of people's behavior are referred to as *alternative movements*. For example, early in the twentieth century the Women's Christian Temperance Union attempted to get people to abstain from drinking alcoholic beverages. Some analysts place "therapeutic social movements" such as Alcoholics Anonymous in this category; however, others do not because of their belief that people must change their lives completely in order to overcome alcohol abuse. More recently, a variety of "New Age" movements have directed people's behavior by emphasizing spiritual consciousness combined with a belief in reincarnation and astrology. Such practices as vegetarianism, meditation, and holistic medicine are often included in the self-improvement category. Some alternative movements have included the practice of yoga (usually without its traditional background in the Hindu religion) as a means by which the self can be liberated and union can be achieved with the supreme spirit or universal soul.

**Resistance Movements** Also referred to as *regressive movements*, *resistance movements* seek to prevent change or to undo change that has already occurred. Virtually all of the social movements previously discussed face resistance from one or more reactive movements that hold opposing viewpoints and want to foster public policies that reflect their own beliefs. Examples of resistance movements are groups organized to oppose gay marriage, abortion, and gun control legislation.

## Stages in Social Movements

Do all social movements go through similar stages? Not necessarily, but there appear to be identifiable stages in virtually all movements that are able to succeed beyond their initial phase of development (see Figure 16.8).

**FIGURE 16.8** STAGES IN SOCIAL MOVEMENTS
© Cengage Learning

In the *preliminary* (or *incipiency*) *stage*, widespread unrest is present as people begin to become aware of a problem. At this stage, leaders emerge to agitate others into taking action. In the *coalescence stage*, people begin to organize and to publicize the problem. At this stage, some movements become formally organized at local and regional levels. In the *institutionalization* (or *bureaucratization*) *stage*, an organizational structure develops, and a paid staff (rather than volunteers) begins to lead the group. When the movement reaches this stage, the initial zeal and idealism of members may diminish as administrators take over management of the organization. Early grassroots supporters may become disillusioned and drop out; they may also start another movement to address some as-yet-unsolved aspect of the original problem. For example, some national environmental organizations—such as the Sierra Club, the National Audubon Society, and the National Parks and Conservation Association—that started as grassroots conservation movements are currently viewed by many people as being unresponsive to local environmental problems. As a result, new movements have arisen.

## Social Movement Theories

**LO7** What conditions are most likely to produce social movements? Why are people drawn to these movements? Sociologists have developed several theories to answer these questions.

### Relative Deprivation Theory

According to *relative deprivation theory*, people who are satisfied with their present condition are less likely to seek social change. Social movements arise as a response to people's perception that they have been deprived of what they consider to be their fair share. Thus, people who suffer relative deprivation are more likely to feel that change is necessary and to join a social movement in order to bring about that change. *Relative deprivation* refers to the discontent that people may feel when they compare their achievements with those of similarly situated persons and find that they have less than they think they deserve. Karl Marx captured the idea of relative deprivation in this description: "A house may be large or small; as long as the surrounding houses are small it satisfies all social demands for a dwelling. But let a palace arise beside the little house, and it shrinks from a little house to a hut" (qtd. in Ladd, 1966: 24). Movements based on relative deprivation are most likely to occur when an upswing in the standard of living is followed by a period of decline, such that people have *unfulfilled rising expectations*—newly raised hopes of a better lifestyle that are not fulfilled as rapidly as the people expected or are not realized at all.

Although most of us can relate to relative deprivation theory, it does not fully account for why people experience social discontent but fail to join a social movement. Even though discontent and feelings of deprivation may be necessary to produce certain types of social movements, they are not sufficient to bring movements into existence.

## Value-Added Theory

The *value-added theory* developed by sociologist Neil Smelser (1963) is based on the assumption that certain conditions are necessary for the development of a social movement. Smelser called his theory the "value-added" approach based on the concept (borrowed from the field of economics) that each step in the production process adds something to the finished product. For example, in the process of converting iron ore into automobiles, each stage "adds value" to the final product (Smelser, 1963). Similarly, Smelser asserted, six conditions are necessary and sufficient to produce social movements when they combine or interact in a particular situation:

■ *Structural conduciveness.* People must become aware of a significant problem and have the opportunity to engage in collective action. According to Smelser, movements are more likely to occur when a person, class, or agency can be singled out as the source of the problem; when channels for expressing grievances either are not available or fail; and when the aggrieved have a chance to communicate among themselves.

■ *Structural strain.* When a society or community is unable to meet people's expectations that something should be done about a problem, strain occurs in the system. The ensuing tension and conflict contribute to the development of a social movement based on people's belief that the problem would not exist if authorities had done what they were supposed to do.

■ *Spread of a generalized belief.* For a movement to develop, there must be a clear statement of the problem and a shared view of its cause, effects, and possible solution.

■ *Precipitating factors.* To reinforce the existing generalized belief, an inciting incident or dramatic event must occur. With regard to technological disasters, some (including Love Canal) gradually emerge from a long-standing environmental threat, whereas others (including the meltdown of a nuclear power plant) involve a suddenly imposed problem.

■ *Mobilization for action.* At this stage, leaders emerge to organize others and give them a sense of direction.

■ *Social control factors.* If there is a high level of social control on the part of law enforcement officials, political leaders, and others, it becomes more difficult to develop a social movement or engage in certain types of collective action.

■ Value-added theory takes into account the complexity of social movements and makes it possible to test Smelser's assertions regarding the necessary and sufficient conditions that produce such movements. However, critics note that the approach is rooted in the functionalist tradition and views structural strains as disruptive to society.

## Resource Mobilization Theory

Smelser's value-added theory tends to underemphasize the importance of resources in social movements. By contrast, *resource mobilization theory* focuses on the ability of members of a social movement to acquire resources and mobilize people in order to advance their cause. Resources include money, people's time and skills, access to the media, and material goods, such as property and equipment. Assistance from outsiders is essential for social movements. For example, reform movements are more likely to succeed when they gain the support of political and economic elites.

Resource mobilization theory is based on the assumption that participants in social movements are rational people. From this perspective, social movements are formed and dissolved, mobilized and deactivated, based on rational decisions about the goals of the group, available resources, and the cost of mobilization and collective action. Resource mobilization theory also assumes that participants must have some degree of economic and political resources to make the movement a success. In other words, widespread discontent alone cannot produce

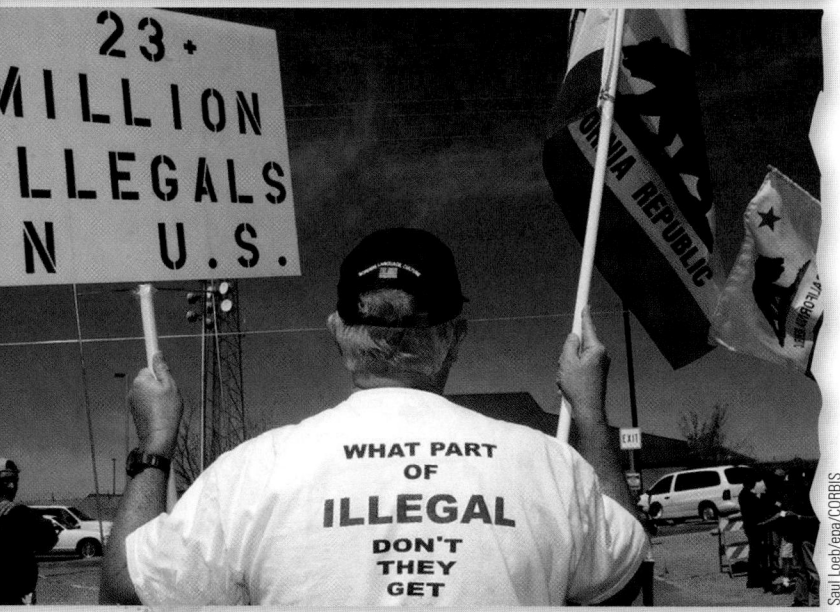

**FIGURE 16.9** How is the issue of immigration framed in these photos? Research based on frame analysis often investigates how social issues are framed and what names they are given.

a social movement; adequate resources and motivated people are essential to any concerted social action.

In the twenty-first century, scholars continue to modify resource mobilization theory and to develop new approaches for investigating the diversity of movements. Emerging perspectives based on resource mobilization theory emphasize the ideology and legitimacy of movements as well as material resources.

Additional perspectives are also needed on social movements in other nations to determine how activists in those countries acquire resources and mobilize people to advance causes such as environmental protection (see "Sociology in Global Perspective").

## Social Constructionist Theory: Frame Analysis

**LO8** Theories based on a symbolic interactionist perspective focus on the importance of the symbolic presentation of a problem to both participants and the general public. *Social constructionist theory* is based on the assumption that a social movement is an interactive, symbolically defined, and negotiated process that involves participants, opponents, and bystanders.

Research based on this perspective often investigates how problems are framed and what names they are given (see Figure 16.9). This approach reflects the influence of the sociologist Erving Goffman's *Frame Analysis* (1974), in which he suggests that our interpretation of the particulars of events and activities is dependent on the framework from which we perceive them. According to Goffman (1974: 10), the purpose of frame analysis is "to try to isolate some of the basic frameworks of understanding available in our society for making sense out of events and to analyze the special vulnerabilities to which these frames of reference are subject." In other words, various "realities" may be simultaneously occurring among participants engaged in the same set of activities. When people come together in a social movement, they assign meanings to their activities in such a way that they build a framework for interacting and socially constructing their grievances so that they can more effectively voice them and know what resolution they want for these issues.

Sociologists have identified at least three ways in which grievances are framed. First, *diagnostic framing* identifies a problem and attributes blame or causality to some group or entity so that the social movement has a target for its actions. Second, *prognostic framing* pinpoints possible solutions or remedies, based on the target previously identified. Third, *motivational framing* provides a vocabulary of motives that compel people to take action. When successful framing occurs, the individual's vague dissatisfactions are turned into well-defined grievances, and people are compelled to join the movement in an effort to reduce or eliminate those grievances.

Beyond motivational framing, additional frame alignment processes are necessary in order to supply a continuing sense of urgency to the movement. *Frame alignment* is the linking together of interpretive orientations of individuals and social movement organizations so that there is congruence between individuals' interests, beliefs, and values and the movement's

# SOCIOLOGY *in Global Perspective*

## Old Environmental Pollution with New Social Pressures in China

- Up to 40 percent of China's rivers are seriously polluted after 75 billion tons of sewage and wastewater were discharged into them.
- Twenty percent of rivers are so polluted that their water quality was rated too toxic for people to come into contact with the rivers' water.
- Nearly 300 million rural residents lack access to drinking water.
- About two-thirds of the cities in China lack sufficient water. (*People's Daily Online*, 2012)

For many years, environmental problems in China have been in the news around the world. The Internet and social media have intensified and greatly sped up this coverage and made environmental activists worldwide more aware of the problems faced by people in that rapidly growing region of the world. For example, it is a well-known fact that cancer is now the leading cause of death in China, partly because of air pollution, water pollution, and other environmental contaminants, some of which may be attributed to factories related to production of goods for high-income nations such as the United States (Larsen, 2011).

What, if anything, is being done about this situation? Some social protests and the beginnings of social movements

Workers in plants in China and other emerging nations typically do not have the same legal protection as U.S. workers. If U.S. companies with factories in these nations become an active force in requiring higher environmental standards, will this improve the quality of life in those nations as well?

are now being found in various provinces of China. Here are a few examples:

- Approximately 12,000 protesters demonstrated in Dalian, China, in 2011 after a storm damaged a paraxylene (PX)

ideologies, goals, and activities. Four distinct frame alignment processes occur in social movements: (1) *frame bridging* is the process by which movement organizations reach individuals who already share the same worldview as the organization, (2) *frame amplification* occurs when movements appeal to deeply held values and beliefs in the general population and link those to movement issues so that people's preexisting value commitments serve as a "hook" that can be used to recruit them, (3) *frame extension* occurs when movements enlarge the boundaries of an initial frame to incorporate other issues that appear to be of importance to potential participants, and (4) *frame transformation* refers to the process whereby the creation and maintenance of new values, beliefs, and meanings induce movement participation by redefining activities and events in such a manner that people believe they must become involved in collective action. Some or all of these frame alignment processes are used by social movements as they seek to define grievances and recruit participants.

Frame analysis provides new insights on how social movements emerge and grow when people are faced with problems such as technological disasters, about which greater ambiguity typically exists, and when people are attempting to "name" the problems associated with things such as nuclear or chemical contamination. However, frame analysis has been criticized for its "ideational biases" (McAdam, 1996). According to the sociologist Doug McAdam (1996), frame analyses of social movements have looked almost exclusively at ideas and their formal expression, whereas little attention has been paid to other significant factors, such as movement tactics, mobilizing structures, and changing political opportunities that influence the signifying work of movements. In this context, *political opportunity* means government structure, public policy, and political conditions that set the boundaries for change and political action. These boundaries are crucial variables in explaining why various social movements have different outcomes.

factory. PX is a toxic chemical used to make polyester. The protesters were mobilized by cell phones and the Internet. Officials announced that the plant would be closed, which was no small feat because $1.5 billion had been invested in a joint venture between a state-owned chemical company and a private developer (Economist.com, 2011).

- Apple announced in 2012 that it would allow independent environmental reviews of factories that supply parts for Apple products. Apple has faced rising criticism about toxic pollution and factory injuries of workers in suppliers' factories in China and other countries. Among the environmental problems cited are hazardous-waste leaks and the use of toxic chemicals that might create health risks not only for workers but also for neighboring communities. On its website, Apple stated that it insists that suppliers "provide safe working conditions, treat workers with dignity and respect, and use environmentally responsible manufacturing processes" (qtd. in Chu, 2012).

Are other Chinese activists and environmental movements trying to bring about environmental conservation in China? Some environmental activists are indeed attempting to highlight the causes and consequences of the various forms of pollution that are assaulting their nation. In the past, other social movement organizers in China found that they risked arrest and prosecution if they publicized their concerns and tried to mobilize others for causes. However, given the pressure that activists from China and other nations can bring to bear through communication on the Internet and Facebook,

Twitter, and a growing number of other social media sites, it appears that Chinese officials are concerned that another Arab Spring (the revolutionary wave of protests that occurred in Tunisia, Egypt, Yemen, Libya, and Syria in 2010–2011) might take place in China if sufficient numbers of people express their discontent about the current state of affairs in their nation.

What will the future hold for environmental protection in China? According to resource mobilization theory, widespread discontent alone cannot produce a social movement: Adequate resources and motivated people are essential for any concerned social action. Some analysts believe that environmental leaders will be able to mobilize people for change because cell phones, the Internet, and social media make it possible for people to organize quickly and demand governmental intervention on pressing problems. If companies such as Apple that do business in China become an active force in requiring environmental reviews of suppliers, this action might increase pressure for improvement by other Chinese suppliers and businesses. Even if this is just the tip of the iceberg, it is a start because many transnational companies currently have contract plants in China.

## reflect & analyze

Do you believe that U.S. companies can play a role in environmental movements in China? Should corporations also be concerned about environmental issues in the United States? Why or why not?

## Political Opportunity Theory

Why do social protests occur? According to political opportunity theorists, the origins of social protests cannot be explained solely by the fact that people possess a variety of grievances or that they have resources available for mobilization. Instead, social protests are directly related to the political opportunities that potential protesters and movement organizers believe exist within the political system at any given point in time. Political opportunity theory is based on the assumption that social protests that take place *outside* of mainstream political institutions are deeply intertwined with more-conventional political activities that take place *inside* these institutions. As used in this context, *opportunity* refers to "options for collective action, with chances and risks attached to them that depend on factors outside the mobilizing group" (Koopmans, 1999: 97). *Political opportunity theory* states that people will choose those options for collective action that are most readily available to

them and those options that will produce the most favorable outcome for their cause.

What are some specific applications of political action theory? Urban sociologists and social movement analysts have found that those cities that provided opportunities for people's protests to be heard within urban governments were less likely to have extensive protests or riots in their communities because aggrieved people could use more-conventional means to make their claims known. By contrast, urban riots were more likely to occur when activists believed that all conventional routes to protest were blocked. Changes in demography, migration, and the political economy in the United States (factors that were seemingly external to the civil rights movement) all contributed to a belief on the part of African Americans in the late 1960s and early 1970s that they could organize collective action and that their claims regarding the need for racial justice might be more readily heard by government officials.

Political opportunity theory has grown in popularity among sociologists who study social movements because this approach highlights the interplay of opportunity, mobilization, and political influence in determining when certain types of behavior may occur. However, like other perspectives, this theory has certain limitations, including the fact that social movement organizations may not always be not completely distinct from, or external to, the existing political system. For example, it is difficult to classify the Tea Party movement, which emerged in the aftermath of the election of President Barack Obama. Some supporters were outside the political mainstream and felt like they had no voice in what was happening in Washington. Keli Carender, who is credited with being one of the first Tea Party campaigners, complained that she tried to call her senators to urge them to vote against the $787 billion stimulus bill but constantly found that their mailboxes were full. As a result, she decided to protest against "porkulus"; in her words, "I basically thought to myself: 'I have two courses. I can give up, go home, crawl into bed and be really depressed and let it happen, or I can do something different, and I can find a new avenue to have my voice get out'" (qtd. in Zernike, 2010: A1).

By contrast, other active supporters of the Tea Party movement are players in the mainstream political process. An example is Sarah Palin, the former Alaska governor and Republican vice presidential candidate, who was a frequent spokesperson at Tea Party rallies across the United States. In political movements, social activists typically *create* their own opportunities rather than wait for them to emerge, and activists are often political entrepreneurs in their own right, much like the state and federal legislators and other governmental officials whom they seek to influence on behalf of their social cause. Political opportunity theory calls our attention to how important the degree of openness of a political system is to the goals and tactics of persons who organize social movements.

## New Social Movement Theory

*New social movement theory* looks at a diverse array of collective actions and the manner in which those actions are based on politics, ideology, and culture. It also incorporates factors of identity, including race, class, gender, and sexuality, as sources of collective action and social movements. Examples of "new social movements" include ecofeminism and environmental justice movements.

Ecofeminism emerged in the late 1970s and early 1980s out of the feminist, peace, and ecology movements. Prompted by the near-meltdown at the Three Mile Island nuclear power plant, ecofeminists established World Women in Defense of the Environment. *Ecofeminism* is based on the belief that patriarchy is a root cause of environmental problems. According to ecofeminists, patriarchy not only results in the domination of women by men but also contributes to a belief that nature is to be possessed and dominated, rather than treated as a partner.

Another "new social movement" focuses on environmental justice and the intersection of race and class in the environmental struggle (see "Sociology Works!"). Sociologist Stella M. Capek (1993) investigated a contaminated landfill in the Carver Terrace neighborhood of Texarkana, Texas, and found that residents were able to mobilize for change and win a federal buyout and relocation by symbolically linking their issue to a larger *environmental justice* framework. Since the 1980s, the emerging environmental justice movement has focused on the issue of **environmental racism—the belief that a disproportionate number of hazardous facilities (including industries such as waste disposal/treatment and chemical plants) are placed in low-income areas populated primarily by people of color** (Figure 16.10) (Bullard and Wright, 1992). These areas have been left out of most of the environmental cleanup that has taken place. Capek concludes that linking Carver Terrace with environmental justice led to it being designated as a cleanup site. She also views this as an important turning point in new social movements: "Carver Terrace is significant not only as a federal buyout and relocation of a minority community, but also as a marker of the emergence of environmental racism as a major new component of environmental social movements in the United States" (Capek, 1993: 21).

Sociologist Steven M. Buechler (2000: 11) has argued that theories pertaining to twenty-first-century social movements should be oriented toward the structural, macrolevel contexts in which movements arise. These theories should incorporate both political and cultural dimensions of social activism:

Social movements are historical products of the age of modernity. They arose as part of a sweeping social, political, and intellectual change that led a significant number of people to view society as a social construction that was susceptible to social reconstruction through concerted collective effort. Thus, from their inception, social movements have had a dual focus. Reflecting the political, they have always involved some form of challenge to prevailing forms of authority. Reflecting the cultural, they have always operated as symbolic

# Sociology WORKS!

## Update on Theories and Data Gathering on Environmental Racism

Throughout *Sociology in Our Times*, we have examined social theories that help us understand the interplay of factors such as race, class, and gender in the everyday lives of millions of people. In the "Sociology Works!" feature, we have focused on specific theories and how applications of those theories can help us understand the world and sometimes make it a better place in which to live.

In this chapter we have looked at the work of new social movement theorists who have demonstrated the intersection of environmental justice with race and class: the belief that hazardous-waste treatment, storage, and disposal facilities are more likely to be located near low-income, nonwhite neighborhoods than near higher-income, predominantly white neighborhoods. This is an important issue because of the potential health risks that such sites may pose for people who live nearby. However, critics have scoffed at the suggestion that race- or class-based discrimination is involved in decisions about where hazardous-waste-materials facilities are located. Can more-accurate data be gathered to help determine the nature and extent to which environmental racism exists?

In the past, national-level studies documenting the location of waste sites and other polluting industrial facilities often used "unit-hazard coincidence" methodology. Based on this approach, researchers selected a predefined geographic unit (such as certain ZIP code areas or census tracts). Then they identified subsets of the units that had, or did not have, the hazard present. The researchers then compared the demographic characteristics of people living within each of the subsets to see if a larger minority population was present near the hazardous facility (see Mohai and Saha, 2007). Unit-hazard coincidence methodology assumes that the people who live within the predefined geographic units included in a study are located closer to the hazard than those individuals who do not live in those geographic units (Mohai and Saha, 2007). The problem with this approach is that the hazardous site is usually not located at the center of the ZIP code or census tract and that the geographic area being examined may be large or small, making it difficult to know for sure the racial and class characteristics of the people who live the closest to the waste facility.

Today, sociologists and other social scientists use GIS (a computer system for capturing, storing, checking, integrating, manipulating, analyzing, and displaying data related to positions on the Earth's surface) to more adequately determine the distance between environmentally hazardous sites and nearby populations. By using distance-based methods to control for proximity around environmentally hazardous sites, those researchers have demonstrated that nonwhites constitute a large percentage of the population living within one mile of hazardous-waste facilities, meaning that racial disparities in the distribution of hazardous sites are much greater than what had been previously reported. According to social scientists Paul Mohai and Robin Saha (2007: 343), "We [find that] these disparities persist even when controlling for economic and sociopolitical variables, suggesting that factors uniquely associated with race, such as racial targeting, housing discrimination, or other race-related factors, are associated with the location of the nation's hazardous waste facilities."

Sociological theories and research pertaining to environmental racism have raised public awareness that the location of hazardous facilities is not purely coincidental in communities throughout our nation. Clearly, proximity to hazardous sites is related to the cost of the land on which the facilities are located, but the issue of proximity based on the racial–ethnic composition of residents raises an even more challenging social and ethical dilemma. But it is also clear that the vast quantity of data available today—and the methods for obtaining those data—make it possible for us to fine-tune previous theories and obtain a better understanding of the social world in which we live.

## reflect & analyze

New technology can improve people's lives but also cause problems for society—for example, a nuclear power plant can produce energy but also be dangerous in a tsunami. How does the location of a new nuclear power plant or other technology affect people based on race and class?

---

laboratories in which reflexive actors pose questions of meaning, purpose, identity, and change.

As we have seen, social movements may be an important source of social change. Throughout this text we have examined a variety of social problems that have been the focus of one or more social movements. For this reason, many groups focus on preserving their gains while simultaneously fighting for changes that they believe are still necessary.

This chapter's Concept Quick Review summarizes the main theories of social movements.

**environmental racism** the belief that a disproportionate number of hazardous facilities (including industries such as waste disposal/treatment and chemical plants) are placed in low-income areas populated primarily by people of color.

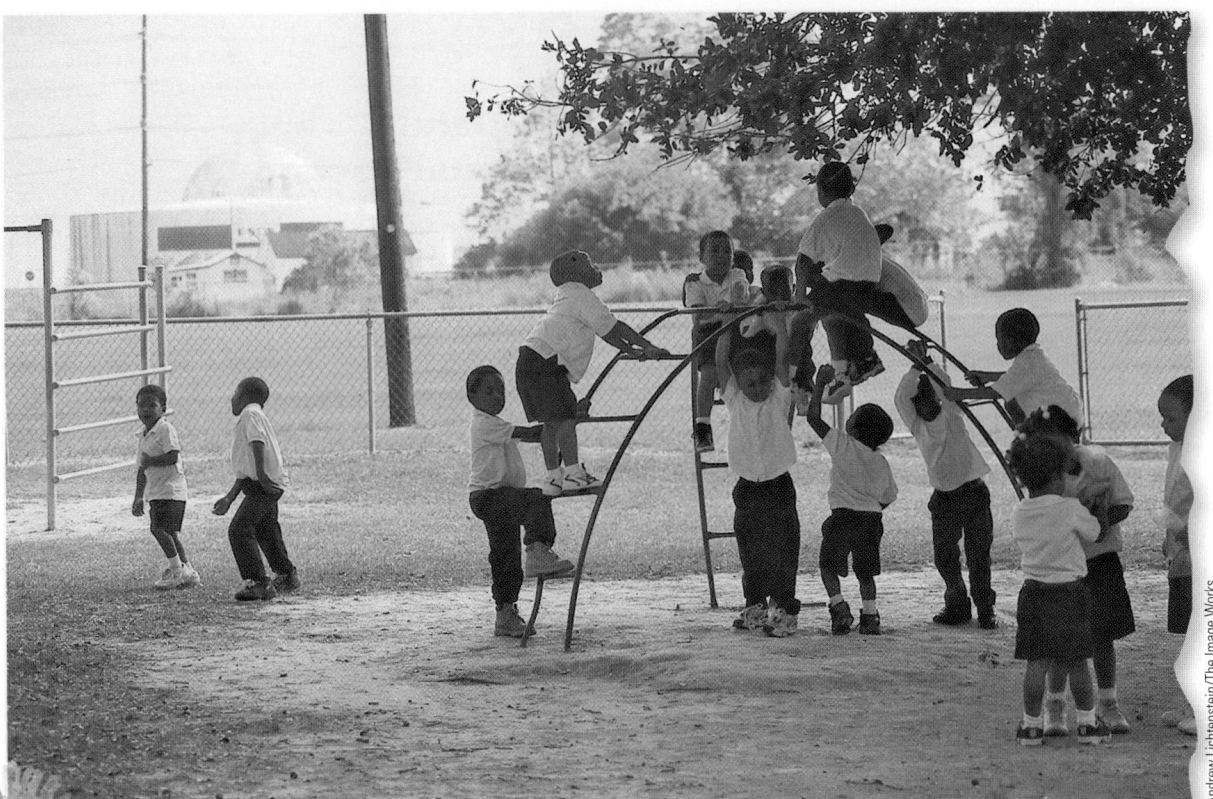

Andrew Lichtenstein/The Image Works

**FIGURE 16.10** Referred to as "Cancer Alley," this area of Baton Rouge, Louisiana, is home to a predominantly African American population and also to many refineries that heavily pollute the region. Sociologists suggest that environmental racism is a significant problem in the United States and other nations. What do you think?

## [ concept quick review ]

### Social Movement Theories

| | Key Components |
|---|---|
| **Relative Deprivation** | People who are discontent when they compare their achievements with those of others consider themselves relatively deprived and join social movements in order to get what they view as their "fair share," especially when there is an upswing in the economy followed by a decline. |
| **Value-Added** | Certain conditions are necessary for a social movement to develop: (1) structural conduciveness, such that people are aware of a problem and have the opportunity to engage in collective action; (2) structural strain, such that society or the community cannot meet people's expectations for taking care of the problem; (3) growth and spread of a generalized belief about causes and effects of and possible solutions to the problem; (4) precipitating factors, or events that reinforce the beliefs; (5) mobilization of participants for action; and (6) social control factors, such that society comes to allow the movement to take action. |
| **Resource Mobilization** | A variety of resources (money, members, access to media, and material goods such as equipment) are necessary for a social movement; people participate only when they believe that the movement has access to these resources. |
| **Social Constructionist Theory: Frame Analysis** | Based on the assumption that social movements are an interactive, symbolically defined, and negotiated process involving participants, opponents, and bystanders, frame analysis is used to determine how people assign meaning to activities and processes in social movements. |
| **Political Opportunity** | People will choose the options for collective action (i.e., "opportunities") that are most readily available to them and those options that will produce the most favorable outcome for their cause. |
| **New Social Movement** | The focus is on sources of social movements, including politics, ideology, and culture. Race, class, gender, sexuality, and other sources of identity are also factors in movements such as ecofeminism and environmental justice. |

# Social Change in the Future

In this chapter we have focused on collective behavior and social movements as potential forces for social change in contemporary societies. For example, the Love Canal disaster and the social movement it produced were, to a large extent, responsible for passage of the Comprehensive Environmental Response, Compensation, and Liability Act, which led to the formation of the Agency for Toxic Substances and Disease Registry. The BP oil spill and the collective behavior following the disaster were responsible for new legislation and guidelines for offshore drilling and emergency preparedness by corporations and the federal government. Will Occupy Wall Street and similar inequality-based protests generate long-term effects in our society? It is too soon to tell because a number of other factors also contribute to social change, including the physical environment, population trends, technological development, and social institutions.

## The Physical Environment and Change

Changes in the physical environment often produce changes in the lives of people (Figure 16.11); in turn, people can make dramatic changes in the physical environment, over which we have only limited control. Throughout history, natural disasters have taken their toll on individuals and societies. Major natural disasters—including tsunamis/hurricanes, floods, tornadoes, and earthquakes—can devastate an entire population. In the twenty-first century, earthquakes have affected India, Pakistan, El Salvador, Iran, China, Italy, Haiti, Chile, New Zealand, Japan, and the United States, among other nations. Hurricanes, tsunamis, and floods have devastated portions of Pakistan, Australia, and the United States. Tornadoes have plagued the United States, with one of the most devastating being the 2013 EF5 tornado that hit the city of Moore, south of Oklahoma City, with winds of 215 miles per hour that left a path of destruction more than a mile wide. Even comparatively "small" natural disasters change the lives of many people. As the sociologist Kai Erikson (1976, 1994) has suggested, the trauma that people experience from disasters may outweigh the actual loss of physical property—memories of such events can haunt people for many years.

Some natural disasters are exacerbated by human decisions. For example, floods are viewed as natural disasters, but excessive development may contribute to a flood's severity. As office buildings, shopping malls, industrial plants, residential areas, and highways are developed, less land remains as ground-cover to absorb rainfall. When heavier-than-usual rains occur, flooding becomes inevitable; some regions of the United States—such as in and around New Orleans—had excessive water on the streets for days or even weeks after Hurricane Katrina, in 2005. Clearly, humans cannot control the rain, but human decisions can worsen the consequences. If Hurricane Katrina's first wave was the storm itself, the second wave was a *human-made disaster* resulting in part from decisions related to planning and budgetary priorities, allocation of funds for maintaining infrastructure, and the importance of emergency preparedness.

*Infrastructure* refers to a framework of systems, such as transportation and utilities, that makes it possible to have specific land uses (commercial, residential, and recreational, for example) and a built environment (buildings, houses, and highways) that support people's daily activities and the nation's economy. It takes money and commitment to make sure that the components of the infrastructure remain strong so that cities can withstand natural disasters and other concerns such as climate change. For example, consider that the city of Chicago has taken a proactive stance on dealing with future increases in temperature and climate conditions that will make Chicago's weather feel

**FIGURE 16.11** Changes in the physical environment often produce changes in the lives of people, a factor that can contribute to social change. What changes may hurricanes such as Hurricane Sandy bring about?

© Leonard Zhukovsky/Shutterstock.com

more like Baton Rouge, Louisiana, than a northern city. To cope with this change, Chicago city officials are already repaving alleyways with water-permeable materials and planting swamp oak and sweet gum trees from the South rather than the more indigenous white oak, which is the state tree of Illinois (Kaufman, 2011). Long-range planning such as this to cope with changes in the physical environment may seem far-fetched, but when the time comes, between 50 and 100 years from now, such endeavors may seem farsighted instead. In the words of one Chicago city official, "Cities adapt or they go away" (qtd. in Kaufman, 2011).

The changing environment is one of many reasons why experts are also concerned about availability of water in the future. Water is a finite resource that is necessary for both human survival and the production of goods. However, water is being wasted and polluted, and the supply of *potable* (drinkable) water is limited. People are causing—or at least contributing to—that problem.

People also contribute to changes in the Earth's physical condition. Through soil erosion and other degradation of grazing land, often at the hands of people, more than 25 billion tons of topsoil is lost annually. As people clear forests to create farmland and pastures and to acquire lumber and firewood, the Earth's tree cover continues to diminish. As millions of people drive motor vehicles, the amount of carbon dioxide in the environment continues to rise each year, contributing to global warming.

Just as people contribute to changes in the physical environment, human activities must also be adapted to changes in the environment. For example, we are being warned to stay out of the sunlight because of increases in ultraviolet rays, a cause of skin cancer, as a result of the accelerating depletion of the ozone layer. If the ozone warnings are accurate, the change in the physical environment will dramatically affect those who work or spend their leisure time outside.

## Population and Change

Changes in population size, distribution, and composition affect the culture and social structure of a society and change the relationships among nations. As discussed in Chapter 15, the countries experiencing the most-rapid increases in population have a less developed infrastructure to deal with those changes. How will nations of the world deal with population growth as the global population continues to move toward eight billion? Only time will provide a response to this question.

In the United States a shift in population distribution from central cities to suburban and exurban areas has produced other dramatic changes. In some areas, central cities have experienced a shrinking tax base as middle-income and upper-middle-income residents and businesses have moved to suburban and outlying areas. In other areas, the largest metropolitan areas have become enclaves for the wealthiest transnational residents who can afford to own homes in many of the world's most expensive cities. Likewise, some suburban areas thrive with McMansion-style residential properties and security-gated communities, while other suburbs have run-down housing, high rates of poverty, and problems similar to the less-fortunate central cities, including decaying infrastructure, low-performing schools, inadequate transportation, and rising rates of crime, to name only a few concerns. The changing composition of the U.S. population has resulted in children from more-diverse cultural backgrounds entering school, producing a demand for new programs and changes in curricula. An increase in the number of single mothers and of women employed outside the household has created a need for more child care; an increase in the older population has created a need for services such as medical care and placed greater stress on programs such as Social Security. Population growth will affect many regions of the country and intensify existing social problems.

## Technology and Change

Technology is an important force for change; in some ways, technological development has made our lives much easier. Advances in communication and transportation have made instantaneous worldwide communication possible but have also brought old belief systems and the status quo into question as never before. Today, we are increasingly moving information instead of people—and doing it instantly. Advances in science and medicine have made significant changes in people's lives in high-income countries.

Scientific advances will continue to affect our lives, from the foods we eat to our reproductive capabilities. Biotechnology is the process by which organisms or their components, such as enzymes, are used to make products such as yogurt, cheese, beer, and wine. Genetically modified products today include medicines and vaccines, as well as foods and food ingredients. Of course, these new technologies are not without controversy; they pose some risks (genomics.energy .gov, 2012). Advances in medicine have made it possible for those formerly unable to have children to procreate; women well beyond menopause are now able to become pregnant with the assistance of medical technology (Figure 16.12). Advances in medicine have also increased the human life span, especially for white and middle- or upper-class individuals in high-income nations; medical advances have also contributed to the declining death rate in low-income nations, where birth rates have not yet been curbed.

Just as technology has brought about improvements in the quality and length of life for many, it has also created the potential for new disasters, ranging from global warfare to localized technological disasters at toxic waste sites. As the sociologist William Ogburn (1966) suggested, when a change in the material culture occurs in society, a period of *cultural lag* follows in which the nonmaterial (ideological) culture has not yet caught up with material development. The rate of technological advancement at the level of material culture today is mind-boggling. Many of us can never hope to understand technological advances in the areas of artificial intelligence, holography, virtual reality, biotechnology, cold fusion, and robotics.

One of the ironies of twenty-first-century high technology is the increased vulnerability that results from the increasing complexity of such systems. We have already seen this in situations ranging from jetliners that are used as terrorist weapons to identity theft and fraud on the Internet.

**FIGURE 16.12** How might advances in medicine—such as the ability of postmenopausal women to bear children—create social change?

Cultura Creative/Alamy

## Social Institutions and Change

Many changes occurred in the family, religion, education, the economy, and the political system during the twentieth century and early in the twenty-first century. The size and composition of families in the United States changed with the dramatic increase in the number of single-person and single-parent households. Changes in families produced changes in the socialization of children, many of whom spend large amounts of time playing video games, texting friends, posting their daily activities on Facebook or Twitter, or spending time in child-care facilities outside their own home. Although some political and religious leaders advocate a return to "traditional" family life, numerous scholars argue that such families never worked quite as well as some might wish to believe.

Public education changed dramatically in the United States during the last century. This country was one of the first to provide "universal" education for students regardless of their ability to pay. As a result, at least until recently, the United States has had one of the most highly educated populations in the world. Today, the United States still has one of the best public education systems in the world for the top 15 percent of the students, but it badly fails the bottom 25 percent. As the nature of the economy changes, schools almost inevitably will have to change, if for no other reason than demands from leaders in business and industry for an educated workforce that allows U.S. companies to compete in a global economic environment. Many business and political leaders believe that education is the single most important factor in the future of the United States. However, in difficult economic times when local, state, and federal budgets are strained, public education is one of the first institutions to undergo the axe as teachers are let go, school buildings are allowed to further decay, and students are not provided with the necessary physical setting and learning tools.

Although we have examined changes in the physical environment, population, technology, and social institutions separately, they all operate together in a complex relationship, sometimes producing large, unanticipated consequences (Figure 16.13). As we move further into the twenty-first century, we need new ways of conceptualizing social life at both the macrolevel and the microlevel. The sociological imagination helps us think about how personal troubles—regardless of our race, class, gender, age, sexual orientation, or physical abilities and disabilities—are intertwined with the public issues of our society and the global community of which we are a part. As one analyst noted regarding Lois Gibbs and Love Canal,

If Love Canal has taught Lois Gibbs—and the rest of us—anything, it is that ordinary people become very smart very quickly when their lives are threatened. They become adept at detecting absurdity, even when it is concealed in bureaucratese and scientific jargon. Lois Gibbs learned

**531**

**FIGURE 16.13** Pollution of lakes, rivers, and other bodies of water has an adverse effect on food supplies, air quality, and the entire environment. What influence does a "business as usual" approach have on environmental quality in your area?

Brent Beach/Alamy

that one cannot always rely on government to act in the best interests of ordinary citizens—at least, not without considerable prodding. She determined that she would prod them until her objectives were attained. She led one of the most successful, single-purpose grassroots efforts of our time. (Levine, 1982: xv)

And, as we learned in the narrative that opened this chapter, vigilance and persistence are important when individuals or groups are trying to bring about positive change. After one problem appears to be alleviated, additional problems crop up, as has been the case for Lois Gibbs throughout her career as an environmental activist and social movement organizer, and she has continued to rise to the challenge.

## A Few Final Thoughts

In this text we have covered a substantial amount of material, examined different perspectives on a wide variety of social issues, and suggested different methods by which to deal with them. The purpose of this text is not to encourage you to take any particular point of view; rather, it is to allow you to understand different viewpoints and ways in which they may be helpful to you and to society in dealing with the issues of the twenty-first century. Possessing that understanding, we can hope that the future will be something we can all look forward to—producing a better way of life, not only in this country but worldwide as well.

# chapter REVIEW

## LO1 What is collective behavior, and what factors contribute to it?

Social change—the alteration, modification, or transformation of public policy, culture, or social institutions over time—is usually brought about by *collective behavior*, which is defined as relatively spontaneous, unstructured activity that typically violates established social norms. Collective behavior occurs when some common influence or stimulus produces a response from a relatively large number of people.

## LO2 What are the most common types of crowd behavior?

A crowd is a relatively large number of people in one another's immediate presence. Sociologist Herbert Blumer divided crowds into four categories:

(1) casual crowds, (2) conventional crowds, (3) expressive crowds, and (4) acting crowds (including mobs, riots, and panics). A fifth type of crowd is a protest crowd.

## LO3 What are some explanations of crowd behavior?

Social scientists have developed several theories to explain crowd behavior. Contagion theory asserts that a crowd takes on a life of its own as people are transformed from rational beings into part of an organism that acts on its own. A variation on this is social unrest and circular reaction—people express their discontent to others, who communicate back similar feelings, resulting in a conscious effort to engage in the crowd's behavior. Convergence theory asserts that people with similar attributes find other like-minded persons with whom they can release underlying personal tendencies. Emergent norm theory asserts that as a crowd develops, it comes up with its own norms that replace more-conventional norms of behavior.

## LO4 How does mass behavior differ from other forms of collective behavior?

Mass behavior is collective behavior that occurs when people respond to the same event in the same way even if they are not in geographic proximity to one another. Rumors, gossip, mass hysteria, fads and fashions, and public opinion are forms of mass behavior.

## LO5 How do fads and fashions compare, and how does the "trickle-down" approach explain the origin of contemporary fashions?

A fad is a temporary but widely copied activity enthusiastically followed by large numbers of people. Fashion is defined as a currently valued style of behavior, thinking, or appearance. Trickle-down theory describes the process by which members of the lower classes emulate the fashions of the upper class.

## LO6 What are the major types of social movements?

A social movement is an organized group that acts consciously to promote or resist change through collective action. Reform, revolutionary, religious, and alternative movements are the major types identified by sociologists. Reform movements seek to improve society by changing some specific aspect of the social structure. Revolutionary movements seek to bring about a total change in society—sometimes by the use of terrorism. Religious movements seek to produce radical change in individuals based on spiritual or supernatural belief systems. Alternative movements seek limited change of some aspect of people's behavior. Resistance movements seek to prevent change or to undo change that has already occurred.

## LO7 How do relative deprivation theory, value-added theory, and resource mobilization theory explain social movements?

Relative deprivation theory asserts that if people are discontent when they compare their accomplishments with those of others similarly situated, they are more likely to join a social movement than are people who are relatively content with their status. According to value-added theory, six conditions are required for a social movement: (1) a perceived problem, (2) a perception that the authorities are not resolving the problem, (3) a spread of the belief to an adequate number of people, (4) a precipitating incident, (5) mobilization of other people by leaders, and (6) a lack of social control. By contrast, resource mobilization theory asserts that successful social movements can occur only when they gain the support of political and economic elites, who provide access to the resources necessary to maintain the movement.

## LO8 How does research based on social constructionist theory—frame analysis, political opportunity theory, and new social movement theory—explain the ways in which people are drawn into social movements?

Research based on social constructionist theory: frame analysis often focuses on how people use social interactions to socially construct their grievances and to determine what they think should be done to resolve them. By contrast, research based on political opportunity theory identifies reasons why social protests are most likely to occur when potential protesters and movement organizers perceive that limited opportunities exist for them within the political system. Finally, research based on new social movement theory looks at various identity factors of potential protesters, including their race, class, gender, and sexuality, to help explain why collective action and social movements emerge (for example, people who believe that they have experienced environmental racism because of their race/ethnicity and class).

## key TERMS

civil disobedience 514

collective behavior 509

crowd 512

environmental racism 526

gossip 516

mass 512

mass behavior 516

mob 513

panic 513

propaganda 518

public opinion 518

riot 513

rumor 516

social change 508

social movement 519

# questions for CRITICAL THINKING

1. What types of collective behavior in the United States do you believe are influenced by inequalities based on race/ethnicity, class, gender, age, or disabilities? Why?

2. Which of the four explanations of crowd behavior (contagion theory, social unrest and circular reaction, convergence theory, and emergent norm theory) do you believe best explains crowd behavior? Why?

3. In the text the Love Canal environmental movement is analyzed in terms of the value-added theory. How would you analyze that movement under (a) the relative deprivation theory and (b) the resource mobilization theory?

4. Using the sociological imagination that you have gained in this course, what are some positive steps that you believe might be taken in the United States to make our society a better place for everyone? What types of collective behavior and/or social movements might be required in order to take those steps?

# Glossary

**absolute poverty** a level of economic deprivation that exists when people do not have the means to secure the most basic necessities of life.

**achieved status** a social position that a person assumes voluntarily as a result of personal choice, merit, or direct effort.

**acute diseases** illnesses that strike suddenly and cause dramatic incapacitation and sometimes death.

**ageism** prejudice and discrimination against people on the basis of age, particularly against older persons.

**agents of socialization** the persons, groups, or institutions that teach us what we need to know in order to participate in society.

**aggregate** a collection of people who happen to be in the same place at the same time but share little else in common.

**agrarian societies** societies that use the technology of large-scale farming, including animal-drawn or energy-powered plows and equipment, to produce their food supply.

**alienation** a feeling of powerlessness and estrangement from other people and from oneself.

**animism** the belief that plants, animals, or other elements of the natural world are endowed with spirits or life forces having an effect on events in society.

**anomie** Emile Durkheim's term for a condition in which social control becomes ineffective as a result of the loss of shared values and of a sense of purpose in society.

**anticipatory socialization** the process by which knowledge and skills are learned for future roles.

**ascribed status** a social position conferred at birth or received involuntarily later in life, based on attributes over which the individual has little or no control, such as race/ethnicity, age, and gender.

**assimilation** a process by which members of subordinate racial and ethnic groups become absorbed into the dominant culture.

**authoritarian leaders** leaders who make all major group decisions and assign tasks to members.

**authoritarian personality** a personality type characterized by excessive conformity, submissiveness to authority, intolerance, insecurity, a high level of superstition, and rigid, stereotypic thinking.

**authoritarianism** a political system controlled by rulers who deny popular participation in government.

**authority** power that people accept as legitimate rather than coercive.

**beliefs** the mental acceptance or conviction that certain things are true or real.

**bilateral descent** a system of tracing descent through both the mother's and father's sides of the family.

**blended family** a family consisting of a husband and wife, children from previous marriages, and children (if any) from the new marriage.

**body consciousness** how a person perceives and feels about his or her body.

**bureaucracy** an organizational model characterized by a hierarchy of authority, a clear division of labor, explicit rules and procedures, and impersonality in personnel matters.

**bureaucratic personality** a psychological construct that describes those workers who are more concerned with following correct procedures than they are with getting the job done correctly.

**capitalism** an economic system characterized by private ownership of the means of production, from which personal profits can be derived through market competition and without government intervention.

**capitalist class (or *bourgeoisie*)** Karl Marx's term for those who own the means of production.

**caste system** a system of social inequality in which people's status is permanently determined at birth based on their parents' ascribed characteristics.

**category** a number of people who may never have met one another but share a similar characteristic, such as education level, age, race, or gender.

**charismatic authority** power legitimized on the basis of a leader's exceptional personal qualities or the demonstration of extraordinary insight and accomplishment that inspire loyalty and obedience from followers.

**chronic diseases** illnesses that are long term or lifelong and that develop gradually or are present from birth.

**church** a large, bureaucratically organized religious organization that tends to seek accommodation with the larger society in order to maintain some degree of control over it.

**civil disobedience** nonviolent action that seeks to change a policy or law by refusing to comply with it.

**civil religion** the set of beliefs, rituals, and symbols that makes sacred the values of the society and places the nation in the context of the ultimate system of meaning.

**class conflict** Karl Marx's term for the struggle between the capitalist class and the working class.

**class system** a type of stratification based on the ownership and control of resources and on the type of work that people do.

**cohabitation** a situation in which two people live together, and think of themselves as a couple, without being legally married.

**collective behavior** voluntary, often spontaneous activity that is engaged in by a large number of people and typically violates dominant-group norms and values.

**comparable worth (or *pay equity*)** the belief that wages ought to reflect the worth of a job, not the gender or race of the worker.

**conflict perspectives** the sociological approach that views groups in society as engaged in a continuous power struggle for control of scarce resources.

**conformity** the process of maintaining or changing behavior to comply with the norms established by a society, subculture, or other group.

**conglomerate**   a combination of businesses in different commercial areas, all of which are owned by one holding company.

**content analysis**   the systematic examination of cultural artifacts or various forms of communication to extract thematic data and draw conclusions about social life.

**contingent work**   part-time work, temporary work, or subcontracted work that offers advantages to employers but that can be detrimental to the welfare of workers.

**control group**   the group that contains the subjects who are not exposed to the independent variable.

**core nations**   according to world systems theory, nations that are dominant capitalist centers characterized by high levels of industrialization and urbanization.

**corporate crime**   illegal acts committed by corporate employees on behalf of the corporation and with its support.

**corporations**   organizations that have legal powers, such as the ability to enter into contracts and buy and sell property, separate from their individual owners.

**correlation**   a relationship that exists when two variables are associated more frequently than could be expected by chance.

**counterculture**   a group that strongly rejects dominant societal values and norms and seeks alternative lifestyles.

**credentialism**   a process of social selection in which class advantage and social status are linked to the possession of academic qualifications.

**crime**   behavior that violates criminal law and is punishable with fines, jail terms, and/or other negative sanctions.

**criminal justice system**   the local, state, and federal agencies that enforce laws, adjudicate crimes, and treat and rehabilitate criminals.

**criminology**   the systematic study of crime and the criminal justice system, including the police, courts, and prisons.

**cross-dresser**   a male who dresses as a woman or a female who dresses as a man but does not alter his or her genitalia.

**crowd**   a relatively large number of people who are in one another's immediate vicinity.

**crude birth rate**   the number of live births per 1,000 people in a population in a given year.

**crude death rate**   the number of deaths per 1,000 people in a population in a given year.

**cult**   a loosely organized religious group with practices and teachings outside the dominant cultural and religious traditions of a society.

**cultural capital**   Pierre Bourdieu's term for people's social assets, including values, beliefs, attitudes, and competencies in language and culture.

**cultural imperialism**   the extensive infusion of one nation's culture into other nations.

**cultural lag**   William Ogburn's term for a gap between the technical development of a society (material culture) and its moral and legal institutions (nonmaterial culture).

**cultural relativism**   the belief that the behaviors and customs of any culture must be viewed and analyzed by the culture's own standards.

**cultural universals**   customs and practices that occur across all societies.

**culture**   the knowledge, language, values, customs, and material objects that are passed from person to person and from one generation to the next in a human group or society.

**culture shock**   the disorientation that people feel when they encounter cultures radically different from their own and believe they cannot depend on their own taken-for-granted assumptions about life.

**deinstitutionalization**   the practice of rapidly discharging patients from mental hospitals into the community.

**demedicalization**   the process whereby a problem ceases to be defined as an illness or a disorder.

**democracy**   a political system in which the people hold the ruling power either directly or through elected representatives.

**democratic leaders**   leaders who encourage group discussion and decision making through consensus building.

**democratic socialism**   an economic and political system that combines private ownership of some of the means of production, governmental distribution of some essential goods and services, and free elections.

**demographic transition**   the process by which some societies have moved

from high birth and death rates to relatively low birth and death rates as a result of technological development.

**demography**   a subfield of sociology that examines population size, composition, and distribution.

**denomination**   a large, organized religion characterized by accommodation to society but frequently lacking in the ability or intention to dominate society.

**dependency theory**   the belief that global poverty can at least partially be attributed to the fact that the low-income countries have been exploited by the high-income countries.

**dependent variable**   in an experiment, the variable assumed to be caused by the independent variable(s).

**deviance**   any behavior, belief, or condition that violates significant social norms in the society or group in which it occurs.

**differential association theory**   the proposition that people have a greater tendency to deviate from societal norms when they frequently associate with persons who are more favorable toward deviance than conformity.

**disability**   a physical or mental impairment that substantially limits one or more major activities that a person would normally do at a given stage of life and that may result in stigmatization or discrimination against the person with the disability.

**discrimination**   actions or practices of dominant-group members (or their representatives) that have a harmful effect on members of a subordinate group.

**division of labor**   how the various tasks of a society are divided up and performed.

**domestic partnerships**   household partnerships in which an unmarried couple lives together in a committed, sexually intimate relationship and is granted some of the same rights and benefits as those accorded to married heterosexual couples.

**dominant group**   a racial or ethnic group that has the greatest power and resources in a society.

**dramaturgical analysis**   Erving Goffman's term for the study of social interaction that compares everyday life to a theatrical presentation.

**drug**   any substance—other than food and water—that, when taken into the body, alters its functioning in some way.

**dual-earner marriages**   marriages in which both spouses are in the labor force.

**dyad**   a group composed of two members.

**ecclesia** a religious organization that is so integrated into the dominant culture that it claims as its membership all members of a society.

**economy** the social institution that ensures the maintenance of society through the production, distribution, and consumption of goods and services.

**education** the social institution responsible for the systematic transmission of knowledge, skills, and cultural values within a formally organized structure.

**egalitarian family** a family structure in which both partners share power and authority equally.

**ego** Sigmund Freud's term for the rational, reality-oriented component of personality that imposes restrictions on the innate pleasure-seeking drives of the id.

**elite model** an analysis of political systems that views power in political systems as being concentrated in the hands of a small group of elites, and the masses are relatively powerless.

**endogamy** the practice of marrying within one's own group.

**environmental racism** the belief that a disproportionate number of hazardous facilities (including industries such as waste disposal/treatment and chemical plants) are placed in low-income areas populated primarily by people of color.

**ethnic group** a collection of people distinguished, by others or by themselves, primarily on the basis of cultural or nationality characteristics.

**ethnic pluralism** the coexistence of a variety of distinct racial and ethnic groups within one society.

**ethnocentrism** the practice of judging all other cultures by one's own culture.

**ethnography** a detailed study of the life and activities of a group of people by researchers who may live with that group over a period of years.

**ethnomethodology** the study of the commonsense knowledge that people use to understand the situations in which they find themselves.

**exogamy** the practice of marrying outside one's own group.

**experiment** a carefully designed situation in which the researcher studies the impact of certain variables on subjects' attitudes or behavior.

**experimental group** the group that contains the subjects who are exposed to an independent variable (the experimental condition) to study its effect on them.

**expressive leadership** leadership that provides emotional support for members.

**extended family** a family unit composed of relatives in addition to parents and children who live in the same household.

**face-saving behavior** Erving Goffman's term for the strategies people use to rescue their performance when they experience a potential or actual loss of face.

**families** relationships in which people live together with commitment, form an economic unit and care for any young, and consider their identity to be significantly attached to the group.

**family of orientation** the family into which a person is born and in which early socialization usually takes place.

**family of procreation** the family that a person forms by having or adopting children.

**feminism** the belief that men and women are equal and should be valued equally and have equal rights.

**feminization of poverty** the trend in which women are disproportionately represented among individuals living in poverty.

**fertility** the actual level of childbearing for an individual or a population.

**folkways** informal norms or everyday customs that may be violated without serious consequences within a particular culture.

**formal organization** a highly structured group formed for the purpose of completing certain tasks or achieving specific goals.

**functionalist perspectives** the sociological approach that views society as a stable, orderly system.

**fundamentalism** a traditional religious doctrine that is conservative, is typically opposed to modernity, and rejects "worldly pleasures" in favor of otherworldly spirituality.

***Gemeinschaft*** **(guh-MINE-shoft)** a traditional society in which social relationships are based on personal bonds of friendship and kinship and on intergenerational stability.

**gender** the culturally and socially constructed differences between females and males found in the meanings, beliefs, and practices associated with "femininity" and "masculinity."

**gender bias** behavior that shows favoritism toward one gender over the other.

**gender identity** a person's perception of the self as female or male.

**gender role** the attitudes, behavior, and activities that are socially defined as appropriate for each sex and are learned through the socialization process.

**gender socialization** the aspect of socialization that contains specific messages and practices concerning the nature of being female or male in a specific group or society.

**gendered racism** the interactive effect of racism and sexism on the exploitation of women of color.

**generalized other** George Herbert Mead's term for the child's awareness of the demands and expectations of the society as a whole or of the child's subculture.

**genocide** the deliberate, systematic killing of an entire people or nation.

**gentrification** the process by which members of the middle and upper-middle classes, especially whites, move into a central-city area and renovate existing properties.

***Gesellschaft*** **(guh-ZELL-shoft)** a large urban society in which social bonds are based on impersonal and specialized relationships, with little long-term commitment to the group or consensus on values.

**global stratification** the unequal distribution of wealth, power, and prestige on a global basis, resulting in people having vastly different lifestyles and life chances both within and among the nations of the world.

**goal displacement** a process that occurs in organizations when the rules become an end in themselves rather than a means to an end, and organizational survival becomes more important than achievement of goals.

**gossip** rumors about the personal lives of individuals.

**government** the formal organization that has the legal and political authority to regulate the relationships among members of a society and between the society and those outside its borders.

**groupthink** the process by which members of a cohesive group arrive at a decision that many individual members privately believe is unwise.

**537**

**health**   a state of complete physical, mental, and social well-being.

**health care**   any activity intended to improve health.

**health maintenance organizations (HMOs)**   companies that provide, for a set monthly fee, total care with an emphasis on prevention to avoid costly treatment later.

**hidden curriculum**   the transmission of cultural values and attitudes, such as conformity and obedience to authority, through implied demands found in rules, routines, and regulations of schools.

**high culture**   classical music, opera, ballet, live theater, and other activities usually patronized by elite audiences.

**high-income countries**   nations with highly industrialized economies; technologically advanced industrial, administrative, and service occupations; and relatively high levels of national and personal income.

**holistic medicine**   an approach to health care that focuses on prevention of illness and disease and is aimed at treating the whole person—body and mind—rather than just the part or parts in which symptoms occur.

**homogamy**   the pattern of individuals marrying those who have similar characteristics, such as race/ethnicity, religious background, age, education, or social class.

**homophobia**   extreme prejudice and sometimes discriminatory actions directed at gays, lesbians, bisexuals, and others who are perceived as not being heterosexual.

**horticultural societies**   societies based on technology that supports the cultivation of plants to provide food.

**hunting and gathering societies**   societies that use simple technology for hunting animals and gathering vegetation.

**hypothesis**   a statement of the expected relationship between two or more variables.

**id**   Sigmund Freud's term for the component of personality that includes all of the individual's basic biological drives and needs that demand immediate gratification.

**ideal type**   an abstract model that describes the recurring characteristics of some phenomenon.

**illegitimate opportunity structures**   circumstances that provide an opportunity for people to acquire through illegitimate activities what they cannot achieve through legitimate channels.

**impression management (presentation of self)**   Erving Goffman's term for people's efforts to present themselves to others in ways that are most favorable to their own interests or image.

**income**   the economic gain derived from wages, salaries, income transfers (governmental aid), and ownership of property.

**independent variable**   in an experiment, the variable assumed to be the cause of the relationship between variables.

**individual discrimination**   behavior consisting of one-on-one acts by members of the dominant group that harm members of the subordinate group or their property.

**industrial societies**   societies based on technology that mechanizes production.

**industrialization**   the process by which societies are transformed from dependence on agriculture and handmade products to an emphasis on manufacturing and related industries.

**infant mortality rate**   the number of deaths of infants under 1 year of age per 1,000 live births in a given year.

**informal side of a bureaucracy**   those aspects of participants' day-to-day activities and interactions that ignore, bypass, or do not correspond with the official rules and procedures of the bureaucracy.

**ingroup**   a group to which a person belongs and with which the person feels a sense of identity.

**institutional discrimination**   the day-to-day practices of organizations and institutions that have a harmful effect on members of subordinate groups.

**instrumental leadership**   goal- or task-oriented leadership.

**intergenerational mobility**   the social movement experienced by family members from one generation to the next.

**interlocking corporate directorates**   members of the board of directors of one corporation who also sit on the board(s) of other corporations.

**internal colonialism**   according to conflict theorists, a practice that occurs when members of a racial or ethnic group are conquered or colonized and forcibly placed under the economic and political control of the dominant group.

**Internet crime**   illegal acts committed on the Internet, including FBI-related scams, identity theft, advance fee fraud, nonauction/nondelivery of merchandise, and overpayment fraud.

**intersexed person**   an individual who is born with a reproductive or sexual anatomy that does not correspond to typical definitions of male or female; in other words, the person's sexual differentiation is ambiguous.

**interview**   a data-collection encounter in which an interviewer asks the respondent questions and records the answers.

**intragenerational mobility**   the social movement of individuals within their own lifetime.

**invasion**   the process by which a new category of people or type of land use arrives in an area previously occupied by another group or type of land use.

**iron law of oligarchy**   according to Robert Michels, the tendency of bureaucracies to be ruled by a few people.

**job deskilling**   a reduction in the proficiency needed to perform a specific job.

**juvenile delinquency**   a violation of law or the commission of a status offense by young people.

**kinship**   a social network of people based on common ancestry, marriage, or adoption.

**labeling theory**   the proposition that deviance is a socially constructed process in which social control agencies designate certain people as deviants and they, in turn, come to accept the label placed upon them and begin to act accordingly.

**laissez-faire leaders**   leaders who are only minimally involved in decision making and encourage group members to make their own decisions.

**language**   a set of symbols that expresses ideas and enables people to think and communicate with one another.

**latent functions**   unintended functions that are hidden and remain unacknowledged by participants.

**laws**   formal, standardized norms that have been enacted by legislatures and are enforced by formal sanctions.

**leadership**   the ability to influence what goes on in a group or social system.

**life chances** Max Weber's term for the extent to which individuals have access to important societal resources such as food, clothing, shelter, education, and health care.

**life expectancy** an estimate of the average lifetime of people born in a specific year.

**looking-glass self** Charles Horton Cooley's term for the way in which a person's sense of self is derived from the perceptions of others.

**low-income countries** primarily agrarian nations with little industrialization and low levels of national and personal income.

**macrolevel analysis** an approach that examines whole societies, large-scale social structures, and social systems instead of looking at important social dynamics in individuals' lives.

**managed care** any system of cost containment that closely monitors and controls health care providers' decisions about medical procedures, diagnostic tests, and other services that should be provided to patients.

**manifest functions** functions that are intended and/or overtly recognized by the participants in a social unit.

**marginal jobs** jobs that differ from the employment norms of the society in which they are located.

**marriage** a legally recognized and/or socially approved arrangement between two or more individuals that carries certain rights and obligations and usually involves sexual activity.

**mass** a number of people who share an interest in a specific idea or issue but who are not in one another's immediate vicinity.

**mass behavior** collective behavior that takes place when people (who are often geographically separated from one another) respond to the same event in much the same way.

**mass media** large-scale organizations that use print or electronic means (such as radio, television, film, and the Internet) to communicate with large numbers of people.

**master status** the most important status that a person occupies.

**material culture** the physical or tangible creations that members of a society make, use, and share.

**matriarchal family** a family structure in which authority is held by the eldest female (usually the mother).

**matriarchy** a hierarchical system of social organization in which cultural, political, and economic structures are controlled by women.

**matrilineal descent** a system of tracing descent through the mother's side of the family.

**matrilocal residence** the custom of a married couple living in the same household (or community) as the wife's parents.

**mechanical solidarity** Emile Durkheim's term for the social cohesion of preindustrial societies, in which there is minimal division of labor and people feel united by shared values and common social bonds.

**medical–industrial complex** local physicians, local hospitals, and global health-related industries such as insurance companies and pharmaceutical and medical supply companies.

**medicalization** the process whereby nonmedical problems become defined and treated as illnesses or disorders.

**medicine** an institutionalized system for the scientific diagnosis, treatment, and prevention of illness.

**meritocracy** a hierarchy in which all positions are rewarded based on people's ability and credentials.

**microlevel analysis** an approach that focuses on small groups rather than large-scale social structures.

**middle-income countries** nations with industrializing economies, particularly in urban areas, and moderate levels of national and personal income.

**migration** the movement of people from one geographic area to another for the purpose of changing residency.

**military–industrial complex** the mutual interdependence of the military establishment and private military contractors.

**mixed economy** an economic system that combines elements of a market economy (capitalism) with elements of a command economy (socialism).

**mob** a highly emotional crowd whose members engage in, or are ready to engage in, violence against a specific target—a person, a category of people, or physical property.

**modernization theory** a perspective that links global inequality to different levels of economic development and suggests that low-income economies can move to middle- and high-income economies by achieving self-sustained economic growth.

**monarchy** a political system in which power resides in one person or family and is passed from generation to generation through lines of inheritance.

**monogamy** a marriage between two partners, usually a woman and a man.

**mores** strongly held norms with moral and ethical connotations that may not be violated without serious consequences in a particular culture.

**mortality** the incidence of death in a population.

**neolocal residence** the custom of a married couple living in their own residence apart from both the husband's and the wife's parents.

**network** a web of social relationships that links one person with other people and, through them, with other people they know.

**new international division of labor theory** the perspective that views commodity production as being split into fragments that can be assigned to whichever part of the world can provide the most profitable combination of capital and labor.

**nonmaterial culture** the abstract or intangible human creations of society that influence people's behavior.

**nonverbal communication** the transfer of information between persons without the use of words.

**norms** established rules of behavior or standards of conduct.

**nuclear family** a family composed of one or two parents and their dependent children, all of whom live apart from other relatives.

**occupational (white-collar) crime** illegal activities committed by people in the course of their employment or financial affairs.

**occupations** categories of jobs that involve similar activities at different work sites.

**official poverty line** the income standard that is based on what the federal government considers to be the minimum amount of money required for living at a subsistence level.

**oligopoly** an industry dominated by just a few companies.

**organic solidarity** Emile Durkheim's term for the social cohesion found in industrial (and perhaps postindustrial) societies, in which people perform very specialized tasks and feel united by their mutual dependence.

**organized crime** a business operation that supplies illegal goods and services for profit.

**outgroup** a group to which a person does not belong and toward which the person may feel a sense of competitiveness or hostility.

**panic** a form of crowd behavior that occurs when a large number of people react to a real or perceived threat with strong emotions and self-destructive behavior.

**participant observation** a research method in which researchers collect systematic observations while being part of the activities of the group being studied.

**pastoral societies** societies based on technology that supports the domestication of large animals to provide food.

**patriarchal family** a family structure in which authority is held by the eldest male (usually the father).

**patriarchy** a hierarchical system of social organization in which cultural, political, and economic structures are controlled by men.

**patrilineal descent** a system of tracing descent through the father's side of the family.

**patrilocal residence** the custom of a married couple living in the same household (or community) as the husband's parents.

**pay gap** the disparity between women's and men's earnings.

**peer group** a group of people who are linked by common interests, equal social position, and (usually) similar age.

**peripheral nations** according to world systems theory, nations that are dependent on core nations for capital, have little or no industrialization (other than what may be brought in by core nations), and have uneven patterns of urbanization.

**personal space** the immediate area surrounding a person that the person claims as private.

**pink-collar occupations** relatively low-paying, nonmanual, semiskilled positions primarily held by women.

**pluralist model** an analysis of political systems that views power as being widely dispersed throughout many competing interest groups.

**political action committees** organizations of special interest groups that solicit contributions from donors and fund campaigns to help elect (or defeat)

candidates based on their stances on specific issues.

**political crime** illegal or unethical acts involving the usurpation of power by government officials or illegal/unethical acts perpetrated against the government by outsiders seeking to make a political statement, undermine the government, or overthrow it.

**political party** an organization whose purpose is to gain and hold legitimate control of government.

**political socialization** the process by which people learn political attitudes, values, and behavior.

**politics** the social institution through which power is acquired and exercised by some people and groups.

**polyandry** the concurrent marriage of one woman with two or more men.

**polygamy** the concurrent marriage of a person of one sex with two or more members of the opposite sex.

**polygyny** the concurrent marriage of one man with two or more women.

**popular culture** activities, products, and services that are assumed to appeal primarily to members of the middle and working classes.

**population composition** the biological and social characteristics of a population, including age, sex, race, marital status, education, occupation, income, and size of household.

**population pyramid** a graphic representation of the distribution of a population by sex and age.

**positivism** a belief that the world can best be understood through scientific inquiry.

**postindustrial societies** societies in which technology supports a service- and information-based economy.

**postmodern perspectives** the sociological approach that attempts to explain social life in contemporary societies that are characterized by postindustrialization, consumerism, and global communications.

**power** according to Max Weber, the ability of persons or groups to achieve their goals despite opposition from others.

**prejudice** a negative attitude based on faulty generalizations about members of specific racial, ethnic, or other groups.

**prestige** the respect or regard that a person or status position is given by others.

**primary deviance** the initial act of rule breaking.

**primary group** a small, less specialized group in which members engage in face-to-face, emotion-based interactions over an extended period of time.

**primary labor market** the sector of the labor market that consists of high-paying jobs with good benefits that have some degree of security and the possibility of future advancement.

**primary sector production** the sector of the economy that extracts raw materials and natural resources from the environment.

**primary sex characteristics** the genitalia used in the reproductive process.

**primary socialization** the process of learning that begins at birth and occurs in the home and family.

**profane** the everyday, secular, or "worldly" aspects of life.

**professions** high-status, knowledge-based occupations.

**propaganda** information provided by individuals or groups that have a vested interest in furthering their own cause or damaging an opposing one.

**property crimes** burglary (breaking into private property to commit a serious crime), motor vehicle theft, larceny-theft (theft of property worth $50 or more), and arson.

**public opinion** the attitudes and beliefs communicated by ordinary citizens to decision makers.

**punishment** any action designed to deprive a person of things of value (including liberty) because of some offense the person is thought to have committed.

**qualitative research** sociological research methods that use interpretive descriptions (words) rather than statistics (numbers) to analyze underlying meanings and patterns of social relationships.

**quantitative research** sociological research methods that are based on the goal of scientific objectivity and that focus on data that can be measured numerically.

**questionnaire** a printed research instrument containing a series of items to which subjects respond.

**race** a category of people who have been singled out as inferior or superior, often on the basis of real or alleged physical characteristics such as skin color, hair texture, eye shape, or other subjectively selected attributes.

**racial socialization** the aspect of socialization that contains specific messages and practices concerning the nature of one's racial or ethnic status as it relates to identity, interpersonal relationships, and location in the social hierarchy.

**racism** a set of attitudes, beliefs, and practices that is used to justify the superior treatment of one racial or ethnic group and the inferior treatment of another racial or ethnic group.

**rational choice theory of deviance** the proposition that deviant behavior occurs when a person weighs the costs and benefits of nonconventional or criminal behavior and determines that the benefits will outweigh the risks involved in such actions.

**rational–legal authority** power legitimized by law or written rules and regulations.

**rationality** the process by which traditional methods of social organization, characterized by informality and spontaneity, are gradually replaced by efficiently administered formal rules and procedures.

**reciprocal socialization** the process by which the feelings, thoughts, appearance, and behavior of individuals who are undergoing socialization also have a direct effect on those agents of socialization who are attempting to influence them.

**reference group** a group that strongly influences a person's behavior and social attitudes, regardless of whether that individual is an actual member.

**relative poverty** a level of economic deprivation that exists when people may be able to afford basic necessities but are still unable to maintain an average standard of living.

**reliability** the extent to which a study or research instrument yields consistent results when applied to different individuals at one time or to the same individuals over time.

**religion** a social institution composed of a unified system of beliefs, symbols, and rituals—based on some sacred or supernatural realm—that guides human behavior, gives meaning to life, and unites believers into a community.

**representative democracy** a form of democracy whereby citizens elect representatives to serve as bridges between themselves and the government.

**research methods** specific strategies or techniques for systematically conducting research.

**resocialization** the process of learning a new and different set of attitudes, values, and behaviors from those in one's background and previous experience.

**riot** violent crowd behavior that is fueled by deep-seated emotions but not directed at one specific target.

**role** a set of behavioral expectations associated with a given status.

**role conflict** a situation in which incompatible role demands are placed on a person by two or more statuses held at the same time.

**role exit** a situation in which people disengage from social roles that have been central to their self-identity.

**role expectation** a group's or society's definition of the way that a specific role ought to be played.

**role performance** how a person actually plays a role.

**role strain** a condition that occurs when incompatible demands are built into a single status that a person occupies.

**role-taking** the process by which a person mentally assumes the role of another person or group in order to understand the world from that person's or group's point of view.

**routinization of charisma** the process by which charismatic authority is succeeded by a bureaucracy controlled by a rationally established authority or by a combination of traditional and bureaucratic authority.

**rumor** an unsubstantiated report on an issue or subject.

**sacred** those aspects of life that are extraordinary or supernatural.

**sanctions** rewards for appropriate behavior or penalties for inappropriate behavior.

**Sapir–Whorf hypothesis** the proposition that language shapes the view of reality of its speakers.

**scapegoat** a person or group that is incapable of offering resistance to the hostility or aggression of others.

**second shift** Arlie Hochschild's term for the domestic work that employed women perform at home after they complete their workday on the job.

**secondary analysis** a research method in which researchers use existing material and analyze data that were originally collected by others.

**secondary deviance** deviance that occurs when a person who has been labeled a deviant accepts that new identity and continues the deviant behavior.

**secondary group** a larger, more specialized group in which members engage in more-impersonal, goal-oriented relationships for a limited period of time.

**secondary labor market** the sector of the labor market that consists of low-paying jobs with few benefits and very little job security or possibility for future advancement.

**secondary sector production** the sector of the economy that processes raw materials (from the primary sector) into finished goods.

**secondary sex characteristics** the physical traits (other than reproductive organs) that identify an individual's sex.

**secondary socialization** the process of learning that takes place outside the home—in settings such as schools, religious organizations, and the workplace—and helps individuals learn how to act in appropriate ways in various situations.

**sect** a relatively small religious group that has broken away from another religious organization to renew what it views as the original version of the faith.

**secularization** the process by which religious beliefs, practices, and institutions lose their significance in sectors of society and culture.

**segregation** the spatial and social separation of categories of people by race, ethnicity, class, gender, and/or religion.

**self-concept** the totality of our beliefs and feelings about ourselves.

**self-fulfilling prophecy** a situation in which a false belief or prediction produces behavior that makes the originally false belief come true.

**semiperipheral nations** according to world systems theory, nations that are more developed than peripheral nations but less developed than core nations.

**sex** the biological and anatomical differences between females and males.

**sex ratio** the number of males for every hundred females in a given population.

**sexism** the subordination of one sex, usually female, based on the assumed superiority of the other sex.

**sexual orientation** an individual's preference for emotional–sexual relationships with members of the different sex (heterosexuality), the same sex (homosexuality), or both (bisexuality).

**sexualization** the acts or processes whereby an individual or group is seen as sexual in nature or persons become aware of their sexuality.

**shared monopoly** a situation that occurs when four or fewer companies supply 50 percent or more of a particular market.

**sick role** the set of patterned expectations that defines the norms and values appropriate for individuals who are sick and for those who interact with them.

**significant others** those persons whose care, affection, and approval are especially desired and who are most important in the development of the self.

**slavery** an extreme form of stratification in which some people are owned or controlled by others for the purpose of economic or sexual exploitation.

**small group** a collectivity small enough for all members to be acquainted with one another and to interact simultaneously.

**social bond theory** the proposition that the probability of deviant behavior increases when a person's ties to society are weakened or broken.

**social change** the alteration, modification, or transformation of public policy, culture, or social institutions over time.

**social construction of reality** the process by which our perception of reality is largely shaped by the subjective meaning that we give to an experience.

**social control** systematic practices that social groups develop in order to encourage conformity to norms, rules, and laws and to discourage deviance.

**social Darwinism** Herbert Spencer's belief that those species of animals, including human beings, best adapted to their environment survive and prosper, whereas those poorly adapted die out.

**social devaluation** a situation in which a person or group is considered to have less social value than other individuals or groups.

**social epidemiology** the study of the causes and distribution of health, disease, and impairment throughout a population.

**social facts** Emile Durkheim's term for patterned ways of acting, thinking, and feeling that exist outside any one individual but that exert social control over each person.

**social group** a group that consists of two or more people who interact frequently and share a common identity and a feeling of interdependence.

**social institution** a set of organized beliefs and rules that establishes how a society will attempt to meet its basic social needs.

**social interaction** the process by which people act toward or respond to other people; the foundation for all relationships and groups in society.

**social mobility** the movement of individuals or groups from one level in a stratification system to another.

**social movement** an organized group that acts consciously to promote or resist change through collective action.

**social script** a "playbook" that "actors" use to guide their verbal replies and overall performance to achieve the desired goal of the conversation or fulfill the role they are playing.

**social stratification** the hierarchical arrangement of large social groups based on their control over basic resources.

**social structure** the complex framework of societal institutions (such as the economy, politics, and religion) and the social practices (such as rules and social roles) that make up a society and that organize and establish limits on people's behavior.

**socialism** an economic system characterized by public ownership of the means of production, the pursuit of collective goals, and centralized decision making.

**socialization** the lifelong process of social interaction through which individuals acquire a self-identity and the physical, mental, and social skills needed for survival in society.

**socialized medicine** a health care system in which the government owns the medical care facilities and employs the physicians.

**society** a large social grouping that shares the same geographical territory and is subject to the same political authority and dominant cultural expectations.

**sociobiology** the systematic study of "social behavior from a biological perspective."

**socioeconomic status (SES)** a combined measure that attempts to classify individuals, families, or households in terms of factors such as income, occupation, and education to determine class location.

**sociological imagination** C. Wright Mills's term for the ability to see the relationship between individual experiences and the larger society.

**sociology** the systematic study of human society and social interaction.

**sociology of family** the subdiscipline of sociology that attempts to describe and explain patterns of family life and variations in family structure.

**split labor market** the division of the economy into two areas of employment, a primary sector or upper tier composed of higher-paid (usually dominant-group) workers in more-secure jobs, and a secondary sector or lower tier composed of lower-paid (often subordinate-group) workers in jobs with little security and hazardous working conditions.

**state** the political entity that possesses a legitimate monopoly over the use of force within its territory to achieve its goals.

**status** a socially defined position in a group or society characterized by certain expectations, rights, and duties.

**status set** all the statuses that a person occupies at a given time.

**status symbol** a material sign that informs others of a person's specific status.

**stereotypes** overgeneralizations about the appearance, behavior, or other characteristics of members of particular categories.

**strain theory** the proposition that people feel strain when they are exposed to cultural goals that they are unable to obtain because they do not have access to culturally approved means of achieving those goals.

**subcontracting** an agreement in which a corporation contracts with other (usually smaller) firms to provide specialized components, products, or services to the larger corporation.

**subculture** a category of people who share distinguishing attributes, beliefs, values, and/or norms that set them apart in some significant manner from the dominant culture.

**subordinate group** a group whose members, because of physical or cultural characteristics, are disadvantaged and subjected to unequal treatment and discrimination by the dominant group.

**succession** the process by which a new category of people or type of land use gradually predominates in an area formerly dominated by another group or type of land use.

**superego** Sigmund Freud's term for the conscience, consisting of the moral and ethical aspects of personality.

**survey** a poll in which the researcher gathers facts or attempts to determine the relationships among facts.

**symbol** anything that meaningfully represents something else.

**symbolic interactionist perspectives** the sociological approach that views society as the sum of the interactions of individuals and groups.

**taboos** mores so strong that their violation is considered to be extremely offensive and even unmentionable.

**technology** the knowledge, techniques, and tools that allow people to transform resources into usable forms, and the knowledge and skills required to use them after they are developed.

**terrorism** the calculated, unlawful use of physical force or threats of violence against persons or property in order to intimidate or coerce a government, organization, or individual for the purpose of gaining some political, religious, economic, or social objective.

**tertiary deviance** deviance that occurs when a person who has been labeled a deviant seeks to normalize the behavior by relabeling it as nondeviant.

**tertiary sector production** the sector of the economy that is involved in the provision of services rather than goods.

**tertiary socialization** the process of learning that takes place when adults move into new settings where they must accept certain ideas or engage in specific behaviors that are appropriate to that specific setting.

**theory** a set of logically interrelated statements that attempts to describe, explain, and (occasionally) predict social events.

**theory of racial formation** the idea that actions of the government substantially define racial and ethnic relations in the United States.

**total institution** Erving Goffman's term for a place where people are isolated from the rest of society for a set period of time and come under the control of the officials who run the institution.

**totalitarianism** a political system in which the state seeks to regulate all aspects of people's public and private lives.

**tracking** the assignment of students to specific curriculum groups and courses on the basis of their test scores, previous grades, or other criteria.

**traditional authority** power that is legitimized on the basis of longstanding custom.

**transgendered person** an individual whose gender identity (self-identification as woman, man, neither, or both) does not match the person's assigned sex (identification by others as male, female, or intersex based on physical/genetic sex).

**transnational corporations** large corporations that are headquartered in one country but sell and produce goods and services in many countries.

**triad** a group composed of three members.

**underclass** those who are poor, seldom employed, and caught in long-term deprivation that results from low levels of education and income and high rates of unemployment.

**unemployment rate** the percentage of unemployed persons in the labor force actively seeking jobs.

**universal health care** a health care system in which all citizens receive medical services paid for by tax revenues.

**urbanization** the process by which an increasing proportion of a population lives in cities rather than in rural areas.

**validity** the extent to which a study or research instrument accurately measures what it is supposed to measure.

**value contradictions** values that conflict with one another or are mutually exclusive.

**values** collective ideas about what is right or wrong, good or bad, and desirable or undesirable in a particular culture.

**variable** any concept with measurable traits or characteristics that can change or vary from one person, time, situation, or society to another.

**victimless crimes** crimes involving a willing exchange of illegal goods or services among adults.

**violent crime** actions—murder, rape, robbery, and aggravated assault—involving force or the threat of force against others.

**wealth** the value of all of a person's or family's economic assets, including income, personal property, and income-producing property.

**welfare state** a state in which there is extensive government action to provide support and services to the citizens.

**working class (or *proletariat*)** Karl Marx's term for those who must sell their labor to the owners in order to earn enough money to survive.

**world systems theory** a perspective that examines the role of capitalism, and particularly the transnational division of labor, in a truly global system held together by economic ties.

**zero population growth** the point at which no population increase occurs from year to year.

# References

AARP.org. 2012. "Social Security and Older Minorities" (Nov. 30). Retrieved Jan. 13, 2013. Online: www.aarp.org/work/social-security/info-11-2012/social-security-older-minorities.html

AAUW (American Association of University Women). 1995. *How Schools Shortchange Girls/The AAUW Report: A Study of Major Findings on Girls and Education.* New York: Marlowe.

—. 2008. "Where the Girls Are: The Facts About Gender Equity in Education." Retrieved Apr. 2, 2010. Online: www.aauw.org/research/upload/whereGirlsARe.pdf

—. 2013. "The Simple Truth About the Gender Pay Gap." Retrieved Feb. 11, 2014. Online: www.aauw.org/files/2013/03/The-Simple-Truth-Fall-2013.pdf

ABC Family. 2010. "Greek: Recaps." Retrieved Dec. 22, 2010. Online: www.abcfamily.go.com/shows/greek/recaps

ABCNews.com. 2006. "Kids Getting 'Thinspiration' from Dangerously Skinny Stars." *Good Morning America* (July 12). Retrieved June 15, 2011. Online: http://abcnews.go.com/GMA/Diet/story?id=2182068&page=1

Adorno, Theodor W., Else Frenkel-Brunswick, Daniel J. Levinson, and R. Nevitt Sanford. 1950. *The Authoritarian Personality.* New York: Harper & Row.

AFP. 2013. "Wombs for Rent: Commercial Surrogacy Big Business in India." *The Express Tribune with the International Herald Tribune* (Feb. 25). Retrieved July 7, 2013. Online: http://tribune.com.pk/story/512264/wombs-for-rent-commercial-surrogacy-big-business-in-india

AIDS.gov. 2013. "Global AIDS Overview: The Global HIV/AIDS Crisis Today." Retrieved Mar. 29, 2014. Online: http://aids.gov/federal-resources/around-the-world/global-aids-overview

Akers, Ronald L. 1998. *Social Learning and Social Structure: A General Theory of Crime and Deviance.* Boston: Northeastern University Press.

Albanese, Catherine L. 2007. *America: Religions and Religion* (4th ed.). Belmont, CA: Wadsworth.

Albas, Daniel, and Cheryl Albas. 1988. "Aces and Bombers: The Post-Exam Impression Management Strategies of Students." *Symbolic Interaction,* 11 (Fall): 289–302.

—. 2011. "Aces and Bombers: The Post-Exam Impression Management. Strategies of Students." *Symbolic Interaction,* 11 (2): 289–302. Retrieved Nov. 8, 2012. Online: http://onlinelibrary.wiley.com/doi/10.1525/si.1988.11.2.289/abstract

Albrecht, Gary L., Katherine D. Seelman, and Michael Bury (Eds.). 2001. *The Handbook of Disability Studies.* Thousand Oaks, CA: Sage.

Altemeyer, Bob. 1981. *Right-Wing Authoritarianism.* Winnipeg: University of Manitoba Press.

—. 1988. *Enemies of Freedom: Understanding Right-Wing Authoritarianism.* San Francisco: Jossey-Bass.

Alvarez, Lizette. 2013. "Girl's Suicide Points to Rise in Apps Used by Cyberbullies." *New York Times* (Sept. 13). Retrieved Dec. 7, 2013. Online: www.nytimes.com/2013/09/14/us/suicide-of-girl-after-bullying-raises-worries-on-web-sites.html?_r=0

Amato, Paul R. 2005. "The Impact of Family Formation Change on the Cognitive, Social, and Emotional Well-Being of the Next Generation." *Marriage and Child Wellbeing* (Fall). Retrieved Feb. 12, 2012. Online: www.princeton.edu/futureofchildren/publications/docs/15_02_05.pdf

Amelie, Mdmse. (2010). "Tips for Surviving Your First Day of Higher Education." Retrieved Dec. 22, 2013. Online: www.educationspace360.com/index.php/first-day-college-first-day-university-higher-education-success-good-impression-22446

American Association of Community Colleges. 2013. "Community College Fact Sheet." Retrieved Apr. 14, 2013. Online: www.aacc.nche.edu/AboutCC/Documents/2013facts_print_revised.pdf

American Indian College Fund. 2011. "Tribal Colleges." Retrieved Apr. 28, 2012. Online: www.collegefund.org/tribal_colleges_listing

American Psychiatric Association. 2013. "DSM-5 Implementation and Support: What's New?" Retrieved May 26, 2013. Online: www.dsm5.org/Pages/Default.aspx

American Psychological Association. 2010. "Report of the APA Task Force on the Sexualization of Girls." Retrieved Feb. 22, 2013. Online: www.apa.org/pi/women/programs/girls/report-full.pdf

American Society for Aesthetic Plastic Surgery. 2010. "Quick Facts: Highlights of the ASAPS 2010 Statistics on Cosmetic Surgery." Retrieved Feb. 5, 2012. Online: www.surgery.org/sites/default/files/2010-quickfacts_0.pdf

American Sociological Association. 1999. "Code of Ethics and Policies and Procedures of the ASA Committee on Professional Ethics." Retrieved Apr. 4, 2014. Online: www.asanet.org/images/asa/docs/pdf/CodeofEthics.pdf

Andersen, Margaret L., and Patricia Hill Collins (Eds.). 2010. *Race, Class, and Gender: An Anthology* (7th ed.). Belmont, CA: Wadsworth.

Angier, Natalie. 1993. "'Stopit!' She Said. 'Nomore!'" *New York Times Book Review* (Apr. 25): 12.

Applebome, Peter, and Brian Stelter. 2012. "Media Spotlight Seen as a Blessing, or a Curse, in a Grieving Town." *New York Times* (Dec. 16). Retrieved Dec. 26, 2012. Online: www.nytimes.com/2012/12/17/business/media/newtown-has-mixed-feelings-about-the-media-horde-in-its-midst.html

Appleton, Lynn M. 1995. "The Gender Regimes in American Cities." In Judith A. Garber and Robyne S. Turner (Eds.), *Gender in Urban Research.* Thousand Oaks, CA: Sage, pp. 44–59.

Asch, Adrienne. 1986. "Will Populism Empower Disabled People?" In Harry G. Boyle and Frank Reissman (Eds.), *The New Populism: The Power of*

*Empowerment.* Philadelphia: Temple University Press, pp. 213–228.

—. 2001. "Critical Race Theory, Feminism, and Disability: Reflections on Social Justice and Personal Identity." *Ohio State Law Journal* (62). Retrieved May 14, 2011. Online: http://moritzlaw.osu.edu/lawjournal/issues/volume62/number1/asch.pdf

Asch, Solomon E. 1955. "Opinions and Social Pressure." *Scientific American,* 193 (5): 31–35.

—. 1956. "Studies of Independence and Conformity: A Minority of One Against a Unanimous Majority." *Psychological Monographs,* 70 (9) (Whole No. 416).

Ashe, Arthur R., Jr. 1988. *A Hard Road to Glory: A History of the African-American Athlete.* New York: Warner.

Asi, Maryam, and Daniel Beaulieu. 2013. "Arab Households in the United States: 2006–2010." U.S. Census Bureau, American Community Survey Briefs (May). Retrieved Feb. 10, 2014. Online: www.census.gov/prod/2013pubs/acsbr10-20.pdf

Associated Press. 2012. "Man Stabs 22 Children in China." New York Times.com (Dec. 14). Retrieved Dec. 27, 2012. Online: www.nytimes.com/2012/12/15/world/asia/man-stabs-22-children-in-china.html?_r=0

Aulette, Judy Root. 1994. *Changing Families.* Belmont, CA: Wadsworth.

Aydelotte, Becky. 2010. "The Normal Paranormal." *Baylor Arts & Sciences* (Fall): 6–7.

abcnews.go.com. 2011. "Gingrich Says Poor Children Have No Work Habits." Retrieved Dec. 26, 2011. Online: http://abcnews.go.com/blogs/politics/2011/12/gingrich-says-poor-children-have-no-work-ethic

Babbie, Earl. 2013. *The Practice of Social Research* (13th ed.). Belmont, CA: Cengage/Wadsworth.

Ballantine, Jeanne H., and Floyd M. Hammack. 2012. *The Sociology of Education: A Systematic Analysis* (7th ed.). Upper Saddle River, NJ: Prentice-Hall/Pearson.

Barkan, Steven E. 2012. *Criminology.* Upper Saddle River, NJ: Prentice-Hall.

Basu, Moni. 2013. "Websites Become Tool for Stopping Veteran Suicides." CNN.com (Sept. 21). Retrieved Dec. 8, 2013. Online: www.cnn.com/2013/09/21/us/facebook-suicide

Baudrillard, Jean. 1983. *Simulations.* New York: Semiotext.

—. 1998. *The Consumer Society: Myths and Structures.* London: Sage (orig. pub. 1970).

Beaulieu, Catherine MJ. 2004. "Intercultural Study of Personal Space: A Case Study." *Journal of Applied Social Psychology,* 34 (4): 794–805.

Becker, Howard S. 1963. *Outsiders: Studies in the Sociology of Deviance.* New York: Free Press.

Belkin, Lisa. 2009. "Do Girls Have More Chores Than Boys?" *New York Times* (Oct. 5). *Motherlode: Adventures in Parenting.* Retrieved Feb. 5, 2012. Online: http://parenting.blogs.nytimes.com/2009/10/05/do-girls-have-more-chores-than-boys

Berger, Peter. 1963. *Invitation to Sociology: A Humanistic Perspective.* New York: Anchor.

—. 1967. *The Sacred Canopy: Elements of a Sociological Theory of Religion.* New York: Doubleday.

Berger, Peter, and Hansfried Kellner. 1964. "Marriage and the Construction of Reality." *Diogenes,* 46: 1–32.

Berger, Peter, and Thomas Luckmann. 1967. *The Social Construction of Reality: A Treatise in the Sociology of Knowledge.* Garden City, NY: Anchor.

Bernard, Jessie. 1982. *The Future of Marriage.* New Haven, CT: Yale University Press (orig. pub. 1973).

Bernstein, Nina. 2012. "Chefs, Butlers, Marble Baths: Hospitals Vie for the Affluent." *New York Times* (Jan. 21). Retrieved Mar. 5, 2012. Online: www.nytimes.com/2012/01/22/nyregion/chefs-butlers-and-marble-baths-not-your-average-hospital-room.html?_r=1&pagewanted=all

Best, Joel. 1999. *Random Violence: How We Talk About New Crimes and New Victims.* Berkeley: University of California Press.

Better Health Channel. 2007. "Food and Celebrations." Retrieved Feb. 11, 2007. Online: www.betterhealth.vic.gov.au

Biblarz, Arturo, R. Michael Brown, Dolores Noonan Biblarz, Mary Pilgram, and Brent F. Baldree. 1991. "Media Influence on Attitudes Toward Suicide." *Suicide and Life-Threatening Behavior,* 21 (4): 374–385.

Bindley, Katherine. 2012. "The Dangerous Mistake We Make: Bullying and Suicide" (Feb. 8). Retrieved Oct. 13, 2012. Online: www.huffingtonpost.com/2012/02/08/bullying-suicide-teens-depression_n_1247875.html?1328712078

Bishaw, Alemayehu. 2013. *Poverty: 2000 to 2012.* U.S. Census Bureau (Sept.). Retrieved Jan. 5, 2013. Online: www.census.gov/prod/2013pubs/acsbr12-01.pdf

Bizshifts-Trends. 2011. "Management Styles: U.S., Europe, Japan, China, India, Brazil, Russia." Retrieved Mar. 13, 2013. Online: http://bizshifts-trends.com/2011/01/10/management-styles-u-s-europe-japan-china-india-brazil-russia

Bjornstrom, Eileen E., Robert L. Kaufman, Ruth D. Peterson, and Michael D. Slater. 2010. "Race and Ethnic Representations of Lawbreakers and Victims in Crime News: A National Study of Television Coverage." *Social Problems,* 57 (2): 269–293.

Blauner, Robert. 1972. *Racial Oppression in America.* New York: Harper & Row.

Block, Fred, Anna C. Korteweg, and Kerry Woodward, with Zach Schiller and Imrul Mazid. 2008. "The Compassion Gap in American Poverty Policy." In Jeff Goodwin and James M. Jasper (Eds.), *The Contexts Reader.* New York: Norton, pp. 166–175. Originally appeared in *Contexts,* a journal published by the American Sociological Association (Spring 2006).

Blumer, Herbert G. 1946. "Collective Behavior." In Alfred McClung Lee (Ed.), *A New Outline of the Principles of Sociology.* New York: Barnes & Noble, pp. 167–219.

—. 1969. *Symbolic Interactionism: Perspective and Method.* Englewood Cliffs, NJ: Prentice Hall.

Blynn, Jamie. 2013. "Inside the 2013 Presidential Inauguration." Seventeen.com (Jan. 22). Retrieved Mar. 18, 2013. Online: www.seventeen.com/college/presidential-election-blog/inauguration-2013

Bonacich, Edna. 1972. "A Theory of Ethnic Antagonism: The Split Labor Market." *American Sociological Review,* 37: 547–549.

—. 1976. "Advanced Capitalism and Black–White Relations in the United States: A Split Labor Market Interpretation." *American Sociological Review,* 41: 34–51.

Bonvillain, Nancy. 2001. *Women & Men: Cultural Constructs of Gender* (3rd ed.). Upper Saddle River, NJ: Prentice Hall.

Bourdieu, Pierre. 1984. *Distinction: A Social Critique of the Judgement of Taste.* Trans. Richard Nice. Cambridge, MA: Harvard University Press.

Bourdieu, Pierre, and Jean-Claude Passeron. 1990. *Reproduction in Education, Society and Culture.* Newbury Park, CA: Sage.

Boyes, William, and Michael Melvin. 2011. *Economics* (8th ed.). Stamford, CT: Cengage.

Bradford, Reuben F. 2013. "Cover Letter on the Sandy Hook Report from Commissioner Bradford." State of Connecticut, Department of Emergency Services and Public Protection, Office of the Commissioner (Dec. 27). Retrieved Dec. 28, 2013. Online: http://cspsandyhookreport.ct.gov /cspshr/Sandy_Hook_Cover_Letter .pdf

Branigan, Tania. 2012. "China's 'Golden Decade' Brings Some Relief to Rural Poor." *Guardian* (Nov. 5). Retrieved Feb. 3, 2013. Online: www.guardian .co.uk/world/2012/nov/05/china -golden-decade-rural-poor

Brault, Matthew W. 2012. "Americans with Disabilities: 2010." United States Census Bureau (July). Retrieved May 11, 2013. Online: www.census .gov/prod/2012pubs/p70-131.pdf

Brill, Steven. 2013. "Bitter Pill: How Outrageous Pricing and Egregious Profits Are Destroying Our Health Care." *Time* (Mar. 4): 16–55.

Brooks, Karen. 2011. "Pageant Parents Set Ugly Example." Couriermail .com (Mar. 30). Retrieved Feb. 19, 2013. Online: www.couriermail .com.au/spike/columnists/pageant -parents-set-ugly-example/story -e6frerfo-1226030268594

Brown, E. Richard. 1979. *Rockefeller Medicine Men.* Berkeley: University of California Press.

Brown, Patrick R., Andy Alaszewski, Trish Swift, and Andy Nordin. 2011. "Actions Speak Louder Than Words: The Embodiment of Trust by Healthcare Professionals in Gynae-oncology." *Sociology of Health & Illness* 33 (2): 280–295.

Brown, Robert W. 1954. "Mass Phenomena." In Gardner Lindzey (Ed.), *Handbook of Social Psychology* (vol. 2). Reading, MA: Addison-Wesley, pp. 833–873.

Bryan, Craig J., and Ann Marie Hernandez. 2013. "The Functions of Social Support as Protective Factors for Suicidal Ideation in a Sample of Air Force Personnel." *Suicide and Life-Threatening Behavior* 45 (Oct.): 562–573.

Buechler, Steven M. 2000. *Social Movements in Advanced Capitalism: The Political Economy and Cultural Construction of Social Activism.* New York: Oxford University Press.

Bullard, Robert B., and Beverly H. Wright. 1992. "The Quest for Environmental Equity: Mobilizing the African-American Community for Social Change." In Riley E. Dunlap and Angela G. Mertig (Eds.), *American Environmentalism: The U.S. Environmental Movement, 1970–1990.* New York: Taylor & Francis, pp. 39–49.

Bumiller, Elisabeth. 2012. "Active-Duty Soldiers Take Their Own Lives at Record Rate." *New York Times* (Jan. 19). Retrieved Sept. 23, 2012. Online: www.nytimes.com/2012/01/20/us /active-duty-army-suicides-reach -record-high.html

Burger, Jerry M. 2009. "Replicating Milgram: Would People Still Obey Today?" *American Psychologist* (Jan.): 1–11. Retrieved Dec. 27, 2013. Online: www.apa.org/pubs/journals/releases /amp-64-1-1.pdf

Burgess, Ernest W. 1925. "The Growth of the City." In Robert E. Park and Ernest W. Burgess (Eds.), *The City.* Chicago: University of Chicago Press, pp. 47–62.

Burgess-Proctor, Amanda. 2006. "Intersections of Race, Class, Gender and Crime: Future Directions for Feminist Criminology." *Feminist Criminology,* 1: 27–47. Retrieved Dec. 24, 2012. Online: http://fcx.sagepub.com /content/1/1/27.full.pdf+html

Caldwell, Leigh Ann. 2012. "Obama Campaign's 'Julia' Turns into Republican Fodder" (May 4). CBSnews .com. Retrieved Mar. 30, 2013. Online: www.cbsnews.com/8301 -503544_162-57428336-503544 /obama-campaigns-julia-turns-into -republican-fodder

*Campus Times.* 2008. "Survey Analyzes Students' Stress Levels" (Aug. 10). Retrieved Feb. 8, 2010. Online: www .campustimes.org/2.4981/survey -analyzes-students-stress-levels-1.492569

Cantril, Hadley. 1941. *The Psychology of Social Movements.* New York: Wiley.

Capek, Stella M. 1993. "The 'Environmental Justice' Frame: A Conceptual Discussion and Application." *Social Problems,* 40 (1): 5–23.

Carson, Rachel. 1962. *Silent Spring.* Boston: Houghton Mifflin.

Carter, Stephen L. 1994. *The Culture of Disbelief: How American Law and Politics Trivializes Religious Devotion.* New York: Anchor/Doubleday.

Castells, Manuel. 1977. *The Urban Question.* London: Edward Arnold (orig. pub. 1972 as *La Question Urbaine,* Paris).

Catalyst. 2013. "2013 Catalyst Census: Fortune 500." Retrieved Feb. 11, 2014. Online: www.catalyst.org/knowledge /2013-catalyst-census-fortune-500 -women-board-directors

*CBS News.* 2007. "Outsourced 'Wombs-for-Rent' in India." Retrieved Feb. 9, 2008. Online: www.cbsnews.com /stories/2007/12/31/health/main 3658750.shtml

*CDC Morbidity and Mortality Weekly Report.* 2013. "CDC Health Disparities and Inequalities Report—United States, 2013" (Nov. 22). Retrieved Feb. 9, 2014. Online: www.cdc.gov /mmwr/pdf/other/su6203.pdf

Center for Responsive Politics. 2013. "Super PACs." OpenSecrets.org. Retrieved May 27, 2014. Online: www .opensecrets.org/pacs/superpacs.php

Chagnon, Napoleon A. 1992. *Yanomamo: The Last Days of Eden.* New York: Harcourt (rev. from 4th ed., *Yanomamo: The Fierce People,* published by Holt, Rinehart & Winston).

Chambliss, William J. 1973. "The Saints and the Roughnecks." *Society,* 11: 24–31.

Cherlin, A. J. 2010. "Demographic Trends in the United States: A Review of Research in the 2000s." *Journal of Marriage and Family,* 72: 403–419.

chiefexecutive.com. 2010. "Changing Management Practice for the 21st Century." Retrieved Mar. 8, 2010. Online: www.the-chiefexecutive.com /features/feature54442

Child Statistics. 2012. "America's Children in Brief: Key National Indicators of Well-Being, 2012." Federal Interagency Forum on Child and Family Statistics. Retrieved Feb. 23, 2013. Online: http://childstats.gov/pdf /ac2012/ac_12.pdf

Child Trends Data Bank. 2012. "Indicators by Topic Area." Retrieved Apr. 4, 2014. Online: www.childtrends.org /databank/indicators-by-topic-area

Child Welfare Information Gateway. 2008. "Long-Term Consequences of Child Abuse and Neglect." Fact Sheet (Apr.). Retrieved Dec. 23, 2011. Online: www.childwelfare.gov/pubs/factsheets /long_term_consequences.pdf

Children's Bureau. 2013. "Child Welfare Outcomes 2008–2011: Report to

Congress." U.S. Department of Health and Human Services. Retrieved Feb. 24, 2014. Online: http://childwelfare.co/wp-content/uploads/2013/09/cwo08_11.pdf

Children's Defense Fund. 2011. "Children in the United States: Fact Sheet." Retrieved Feb. 21, 2011. Online: www.childrensdefense.org/child-research-data-publications/data/state-data-repository/cits/2011/children-in-the-states-2011-united-states.pdf

—. 2012a. "Each Day in America." Retrieved Mar. 3, 2013. Online: www.childrensdefense.org/child-research-data-publications/each-day-in-america.html

—. 2012b. *The State of America's Children: Handbook, 2012.* Retrieved Jan. 12, 2013. Online: www.childrensdefense.org/child-research-data-publications/data/soac-2012-handbook.pdf

—. 2013. "Child Poverty in America 2012: National Analysis." Retrieved Jan. 20, 2014. Online: www.childrensdefense.org/child-research-data-publications/data/child-poverty-in-america-2012.pdf

*Chronicle of Higher Education.* 2011. "Almanac Issue: 2011–12" (Aug. 26).

—. 2012. "Almanac." Online: http://chronicle.com/article/Almanac-2012-Students/133763

—. 2013. "Almanac." Online: http://chronicle.com/article/Almanac-2013-Students/140795

Chu, Kathy. 2012. "Apple Plans Environmental Audits of China Suppliers." *USA Today* (Feb. 20). Retrieved Feb. 17, 2012. Online: www.usatoday.com/tech/news/story/2012-02-20/apple-china-environmental-audits/53167970/1

*CIA Factbook.* 2013. "Country Comparisons: Life Expectancy at Birth." Retrieved July 8, 2013. Online: www.cia.gov/library/publications/the-world-factbook/rankorder/2102rank.html

—. 2014. "The World Factbook." Retrieved Apr. 6, 2014. Online: www.cia.gov/library/publications/the-world-factbook/fields/2102.html

cia.gov. 2013. "World Factbook—Country Comparisons: Net Migration Rate." Retrieved July 9, 2013. Online: www.cia.gov/library/publications/the-world-factbook/rankorder/2112rank.html

Clayman, Steven E. 1993. "Booing: The Anatomy of a Disaffiliative Response." *American Sociological Review,* 58 (1): 110–131.

Clinton, Hillary Rodham. 2011. "Empowering Women Helps Fuel Global Economic Growth." Bloomberg.com (Mar. 8). Retrieved Mar. 14, 2011. Online: www.bloomberg.com/news/print/2011-03-08/empowering-women-helps-global-growth-commentary-by-hillary-rodham-clinton.html

Cloward, Richard A., and Lloyd E. Ohlin. 1960. *Delinquency and Opportunity: A Theory of Delinquent Gangs.* New York: Free Press.

CNN.com. 2012a. "At a Glance: Supreme Court Decision on Arizona's Immigration Law." Retrieved May 14, 2013. Online: www.cnn.com/interactive/2012/06/us/scotus.immigration/index.html?pos=canon

—. 2012b. "Knife Attack at Chinese School Wounds 22 Children" (Dec. 14). Retrieved Dec. 27, 2012. Online: www.cnn.com/2012/12/14/world/asia/china-knife-attack/index.html

—. 2013. "Milan's Boateng Takes Stand over Racism" (Jan. 4). Retrieved Feb. 9, 2013. Online: http://edition.cnn.com/2013/01/03/sport/football/football-ac-milan-racism-boateng

Coakley, Jay. 2009. *Sports in Society: Issues and Controversies* (10th ed.). New York: McGraw-Hill.

Coleman, Richard P., and Lee Rainwater. 1978. *Social Standing in America: New Dimensions of Class.* New York: Basic.

Coleman-Jensen, Alisha, Mark Nord, and Anita Singh. 2013. "Household Food Security in the United States in 2012." United States Department of Agriculture (Sept.). Retrieved Jan. 5, 2014. Online: www.ers.usda.gov/publications/err-economic-research-report/err155.aspx

Colford, Paul. 2013. "Illegal Immigrant No More." The Definitive Source (Apr. 2). Retrieved May 16, 2013. Online: http://blog.ap.org/2013/04/02/illegal-immigrant-no-more

Collins, Patricia Hill. 2000. *Black Feminist Thought: Knowledge, Consciousness, and the Politics of Empowerment* (2nd ed.). New York: Routledge.

Collins, Randall. 1979. *The Credential Society: An Historical Sociology of Education.* New York: Academic Press.

—. 1982. *Sociological Insight: An Introduction to Non-Obvious Sociology.* New York: Oxford University Press.

Collins, Rebecca L. 2011. "Content Analysis of Gender Roles in Media: Where Are We Now and Where Should We Go?" *Sex Roles,* 64 (3/4): 290–329.

Coltrane, Scott. 2010. "Fathering: Paradoxes, Contradictions, and Dilemmas." In Michael S. Kimmel and Michael A. Messner (Eds.), *Men's Lives* (8th ed.). Boston: Allyn & Bacon, pp. 432–449.

Congressional Budget Office. 2011. "A Description of the Immigrant Population: An Update." Retrieved Feb. 5, 2012. Online: www.cbo.gov/ftpdocs/121xx/doc12168/06-02-Foreign-BornPopulation.pdf

Cooley, Charles Horton. 1963. *Social Organization: A Study of the Larger Mind.* New York: Schocken (orig. pub. 1909).

—. 1998. "The Social Self—the Meaning of 'I.'" In Hans-Joachim Schubert (Ed.), *On Self and Social Organization—Charles Horton Cooley.* Chicago: University of Chicago Press, pp. 155–175. Reprinted from Charles Horton Cooley, *Human Nature and the Social Order.* New York: Schocken, 1902.

Coontz, Stephanie. 2000. *The Way We Never Were: American Families and the Nostalgia Trap.* New York: Basic.

Cooper, Elizabeth. 2011. "Citizen's Voice: Why Intelligent Design Doesn't Qualify as Science." Knoxnews.com (Apr. 9). Retrieved May 2, 2011. Online: www.knoxnews.com/news/2011/apr/09/why-intelligent-design-doesnt-qualify-as-science/?print=1

Copen, Casey E., Kimberly Daniels, and William D. Mosher. 2013. "First Premarital Cohabitation in the United States: 2006–2010 National Survey of Family Growth." National Health Statistics Report (Apr. 4). Centers for Disease Control and Prevention, U.S. Department of Health and Human Services. Retrieved Feb. 19, 2014. Online: www.cdc.gov/nchs/data/nhsr/nhsr064.pdf

Corsaro, William A. 2011. *The Sociology of Childhood* (3rd ed.). Thousand Oaks, CA: Pine Forge.

Cortese, Anthony J. 2004. *Provocateur: Images of Women and Minorities in Advertising* (2nd ed.). Latham, MD: Rowman and Littlefield.

Cox, Oliver C. 1948. *Caste, Class, and Race.* Garden City, NY: Doubleday.

Cronin, John, and Robert F. Kennedy Jr. 1999. *The Riverkeepers: Two Activists Fight to Reclaim Our Environment as a Basic Human Right.* New York: Touchstone.

Curtiss, Susan. 1977. *Genie: A Psycholinguistic Study of a Modern Day "Wild Child."* New York: Academic Press.

Cyrus, Virginia. 1993. *Experiencing Race, Class, and Gender in the United States.* Mountain View, CA: Mayfield.

Dahl, Robert A. 1961. *Who Governs?* New Haven, CT: Yale University Press.

Daly, Mary. 1973. *Beyond God the Father.* Boston: Beacon.

Daniels, Jessie. 2009. *Cyber Racism: White Supremacy Online and the New Attack on Civil Rights.* Lanham, MD: Rowman & Littlefield.

—. 2010. "Cyber Racism on College Campuses." *Race-Talk* (May 4). Retrieved Apr. 29, 2011. Online: http://blogs.alternet.org/speakeasy/2010/05/04/cyber-racism-on-college-campuses

Daniels, Roger. 1993. *Prisoners Without Trial: Japanese-Americans in World War II.* New York: Hill & Wang.

Davidson, Jacob. 2013. "Two Years Later, Portland's 'Last Occupier' Still Going Strong." *Time* (Sept. 16). Retrieved Dec. 28, 2013. Online: http://newsfeed.time.com/2013/09/16/two-years-later-portlands-last-occupier-still-going-strong/print

Davidson, Janet, and Meda Chesney-Lind. 2009. "Discounting Women: Context Matters in Risk and Need Assessment." *Critical Criminology,* 17 (4): 221–245.

Davis, Fred. 1992. *Fashion, Culture, and Identity.* Chicago: University of Chicago Press.

Davis, Kingsley. 1940. "Extreme Social Isolation of a Child." *American Journal of Sociology,* 45 (4): 554–565.

Davis, Kingsley, and Wilbert Moore. 1945. "Some Principles of Stratification." *American Sociological Review,* 7 (April): 242–249.

Davis-Delano, Laurel R., April Pollock, and Jennifer Ellsworth Vose. 2009. "Apologetic Behavior Among Female Athletes: A New Questionnaire and Initial Results." *International Review for the Sociology of Sport,* 44: 131–150.

Death Penalty Information Center. 2013. "Facts About the Death Penalty." Retrieved Dec. 27, 2013. Online: www.deathpenaltyinfo.org/documents/FactSheet.pdf

DeCarlo, Scott. 2013. "The World's Biggest Companies." Forbes.com (Apr. 17, 2013). Retrieved Apr. 19, 2013. Online: www.forbes.com/sites/scottdecarlo/2013/04/17/the-worlds-biggest-companies-2

Delgado, Richard. 1995. "Introduction." In Richard Delgado (Ed.), *Critical Race Theory: The Cutting Edge.* Philadelphia: Temple University Press, pp. xiii–xvi.

DeNavas-Walt, Carmen, Robert W. Cleveland, and Bruce H. Webster Jr. 2003. "Income in the United States: 2002." U.S. Census Bureau, Current Population Reports, P60–221. Washington, DC: U.S. Government Printing Office.

DeNavas-Walt, Carmen, Bernadette D. Proctor, and Jessica C. Smith. 2012. *Income, Poverty, and Health Insurance Coverage in the United States: 2011.* U.S. Census Bureau (Sept.). Retrieved May 10, 2013. Online: www.census.gov/prod/2012pubs/p60-243.pdf

—. 2013. *Income, Poverty, and Health Insurance Coverage in the United States, 2012.* U.S. Census Bureau, Current Population Reports, P60-245. Washington, DC: U.S. Government Printing Office. Retrieved Jan. 5, 2013. Online: www.census.gov/prod/2013pubs/p60-245.pdf

Derber, Charles. 1983. *The Pursuit of Attention: Power and Individualism in Everyday Life.* New York: Oxford University Press.

Dolan, Andy, and Martin Robinson. 2013. "Schoolgirl Is 'Trolled to Death': Parents' Agony as Daughter, 14, 'Hangs Herself' After Horrific Abuse from Bullies on Website Ask.fm." *Daily Mail Online* (Aug. 5). Retrieved Dec. 7, 2013. Online: www.dailymail.co.uk/news/article-2384866/Schoolgirl-Hannah-Smith-trolled-death-bullies-Ask-fm-website.html

Dollard, John, Neal E. Miller, Leonard W. Doob, O. H. Mowrer, and Robert R. Sears. 1939. *Frustration and Aggression.* New Haven, CT: Yale University Press.

Dolnick, Sam. 2010. "The Obesity–Hunger Paradox." *New York Times* (Mar. 14): A27.

Domhoff, G. William. 2002. *Who Rules America? Power and Politics* (4th ed.). New York: McGraw-Hill.

—. 2005. "The Class-Domination Theory of Power." Retrieved Mar. 23, 2013. Online: www2.ucsc.edu/whorulesamerica/power/class_domination.html

—. 2014. "Power in America: The Class-Domination Theory of Power." Retrieved May 27, 2014. Online: www2.ucsc.edu/whorulesamerica/power/class_domination.html

Driskell, Robyn Bateman, and Larry Lyon. 2002. "Are Virtual Communities True Communities? Examining the Environments and Elements of Community." *City & Community,* 1 (4): 1–18.

Du Bois, W. E. B. 1967. *The Philadelphia Negro: A Social Study.* New York: Schocken (orig. pub. 1899).

Dunbar, Polly. 2007. "Wombs to Rent: Childless British Couples Pay Indian Women to Carry Their Babies." *Daily Mail* (Dec. 8). Retrieved Feb. 9, 2008. Online: www.dailymail.co.uk/pages/live/articles/news/worldnews.html?in_article_id=500601

Durkheim, Emile. 1933. *The Division of Labor in Society.* Trans. George Simpson. New York: Free Press (orig. pub. 1893).

—. 1956. *Education and Sociology.* Trans. Sherwood D. Fox. Glencoe, IL: Free Press.

—. 1964a. *The Rules of Sociological Method.* Trans. Sarah A. Solovay and John H. Mueller. New York: Free Press (orig. pub. 1895).

—. 1964b. *Suicide.* Trans. John A. Sparkling and George Simpson. New York: Free Press (orig. pub. 1897).

—. 1995. *The Elementary Forms of Religious Life.* Trans. Karen E. Fields. New York: Free Press (orig. pub. 1912).

Dye, Thomas R., Harmon Zeigler, and Louis Schubert. 2012. *The Irony of Democracy: An Uncommon Introduction to American Politics* (15th ed.). Boston: Wadsworth.

Early, Kevin E. 1992. *Religion and Suicide in the African-American Community.* Westport, CT: Greenwood.

Ebaugh, Helen Rose Fuchs. 1988. *Becoming an EX: The Process of Role Exit.* Chicago: University of Chicago Press.

Eckholm, Erik. 2009. "Gang Violence Grows on an Indian Reservation." *New York Times* (Dec. 14): A13.

—. 2013. "With Police in Schools, More Children in Court." *New York Times* (Apr. 12). Retrieved Apr. 13, 2013. Online: www.nytimes.com/2013/04/12/education/with-police-in-schools-more-children-in-court.html?_r=0

Economist.com. 2011. "Environmental Activism in China: Poison Protests—A Huge Demonstration Over a Chemical Factory Unnerves Officials" (Aug. 20). Retrieved Mar. 17, 2012. Online: www.economist.com/node/21526417

edivorcestatistics.com. 2013. "Remarriage After Divorce Statistics." Retrieved Apr. 8, 2013. Online:

www.edivorcestatistics.com/remarriage.html

Edwards, Kathryn A., Anna Turner, and Alexander Hertel-Fernandez. 2011. *A Young Person's Guide to Social Security*. Economic Policy Institute. Retrieved Dec. 17, 2011. Online: www.epi.org

Egley, Arlen, Jr., and James C. Howell. 2012. "Highlights of the 2010 National Youth Gang Survey" (Apr.). U.S. Department of Justice, Office of Juvenile Justice and Delinquency Prevention. Retrieved Dec. 28, 2013. Online: www.ojjdp.gov/pubs/237542.pdf

Ehrenreich, Barbara. 2001. *Nickel and Dimed: On (Not) Getting by in America*. New York: Metropolitan.

—. 2011. *Nickel and Dimed: On (Not) Getting by in America* (10th anniversary edition). New York: Picador.

Ehrlich, Paul R., and Anne H. Ehrlich. 2009. "The Population Bomb Revisited." *Electronic Journal of Sustainable Development*, 1 (3). Retrieved May 22, 2011. Online: http://fragette.free.fr/demography/The_Population_Bomb_Revisited.pdf

Eighner, Lars. 1993. *Travels with Lizbeth*. New York: St. Martin's.

Eisenstein, Zillah R. 1994. *The Color of Gender: Reimaging Democracy*. Berkeley: University of California Press.

Elkind, David. 1995. "School and Family in the Postmodern World." *Phi Delta Kappan* (September): 8–21.

Engels, Friedrich. 1970. *The Origins of the Family, Private Property, and the State*. New York: International (orig. pub. 1884).

Ennis, Sharon, Merarys Rios-Vargas, and Nora G. Albert. 2011. "The Hispanic Population: 2010." 2010 Census Briefs (May). Retrieved Dec. 21, 2011. Online: www.census.gov/prod/cen2010/briefs/c2010br-04.pdf

EPA.gov. 2013. "Creating Equitable, Healthy, and Sustainable Communities" (Feb.). Retrieved May 16, 2013. Online: www.epa.gov/smartgrowth/pdf/equitable-dev/equitable-development-report-508-011713b.pdf

Erikson, Kai T. 1962. "Notes on the Sociology of Deviance." *Social Problems*, 9: 307–314.

—. 1964. "Notes on the Sociology of Deviance." In Howard S. Becker (Ed.), *The Other Side: Perspectives on Deviance*. New York: Free Press, pp. 9–21.

—. 1976. *Everything in Its Path: Destruction of Community in the Buffalo Creek Flood*. New York: Simon & Schuster.

—. 1991. "A New Species of Trouble." In Stephen Robert Couch and J. Stephen Kroll-Smith (Eds.), *Communities at Risk: Collective Responses to Technological Hazards*. New York: Land, pp. 11–29.

—. 1994. *A New Species of Trouble: Explorations in Disaster, Trauma, and Community*. New York: Norton.

Essed, Philomena. 1991. *Understanding Everyday Racism*. Newbury Park, CA: Sage.

Etzioni, Amitai. 1975. *A Comparative Analysis of Complex Organizations: On Power, Involvement, and Their Correlates* (rev. ed.). New York: Free Press.

farenet.org. 2011. "Star Tells of Sadness at Racist Taunts." FARE. Retrieved May 23, 2011. Online: www.farenet.org/default.asp?intPageID=7&intArticleID=714

Feagin, Joe R. 1991. "The Continuing Significance of Race: Antiblack Discrimination in Public Places." *American Sociological Review*, 56 (February): 101–116.

Feagin, Joe R., and Clairece Booher Feagin. 1994. *Social Problems: A Critical Power–Conflict Perspective* (4th ed.). Englewood Cliffs, NJ: Prentice Hall.

—. 2012. *Race and Ethnic Relations, Census Update* (9th ed.). Upper Saddle River, NJ: Pearson.

Feagin, Joe R., and Robert Parker. 2002. *Building American Cities: The Urban Real Estate Game* (2nd ed.). Hopkins, MN: Beard.

Feagin, Joe R., and Hernán Vera. 1995. *White Racism: The Basics*. New York: Routledge.

Federal Bureau of Investigation (FBI). 2012. *Crime in the United States, 2011*. Retrieved Dec. 24, 2012. Online: www.fbi.gov/about-us/cjis/ucr/crime-in-the-u.s/2011/crime-in-the-u.s.-2011/violent-crime/violentcrimemain_final.pdf

—. 2013. *Crime in the United States 2012*. Federal Bureau of Investigation. Retrieved Dec. 31, 2013. Online: www.fbi.gov/about-us/cjis/ucr/crime-in-the-u.s/2012/crime-in-the-u.s.-2012/resource-pages/download-printable-files

Federal Interagency Forum on Child and Family Statistics. 2013. "America's Children: Key National Indicators of Well-Being, 2013." National Institute of Health. Retrieved Feb. 22, 2014. Online: www.nichd.nih.gov/publications/pubs/Documents/Americas_Children_2013_DRAFT.pdf

*Federal Register*. 2013. "Annual Update on HHS Poverty Guidelines" (Jan. 24). Retrieved Jan. 4, 2014. Online: www.federalregister.gov/articles/2013/01/24/2013-01422/annual-update-of-the-hhs-poverty-guidelines#t-1

Feiler, Bruce. 2013. "The Stories That Bind Us." *New York Times* (Mar. 15). Retrieved Apr. 1, 2013. Online: www.nytimes.com/2013/03/17/fashion/the-family-stories-that-bind-us-this-life.html?ref=fashion&_r=1&

File, Thom. 2013. "Computer and Internet Use in the United States" (May). Retrieved Dec. 26, 2013. Online: www.census.gov/prod/2013pubs/p20-569.pdf

Findlay, Deborah A., and Leslie J. Miller. 1994. "Through Medical Eyes: The Medicalization of Women's Bodies and Women's Lives." In B. Singh Bolaria and Harley D. Dickinson (Eds.), *Health, Illness, and Health Care in Canada* (2nd ed.). Toronto: Harcourt, pp. 276–306.

Firestone, Shulamith. 1970. *The Dialectic of Sex*. New York: Morrow.

Flexner, Abraham. 1910. *Medical Education in the United States and Canada*. New York: Carnegie Foundation.

*Forbes*. 2012. "Average America vs. the One Percent" (Mar. 21). Retrieved Jan. 4, 2013. Online: www.forbes.com/sites/moneywisewomen/2012/03/21/average-america-vs-the-one-percent

—. 2013. "The World's Billionaires" (Mar.). Retrieved Jan. 4, 2014. Online: www.forbes.com/billionaires/list

Foucault, Michel. 1979. *Discipline and Punish: The Birth of the Prison*. New York: Vintage.

—. 1988. *Madness and Civilization: A History of Insanity in the Age of Reason*. New York: Vintage (orig. pub. 1961).

—. 1994. *The Birth of the Clinic: An Archeology of Medical Perception*. New York: Vintage (orig. pub. 1963).

Frank, Marc. 2014. "Cuba, United States Meet on Migration in Latest Sign of a Thaw." Reuters.com (Jan. 9). Retrieved Feb. 10, 2014. Online: www.reuters.com/article/2014/01/09/us-cuba-usa-idUSBREA0810J20140109

Franklin, John Hope. 1980. *From Slavery to Freedom: A History of Negro Americans*. New York: Vintage.

Freidson, Eliot. 1965. "Disability as Social Deviance." In Marvin B. Sussman (Ed.), *Sociology and Rehabilitation*.

Washington, DC: American Sociology Association, pp. 71–99.

—. 1970. *Profession of Medicine.* New York: Dodd, Mead.

—. 1986. *Professional Powers.* Chicago: University of Chicago Press.

French, Sally. 1999. "The Wind Gets in My Way." In Mairian Corker and Sally French (Eds.), *Disability Discourse.* Buckingham, England: Open University Press, pp. 21–27.

Freud, Sigmund. 1924. *A General Introduction to Psychoanalysis* (2nd ed.). New York: Boni & Liveright.

Frey, William H. 2011. "Melting Pot Cities and Suburbs: Racial and Ethnic Changes in Metro America in the 2000s." Metropolitan Policy Program at Brookings. Retrieved Mar. 15, 2012. Online: www.brookings.edu/~/media /Files/rc/papers/2011/0504_census _ethnicity_frey/0504_census_ethnicity _frey.pdf

Frisbie, W. Parker, and John D. Kasarda. 1988. "Spatial Processes." In Neil Smelser (Ed.), *The Handbook of Sociology.* Newbury Park, CA: Sage, pp. 629–666.

Frome, Pamela M., Corinne J. Alfeld, Jacquelynne S. Eccles, and Bonnie L. Barber. 2006. "Why Don't They Want a Male-Dominated Job: An Investigation of Young Women Who Changed Their Occupational Aspirations." *Educational Research and Evaluation,* 12 (4): 359–372.

Fry, Richard, and Paul Taylor. 2013. "An Uneven Recovery, 2009–2011: A Rise in Wealth for the Wealthy; Declines for the Lower 93%" (Apr. 23). Pew Research Social & Demographic Trends. Retrieved Jan. 13, 2014. Online: www .pewsocialtrends.org/files/2013/04 /wealth_recovery_final.pdf

Galbraith, John Kenneth. 1985. *The New Industrial State* (4th ed.). Boston: Houghton Mifflin.

Gallup Organization. 2011a. "Immigration." Retrieved Feb. 5, 2012. Online: www.gallup.com

—. 2011b. "Race Relations." Retrieved Dec. 27, 2011. Online: www.gallup .com/poll/1687/race-relations.aspx

Gambino, Richard. 1975. *Blood of My Blood.* New York: Doubleday/Anchor.

Gandara, Ricardo. 1995. "*Dichos de la Vida:* Homespun Proverbs Link Hispanic Culture's Past with the Present." *Austin American-Statesman* (Jan. 21): E1, E10.

Gans, Herbert. 1982. *The Urban Villagers: Group and Class in the Life of Italian Americans* (updated and expanded ed.; orig. pub. 1962). New York: Free Press.

Garfinkel, Harold. 1967. *Studies in Ethnomethodology.* Englewood Cliffs, NJ: Prentice Hall.

Garreau, Joel. 1991. *Edge City: Life on the New Frontier.* New York: Doubleday.

Gates, Gary J. 2011. "How Many People Are Lesbian, Gay, Bisexual and Transgender?" Williams Institute at the University of California School of Law (Apr.). Retrieved Feb. 10, 2012. Online: http:// williamsinstitute.law.ucla.edu/research /census-lgbt-demographics-studies /how-many-people-are-lesbian-gay -bisexual-and-transgender

Gates, Gary J., and Frank Newport. 2012. "Special Report: 3.4% of U.S. Adults Identify as LGBT" (Oct. 18). Retrieved Feb. 21, 2013. Online: www .gallup.com/poll/158066/special -report-adults-identify-lgbt.aspx

Gavora, Jessica. 2012. "Obama's 'Julia' Ad and the New Hubby State" (May 11). Washingtonpost.com. Retrieved Mar. 30, 2013. Online: http://articles .washingtonpost.com/2012-05-11 /opinions/35455415_1_single-women -unmarried-women-women-favor

Gawande, Atul. 2002. *Complications: A Surgeon's Notes on an Imperfect Science.* New York: Picador.

Geertz, Clifford. 1966. "Religion as a Cultural System." In Michael Banton (Ed.), *Anthropological Approaches to the Study of Religion.* London: Tavistock, pp. 1–46.

General Motors. 2013. "About GM: Board of Directors." Retrieved July 2, 2013. Online: www.gm.com /company/board-of-directors

genomics.energy.gov. 2007. "Human Genome Project Information: Minorities, Race, and Genomics." Retrieved Feb. 9, 2013. Online: www.ornl.gov /sci/techresources/Human_Genome /elsi/minorities.shtml

—. 2012. "Genetically Modified Food and Organisms." Retrieved Mar. 17, 2012. Online: www.ornl.gov/sci /techresources/Human_Genome/elsi /gmfood.shtml

Gerth, Hans H., and C. Wright Mills. 1946. *From Max Weber: Essays in Sociology.* New York: Oxford University Press.

Gibbs, Lois. 2013. "Backyard Talk: A Circle of Poison and Poverty." Center for Health, Environment and Justice (May 16). Retrieved May 16, 2013. Online: http://chej.org/backyard-talk

Gibbs, Lois Marie, as told to Murray Levine. 1982. *Love Canal: My Story.* Albany: SUNY Press.

Gibson, Campbell, and Kay Jung. 2006. "Historical Census Statistics on the Foreign-Born Population of the United States: 1850 to 2000." U.S. Census Bureau. Retrieved Feb. 5, 2012. Online: www.census.gov /population/www/documentation /twps0081/twps0081.html#trends

Gilbert, Dennis. 2010. *The American Class Structure in an Age of Growing Inequality* (8th ed.). Newbury Park, CA: Pine Forge/Sage.

—. 2011. *The American Class Structure in an Age of Growing Inequality* (8th ed.). Thousand Oaks, CA: Sage.

Gill, Nikhila. 2012. "Extreme Poverty Drops Worldwide." *New York Times* (Mar. 4). Retrieved Feb. 3, 2013. Online: india.blogs.nytimes .com/2012/03/04/extreme-poverty -drops-worldwide/?pagewanted=print

Gilligan, Carol. 1982. *In a Different Voice: Psychological Theory and Women's Development.* Cambridge, MA: Harvard University Press.

Gilman, Charlotte Perkins. 1976. *His Religion and Hers.* Westport, CT: Hyperion (orig. pub. 1923).

Giuliano, Traci A., Kathryn E. Popp, and Jennifer L. Knight. 2000. "Footballs Versus Barbies: Childhood Play Activities as Predictors of Sports Participation by Women." *Sex Roles,* 42: 159–181.

Glaeser, Edward. 2011. *Triumph of the City.* New York: Penguin.

Goffman, Erving. 1959. *The Presentation of Self in Everyday Life.* Garden City, NY: Doubleday.

—. 1961a. *Asylums: Essays on the Social Situation of Mental Patients and Other Inmates.* Chicago: Aldine.

—. 1961b. *Encounters: Two Studies in the Sociology of Interaction.* London: Routledge and Kegan Paul.

—. 1963a. *Behavior in Public Places: Notes on the Social Structure of Gatherings.* New York: Free Press.

—. 1963b. *Stigma: Notes on the Management of Spoiled Identity.* Englewood Cliffs, NJ: Prentice Hall.

—. 1967. *Interaction Ritual: Essays on Face to Face Behavior.* Garden City, NY: Anchor.

—. 1974. *Frame Analysis: An Essay on the Organization of Experience.* Boston: Northeastern University Press.

Gold, Matea, and Melanie Mason. 2012. "NRA Calls for Armed Guards in Schools to Prevent Gun Violence."

*Los Angeles Times* (Dec. 22). Retrieved Dec. 27, 2012. Online: http://articles.latimes.com/print/2012/dec/22/nation/la-na-guns-nra-20121222

Gordon, Milton. 1964. *Assimilation in American Life: The Role of Race, Religion, and National Origins.* New York: Oxford University Press.

Gottdiener, Mark. 1985. *The Social Production of Urban Space.* Austin: University of Texas Press.

Gottfredson, Michael R., and Travis Hirschi. 1990. *A General Theory of Crime.* Palo Alto, CA: Stanford University Press.

Green, Erica L. 2011. "Prayer Service at City School Called Improper." *Baltimore Sun* (Mar. 13). Retrieved May 8, 2011. Online: http://articles.baltimoresun.com/2011-03-13/news/bs-md-ci-msa-prayer-service-20110313_1_prayer-service-prayer-in-public-schools-voluntary-student-prayer

Green, Rick. 2010. "We Need to Pay More Attention to Bullying at Schools." Courant.com (Apr. 2). Retrieved Apr. 2, 2010. Online: www.courant.com/news/opinion/courant-columnists/hc-green,0,3979875.columnist

GreenParty.org. 2013. "Another U.S. Is Possible—Another Party Is Necessary." Retrieved Mar. 29, 2013. Online: www.gp.org/history/php

Griswold, Eliza. 2012. "How 'Silent Spring' Ignited the Environmental Movement." *New York Times* (Sept. 21). Retrieved May 16, 2013. Online: www.nytimes.com/2012/09/23/magazine/how-silent-spring-ignited-the-environmental-movement.html?pagewanted=all&_r=0

Grogan, Sarah. 2007. *Body Image: Understanding Body Dissatisfaction in Men, Women and Children.* New York: Routledge.

Gross, Daniel. 2011. "The Real China: Urban Wealth, Rural Poverty." Yahoo! Finance (Nov. 7). Retrieved Feb. 3, 2013. Online: http://finance.yahoo.com/blogs/daniel-gross/real-china-urban-wealth-rural-poverty-124416176.html

Gryn, Thomas A., and Luke J. Larsen. 2010. "Nativity Status and Citizenship in the United States: 2009." American Community Survey Briefs, U.S. Census Bureau. Retrieved Feb. 5, 2012. Online: www.census.gov/prod/2010pubs/acsbr09-16.pdf

Guerriero, John. 2012. "Homeless in Tents: 'Distressingly Common.' " *Erie Times-News* (Sept. 16). Retrieved Nov. 9, 2012. Online: www.goerie.com/article/20120916/NEWS02/309169921/Homeless-in-tents%3A-%22Distressingly-common%27%27

Guha, Ramachandra. 2004. "The Sociology of Suicide." Retrieved Dec. 20, 2007. Online: www.indiatogether.org/2004/aug/rgh-suicide.htm

Hahn, Harlan. 1987. "Civil Rights for Disabled Americans: The Foundation of a Political Agenda." In Alan Gartner and Tom Joe (Eds.), *Images of the Disabled, Disabling Images.* New York: Praeger, pp. 181–203.

—. 1997. "Advertising the Acceptably Employable Image." In Lennard J. Davis (Ed.), *The Disability Studies Reader.* New York: Routledge, pp. 172–186.

Hall, Edward. 1966. *The Hidden Dimension.* New York: Anchor/Doubleday.

Hall, Edward T. 1959. *The Silent Language.* New York: Anchor/Doubleday.

Hall, Judy A., Jason D. Carter, and Terrence G. Horgan. 2000. "Gender Differences in Nonverbal Communication of Emotion." In Agneta H. Fischer (Ed.), *Gender and Emotion: Social Psychological Perspectives.* New York: Cambridge University Press, pp. 97–117.

Hallinan, Maureen. 2005. "Should Your School Eliminate Tracking? The History of Tracking and Detracking in America's Schools." *Education Matters,* 1–2.

Hamilton, Brady E., and Paul D. Sutton. 2013. "Recent Trends in Births and Fertility Rates Through June 2013." U.S. Centers for Disease Control and Prevention (December 6). Retrieved July 8, 2014. Online: http://www.cdc.gov/nchs/data/hestat/births_fertility_june_2013/births_june_2013.htm#figures

Harcourt, Bernard E. 2011. "Occupy Wall Street's 'Political Disobedience.' " *New York Times* (Oct. 13). Retrieved Mar. 16, 2012. Online: http://opinionator.blogs.nytimes.com/2011/10/13/occupy-wall-streets-political-disobedience/?pagemode=print

Harlow, Harry F., and Margaret Kuenne Harlow. 1962. "Social Deprivation in Monkeys." *Scientific American,* 207 (5): 137–146.

—. 1977. "Effects of Various Mother–Infant Relationships on Rhesus Monkey Behaviors." In Brian M. Foss (Ed.), *Determinants of Infant Behavior* (vol. 4). London: Methuen, pp. 15–36.

Harris, Chauncey D., and Edward L. Ullman. 1945. "The Nature of Cities." *Annals of the Academy of Political and Social Sciences* (November): 7–17.

Harris, Marvin. 1974. *Cows, Pigs, Wars, and Witches.* New York: Random House.

—. 1985. *Good to Eat: Riddles of Food and Culture.* New York: Simon & Schuster.

Hatton, Erin, and Mary Nell Trautner. 2011. "Equal Opportunity Objectification? The Sexualization of Men and Women on the Cover of *Rolling Stone.*" *Sexuality & Culture,* 15: 256–278.

Haun, Daniel B. M., and Michael Tomasello. 2011. "Conformity to Peer Pressure in Preschool Children." *Child Development,* 82 (6): 1759–1767.

Haviland, William A., Harald E. L. Prins, Bunny McBride, and Dana Walrath. 2013. *Cultural Anthropology: The Human Challenge* (14th ed.). Stamford, CT: Cengage Learning.

Hawley, Amos. 1950. *Human Ecology.* New York: Ronald.

—. 1981. *Urban Society* (2nd ed.). New York: Wiley.

Hayes, Liz. 2011. "Black Officials Scarce, Despite Population Gains." *Valley News Dispatch* (Dec. 4). Retrieved Dec. 29, 2011. Online: www.pittsburghlive.com/x/valleynewsdispatch/s_770348.html

Health Affairs. 2012. "Differences in Life Expectancy Due to Race and Educational Differences Are Widening, and Many May Not Catch Up." Retrieved Apr. 6, 2014. Online: http://content.healthaffairs.org/content/31/8/1803

Heritage, John. 1984. *Garfinkel and Ethnomethodology.* Cambridge, MA: Polity.

Hermsen, Joan M., David Cotter, and Reeve Vanneman. 2011. "The End of the Gender Revolution? Gender Role Attitudes from 1977 to 2008." *American Journal of Sociology* 117 (1): 259–289.

Hiebert, Paul G. 1983. *Cultural Anthropology* (2nd ed.). Grand Rapids, MI: Baker.

Hipple, Steven F. 2010. "Self-Employment in the United States." *Monthly Labor Review* (Sept.): 17–32.

Hirschi, Travis. 1969. *Causes of Delinquency.* Berkeley: University of California Press.

Hochschild, Arlie Russell. 1983. *The Managed Heart: Commercialization of Human Feeling.* Berkeley: University of California Press.

—. 1997. *The Time Bind: When Work Becomes Home and Home Becomes Work.* New York: Metropolitan.

—. 2003. *The Commercialization of Intimate Life: Notes from Home and Work.* Berkeley: University of California Press.

—. 2012. *The Outsourced Self: Intimate Life in Market Times.* New York: Metropolitan.

Hochschild, Arlie Russell, with Ann Machung. 2012. *The Second Shift: Working Families and the Revolution at Home.* New York: Penguin.

Hoffman, Jan. 2009. "Binging on Celebrity Weight Battles." *New York Times* (May 31). Retrieved Mar. 11, 2011. Online: www.nytimes.com/2009/05/31/fashion/31fat.html?_r=2&pagewanted=1

—. 2011. "Fighting Teenage Pregnancy with MTV Stars as Exhibit A." *New York Times* (Apr. 10): ST1, ST11.

Holland, Dorothy C., and Margaret A. Eisenhart. 1990. *Educated in Romance: Women, Achievement, and College Culture.* Chicago: University of Chicago Press.

Holyk, Greg. 2012. "Lackluster Popularity Dogs the Political Parties." ABCnews.com (Oct. 3). Retrieved Mar. 29, 2013. Online: http://abcnews.go.com/blogs/politics/2012/10/lackluster-popularity-dogs-the-political-parties

Hondagneu-Sotelo, Pierrette. 2007. *Doméstica: Immigrant Workers Cleaning and Caring in the Shadows of Affluence.* Berkeley: University of California Press.

Howie, John R. R. 2010. "A Final Word." *Blog-Abroad.* StudyAbroad.com. Retrieved Feb. 12, 2010. Online: www.studyabroad.com/blog-abroad/howie

Hoyt, Homer. 1939. *The Structure and Growth of Residential Neighborhoods in American Cities.* Washington, DC: Federal Housing Administration.

Hughes, Everett C. 1945. "Dilemmas and Contradictions of Status." *American Journal of Sociology,* 50: 353–359.

Humes, Karen R., Nicholas A. Jones, and Roberto R. Ramirez. 2011. "Overview of Race and Hispanic Origin: 2010." U.S. Census Bureau (Mar. 2011). Retrieved June 25, 2013. Online: www.census.gov/prod/cen2010/briefs/c2010br-02.pdf

Hurtado, Aida. 1996. *The Color of Privilege: Three Blasphemies on Race and Feminism.* Ann Arbor: University of Michigan Press.

Hussey, Kristin, and John Leland. 2013. "After Boy's Suicide, Questions About Missed Signs." *New York Times* (Aug. 30): A13.

Institute of International Education. 2011. "Profile of U.S. Study Abroad Students." *Open Door Report on International Educational Exchange.* Retrieved Nov. 24, 2012. Online: www.ile.org/opendoors

International Foundation for Electoral Systems. 2011. "Gender Issues." Retrieved Mar. 12, 2011. Online: www.ifes.org/Content/Topics/Gender-Issues.aspx

International Labor Organization. 2013. "Global Employment Trends 2013." Retrieved Jan. 24, 2014. Online: www.ilo.org/wcmsp5/groups/public/—dgreports/—dcomm/—publ/documents/publication/wcms_202326.pdf

Internet Crime Complaint Center. 2013. "Internet Crime Report, 2012." Retrieved Dec. 31, 2013. Online: www.ic3.gov/media/annualreport/2012_ic3report.pdf

Intersex Society of North America. 2011. "What Is Intersex?" Retrieved Mar. 11, 2011. Online: www.isna.org/faq/what_is_intersex

Jacobs, Andrew. 2011. "Where 'Jasmine' Means Tea, Not a Revolt." *New York Times* (Apr. 3): WK4.

—. 2013. "Leader's Visit Lifts a Village, yet Lays Bare China's Woes." *New York Times* (Jan. 26). Retrieved Feb. 3, 2013. Online: www.nytimes.com/2013/01/27/world/asia/chinas-xi-jinping-faces-problem-of-rural-poverty.html?pagewanted=all&_r=0

Janis, Irving. 1972. *Victims of Groupthink.* Boston: Houghton Mifflin.

—. 1989. *Crucial Decisions: Leadership in Policymaking and Crisis Management.* New York: Free Press.

Japsen, Bruce. 2012. "Drug Makers Dial Down TV Advertising." *New York Times* (Feb. 2). Retrieved Mar. 5, 2012. Online: http://prescriptions.blogs.nytimes.com/2012/02/02/drug-makers-dial-down-tv-advertising

John Curley Center for Sports Journalism at Penn State. 2007. "Jason Whitlock and the New Racism." *Sports, Media and Society* (Oct. 24). Retrieved Mar. 8, 2010. Online: http://sportsmediasociety.blogspot.com/2007/10/jason-whitlock-and-new-racism.html

Johnson, Earvin "Magic," with William Novak. 1992. *My Life.* New York: Fawcett Crest.

Johnson, Guy. 2006. On the Move: A Longitudinal Study of Pathways in and out of Homelessness. Ph.D. Dissertation, submitted to the School of Global Studies, Social Science and Planning, RMIT University, Australia (May). Retrieved Feb. 26, 2010. Online: www.ahuri.edu/au/downloads/Research_Training/PhDs/Johnson_Onthemove.pdf

Kaiser Family Foundation. 2012. "The Global HIV/AIDS Epidemic: Fact Sheet" (Dec.). Retrieved May 12, 2013. Online: http://kaiserfamilyfoundation.files.wordpress.com/2013/01/3030-17.pdf

—. 2013. "Snapshots: Premiums, Cost-Sharing and Coverage at Public, Private and Non-Profit Firms." Retrieved May 9, 2013. Online: http://kff.org/health-costs/issue-brief/snapshots-premiums-cost-sharing-and-coverage-at-public-private-and-non-profit-firms

Kaklenberg, Susan, and Michelle Hein. 2010. "Progression on Nickelodeon? Gender-Role Stereotypes in Toy Commercials." *Sex Roles,* 62 (11): 830–847.

Kandel, William A. 2011. "The U.S. Foreign-Born Population: Trends and Selected Characteristics." Congressional Research Service. Retrieved Feb. 5, 2011. Online: www.crs.gov

Kansas State University. 2010. "Timeline for Transition." Retrieved Dec. 22, 2010. Online: www.k-state.edu/parentsandfamily/resources/timeline.htm

Kanter, Rosabeth Moss. 1983. *The Change Masters: Innovation and Entrepreneurship in the American Corporation.* New York: Simon & Schuster.

—. 1985. "All That Is Entrepreneural Is Not Gold." *Wall Street Journal* (July 22): 18.

—. 1993. *Men and Women of the Corporation.* New York: Basic (orig. pub. 1977).

Kasarda, John D., and Greg Lindsay. 2011. *Aerotropolis: The Way We'll Live Next.* New York: Farrar, Straus and Giroux.

Katzer, Jeffrey, Kenneth H. Cook, and Wayne W. Crouch. 1991. *Evaluating Information: A Guide for Users of Social Science Research.* New York: McGraw-Hill.

Kaufman, Leslie. 2011. "City Prepares for a Warm Long-Term Forecast." *New York Times* (May 23): A1, A14.

Keen, Judy. 2010. "Wis. Law Lets Residents Challenge Race-Based Mascots."

*USA Today* (Oct. 7). Retrieved Dec. 25, 2011. Online: www.usatoday.com/news/nation/2010-10-06-Kewaunee-school-mascot_N.htm

Kendall, Diana. 1980. Square Pegs in Round Holes: Non-Traditional Students in Medical Schools. Unpublished doctoral dissertation, Department of Sociology, the University of Texas at Austin.

—. 2002. *The Power of Good Deeds: Privileged Women and the Social Reproduction of the Upper Class.* Lanham, MD: Rowman & Littlefield.

—. 2008. *Members Only: Elite Clubs and the Process of Exclusion.* Lanham, MD: Rowman & Littlefield.

—. 2011. *Framing Class: Media Representations of Wealth and Poverty in the United States* (2nd ed.). Lanham, MD: Rowman & Littlefield.

Kenyon, Kathleen. 1957. *Digging Up Jericho.* London: Benn.

Kerbo, Harold R. 2000. *Social Stratification and Inequality: Class Conflict in Historical, Comparative, and Global Perspective* (4th ed.). New York: McGraw-Hill.

Kerkman, Maggie. 2011. "Educating Our Children: The Evolution of Home Schooling." Foxnews.com (Feb. 9). Retrieved Apr. 29, 2011. Online: www.foxnews.com/us/2011/02/09/educating-children-evolution-home-schooling

Kimmell, Michael S., and Michael A. Messner. 2012. *Men's Lives* (9th ed.). Upper Saddle River, NJ: Pearson.

King, Ryan D., Kecia R. Johnson, and Kelly McGeever. 2010. "Demography of the Legal Profession and Racial Disparities in Sentencing." *Law & Society Review,* 44 (1): 1–31.

Kitsuse, John I. 1980. "Coming Out All Over: Deviance and the Politics of Social Problems." *Social Problems,* 28: 1–13.

Klein, Ezra. 2012. "Twelve Facts About Guns and Mass Shootings in the United States." *Washington Post* (Dec. 14). Retrieved June 11, 2013. Online: www.washingtonpost.com/blogs/wonkblog/wp/2012/12/14/nine-facts-about-guns-and-mass-shootings-in-the-united-states

Kohl, Beth. 2007. "On Indian Surrogates." *The Huffington Post* (Oct. 30). Retrieved Feb. 9, 2008. Online: www.huffingtonpost.com/beth-kohl/on-indian-surrogates_b_70425.html

Kohlberg, Lawrence. 1969. "Stage and Sequence: The Cognitive–Developmental Approach to Socialization." In David A. Goslin (Ed.), *Handbook of Socialization Theory and Research.* Chicago: Rand McNally, pp. 347–480.

—. 1981. *The Philosophy of Moral Development: Moral Stages and the Idea of Justice,* vol. 1: *Essays on Moral Development.* San Francisco: Harper & Row.

Koopmans, Ruud. 1999. "Political. Opportunity. Structure. Some Splitting to Balance the Lumping." *Sociological Forum* (Mar.): 93–105.

Korsmeyer, Carolyn. 1981. "The Hidden Joke: Generic Uses of Masculine Terminology." In Mary Vetterling-Braggin (Ed.), *Sexist Language: A Modern Philosophical Analysis.* Totowa, NJ: Littlefield, Adams, pp. 116–131.

Kozol, Jonathan. 1991. *Savage Inequalities: Children in America's Schools.* New York: Crown.

Kramer, Andrew E. 2010. "Philip Morris Is Said to Benefit from Child Labor." *New York Times* (July 14). Retrieved Mar. 6, 2011. Online: www.nytimes.com/2010/07/14/business/global/14smoke.html?_r=2&ref=childlabor

—. 2011. "Russia Cashes in on Anxiety Over Supply of Middle East Oil." *New York Times* (Mar. 7). Retrieved Mar. 28, 2011. Online: www.nytimes.com/2011/03/08/business/global/08oil.html

Kreider, Rose M., and Diana B. Elliott. 2009. *America's Families and Living Arrangements: 2007.* U.S. Census Bureau. Retrieved Apr. 2, 2010. Online: www.census.gov/population/www/socdemo/hh-fam/p20-561.pdf

Kristof, Nicholas D. 2006. "Looking for Islam's Luthers." *New York Times* (Oct. 15): A22.

Kristof, Nicholas D., and Sheryl WuDunn. 2009. *Half the Sky: Turning Oppression into Opportunity for Women Worldwide.* New York: Vintage.

Kurtz, Lester. 1995. *Gods in the Global Village: The World's Religions in Sociological Perspective.* Thousand Oaks, CA: Sage.

Lachance-Grzela, Mylene, and Genevieve Bouchard. 2010. "Why Do Women Do the Lion's Share of Housework? A Decade of Research." *Sex Roles,* 63 (11–12): 767–780.

Ladd, E. C., Jr. 1966. *Negro Political Leadership in the South.* Ithaca, NY: Cornell University Press.

LaFrance, Marianne, and Marvin Hecht. 2000. "Gender and Smiling: A Meta-Analysis." In Agneta H. Fischer (Ed.), *Gender and Emotion: Social Psychological Perspectives.* New York: Cambridge University Press, pp. 118–142.

Lakeman, Richard, and Mary Fitzgerald. 2009. "Ethical Suicide Research: A Survey of Researchers." *International Journal of Mental Health Nursing* 18 (1): 10–17.

Lamanna, Mary Ann, and Agnes Riedmann. 2012. *Marriages, Families, and Relationships: Making Choices in a Diverse Society* (11th ed.). Belmont, CA: Wadsworth/Cengage.

*Lancet.* 2012. "Suicide Mortality in India: A Nationally Representative Survey" (June). Retrieved Oct. 12, 2012. Online: www.thelancet.com/journals/lancet/article/PIIS0140-6736(12)60606-0/abstract

Landale, Nancy S., and Ralph S. Oropesa. 2007. "Hispanic Families: Stability and Change." *Annual Review of Sociology,* 33: 381–405.

Langhout, Regina D., and Cecily A. Mitchell. 2008. "Engaging Contexts: Drawing the Link Between Student and Teacher Experiences of the Hidden Curriculum." *Journal of Community & Applied Social Psychology,* 18 (6): 593–614.

Lapchick, Richard. 2010. "The 2010 Racial and Gender Report Card: College Sport." University of Central Florida: The Institute for Diversity and Ethics in Sport. Retrieved Mar. 7, 2011. Online: www.tidesport.org/RGRC/2010/2010_College_RGRC_FINAL.pdf

Lapchick, Richard (with Devin Beahm and Jason Robinson). 2014. "Small Progress Throughout Collegiate Athletic Leadership: Assessing Diversity Among Campus and Conference Leaders for Football Bowl Subdivision (FBS) Schools in the 2013–14 Academic Year." Institute for Diversity and Ethnics in Sports, University of Central Florida (Jan. 2). Retrieved Feb. 3, 2014. Online: http://tidesport.org/Grad%20Rates/2013%20D1%20Leadership%20Report.pdf

Lapham, Lewis H. 1988. *Money and Class in America: Notes and Observations on Our Civil Religion.* New York: Weidenfeld & Nicolson.

Lareau, Annette. 2011. *Unequal Childhoods: Class, Race, and Family Life* (2nd ed.). Berkeley: University of California Press.

Larsen, Janet. 2011. "Cancer Now Leading Cause of Death in China." Earth

Policy Institute (May 25). Retrieved Mar. 17, 2012. Online: www.earth-policy.org/plan_b_updates/2011/update96

Larson, Magali Sarfatti. 1977. *The Rise of Professionalism: A Sociological Analysis.* Berkeley: University of California Press.

Laumann, Edward O., John H. Gagnon, Robert T. Michael, and Stuart Michaels. 1994. *The Social Organization of Sexuality.* Chicago: University of Chicago Press.

Lauzen, Martha M., and David M. Dozier. 2005. "Maintaining the Double Standard: Portrayals of Age and Gender in Popular Films." *Sex Roles,* 52 (7/8): 437–446.

Le, C. N. 2013. "The Model Minority Image." *Asian-Nation: The Landscape of Asian America.* Retrieved Apr. 4, 2013. Online: www.asian-nation.org/model-minority.shtml

Le Bon, Gustave. 1960. *The Crowd: A Study of the Popular Mind.* New York: Viking (orig. pub. 1895).

Leenaars, Antoon A. 1988. *Suicide Notes: Predictive Clues and Patterns.* New York: Human Sciences Press.

Lemert, Edwin M. 1951. *Social Pathology.* New York: McGraw-Hill.

Leonard, Devin. 2012. "Medicare and Medicaid Must Be Cut. Period." *Bloomberg Businessweek* (Nov. 8). Retrieved May 9, 2013. Online: www.businessweek.com/printer/articles/81414-medicare-and-medicaid-must-be-cut-dot-period

Lerner, Gerda. 1986. *The Creation of Patriarchy.* New York: Oxford University Press.

Levine, Murray. 1982. "Introduction." In Lois Marie Gibbs, *Love Canal: My Story.* Albany: SUNY Press.

Levine, Peter. 1992. *Ellis Island to Ebbets Field: Sport and the American Jewish Experience.* New York: Oxford University Press.

Lewis, Thabiti. 2010. *Ballers of the New School: Race and Sports in America.* Chicago: Third World.

Lipka, Michael. 2013. "Abercrombie Hijab Firing Highlights Muslim Concern About Discrimination." Pew Research Center (Sept.). Retrieved Feb. 10, 2014. Online: www.pewresearch.org/fact-tank/2013/09/11/abercrombie-hijab-firing-highlights-muslim-concern-about-discrimination

Littal, Robert. 2010. "Why Is This Site Called Black Sports Online?" (Oct. 2). Retrieved June 25, 2013. Online: http://blacksportsonline.com/home/why-is-the-site-called-blacksportsonline

Lofland, John. 1993. "Collective Behavior: The Elementary Forms." In Russell L. Curtis Jr., and Benigno E. Aguirre (Eds.), *Collective Behavior and Social Movements.* Boston: Allyn & Bacon, pp. 70–75.

Lorber, Judith. 1994. *Paradoxes of Gender.* New Haven, CT: Yale University Press.

Lorber, Judith (Ed.). 2005. *Gender Inequality: Feminist Theories and Politics* (3rd ed.). Los Angeles: Roxbury.

Low, Setha. 2003. *Behind the Gates: Life, Security, and the Pursuit of Happiness in Fortress America.* New York: Routledge.

Lundberg, Ferdinand. 1988. *The Rich and the Super-Rich: A Study in the Power of Money Today.* Secaucus, NJ: Lyle Stuart.

Luo, Michael, and Megan Thee-Brenan. 2009. "Emotional Havoc Wreaked on Workers and Families." *New York Times* (Dec. 15): A1, A26.

Lupton, Deborah. 1997. "Foucault and the Medicalisation Critique." In Alan Petersen and Robin Bunton (Eds.), *Foucault: Health and Medicine.* London: Routledge, pp. 94–110.

Lusca, Emanuel. 2008. "Race as a Social Construct." Anthropology.net (Oct. 1). Retrieved Feb. 9, 2013. Online: http://anthropology.net/2008/10/01/race-as-a-social-construct

Mackin, Deborah. 2010. "BP Oil Spill, the Challenger and Columbia Explosions, and the Deaths on Mt. Everest: Key Decision Making Tragedies and What We Can Learn from Them." Retrieved Dec. 27, 2011. Online: www.newdirectionsconsulting.com/2010/06/bp-oil-spill-the-challenger-and-columbia-explosions-and-the-deaths-on-mt-everest-key-decision-making-tragedies-and-what-we-can-learn-from-them

Malinowski, Bronislaw. 1922. *Argonauts of the Western Pacific.* New York: Dutton.

Malthus, Thomas R. 1965. *An Essay on Population.* New York: Augustus Kelley (orig. pub. 1798).

Mangione, Jerre, and Ben Morreale. 1992. *La Storia: Five Centuries of the Italian American Experience.* New York: HarperPerennial.

Mann, Patricia S. 1994. *Micro-Politics: Agency in a Postfeminist Era.* Minneapolis: University of Minnesota Press.

Manning, Jennifer E. 2011. "Membership of the 112th Congress: A Profile." Congressional Research Service (Mar. 1).

Retrieved Nov. 10, 2012. Online: www.senate.gov/reference/resources/pdf/R41647.pdf

Marger, Martin N. 1994. *Race and Ethnic Relations: American and Global Perspectives.* Belmont, CA: Wadsworth.

—. 2012. *Race & Ethnic Relations: American and Global Perspectives* (9th ed.). Belmont, CA: Wadsworth, Cengage.

Marshall, Gordon. 1998. *A Dictionary of Sociology* (2nd ed.). New York: Oxford University Press.

Martineau, Harriet. 1962. *Society in America* (edited, abridged). Garden City, NY: Doubleday (orig. pub. 1837).

Martinez, Michael. 2014. "10 Things to Know About Nation's First Recreational Marijuana Shops in Colorado." CNN.com (Jan. 1). Retrieved Mar. 22, 2014. Online: www.cnn.com/2013/12/28/us/10-things-colorado-recreational-marijuana

Marx, Karl. 1967. *Capital: A Critique of Political Economy.* Ed. Friedrich Engels. New York: International (orig. pub. 1867).

Marx, Karl, and Friedrich Engels. 1967. *The Communist Manifesto.* New York: Pantheon (orig. pub. 1848).

—. 1970. *The German Ideology,* Part 1. Ed. C. J. Arthur. New York: International (orig. pub. 1845–1846).

Massey, Douglas J., G. Hugo Arango, A. Kowasuci, A. Pellegrino, and J. E. Taylor. 1993. "Theories of International Migration: A Review and Appraisal." *Population and Development Review,* 19: 431–466.

Matthews, Christopher. 2013. "How Does One Fake Tweet Cause a Stock Market Crash?" *Time* (Apr. 24). Retrieved May 23, 2013. Online: http://business.time.com/2013/04/24/how-does-one-fake-tweet-cause-a-stock-market-crash/#ixzz2U8tGJinY

Mattingly, Beth, and Wendy A. Walsh. 2010. "Rural Families with a Child Abuse Report Are More Likely Headed by a Single Parent and Endure Economic and Family Stress." Carsey Institute Issue Brief no. 10 (Winter). Online: www.carseyinstitute.unh.edu/publications/FS-Mattingly-Childabuse.pdf

Mayo Clinic. 2011. "Infertility: Causes" (Sept. 9). Retrieved Jan. 5, 2012. Online: www.mayoclinic.com/health/infertility/DS00310/DSECTION=causes

McAdam, Doug. 1996. "Conceptual Origins, Current Problems, Future Directions." In Doug McAdam, John D. McCarthy, and Meyer N. Zald (Eds.), *Comparative Perspectives on Social Movements.* New York: Cambridge University Press, pp. 23–40.

McClure, Robin. 2010. "Afterschool Child Care—Number of Kids Home Alone After School Has Risen." Retrieved Apr. 2, 2010. Online: http://childcare.about.com/od /schoolagetopics/a/homealone.htm

McHale, Susan M., Ann C. Crouter, Ji-Yeon Kim, Linda M. Burton, Kelly D. Davis, Aryn M. Dotterer, and Dena P. Swanson. 2006. "Mothers' and Fathers' Racial Socialization in African American Families: Implications for Youth." *Child Development, 77* (5): 1387–1402.

McKenzie, Roderick D. 1925. "The Ecological Approach to the Study of the Human Community." In Robert Park, Ernest Burgess, and Roderick D. McKenzie, *The City.* Chicago: University of Chicago Press.

McPhail, Clark, and Ronald T. Wohlstein. 1983. "Individual and Collective Behavior Within Gatherings, Demonstrations, and Riots." In Ralph H. Turner and James F. Short, Jr. (Eds.), *Annual Review of Sociology*, vol. 9. Palo Alto, CA: Annual Reviews, pp. 579–600.

McPherson, Barry D., James E. Curtis, and John W. Loy. 1989. *The Social Significance of Sport: An Introduction to the Sociology of Sport.* Champaign, IL: Human Kinetics.

Mead, George Herbert. 1934. *Mind, Self, and Society.* Chicago: University of Chicago Press.

Merton, Robert King. 1938. "Social Structure and Anomie." *American Sociological Review,* 3 (6): 672–682.

—. 1949. "Discrimination and the American Creed." In Robert M. MacIver (Ed.), *Discrimination and National Welfare.* New York: Harper & Row, pp. 99–126.

—. 1968. *Social Theory and Social Structure* (enlarged ed.). New York: Free Press.

Messenger, David. 2009. "Studies Show People More Stressed as Students Than at Other Stages of Life Due to Work, Relationships." Retrieved Feb. 8, 2010. Online: www.studlife.com /news/2009/10/30/studies-shows -people-more-stressed-as-students -than-at-other-stages-of-life-due-to -work-relationships

Michael, Robert T., John H. Gagnon, Edward O. Laumann, and Gina Kolata. 1994. *Sex in America.* Boston: Little, Brown.

Michels, Robert. 1949. *Political Parties.* Glencoe, IL: Free Press (orig. pub. 1911).

Milgram, Stanley. 1963. "Behavioral Study of Obedience." *Journal of Abnormal and Social Psychology,* 67: 371–378.

—. 1967. "The Small World Problem." *Psychology Today,* 1: 61–67.

—. 1974. *Obedience to Authority.* New York: Harper & Row.

Miller, Casey, and Kate Swift. 1991. *Words and Women: New Language in New Times* (updated ed.). New York: HarperCollins.

Miller, Lisa. 2013. "The Retro Wife." *New York* (Mar. 25): 20–25, 78–79.

Mills, C. Wright. 1956. *White Collar.* New York: Oxford University Press.

—. 1959a. *The Power Elite.* Fair Lawn, NJ: Oxford University Press.

—. 1959b. *The Sociological Imagination.* London: Oxford University Press.

—. 1976. *The Causes of World War Three.* Westport, CT: Greenwood.

Mills, Iain. 2010. "China Puts Healthcare Cart Before the Horse." *Asia Times Online* (Apr. 21). Retrieved May 13, 2011. Online: www.atimes .com/atimes/China/LD21Ad01.html

Minino, Arialdi M. 2013. "Death in the United States." NCHS Data Brief, No. 115 (Mar. 2013). U.S. Department of Health and Human Services. Retrieved May 7, 2013. Online: www .cdc.gov/nchs/data/databriefs/db115 .pdf

Mishler, Elliot G. 1984. *The Discourse of Medicine: Dialectics of Medical Interviews.* Norwood, NJ: Ablex.

—. 2005. "The Struggle Between the Voice of Medicine and the Voice of the Lifeworld." In Peter Conrad (Ed.), *The Sociology of Health and Illness: Critical Perspectives* (7th ed.). New York: Worth, pp. 319–330.

Mixedfolks.com. 2011. "Well Known Mixed Athletes." Retrieved Mar. 23, 2011. Online: www.mixedfolks.com /mfc/Athletes.html

Mohai, Paul, and Robin Saha. 2007. "Racial Inequality in the Distribution of Hazardous Waste: A National-Level Reassessment." *Social Problems,* (August): 343–370.

Mollborn, Stefanie, and Jeff A. Dennis. 2011. "Investigating the Life Situations and Development of Teenage Mothers' Children: Evidence from the ECLS-B." *Population Research and Policy Review,* 31: 31–66.

Mollborn, Stefanie, and Peter J. Lovegrove. 2010. "How Teenage Fathers Matter for Children." *Journal of Family Issues,* 32: 3–30.

Molnar, Alex, Gary Miron, and Jessica L. Urschel. 2010. "Profiles of For-Profit Education Management Organizations, 2009–2010." National Education Policy Center, University of Colorado at Boulder. Retrieved Apr. 29, 2011. Online: http://nepc .colorado.edu/files/EMO-FP-09 -10.pdf

Monger, Randall, and James Yankay. 2013. "U.S. Legal Permanent Residents: 2012." U.S. Department of Homeland Security (Mar.). Retrieved May 13, 2013. Online: www.dhs .gov/sites/default/files/publications /ois_lpr_fr_2012_2.pdf

Mooney, Karen. 2011. "Raleigh Marchers Fight 'Re-segregation' Plan." abc.news.com (Feb. 12). Retrieved Apr. 25, 2011. Online: http://abcnews .go.com/Politics/raleigh-nc -marchers-fight-segregation-plan /story?id=12904201

Moore, Patricia, with C. P. Conn. 1985. *Disguised.* Waco, TX: Word.

Morselli, Henry. 1975. *Suicide: An Essay on Comparative Moral Statistics.* New York: Arno (orig. pub. 1881).

Muehlfried, Florian. 2007. "Sharing the Same Blood—Culture and Cuisine in the Republic of Georgia." *Anthropology of Food* (Dec.). Retrieved Oct. 13, 2012. Online: http://aof .revues.org/2342?&id=2342#text

Murdock, George P. 1945. "The Common Denominator of Cultures." In Ralph Linton (Ed.), *The Science of Man in the World Crisis.* New York: Columbia University Press, pp. 123–142.

Murphy, Andrea. 2012. "America's Largest Private Companies 2012." Forbes.com (Nov. 28). Retrieved Apr. 19, 2013. Online: www.forbes .com/sites/andreamurphy/2012/11/28 /americas-largest-private-companies -2012

Murray, Jane, Rick Linden, and Diana Kendall. 2012. *Sociology in Our Times: The Essentials* (5th Canadian ed.). Toronto, CA: Nelson/Cengage.

National Alliance for Public Charter Schools. 2013. "Back to School Tallies: Estimated Number of Public Charter Schools and Students, 2012–2013" (Jan.). Retrieved Apr. 11, 2013.

Online: www.publiccharters .org/data/files/Publication_docs /NAPCS%202011-12%20New%20 and%20Closed%20Charter%20 Schools_20111206T125251.pdf

National Center for Educational Statistics. 2012a. *The Condition of Education 2012* (May). Retrieved Apr. 13, 2013. Online: http://nces.ed.gov /pubs2012/2012045.pdf

—. 2012b. *Indicators of School Crime and Safety: 2011* (Feb.). Retrieved Apr. 13, 2013. Online: www.bjs.gov /content/pub/pdf/iscs11.pdf

—. 2014. "The Condition of Education: Status Dropout Rates, 2012." Retrieved Feb. 27, 2014. Online: nces .ed.gov/programs/coe/indicator_coj .asp

National Center for Health Statistics. 2011. *Health, United States, 2010 with Special Feature on Death and Dying.* Hyattsville, MD: U.S. Government Printing Office.

—. 2012. "Prevalence of Obesity in the United States, 2009–2010" (Jan.). Retrieved May 12, 2013. Online: www .cdc.gov/nchs/data/databriefs/db82 .pdf

National Coalition for the Homeless. 2012. "National Homeless Civil Rights Organizing Project." Retrieved Nov. 9, 2012. Online: www.nationalhomeless .org/projects/nhcrop.html

National Collegiate Athletic Association (NCAA). 2012. "Estimated Probability of Competing in Athletics Beyond the High School Interscholastic Level." Retrieved Apr. 28, 2012. Online: www.ncaa.org/wps/wcm/connect /public/test/issues/recruiting /probability+of+going+pro

National Council of Churches. 2012. *Yearbook of American and Canadian Churches: 2012.* New York: National Council of Churches.

National Education Association. 2012. "Rankings and Estimates: Rankings of the States 2012 and Estimates of School Statistics 2013" (Dec.). Retrieved Apr. 11, 2013. Online: www.nea.org/assets /img/content/NEA_Rankings_And _Estimates-2013_(2).pdf

National Institute of Justice. 2011. "Transnational Organized Crime." Retrieved Feb. 9, 2011. Online: www .ojp.usdoj.gov/nij/topics/crime /transnational-organized-crime /welcome.htm

National League of Cities. 2010. "Significant Budget Shortfalls Could Mean More Job Losses" (May 24). National League of Cities. Retrieved June 10, 2010. Online: www.ncl.org.PRESS ROOM/PRESSRELEASEITEMS /SoACJobsEcon5.10.aspx

National Marriage Project. 2010. "When Marriage Disappears: The New Middle America." Retrieved Apr. 18, 2011. Online: www.virginia .edu/marriageproject/pdfs/Union_11 _12_10.pdf

National Vital Statistics Reports. 2013a. "Births: Final Data for 2012" (Dec. 30). Retrieved Feb. 24, 2014. Online: www.cdc.gov/nchs/data/nvsr /nvsr62/nvsr62_09.pdf

—. 2013b. "Deaths: Leading Causes." Retrieved Apr. 6, 2014. Online: www .cdc.gov/nchs/data/nvsr/nvsr62 /nvsr62_06.pdf

Navarrette, Ruben Jr. 1997. "A Darker Shade of Crimson." In Diana Kendall (Ed.), *Race, Class, and Gender in a Diverse Society.* Boston: Allyn & Bacon, 1997: 274–279. Reprinted from Ruben Navarrette, Jr., *A Darker Shade of Crimson.* New York: Bantam, 1993.

nbcnews.com. "Presidential Election Results by State." Retrieved May 27, 2014. Online: http://elections .nbcnews.com/ns/politics/2012/all /president/#.U4T8qrlOUfQ

NDTV.com. 2012. "Suicide Rates in India Are Highest in the 15–29 Age Group: Report." Retrieved Oct. 12, 2012. Online: www.ndtv.com/article /india/suicide-rates-in-india-are -highest-in-the-15-29-age-group -report-234986

Newport, Frank. 2011. "For First Time, Majority of Americans Favor Legal Gay Marriage." Gallup Organization (May 20). Retrieved Jan. 1, 2012. Online: www.gallup.com/poll/147662 /first-time-majority-americans-favor -legal-gay-marriage.aspx

Ng, Christina. 2013. "Rebecca Sedwick Suicide Part of 'Cool to be Cruel' Cyber Culture." ABCnews.com (Oct. 16). Retrieved Dec. 9, 2013. Online: http://abcnews.go.com /US/rebecca-sedwick-suicide -part-cool-cruel-cyber-culture /story?id=20588017

Niebuhr, H. Richard. 1929. *The Social Sources of Denominationalism.* New York: Meridian.

NIH. 2012. "Regular Marijuana Use by Teens Continues to Be a Concern." National Institute of Health (Dec. 19). Retrieved May 7, 2013. Online: www .nih.gov/news/health/dec2012/nida -19.htm

Nolan, Patrick, and Gerhard E. Lenski. 2010. *Human Societies: An Introduction to Macrosociology* (11th ed.). New York: Oxford University Press.

NowPublic.com. 2011. "Paisley Dickey Dressed as Prostitute on *Toddlers & Tiaras*" (Sept. 7). Retrieved Feb. 24, 2013. Online: www.nowpublic.com /culture/paisley-dickey-dressed -prostitute-toddlers-tiaras-video -2834728.html#ixzz2Lq5ldpMf

OECD. 2013. "OECD Health Data 2013: How Does the United States Compare?" Retrieved Mar. 23, 2014. Online: www.oecd.org/unitedstates /Briefing-Note-USA-2013.pdf

Ogburn, William F. 1966. *Social Change with Respect to Culture and Original Nature.* New York: Dell (orig. pub. 1922).

Ogden, Jane, and Talin Avades. 2011. "Being Homeless and the Use and Nonuse of Services: A Qualitative Study." *Journal of Community Psychology,* 39 (4): 499–505.

Ohio State University. 2007. "Cultural Diversity: Eating in America." Ohio State University Extension Fact Sheet, Family and Consumer Sciences Series. Retrieved Feb. 11, 2007. Online: www.ohioline.ag.ohio-state.edu

Oliffe, John L., Christina S. E. Han, John S. Ogrodniczuk, J. Craig Phillips, and Philippe Roy. 2011. "Suicide from the Perspectives of Older Men Who Experience Depression: A Gender Analysis." *American Journal of Men's Health* 5 (5): 444–454.

Omi, Michael, and Howard Winant. 1994. *Racial Formation in the United States: From the 1960s to the 1990s.* New York: Routledge.

—. 2013. *Racial Formation in the United States: From the 1960s to the 1990s* (3rd ed.). New York: Routledge.

Opportunity Agenda. 2013. "The State of Media Coverage of Immigration: 2012–2013." Retrieved May 16, 2013. Online: http://opportunityagenda .org/files/field_file/2013_imm _media_scan.pdf

Orfield, Gary, John Kucsera, and Genevieve Siegel-Hawley. 2012. "E Pluribus . . . Separation: Deepening Double Segregation for More Students" (Sept.). Retrieved Apr. 13, 2013. Online: http:// civilrightsproject.ucla.edu/research /k-12-education/integration -and-diversity/mlk-national/e -pluribus...separation-deepening -double-segregation-for-more

-students/orfield_epluribus
_revised_omplete_2012.pdf

Orr, Matthew, and Susan Saulny. 2011. "Young and Mixed in America" (video). Retrieved Mar. 7, 2011. Online: www.nytimes.com

Ott, Thomas. 2011. "Cleveland Students Hold Their Own with Voucher Students on State Tests." Cleveland.com (Feb. 22). Retrieved Apr. 27, 2011. Online: http://blog.cleveland.com /metro/2011/02/cleveland_students _hold_own_wi.html

Oxendine, Joseph B. 2003. *American Indian Sports Heritage* (rev. ed.). Lincoln: University of Nebraska Press.

Padilla, Felix M. 1993. *The Gang as an American Enterprise.* New Brunswick, NJ: Rutgers University Press.

Page, Charles H. 1946. "Bureaucracy's Other Face." *Social Forces,* 25 (October): 89–94.

Pampel, Fred. 2011. "Cohort Change, Diffusion, and Support for Gender Egalitarianism in Cross-National Perspective." *Demographic Research,* 25: 667–694. Retrieved Feb. 11, 2012. Online: www.demographic-research .org/volumes/vol25/21/25-21.pdf

Parenti, Michael. 1998. *America Besieged.* San Francisco: City Lights.

Park, Robert E. 1915. "The City: Suggestions for the Investigation of Human Behavior in the City." *American Journal of Sociology,* 20: 577–612.

—. 1928. "Human Migration and the Marginal Man." *American Journal of Sociology,* 33.

—. 1936. "Human Ecology." *American Journal of Sociology,* 42: 1–15.

Park, Robert E., and Ernest W. Burgess. 1921. *Human Ecology.* Chicago: University of Chicago Press.

Parker, Ashley. 2013. "Immigration Overhaul Proposal Is Likely to Ignite Fierce Debate." *New York Times* (Apr. 16). Retrieved May 14, 2013. Online: www.nytimes.com/2013 /04/16/us/politics/senate-immig ration-overhaul-likely-to-set-off -fight.html?ref=us

Parker, Kim. 2011. "A Portrait of Stepfamilies." Pew Research Center (Jan. 13). Retrieved Jan. 1, 2012. Online: www.pewsocialtrends.org /2011/01/13/a-portrait-of-step families

—. 2013. "Modern Parenthood; Roles of Moms and Dads Converge as They Balance Work and Family." Pew Research Social and Demographic Trends (Mar. 14). Retrieved Feb. 11,

2014. Online: www.pewsocialtrends .org/2013/03/14/modern-parenthood -roles-of-moms-and-dads-converge -as-they-balance-work-and-family

Parsons, Talcott. 1951. *The Social System.* Glencoe, IL: Free Press.

—. 1955. "The American Family: Its Relations to Personality and to the Social Structure." In Talcott Parsons and Robert F. Bales (Eds.), *Family, Socialization and Interaction Process.* Glencoe, IL: Free Press, pp. 3–33.

Passel, Jeffrey S., and D'Vera Cohn. 2011. "Unauthorized Immigrant Population: National and State Trends, 2010." Pew Hispanic Research Center. Retrieved Feb. 5, 2012. Online: www .pewhispanic.org

Passel, Jeffrey S., D'Vera Cohn, and Ana Gonzalez-Barrera. 2012. "Net Migration from Mexico Falls to Zero—and Perhaps Less" (May 3). Retrieved May 13, 2013. Online: www.pewhispanic.org/2012/04/23 /net-migration-from-mexico-falls-to -zero-and-perhaps-less

Passel, Jeffrey S., Wendy Wang, and Paul Taylor. 2010. "Marrying Out: One-in-Seven New U.S. Marriages Is Interracial or Interethnic." Pew Research Center (June 4). Retrieved Dec. 22, 2011. Online: http:// pewresearchcenter.org

PBS. 2005. "The Meaning of Food: Food & Culture." Retrieved Feb. 11, 2007. Online: www.pbs.org/opb /meaningoffood

pbs.org. 2012. "Health Costs: How the U.S. Compares with Other Countries" (Oct. 22). Retrieved July 8, 2013. Online: www.pbs.org/newshour/rundown /2012/10/health-costs-how-the-us -compares-with-other-countries.html

Pearce, Diana. 1978. "The Feminization of Poverty: Women, Work, and Welfare." *Urban and Social Change Review,* 11 (1/2): 28–36.

*People's Daily Online.* 2012. "China's River Pollution a Threat to People's Lives" (Feb. 17). Retrieved Mar. 17, 2012. Online: http://english.people .com.cn/90882/7732438.html

PerfSpot.com. 2011. "The Enjoyment of Chat Groups." Retrieved Feb. 7, 2011. Online: www.perfspot.com/articles /groups_for_chat.asp

Perry, Gina. 2013. *Behind the Shock Machine: The Untold Story of the Notorious Milgram Psychology Experiments.* New York: New Press.

Peterson, Robert. 1992. *Only the Ball Was White: A History of Legendary Black Players and All-Black*

*Professional Teams.* New York: Oxford University Press (orig. pub. 1970).

Petrecca, Laura. 2010. "More Employers Use Tech to Track Workers." USAToday.com (Mar. 17). Retrieved Nov. 10, 2012. Online: http://usatoday30 .usatoday.com/money/workplace/2010 -03-17-workplaceprivacy15_CV_N .htm?csp=usat.me

Pew Charitable Trusts. 2013. "Cities Squeezed by Pension and Retiree Health Care Shortfalls" (Mar.). Retrieved May 16, 2013. Online: www .pewstates.org/uploadedFiles/PCS _Assets/2013/Pew_city_pensions _brief.pdf

Pew Forum on Religion and Public Life. 2007a. "Religion in the Public Schools" (May). Retrieved Feb. 18, 2012. Online: www.pewforum.org /uploadedfiles/Topics/Issues/Church -State_Law/religion-public-schools.pdf

—. 2007b. "U.S. Religious Landscape Survey." Retrieved Feb. 18, 2012. Online: http://religions.pewforum.org /pdf/report-religious-landscape -study-key-findings.pdf

—. 2008. "U.S. Religious Landscape Survey: Religious Affiliation— Diverse and Dynamic." Retrieved May 8, 2011. Online: http://religions .pewforum.org/pdf/report-religious -landscape-study-full.pdf

—. 2012b. "'Nones' on the Rise" (Oct. 9). Retrieved Apr. 19, 2013. Online: www .pewforum.org/uploadedFiles/Topics /Religious_Affiliation/Unaffiliated /NonesOnTheRise-full.pdf

Pew Research Center. 2012a. "Hispanics of Puerto Rican Origin in the United States, 2012" (June 27). Retrieved Apr. 6, 2013. Online: www.pewhispanic .org/2012/06/27/hispanics-of-puerto -rican-origin-in-the-united-states-2010

—. 2012b. "Internet and American Life: Social Media and Political Engagement" (Oct. 19). Retrieved May 27, 2014. Online: www.pewinternet .org/2012/10/19/social-media-and -political-engagement

—. 2012c. "Majority of Public Agrees Labor Unions Needed" (Aug. 30). Retrieved Mar. 11, 2013. Online: www.pewresearch.org/daily-number /majority-of-public-agrees-labor -unions-needed

—. 2012d. "The Rise of Asian Americans" (July 12). Retrieved Apr. 6, 2013. Online: www.pewsocialtrends.org/files /2013/04/Asian-Americans-new-full -report-04-2013.pdf

—. 2013a. "Gay Marriage: Key Data Points from Pew Research." Retrieved

Apr. 7, 2013. Online: www
.pewresearch.org/2013/03/21/gay
-marriage-key-data-points-from
-pew-research

—. 2013b. "Protecting the Environment
Ranks in the Middle of Public's
Priorities for 2013" (Apr. 22). Retrieved
May 23, 2013. Online: www.pew
research.org/daily-number/protecting
-the-environment-ranks-in-the-middle
-of-publics-priorities-for-2013

—. 2013c. "The State of the News
Media: The Media and Campaign
2012." Retrieved May 27, 2014. Online: http://stateofthemedia.org/2013
/special-reports-landing-page
/the-media-and-campaign-2012

—. 2014. "Deficit Reduction Declines as
Policy Priority" (Jan.). Retrieved June
1, 2014. Online: www.people-press.org
/files/legacy-pdf/01-27-14%20Policy
%20Priorities%20Release%202.pdf

Pew Research Center Global Attitudes
Project. 2011. "The American–Western
European Values Gap" (Nov. 17).
Retrieved Apr. 4, 2012. Online: www
.pewglobal.org/2011/11/17/the
-american-western-european-values
-gap

Pew Research Hispanic Trends Project.
2013. "Hispanics of Cuban Origin in
the United States, 2011." Retrieved Feb.
10, 2014. Online: www.pewhispanic
.org/2013/06/19/hispanics-of-cuban
-origin-in-the-united-states-2011

Pew Research Religion and Public Life
Project. 2014. "Religious Landscape
Survey." Retrieved Feb. 27, 2014. Online: http://religions.pewforum.org
/affiliations

Pew Research Social & Demographic
Trends. 2013a. "On Pay Gap, Millennial Women Near Parity—for Now."
Retrieved Jan. 20, 2014. Online: www
.pewsocialtrends.org/files/2013/12
/gender-and-work_final.pdf

—. 2013b. "Race in America: Tracking
50 Years of Demographic Trends."
Retrieved Jan. 19, 2014. Online: www
.pewsocialtrends.org/2013/08/22
/race-demographics

Phillips, John C. 1993. *Sociology of
Sport.* Boston: Allyn & Bacon.

Piaget, Jean. 1954. *The Construction of
Reality in the Child.* Trans. Margaret
Cook. New York: Basic.

Picca, Leslie Houts, and Joe R. Feagin.
2007. *Two-Faced Racism: Whites
in the Backstage and Frontstage.*
New York: Routledge.

Pillsbury, George, and Julian
Johannesen. 2012. "America Goes to
the Polls 2012: A Report on Voter

Turnout in the 2012 Presidential
Election." Retrieved Mar. 24, 2013.
Online: www.nonprofitvote.org/voter
-turnout.html

Pines, Maya. 1981. "The Civilizing
of Genie." *Psychology Today,* 15
(September): 28–29, 31–32, 34.

Plunkett Research. 2013. "Health Expenditures and Services in the U.S."
Retrieved May 11, 2013. Online:
www.plunkettresearch.com/health
-care-medical-market-research
/industry-and-business-data

Population Reference Bureau. 2013.
"2013 World Population Data Sheet."
Retrieved Mar. 31, 2014. Online:
www.prb.org/pdf13/2013-population
-data-sheet_eng.pdf

Porter, Eduardo. 2014. "Old Forecast
of Famine May Yet Come True." *New
York Times* (Apr. 1, 2014). Retrieved
Apr. 10, 2014. Online: www
.nytimes.com/2014/04/02/business
/energy-environment/a-200-year
-old-forecast-for-food-scarcity-may
-yet-come-true.html

Powell, Michael. 2004. "Evolution
Shares a Desk with 'Intelligent
Design.'" *Washington Post*
(Dec. 26): A1.

Preston, Julia. 2012. "Immigration Decreases, But Tensions Remain High."
*New York Times* (Mar. 10). Retrieved
Mar. 15. 2012. Online: www.nytimes
.com/2012/01/07/us/path-to-green
-card-for-illegal-immigrant-family
-members-of-americans.html?_r=2

—. 2013. "Beside a Path to Citizenship,
a New Path on Immigration." *New
York Times* (Apr. 18). Retrieved May
13, 2013. Online: www.nytimes
.com/2013/04/17/us/senators
-set-to-unveil-immigration-bill
.html?pagewanted=1

ProQuest. 2014. *ProQuest Statistical
Abstract of the United States 2014: The
National Data Book.* Lanham, MD:
Bernan Press/Rowman & Littlefield.

Puette, William J. 1992. *Through
Jaundiced Eyes: How the Media View
Organized Labor.* Ithaca, NY: ILR.

Purcell, Kristen. 2012. "Teens 2012:
Truth, Trends, and Myths About
Teen Online Behavior." Pew Internet
& American Life Project. Retrieved
Mar. 7, 2013. Online: http://
pewinternet.org/~/media//Files
/Presentations/2012/July
/KPurcell%20ACT%20Conf_PDF.pdf

Putnam, Robert D., and David E.
Campbell. 2010. *American Grace:
How Religion Divides and Unites Us.*
New York: Simon & Schuster.

Quinney, Richard. 2001. *Critique of the
Legal Order.* Piscataway, NJ: Transaction (orig. pub. 1974).

Reckless, Walter C. 1967. *The Crime
Problem.* New York: Meredith.

Rehavi, M. Marit, and Sonia B. Starr.
2012. "Racial Disparity in Federal
Criminal Charging and Its Sentencing Consequences." University of
Michigan Law School, Empirical
Legal Studies Center Paper #12-002.
Retrieved May 13, 2013. Online:
http://papers.ssrn.com/sol3/papers
.cfm?abstract_id=1985377

Reich, Robert B. 2010. *Aftershock: The
Next Economy and America's Future.*
New York: Knopf.

Reiman, Jeffrey, and Paul Leighton. 2010.
*The Rich Get Richer and the Poor Get
Prison: Ideology, Class, and Crimi-
nal Justice.* Upper Saddle River, NJ:
Pearson.

Reiner, Robert. 2007. "Media Made
Criminality: The Representation
of Crime in the Mass Media." In
Mike Maguire, Rodney Morgan, and
Robert Reiner (Eds.), *The Oxford
Handbook of Criminology.* New
York: Oxford University Press,
pp. 376–416.

Relman, Arnold S. 1992. "Self-
Referral—What's at Stake?" *New
England Journal of Medicine,* 327
(Nov. 19): 1522–1524.

Reuters News Service. 2008. "New
mtvU and Associated Press Poll
Shows How Stress, War, the
Economy, and Other Factors Are
Affecting College Students' Mental
Health." Retrieved Feb. 8, 2010.
Online: www.reuters.com/article
/idUS173716+19-Mar2008+PRN
-20080319

Rideout, Victoria J., Ulla G. Foehr, and
Donald F. Roberts. 2010. *Generation M2: Media in the Lives of 8- to
18-Year-Olds* (a Kaiser Family Foundation study). Retrieved Dec. 21,
2011. Online: www.kff.org/entmedia
/upload/8010.pdf

Ridgeway, Greg. 2006. "Assessing the
Effect of Race Bias in Post-Traffic
Stop Outcomes Using Propensity
Scores." *Journal of Quantitative Crimi-
nology,* 22 (1): 1–29.

Rigler, David. 1993. "Letters: A Psychol-
ogist Portrayed in a Book About an
Abused Child Speaks Out for the First
Time in 22 Years." *New York Times
Book Review* (June 13): 35.

Ritzer, George. 2011. *Sociological Theory* (8th ed.). New York:
McGraw-Hill.

—. 2013. *The McDonaldization of Society,* 20th anniversary edition. Thousand Oaks, CA: Sage.

Robbins, Alexandra. 2004. *Pledged: The Secret Life of Sororities.* New York: Hyperion.

Roberts, John. 2011. "Marine Corps Steps Up Suicide Prevention Efforts to Halt Deadly Trend" (Mar. 30). Retrieved Oct. 13, 2012. Online: www .foxnews.com/us/2011/03/30/marine -corps-steps-suicide-prevention -efforts-halt-deadly-trend

Roberts, Sam. 2010. "Study Finds Cohabiting Doesn't Make a Union Last." *New York Times* (Mar. 2). Retrieved Feb. 11, 2012. Online: www .nytimes.com/2010/03/03/us/03 marry.html

Robnett, Rachael D., and Joshue E. Susskind. 2010. "Who Cares About Being Gentle? The Impact of Social Identity and the Gender of One's Friends on Children's Display of Same-Gender Favoritism." *Sex Roles,* 64 (1/2): 90–102.

Rocca, Mo. 2007. "TV Drug Ads' Side Effects." CBS News *Sunday Morning* (Oct. 14). Retrieved Feb. 24, 2008. Online: www.cbsnews .com/stories/2007/10/14/sunday /main3365346.shtml

Rogers, Dexter. 2011. "White Heterosexual Men Dominate the Sports Journalism Arena but Are They Objective About Race Issues?" *Color- lines Magazine* (Jan. 3). Retrieved Mar. 8, 2010. Online: http://gayblack canadianman.com/2011/01/03 /colorlines-magazine-article-white -heterosexual-men-dominate-the -sports-journalism-arena-but-are -they-objective-about-race-issues

Rollins, Judith. 1985. *Between Women: Domestics and Their Employers.* Phila- delphia: Temple University Press.

Ropers, Richard H. 1991. *Persistent Poverty: The American Dream Turned Nightmare.* New York: Plenum.

Rose, Stephen J., and Heidi Hartman. 2008. "Still a Man's Labor Market: The Long-Term Earnings Gap." IWPR #C366 (Feb.). New York: Institute for Women's Policy Research.

Rosenbaum, Dorothy. 2013. "SNAP Costs Leveling Off, Almost Certain to Fall Next Year." Center on Budget and Policy Priorities (Nov. 20). Retrieved Jan. 5, 2014. Online: www.cbpp.org /files/11-20-13fa.pdf

Ross, Dorothy. 1991. *The Origins of American Social Science.* Cambridge, England: Cambridge University Press.

Ross, Sally R., and Kimberly J. Shinew. 2008. "Perspectives of Women Col- lege Athletes on Sport and Gender." *Sex Roles,* 58 (1): 40–57.

Rostow, Walt W. 1971. *The Stages of Economic Growth: A Non-Communist Manifesto* (2nd ed.). Cambridge: Cambridge University Press (orig. pub. 1960).

—. 1978. *The World Economy: History and Prospect.* Austin: University of Texas Press.

Rothman, Ellen Lerner. 1999. *White Coat: Becoming a Doctor at Harvard Medical School.* New York: Morrow.

Rousseau, Ann Marie. 1981. *Shopping Bag Ladies: Homeless Women Speak About Their Lives.* New York: Pilgrim.

Roy, Jessica. 2013. "This Is the Age You'll Get Married Based on Where You Live." *Time Magazine: Newsfeed* (Nov. 22). Retrieved Feb. 19, 2014. Online: http://newsfeed.time .com/2013/11/22/this-is-the-age -youll-get-married-based-on-where -you-live

Rural Policy Research Institute. 2004. "Place Matters: Addressing Rural Pov- erty." Rural Poverty Research Insti- tute. Retrieved Jan. 21, 2008. Online: www.rprconline.org/synthesis.pdf

Ryan, Camille. 2013. "Language Use in the United States: 2011." U.S. Census Bureau (Aug.). Retrieved Dec. 16, 2013. Online: www.census.gov/prod /2013pubs/acs-22.pdf

Rymer, Russ. 1993. *Genie: An Abused Child's Flight from Silence.* New York: HarperCollins.

Saad, Lydia. 2011. "Doctor-Assisted Suicide Is Moral Issue Dividing Americans Most." Gallup Organiza- tion (May 31). Retrieved Jan. 3, 2012. Online: www.gallup.com/poll/147842 /Doctor-Assisted-Suicide-Moral -Issue-Dividing-Americans.aspx

—. 2012. "Majority in U.S. Drink Alco- hol, Averaging Four Drinks a Week." Gallup.com (Aug. 17). Retrieved May 7, 2013. Online: www.gallup .com/poll/156770/Majority-Drink -Alcohol-Averaging-Four-Drinks -Week.aspx

Sadker, David, and Karen Zittleman. 2009. *Still Failing at Fairness: How Gender Bias Cheats Boys and Girls in Schools.* New York: Simon & Schuster.

SAMHSA. 2012. "Twenty Percent of U.S. Adults Experienced Mental Ill- ness in the Past Year, Report Says." Substance Abuse and Mental Health Services Administration (Nov. 27).

Retrieved May 10, 2013. Online: www.samhsa.gov/newsroom /advisories/1211273220.aspx

Sandals, Leah. 2007. "'Public Space Protection'—But for Which 'Public'?" Retrieved Feb. 24, 2007. Online: http://spacing.ca/wire/?p=1466

Sapir, Edward. 1961. *Culture, Language and Personality.* Berkeley: University of California Press.

Saulny, Susan. 2011a. "Black? White? Asian? More Young Americans Choose All of the Above." *New York Times* (Jan. 29). Retrieved Mar. 7, 2011. Online: www.nytimes.com /2011/01/30/us/30mixed.html

—. 2011b. "In a Multiracial Nation, Many Ways to Tally." *New York Times* (Feb. 10): A1, A17.

Schaefer, Richard T., and William W. Zellner. 2010. *Extraordinary Groups: An Examination of Unconventional Lifestyles* (9th ed.). New York: Worth.

Scherer, Michael. 2012a. "Friended: How the Obama Campaign Con- nected with Young Voters" (Nov. 20). Retrieved Mar. 30, 2013. Online: http://swampland.time.com/2012 ./11/20/friended-how-the-obama -campaign-connected-with-young -voters

—. 2012b. "'Julia' Becomes Vehicle for Obama's Messaging" (May 3). Retrieved Mar. 30, 2013. Online: http://thecaucus.blogs.nytimes .com/2012/05/03/julia-becomes -vehicle-for-obamas-messaging

*Scholastic Parent & Child.* 2007. "How and When to Praise." Retrieved Jan. 3, 2008. Online: http://content.scholastic .com/browse/article.jsp?id=2064

Schur, Edwin M. 1983. *Labeling Women Deviant: Gender, Stigma, and Social Control.* Philadelphia: Temple Univer- sity Press.

Schwarz, John E., and Thomas J. Volgy. 1992. *The Forgotten Americans.* New York: Norton.

Sciencedaily.com. 2008. "Three Out of Four American Women Have Disordered Eating, Survey Suggests" (Apr. 22). Retrieved Jan. 30, 2012. Online: www.sciencedaily.com /releases/2008/04/080422202514.htm

—. 2011. "Educational Development Stunted by Teenage Fatherhood" (Mar. 30). Retrieved Feb. 12, 2012. Online: www.sciencedaily.com /releases/2011/03/110330094353.htm

Shapiro, Thomas, Tatjana Meschede, and Sam Osoro. 2013. "The Roots of the Widening Racial Wealth Gap: Ex- plaining the Black–White Economic

Division." Institute on Assets and Social Policy (Feb.). Retrieved Jan. 13, 2014. Online: http://iasp.brandeis .edu/pdfs/Author/shapiro-thomas-m /racialwealthgapbrief.pdf

Shawver, Lois. 1998. "Notes on Reading Foucault's *The Birth of the Clinic*." Retrieved Oct. 2, 1999. Online: www .california.com/~rathbone/foucbc .htm

Shevky, Eshref, and Wendell Bell. 1966. *Social Area Analysis: Theory, Illustrative Application and Computational Procedures*. Westport, CT: Greenwood.

Shierholz, Heidi. 2010. "Immigration and Wages—Methodological Advancements Confirm Modest Gains for Native Workers." EPI Briefing Paper #255 (Feb. 4). Retrieved Feb. 5, 2012. Online: www.epi.org

Short, Elizabeth, Damien W. Riggs, Amaryll Perlesz, Rhonda Brown, and Graeme Kane. 2007. "Lesbian, Gay, Bisexual and Transgender (LGBT) Parented Families" (Aug.). Australian Psychological Society. Retrieved Apr. 17, 2011. Online: www .psychology.org.au/Assets/Files /LGBT-Families-Lit-Review.pdf

Siegel, Larry J. 2012. *Criminology*, 11th ed. Belmont, CA: Cengage.

Simmel, Georg. 1950. *The Sociology of Georg Simmel*. Trans. Kurt Wolff. Glencoe, IL: Free Press (orig. written in 1902–1917).

—. 1957. "Fashion." *American Journal of Sociology*, 62 (May 1957): 541–558. Orig. pub. 1904.

—. 1990. *The Philosophy of Money*. Ed. David Frisby. New York: Routledge (orig. pub. 1907).

Simon, David R. 2012. *Elite Deviance* (10th ed.). Upper Saddle River, NJ: Pearson.

Sjoberg, Gideon. 1965. *The Preindustrial City: Past and Present*. New York: Free Press.

Smelser, Neil J. 1963. *Theory of Collective Behavior*. New York: Free Press.

—. 1988. "Social Structure." In Neil J. Smelser (Ed.), *Handbook of Sociology*. Newbury Park, CA: Sage, pp. 103–129.

Smith, Adam. 1976. *An Inquiry into the Nature and Causes of the Wealth of Nations*. Ed. Roy H. Campbell and Andrew S. Skinner. Oxford, England: Clarendon (orig. pub. 1776).

Smith, Earl, and Angela J. Hatter. 2011. "Wrongful Convictions, and Exonerations." *Journal of African American Studies*, 15: 74–79.

Snow, David A., and Leon Anderson. 1993. *Down on Their Luck: A Case Study of Homeless Street People*. Berkeley: University of California Press.

Snow, David A., Louis A. Zurcher, and Robert Peters. 1981. "Victory Celebrations as Theater: A Dramaturgical Approach to Crowd Behavior." *Symbolic Interaction*, 4 (1): 21–41.

Society for Human Resource Management. 2010. "Workers with Disabilities Face Steep Occupational Obstacles" (Dec. 8). Retrieved Apr. 2, 2011. Online: www.shrm.org/hrdisciplines/Diversity /Articles/Pages/FaceSteepOccupational Obstacles.aspx

Solomon, Jon. 2012. "College Football's Diversity Quietly Improves Among Head-Coaching Hires" (Jan. 16). Retrieved Apr. 28, 2012. Online: www .al.com/sports/index.ssf/2012/01 /diversity_quietly_improves_in.html

Stainback, Kevin, and Donald Tomaskovic-Devey. 2012. *Documenting Discrimination: Racial and Gender Segregation in Private-Sector Employment Since the Civil Rights Act*. New York: Russell Sage Foundation.

Stanley, Alessandra. 2009. "A Wink at Colleges and a Nod to Cliches." *New York Times* (Sept. 17). Retrieved Dec. 21, 2010. Online: http:// tv.nytimes.com/2009/09/17/arts /television/17community.html

Stark, Rodney, and William Sims Bainbridge. 1981. "American-Born Sects: Initial Findings." *Journal for the Scientific Study of Religion*, 20: 130–149.

—. 1985. *The Future of Religion: Secularization, Revival and Cult Formation*. Berkeley, CA: University of California Press.

Stark, Rodney, and Gary Tobin. 2008. "Guest Commentary: Competition and the American Religious Marketplace." JCR: Institute for Jewish and Community Research (Mar.). Retrieved May 2, 2011. Online: www .jewishresearch.org/v2/2008/articles /demography/03_08.htm

St. John, Warren. 2007. "A Laboratory for Getting Along." *New York Times* (Dec. 25): A1, A14.

Strohmeyer, Robert. 2011. "How to Monitor Your Employees' PCs Without Going Too Far." PCWorld.com (Mar. 22). Retrieved Nov. 10, 2012. Online: www .pcworld.com/article/222169/how _to_monitor_your_employees _without_going_too_far.html

Sugrue, Thomas J. 2011. "A Dream Still Deferred." *New York Times* (Mar. 26). Retrieved May 23, 2011. Online: www.nytimes.com/2011 /03/27/opinion/27Sugrue.html?ref =michigan&pagewanted=print

Sullivan, Oriel. 2011. "Gender Deviance Neutralization Through Housework." *Journal of Family Theory & Review*, 3 (Mar.): 27–31.

Sumner, William G. 1959. *Folkways*. New York: Dover (orig. pub. 1906).

Sutherland, Edwin H. 1939. *Principles of Criminology*. Philadelphia: Lippincott.

—. 1949. *White Collar Crime*. New York: Dryden.

Swidler, Ann. 1986. "Culture in Action: Symbols and Strategies." *American Sociological Review*, 51 (April): 273–286.

Szasz, Thomas S. 1984. *The Myth of Mental Illness: Foundations of a Theory of Personal Conduct*. New York: HarperCollins.

Takaki, Ronald. 1993. *A Different Mirror: A History of Multicultural America*. Boston: Little, Brown.

Tannen, Deborah. 1993. "Commencement Address, State University of New York at Binghamton." Reprinted in *Chronicle of Higher Education* (June 9): B5.

Tarbell, Ida M. 1925. *The History of Standard Oil Company*. New York: Macmillan (orig. pub. 1904).

Tavris, Carol. 2013. "Book Review: 'Behind the Shock Machine' by Gina Perry." Retrieved Dec. 27, 2013. Online: http://online.wsj.com/news /articles/SB100014241278873233249 04579040672110673420

Taylor, Paul. 2012. "The Growing Electoral Clout of Blacks in Drive by Turnout, Not Demographics." Pew Research Social and Demographic Trends (Dec. 26). Retrieved Mar. 13, 2013. Online: www.pewsocialtrends .org/files/2013/01/2012_Black_Voter _Project_revised_1-9.pdf

Taylor, Steve. 1982. *Durkheim and the Study of Suicide*. New York: St. Martin's.

Teicher, Stacy A. 2006. "Researchers Say That Middle-School Bullying Could Be Curbed by Showing That It's Not Normal." *Christian Science Monitor* (Aug. 17). Retrieved Feb. 12, 2008. Online: www.csmonitor com./2006 /0817/p15s02-legn.html

thetaskforce.org. 2011. "Injustice at Every Turn: A Report on the National Transgender Discrimination Survey." Retrieved May 2, 2012. Online: www .thetaskforce.org/downloads/reports /reports/ntds_full.pdf

Tilly, Charles. 1975. *The Formation of National States in Western Europe.* Princeton, NJ: Princeton University Press.

*Toddlers & Tiaras.* 2009. "Marleigh's Mom." Learning Channel (Apr. 8). Retrieved Feb. 22, 2013. Online: www.youtube.com/watch?v=MRZyzccF02I

Toft, Monica Duffy, Daniel Philpott, and Timothy Samuel Shah. 2011. "God's Partisans Are Back." *Chronicle Review* (Apr. 22): B4, B5.

Tönnies, Ferdinand. 1940. *Fundamental Concepts of Sociology* (Gemeinschaft *und* Gesellschaft). Trans. Charles P. Loomis. New York: American Book Company (orig. pub. 1887).

—. 1963. *Community and Society* (Gemeinschaft *and* Gesellschaft). New York: Harper & Row (orig. pub. 1887).

Toobin, Jeffrey. 2011. "Money Talks." *New Yorker* (Apr. 11). Retrieved Mar. 9, 2011. Online: www.newyorker.com/talk/comment/2011/04/11/110411taco_talk_toobin

Trounson, Rebecca. 2012. "Most Young People OK with Living at Home in Tough Times." *Los Angeles Times* (Mar. 16). Retrieved Apr. 1, 2013. Online: http://articles.latimes.com/2012/mar/16/local/la-me-0316-boomerang-20120316

Truman, Jennifer, Lynn Langton, and Michael Planty. 2013. *Criminal Victimization, 2012.* U.S. Department of Justice (Oct.). Retrieved Dec. 31, 2013. Online: www.bjs.gov/content/pub/pdf/cv12.pdf

Truman, Jennifer L., and Michael Planty. 2012. "Crime Victimization 2011." U.S. Department of Justice (Oct.). Retrieved Nov. 11, 2012. Online: www.bjs.gov/content/pub/pdf/cv11.pdf

Turkle, Sherry. 2011. *Alone Together: Why We Expect More from Technology and Less from Each Other.* New York: Basic.

Turner, Jonathan, Leonard Beeghley and Charles H. Powers. 2002. *The Emergence of Sociological Theory* (5th ed.). Belmont, CA: Wadsworth.

—. 2007. *The Emergence of Sociological Theory* (6th ed.). Belmont, CA: Wadsworth.

Turner, Jonathan H., Royce Singleton Jr., and David Musick. 1984. *Oppression: A Socio-History of Black–White Relations in America.* Chicago: Nelson-Hall (reprinted 1987).

Turner, Ralph H., and Lewis M. Killian. 1993. "The Field of Collective Behavior." In Russell L. Curtis, Jr., and Benigno E. Aguirre (Eds.), *Collective Behavior and Social Movements.* Boston: Allyn & Bacon, pp. 5–20.

UNAIDS. 2013. "2013 Global Fact Sheet." Retrieved Mar. 23, 2014. Online: www.unaids.org/en/media/unaids/contentassets/documents/epidemiology/2013/gr2013/20130923_FactSheet_Global_en.pdf

UNFPA (United Nations Population Fund). 2011. *The State of World Population 2011.* Retrieved Feb. 7, 2012. Online: http://foweb.unfpa.org/SWP2011/reports/EN-SWOP2011-FINAL.pdf

unicef.org. 2011. "Child Protection from Violence, Exploitation and Abuse: Child Labour." Retrieved Mar. 6, 2011. Online: www.unicef.org/protection/index_childlabour.html

*Unification Church News.* 2011. "Who We Are." Retrieved May 10, 2011. Online: www.familyfed.org/about

United Nations. 1948. "The Universal Declaration of Human Rights." Retrieved Apr. 14, 2011. Online: www.un.org/en/documents/udhr/index.shtml

—. 2012. "Highlights." *World Urbanization Prospects: The 2011 Revision.* Retrieved July 9, 2013. Online: http://esa.un.org/unup/pdf/WUP2011_Highlights.pdf

—. 2013. "The Millennium Development Goals Report 2013." Retrieved Jan. 26, 2014. Online: www.un.org/millenniumgoals/pdf/report-2013/mdg-report-2013-english.pdf

United Nations (Department of Economic and Social Affairs, Population Division [DESAPD]). 2010. *World Urbanization Prospects: The 2009 Revision.* Retrieved Feb. 7, 2012. Online: http://esa.un.org/unpd/wup/index.htm

United Nations Development Programme. 2010. *Human Development Report: 2010.* Retrieved Feb. 25, 2010. Online: http://hdr.undp.org/en/media/HDR_2010_EN_Complete_reprint.pdf

—. 2011. *Human Development Report 2011: Sustainability and Equity: A Better Future for All.* Retrieved Mar. 4, 2012. Online: http://hdr.undp.org/en/media/HDR_2011_EN_Complete.pdf

—. 2013. "Human Development Report 2013." Retrieved Jan. 24, 2014. Online: http://hdr.undp.org/sites/default/files/reports/14/hdr2013_en_complete.pdf

United Nations Office of Drugs and Crime. 2014. "Transnational Organized Crime—the Globalized Illegal Economy." Retrieved May 14, 2014. Online: www.unodc.org/toc/en/crimes/organized-crime.html

United Nations Population Division. 2011. *World Population Prospects: The 2010 Revision.* Accessed Feb. 7, 2011. Online: http://esa.un.org/unpd/wpp/index.htm

United Nations Regional Information Center for Western Europe. 2014. "The Evolving Definition of Literacy." Retrieved Jan. 26, 2014. Online: www.unric.org/en/literacy/27791-the-evolving-definition-of-literacy

United Nations World Urbanization Prospects On-Line Data. 2012. "Urban Agglomerations." Retrieved May 13, 2013. Online: http://esa.un.org/unpd/wup/unup/index_panel2.html

United Press International. 2007. "Indian Women Carry Children for Foreigners." *United Press International.com* (Nov. 11). Retrieved Feb. 9, 2008. Online: www.upi.com/NewsTrack/Science/2007/11/11/indian_women_carry_children_for_foreigners/2909

University of Michigan. 2007. "Boys Mow Lawns, Girls Do Dishes: Are Parents Perpetuating the Chore Wars?" Institute for Social Research. Retrieved May 2, 2012. Online: www.isr.umich.edu/home/news/archive.html

UPI.com. 2010. "Philip Morris Vows to End Child Labor." UPI.com (July 14). Retrieved Mar. 6, 2011. Online: www.upi.com/Top_News/US/2010/07/14/Philip-Morris-vows-to-end-child-labor/UPI-61821279121448

U.S. Army. 2012. "Army 2020: Generating Health & Discipline in the Force Ahead of the Strategic Reset: Report 2012." Retrieved Sept. 23, 2012. Online: http://usarmy.vo.llnwd.net/e2/c/downloads/232541.pdf

U.S. Bureau of Labor Statistics. 2012a. "Consumer Price Index, Nov. 2012." Retrieved Jan. 8, 2012. Online: www.bls.gov/news.release/cpi.t01.htm

—. 2012b. "Employment and Wages for the Highest and Lowest Paying Occupations, May 2011." Retrieved Mar. 10, 2013. Online: www.bls.gov/oes/current/high_paying_occs.htm

—. 2012c. "Employment Situation Summary." Retrieved Jan. 13, 2013.

Online: www.bls.gov/news.release
/empsit.nr0.htm

—. 2012d. "Fastest Growing Occupations." Retrieved Mar. 9, 2013. Online: www.bls.gov/ooh/fastest-growing.htm

—. 2012e. "Highlights of Women's Earnings in 2011" (Oct.). Retrieved Feb. 22, 2013. Online: www.bls.gov/cps/cpswom2011.pdf

—. 2013a. "The Employment Situation—February 2013." Retrieved Mar. 10, 2013. Online: www.bls.gov/news.release/pdf/empsit.pdf

—. 2013b. "Employment Situation Summary" (June 7). Retrieved June 14, 2013. Online: www.bls.gov/news.release/empsit.nr0.htm

—. 2013c. "Major Work Stoppages in 2012." Retrieved Mar. 10, 2013. Online: www.bls.gov/news.release/pdf/wkstp.pdf

—. 2013d. "Union Members Summary." Retrieved Mar. 10, 2013. Online: http://data.bls.gov/cgi-bin/print.pl/news.release/union2.nr0.htm

—. 2013e. "Usual Weekly Earnings Summary" (Jan. 18). Retrieved Apr. 13, 2013. Online: http://data.bls.gov/cgi-bin/print.pl/news.release/wkyeng.nr0.htm

—. 2013f. "Women in the Labor Force: A Databook." BLS Report 1040 (Feb.). Retrieved May 3, 2014. Online: www.bls.gov/cps/wlf-databook-2012.pdf

—. 2014a. "Employment Characteristics of Families—2012" (Apr. 26). Retrieved Feb. 22, 2014. Online: www.bls.gov/news.release/pdf/famee.pdf

—. 2014b. "The Employment Situation—February 2014." Retrieved Mar. 17, 2014. Online: www.bls.gov/news.release/pdf/empsit.pdf

—. 2014c. "Employment Situation Summary." Retrieved Jan. 19, 2014. Online: http://data.bls.gov/cgi-bin/print.pl/news.release/empsit.nr0.htm

—. 2014d. "Employment Status of the Civilian Population by Sex, Age, and Disability Status." Retrieved Mar. 17, 2014. Online: www.bls.gov/news.release/empsit.t06.htm

—. 2014e. "Union Members—2013." Retrieved Mar. 17, 2014. Online: www.bls.gov/news.release/pdf/union2.pdf

U.S. Census Bureau. 2004. "Ancestry: 2000" (C2KBR-35). Retrieved Apr. 7, 2014. Online: www.census.gov/prod/2004pubs/c2kbr-35.pdf

—. 2011a. "American Indian and Alaska Native Heritage Month: November 2011." U.S. Census Bureau News, Facts for Features. Retrieved Dec. 29, 2011. Online: www.census.gov/newsroom/releases/pdf/cb11ff-22_aian.pdf

—. 2011b. *America's Families and Living Arrangements: 2011*. Retrieved Jan. 4, 2012. Online: www.census.gov/population/www/socdemo/hh-fam/cps2011.html

—. 2011c. "Asian/Pacific American Heritage Month: May 2011." U.S. Census Bureau News, Facts for Features (May). Retrieved Dec. 29, 2011. Online: www.census.gov/newsroom/releases/archives/facts_for_features_special_editions/cb11-ff06.html

—. 2011d. "Census Bureau Releases Estimate of Same-Sex Married Couples." Retrieved Jan. 5, 2011. Online: http://2010.census.gov/news/releases/operations/cb11-cn181.html

—. 2011e. "Current Population Survey: Definitions and Explanations." Retrieved Apr. 14, 2011. Online: www.census.gov/population/www/cps/cpsdef.html

—. 2011f. *Historical Income Tables: Families*. Table F-3: Mean Income Received by Each Fifth and Top 5 Percent of Families, All Races: 1966 to 2010. Retrieved Dec. 12, 2011. Online: www.census.gov/hhes/www/income/data/historical/families/index.html

—. 2012a. "American Families and Living Arrangements: 2012." Retrieved Apr. 1, 2013. Online: www.census.gov/hhes/families/data/cps2012.html

—. 2012b. "Language Spoken at Home by Ability to Speak English for the Population 5 Years and Over." Retrieved Oct. 14, 2012. Online: http://factfinder2.census.gov/faces/tableservices/jsf/pages/productview.xhtml?pid=ACS_11_1YR_B16001&prodType=table

—. 2013a. "America's Families and Living Arrangements: 2013: Children (C table series)." Retrieved Feb. 17, 2014. Online: www.census.gov/hhes/families/data/cps2013C.html

—. 2013b. "Disability Employment Tabulation." Retrieved Mar. 22, 2014. Online: www.census.gov/people/disabilityemptab/data

"U.S. Census Bureau Facts on American Indian and Alaska Native Heritage Month." 2013. Retrieved Feb. 9, 2014. Online: www.census.gov/newsroom/releases/pdf/cb13ff-26_aian.pdf

"U.S. Census Bureau Facts on Asian/Pacific American Heritage Month." 2013. Retrieved Feb. 10, 2014. Online: www.census.gov/newsroom/releases/pdf/cb13ff-09_asian.pdf

"U.S. Census Bureau Facts on Hispanic Heritage Month." 2013. Retrieved Feb. 10, 2014. Online: www.census.gov/newsroom/releases/pdf/cb13ff-19_hispanicheritage.pdf

U.S. Census Bureau Historical Income Tables. 2012. "Historical Income Tables: Households." Retrieved Jan. 10, 2013. Online: www.census.gov/hhes/www/income/data/historical/household

U.S. Census Bureau International Data Base. 2011. "World Vital Events Per Time Unit: 2011." Retrieved May 15, 2011. Online: www.census.gov/cgi-bin/ipc/pcwe

U.S. Census Bureau News. 2013. "American Indian and Alaska Native Heritage Month: November 2013." Retrieved Jan. 20, 2014. Online: www.census.gov/newsroom/releases/pdf/cb13ff-26_aian.pdf

U.S. Centers for Disease Control and Prevention (U.S. CDC). 2011a. "CDC Report: Mental Illness Surveillance Among Adults in the United States." Retrieved Jan. 31, 2012. Online: www.cdc.gov/mentalhealthsurveillance/fact_sheet.html

—. 2011b. "Preventing Teen Pregnancy in the US." *Vital Signs*. Retrieved Jan. 1, 2012. Online: www.cdc.gov/vitalsigns/TeenPregnancy/index.html

—. 2011c. "Sexually Transmitted Disease Surveillance: 2010." Retrieved Mar. 5, 2012. Online: www.cdc.gov/std/stats10/surv2010.pdf

—. 2011d. "Teen Birth Rates Declined Again in 2009." Retrieved Jan. 1, 2012. Online: www.cdc.gov/features/dsteenpregnancy

—. 2012a. "African Americans and Sexually Transmitted Diseases" (Dec.). Retrieved May 8, 2013. Online: www.cdc.gov/nchhstp/newsroom/docs/AAs-and-STD-Fact-Sheet-042011.pdf

—. 2012b. "Binge Drinking." Retrieved Mar. 5, 2012. Online: www.cdc.gov/VitalSigns/pdf/2012-01-vitalsigns.pdf

—. 2012c. "HIV in the United States." Fact Sheet. Retrieved Jan. 29, 2012. Online: www.cdc.gov/hiv/resources/factsheets/us.htm

—. 2012d. "Suicide: Facts at a Glance." Retrieved Oct. 12, 2012. Online: www.cdc.gov/ViolencePrevention/pdf/Suicide_DataSheet_2012-a.pdf

—. 2012e. "Web-Based Injury Statistics Query and Reporting System (WISQARS)." Retrieved Oct. 12,

2012. Online: www.cdc.gov/injury/wisqars/index.html

—. 2013a. "Adult Obesity: Data and Statistics." Retrieved Mar. 30, 2014. Online: www.cdc.gov/obesity/data/adult.html

—. 2013b. "HIV in the United States: At a Glance." Retrieved Mar. 30, 2014. Online: www.cdc.gov/hiv/pdf/statistics_basics_factsheet.pdf

—. 2013c. "Secondhand Smoke (SHS) Facts." Retrieved July 8, 2013. Online: www.cdc.gov/tobacco/data_statistics/fact_sheets/secondhand_smoke/general_facts

—. 2014a. "2012 Sexually Transmitted Diseases Surveillance: Chlamydia." Retrieved Mar. 29, 2014. Online: www.cdc.gov/std/stats12/chlamydia.htm

—. 2014b. "2012 Sexually Transmitted Diseases Surveillance: Gonorrhea." Retrieved Mar. 29, 2014. Online: www.cdc.gov/std/stats12/gonorrhea.htm

—. 2014c. "2012 Sexually Transmitted Diseases Surveillance: Syphilis." Retrieved Mar. 29, 2014. Online: www.cdc.gov/std/stats12/syphilis.htm

U.S. Citizenship and Immigration Services. 2011. *A Guide to Naturalization.* Retrieved Feb. 6, 2012. Online: www.uscis.gov/files/article/M-476.pdf

U.S. Conference of Mayors. 2011. "Hunger and Homeless Survey" (Dec.). Retrieved Dec. 24, 2011. Online: http://usmayors.org/pressreleases/uploads/2011-hhreport.pdf

—. 2012. "U.S. Metro Economies: Key Findings" (Jan.). Retrieved Mar. 15, 2012. Online: http://usmayors.org/pressreleases/uploads/2012/MetroEconomiesKeyFindings_011812.pdf

—. 2013. *Hunger and Homelessness Survey: A Status Report of Hunger and Homelessness in America's Cities. A 25-City Survey* (Dec.). Retrieved Dec. 26, 2013. Online: www.usmayors.org/pressreleases/uploads/2013/1210-report-HH.pdf

U.S. Department of Health and Human Services. 2008a. "Adoption Experiences of Women and Men and Demand for Children to Adopt by Women 18–44 Years of Age in the United States, 2002." National Centers for Disease Control and Prevention. Retrieved Apr. 2, 2010. Online: www.cdc.gov/nchs/data/series/sr_23/sr23_027.pdf

—. 2008b. "Stop Bullying Now." Retrieved Feb. 12, 2008. Online: www.stopbullyingnow.hrsa.gov

—. 2013a. "Adoptive Family Structure (July 2013)." Retrieved Feb. 22, 2014. Online: www.acf.hhs.gov/sites/default/files/cb/family_structure2012.pdf

—. 2013b. *Child Health USA, 2012.* Retrieved Dec. 22, 2013. Online: http://mchb.hrsa.gov/chusa12

U.S. Department of Housing and Urban Development. 2011. "Affordable Housing." Retrieved Dec. 15, 2011. Online: www.hud.gov/offices/cpd/affordablehousing

U.S. Department of State. 2012. "A Day in Your Life: Touched by Modern Slavery." Washington, DC: Office to Monitor and Combat Tracking of Persons. Retrieved Jan. 6, 2013. Online: www.state.gov/j/tip/rls/fs/2012/194712.htm

—. 2013. "Independent States of the World: Fact Sheet" (Dec. 9). Retrieved Mar. 13, 2014. Online: www.state.gov/s/inr/rls/4250.htm

U.S. Equal Employment Opportunity Commission. 2012. "SunTrust Bank Sued by EEOC in Class Sexual Harassment Lawsuit" (June 14). Retrieved Apr. 7, 2013. Online: www.eeoc.gov/eeoc/newsroom/release/6-14-12.cfm

U.S. Office of Personnel Management. 2013. "Data Analysis and Documentation: Profile of Federal Civilian Non-Postal Employees." Retrieved Mar. 29, 2013. Online: www.opm.gov/policy-data-oversight/data-analysis-documentation/federal-employment-reports/reports-publications/profile-of-federal-civilian-non-postal-employees

Van Ausdale, Debra, and Joe R. Feagin. 2001. *The First R: How Children Learn Race and Racism.* Lanham, MD: Rowman & Littlefield.

Van Gundy, Alana, and Victor E. Kappeler. 2013. *Feminist Theory, Crime, and Social Justice.* New York: Anderson/Elsevier.

Vargas, Jose Antonio. 2011. "My Life as an Undocumented Immigrant." *New York Times* (June 22). Retrieved Mar. 13, 2011. Online: www.nytimes.com/2011/06/26/magazine/my-life-as-an-undocumented-immigrant.html?pagewanted=all

—. 2012. "About Jose." defineamerican.com. Retrieved Mar. 14, 2012. Online: www.defineamerican.com/page/about/about-jose

Veblen, Thorstein. 1967. *The Theory of the Leisure Class.* New York: Viking (orig. pub. 1899).

Venkatesh, Sudhir Alladi. 2006. *Off the Books: The Underground Economy of the Urban Poor.* Cambridge: Harvard University Press.

—. 2013. *Floating City: A Rogue Sociologist Lost and Found in New York's Underground Economy.* New York: Penguin.

Vespa, Jonathan, Jamie M. Lewis, and Rose M. Kreider. 2013. "America's Families and Living Arrangements: 2012." U.S. Census Bureau (Aug.). Retrieved May 23, 2014. Online: www.census.gov/prod/2013pubs/p20-570.pdf

Wade, Lisa. 2012. "Women, Education, and Trends in Childlessness" (Sept. 24). Retrieved Apr. 8, 2013. Online: http://thesocietypages.org/socimages/2012/09/24/women-education-and-trends-in-childlessness

Wadsworth, Tim, and Charis E. Kubrin. 2007. "Hispanic Suicide in U.S. Metropolitan Areas: Examining the Effects of Immigration, Assimilation, Affluence, and Disadvantage." *American Journal of Sociology* 112 (6): 1848–1885.

Wadud, Amina. 2002. "A'ishah's Legacy: Amina Wadud Looks at the Struggle for Women's Rights Within Islam." *New Internationalist* (May). Retrieved Mar. 18, 2007. Online: http://newint.org/features/2002/05/01/aishahs-legacy

Wald, Matthew L. 2011. "Japan Orders Evacuation Near 2nd Nuclear Plant." *New York Times* (Mar. 11). Retrieved May 23, 2011. Online: www.nytimes.com/2011/03/12/world/asia/12nuclear.html?_r=1&scp=3&sq=evacuation%20near%20nuclear%20reaction%20in%20Japan?&st=cse

Wallace, Walter L. 1971. *The Logic of Science in Sociology.* New York: Aldine de Gruyter.

Wallerstein, Immanuel. 1979. *The Capitalist World-Economy.* Cambridge, England: Cambridge University Press.

—. 1984. *The Politics of the World Economy.* Cambridge, England: Cambridge University Press.

—. 1991. *Unthinking Social Science: The Limits of Nineteenth-Century Paradigms.* Cambridge, England: Polity.

—. 2011. *The Modern World System I–IV.* Berkeley: University of California Press.

Warner, W. Lloyd, and Paul S. Lunt. 1941. *The Social Life of a Modern Community.* New Haven, CT: Yale University Press.

Watson, Tracey. 1987. "Women Athletes and Athletic Women: The Dilemmas and Contradictions of Managing Incongruent Identities." *Sociological Inquiry,* 57 (Fall): 431–446.

Weber, Max. 1968. *Economy and Society: An Outline of Interpretive Sociology.* Trans. G. Roth and G. Wittich. New York: Bedminster (orig. pub. 1922).

—. 1976. *The Protestant Ethic and the Spirit of Capitalism.* Trans. Talcott Parsons. Introduction by Anthony Giddens. New York: Scribner (orig. pub. 1904–1905).

Weeks, John R. 2012. *Population: An Introduction to Concepts and Issues* (11th ed.). Belmont, CA: Wadsworth.

Weigel, Russell H., and P. W. Howes. 1985. "Conceptions of Racial Prejudice: Symbolic Racism Revisited." *Journal of Social Issues,* 41: 124–132.

Weiss, Gregory L., and Lynne E. Lonnquist. 2012. *Sociology of Health, Healing, and Illness* (7th ed.). Upper Saddle River, NJ: Prentice Hall.

Weiss, Jeffrey. 2004. "Beliefs in UFOs and Ritual Abuse Are Studied as New Religious Movements." *Dallas Morning News* (Aug. 28). Retrieved Jan. 11, 2008. Online: www.dallasnews.com

Weitz, Rose. 2004. *The Sociology of Health, Illness, and Health Care* (3rd ed.). Belmont, CA: Wadsworth.

—. 2013. *The Sociology of Health, Illness, & Health Care: A Critical Approach* (6th ed.). Boston: Wadsworth/Cengage.

Wellman, Barry. 2001. "Physical Place and Cyberplace: The Rise of Personalized Networking." *International Journal of Urban and Regional Research,* 22 (2): 227–252.

Werhane, Veronica. 2011. "Food Trucks: Spreading Culture Through Food." Neontommy.com (Apr. 12). USC Annenberg School for Communication and Journalism. Retrieved Dec. 11, 2013. Online: www .neontommy.com/news/2011/04 /food-and-food-trucks-using-food -spread-cultural-awareness

White House. 2013. "Education Blueprint: An Economy Built to Last." Retrieved Apr. 14, 2013. Online: www .whitehouse.gov/sites/default/files /cantwait/final_-_education_blueprint_-_an_economy_built_to_last.pdf

whitehouse.gov. 2011. "Strategy to Combat Transnational Organized Crime" (July). Retrieved Jan. 2, 2012. Online: www.whitehouse.gov/sites/default /files/microsites/2011-strategy-combat -transnational-organized-crime.pdf

—. 2012. "President Obama: 'Words Need to Lead to Action' on Gun Violence." Retrieved Dec. 27, 2012.

Online: www.whitehouse.gov /blog/2012/12/19/president-obama -words-need-lead-action-gun-violence

Whorf, Benjamin Lee. 1956. *Language, Thought and Reality.* Ed. John B. Carroll. Cambridge, MA: MIT Press.

Williams, Robin M., Jr. 1970. *American Society: A Sociological Interpretation* (3rd ed.). New York: Knopf.

Wilson, David Sloan, and Edward O. Wilson. 2007. "Rethinking the Theoretical Foundation of Sociobiology." *Quarterly Review of Biology,* 82 (4): 327–348.

Wilson, William Julius. 1978. *The Declining Significance of Race: Blacks and Changing American Institutions.* Chicago: University of Chicago Press.

Wilson, William Julius, and Richard P. Taub. 2007. *There Goes the Neighborhood: Racial, Ethnic and Class Tension in Four Chicago Neighborhoods and Their Meaning for America.* New York: Vintage.

Wintour, Patrick. 2013. "UN Goals Must Aim to Eradicate Extreme Poverty." *Guardian* (Feb. 1). Retrieved Feb. 3, 2013. Online: www.guardian .co.uk/world/2013/feb/01/un-goals -extreme-poverty-david-cameron /print

Wirth, Louis. 1938. "Urbanism as a Way of Life." *American Journal of Sociology,* 40: 1–24.

Wischnowsky, Dave. 2005. "Small Town a Bastion of Bigfoot Belief." *Chicago Tribune* (Oct. 10): 2.

Woodard, Stephanie. 2012. "Suicide Is Epidemic for American Indian Youth: What More Can Be Done?" NBCNews.com (Oct. 10). Retrieved Dec. 7, 2013. Online: http://investiga tions.nbcnews.com/_news/2012 /10/10/14340090-suicide-is-epidemic -for-american-indian-youth-what -more-can-be-done

World Bank. 2012. "World Development Indicators: 2012." Retrieved Jan. 14, 2013. Online: www.oecd -ilibrary.org/docserver/download /3011041ec020.pdf?expires=1359239 194&id=id&accname=guest&checks um=254AE45119DAF1066B9D9EEE 031C65DD

—. 2013. "New Country Classifications." Retrieved Jan. 24, 2014. Online: http://data.worldbank.org /news/new-country-classifications

World Health Organization. 2012. *World Health Statistics: A Snapshot of Global Health.* Retrieved May 6, 2013. Online: http://apps.who.int/iris /bitstream/10665/70889/1/WHO _IER_HSI_12.1_eng.pdf

—. 2013. "Tobacco: Fact Sheet." Retrieved Jan. 26, 2014. Online: www.who.int /mediacentre/factsheets/fs339/en

Wright, Erik Olin. 1978. "Race, Class, and Income Inequality." *American Journal of Sociology,* 83 (6): 1397.

—. 1979. *Class Structure and Income Determination.* New York: Academic Press.

—. 1985. *Class.* London: Verso.

—. 1997. *Class Counts: Comparative Studies in Class Analysis.* Cambridge, England: Cambridge University Press.

—. 2010. *Envisioning Real Utopias.* London: Verso.

*Yale News.* 2013. "Students Map Diversity and Culture of New Haven Through Food Trucks." Yale University (Dec. 4). Retrieved Dec. 11, 2013. Online: http://news.yale.edu/2013/12 /04/students-map-diversity-and -culture-new-haven-through-food -trucks

Yong, Ed. 2011. "New Evidence That IQ Is Not Set in Stone." cbsnews .com (Apr. 26). Retrieved Apr. 28, 2011. Online: www.cbsnews .com/stories/2011/04/26/scitech /main20057536.shtml

Zellner, William M. 1978. Vehicular Suicide: In Search of Incidence. Unpublished M.A. thesis, Western Illinois University, Macomb. Quoted in Richard T. Schaefer and Robert P. Lamm. 1992. *Sociology* (4th ed.). New York: McGraw-Hill, pp. 54–55.

Zernike, Kate. 2010. "A Young and Unlikely Activist Who Got to the Tea Party Early." *New York Times* (Feb. 28): A1, A19.

Zernike, Kate, and Megan Thee-Brenan. 2010. "Discontent's Demography: Who Backs the Tea Party?" *New York Times* (Apr. 15): A1, A17.

Zoroya, Gregg. 2012. "Suicides No. 2 Cause of Death in Military." *USA Today* (June 14). Retrieved Sept. 24, 2012. Online: www.usatoday.com /news/military/story/2012-06-13 /military-suicides/55585182/1

Zuckerman, Mortimer B. 2011. "The Great Recession Goes on." *USNews* (Feb. 11). Retrieved Jan. 2, 2012. Online: www.usnews.com/opinion /mzuckerman/articles/2011/02/11 /the-great-jobs-recession-goes-on

Zurbriggen, Eileen L., and Tomi-Ann Roberts (Eds.). 2012. *The Sexualization of Girls and Girlhood: Causes, Consequences, and Resistance.* New York: Oxford University Press.

# Name Index

# Subject Index

Normative organization, 150–151
Norms, 54–56
Norwegian Constitution Day, 61
Nuclear family, 333

Obamacare. *See* Affordable Care Act
Obedience, 147
Obesity, 305, 448–449
Object fad, 63
Object permanence, 81
Objectification, 297
Objectification of women, 298
Obsessive-compulsive disorder, 464
Occupational crime, 179
Occupational socialization, 96
Occupations, 422, 428
Occupy movement, 168, 508–509, 514
Occupy Wall Street, 60, 181, 508
"Off the books," 429
Office of Management and Budget, 418–419
Official English laws, 51
Official poverty line, 221
Old Age, Survivors, Disability, and Health
    Insurance, 222
Old-boy social network, 113, 114
"Old money," 210
Old-old, 96
Old Order Amish, 58–59
"Old Pat Moore," 98
Oldest-old, 96
Oligarchy, 156–157
Oligopoly, 423–424
Online communities, 141
Online matching services, 340
Open-air markets, 118
Open system, 202
Opportunity theory, 169–170
Orb web model, 85
Organic solidarity, 121, 489
Organization. *See* Groups and organizations
Organized crime, 182–183
Othermothers, 92–93
"Outdated corporate culture," 306
Outer containments, 174
Outgroup, 141–142
Outlet malls, 502
Overt racism, 265
Overview. *See* Concept quick review
OWN, 18
Ozone layer, 530

Pacific Islanders. *See* Asian Americans
PACs, 409–410
Paid work and family work, 319–320
Panic, 513, 516–517
Panopticon, 177
Paraphilic disorders, 464
*Parens patriae*, 190
*Part of History: College Students and the 2013
    Inauguration, A* (Kucharshi), 401
Part-time workers, 428
Participant observation, 31
Pastoral societies, 116
Pastoralism, 308
Patriarchal family, 335
Patriarchy, 18, 306, 307
Patrilineal descent, 334
Patrilocal residence, 336
Patriot Act, 289
Patriot Sunsets Extension Act, 289
Patterns of descent and inheritance, 334–335
Pay equity, 318–319
Pay gap, 318
Payday loan scam, 182
Peer culture, 85
Peer groups, 90
Pell Grant, 378

Pelvic inflammatory disease (PID), 446
Peripheral nations, 250, 498
Permanent government (Washington), 418
Permeable family, 338
Persian Americans, 288
Personal distance, 131
Personal space, 131–132
Personal troubles, 7
Personality fad, 63
Persons with disabilities, 494–496
Persons with disabilities, 430–431, 465–468, 469
Perspective, 16. *See also* Theoretical
    perspectives
PetSmart, 502
Pharmaceutical companies, 438–439
*Philadelphia Negro: A Social Study, The*
    (Du Bois), 15
Philip Morris, 241
Physical environment and change, 529–530
Physical self, 82
Physician–patient relationship, 462
Picketing, 430
Pink-collar occupations, 211
Platform, 413
Play stage, 84
Plea bargaining, 190
Pluralist model, 407–410, 411
Police, 188–189
Political action committees (PACs), 409–410
Political crime, 183–184
Political economy models (cities), 491–493, 497
Political opportunity, 524
Political opportunity theory, 525–526, 528
Political participation, 416–417
Political parties, 413–416
Political preferences, 417–418
Political science, 402
Political socialization, 416
Political sociology, 402
Politics and government, 399–433
    authoritarianism, 406
    conflict perspectives, 410–412
    definitions, 401
    democracy, 407
    elite models, 410–412
    federal civilian employee, 420
    functionalist perspectives, 407–410
    future directions, 431
    governmental bureaucracy, 418–420
    military–industrial complex, 419–420
    monarchy, 405–406
    PACs, 409–410
    pluralist model, 407–410, 411
    political participation, 416–417
    political parties, 413–416
    political preferences, 417–418
    power and authority, 403–404, 405
    power elite, 410–411
    quiz, 401, 402
    ruling class, 411
    special interest groups, 408–409
    super PACs, 409–410
    totalitarianism, 406–407
Pollution, 532
Polyandry, 334
Polygamy, 334, 335
Polygyny, 334
Polytheism, 382
Popular culture, 62–63
Population and urbanization, 473–505
    capitalism and urban growth, 492
    conflict perspectives, 491–493, 497
    demographic transition, 485
    demographic transition theory, 485–486
    divided interests (cities and suburbs),
        498–499
    ecological models, 489–491, 497

    edge cities, 500
    emergence/evolution of the city, 487
    exurban areas, 500–501
    fertility, 476–477, 486, 503
    fiscal crises in cities, 501
    fortressing of America, 493
    functionalist perspectives, 489–491, 497
    future directions, 502–503
    Gans's urban villagers, 494, 495
    gated communities, 492–493
    gender regimes, 493
    gentrification, 490, 491
    global cities, 496–498
    industrial cities, 488
    international migration theories, 486–487
    Malthusian perspective, 482–484
    Marxist perspective, 484
    migration, 478–482. *See also* Immigration
    mortality, 477–478
    neo-Malthusian perspective, 484–485
    "new cities," 503
    persons with disabilities, 494–496
    political economy models, 491–493, 497
    population composition, 482
    population growth, 475, 502, 503
    population pyramid, 482, 483
    postindustrial cities, 488–489
    preindustrial cities, 487–488
    quiz, 475, 477
    rural community issues, 501–502
    rural-to-urban migration, 498
    Simmel's view of city life, 493–494
    Smart Growth, 503
    social change, 530
    suburbs, 499–500
    symbolic interactionist perspectives,
        493–496, 497
    urbanism, 494
    value of urban space, 492
    world's largest agglomerations, 497
*Population Bomb, The* (Ehrlich), 484
"Population Bomb Revisited, The" (Ehrlich/
    Ehrlich), 484
Population composition, 482
Population Connection, 408
Population growth, 475, 502, 503
Population pyramid, 482, 483
Populist Party, 413
"Porkulus," 526
Positive checks (population), 483
Positive ethnocentrism, 61
Positive sanctions, 55
Positivism, 10
Post-racial and post-homophobia framing, 91
Postal jobs, 225
Postconventional stage, 82
Postindustrial cities, 488–489
Postindustrial economies, 420–421
Postindustrial societies, 20, 116, 119–120,
    309–310
Postindustrialization, 485
Postmodern societies, 20
Postmodernist perspectives
    culture, 66–67
    defined, 20
    families and intimate relationships, 338–339
    groups and organizations, 144
    health and health care, 462–463
    overview (concept quick review), 21
    suicide, 20
Poverty, 221–225. *See also* Global stratification
    absolute/relative, 222
    age, 222–223
    economic and structural sources, 224–225
    gender, 223
    health, 218
    high school graduation rates, 220

War veterans, 468
Warner Music Group, 424
WASPs, 279
Water, 530
Watergate, 404
Wealth, 208
Wealth inequality, 217–218
*Wealth of Nations* (Smith), 425
Weber, Max
  authority, 403–404, 405
  bureaucracy, 152–153
  religion, 386
  social stratification, 207–209
Weberian model of U.S. class structure, 210–212
Weddings, 48
Welfare programs, 225
Welfare state, 426
Western Wall, 381
"What it would be like?" framing, 350
White Anglo-Saxon Protestants (WASPs), 279
White coat ceremony (Harvard Medical
    School), 74
White-collar crime, 179
White-collar workers, 214
White ethnic Americans, 282–283
White Hat Management, 374–375

"White men can't jump" assumption, 260
WikiLeaks, 183
Wombs-for-rent (India), 346–347
Women. *See also* Gender; Sex, gender, and
    sexuality
  bodybuilders, 299
  cohabitation, 341
  female/athlete paradox, 111
  feminist viewpoint. *See* Feminist approaches/
    perspectives
  feminization of poverty, 223
  gender-segregated work, 316
  objectification of, 298
  occupations, 317
  police, 189
  richest people in the world, 214
  second shift, 320, 343, 344
  sex discrimination, 306
  street encounters, 123
  U.S. senators, 322
*Women and Crime* (Simon), 172
Women's Christian Temperance Union, 521
Women's suffrage movement, 519
Work culture, 153–154
Work ethic, 247, 386
Worker activism, 430

Working class
  Marx, 206
  Weberian model, 211
  Wright, 214
Working poor, 211–212, 224
Workplace socialization, 96
World Food Program, 248
World systems theory, 248, 249–250, 486, 498
*World Urbanization Prospects*, 503
World Women in Defense of the Environment,
  526
World's largest agglomerations, 497
World's largest corporations, 423
Worldwide racial and ethnic struggles, 289
Wright's model of class structure, 213–214

Xanith, 300

Yahoo!, 420
Yanomamö, 60–61
Young adults, 96, 312
Young-old, 96

*Zelman v. Simmons-Harris*, 374
Zero population growth, 485